The interplay between probability, physics, and geometry is at the frontier of current studies of river basins. This book considers river basins and drainage networks in light of their scaling and multiscaling properties and the dynamics responsible for their development. The hydrology of river basins and prediction of their growth demands knowledge of a range of temporal and spatial scales. At the core of *Fractal River Basins* is the search for the hidden order of these temporal and spatial variabilities in river basins, despite variations in size, climate, and geology. The search concentrates on the detection and dynamic origins of fractal features and the crucial role of self-organization.

In engaging style, Rodriguez-Iturbe and Rinaldo provide a theoretical basis for the arrangement of branching networks of river basins. The commonality of branching networks to other natural phenomena makes this book applicable to a wide range of disciplines. Hydrologists and geomorphologists will find that this book opens up the important topic of the fractal structure of networks at an accessible level. Mathematicians and physicists will appreciate the application of the theory to this aspect of the earth sciences. Comprehensive, well illustrated, and with many real-world examples, *Fractal River Basins* will be useful to researchers and students alike.

FRACTAL RIVER BASINS

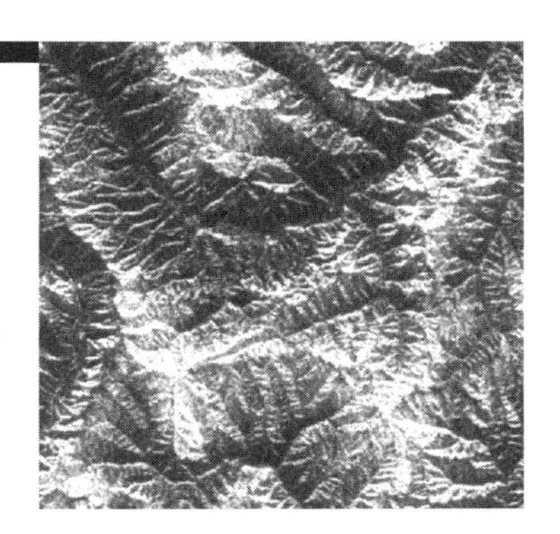

FRACTAL RIVER BASINS

CHANCE AND SELF-ORGANIZATION

Ignacio Rodriguez-Iturbe
Princeton University

Andrea Rinaldo
University of Padua, Italy

PUBLISHED BY THE PRESS SYNDICATE OF THE UNIVERSITY OF CAMBRIDGE
The Pitt Building, Trumpington Street, Cambridge, United Kingdom

CAMBRIDGE UNIVERSITY PRESS
The Edinburgh Building, Cambridge CB2 2RU, UK
40 West 20th Street, New York, NY 10011-4211, USA
10 Stamford Road, Oakleigh, Melbourne 3166, Australia
Ruiz de Alarcón 13, 28014 Madrid, Spain
Dock House, The Waterfront, Cape Town 8001, South Africa

http://www.cambridge.org

First published 1997
First paperback edition 2001

Printed in the United States of America

Typeface Melior 10/12 pt.

A catalog record for this book is available from the British Library

Library of Congress Cataloging in Publication data is available

ISBN 0 521 47398 5 hardback
ISBN 0 521 00405 5 paperback

Para mis hijos y sus hijos,
a la memoria constante de mi padre y de mi madre y de Maria L. Iturbe...
pero sobre todo para Mercedes con el mismo amor
y la misma ilusión de hace 35 años

IRI

per Daniele, Carlotta e Tobia,
per cui tutto e' troppo poco,
per i miei genitori con gratitudine e infinito affetto,
per Alberico Putti che onoro e ricordo sempre,
soprattutto per Caterina, con la stessa ammirazione
e lo stesso entusiasmo di 18 anni fa

AR

Contents

Foreword

Under the driving force of gravity, river basins take in moisture from the atmosphere in a distributed and intermittent manner, store it in the soil, and release it gradually into a network of stream channels where it is concentrated and delivered more-or-less continuously to the outlet. Along the way this water interacts with the basin, putting mineral constituents into solution, nourishing animal and vegetal communities as well as endangering them, and reshaping the basin through entrainment and redepostition of its soil. Their life-support and life-threatening roles have made river basins a subject of intense interest and study by agriculturalists and engineers for over 2,000 years, while their physical and chemical structure has more recently provided earth scientists with insights into Earth's early history. Each of us lives in a river basin, and just as we differ from each other by virtue of our size, shape, age, and metabolism, so do our river basins vary in the same measures. In either case, collective understanding is often obscured by individual differences!

In the heart of this important book, Professors Rodriguez-Iturbe and Rinaldo bring a fundamental advance to both geomorphic science and hydrologic science by uncovering and exploiting "the deep statistical symmetry" inherent in the scale-free fractal form which unifies the characterization of river networks despite their extraordinary individual diversity. Motivated to seek the dynamic origin of this fractal structure, the authors propose and verify a condition of minimum energy expenditure for the entire fractal network which leads to their definition of "Optimal Channel Networks." In so doing they establish for the first time the connection between optimality and fractal growth, and offer fascinating speculation on the difference between network structures based upon minimum energy expenditure and those resulting in maximum entropy. Finally, they demonstrate that Optimum Channel Networks are spatial examples of large, forced dynamical systems which self-organize into a critical state.

A concluding chapter explores the practical utilization of these findings by relating the hydrologic response of the river basin to its geomorphology through the fractal structure of the stream channel network.

In summary, the authors discover and demonstrate "profound order" in the large, coordinated, dissipative systems we call river basins and show that we need not know all the details of the processes to explain the generic properties of that order. In so doing, they exhibit a rare conjunction of theoretical sophistication, physical intuition, and original thought. This monumental work, which makes a coherent story from over 15 years' work by the authors, their colleagues, and their students, will surely take a position alongside the pathfinding contributions of Mandelbrot (fractals) and Bak (Self-Organized Criticality) which inspired it.

The importance of this book to the development of hydrology as a geoscience cannot be overestimated and will be long-lasting. The authors have demonstrated to their fellow hydrologists that advances in basic understanding of hydrologic science can be achieved through transferral of ideas from the frontiers of sister sciences, and more importantly perhaps, they have shown readers from these sciences that hydrology has challenging problems worthy of their own finest efforts. It is through work such as this book that a science is built.

Peter S. Eagleson
Edmund K. Turner Professor of Civil and Environmental Engineering, Emeritus, and Professor of Earth, Atmospheric and Planetary Sciences, Emeritus
Massachusetts Institute of Technology

Preface

In the introduction of his beautiful book on the origins of order, Stuart Kauffman observes that "*. . . there are times in any science when one senses that a transformation to deeper understanding is pressing upward in some as yet poorly articulate form*". Although Kauffman refers to evolutionary biology, we may very well be in such a period in hydrology.

Many factors must come together to effect such a transformation. The first factor is undoubtedly the appearance on the scientific scene of Benoit Mandelbrot's ideas on fractal geometry. The recognition and the implications of the fractal geometry of nature have radically changed the way we perceive and measure hydrologic phenomena. The second factor, the theory of self-organized criticality by Per Bak and collaborators, has provided general foundations for studying the linkages between fractal geometry and the dynamics behind its growth. The third factor, the accessibility of large data sets, objectively collected, of acceptable precision and spanning the natural phenomena over a wide ranges of scales, has provided hydrologists with a unique opportunity for the analysis and testing of otherwise empirical assumptions. As we now fully realize, many geomorphological relationships empirically known – some dating from the last century – carry the signatures of fractal growth and of critical self-organization.

A good amount of hydrologic research has become of interest to many disciplines, which up to now had very little to share with hydrology. River networks play a central role in this transfer, partly due to the outstanding experimental evidence available and partly because of the widespread interest in physics for tree-like structures like percolation clusters, diffusion-limited aggregates, or molecular chains. As such, hydrology finds itself playing a role of increasing importance in the arena of physical sciences.

This book has a few ancestors. One is certainly Peter S. Eagleson's *Dynamic Hydrology*, not strictly because of its contents, but rather because it strives to link different subjects within the common denominator of hydrologic research. Kauffman's book, *Origins of Order,* is indeed a model for its powerful attempt to focus attention on new themes in evolutionary biology that bridge the gap between physics and biology through cross-fertilization. *Opportunities in the Hydrological Sciences*, by the U.S. National Research Council Committee of the same name, provided stimulus to try to strengthen the scientific basis of the important branch of hydrology that is the subject of this book.

The interplay between probability, physics, and geometry has always been in the background of hydrologic theories and practice. In the case of river basins this interplay is now at the frontier of contemporary science. We are dealing with branching networks and searching for scaling and multiscaling properties and searching even more for the dynamics responsible for the growth of such structures. This has implications of both a scientific and practical character. As described in *Opportunities in the Hydrologic Sciences*:

"*. . . Modern approaches to problems in hydrology are moving toward scaling theories as much out of pragmatic necessity as out of pure scientific curiosity and rigor. This is true whether one looks at current theoretical efforts dealing with water in the atmosphere, on the surface, or beneath the ground. The pragmatic reason is that hydrologic understanding and predictions are needed over a broad range of scales, ranging from 100 m to 10,000 km in space and from a few minutes to many years in time. Over such a range, measurements are hard to make and hard to follow because of noise and nonlinearity. Therefore it is all the more important to make theoretically meaningful observations on such natural systems that are subject to the paradox of measurement. The purely scientific reason happens to be the same one. If the spatial and/or temporal variabilities embody a fundamental hidden order that manifests itself across a wide range of scales as an invariance property, then it must be formulated mathematically and tested empirically, for the presence of such a property must be obeyed by more specific mathematical models.*"

We also agree with Kauffman's statement that in writing a scientific book, ideas follow unsuspected paths, mature in unpredictable ways, and mingle with their own logic. Nothing, though, neither the exciting new phases of research nor the linkages across different fields, can match the pleasure we had in writing this book. In a sense the writing strengthened our feelings for what we might term the fractality of the cognitive process and of knowledge itself: As we zeroed in what seemed to be details, they often revealed unexpected depth, and new particulars or technical problems surfaced under the new resolution so that the deepening process became endless. This book is the byproduct of a cutoff in the endless dissection of particulars, dictated in equal parts by chance and necessity: chance, because we have moved into (or out of) some problems by pure randomness; and necessity, because the dissection stops with the current state of knowledge.

Nevertheless, the pleasure of scientific research hinges more on the challenge posed by the questions it raises rather than on the answers – imperfect and limited as they always are – it contrives. Again in Kauffman's words, "what a pleasure it is to seek." In our case, our scientific interests and friendship afforded us a very special pleasure in seeking together. We wish to ask the reader forgiveness when we get carried away in transmitting our feelings regarding the importance of some topics and ideas. Somehow the pleasure of writing was inseparable from the excitement about the problems and the progress we believe has recently been made.

The subject and the results that this book covers are by no means exhaustive, nor are they conclusive, as many contributions are appearing covering new and unexpected facts. However, it seemed appropriate at this point to order the existing material and to place the results in perspective, facilitating future work. Of course we do not claim that all natural forms appearing in the river basins are fractal. Nevertheless, many of them are. In this book we focus our interest only in the processes that either create or interfere with the growth of fractal forms. The complexity of the processes leading to fractal and multifractal structures in river basins favors a distinctive modeling perspective, not based on systems of nonlinear differential balance equations but on conceptually different approaches that are capable of explaining the dynamics of fractal forms. Complex results do not need a complex model in order to be generated, and theories of self-organization provide a perfect example for this basic claim. This modeling perspective permeates the structure of the book and by necessity limits the type of problems and approaches that we review or examine. We have not been strict in trying to avoid some repetition of concepts in different parts of this book. We feel that some of this contributes to the understanding of the ideas.

This book does not attempt to solve any urgent problems of engineering hydrology. This does not mean that they should not be solved and should not receive all the attention they deserve. But somehow, it is our belief, throughout the years hydrology lost touch with the difference between the important and the urgent. Antoine de St. Exupery says, in his *Cittadelle:*

"*...but I have long learned to distinguish that which is important from that which is urgent. True, it is urgent that man should eat, for else he cannot live and death abides no question. Yet love, the sense of life and the quest of God are more important ...*"

Hydrologists indeed need to solve urgent problems, but it is also true that we should seek the exciting and the interesting, keeping the flavor and the rigor of science in our disciplines, which will then emerge as a distinct science motivated by the desire to understand and, as a consequence, will bring the capability to solve.

We owe recognition to institutions and individuals. Our main institutions, the Department of Civil Engineering of Texas A & M University and the Istituto di Idraulica "G.Poleni" of the University of Padua, Italy, are gratefully acknowledged for their continuing support. Through the years of our enduring collaboration we have served on the faculties of the Instituto Internacional de Estudios Avanzados and the Universidad Simon Bolivar, in Caracas, Venezuela (IRI) and the Universita' di Trento (AR) to which we are indebted and truly attached. Throughout the years IRI and recently AR have been associated with the Parsons Laboratory of the Massachusetts Institute of Technology (MIT) in Cambridge, which we are happy to acknowledge. Istituto Veneto di Scienze, Lettere ed Arti in Venice, Italy, and the University of Genova, Italy, were instrumental in the initial phase of our collaboration. The International Centre for Hydrology "Dino Tonini" of the University of Padua provided different kinds of support throughout the writing stage. The support and encouragement of all these institutions are gratefully acknowledged.

We are also indebted to many individuals. Peter S. Eagleson from M.I.T. has deeply influenced our own approach toward research through personal example. His friendship and academic style have been a source of guidance for IRI. Rafael L. Bras from M.I.T. has been a special companion in research for many years for IRI and more recently for AR. It is a particular pleasure to acknowledge his help, support and friendship. AR owes much to Claudio Datei from the University of Padua, who provided inspiration, guidance, and a vital support in the pursuit of AR's academic interests. Donald R.F. Harleman from M.I.T. has been a close friend and an academic model throughout the years for IRI and recently for AR. Alessandro Marani from the University of Venice has been a close friend to both AR and IRI and a continuous cultural reference for AR. William E. Dietrich of the University of California at Berkeley provided many useful insights into the topics of this book and a stimulating partnership in research. Enrico Marchi, from the University of Genova, distinctively influenced us through his friendship and academic standards. Amos Maritan, from the Scuola Internazionale Superiore di Studi Avanzati in Trieste, provided many comments on drafts of key parts of this book and a challenging partnership in research. The late Augusto Ghetti and Aldo Giorgini, formerly from the University of Padua and Purdue University, were instrumental in AR's education and are reference scholars who will be sorely missed.

Juan Valdes (Texas A & M University), Dara Entekhabi (M.I.T.), Per Bak (Brookhaven National Laboratory), Franco Siccardi and Giovanni Seminara (University of Genova), Alessandro Franchini (Istituto Veneto di Scienze, Lettere ed Arti), José Córdova and Marcelo Gonzáles (Universidad Simon Bolivar), Vijay Gupta (University of Colorado), Ed Waymire (Oregon State University), Garrison Sposito (University of California at Berkeley), Gedeon Dagan (Tel Aviv University), Luigi Da Deppo, Giuseppe Gambolati and Mario Putti (University of Padova), and Jayanth R. Banavar (Pennsylvania State University) are colleagues who in more than one way have influenced this book through interaction with both authors. We thank them.

It is a pleasure to acknowledge the generosity of William E. Dietrich and Dr. Ronald Blom (Jet Propulsion Laboratory, California Institute of Technology) in providing most of the photographs used in this book. The National Aeronautics and Space Administration (NASA) was also instrumental in facilitating some photographs from the space program.

Our research in the subject matter of this book would have been impossible without the dedicated effort of many of our former (or current) doctoral students. We are pleased to acknowledge the help and the contribution of David Tarboton, Garry Willgoose, and Ede Ijjasz-Vasquez from M.I.T., Riccardo Rigon and Alberto Bellin from the University of Trento, Gregor Vogel from both the University of Trento and Texas A & M University, Luis Parra from Texas A & M University, Paolo D'Odorico and Marco Marani from the University of Padua. Special thanks have to be expressed to Riccardo Rigon and Marco Marani for their dedication and companionship.

Ignacio Rodríguez-Iturbe
Andrea Rinaldo

College Station, Padova
March 1997

CHAPTER 1

A View of River Basins

> River basins are the fundamental natural system of many hydrologic phenomena. Hydrology as an earth science covers an extremely wide range of temporal and spatial scales and a myriad of different processes. This introductory chapter reviews the fluvial system in the search for observational evidence of scale invariance. We also review geomorphological concepts and tools, as well as the relevant characters of stream channel and basin landscape geometries.

1.1 Introduction

Similarities can be observed among all kinds of river basins. These similarities respond to general operating criteria that control the manner in which river basins work. The study of these similarities and the reasons behind their existence are at the core of hydrology as an earth science, and they are the main subject of this book.

It is important to clarify, first of all, what we call a river basin. The whole picture of a river system may be divided in three loosely separated, but distinct, regions. These are shown in Figure 1.1, and according to their main working purpose they are called, following Schumm [1977], the *production zone*, the *transportation or transfer zone*, and the *delivery or deposition zone*.

The production zone is what we call the river basin or watershed. It originates most of the water and sediments that are then transported through the plains for their delivery to the oceans through deltas and estuaries. Although each of these sections has its own peculiar properties, it is in the river basin where we perceive the greatest challenges and fascinating phenomena from the hydrologic point of view.

We will not deal with all the aspects of a river basin; in fact, we will skip most issues related to vegetation and the soil system and concentrate mainly on the structure of the drainage network and on the three-dimensional landscape the network drains. Not that the network is unrelated to the soil and vegetation characteristics; on the contrary, its structure reflects the general properties of the soil and vegetation system that, when linked to a particular climate, yield the basin runoff and the sediment load that the network collects and transports to the outlet of the basin.

The river basin is made up of two interrelated systems: the drainage network and the hillslopes. The hillslopes control the production of storm water runoff, which, in turn, is transported through the channel network toward the basin outlet. The runoff-contributing areas of the hillslopes are both a cause and an effect of the drainage network's growth and development. This cause and effect relationship may be visualized through the following considerations. In short and intense rainfall storms, most of the runoff is generally contributed either by Hortonian overland flow (in which the rainfall intensity exceeds the infiltration capacity of the soil) or by direct precipitation into the saturated areas around the channels. Which of

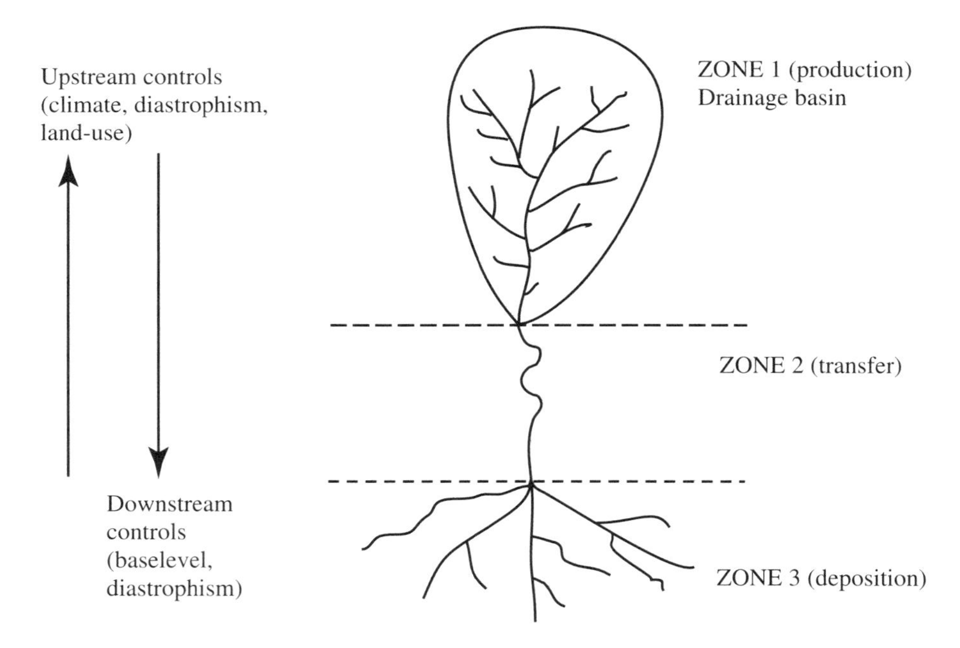

Figure 1.1. Idealized fluvial system [after Schumm, 1977].

these two mechanisms prevails in a given climate depends on the soil and vegetation characteristics regulating the infiltration capacity. The second mechanism produces what is called the saturated overland flow. When rainfall is not very intense but has a long duration, contributions of subsurface storm flow occur mainly from areas close to the drainage network. In such long storms the contributions of saturated overland flow are also very important and may increase more rapidly than those of the subsurface stormflow. The saturated areas expand rapidly from poorly drained soils into areas of initially better drained soils and steeper topography. As explained by Dunne [1978], hollows are preferred avenues for such expansions, partly because of their role in concentrating downslope surface and subsurface flow and also, to a much greater extent, because their concave profiles cause the water table to be closer to the surface at the beginning of the storm. The convergence of overland flow is a major cause of the growth upslope of the channels but, on the other hand, hollows are a cause of the occurrence of overland flow. From this point of view the drainage network itself may be viewed as a reflection of the runoff-producing mechanisms occurring in a basin. Indeed, there has been a continuous advancement in the understanding of the physical processes that control the response of a given hillslope to a precipitation input under certain simplifying assumptions of homogeneity in soil and vegetation characteristics [Kirkby, 1971]. Nevertheless, basins are made up of a very large number of hillslopes, and the hillslopes can display a large overall variability of soil and vegetation properties. The knowledge of the principles that control the behavior of a single hillslope – which is both necessary and valuable – by itself cannot be expected to lead to an understanding of the role of the hillslope system in the hydrologic response of a basin or of its impact on landscape evolution.

From the previous discussion the existence of circuits of reciprocal control between the system of hillslopes and the drainage network of a basin is apparent. One expects to find a whole interlocking system that commands growth and differentiation in the drainage network. The importance of these reciprocal controls lies at the very heart of hydrology. Every branch of the network is linked to a downstream branch for the transportation of water and sediment, but it is also linked – through the hillslope system – to another branch and, in a sense, to every other branch. The second linkage through the hillslope system results from the fact that, in order for a stream channel to be maintained, the slope-dependent supporting area where runoff is generated must not fall below a certain minimum.

In the above interlocking system the drainage network should be seen as the pattern that connects the different parts of the basin to each other. As a whole, and together with the hillslope system, it relates the precipitation input into the basin to the surface runoff at the outlet. This concept is embedded in the work of Horton [1945], in our opinion an even greater accomplishment for hydrology than are his specific laws. He provided hydrologists and geomorphologists with the ladder of an ordering system, a pattern that connects.

The nature of the above pattern, as well as the features of the individual components of the general structure, were traditionally described through a framework dependent on the existence of fundamental scales. This type of description changed dramatically after the introduction of *fractal geometry* by Mandelbrot in the late sixties [e.g., 1967]. *Fractals* brought a completely new and different perspective into the analysis of river basins. New relationships were found, in both the planar and the altitude dimensions, which spanned a wide range of scales in the form of power laws. The sinuosity of individual river channels as well as the branching characteristics of the drainage tree are now seen through a fractal lens. Even the response function of the basin to a rainfall input acquires new and important features. In this introductory chapter we will use the term fractal without definition; a definition will be formally given in Chapter 2.

This chapter deals mainly with the more traditional view of river networks. It is not meant to be an exhaustive literature review, but rather something more than a superficial overview of the main results of fluvial geomorphology, which later on will be shown to be intimately linked to fractals.

The infinite variety of patterns in natural river basins, with an underlying unity with the fractal geometry that describes them (as will be seen in Chapter 2), suggests the existence of a basic unifying evolutionary dynamic that is responsible for pattern formation. The search for such a basic dynamic is a major constituent of this book. Indeed, this frontier, which goes beyond the description of fluvial fractals, permeates many of the modern scientific disciplines in which many processes are shown to exhibit fractal signatures. A major advance in this direction was recently made by Bak and coworkers [e.g., 1987], who studied the behavior of large dissipative dynamical systems with many degrees of freedom. Initially, Bak and his collaborators worked on computer simulations of coupled torsion pendula. They attempted to simulate the behavior of a complex dynamical system consisting of many (simple) parts, driven by the constant feeding of energy into the system. The pendula behaved in a very concerted cooperative manner that differed radically from a state of equilibrium where all pendula are in the downward position. Bak speculated that the system would evolve to a critical state if properly driven, like a chain reaction. Indeed, computer simulations confirmed that this was the case.

The above proved to be an important result. In fact, the scientific community has been puzzled for a long time by two difficult problems. First, the deep dynamic reason discussed above for the ubiquity of Mandelbrot's fractal forms in nature, such as river basins, mountain ranges, and coastlines, which look alike on all length scales. Second, the widespread phenomenon of *1/f noise*, that is, a signal emitted from a variety of sources ranging from quasars to river flows that has components of all time scales. Spatial and temporal scale-free behaviors are the fingerprints of both fractal structures and critical processes. Bak therefore concluded that the abundance of such signals in nature indicates a universal tendency of large, driven dynamical systems to self-organize into a critical state far away from equilibrium. This constitutes what is now called the *theory of self-organized criticality*.

The word *critical* means that the state that is naturally reached by the system is one that may undergo a full spectrum of changes (from small to very large) when subjected to minimal perturbations. However, thermodynamic properties of spanning networks will suggest a somewhat different interpretation of criticality in self-organizing dynamical contexts. Moreover, the 'critical' state is characterized by large periods of stasis interrupted by intermittent bursts of activity of all sizes. Earthquakes follow precisely such a pattern, which is known as the Gutemberg–Richter law [Bak and Tang, 1989]. Mandelbrot [1963] suggested that fluctuations in economics behave much in the same way. Also, biological evolution itself is not gradual, but rather exhibits *punctuated equilibria* that describe the intermittent behavior described above [Gould and Eldredge, 1993; Kauffman, 1993], prompting Bak to boldly suggest that life itself could indeed be a self-organized critical process [Bak and Sneppen, 1993]. In fact, Darwinian biological evolution is currently being revisited in view of *critical evolution*, which is much faster than noncooperative scenarios as no large and coordinated – hence prohibitively unlikely – mutations are necessary.

The above arguments lead us to believe that fractal river basins are best studied within this framework. The theory, and its extensions, provide many of the needed tools to firmly link fractal growth and processes for river basins, as we will discuss throughout this book.

This chapter discusses geomorphologic material of interest and of use in the subsequent revisitation of basin geomorphology through the fractal perspective.

1.2 River Basin Geomorphology: A Brief Review

1.2.1 Ordering of the Channel Network

Ordering systems are used to group or characterize the parts that constitute a drainage network. These systems can work through the network starting from the outlet and moving in the upstream direction or from each source and moving downstream. The most successful have been the downstream-moving ordering systems, the first of which was proposed by Horton [1932,

1945]. Strahler [1952, 1957] revised Horton's scheme to avoid some ambiguities, and the so-called Strahler system or Horton–Strahler ordering system is now the most commonly used in hydrogeomorphology. This ordering procedure analyzes networks as follows:

- channels that originate at a source – have no tributaries – are defined to be first-order streams;
- when two streams of order ω join, a stream of order $\omega + 1$ is created;
- when two streams of different order join, the channel segment immediately downstream has the higher order of the two combining streams (see Figure 1.2).

The whole network embodies a deep sense of regularity, not the trivial regularity of size, but the much deeper regularity of formal relations between the parts. This deeper regularity was first observed by Horton [1945] in the planar projection of the drainage network.

Horton's law of stream numbers is expressed as

$$\frac{N(\omega)}{N(\omega + 1)} = R_B \tag{1.1}$$

where $N(\omega)$ is the number of streams of order ω and R_B is called the bifurcation ratio. $N(\omega)$ is estimated with the Strahler ordering procedure.

Horton's law of stream lengths is expressed as

$$\frac{\bar{L}(\omega + 1)}{\bar{L}(\omega)} = R_L \tag{1.2}$$

where $\bar{L}(\omega)$ is the (arithmetic) average of the length of streams of order ω and R_L is called the length ratio.

Typical values for R_B and R_L are 4 and 2, respectively, with a range between 3 and 5 for R_B and 1.5 and 3.5 for R_L. Horton discovered that R_B

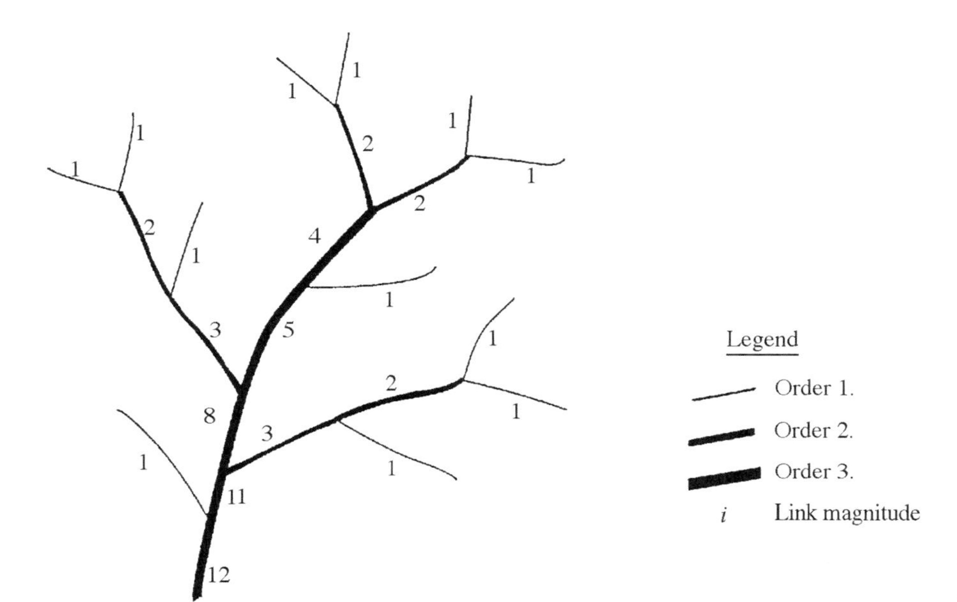

Figure 1.2. Horton–Strahler ordering and link magnitude.

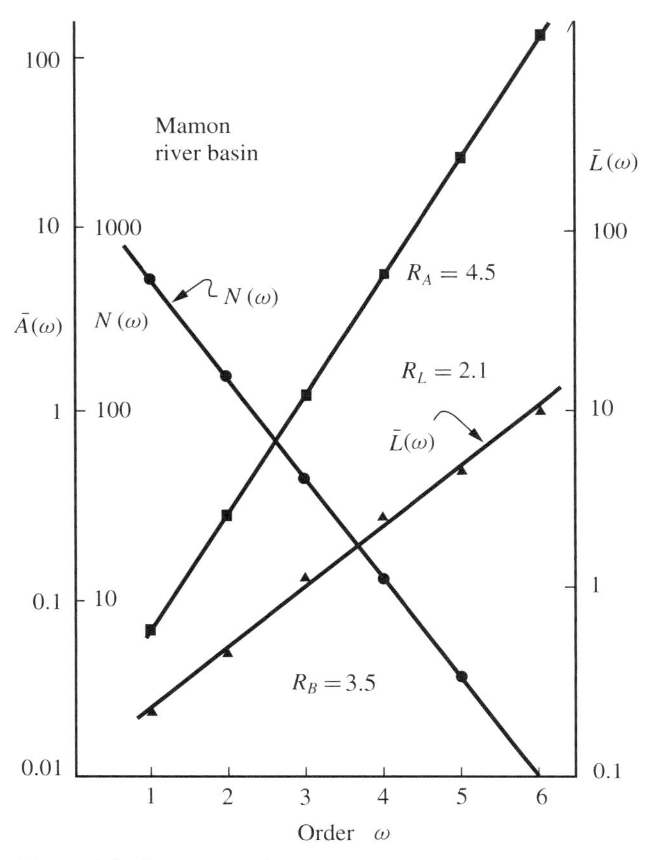

Figure 1.3. Stream numbers, lengths, and areas versus order illustrating Horton's laws for the Mamon Basin, Venezuela [after Valdes et al., 1979].

and R_L were approximately constant through semilog plots of $N(\omega)$ and $\bar{L}(\omega)$ against order ω. The ratio, or *Horton number*, is obtained from the slope of the straight line fit to such plots; the procedure is called a *Horton analysis* (see Figure 1.3).

In the case of R_B the line of best fit must pass through the point $(1, \Omega)$, where Ω is the order of the highest-order stream, also called the order of the basin. In the case of R_L the best fit line is not required to pass through any particular point. Because the ratios are approximately constant, the above geometric descriptions are called "Horton's laws." Although Horton did not specifically include basin areas in his laws of drainage basin composition, he implied that areas should satisfy a geometric series like stream numbers and lengths [Horton, 1945; Shreve, 1969]. The law of basin areas was explicitly stated by Schumm [1956] (law of stream areas):

$$\frac{\bar{A}(\omega)}{\bar{A}(\omega - 1)} = R_A \qquad (1.3)$$

where $\bar{A}(\omega)$ is the mean total area contributing to streams of order ω and R_A is called the area ratio, whose typical value is near 5. The regular geometric relationships contained in Horton's laws and observed in natural drainage networks have been interpreted as the signature of some particular evolutionary criteria in network organization. They have also been interpreted as evidence that drainage networks are topologically random, meaning that chance is the only criteria operating on the organization of the network. Sometimes the previous dichotomy has been at the center of the controversy over the relative merits of deterministic and random models in fluvial geomorphology. Besides the above, it has also been argued that Horton's relations are not specific to any kind of networks as they describe the vast majority of networks, random or not. This topic will be discussed extensively after we describe some models for the planar structure of the drainage network, including the so-called random topology model.

Another ordering system is the link magnitude system owing to Shreve [1966]. Here source streams or links have magnitude 1. At a bifurcation the downstream link takes as its magnitude the sum of the two incoming magnitudes. Thus the magnitude of each link represents the number of sources in the network draining into that link. Figure 1.2 shows, along with the Horton–Strahler ordering system, the concept of link magnitude.

Some terminology of common use in hydrogeomorphology will now be defined. A river network is idealized as a trivalent planted tree, the root of which is the *outlet* or point furthest downstream. *Sources* are points furthest upstream, and a point at which two upstream channels join to form one downstream channel is called a *junction* or *node*. *Exterior links* are the segments of channel between a source and the first junction downstream and *interior links* are the segments of channel between two successive nodes or a node and the outlet. Each link has certain properties: *length* along the stream; *geometric length*, the distance between end points; *height*, or *drop*, the elevation difference between upstream and downstream nodes; *average slope*, height divided by length; *contributing area*, the total area draining through the link measured at the downstream end; and *local* or *directly contributing area*, the area draining directly into a link, that is, not through any other links.

1.2.2 Drainage Density and the Hillslope Scale

The previous section presented the Horton-Strahler ordering system and the concept of link magnitude as measures of the size and scale of a network. These are topological, dimensionless measures of size. They need to be related to physical sizes. This relationship will be established through the drainage density, which is a measure of the degree to which the basin is dissected by channels. Drainage density is closely related to stream and link frequency (defined later in this section), to mean link length, and to mean hillslope length.

The concept of drainage density implies the existence of a fundamental length scale associated with the dissection of the landscape by the river network. Horton [1932, 1945] defines the drainage density $\mathcal{D}$ as

$$\mathcal{D} = \frac{L_T}{A} \tag{1.4}$$

where L_T is the total length of stream channels and A is the total area. An alternative, dimensionless measure of drainage density, will at times be adopted, that is, that defined by the ratio of channelized area (a measure of the number of area units making up the basin occupied by established channels, as opposed to unchannelized or hillslope areas) to the total basin area.

Horton suggested that the average length of overland flow or hillslope length is approximately half the average distance between stream channels and is therefore approximately equal to half the reciprocal of drainage density, $1/2\mathcal{D}$.

Smith [1950] defined a topographic texture ratio: the number of contour crenulations divided by the contour length. He essentially showed that texture is indeed correlated with $\mathcal{D}$ so the notion of a well or poorly drained basin corresponds to the notion of fine or coarse texture.

Horton also introduced the concept of *stream frequency* or channel frequency, F_s, defined as

$$F_s = \frac{N_s}{A} \tag{1.5}$$

where N_s is the number of Strahler streams. Link frequency, F_l, is similarly defined using the number of links. As shown in Figure 1.4 it is possible to construct two hypothetical basins having the same drainage density but different stream frequency, and, on the other hand, it is possible to have two basins with the same stream frequency but with different drainage density. In Figures 1.4(a) and 1.4(b) the basins have the same drainage density but different stream frequencies; in 1.4(c) and 1.4(d) the basins have the same stream frequency but different drainage densities.

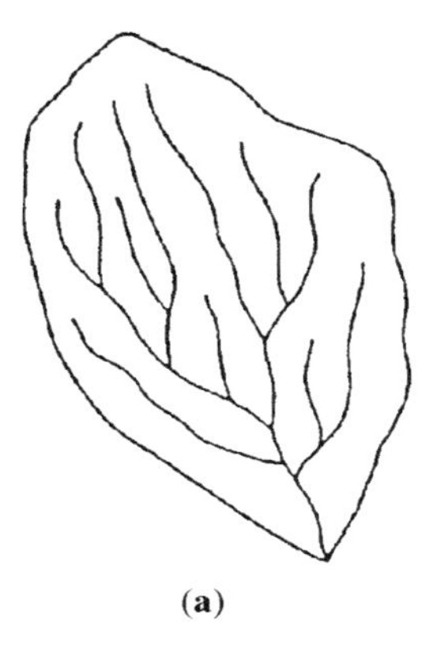

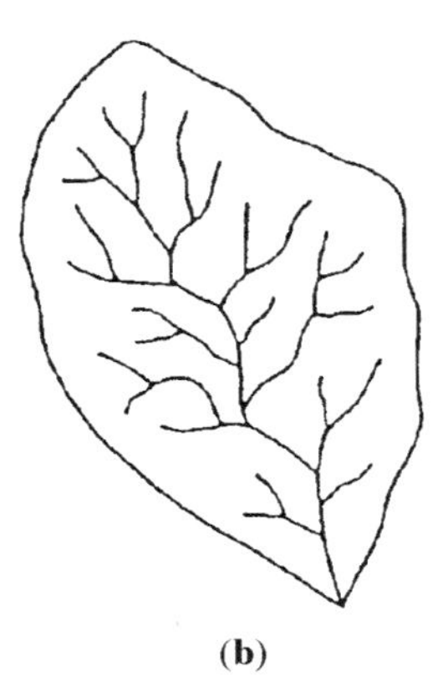

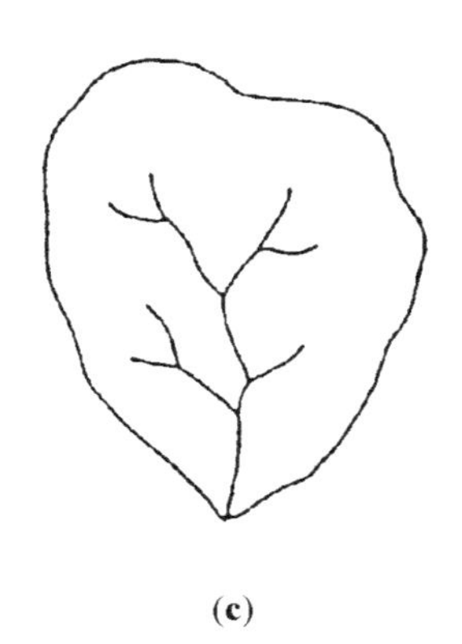

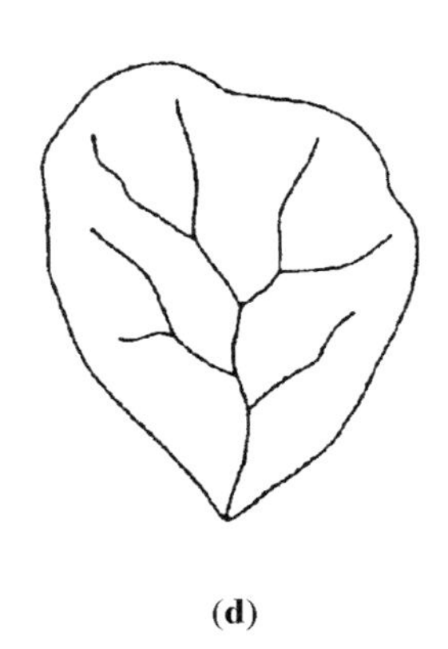

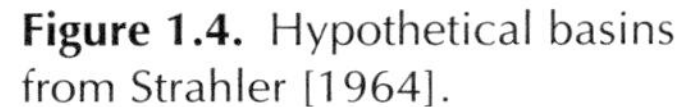

Figure 1.4. Hypothetical basins from Strahler [1964].

Melton [1958] showed that F_s was strongly correlated with drainage density. He plotted F_s versus $\mathcal{D}$ for 156 drainage basins covering a vast range in scale, climate, relief, surface cover, and geologic type. Remarkably, small scatter exists showing that the relationship of F_s versus $\mathcal{D}$ tends to be conserved as a constant in nature through the so-called Melton's law:

$$F_s = 0.694\ \mathcal{D}^2 \tag{1.6}$$

Smart [1978] states that mean link length and mean link frequency are also closely related to drainage density. If the mean area directly draining to a link is κl^2 (suggested by Shreve [1967]), where l is the mean link length, the relationship is

$$\mathcal{D} = F_l\, l = \frac{1}{\kappa l} \tag{1.7}$$

We thus see that drainage density, stream or link frequency, mean link length, hillslope length, and texture are all essentially related to the same measure, the fundamental horizontal length scale associated with how the channel network dissects the landscape. The determination of this scale is generally dependent on the resolution of the map used. Historically, researchers have called on the highest resolution maps and (or) field work to measure these quantities.

Mark [1983] discusses the differences between drainage networks obtained from maps and field surveys and the merits of various procedures such as use of contour crenulations to "extend" the network. He concludes that first-order streams defined from contour crenulations on 1:24,000 maps are identifiable topographic features in the field. However, most first-order basins defined on the map contain more than one fluvial channel in the field. Accordingly, the exterior links drawn by contour crenulations do not always represent unbranched channels.

In the context of scaling and fractals [Mandelbrot, 1983], questions arise as to whether the notion of the existence of fundamental scales is well founded or whether river networks dissect the landscape indefinitely, requiring characterization as a scaling phenomenon (i.e., scale-free, see Chapter 2). This is an important question that was early recognized by Davis [1899, p. 495], who wrote:

"Although the river and hillside waste do not resemble each other at first sight, they are only the extreme members of a continuous series and when this

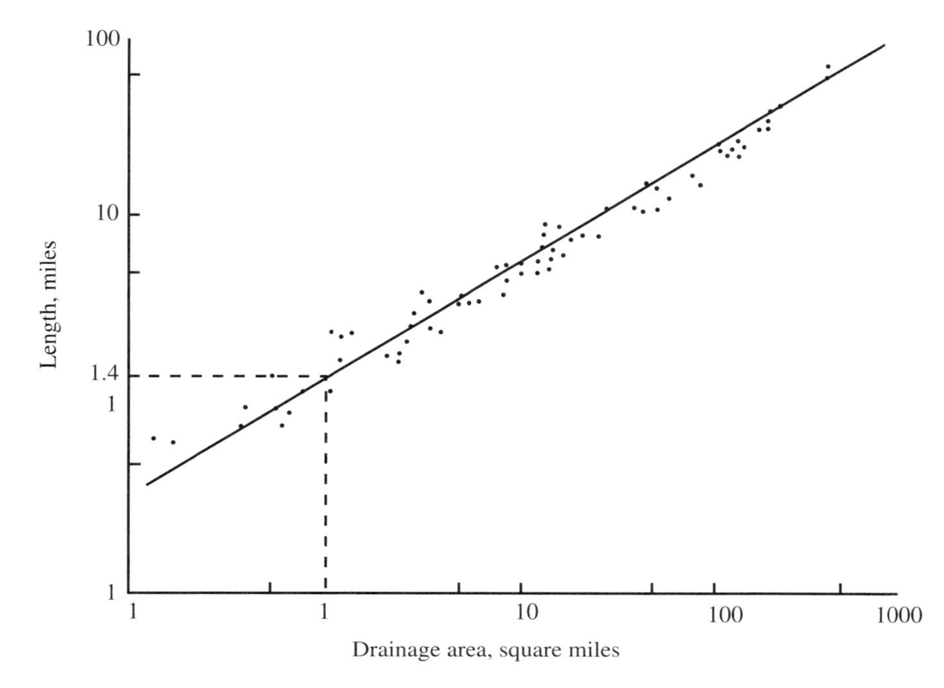

Figure 1.5. Relation of the length of the longest stream to the drainage area for Shenandoah Valley (after Hack [1957]).

generalization is appreciated one may fairly extend the 'river' all over its basin and up to its very divide. Ordinarily treated the river is like the veins of a leaf; broadly viewed it is the entire leaf."

Later on in this chapter (Section 1.2.9) we will extend the concept of drainage density and introduce valley density as a new measure of hillslope scale, distinguishing the channelized part of the basin from that characterized by the unchanneled convergent topography typical of colluvium regions.

1.2.3 Relation of Area to Length

Absolute stream length, measured headward to the divide from a given point on a stream, is functionally related to the area of watershed upstream from a given point. Hack [1957] demonstrated the applicability of a power function relating length and area for streams of the Shenandoah Valley and adjacent mountains in Virginia. He found the equation

$$L = 1.4\, A^{0.6} \tag{1.8}$$

where L is the length of the longest stream, in miles, in the drainage region measured to a point in the divide and A is the area in square miles. He later extended these studies to other basins in the United States and different large rivers of the world, finding that the exponent of Eq. (1.8) remained close to 0.6 (see Figure 1.5).

Hack noted that if geometrical similarity is to be preserved as a drainage basin increases in area downstream, meaning that there is an increase in size but no change in shape, then the exponent in Eq. (1.8) should be 0.5.

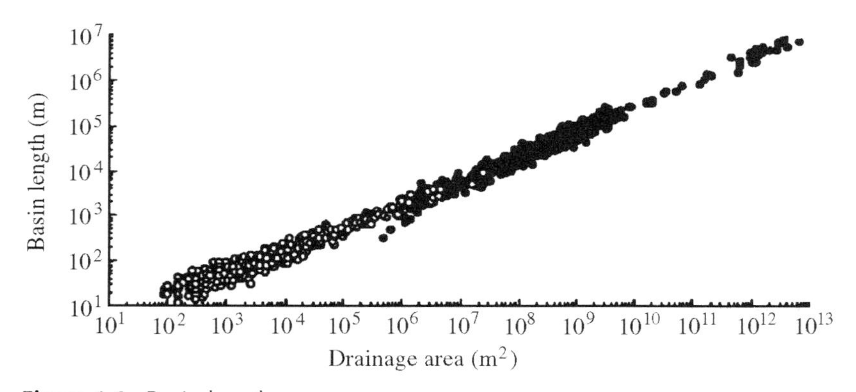

Figure 1.6. Basin length versus area for unchanneled valleys, source areas and low-order channels (empty circles). Solid circles are reported data for large channel networks [after Montgomery and Dietrich, 1992].

Because Hack's equation is

$$\frac{A}{L^2} \approx \frac{1}{2} A^{-0.2} \tag{1.9}$$

it has been interpreted to imply that catchments of all sizes are not entirely similar in shape. Rather, as the area increases, A/L^2 decreases, which indicates a tendency toward elongation of the larger catchments. In other words, Hack's law suggests that basins tend to become longer and narrower as they enlarge. A causal explanation of this tendency is a fascinating problem that will be explored in Chapter 2.

Gray [1961] later perfected an analysis of Hack's law finding the relationship $L \propto A^{0.568}$. We notice that the measure of the watercourse length L is by necessity imperfect without the tools of fractal geometry (see Chapter 2). The difference in the exponents in the range 0.568–0.6 will also be carefully discussed in Chapter 2.

Montgomery and Dietrich [1992] collected data from small drainage basins and observed that Hack-like relationships of characteristic basin lengths versus drainage area tend to also hold for unchanneled valleys, source areas, and low-order channels. Figure 1.6 illustrates the results of data collection for small drainage basins where reported drainage data from larger networks are included. Here the basin length L replaces the mainstream length because it extends to unchanneled regions of convergent topography. It is defined as the length along the main valley axis to the drainage divide. The exponent h of $L \propto A^h$ is estimated with some arbitrariness but, according to Montgomery and Dietrich [1992], could reasonably fit a value significantly lower than in Hack's original analysis, as they found it to be close to 0.5 (Figure 1.6). The composite data set covers basin areas going from 100 m^2 up to 10^7 km^2. Indeed, Hack's law was not the main goal of Montgomery and Dietrich's research; a span of more than 11 orders of magnitude in basin area, from unchanneled hillslope depressions to the world's largest rivers, is, as a whole, not the most adequate to fit when investigating Hack's equation. The investigations of Muller [1973] point at the inconvenience of this.

Nevertheless, the relation observed suggests that there is a basic geometric similarity between drainage basins and the smaller basins they contain that holds down to the finest scale to which the landscape is dissected [Montgomery and Dietrich, 1992]. In Section 1.2.12 we will examine the nature of this limiting scale for the dissection of the landscape, identified by topographically divergent ridges that separate fine-scale valleys. Although a quantitative assessment of the variance in the observed data is difficult because the headward extent of the stream network is often identified through maps of varying scale, it can be seen that landscape dissection results in an integrated network of valleys that capture geometrically similar drainage basins at scales ranging from the largest rivers to the finest-scale valleys. Within this range of scales there appears to be little inherent to the channel network and the corresponding shape of the drainage area that would provide any reference to some *absolute* scale [Montgomery and Dietrich, 1992].

Later in this book (Chapter 6) we will see that unchanneled valleys often bear the signature of patterns from fluvial processes. This may provide an explanation for the extension of Hack's relationship to unchanneled areas.

Finally, we observe that Hack's law, whether in its original or in its extended form, cannot ignore the role and the impact of statistical fluctuations which are not accounted for by the simple, deterministic relationship $L \propto A^h$. These fluctuations we will interpret within a scaling framework in Section 2.10.

1.2.4 Relation of Area to Discharge

Another regular factor among different river basins is the relation between discharge of a given frequency of occurrence and drainage area. Bankfull discharge has a recurrence interval averaging 1.5 years [Leopold, Wolman, and Miller, 1964], and its relation to area is approximately given by an equation of the type

$$Q \propto A^{0.75} \tag{1.10}$$

The mean annual discharge usually fills a river channel to about one-third of its bankfull depth, and it tends to have a similar frequency of occurrence among rivers of different features. This flow is equaled or exceeded on the average about 25% of the time. It roughly represents the discharge exceeded for one day in every four over a large period of time. An important feature of the mean annual discharge is that even though it may not occur at all points in the drainage network on any given day of record, it is nevertheless closely approximated everywhere during many days of the year. For mean annual flows the exponent in the discharge versus area relationship is about 1.0:

$$Q \propto A \tag{1.11}$$

This is frequently the case in basins with relatively minor differences in mean annual rainfall for different locations inside the watershed. The lower exponent– 0.75 – at higher flows is a measure of storage in river valleys and a consequence of the fact that rains of high intensity rarely cover the entire basin, but are instead widely and irregularly spaced [Leopold et al., 1964]. An example of Eq. (1.11) is given in Figure 1.7 which plots mean annual discharge versus area for all gauging stations on the Potomac River Basin.

When studying the spatial distribution of discharge-dependent quantities, for example energy expenditure, it is usually the case that one does not have flow measurements throughout a major part of the links that make up the drainage network. Thus equations such as (1.10) and (1.11) are of great utility because they allow the use of drainage area as a surrogate variable for discharge, whether it be bankfull or mean annual flow.

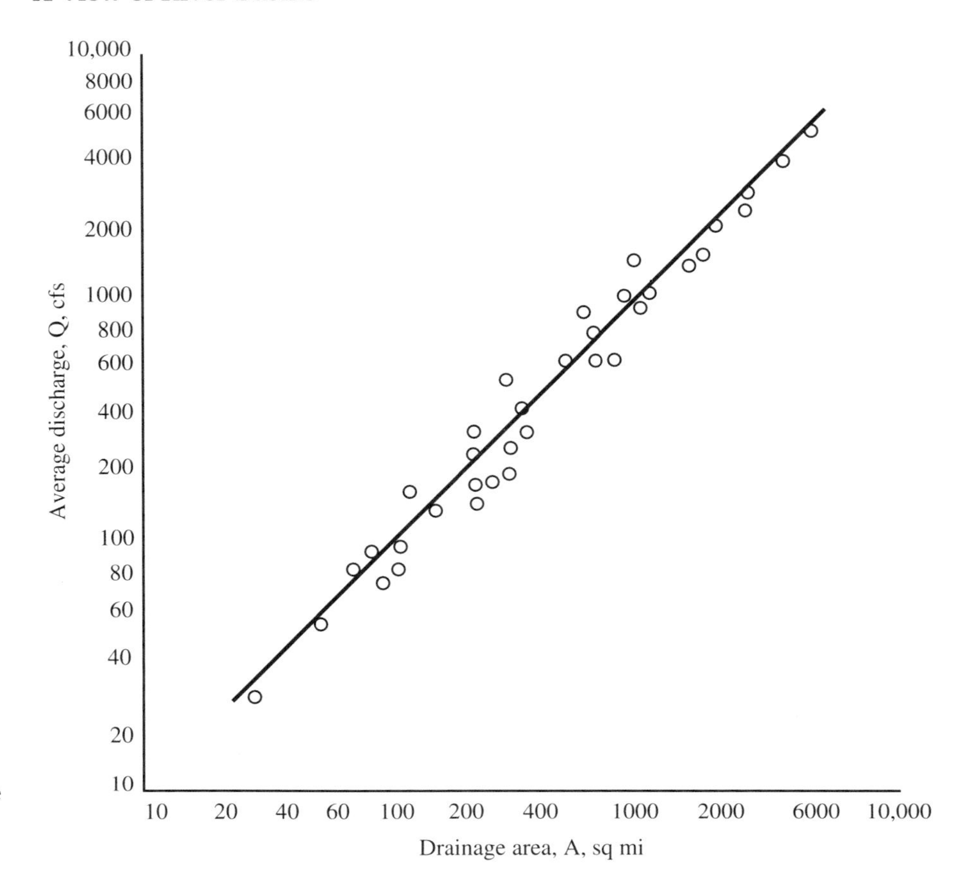

Figure 1.7. Relation of discharge to drainage area for all gaging stations in the Potomac River. The line has slope 1 [after Hack, 1957].

1.2.5 Relation between Magnitude and Area

The magnitude n of a basin (or subbasin) is the number of exterior links in the network, or equivalently, the number of first-order streams. In any network the number of internal links is $n - 1$, and hence the total number of links is $2n - 1$.

There exists an excellent linear relationship between total area and the number of links, or the magnitude. The implication is that each link has an associated area that directly contributes and that does not vary much throughout the basin.

1.2.6 Stream Channel Geometry

In a fundamental study of river hydraulics, Leopold and Maddock [1953] demonstrated how some characteristics of stream channels – depth, width, velocity, and suspended load – vary with discharge as simple power functions at a given river cross section. Power law relationships also exist among the previously discussed variables and discharge when the channel characteristics are measured along the length of the river under the condition that discharge at all points is equal in frequency of occurrence. The velocity v, width w, and depth d of flow Q are given by

$$v = kQ^m \qquad w = aQ^b \qquad d = cQ^f \tag{1.12}$$

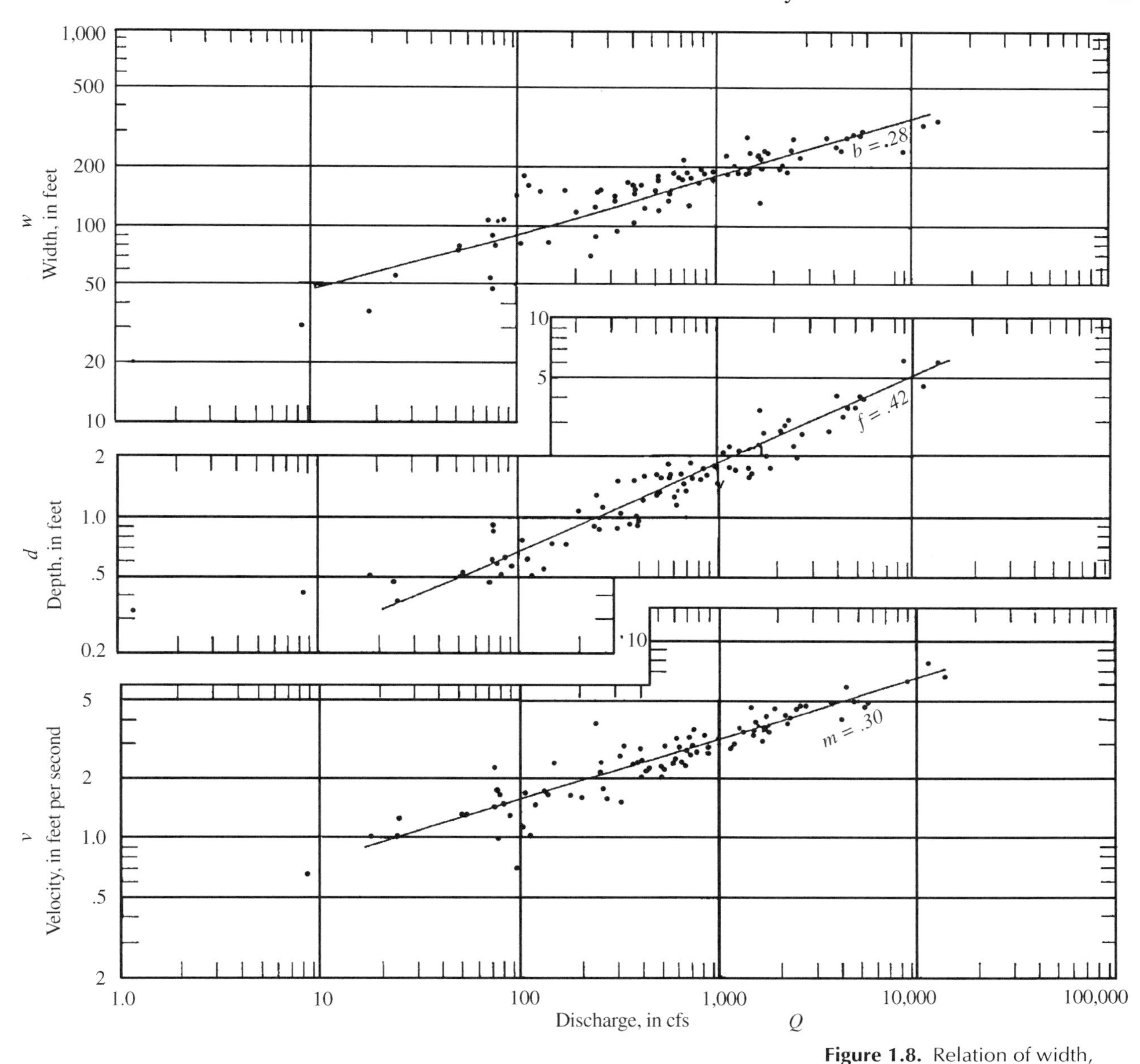

Figure 1.8. Relation of width, depth, and velocity to discharge; Powder River at Locate, Montana [after Leopold and Maddock, 1953].

where Q is the discharge; k, a, and c are proportionality constants; and m, b, and f are the power law exponents.

Because discharge is approximately given by $Q \approx vwd$, Eq. (1.12) implies that $m+b+f=1$. When they analyzed the variation of the hydraulic characteristics in a particular cross section of the river as a function of discharge (*at-a-station* type of analysis), Leopold and Maddock [1953] found average values of $b=0.26$, $f=0.40$, and $m=0.34$.

An example of the power law relationships at a station is shown in Figure 1.8 for the Powder River at Locate, Montana.

A value of 0.4 for f indicates that discharge increases much faster than depth of water in a stream. When the mean depth doubles, the discharge increases proportionally to $2^{2.5}$, that is, it increases nearly 6 times.

The equation $d = cQ^f$ indicates that a rating curve – a graph of water stage versus discharge – should plot approximately as a straight line on log paper and, indeed, such plots are widely used in engineering practice.

In the analysis of hydraulic characteristics in different cross sections along the length of a river, the comparison is valid only under a condition of constant frequency of discharge at all cross sections. This analysis is called *in the downstream direction*, and it is of special interest for interpreting the structure of the drainage network as a pattern that connects the elements of a basin. Leopold and Maddock [1953] showed that similar power law equations are obtained whether one uses the mean discharge or bankfull discharge. The average values of the exponents they obtained are $b = 0.5$, $f = 0.4$, and $m = 0.1$.

An example of the analysis in the downstream direction is shown in Figure 1.9 for the Bighorn River and its tributaries in Wyoming and Montana.

Proceeding downstream in a given river, discharge will increase because of the increasing area (except in very arid regions where there may exist large losses in the downstream direction). The interesting thing is that, along a river channel, such a progressive increase is structurally related to the changes in width and water depth in the channel, regardless of where in the watershed or on what tributary the cross sections may be.

The very low value of the exponent m indicates that the velocity tends to remain constant or increase slightly in the downstream direction as long as the discharge at all points is of similar frequency. This important finding has also been confirmed by a variety of field experiments (e.g., those of Pilgrim [1976, 1977] in Australia; see also Carlston [1969]).

The above indicates that for mean annual discharge everywhere in the basin, the increase in depth compensates – or slightly overcompensates – for the decreasing river slope in the downstream direction, with a net result that velocity is nearly constant everywhere in the basin. The nature of this compensation, in case of uniform flow, may be seen in Manning's equation [e.g., Chow, 1959] relating velocity to depth of flow (d), slope (∇z), and channel roughness (n):

$$v = \frac{1}{n} d^{2/3} \nabla z^{1/2} \tag{1.13}$$

where d approximates the hydraulic radius. One observes that velocity depends on depth to the power 2/3 and on slope to the power 1/2. Mean velocity at a series of gaging stations along the course of the Yellowstone–Missouri–Mississippi river system was obtained by Leopold et al. [1964] from flood measurement data interpolated to represent floods of fifty-year and five-year recurrence intervals. It was found that velocity remains essentially constant in the downstream direction at each flood frequency.

The constant-velocity characteristic everywhere in the network is an important feature that has crucial implications when studying the properties of the hydrologic response of the basin to precipitation inputs (Chapter 7).

It is important to notice that a value of $m = 0$ in Eq. (1.12) suggests reasonable values of 0.5 for the exponents b and f, which control the power law variation of width and depth with respect to discharge.

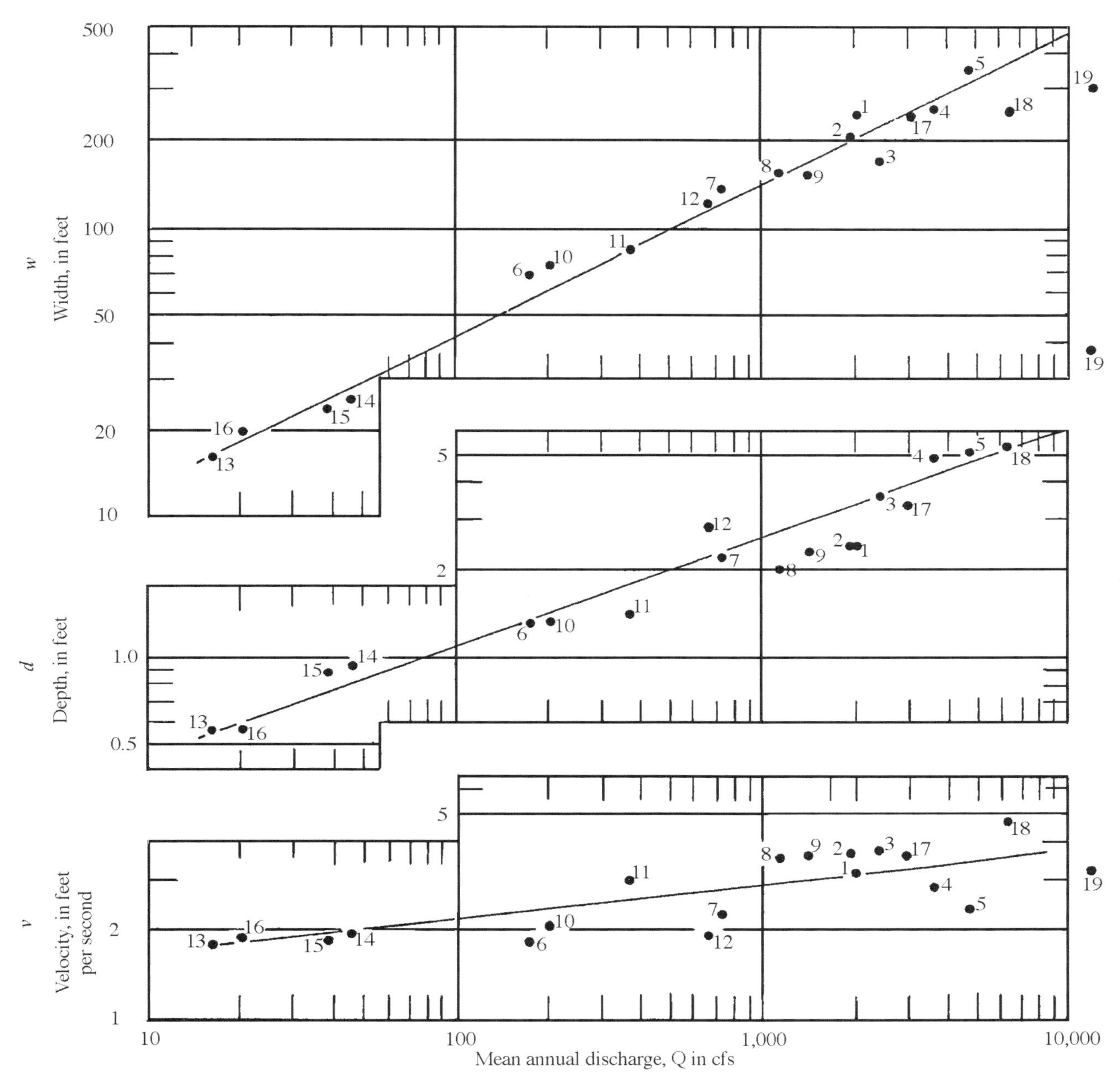

Figure 1.9. Width, depth, and velocity in relation to mean annual discharge as discharge increases downstream, for the Bighorn River and tributaries, Wyoming and Montana, and the Yellowstone River, Montana [after Leopold and Maddock, 1953].

1.2.7 The Width Function

A fundamental property of any drainage network is that there is a unique one-dimensional path connecting any pair of points in the tree. In particular, the flow path from any point to the basin outlet is uniquely determined.

When studying the structural characteristics of a drainage network and, most importantly, the implications of such a structure in the hydrologic response of a basin to any precipitation input, the arrangements of the flow paths from any point of the basin to the outlet are of the most crucial importance. This arrangement is characterized by the so-called width function

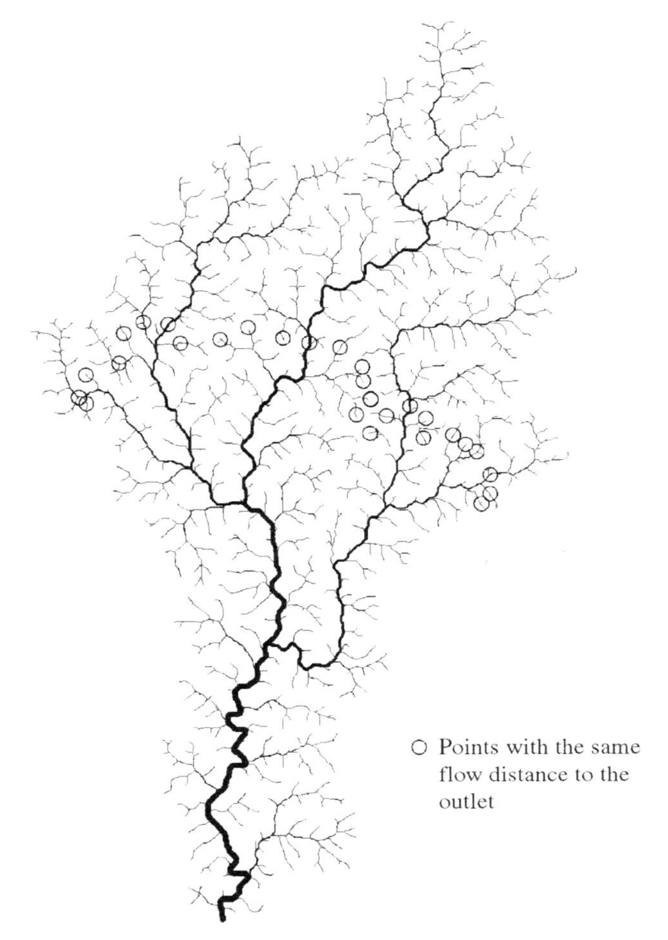

Figure 1.10. Definition of width function.

of the basin, first introduced by Shreve [1969]. The width function $W(x)$ gives the number of links in the network at a flow distance x from the outlet. It is important here to underline the fact that distance is measured along the network rather than, say, radially from the outlet. At this point one should distinguish the notion of geometric link length from that of the flow link length (or distance along the channel). The former definition used by Shreve [1969] denotes the (straight line) distance between two junctions or a junction and a source defining a link. The latter definition denotes the length along a channel between two junctions defining a link or between a junction and a source. Figure 1.10 depicts these two definitions of link length as well as the definition of width function. Except when explicitly stated, the distance used in the width function will always be measured along the channel network.

Figure 1.11 shows examples of width functions from real basins of very different characteristics. They look extremely erratic but, indeed, there are many elements of a common fundamental structure, as will be seen in Section 3.4.

A natural extension of the width function of a basin, defined above through its links, is the *distance-area* function, where the relative proportion of drainage area is organized on the basis of its distance to the outlet. In a system where a link is associated with every unit area used to discretize the catchment surface, the width function and the area-distance functions are equal. If procedures to automatically extract the network employ critical support area concepts, say A_t (through which a channel link is defined only when it drains more than a given cumulated drainage area), then the one-to-one correspondence between width and area functions is not granted. In other words, when the necessary support is the unit area, the network is space filling because every pixel is assigned a link and no holes are allowed in the distribution of pixels covering the total drainage area. Unless otherwise specified, throughout this book the area function will be used to identify the basin width function, making use of all the information derived from the basin elevation field.

The impact of a critical support or threshold area on the characteristics of the width and area functions may be observed in the results of Snell and Sivapalan [1994]. They compared experimental values of width and area functions of two basins in Australia for which very detailed (≈ 1 m resolution) digital elevation maps were available. Figure 1.12 shows that the smoothing operations embedded in the determination of the area function and the effects of the threshold area used to extract the network, as large as $A_t = 23$ pixels in this example, do not substantially change the character of the width function in a basin of 39.6 km^2, where the maximum distance from source to outlet is 13.945 km. Nevertheless, some differences are observed in the area of maximum relative contribution.

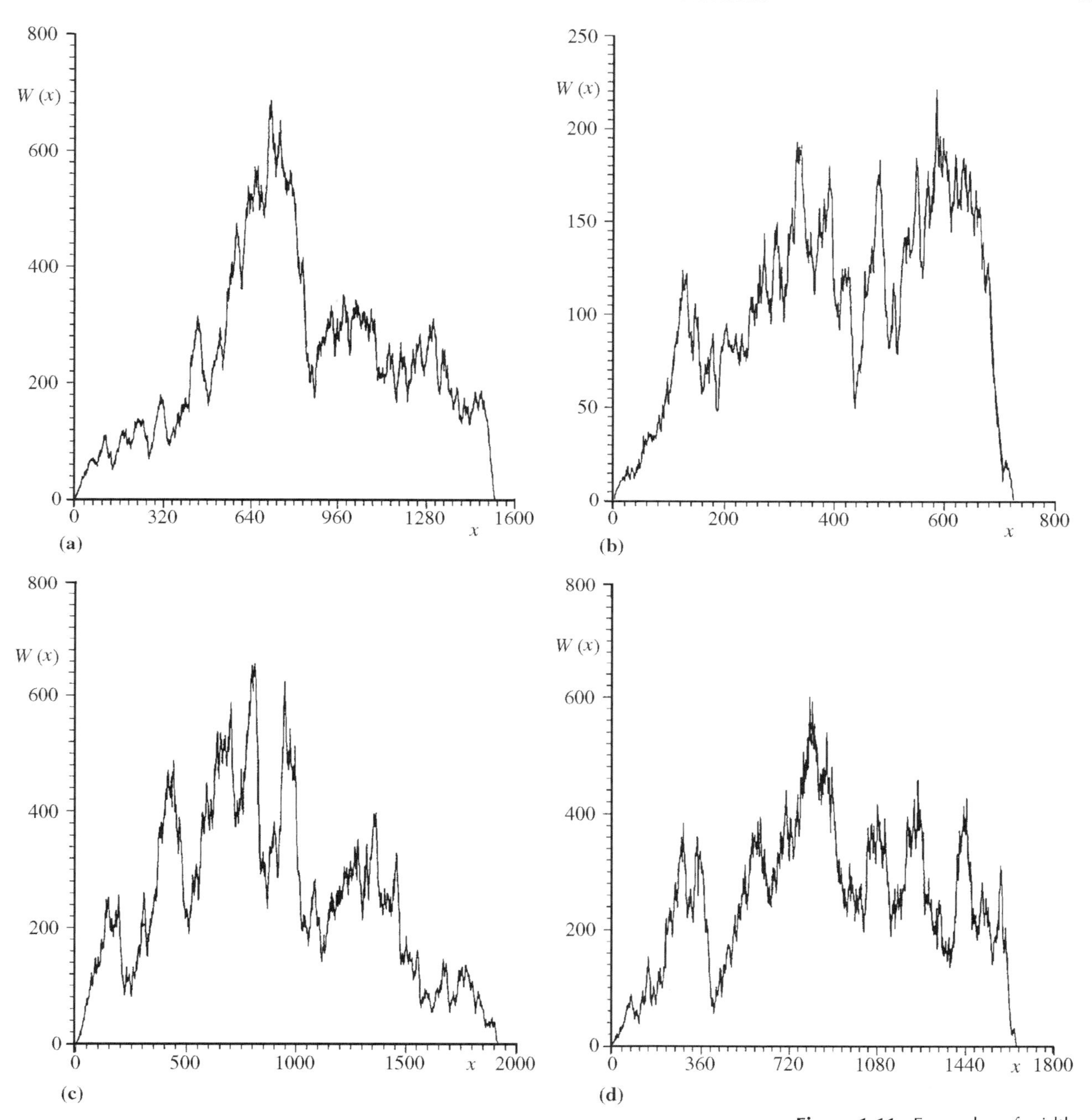

Figure 1.11. Examples of width functions of real basins: (a) Schoharie, (b) Nelk, (c) St. Joe, and (d) Racoon. The main characteristics of these basins are described in Table 1.3. The distances x are expressed in multiples of the unit pixel length.

Finally, the identification of the basin width function with the area function overlooks the role of transitions from areas of divergent topography to areas of convergent topography with or without channelization. When studying scale-invariant properties of the spatial organization of a river network this is a viable approximation. This is so because in our fluvial environments the unchanneled valleys are relicts of channelizations in wetter climates and, as seen in Figure 1.6, their organization has geometrical properties similar to those of larger fluvial basins. Hence, the filling processes associated with nonchanneled sediment transport do not substantially modify the overall shape of the planar structure and, in

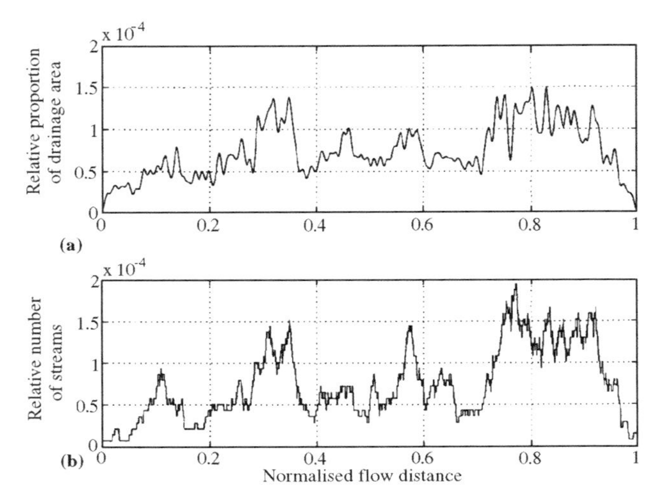

Figure 1.12. Area and width function for Conjurunup catchment from a 1-m digital elevation map (DEM). Here a threshold area of twenty-three pixels has been used to extract the network [after Snell and Sivapalan, 1994]. Ordinates are dimensionless ratios obtained via the contributing area at normalized distance x divided by total area, and via the contributing length at x divided by total network length.

particular, the distribution of spatial curvatures. This will be explicitly studied in Chapter 6. Nevertheless, when studying the hydrologic response of a basin (Chapter 7), such distinction cannot be overlooked because runoff production mechanisms and travel time distributions are strongly related to the nature of hillslope–channel transitions.

1.2.8 The Three-Dimensional Structure of River Basins

In an early contribution toward the understanding of landscape evolution, Leopold et al. [1964] argued that because no single process controls slope forms, as streamflow does within the river channel, any realistically coupled model of hillslope, channel network, and river basin evolution needs to be complex. We will see in Chapter 5, when outlining the foundations of the theory of self-organized criticality, that structural complexity does not need a complex model to be explained. However, the number of factors affecting the three-dimensional structure of the river basin is indeed exceedingly large. As an illustrative model of landform development, Leopold et al. [1964] describe the case in which the rate of erosion is assumed to be a function of shear stress and the form of the land is derived as a function of various lithologies and possible rates and loci of geologic uplift. Thus all kinds of concave and convex forms and combinations of these shapes can be generated by assuming various combinations of erosion and uplift, and reasonable results can consequently be obtained. Leopold et al. [1964] conclude that whatever model is used, the forms derived are not necessarily unique to the conditions assumed. This is indeed the case in river basin morphology, but it is also true that the quest for an underlying unity in the natural landforms is quite relevant.

What is behind the extraordinary complexity and yet the deep structural symmetry that nature exhibits in river basins? Why do certain recurrent geomorphic relationships among the parts of different complex basins hold regardless of geology, altitude, climate, and size of the basin? In this book we will show that a lot more is known now, especially in light of fractal theories of self-organization, than at the time of Leopold et al.'s [1964] book. This is particularly true for the observational evidence related to the scaling properties of the three-dimensional structure of river basins and the related progress in the understanding of landscape-forming processes.

We believe that data from digital elevation maps (DEMs) provide key missing factors of many geomorphological analyses of the past. In fact, until recently most basin geomorphology hinged on planar features of river networks. Although planar features are basically related to elevation fields, describing the landscape through the definition of drainage directions (see Section 1.2.9), we will show in Chapters 4, 5, and 6 that quite unrealistic basins may have realistic planar features. A key role in evaluating the goodness of any geomorphological representation is that played by the statistics related to the third dimension. In this sense, data relative to the elevation field will be crucial for addressing the issue of the interplay of chance and necessity in the evolutionary process leading to basin morphology. This is so because of the statistical inevitability of certain planar properties, for example, the bifurcation structure observed in nature as the outcome of the (exceedingly) most probable state, proposed by Shreve [1967] and discussed in Section 1.3. To discuss the experimental evidence available, we will first provide a brief review of procedures for data analysis.

1.2.9 River Basins from Digital Elevation Models

Deeper insight into the structure – both planar and three-dimensional – of large channel networks has been gained after introduction of digital elevation maps (DEMs). In particular, the analysis of large river networks (say, of the order of hundreds or thousands of square kilometers) obtained from DEMs has made possible a completely new set of statistical analyses aimed at the determination of scaling properties of the observed fields. This section briefly describes the procedures to extract channel networks from aerial topography.

The early procedures for obtaining channel networks were based on work by O'Callaghan and Mark [1985], Band [1986], and Carrara [1988]. O'Callaghan and Mark [1985] restrict their analysis to the most commonly used data structure for DEMs, that of the regular square grid. In such a grid, elevations are available as a matrix of points equally spaced in two orthogonal directions. Spacing in each direction is not necessarily the same, that is, rectangular grids are commonplace. Other data structures have been used for DEMs in hydrologic analyses like triangular irregular networks, contour-based DEMs, and multiple flow direction maps. The reader is referred to Tarboton et al. [1989b] and Moore, O'Loughlin, and Burch [1988] for further details.

The DEMs used by Tarboton et al. [1988; 1989a,b] to extract channel networks consist of elevations, obtained by topography from space, in a grid (square or rectangular) whose spacing is of the order of 30 m. Each grid block is called a pixel, and the first step in determining drainage networks is to assign a drainage direction from each pixel to one of its eight neighbors (including the diagonals). Usually this is done in the direction of steepest descent (multiple flow directions are also employed at times; their implementation is discussed in Chapter 6). In the simplest model, where the slopes ∇z in two or more directions are the same within accuracy, as is typically the case in flat areas, the directions may be assigned arbitrarily, provided no loops are formed.

At this stage, other data anomalies need be resolved, for example, the formation of pits. Pits are anomalies manifested as sites lower than all surrounding pixels. They are usually regarded as data errors within a sufficiently large fluvial landscape, because pits large enough to be resolved on a 30-m grid are rarely seen in practice [Tarboton et al., 1989a]. In DEMs pits are resolved by filling them, that is, increasing their elevation until they drain, meaning that consistent drainage directions with nonnegative slopes can be determined. These adjustments are relatively minor and typically are only required for a small subset ($< 2\%$) of the pixels. A thorough discussion of data sources and accuracy is presented in Tarboton et al. [1989a,b]. The basic result is a field of elevations z_i where i is the two-dimensional spatial position vector of the arbitrary pixel.

Figure 1.13. A three-dimensional plot of the DEM of the Fella River Basin [after Rigon, 1994].

The next step is to count the number of pixels that drain through each pixel. This provides a measure of convergence of flow, with the largest accumulation of area being in the valleys along the streams. This suggests defining streams as those pixels with total drainage area greater than a support area threshold A_t. This is conceptually similar to the notion of channel maintenance [Schumm, 1956]. Carrara [1988] uses a support area threshold as well as minimum source stream length to define streams. In this book the issue of where channels begin will be addressed, particularly because it has been argued [Montgomery and Dietrich, 1988] that support area alone may not be sufficient to determine stream *initiation*.

Setting of directions and accumulation of areas is done for the whole of a rectangular DEM data set. Isolation of the drainage basin consists of identifying those pixels that eventually drain through an outlet pixel. The channel network is defined as the geometric tree of lines along flow directions joining the centers of all pixels with an accumulation area above the support area threshold. The terminology used is fairly standard and basically that of Shreve [1966, 1967] described earlier. Sources are then the points upstream on this network and junctions are points where two (or more) channels join. As seen in Section 1.2, exterior links are defined as segments of the channel network between a source and the first junction downstream, and interior links are segments between two successive junctions or between the outlet and first junction upstream. Also, the magnitude, n, of a network is the number of sources or exterior links. The

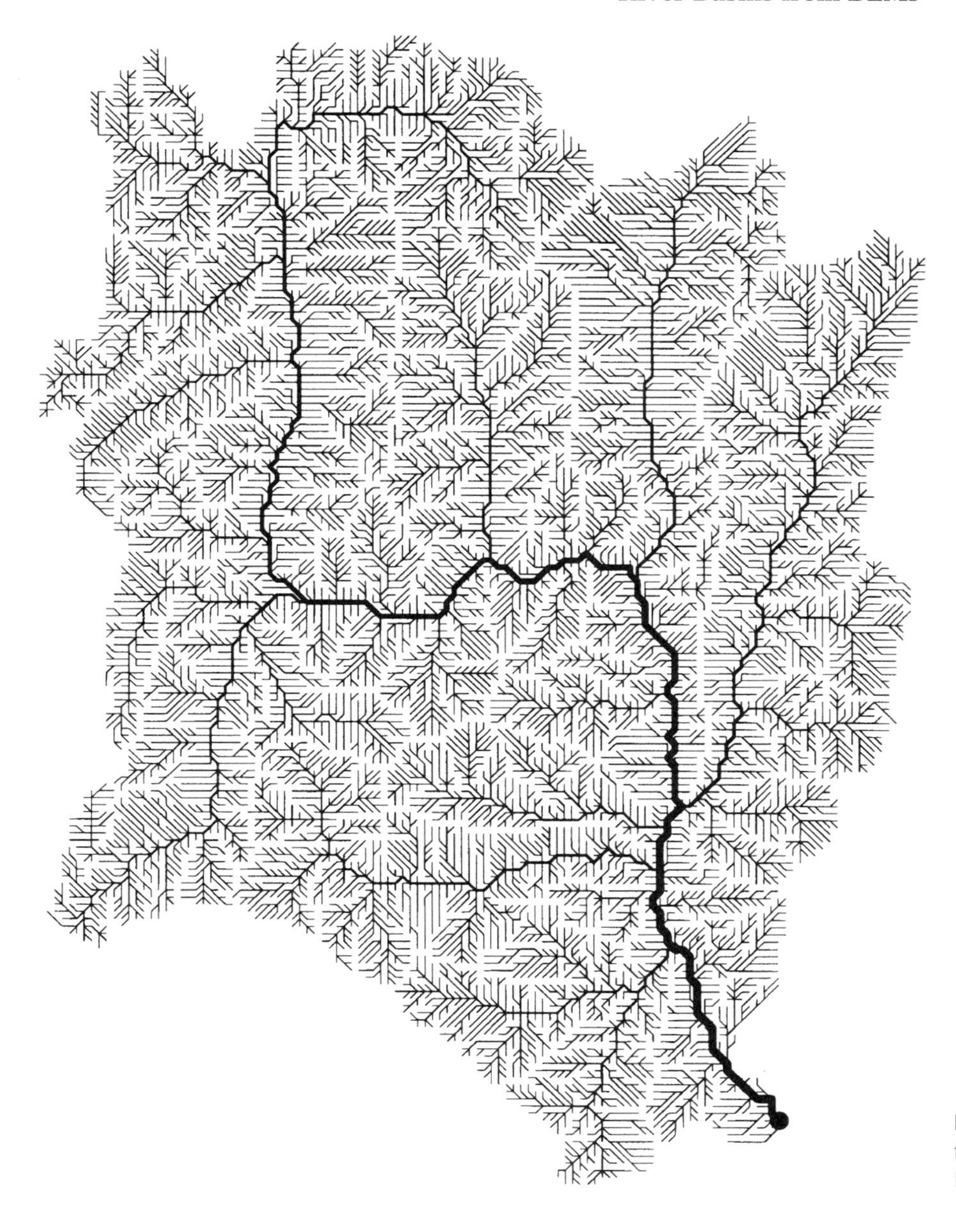

Figure 1.14. Drainage directions for the DEM of the Fella River Basin.

networks can be described by the Horton–Strahler ordering system described in Figure 1.2. In brief, exterior links have order 1. Where two or more links of order $m_1, m_2, m_3, \ldots$ join with $m_1 \geq m_2 \geq m_3, \ldots$ the order of the next downstream link is max $(m_1, m_2 + 1)$. This is a slight modification of the usual system in Figure 1.2 to allow for the fact that, on a discrete grid, junctions of more than two links sometimes occur. A Strahler stream is thus a sequence of links of the same order.

A three-dimensional plot of a DEM and its corresponding plot of drainage directions, irrespective of whether the arbitrary site is channelized or not, are shown in Figures 1.13 and 1.14 for the 706-km^2 Fella River Basin (northern Italy). In this example, DEM data are given on a square grid of 150×150 m^2.

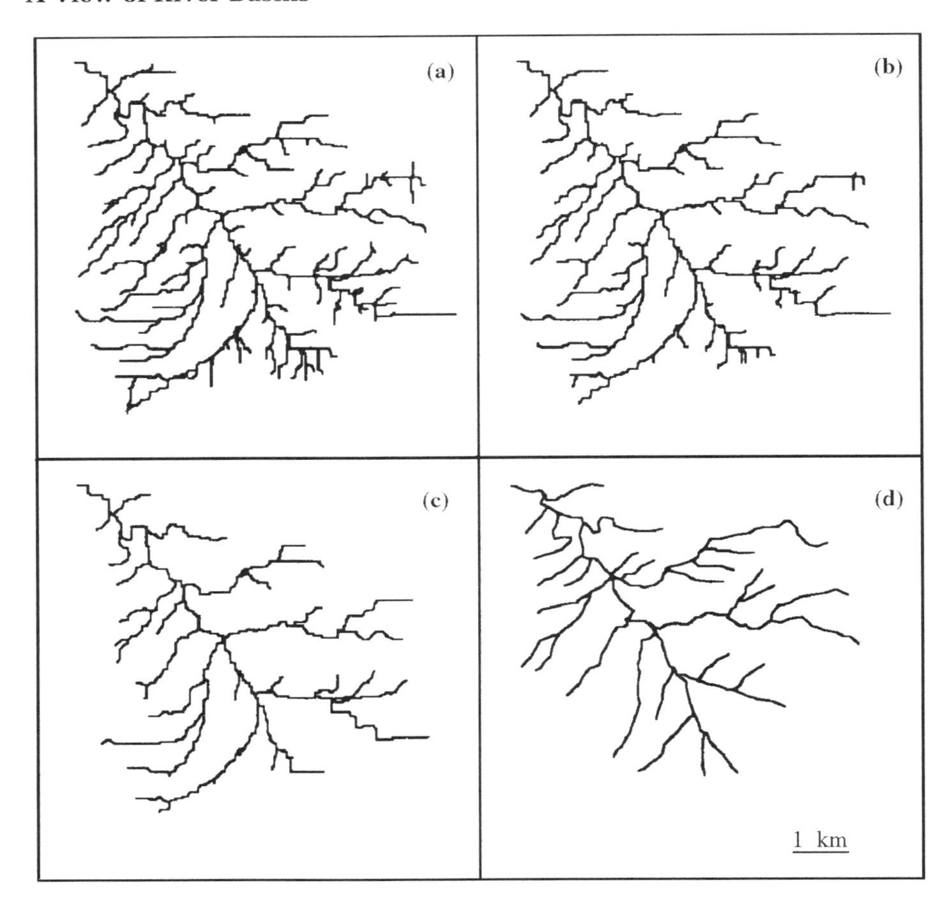

Figure 1.15. (a) to (c) show the the effect of different support areas A_t on automatic network extraction from a DEM; (d) shows the network resulting from the 'blue lines' of a conventional map [after Tarboton et al., 1989a].

The most appropriate choice of support area threshold A_t used to define the network is a matter of debate. Although it is true that DEMs may cloud the correct scales for channel initiation [e.g., Montgomery and Dietrich, 1988], at large enough sizes of the basin such features may lose relevance. As shown in Section 1.2.11, Tarboton et al. [1988] simply use the *constant* support area threshold that gives the correct drainage density for the data set used. An example of channel networks obtained from the same DEM with varying support areas A_t for channel identification is shown in Figure 1.15.

Once the network is identified through drainage directions, computing the properties of each link, that is, magnitude, drop, order, and length, is straightforward.

In the context of this book we frequently focus on the evaluation of slopes and their relationship with drainage area. Slope ∇z_i at the ith link is defined as drop/length, that is, is an average over the link. Total contributing area, A_i, is measured at the downstream end of each link i and is the area of the whole subbasin draining through the outlet of the link.

The basic algorithm for computing total contributing area proceeds as follows. A scalar, A_i, denotes the total drainage area, that is, the number of pixels draining through the ith link, which is defined by a pair of integers (i_1, i_2) denoting its spatial position within a spatial grid. In *pixel units* the total area at i is

$$A_i = \sum_{j \in nn(i)} W_{ij} A_j + 1 \tag{1.14}$$

where

$$W_{ij} = \begin{cases} 1, & \text{if } i, j \text{ are connected, that is, if } j \to i \text{ is a drainage direction} \\ 0, & \text{otherwise} \end{cases}$$

The subindex j spans the eight neighboring ($nn(i)$ stands for nearest neighbors to i) pixels of the arbitrary ith site and W_{ij} is a functional operator. The unit area added in Eq. (1.14) refers to the area of the actual site and represents the area draining directly into the ith link.

An example of the extraction of a drainage network from DEMs through Tarboton et al.'s [1988] constant area threshold criterion is shown in Figure 1.16. The data refer to Big Creek, a tributary of the St. Joe River near Calder, Idaho, and the St. Joe River itself. Big Creek is a 147-km^2 basin covered by a combination of four U.S. Geological Survey 7.5-minute DEMs that give elevations on a 30-m grid. The St. Joe River is a 2, 834-km^2 basin with DEM data from a combination of three DEMs that give elevations on a 3 arc second grid (60 $\times$ 90 m^2 approximately). Figure 1.16 shows the resulting maps.

Table 1.1 shows the characteristics of DEM data sets used by Tarboton et al. [1989a,b; 1990]. These data are a part of the geomorphological evidence used throughout this book.

Contour-based digital terrain models divide the land surface into elements defined by topographic contours and flow lines. They have proved to be useful for the analysis of river basins. Dietrich et al. [1988, 1992, 1993] analyzed erosion thresholds, channel networks, and landscape morphology via accurate terrain models and concluded that prediction of the full extent of the channel network in a landscape requires elevation data of sufficiently high resolution such that the finest-scale source-area basins can be quantitatively analyzed.

An example of network and landform extraction from accurate traditional mapping is shown in Figure 1.17.

A few geomorphologic concepts need to be recalled here. One concerns the distinction of hillslopes, valleys, and channels (Figure 1.18). Hillslopes

Table 1.1. *Digital elevation model data sets*

Name	Location	Pixel size [m^2]	Area [km^2]
St. Joe River	Mont., Idaho	62.2 × 92.2	2834
Schoharie River	N.Y.	68.3 × 92.2	2408
Beaver River	Ohio, Penn., Minn.	70.5 × 92.2	1223
Edel River	N.Y.	68.3 × 92.2	993
Buck River	Califo.	30 × 30	606
Racoon River	Penn.	30 × 30	448
Nelk River	Idaho	62.6 × 92.2	440
Brushy River	Ala.	30 × 30	325
Cald River	Idaho	30 × 30	147

Note: Basin names, approximate locations, pixel sizes, and basin area.
Source: Tarboton et al., 1989a.

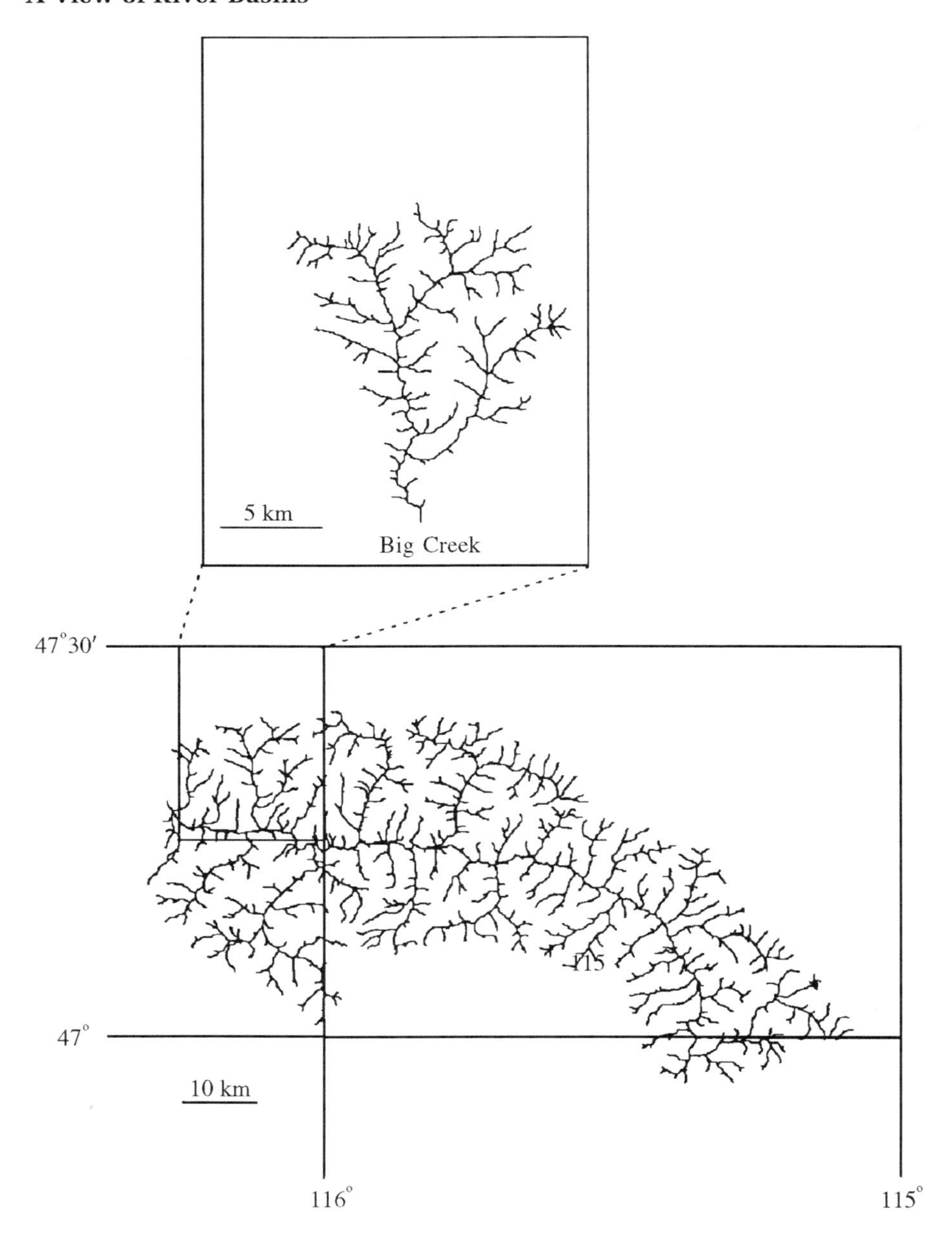

Figure 1.16. St. Joe River and Big Creek location maps [after Tarboton et al., 1989a].

are seen as areas of topographic divergence, and valleys are areas of topographic convergence. In Figure 1.18 the dashed lines indicate the transitions of hillslopes to valleys. Channels appear within areas of topographic convergence but are not defined by curvature alone. In fact, it is difficult to define a channel from DEMs, a morphologic feature defined by a concentration of transport of water and sediment within a defined geometry where, for instance, banks are meaningful concepts.

One interesting feature that can be objectively measured from accurate DEM data is the convergence (or the divergence) of topography at any point

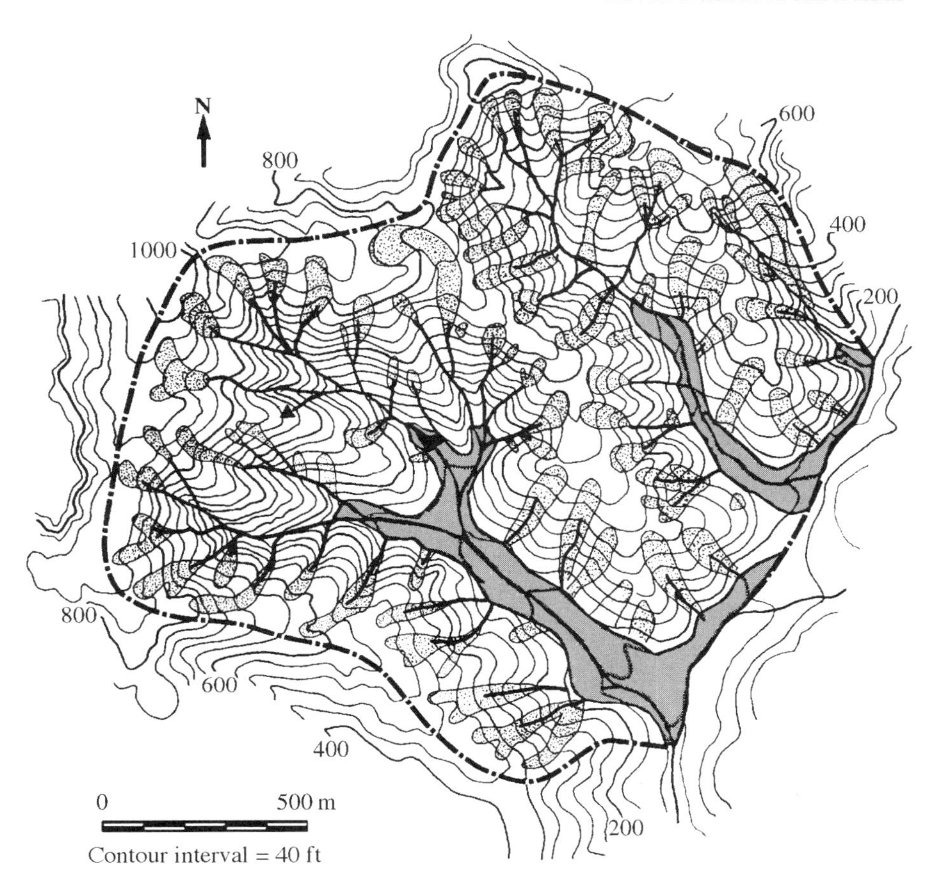

Figure 1.17. Map of the channel network (solid lines), areas of thick colluvium (stippled areas) and valley areas defined from floor deposits of colluvium and alluvium (shaded areas) in the Tennessee Valley study area, Marin County, CA [after Dietrich et al., 1993].

of the landscape. In fact, let $z(\mathbf{x}) = z(x, y)$ be the field of landscape elevations, one can measure regions of convergent topography by the condition that

$$\nabla^2 z(\mathbf{x}) \geq 0 \qquad (1.15)$$

(where $\nabla^2 = \partial/\partial x^2 + \partial/\partial y^2$ is the Laplace operator) and vice versa ($\nabla^2 z < 0$) for regions of divergent topography. An example of the spatial pattern of convergent, divergent, and planar topographic elements of larger (1.2 km^2) subcatchments in Figure 1.17 is shown in Figure 1.19 [after Dietrich et al., 1993].

While the classification is based on the sign of $\nabla^2 z$ for discriminating between convergent or divergent elements (Eq. (1.15)), elements are considered planar when $|\nabla^2 z| < \epsilon$, where ϵ is a suitable cutoff.

The accurate evaluation of the Laplacian in Eq. (1.15) is an interesting numerical problem. However, as discussed in Chapter 6, in assessing the convergent or divergent nature of the topography, one is interested in the *sign* of $\nabla^2 z$ and thus the accurate evaluation of its absolute value is of lesser importance.

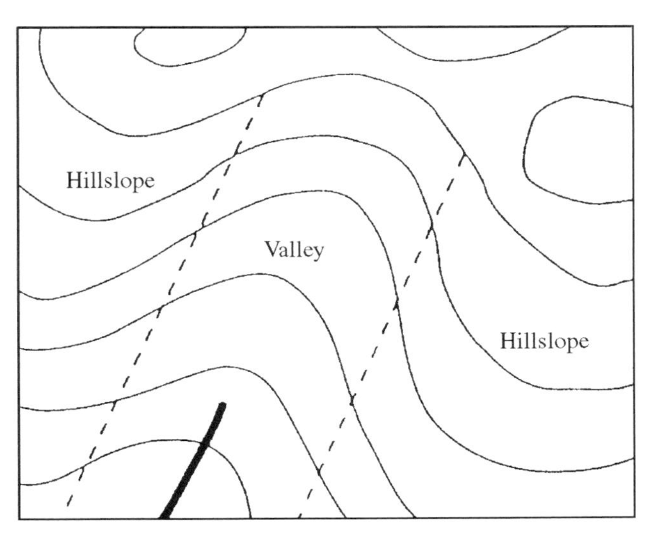

Figure 1.18. Schematic topographic map illustrating distinctions between hillslopes, valleys, and channels [after Montgomery and Foufoula-Georgiou, 1993].

Figure 1.20 shows the results of the analysis of concave sites (black) for the DEM of the Fella River shown in Figure 1.14. The related extraction of

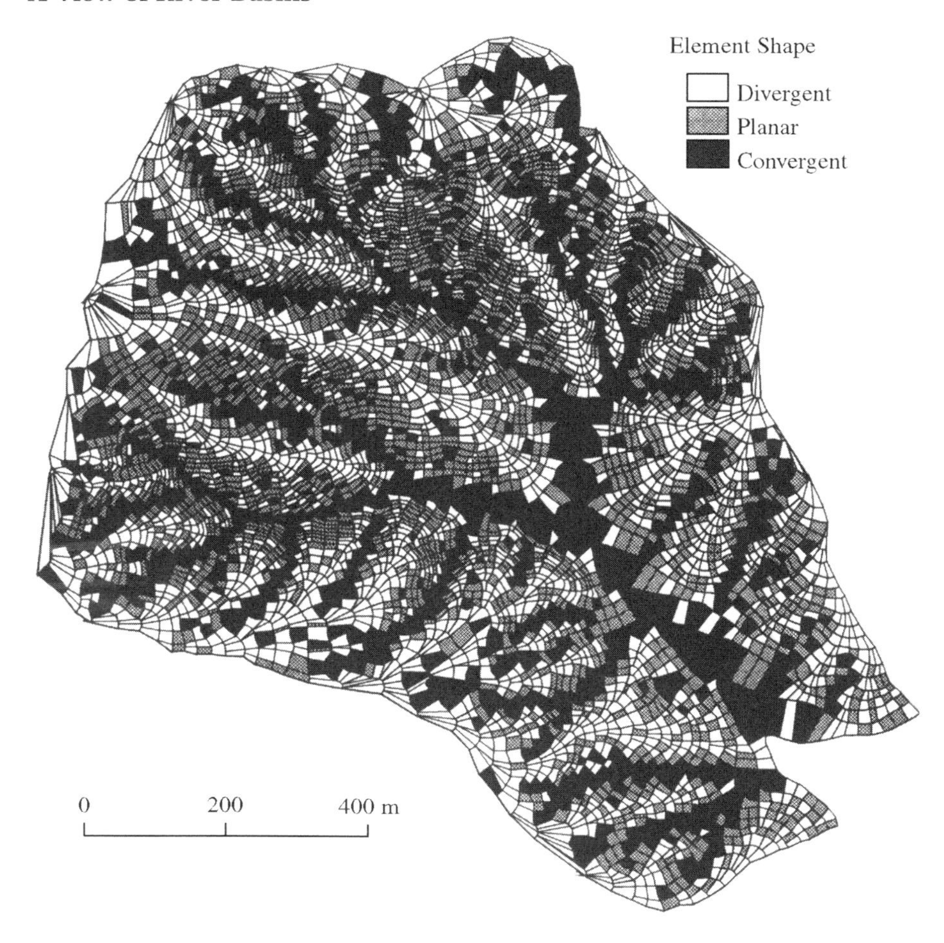

Figure 1.19. Topographic elements classified according to their curvature [after Dietrich et al., 1993].

the channel network based on the connection of concave pixels as dictated by flow in the steepest descent direction and thus based on a criterion different from that of a constant support area is shown in Figure 1.21.

Finally, we observe that DEM data based on regular grid points may be somewhat less flexible in handling divergent topographies because multiple flow directions need to be assigned to pixels placed in a divergent landscape (see, e.g., Cabral and Burges [1994]). In these cases a reformulation of Eq. (1.14) is needed because the total flow from the contributing area at a point (surrogated by the upslope area) should be distributed among more than one downslope neighbor.

1.2.10 Slope–Area Scaling

The first empirical characterization of elevation and slope properties was Horton's [1945] slope law:

$$R_{\nabla z} = \frac{\nabla z(\omega)}{\nabla z(\omega + 1)} \tag{1.16}$$

Here $R_{\nabla z}$ is the slope ratio and $\nabla z(\omega)$, $\nabla z(\omega + 1)$ are the mean slopes of streams of order $\omega, \omega + 1$, respectively. In practice, $R_{\nabla z}$ is obtained from a least squares fit to plots of log $\nabla z(\omega)$ versus ω and typically has values

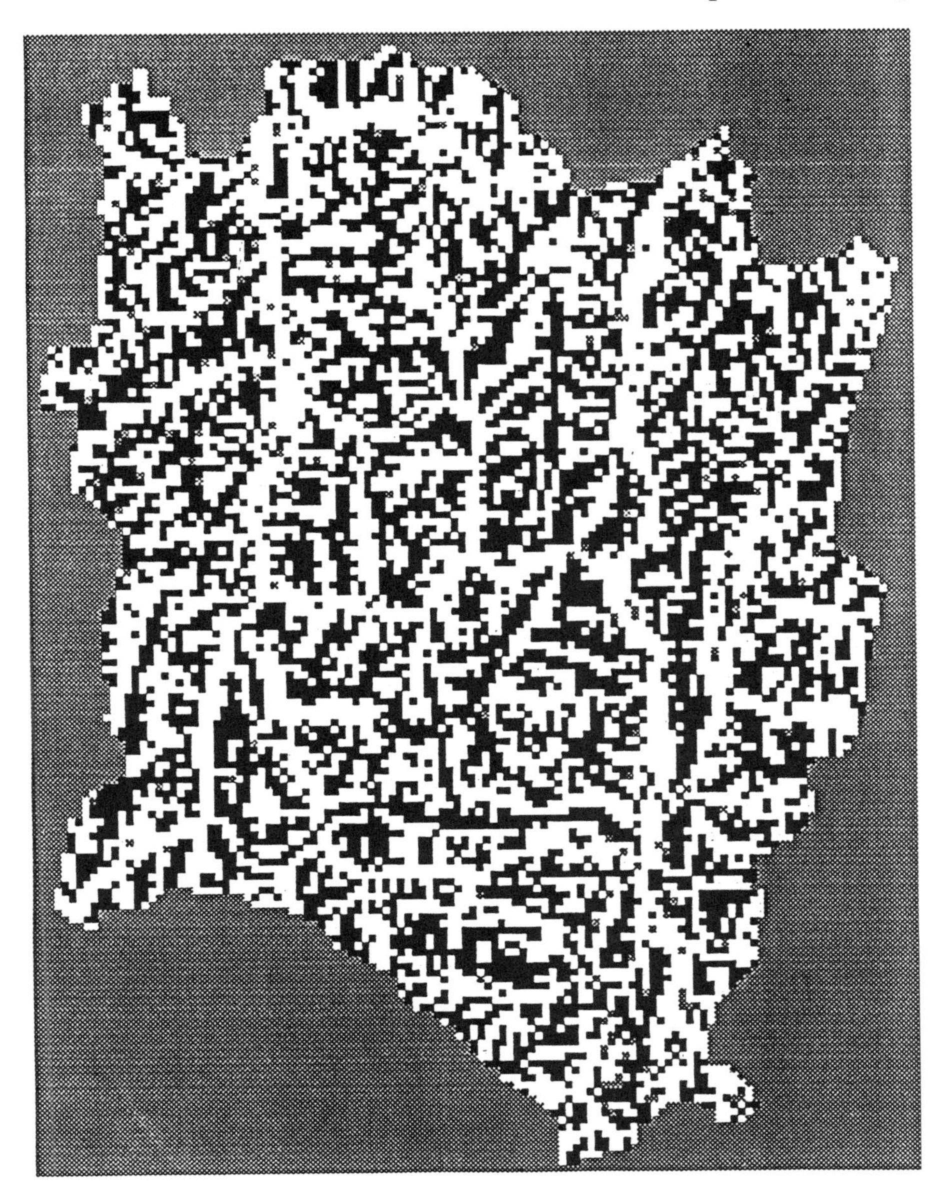

Figure 1.20. Concave (white) sites for the Fella River Basin in Figure 1.14.

between 1.5 and 3. We note that the above so-called slope law has not been as well documented in data analysis as the planar Horton's laws (of bifurcation, length, and area). One reason is that only recently have DEMs been widely available. It also seems that the scatter observed in matching Eq. (1.16) to real data is much larger than that present in the planar cases. The validity of Eq. (1.16) is indeed doubtful, but the topic is not a crucial one because practically all the analyses in the altitude space are now carried out on the basis of channel links and their corresponding magnitude.

Note that the slope law (Eq. (1.16)) implies exponential scaling of mean slope with order:

$$\nabla z(\omega) = (R_{\nabla z} \nabla z(1)) R_{\nabla z}^{-\omega} = (R_{\nabla z} \nabla z(1))\, e^{-\omega \log R_{\nabla z}} \tag{1.17}$$

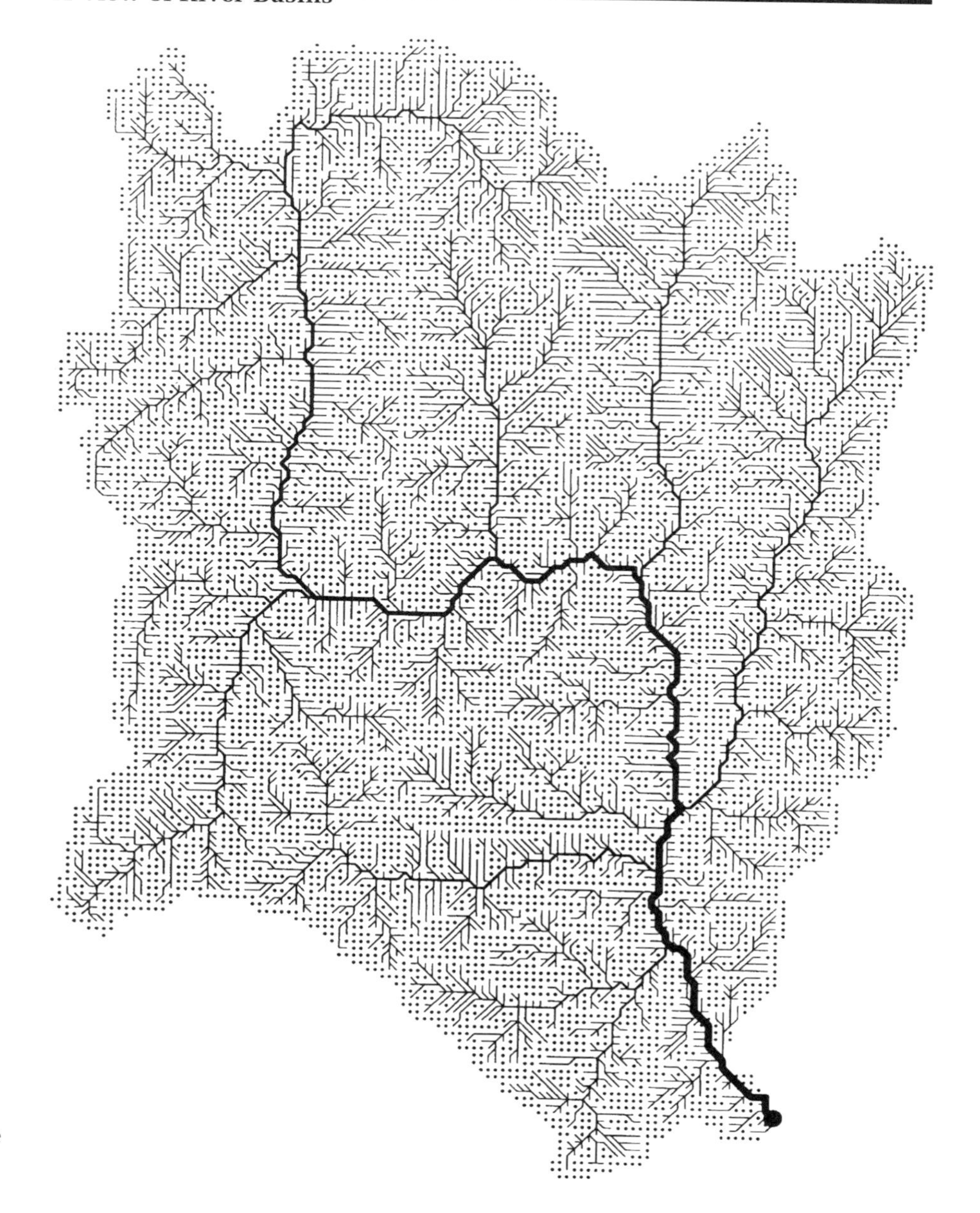

Figure 1.21. Channel network identified by concave sites for the Fella River Basin in Figure 1.20.

Power law relationships have been widely used in the hydrologic and geomorphologic literature to describe the scaling of hydraulic-geometric variables, as seen in Sections 1.2.3 to 1.2.6 [Wolman, 1955; Leopold et al., 1964; Leopold and Miller, 1956; Flint, 1973, 1974]. One additional relation is

$$\nabla z \propto Q^{-\alpha} \tag{1.18}$$

where, as usual, ∇z is slope, Q is discharge, and α is a suitable exponent.

As seen in Section 1.2.4, a common relationship between discharge and area [Flint, 1974; Leopold and Miller, 1956; Leopold et al., 1964] is

$$Q \propto A^{\beta} \tag{1.19}$$

where β is another coefficient. Combining these, Flint [1974] gives

$$\nabla z \propto A^{-\alpha\beta} \tag{1.20}$$

Grouping the coefficients leads to the fundamental scaling described by Gupta and Waymire [1989], that is

$$E[\nabla z(A)] \propto A^{-\theta} \tag{1.21}$$

where the expectation $E[\cdot]$ is used because Gupta and Waymire suggested that slope should be viewed as a random variable. This aspect, and the crucial connection of scaling properties with fractal geometry, will be discussed in Chapter 2.

In Flint's [1974] work the exponent θ takes on values with an average of 0.60 and a range of 0.37–0.83. A connection between the exponent θ and Horton's laws is also provided by Flint [1974]:

$$\theta = \frac{\log R_{\nabla z}}{\log R_A} \tag{1.22}$$

where R_A is defined in Eq. (1.3) as the area ratio. Eq. (1.22) can obtained by using the area law in Eq. (1.21). This connects the exponential scaling with the basin order described by Horton's slope law and the power law scaling with the area present in Eq. (1.21).

A connection with magnitude is seen using Shreve's [1967] relationship between area and magnitude, $A = a(2n - 1)$, where a is the mean area draining directly into a link and $2n - 1$ is the total number of links. In Eq. (1.21) this gives

$$E[\nabla z(A)] = E[\nabla z(n)] \propto (2n - 1)^{-\theta} \tag{1.23}$$

Flint [1974] actually estimated θ by a regression of $\log \nabla z$ versus $\log(2n-1)$. For large n the previous equation can be written as

$$E[\nabla z(n)] \propto n^{-\theta} \tag{1.24}$$

where the -1 term is neglected and the prefactor $2^{-\theta}$ is incorporated in the proportionality constant. This power law scaling of slope with magnitude is the form used by Gupta and Waymire [1989] in their scaling model for landscape elevations, discussed in Chapter 2.

By using Horton's [1945] bifurcation law, Eq. (1.1), we can estimate the number of first-order streams (i.e., the magnitude) in a subbasin of order ω as

$$n = R_B^{\omega-1} \tag{1.25}$$

Using this result in Eq. (1.24) gives

$$\theta = \frac{\log R_{\nabla z}}{\log R_B} \tag{1.26}$$

This connects the exponential scaling with basin order (Eq. (1.17)) and the power law scaling represented by Eq. (1.25). Eq. (1.25) basically shows that log n and order ω are equivalent measures of network scale, as suggested by Shreve [1967].

Tarboton et al. [1989a] observed that there is a discrepancy between Eqs. (1.26) and (1.22) in the fact that R_B cannot equal R_A in a given finite channel network unless a connectivity conjecture [Marani, Rigon and Rinaldo, 1991] is assumed. This discrepancy arises from the approximation $n \approx A$ and some mathematical inconsistencies [Tarboton et al., 1989a]. In fact, taking any two of Horton's planar ratios to hold exactly contradicts the third. This is of no practical importance, as these results are all empirical observations that do not hold exactly in nature. Their deeper theoretical implications in view of fractal geometries is dealt with in Chapter 2.

The slope scaling and concave longitudinal stream profiles are often attributed to notions of dynamic equilibrium [Langbein and Leopold, 1964]. There have been attempts to quantify and explain the scaling of slopes in terms of minimum work or minimum rate of entropy production principles [Langbein, 1964; Leopold and Langbein, 1962; Scheidegger, 1964; Yang, 1971b]. Although the mathematical validity of these results is in doubt [Kennedy, Richardson and Sutera, 1965], we agree that the widespread observation that slopes have a scaling structure should be explained through some general principle. The apparent self-similarity of landscapes [Mandelbrot, 1983] will also be related to these features (Chapters 4 to 6).

Note that the slope scaling as described by Horton's slope law or the power law, Eq. (1.24), cannot extend indefinitely, as that would imply infinite slopes at small scales. A break in scaling at scales smaller than about 0.6 km was observed by Mark and Aronson [1984]. Tarboton, Bras, and Rodriguez-Iturbe [1991] used a break in slope scaling to identify the fundamental scale approximately represented by the inverse of the drainage density of a network. This fundamental scale is viewed as one that sets a finite limit to what would become an infinite dissection of the landscape by the channel network. One may interpret this scale as one in which domination of sediment transport changes from fluvial processes to nonfluvial (i.e., hillslope) processes. Andrle and Abrahams [1989], working at scales of less than 10 m on talus slopes, also identify fundamental scales associated with the size of boulders. Montgomery and Dietrich [1988, 1989] and Dietrich et al. [1992, 1993] have also addressed the issue of channel–hillslope transition. It appears therefore that there are at least two fundamental scales in the landscape:

- particle/boulder size (0.001–10 m); and
- hillslope scale (10–10^3 m).

The precise value of these scales will vary for different landscapes and are dependent on local, geological, and climatic conditions. Between these scales there is a range of scaling characterized by diffusive degradation (see Chapter 6) and above the hillslope scale the landscape scaling is characterized by the stream slope scaling discussed here.

1.2.11 Empirical Evidence

The data presented here refer to the St. Joe River near Calder, Idaho. As described in Section 1.2.9, this data is part of an extensive empirical study over a wide range of areas (see Table 1.1) conducted by Tarboton et al. [1989a]. The results obtained are typical of all the areas studied.

Figure 1.16 illustrates the networks extracted from the DEMs of the St. Joe River.

Figure 1.22 shows link slopes plotted against magnitude, $2n - 1$, and area for these two data sets. In these figures n, $2n - 1$, and area A are for practical purposes interchangeable scaling indices. Area was thus interpreted as being the fundamental scaling index, with n and $2n - 1$ good surrogate measures. In the remainder of this section, n or area A are used interchangeably as our scaling index.

In Figure 1.22 there is considerable scatter in the individual link slopes. However, by ranking the links according to the scale index (A, n, or $2n-1$), the links are grouped into bins containing at least twenty links that cover a narrow range of scale index. The group sample means, which are plotted as circles in Figure 1.22, show power law scaling approximately proportional to $n^{-0.5}$.

The above scaling law is determined once the network properties are extracted from the DEM with a relatively large support area, that is, overcoming the often difficult issue of where channels begin. In the study of this issue, examining DEMs at the highest possible resolution is of great interest. Even though it has been inferred that the correct identification of channel initiation may generally require finer data sets [e.g., Dietrich et al., 1993], the information processed by DEMs is of great importance for a complete understanding of the dominant scales acting in landscape-forming processes.

Figure 1.23 shows a plot of slope versus area, where a small support area was used to extract the network. Area is the total contributing area measured at the downstream end of each link, and slope is mean link slope defined as elevation drop divided by link length. The individual points in Figure 1.23 exhibit considerable scatter, indicating that link slope is highly variable. As in the previous example of Figure 1.22, when many links with similar area are grouped together and averaged (circles in Figure 1.23), the mean slope is seen to follow a fairly smooth trend. The line to the right corresponds to the scaling described earlier. The appearance of a switch point gives the scale at which this scaling breaks and is related to the support area, A_t, that should be used to extract channel networks from DEMs.

Furthermore, Tarboton et al. [1991] suggest using the area where the breakpoint in slope occurs to infer the drainage density that corresponds to the basic scale of the landscape. From dimensional analysis one expects the drainage density, $\mathcal{D}$, defined in Eq. (1.4), to be proportional to such support area via

$$\mathcal{D} \propto \frac{1}{\sqrt{A_t}} \tag{1.27}$$

The issue of the interplay of channel and hillslope processes will be

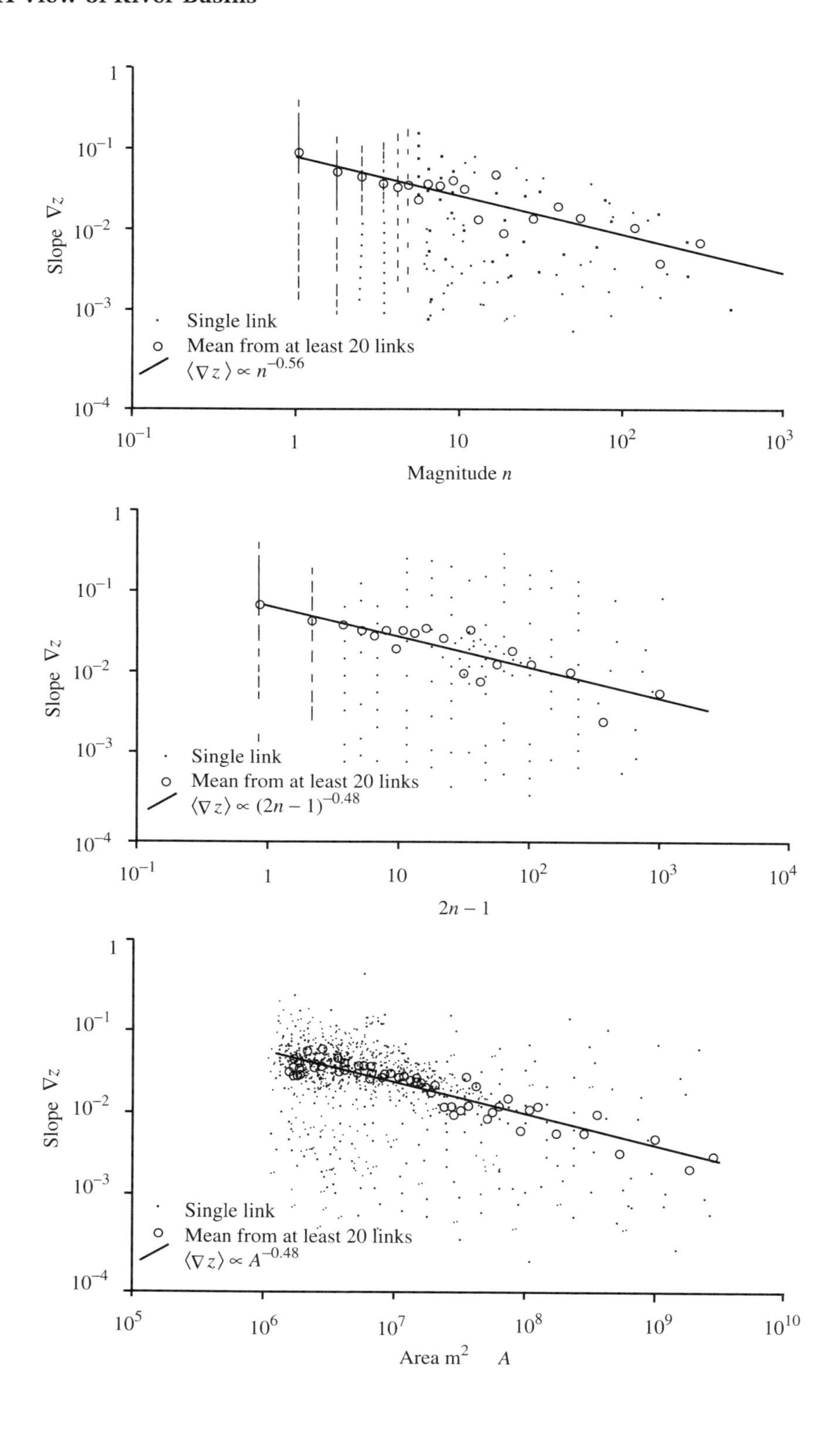

Figure 1.22. Link slopes for the St. Joe River, Idaho [after Tarboton et al., 1989a].

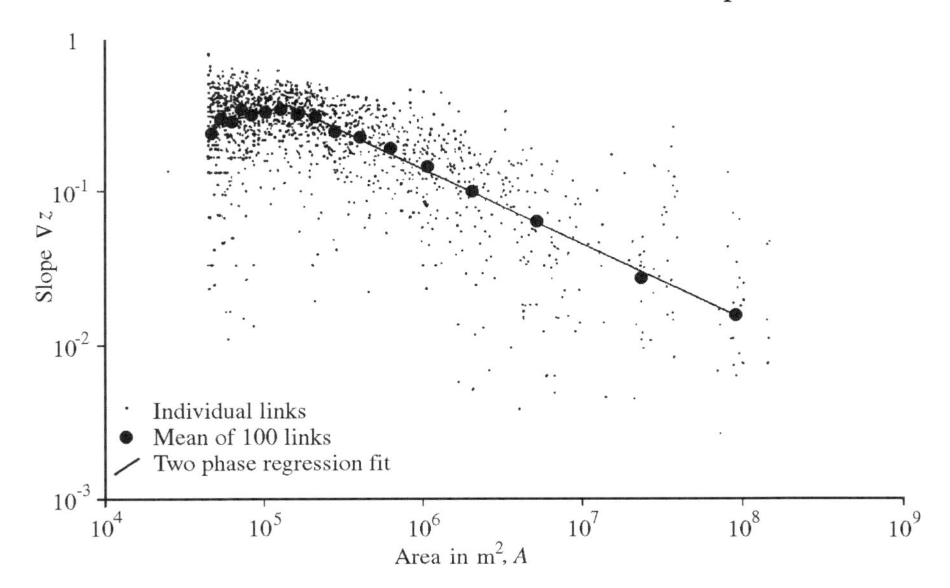

Figure 1.23. Link slopes with support area of 50 pixels used to extract the network. Data are from Big Creek, Idaho, a tributary of the St. Joe River Basin [after Tarboton et al., 1989a].

discussed in Section 1.2.12 through high-resolution DEM data and in Chapters 5 and 6 when the statistics of the type used in Figures 1.22 and 1.23 will be analyzed in simulated landscapes. Additional examples of analyses of the break points in the slope versus area diagrams are shown in Figure 1.24. From the results of Tarboton et al. [1989b] we observe that the break point occurs at different supporting areas. We will return to this point in Section 2.1.12 and in Chapter 6. Also, the interpretation of the behavior of the slope–area relationship before the establishment of the fluvial scaling poses observational and theoretical problems, because there does not seem to exist a common behavior in the DEMs at small contributing areas. Willgoose, Bras, and Rodriguez-Iturbe [1989] suggested that in a slope–area analysis of the kind shown in Figure 1.24, hillslopes should exhibit a systematic *increase* in contributing area with increasing slope (a convex profile) because hillslope form is presumably controlled by slope-dependent transport processes. Because channeled portions of the landscape show systematic *decrease* in slope with increasing drainage area, the inflection point of the curve merging the two regimes could allow for an objective definition of the drainage density. Nevertheless, as discussed in Section 2.1.12, in low-resolution DEM data the inflection point may be an artifact [Dietrich et al., 1993].

It is of great interest to examine in depth the scatter exhibited by slope–area diagrams. In Chapter 2 we will examine the fractal implications of the scaling structure exhibited by this scatter. Figure 1.25 shows the variance among the slopes of the links contained in the different groups, each group being characterized by a contributing area. This procedure is often referred to as binning, meaning that significant sample sizes are obtained by assigning a value of contributing area, say A_i, to all values found for contributing areas in the range $(A_i - \Delta A, A_i + \Delta A)$.

The group sample variances $Var[\nabla z(A)]$ show power-law scaling proportional to $A^{-\theta'}$, where θ' is observed in the range 0.53–0.78. It is of interest to observe that indeed the sample variance scales with area and that the exponent of the scaling relationship varies from basin to basin. This fact

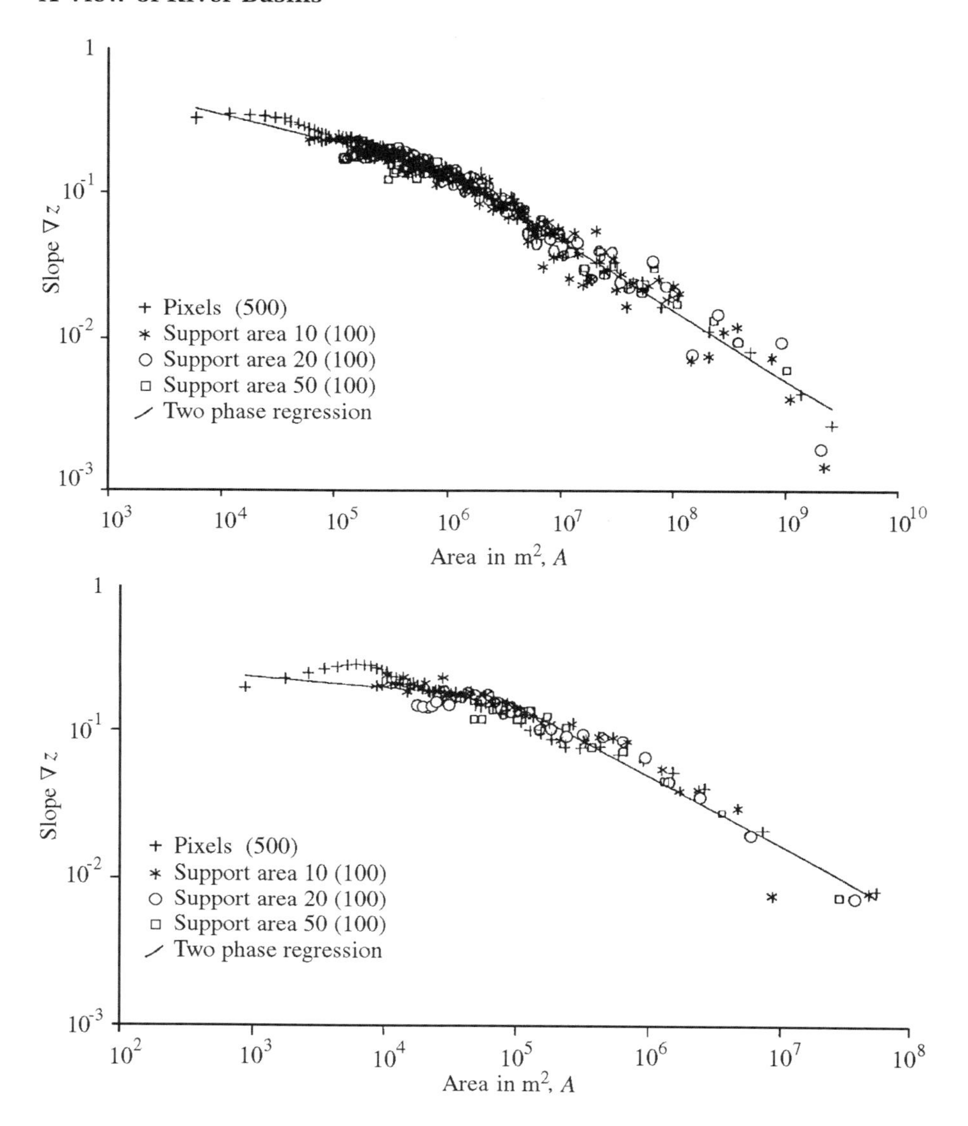

Figure 1.24. Mean slope versus area with different DEM support areas, in the St. Joe basin (St. Joe River, Minnesota and Idaho; square grid 30 × 30 m^2; 380 × 127 grid) [after Tarboton et al., 1989b].

has important theoretical implications because it impairs *simple scaling* models of slopes versus areas, suggesting that instead the behavior is *multiscaling* because different moments scale with different laws. This will be extensively discussed in other parts of this book.

Higher moments of the samples are difficult to estimate due to sample size effects.

Other experimental data come from the statistics of drops in elevation. Tarboton et al. [1989a,b] examined the properties of stream drops and concluded that slopes are to be regarded as the fundamental scaling parameter to characterize the three-dimensional structure of river basins.

1.2.12 Where Do Channels Begin?

As mentioned before, an important issue in the study of the fractal nature of river basins concerns the range of scales allowed for the evolution of

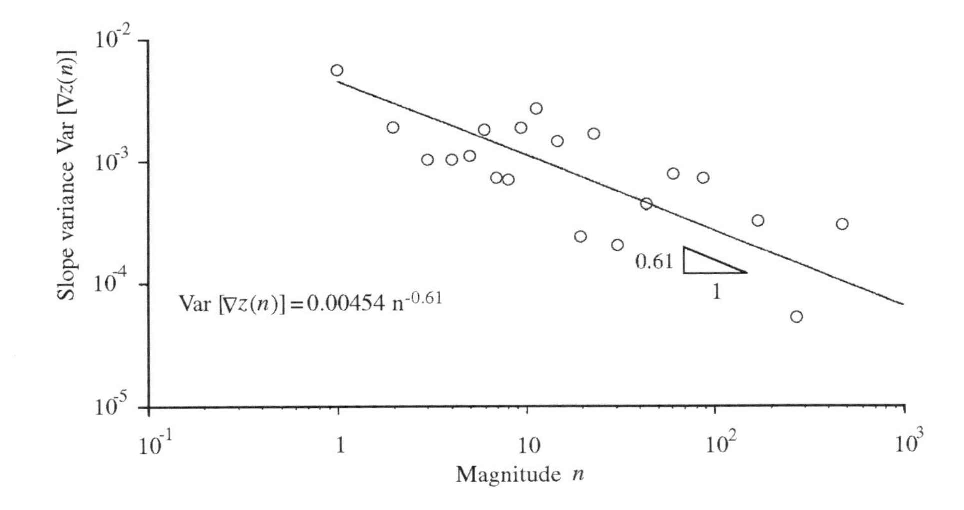

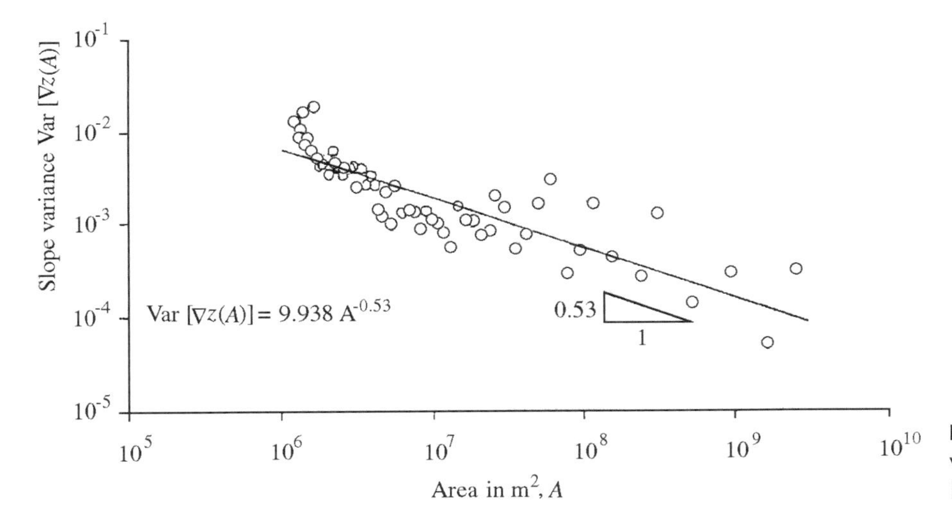

Figure 1.25. Link slope variances, St. Joe River, Idaho [after Tarboton et al., 1989b].

the different natural processes contributing to the formation of the river basin and its embedded network. This requires an understanding of the mechanisms of channel initiation, that is, of the interplay of channel and hillslope evolution.

Figure 1.26, from Montgomery and Dietrich [1992], illustrates the effects of depicting an area of similar topography at different scales. Figures 1.26(a) and 1.26(b) show adjacent basins in the Oregon coast range; Figure 1.26(b) covers an area four times larger than Figure 1.26(a) and has twice the contour interval. The lower two maps illustrate very different landscapes, and detailed mapping was done to resolve the finest scale valleys that determine the extent of landscape dissection. Figure 1.26(c) shows a portion of badlands at Perth Amboy, New Jersey (scale bar represents 2 m, contour interval 0.3 m); Figure 1.26(d) shows a portion of San Gabriel Mountains, southern California (scale bar is 100 m, contour interval is 15 m).

In Figure 1.26 all four maps suggest a limit to the landscape dissection defined by the size of the hillslope separating valleys. Montgomery and

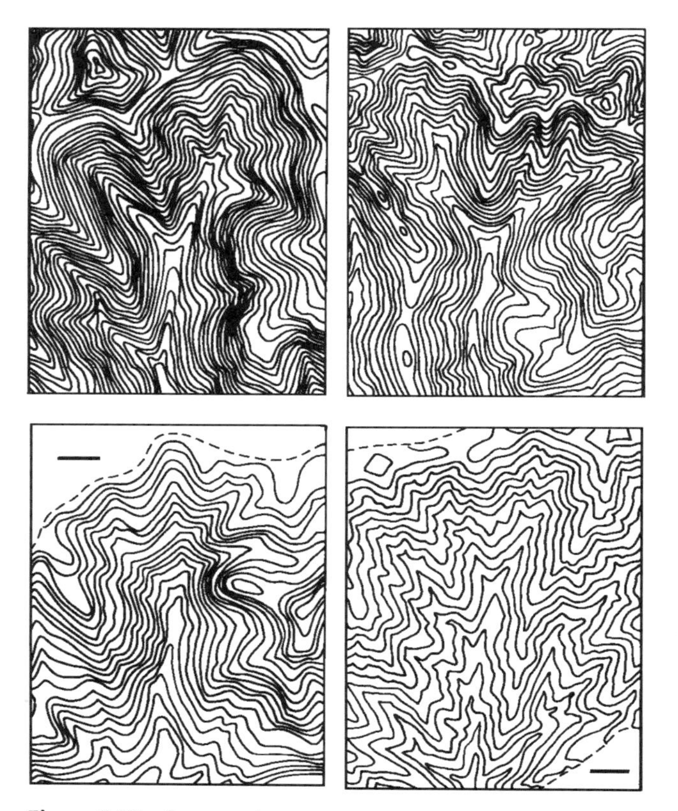

Figure 1.26. Four contour maps at different scales [after Montgomery and Dietrich, 1992].

Dietrich [1992, 1994] noted that this apparent limit only corresponds to the extent of valley dissection definable in the field for the cases of Figures 1.26(c) and 1.26(d). Clearly, finer and finer scales of dissection are found upon closer and closer inspection. Nevertheless, landscape dissection into distinct valleys is limited by a threshold of channelization that sets a finite scale to the landscape. However, it is also true that within a large range of spatial scales without a scale bar it is almost impossible to determine even the approximate scale of a topographic map. The problem is complicated by the fact that in natural river basins, channel initiation is heavily influenced by spatial heterogeneity, which reflects the variations in the exposed lithology of the soil mantle, in the vegetation cover, and in the prevailing erosional mechanism. Thus vegetation, on colonizing an initially bare soil surface, increases by orders of magnitude the erosion threshold by overland flow. A synthesis of the possible mechanisms for channel initiation has been recently given by Dietrich et al. [1992, 1993] and is briefly described here. The complexity of the various intertwined processes will be evident.

Threshold theories for land surface morphology are an active area of geomorphological research. Field studies [Montgomery and Dietrich, 1992] suggest that an empirically defined topographic threshold associated with channel head locations defines the boundary between essentially smooth and undissected slopes and the valley bottoms to which they drain. Figure 1.27 (from Montgomery and Dietrich [1992]) shows the experimental relationship of drainage area versus local slope for channel heads, unchanneled valleys, and low-order channel networks from different study areas. Local slope was measured in the field and drainage area was determined from topographic base maps.

Figure 1.27 clearly shows that channel heads are defined through a topographically defined (i.e., based on total contributing area and slope) threshold between channeled and unchanneled regions of the landscape. The studies of Montgomery and Dietrich [1992] show that threshold-based channel initiation models predict that this central tendency reflects the general environmental controls on channel initiation, typically climate and vegetation, and the form of this transition reflects the different channel initiation processes involved. The variability produced by these controls in the threshold is shown in the shaded region of Figure 1.27(d).

Thresholds for channelization can be thought of as the by-product of subsurface saturation, slope instability through shallow landsliding, and/or erosion by overland flow [see, e.g., Dietrich et al., 1993]. In the case of erosion by overland flow, we will show (Section 1.4.3) that the actual shear stress at the ith site, say τ_i, is proportional to the contributing area and the local slope at i, say $\tau_i \propto \sqrt{A_i}|\nabla z_i|$. The exceedence of a critical stress ($\tau_i > \tau_c$) thus leads to channelization. The contributing area to support a channel head generated by the above mechanism ($A_i = A_t$) can be cast in

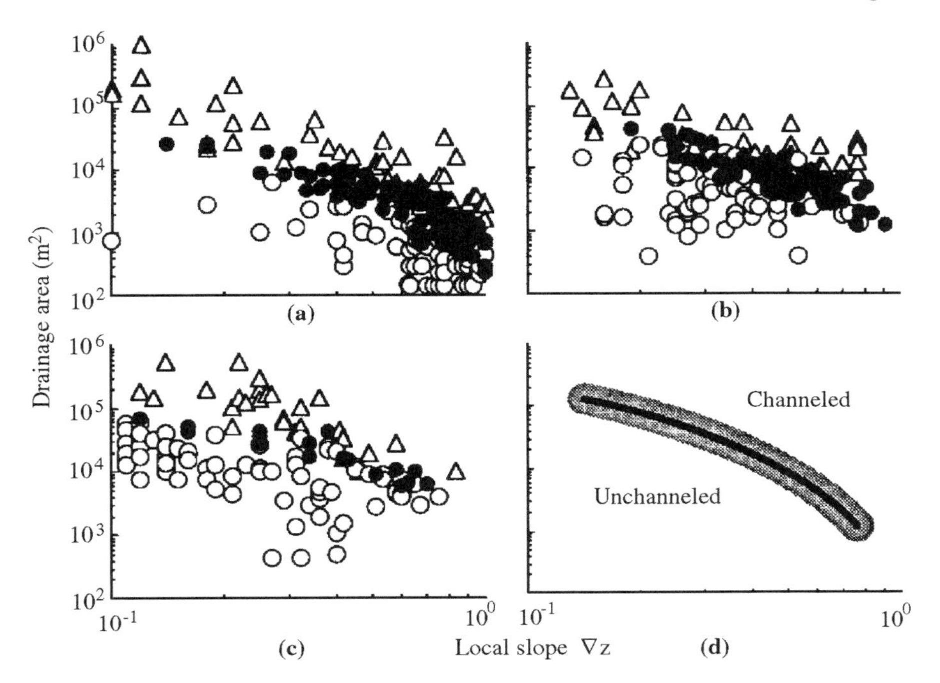

Figure 1.27. Drainage area versus local slope from field data. Observed values of drainage areas versus local slope for channel heads (solid dots), unchannelled valleys (dots), and low-order channel networks (triangles) from study areas in (a) coastal Oregon, (b) northern California, and (c) southern California; (d) shows a schematic of the transition between unchanneled and channeled regions [after Montgomery and Dietrich, 1992].

the form $A_t = C/|\nabla z|^2$, where $C = f(\tau_c)$. Thus smaller drainage areas are needed to initiate a channel on steeper slopes. Later in this section we will show that regardless of the interplay of the different actions that control channelization, slope-dependent area thresholds A_t nicely define the real extent and structure of the channel networks from topographic data.

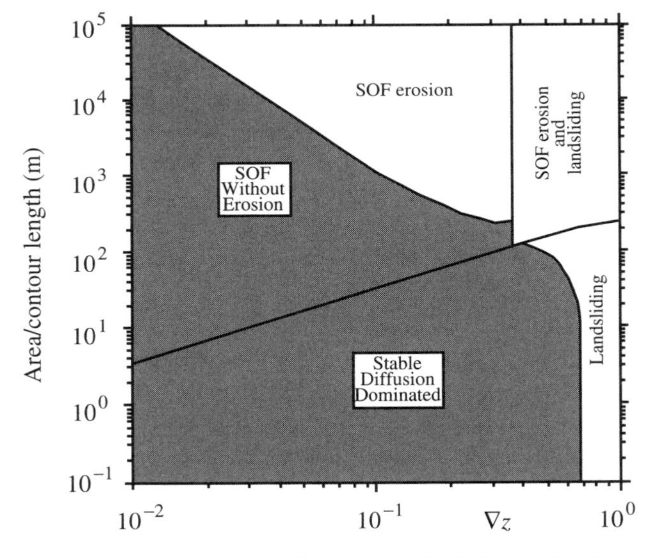

Figure 1.28. Definition of regions for applicability of different channel initiation mechanisms [after Dietrich, Wilson, and Reneau, 1986].

The field observations described by Dietrich and coworkers suggest that any reasonable model for channel initiation at the basin scale should include some degree of spatial heterogeneity. Examination of the different types of sediment transport mechanisms that could be responsible for dominating landscape forms [Dietrich et al., 1992] leads to the conclusion that channel initiation is basically a random process that depends on slope and supporting area. A synthesis of the possible mechanisms for channel initiation has been recently given by Dietrich et al. [1993] (Figure 1.28), where regions of prevailing landsliding, subsurface saturation without erosion, and diffusion-dominated and combined transport mechanisms are observed (note that in Figure 1.28 SOF erosion means erosion by saturation overland flow). Figure 1.28 uses the parameter T/q in characterizations of regions where T, m^2/day, is soil transmissivity and q, m/day, is subsurface runoff per unit area (in the case of Figure 1.28 this parameter is fixed at $T/q = 350$ m). The ratio a/b indicates the total contributing upslope area per unit contour length. In Figure 1.28 the subsurface runoff per unit area is taken as $q = 0.05$ m/day, critical shear stress is equal to

Figure 1.29. Gently sloping landscape mostly made up by convex hillslopes. Ephemeral migrations of the channel heads create the conditions for the disappearance of channels downstream of a source [courtesy of W. E. Dietrich].

160 dyne/cm^2, and the internal friction angle of the material constituting the slope is $\phi \sim 35^o$. The complexity of the various intertwined processes is evident when one observes the shape of the shaded area that implies stable slopes, that is, no channels. Figures 1.29 and 1.30 show two examples of the complex interplay among the different processes that lead to fundamentally different landscapes.

The above studies are important for the interpretation of channel initiation from DEM data. In fact:

- the logic behind a constant critical support area comes from the early observation by Gilbert [1909] that slope-dependent sediment transport on hillslopes gives rise to convex slopes, whereas discharge- and slope-dependent sediment transport in channels gives rise to concave slope profiles. Nevertheless, the transition from convex to concave slopes commonly coincides with the transition from convergent to divergent topography, suggesting that the transition from one dominant transport mode to another is more appropriate in general for representing controls on valley development than on channel initiation [Dietrich and Dunne, 1993];
- in the case of soil-mantled landscapes, field observations generally support the association of channel heads with a change in sediment transport processes at a critical contributing area A_t. The change essentially distinguishes slope-dependent processes upslope of the channel head and discharge- and slope-dependent processes downslope of the channel head. The processes

Figure 1.30. Landscape with hillslope–channel transitions characterized by approximately linear slopes and a very efficient erosion mechanism that pushes the channel head almost up to the divide unless for the effects of vegetation [courtesy of W. E. Dietrich].

defining the critical contributing area A_t may include overland flow, seepage, piping, and landsliding, thus resulting in spatially heterogeneous characterizations.

Observational evidence supports the above conclusions. Figure 1.31 shows the field-mapped channel network and the distribution of the parameter $A|\nabla z|^2$ for a study area in northern California [Dietrich et al., 1993]. Note that the ratio a/b is required in place of A in contour-based DEMs indicating total area per unit contour length. A strong correspondence is observed between the extent of the mapped network and the networks obtained through thresholds in the range $25 \leq A|\nabla z|^2 \leq 200$. Not a single channel extends below this range, strongly suggesting the validity of topographic thresholds of this type for automatic determination of channel initiation. According to Dietrich et al. [1988, 1992, 1993], the related variations implied for the critical shear stress τ_c are relatively small. This observation could support models of network evolution where the threshold for landscape channelization is considered spatially constant. Chapter 6 will study models of this type as well as models where τ_c is considered to be a random field with a spatial correlation structure. We will show that the resulting landscapes from both of the foregoing cases have many common statistical characteristics but that there are also some important differences induced by spatial heterogeneity.

We notice that in the interpretation of DEMs some confusion has sur-

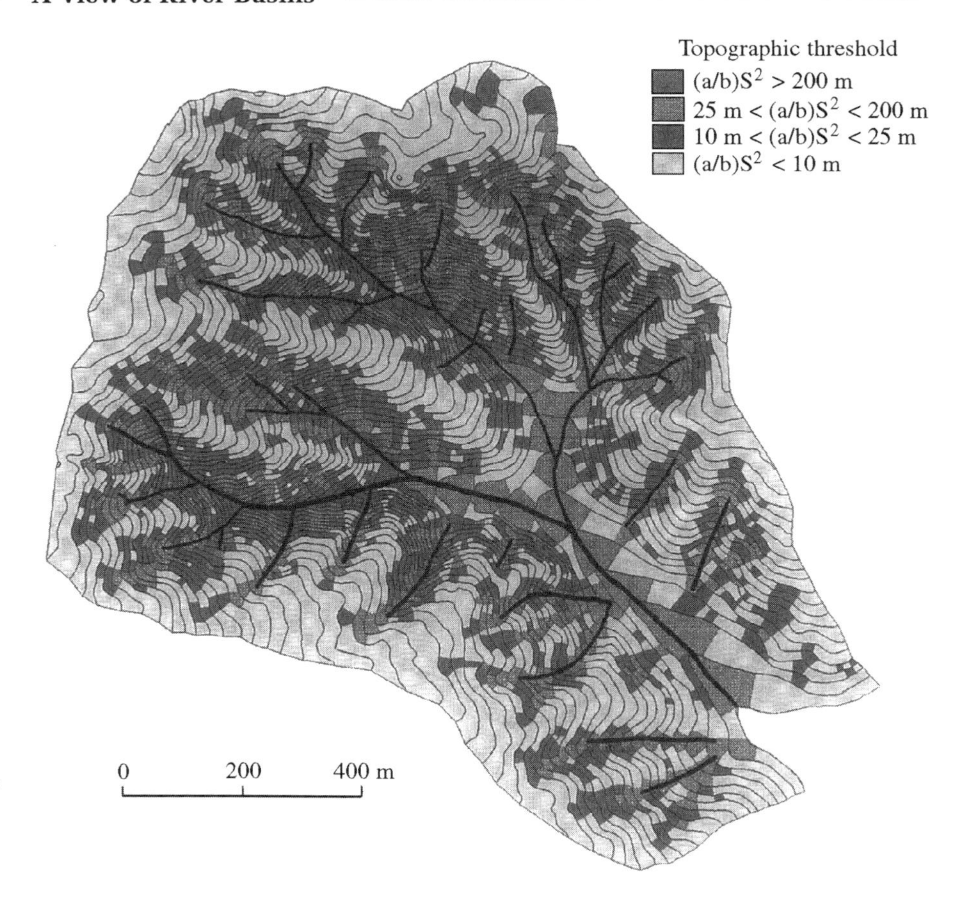

Figure 1.31. Comparison of observed channel network with topographic thresholds defined by $A|\nabla z|^2$. Here $a/b = A$ and $S = \nabla z$ [after Dietrich et al., 1993].

faced in the literature from an unclear distinction between the stream (or channel) network, defined by streams with well-defined banks and sources, and the valley (or drainage) network, which is defined on the basis of basin morphometry [Howard, 1994]. Of course, both networks are related because streams occupy most drainage networks, but channels may expand or contract as a result of short-term climate or land-use changes. Indeed, the factors determining the location and density of valleys and stream channels in drainage basins have a variety of implications for understanding the origin, scale, morphology, and hydrologic response of a basin. Moreover, the expansions and contractions of the channel network, and thus *the interplay of the channel heads with the unchanneled valleys,* play an important role for reading the signatures of past climates, a role we will study in Chapter 6.

Recently, methods for identifying the hillslope scales and important inferences on the crossovers in the scaling relationship of slopes and areas from DEM data have been examined by Montgomery and Foufoula-Georgiou [1993]. As seen earlier in this section, a slope-dependent critical support area is a sound concept for extracting the channelized portion of a DEM. The question is how to detect the threshold *directly* from DEM observations. Montgomery and Foufoula-Georgiou [1993] examined the various channel initiation mechanisms described in this section to test their appropriateness for locating channel network sources. The exami-

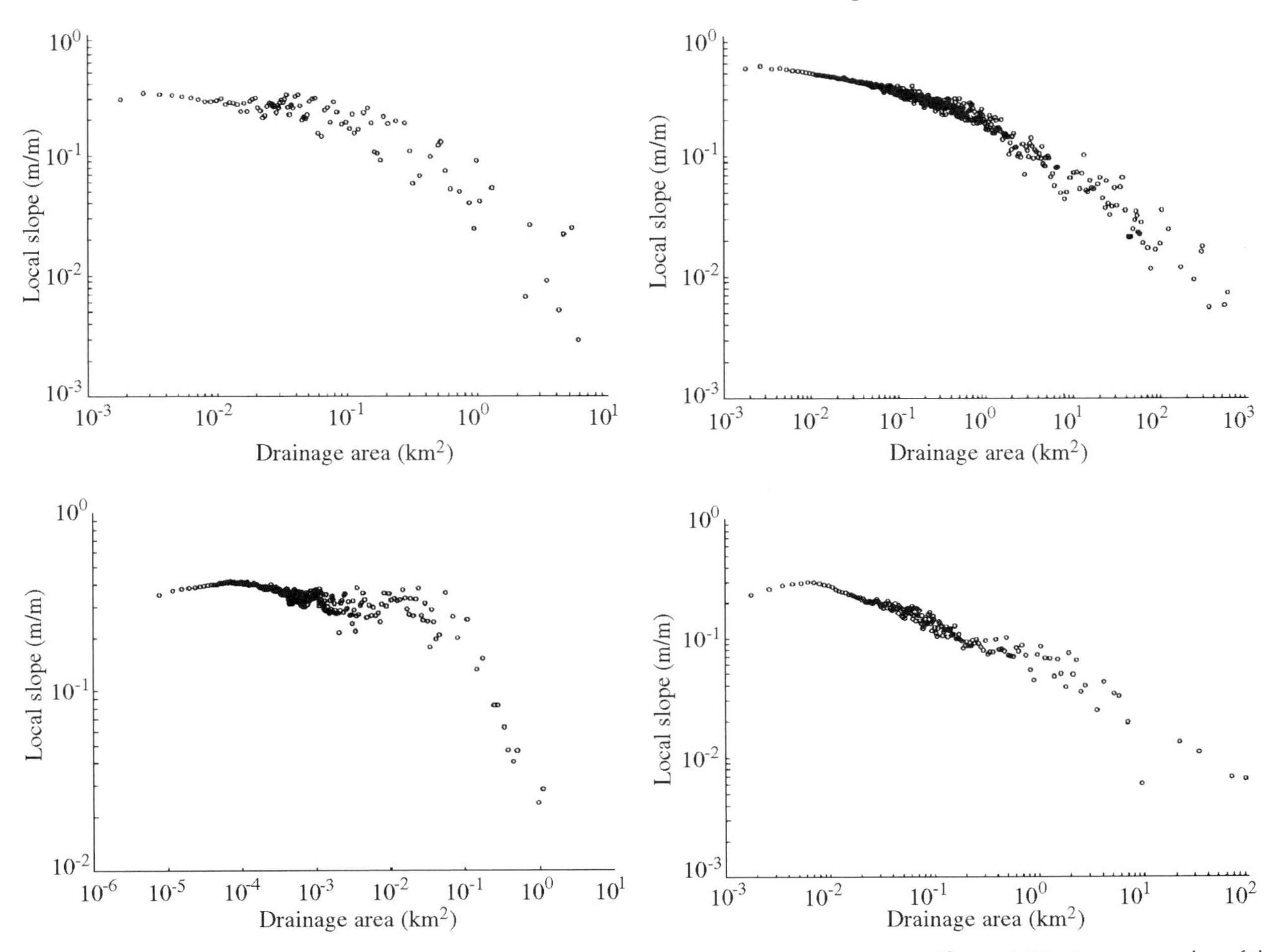

Figure 1.32. Four examples of the relationship between drainage area and local slope for averaged data [after Montgomery and Foufoula-Georgiou, 1993].

nation was carried out using field data and high-resolution DEMs. Some of their results are shown in Figure 1.32, where the source data are (a) USGS 5.7' DEM data for the Tennessee Valley area, Marin County, California; (b) 2-m-resolution DEM data from field survey and low-altitude stereo aerial topography of the area shown in Figure 1.32(a), (c) a USGS 30-m DEM of South Fork Smith River, California, and (d) the same as (c) but for Schoharie Creek, New York.

It appears that two transitions may exist in the drainage area–slope relationship:

- a reversal at very small drainage areas;
- an inflection at local slopes of about 0.2 to 0.3 in the cases given in Figure 1.32.

This helps to understand the limitations of the early observations by Tarboton et al. [1989a], where a constant critical support area was used to extract channel networks from DEMs. The critical support area, as briefly described in Section 1.2.9, was deduced from the inflection point found in the slope–area relationship. However, the method of directing the full flow from one pixel to one of its eight neighbors (e.g., steepest descent criterion) does not allow for the representation of flow in divergent topographies, which matters at small scales [e.g., Cabral and Burges, 1994]. Thus this method appropriately describes the convergent topography of the valleys but not the divergent features of hillslopes. The representation of hillslopes with linear, rather than divergent, flow allows for an interesting check on hillslope scales [Montgomery and Foufoula-Georgiou, 1993]. The inflection points in the link slope plots of Tarboton et al. [1989a,b] (see, e.g., Figure 1.24) occur at drainage areas of the order of 10^5 to 10^6 m^2. For a pixel size of 30 m this implies hillslope lengths of the order of 3.3 to 33 km. Field surveys indicate this value is far too large because hillslope lengths are typically of the order of hundreds of meters or less, approaching O(1) km only for flat semiarid landscapes. Thus the inflection that one can infer from low-resolution DEM data reflects something other than hillslope–valley transitions.

From the results in Figure 1.32 we see that the averaged local slopes plotted versus drainage area exhibit an inflection point at a drainage area of about 0.1 km^2 and a reversal at a drainage area of about 3×10^{-3} km^2. The hillslope length implied from the inflection is about 6 km. The hillslope length implied by the reversal is of the order of 100 m. The field survey indicated that the latter is a reasonable estimate for the hillslope length in the region.

We have already observed that the transition to the alluvial channel regime is marked by a slope-dependent area threshold. Montgomery and Dietrich [1989] report that for thresholds of the type $A_t > C/|\nabla z|^2$ the constant C could be of the type $C \approx 10^6/p$ m^2, where p is the mean annual rainfall expressed in millimeters. The proper identification of the network from DEMs depends on the value of C that controls the spatially varying A_t. Montgomery and Foufoula-Georgiou [1993] propose using as C the smallest value that does not result in a significant number of small channels issuing from the sides of the network. This pattern is shown in Figure 1.33 and is called feathering. Notice the feathering along low-order channels in the lower portions of the catchment in Figures 1.33(c) and 1.33(d), identified by values of $A_t|\nabla z|^2 = 4{,}000$ and 2,000, respectively.

The above concepts have also been applied to the identification of the channel network of the Fella River Basin shown in Figures 1.13 and 1.14. We have extracted the channel network by defining the arbitrary channelized site i through the simultaneous occurrence of

- convergent topography, that is, $\nabla^2 z_i \geq 0$; and
- the exceedence of the critical threshold, that is, $\nabla z_i \sqrt{A_i} \geq \tau_c$,

where, as usual, z_i is the elevation at the ith site and A_i is its total contributing area. Figures 1.34 and 1.35 show the results of the application of the above criterion to the Fella River Basin with values of $\tau_c = 100$ and 400 in pixel units.

Notice that drainage directions in unchanneled sites are represented by dotted lines. Notwithstanding the rather large range of values of τ_c, the network is fairly well defined. Also notice that, as opposed to the case of Figure 1.33, feathering is less evident owing to the coarse scale (150×150 m^2) of the basic DEM.

Other criteria exist for the determination of the channelized part of a

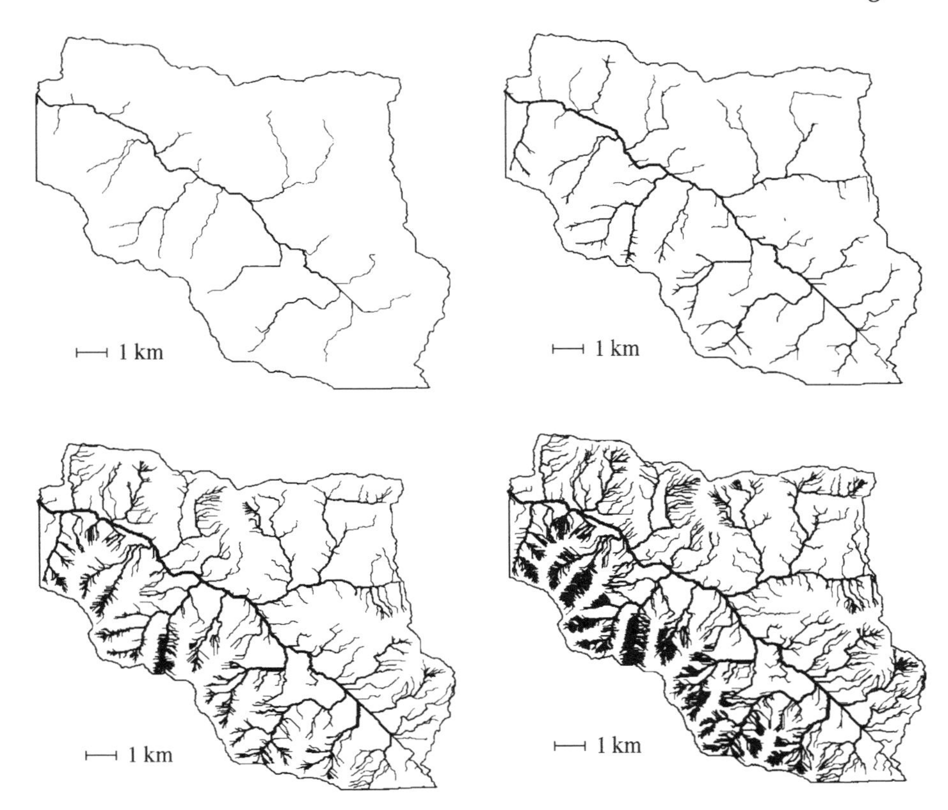

Figure 1.33. Maps of Schoharie Creek channel network defined using $A_t|\nabla z|^2 =$: (a) 64.000 m^2; (b) 16.000 m^2; (c) 4.000 m^2; (d) 2.000 m^2 [after Montgomery and Foufoula-Georgiou, 1993].

DEM. Howard [1994] examined in detail several criteria and found that a criterion based on a threshold, say D_t, for a generalized curvature also works satisfactorily. The criterion is based on the values of the gradient divergence, that is, whenever

$$\nabla^2 z(\mathbf{x}, t) \leq D_t \tag{1.28}$$

lattice sites are defined to have channels extending downgradient from the site to the adjacent downstream site. When one plots drainage density $\mathcal{D}$ that would be defined as a function of D_t for any DEM, a typical shoulder is exhibited at which the rate of change of drainage density with D_t is small. Howard [1994] suggests selecting values of D_t within the range of the shoulder such that the resulting network is not strongly sensitive to modest variations in the defining parameter D_t.

Indeed, Howard [1994] observed that the assumption of a critical D_t defining valley heads produces a relationship between contributing area and slope gradient in first-order valleys similar to that observed by Montgomery and Dietrich [1988, 1992], described above. Thus we assume that a reasonable description of channelized valley sites can be obtained by any one of the criteria defined above.

In Chapter 6 we will return to the various mechanisms of channel initiation in the search for an explanation for the macroscopic characters of the three-dimensional field of elevations of river basins.

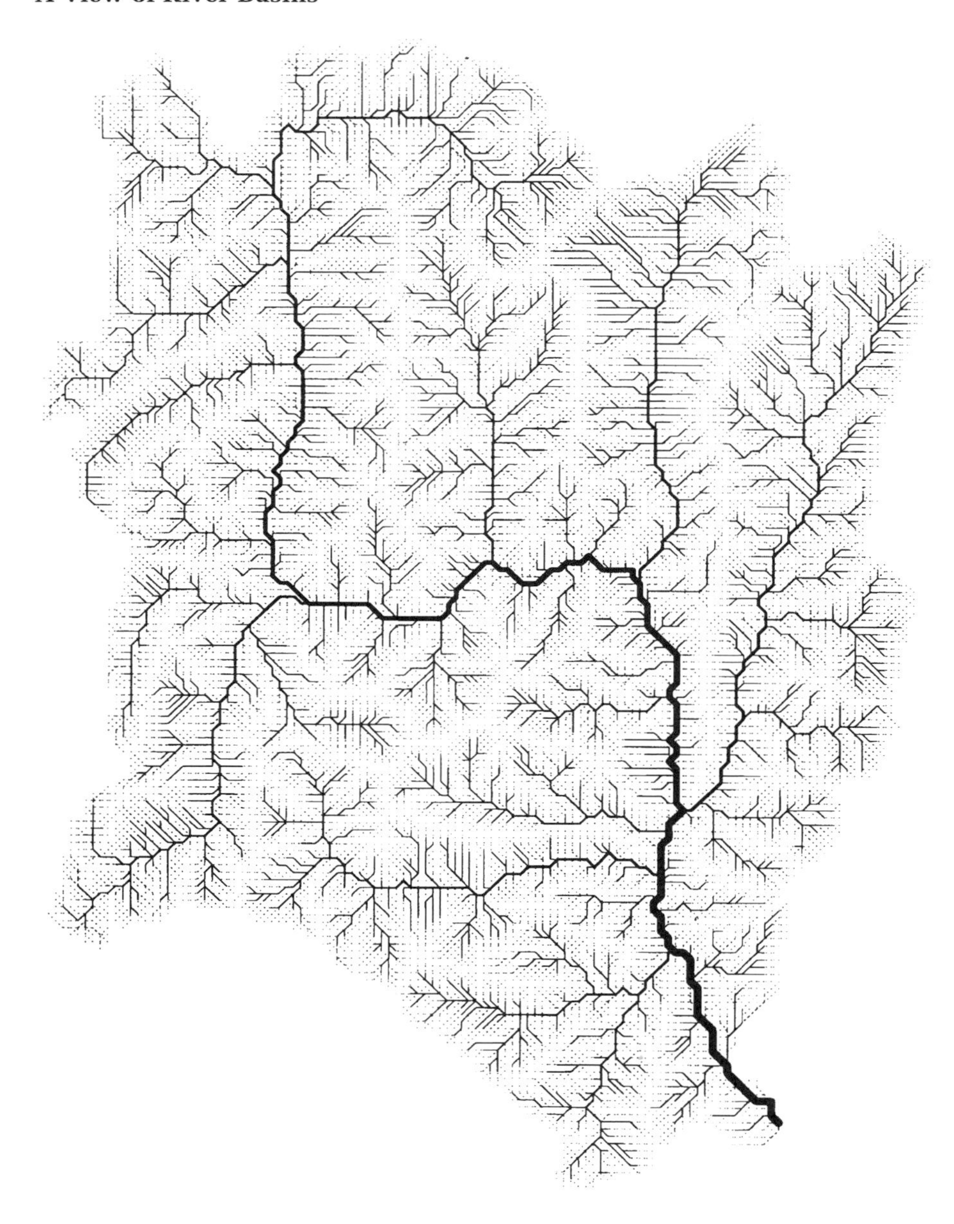

Figure 1.34. Fella River network. The channelized site i has $\nabla^2 z_i \geq 0$ and $\tau_c \geq 100$ (pixel units).

1.2.13 Experimental Fluvial Geomorphology

Following Schumm, Mosley, and Weaver [1987], experimental geomorphology may be defined as the science that studies, under closely monitored and controlled experimental conditions, a physical model of selected geomorphic features. Schumm et al. [1987], as Chorley [1969], Henderson [1966], and Mosley [1972] did earlier, classify physical models as either parts of unscaled reality or as scale models, basically distinguishing whether the study monitors reality (like a river reach) or suitably scaled replicas of natural landforms. Analog models that reproduce forms and functions of natural phenomena by suitably employing forces, materials, and processes different from those acting in nature have also been used.

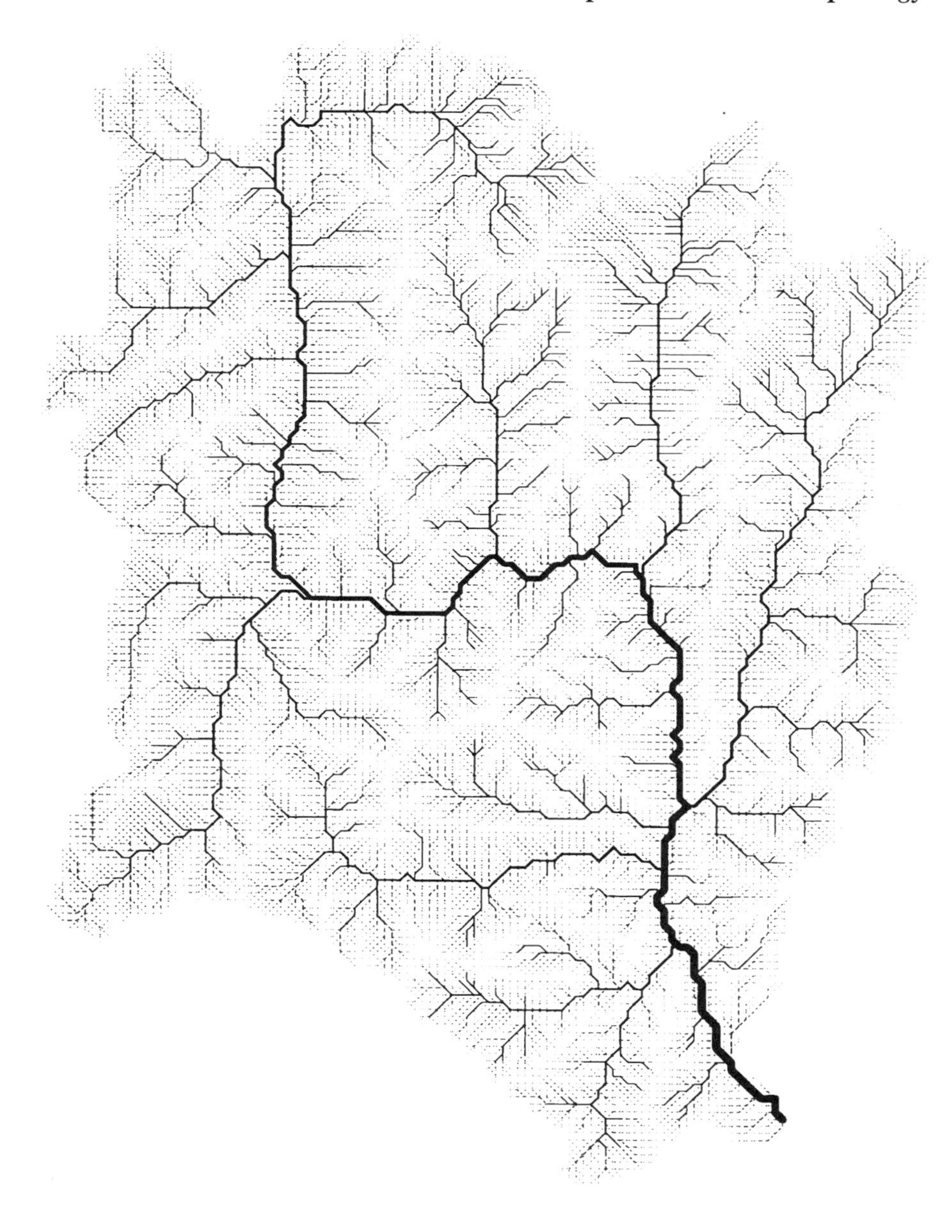

Figure 1.35. Fella River network. The channelized site i has $\nabla^2 z_i \geq 0$ and $\tau_c \geq 400$ (pixel units).

In this book the attention paid to these issues is by necessity limited, although some of the most recent work is significant to fractal river basins and therefore will be described.

The basic advantage of the experimental approach is that it permits the study of evolving geomorphic systems and of the differences between equilibrium and nonequilibrium states. It also allows the testing of various boundary and initial conditions. On the other hand, a strong drawback lies in the fact that these initial and boundary conditions may not be analogous to those in nature. Also, natural interactions may be hidden in the model and some natural processes (like rates of evolution) may not relate at all to those observed in the model. Scale effects may also be important, and a tradeoff between accuracy of measurements (generally improving with the

Figure 1.36. Aerial photographs taken 16 m above the surface of the Colorado experiment. The two panels refer to different experiments [after Schumm et al., 1987].

scale of reduction) and the impact of such effects needs to be assessed. In any case, transferring experimental results by analogy to larger landforms has always proved a major task, and the direct field data acquisition now available through DEMs has stolen researchers' attention from this subject.

Some field observations related to the evolution of natural river networks are available. For instance, Schumm [1956] (who studied badland developments) and Morisawa [1964] (who studied an upraised lake floor) noted that network elongation and the addition of tributaries tend to occur simultaneously, with a concurrent loss of tributaries near the major streams. Earlier, Ruhe [1952] studied drainage patterns on glacial till sheets of different ages and tried to estimate network changes that occurred during the past 40,000 years. He observed a behavior, called Glockian after Glock [1931], characterized by the initial growth (extension) of long first-order streams followed by addition of tributaries. This is a behavior that contrasts with the Hortonian view of network formation via the initial development of parallel rills that are replaced by a dendritic pattern. It also contrasts with development of dendritic patterns through headward growth and branching.

A significant laboratory experiment was constructed by Schumm and coworkers [Schumm, 1977; Schumm and Kahn, 1971; Schumm et al., 1972] at the Rainfall Erosion Facility of Colorado State University. The experimental facility was a 9-m by 15-m container with an outlet flume whose elevation could be adjusted. A system of sprinklers provided relatively uniform and known rates of artificial precipitation. The container was filled with mixtures of clay, sand, and silt of known textures and properties.

Examples of networks obtained in the Colorado experiments are shown in Figure 1.36. The experiments suggested that initial conditions play a major role in network development, although the robustness of the statistics of the resulting networks was not tested.

The most interesting observation from the experimental analyses is that the interplay of the initial slope and the baselevel elevation of the outlet flume could produce two different growth modes: (i) the first characterized by rapid headward elongation of the streams of first-order, later followed by addition of tributaries; and (ii) the second characterized by slow headward growth with full elaboration of the entire texture by streams of all orders. Figure 1.37 shows the marked effects of the initial conditions at equivalent times in two different experiments. The foremost effect lies in the different mode of growth. In experiment 1 the baselevel was lowered before precipitation was applied, and the resulting network developed by extending headward on the initial surface (the so-called expansion mode of growth). In experiment 2 the drainage network developed on a steeper slope and without an initial baselevel lowering. A skeleton of drainage network rapidly formed that spanned the whole watershed (the so-called extension mode).

The extension mode ultimately gave a maximum total channel length less than that produced by headward growth for approximately the same relief. Initially, the network developing on the steeper surface (experiment 2) produced a higher drainage density because the low-order tributaries were longer as they extended farther into the available area. However, at maximum extension the network had a lower drainage density $\mathcal{D}$. These results cannot be considered conclusive because of the difficult assessment of other effects, in particular that of baseline lowering. Furthermore, some of the features of the resulting networks are very different from those exhibited by natural landforms (e.g., Melton's relationship (Eq. (1.6)) in Parker's structures is $F_s \propto \mathcal{D}^{1.15}$, rather than the common relationship that depends on the power 2 of the drainage density).

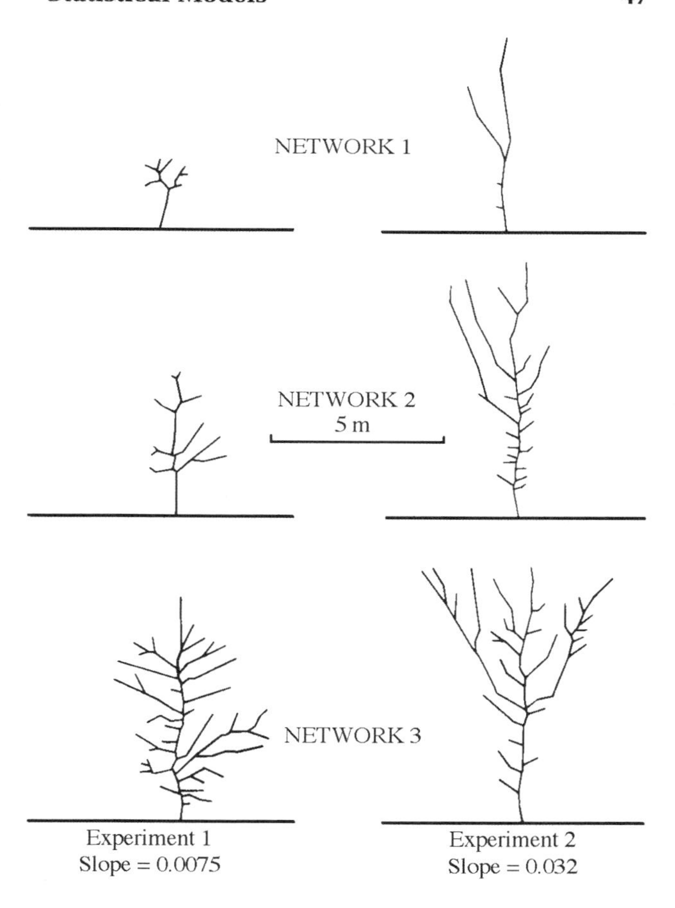

Figure 1.37. Drainage network growth under different initial conditions but at equivalent times in two different experiments by Parker [1977].

An interesting laboratory experiment was recently designed and conducted by Wittmann et al. [1991]. In it a circular, flat 1.4-m-diameter sandbox was watered uniformly by a rotating arm fitted with sprinklers. A sinkhole was placed in the center of the sandbox where water and mobilized sand exited. Once uniformly saturated conditions were established the sink was instantaneously opened, and the erosion of sand carried toward the sink led to the formation of dendritic structures with planar features similar to those of river networks. Wittmann et al. [1991] measured some geometric features of the outer boundary of the developing network (Figure 1.38) in relation to the flow rates of water. They found that the borderline of the capture zone possesses fractal characters.

1.3 Statistical Models of Network Evolution

1.3.1 Introduction

In this section we will examine two broad classes of statistical models for the representation of streamflow networks. Neither of the models

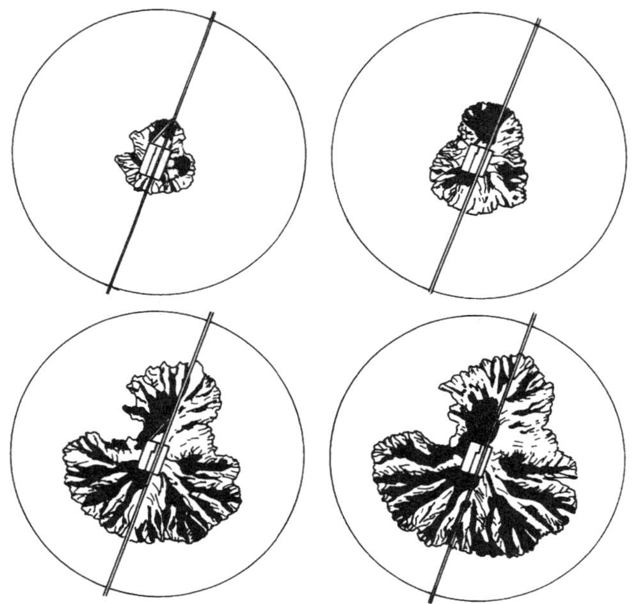

Figure 1.38. A 1.4-m-diameter circular sand-filled container is drained through a sinkhole placed in its center. Different stages of growth are illustrated [after Wittmann et al., 1991].

incorporates the altitude dimension so vital for a valuable description of river basins.

The first of these approaches, based on random walks of various types, introduces the concept of growth and development of the network. The second approach, the random topology model, is a static description of the planar characteristics of the drainage network and as such does not try to incorporate any evolutionary concept.

We will first review the general physical mechanisms that have been proposed for channel network growth. A review of three conceptual models of drainage network evolution is illustrated in Figure 1.39 [after Schumm et al., 1987]. Here (a) represents the Hortonian picture, where parallel rills develop into a dendritic pattern; (b) illustrates headward growth and branching; and (c) illustrates Glockian evolution characterized by the extension of long first-order streams followed by elaboration.

An early scheme is due to Horton [1945], who described a growth process in which a thin sheet of water in uniform flow conditions exceeds a critical shear stress at a distance x downstream from the divide. The critical shear stress is thought of as a threshold for mobilization of bottom material, and thus a system of parallel rills (Figure 1.39(a)) is developed, which rapidly propagates over the entire surface. Divide migration through competition and transverse grading subsequently generate a dendritic pattern. Divide migration refers to the capture of small rills by larger ones and transverse grading refers to the development of side hillslopes, which drain toward the dominant rill through drainage directions established by the maximum gradients. Horton's scheme has the essential ingredients of large-scale network growth but lacks the ability to describe the effects of heterogeneities in the surface structure on the development of the network. Thus the Hortonian picture is more likely to be representative of the processes taking place in relatively small and flat areas.

A second mechanism is headward growth and branching [e.g., Schumm, 1956; Howard, 1971a,b,c; Smart and Moruzzi, 1971a,b]. According to this model, a network is formed as a wave of dissections progressing from the outlet into an unchanneled landscape. Thus channels grow upstream and bifurcate, filling the available drainage area. Whatever the rule for branching (see Section 1.3.2), growing networks may be subjected to a process of stream capture through which large streams migrate sideways, capturing smaller ones (Figure 1.39(b)). As an example of the mechanism for headward growth, Dunne [1980] suggested that channel heads are formed by reemergence of subsurface flows, where the cohesive force of the surface layer is exceeded by the drag of the emerging subsurface flow.

A different conceptual picture of network growth is that proposed by Glock [1931], where the stages of development are classified as follows (Figure 1.39(c)):

- network initiation through the rapid carving of a skeletal pattern;
- network elongation by headward growth up to maximum extension;

- network elaboration through the development of tributaries; and
- a stage of simplification where tributaries disappear owing to the reduced relief.

The Glockian picture is obtained by merging the previous mechanisms and as such is likely to represent large-scale processes. However, although any experimental verification of the individual role of distinct processes is quite difficult, Schumm et al. [1987] observed from laboratory experimentation that, within the same physical setup, the modification of physical constraints (i.e., relief, initial slope) may induce either of the above mechanisms. Therefore it is highly likely that no general rule can be inferred from simple conceptual models. Network growth is instead produced by complex interactions that seldom yield to a simplified description.

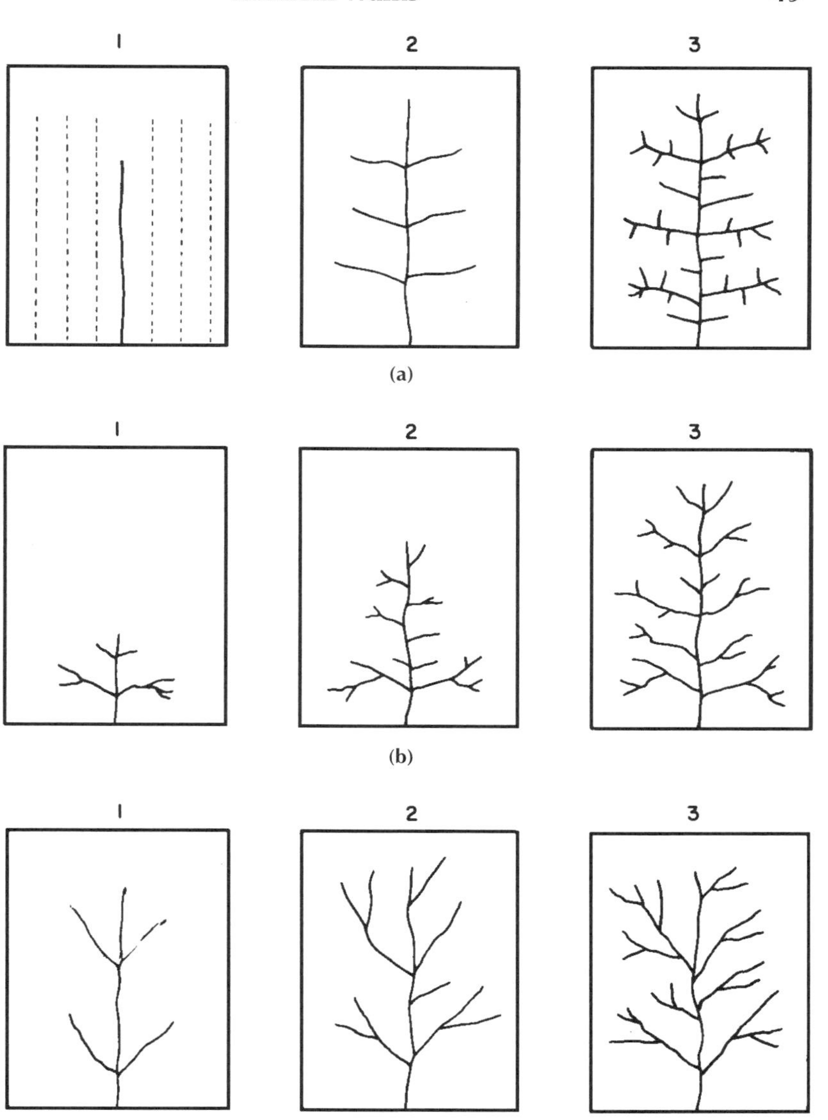

Figure 1.39. Conceptual models of drainage network evolution [after Schumm et al., 1987].

1.3.2 Random-Walk Drainage Basin Models

Leopold and Langbein [1962] first carried out modeling studies of drainage basins by simulating the development of drainage networks through random walks in a rectangular region. In their approach the region to be drained is covered with square tiles of unit area. Every square tile is drained and the drainage direction (i.e., the flow path) has an equal chance of flowing to any of the adjacent tiles. Diagonal connections are not allowed. A further restriction is imposed: once a choice is made, flow in the reverse direction is not possible.

The source of the first stream is determined by selecting a square at random, and a channel is generated by making random moves into adjacent squares. On subsequent moves no flow reversal is allowed. The first stream continues until it goes off the area of simulation and another stream is generated as before. Subsequent streams will finish when they either reach the boundaries of the rectangular region or join a preexisting stream.

The procedure continues until all squares are filled. Arbitrary decisions are needed in some situations (for instance, dealing with a stream looping

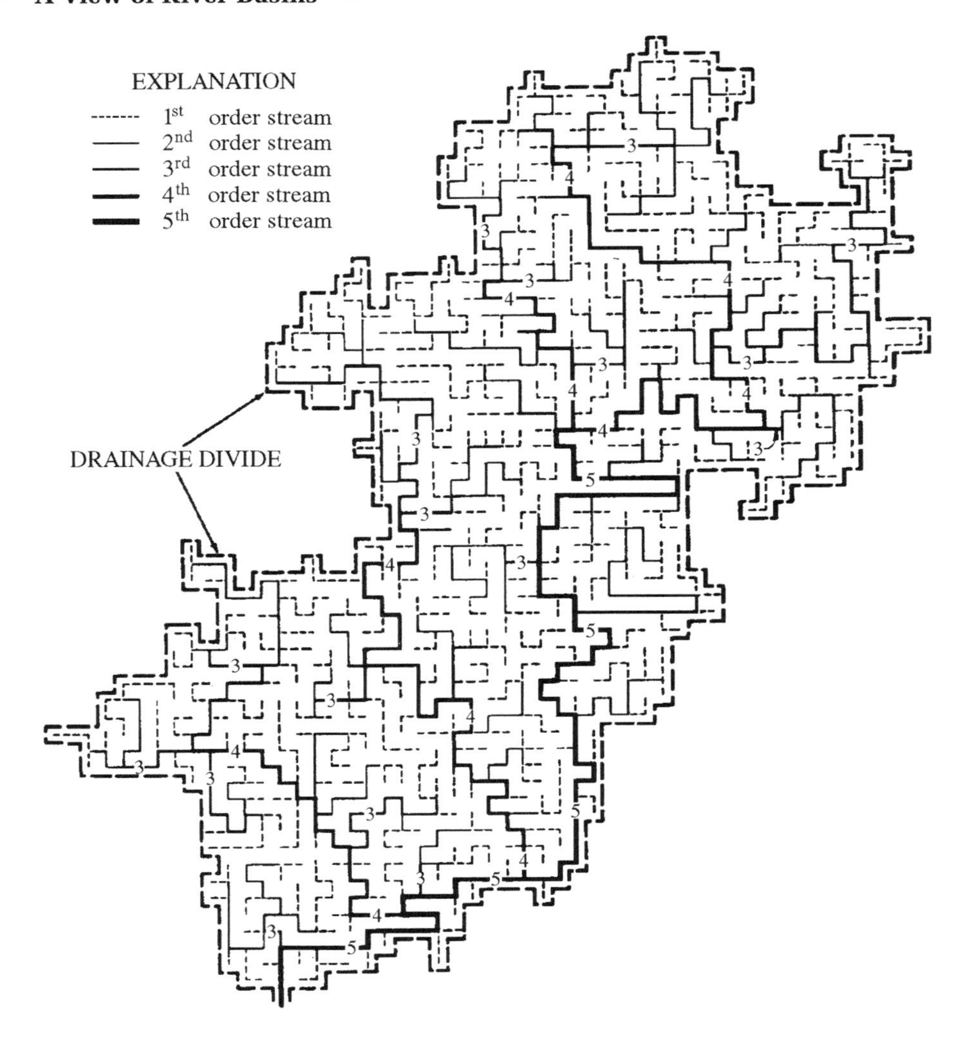

Figure 1.40. Development of a random-walk drainage basin network [from Leopold and Langbein, 1962].

back on itself), and these rules are discussed by Schenck [1963] and by Smart, Surkan, and Cosidine [1967]. Figure 1.40 shows a stream network generated by this procedure. The network exhibits striking similarities to natural drainage nets. Divides are developed and the streams join so as to create rivers of increasing size. The Horton analysis of the network shown in Figure 1.40 is presented in Figure 1.41. Again, one observes that Horton's laws of stream numbers and stream lengths are relatively well fulfilled in the simulated network, with R_B and R_L values close to those observed in real basins.

It is worth observing the much better fit of the law of stream numbers than the fit of the stream lengths law. In our experience with networks generated by random-walk mechanisms we have observed that this behavior holds consistently.

Figure 1.42 shows the relationship between mainstream length and drainage area, that is, Hack's law, which in the particular example of Figure 1.40 follows a well-defined power law with exponent 0.64, significantly different from 0.5.

There have been several further studies along the lines of Leopold and Langbein's planar modeling of drainage networks.

Scheidegger [1967] carried out simulations of drainage patterns through a random-walk process. He was interested in Alpine valleys and rightly considered it necessary to impose conditions such that all the drainage paths would be essentially in the direction of the high gradients between the watershed and the main valley. This preferential flow condition (the 'directed' network) is absent in the model of Leopold and Langbein and results in nontrivial consequences. As shown in Figure 1.43(a) the drainage in a unit time step is always forward (toward the main valley) but it may randomly go one-half space unit to the left or to the right. The drained strip is modeled by a grid of points in which each row is displaced by one-half the lattice distance with regard to the former row. The final step is then to draw the boundaries between the basins generated by the above procedure. An example is shown in Figure 1.43(b). Obviously, the resulting basins are elongated but their quantitative characteristics depend on the rather restrictive assumption of always going one-half space unit either to the left or to the right at each time step with equal probability. The main point is that it is rather artificial to insert preferential drainage directions, for example, the effect of gravity, in the planar model of a drainage network. This crucial effect has to be dealt with through the explicit consideration of the altitude dimension, which is absent by construction in all models of this type.

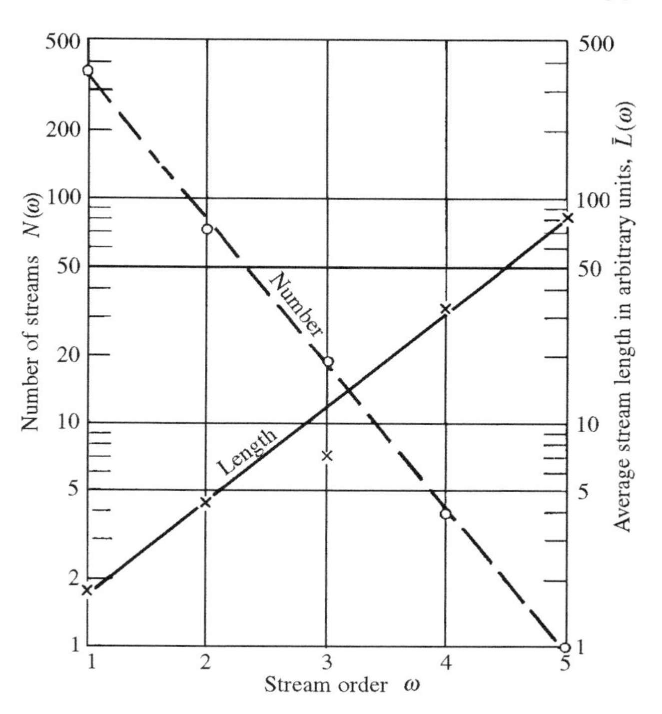

Figure 1.41. Relation of number and average lengths of streams to stream order for the random-walk model [from Leopold and Langbein, 1962].

Interestingly, Huber [1991] has shown that Scheidegger's river model is equivalent to the one-dimensional model of aggregation with injection of mass studied by Takayasu and Takayasu [1989] and Takayasu, Nishikawa, and Tasaki [1988]. This is indeed interesting because fractal properties of aggregation with injection are well known, as will be described in Chapter 2.

The models of Leopold and Langbein [1962] and Scheidegger [1967] deal with the growth of a stream network by coalescence of streams and thus involve generation of a stream network starting from the head of first-order streams. Howard [1971b] used a different type of random-walk approach to the problem of stream network growth and development. The approach involves headward growth and branching in a random fashion. Howard first investigated purely topological headward growth models, where the network growth starts from a single stream and grows through successive generations. In each generation, every active stream may continue unchanged, branch, or become unactive (terminate); the probabilities of these events are parameters of the model. The process may be continued indefinitely until the entire network becomes inactive or until it reaches a predetermined size. A special case of the topological headward growth model occurs when the probability of termination is zero and the simulation becomes a pure growth process. The topological characteristics of

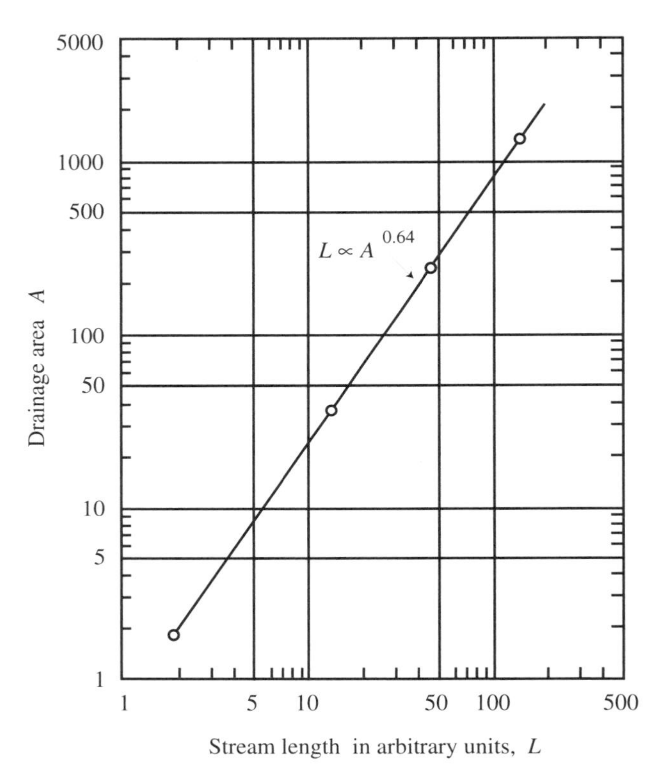

Figure 1.42. Relation between stream length and drainage area for the random-walk model [from Leopold and Langbein, 1962].

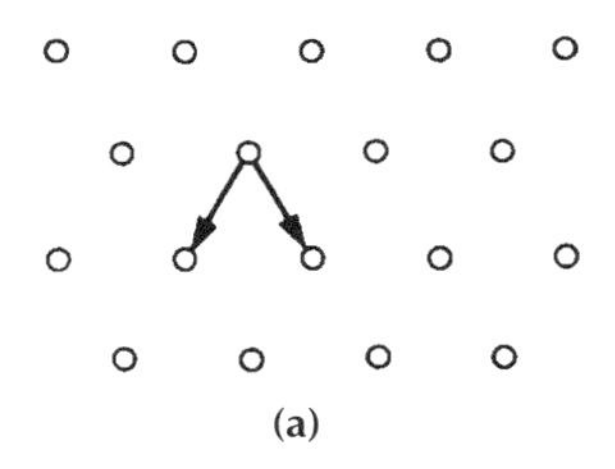

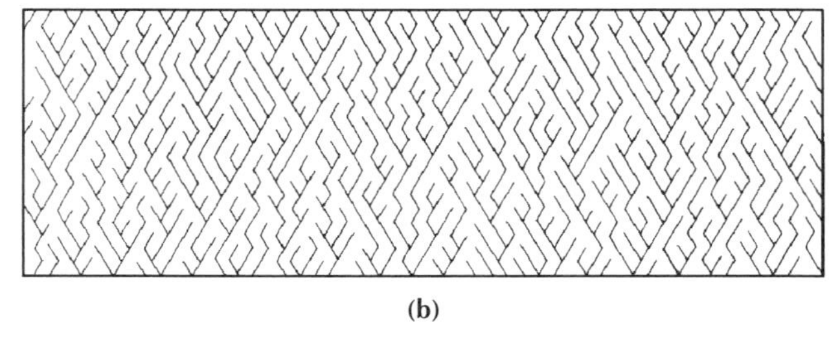

Figure 1.43. (a) The model grid showing the possibilities for drainage at each point; (b) random-generated drainage net [from Scheidegger, 1967].

the obtained networks depend crucially on the probabilities of branching, and furthermore they do not conform to those observed in nature.

If the probability of branching equals unity, the bifurcation ratio equals 2.0 because every stream bifurcates at each generation. If the branching probability is less than 1.0, the bifurcation ratio will be greater than 2.0 because some streams will not change throughout generations and may thus enter a stream of higher order. Random headward growth produces networks with properties quite different from the random topological models discussed later in this chapter [Howard, 1971b]. In fact, irrelevant to the probability of branching, the pure growth model – with zero probability of death or termination of a stream – results in average stream numbers $N(\omega)$ and bifurcation ratios R_B that are much smaller than those found in nature. To correct for the large number of low-order tributaries directly entering high-order streams, one may introduce a finite probability of termination during each generation at each active stream head. Although this procedure improves the situation when compared to the pure growth model, the numbers are nevertheless still lower than those commonly observed in nature [Howard, 1971b,c]. Besides, it seems rather doubtful that the two probabilities involved in the scheme – branching and termination – could be related to other physical mechanisms or processes taking place in the basin.

As pointed out by Howard [1971b], a critical fault of these models is that the probabilities that control the growth remain the same everywhere; this is not the case in nature, where there exist effects of divides and competition among streams that are not felt simultaneously by all stream heads, but rather are felt selectively, depending upon the position of the stream within the network. To deal with this, Howard [1971b] developed the so-called areal growth model, where the probability of growth at any active site depends on the position of the site inside a region where several networks are growing in competition. This type of model was also pursued by Smart and Moruzzi [1971a, b], whose main goal was to replace the arbitrary probabilities for growth and branching by rules that more closely reflect the actual process involved. In their scheme the decision as to whether a given stream segment should be allowed to grow is reached by examining the area immediately upslope. As in other random-walk models, an array of squares covers a rectangular domain where construction takes place. The squares along the bottom row are all possible outlet locations and are considered active sites. The potential contributing area for an active site consists of the three columns of squares immediately above and at each side of it. Figure 1.44 from Smart and Moruzzi [1971a] shows an example of the procedure. One of the active sites is chosen at random and the following V-rule is used to determine if the site becomes an outlet:

V**-rule:** Growth is not allowed if any square in in the central column of the potential contributing area is occupied.

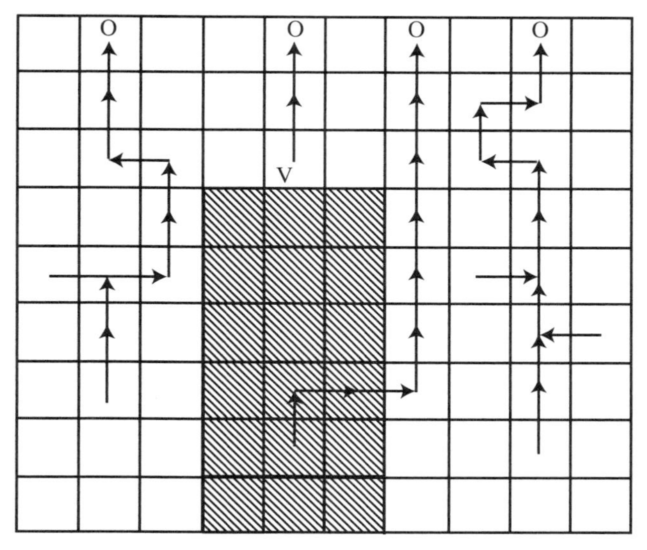

Figure 1.44. Headward growth random-walk simulation: O–outlets; V–active site. The potential contributing area for V is cross-hatched [from Smart and Moruzzi, 1971a].

If the site becomes an outlet, a segment of stream is created from the square above into the outlet. The square above is then added to the active-site list and the outlet location is no longer considered an active site. Another site is now chosen at random from the updated active-site list. If the new site is an outlet location, the procedure is repeated. If the new site is an interior location, the number of streams already entering it is determined and the three neighboring sites (one above, one at each side) are examined to see if they are occupied. The site is removed from the active-site list if it contains the junction of two streams or if all of its neighbours are occupied; then a new random selection is made. Otherwise one of the permitted directions of growth is chosen at random. If the vertical direction is chosen, the V-rule is again used to determine whether growth occurs; if one of the horizontal directions is chosen, the following H-rule is then used:

H**-rule:** Growth to the right (left) is not allowed if any square in the right (left) column of the potential contributing area is occupied or if the active site is on the second row from the top.

Whatever direction is chosen (vertical or horizontal), if growth is not allowed the site is removed from the active list; if growth is allowed, the newly connected site is added to the list of active sites unless it is located in the top row of the array. A new active site is then chosen and the process is repeated until there are no active sites in the region. The above procedure simulates networks which, according to Smart and Moruzzi [1971b] satisfactorily resemble the systems of drainage networks for a homoclinical reach (Clinch Mountain, Virginia).

Many other approaches deal with network-like fractal structures generated by different types of random walks. We will briefly review some examples relevant to river networks, avoiding a deeper and complete review more suitable to books on general fractal growth phenomena [e.g., Vicsek, 1989].

A different type of random walk in the context of river networks has been proposed by Seginer [1969]. Rather than assigning random flow directions to all grid points (i.e., all nodes of the rectangular lattice on which the walkers are allowed to stop), Seginer defined a matrix of random elevations. Flow directions are then defined according to steepest descent,

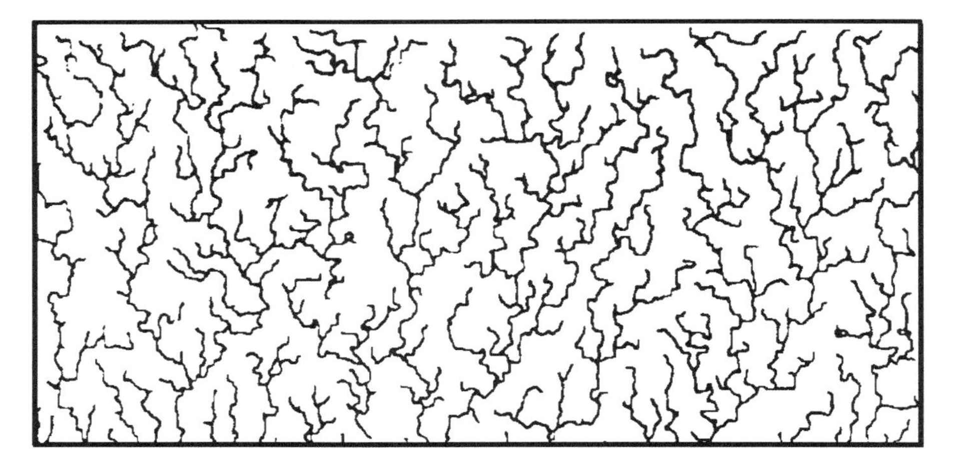

Figure 1.45. Sample network created with self-avoiding percolation on a lattice size 512×256. Streams of fourth and higher order are shown [after Stark, 1991].

that is, from each lattice site to the lowest nearest neighbor. The idea is indeed very appealing because the arbitrariness of the elevation field allows a variety of landscapes to be treated. However, we do not consider this an appropriate way of accounting for the altitude dimension because the a priori imposition of the elevation field (through the random rule, in this case, or otherwise [e.g., Leheny and Nagel, 1993]) fixes the most important structural characteristics of the landscape and thus does not allow researchers any new insight into the geomorphic process responsible for landscape formation nor does it allow for any demanding comparisons with real data.

Percolation models have provided a well-studied field of research for critical phenomena in physics [e.g., Feder, 1988; Vicsek, 1989]. A stochastic model of invasive percolation has been proposed for river network growth by Stark [1991], following rules of the so-called self-avoiding random walk on a discrete lattice. In Stark's approach a grid and a strength field are defined. The strength field mimics some sort of resistance to shearing actions imposed on the surface layer (or substrate) of the developing medium, which is taken to be a random function. Stark's strength field is meant to reflect topographical and hydrological influences, including inhomogeneities in the vegetation cover, although no insight is offered regarding the characterization of these influences. Streams and links grow by headward extension and branching moving from boundaries where the substrate is assumed weakest and hence most likely to fail and become part of the network.

Stark's [1991] model is a self-avoiding invasive percolation whose scaling properties are analogous to those of other lattice percolation clusters [Stauffer, 1985; Feder, 1988]. One example of Stark's networks is shown in Figure 1.45, where the substrate strength was randomly distributed over the lattice with mild anisotropy. This model was successful in reproducing some of the statistics of real networks, such as Horton's ratios and Hack's law ($L \propto A^{0.565}$). Unfortunately, the networks developed through this model have not been subjected to any deep screening even on planar statistics. As discussed in Chapter 2, most random growth models reproduce the statistics found by Stark. Moreover, random growth matches more refined statistics like the power-law distribution of total contributing areas.

We will end our review of the literature on random growth models at this point, although by no means do we claim to have covered the whole subject. Our objective was merely to provide a brief account of some recent work more directly related to river networks. The reader is referred to other texts for complete treatments of this subject [e.g., Feder, 1988; Vicsek, 1989].

As stated before, one purpose of growth models is to replace the assignment of arbitrary probabilities for growth and branching by rules that somehow are thought to more realistically represent the physics of the ongoing evolution. Again, it seems obvious that without explicit consideration of the altitude dimension the theoretical foundation of these efforts is questionable. Furthermore, the comparisons with reality of all planar simulations are highly nondemanding, as they are based on planar statistics which, although some are more strict than others, tend to be a much less stringent test than those involving the third dimension. Some authors [e.g., Kirchner, 1993] have labeled the frequent occurrence of Horton's ratios that are similar to the ones produced in nature by planar network models as a statistical inevitability. However, little consideration has been given to the embedded limitations of the models chosen and to differences in the goodness of fit in Horton's diagrams. The impact of particular assumptions built into common random models (e.g., constant link lengths, fixed number of sources, constant contributing area to each link – to name some of the most common ones) is particularly strong on planar features. Thus the statistical inevitability of Horton's laws in nature has merely been suggested rather than demonstrated.

Chapter 4 and especially Chapters 5 and 6 address the incorporation of altitude space into a comprehensive framework for network and landscape evolution.

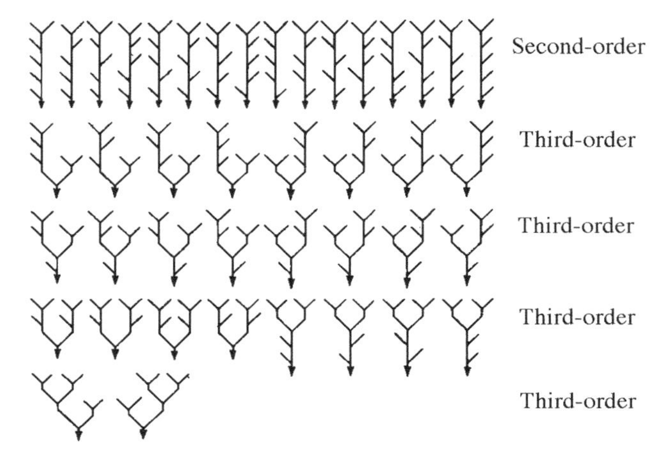

Figure 1.46. Topologically distinct networks having six sources [from Shreve, 1966].

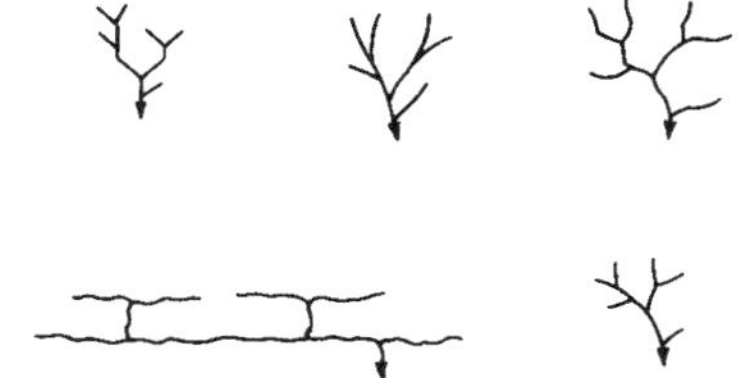

Figure 1.47. Topologically identical networks having different stream patterns [from Shreve, 1966].

1.3.3 The Random-Topology Model

In a series of landmark papers in the 1960s, Shreve introduced the basis of what is now called the random-topology model. Two basic postulates constitute the foundation of this model:

- In the absence of geologic controls, channel networks are topologically random. Shreve [1966] first introduced to geomorphology the concept of a topologically random population of channel networks in which all topologically distinct channel networks (TDCNs) with a given number of sources are equally likely. Topologically distinct networks, illustrated in Figure 1.46, in contrast to topologically identical networks, shown in Figure 1.47, are those whose schematic map projections cannot be continuously deformed and rotated in the plane of projection so as to become congruent.
- For drainage basins developed under comparable environmental conditions, the exterior and interior link lengths are independent random variables with a single common distribution for each type [Smart, 1968].

As pointed out by Smart [1973] the justification for this model is purely pragmatic. Starting with these two postulates, many observed features of drainage basin composition relating to topology and channel lengths can be adequately reproduced.

The number $\mathcal{W}$ of TDCNs with $N(1)$ sources was first obtained by Cayley [1859]:

$$\mathcal{W}(N(1)) = \frac{1}{2N(1) - 1} \binom{2N(1) - 1}{N(1)} \tag{1.29}$$

where the term in parenthesis is the binomial coefficient, that is, $\binom{a}{b} = a!/b!(a-b)!$.

The number of TDCNs having $N(1), N(2), \ldots, N(\Omega - 1), 1$ streams of order $1, 2, \ldots, \Omega - 1, \Omega$, respectively, is given by Shreve [1966]:

$$\mathcal{W}(N(\Omega)) = \prod_{\omega=1}^{\Omega-1} 2^{N(\omega)-2N(\omega+1)} \binom{N(\omega) - 2}{N(\omega) - 2N(\omega + 1)} \tag{1.30}$$

where $N(\Omega)$ is a convenient notation for the set $N(1), N(2), \ldots, N(\Omega - 1), 1$.

The general term of the product in Eq. (1.30) is the number of topologically distinct ways in which the $N(\omega)$ streams of order ω may be arranged as tributaries to the streams of higher order in the network. Of the $N(\omega)$ streams, $2N(\omega + 1)$ must join in pairs to form the $N(\omega + 1)$ streams of order $\omega + 1$, whereas the remaining $N(\omega) - 2N(\omega + 1)$ can be joined in a number of different arrangements to the $2N(\omega + 1) - 1$ links of order $\omega + 1$ or greater. The binomial coefficient gives the number of distinct ways that the $N(\omega) - 2N(\omega + 1)$ streams may be divided among $2N(\omega + 1) - 1$ links, and the factor $2^{N(\omega)-2N(\omega+1)}$ occurs because each tributary may enter either from the right or from the left.

The probability of occurrence of selecting a channel network with a set of stream numbers $N(\Omega)$ from a topologically random population of networks with $N(1)$ sources is given by

$$P\left[N(\Omega); N(1)\right] = \frac{\mathcal{W}(N(\Omega))}{\mathcal{W}(N(1))} \tag{1.31}$$

Shreve [1966] gives a recursive relation to calculate the number $\mathcal{W}_\Omega$ of TDCNs with $N(1)$ sources and a given basin order Ω. He also derives the number N_Ω of different sets of stream numbers with given $N(1)$ and Ω. Table 1.2 from Smart [1972] shows how the astronomical numbers of TDCNs can be grouped according to the above criteria.

From Eq. (1.31), Shreve [1966] proceeded to show that in a topologically random population the most probable networks approximately obey Horton's law of stream numbers but exhibit some minor systematic deviations. He also showed that for networks with a given number of first-order streams the most probable network order is that which makes the geometric mean bifurcation ratio closest to 4. Table 1.3 from Smart [1972] illustrates this for the case $N_1 = 60$. $P[N(\Omega); 60]$ is given by Eq. (1.31). $\sum_P$ stands for the cumulative probability of all networks with equal of higher likelihood of occurrence than the one being considered. The top half of the table lists the ten most probable sets of stream numbers for $N_1 = 60$, all of them being of fourth order and accounting, by nearly half of the possible TDCNs, for sixty

Table 1.2. *Values of the number $\mathcal{W}_\Omega$ of TDCNs for $N(1)$ number of sources and given basin order Ω*

Ω	$N(1)$	$\mathcal{W}_\Omega$	N_Ω
2	6	16	1
3	6	26	2
4	6	0	0
2	24	4.19×10^6	1
3	24	1.35×10^{11}	11
4	24	2.08×10^{11}	25
5	24	1.63×10^8	10
2	40	2.75×10^{11}	1
3	40	4.60×10^{19}	19
4	40	6.10×10^{20}	81
5	40	2.42×10^{19}	84
6	40	1.85×10^{10}	10
2	60	2.88×10^{17}	1
3	60	2.13×10^{30}	29
4	60	3.19×10^{32}	196
5	60	8.45×10^{31}	364
6	60	4.91×10^{26}	140

Note: Also shown are the numbers N_Ω of different sets of stream numbers for given $N(1)$ and Ω.
Source: Smart [1972].

Table 1.3. *Stream number probabilities for $N(1) = 60$*

$N(1)$	$N(2)$	$N(3)$	$N(4)$	$P[N(\Omega); 60]$	$\sum_P$	R_B
60	15	4	1	0.0675	0.0675	3.91
60	14	3	1	0.0594	0.1269	3.79
60	16	4	1	0.0591	0.1860	3.95
60	15	3	1	0.0563	0.2423	3.83
60	14	4	1	0.0554	0.2978	3.87
60	14	3	1	0.0459	0.3436	3.75
60	16	3	1	0.0394	0.3830	3.87
60	17	4	1	0.0375	0.4206	3.98
60	13	4	1	0.0321	0.4527	3.83
60	16	5	1	0.0296	0.4822	4.01
60	19	4	2	0.0015	0.9679	2.60
60	20	4	1	0.0015	0.9694	4.07
60	12	5	1	0.0014	0.9707	3.85
60	20	5	2	0.0013	0.9720	2.65
60	20	6	2	0.0013	0.9733	2.69
60	18	7	2	0.0012	0.9745	2.69
60	13	1	0	0.0011	0.9756	8.59
60	17	7	2	0.0011	0.9767	2.67
60	18	2	1	0.0010	0.9777	3.82
60	12	5	2	0.0010	0.9788	2.52

Source: Smart [1972].

sources. The individual bifurcation ratios among successive orders range between 3 and 5.3, and the overall values of R_B are all near 4. Thus the most probable networks rather closely obey Horton's law of stream number. The bottom half of Table 1.3 lists ten sets of stream numbers of low – but not the lowest – probability of occurrence. The individual bifurcation ratios of the least likely networks range widely between 2 and 13, and moreover these sets do not conform to Horton's law of stream numbers. Also, most R_Bs are quite different from 4, which, as seen in the top half of Table 1.3, is the value to which the most probable networks tend to adjust. In fact, $R_B = 4$ is a general result for the most probable sets of stream numbers for any given $N(1)$.

Shreve [1966] shows in detail one specific example that was obtained through an algorithm for computer generation of the sets of stream numbers in a topologically random population of channel networks with 64 first-order streams. There are 9.430×10^{34} topologically distinct networks with sixty-four first-order streams and the immense majority of them conform closely to Horton's law of stream numbers with bifurcation ratio near 4. Similar results may be obtained for Horton's law of stream lengths by using the second postulate of the random-topology model. Again, the most probable networks conform closely to Horton's law of stream lengths with mean stream length ratio, R_L, close to 2 [Shreve, 1969; Smart, 1972].

In two other papers, Shreve [1967, 1969] introduced and studied the properties of infinite topologically random channel networks. The analytics of infinite networks is much simpler than that for finite networks. The basic idea is that in an infinite topologically random channel network, all TDCNs of a given magnitude occur with equal frequency. Some of the main results obtained by Shreve for infinite topologically random channel networks are given by Smart [1972] as follows:

(1) The probability of draining a link of magnitude μ is $2^{-(2\mu-1)}\mathcal{W}(\mu)$.
(2) The probability of draining a link, a subnetwork, or a basin of order ω is $1/2^{\omega}$.
(3) The average magnitude of a link of order ω is $(2^{2\omega-1}+1)/3$.
(4) The probability of draining a stream of order ω (from the population of streams) is $3/4^{\omega}$. This implies that $N(\omega)/N(\omega+1) = (3/4^{\omega})/(3/4^{\omega+1}) = 4$. Thus in random samples of Strahler streams from an infinite topologically random channel network, the expected value of the number of streams of successive orders approaches a geometric series with $R_B = 4$ as the size of the sample increases.
(5) The average number of links in stream of order ω is $2^{\omega-1}$.
(6) The average number of tributaries to streams of order ω is $2^{\omega-1} - 1$, as required by item (5) above. The average number of these tributaries that have order ω^* $(1 \leq \omega^* \leq \omega - 1)$ is $2^{\omega-\omega^*-1}$.

In finite topologically random channel networks with specified stream numbers, many similar results are available but they are given by very complicated formulas that will not be reported here.

It is also possible to make a simple estimate of stream length behavior for an infinite topologically random channel network [Smart, 1969, 1972]. The expected number of links for streams of order ω is $2^{\omega-1}$. Making the simplifying assumption that all links have the same length, the mean stream length increases with ω as a geometric series with ratio 2, which agrees well with the observations in natural drainage networks.

With respect to the law of basin areas, similar verifications to those of the law of stream lengths have been obtained. Assuming that the first-order areas and the area draining directly overland into individual interior links are independent random variables with a single common distribution for each type, Shreve [1969] has shown that the most probable finite topologically random channel networks conform rather well to the law of stream areas with a geometric-mean area ratio near 4. The random-topology model can also be manipulated to provide a tentative explanation of Melton's law relating stream frequency and drainage density, Eq. (1.6). For this, one begins with an infinite topologically random channel network. The total number of Strahler streams is approximately given by $n + n/4 + n/16 + \cdots = 4n/3$. Assume now that all links have a length ℓ and directly drain a region of area a. Furthermore, assume $a = \ell^2$. This model also has a uniform drainage density $\mathcal{D} = \ell/a$. Then Melton's ratio becomes

$$\frac{F_s}{\mathcal{D}^2} = \frac{4n/3}{A}\frac{1}{a^{-1}} = \frac{4n/3}{(2n-1)a}a \approx \frac{2}{3} \tag{1.32}$$

(where $A = (2n-1)a$ is the total area), in very good agreement with Melton's observed value of 0.69 [Shreve, 1967]. Notice, nevertheless, that several expedient assumptions were used in the previous derivation.

Particularly interesting because of its implications with respect to the structure of the drainage network is the so-called Hack's relationship between the length of the longest stream and the drainage area of a basin, $L \sim A^{0.6}$ (Eq. (1.8)). The explanation of Hack's relationship as a consequence of the random topology model has been attempted by several investigators [Shreve, 1974; Werner and Smart, 1973]. The basic line of these investigations was to simulate, through Monte Carlo techniques, topologically random networks with magnitudes randomly chosen in relatively wide ranges (Shreve used magnitudes up to 10^6, Werner and Smart chose magnitudes in the range 20–200). In the Monte Carlo calculation the link lengths and associated area may be assigned either from some prescribed distributions along the postulates of the random-topology model (using $a = k\ell^2$, with k being a random variable of expected value equal to one) or from tables taken from real data. Shreve [1974] assigned link lengths and associated areas by drawing them at random from two tables, one for the interior links and one for the exterior links. Each table consisted of 390 lengths and corresponding areas measured on U.S. Geological Survey 1:24.000-scale maps for a region in eastern Kentucky. Both Shreve [1974] and Werner and Smart [1973] report excellent agreement between their results and Hack's relationship. Of special interest are the numerical values obtained by Shreve for the exponent in $L \propto A^h$ as a function of the area of the basin. He detected a continuous decrease in h from $h = 0.6$ for basins in the range 1–10^3 km^2 down to $h \approx 0.5$ for the largest basins (10^6 km^2). Furthermore, Shreve [1974] indicated that this is what is found in real basins.

Shreve's numerical results were indeed nicely corroborated by the more recent work of Mesa and Gupta [1987], who derived the theoretical value of Hack's exponent h for the random topology model of channel networks:

$$h(n) = \frac{1}{2}\left(\frac{\pi + (\pi/n)^{1/2}}{\pi - 1/n}\right) \tag{1.33}$$

where n is the basin's magnitude. For $n = 10, 100$, or 500 the slope $h(n)$ is

0.68, 0.530, and 0.513, respectively. When n tends to infinity, h tends to the asymptotic value of 0.5. This result makes quite clear the fundamental importance of the magnitude of the network in the value of the exponent h under the premises of the random-topology model. It also explains some contradictory or not very illuminating results offered by simulations of random topological networks, where Hack's relationship was derived without paying much attention to the magnitude (or, equivalently, area) of the networks being analyzed.

Do the data from natural basins show changes of h as function of the basin's size? The most careful investigations in this direction are those of Muller [1973], who, based on extensive data analysis of several thousand basins, found that the exponent in Hack's equation did not change continuously as the basin increased in size from very small to very large. Surprisingly, the exponent remained rigid at 0.6 for basins less than 8,000 square miles, then suddenly changed to 0.5 for basins between 8,000 and 10^5 square miles, and finally dropped to 0.466 for basins larger than 10^5 square miles. As Mesa and Gupta [1987] point out, Muller's empirical observations are not consistent with the implications of the random topological model as reflected in Eq. (1.33), which implies a continuously decreasing exponent $h(n)$ with an increasing n. Is there any reason why the exponent in the relation $L \propto A^h$ should change from near 0.6 for basins below, say, 10,000 square miles to 0.5 for basins between 10^4 and 10^5 square miles? Is the change from 0.5 to 0.47 for very large basins – above 100,000 square miles – a significant one, or is it perhaps due to a smaller sample size and more uncertainties in measuring L in such very large basins? We believe that an explanation should be found for the first of these questions, but with regard to $h \approx 0.47$ for very large basins, more detailed measurements are needed before arguing that this value is truly different than 0.5. The significant change in the value of h at a threshold of sizes near 10,000 square miles points to an explanation of the elongation of basins that operates up to or around certain basin sizes. After that size – which obviously would not be an exact rigid value – whatever physical mechanism was operating ceases to be valid and other causes and processes control the shape of the basin, which stops its tendency toward elongation. This cannot be explained by the random-topology model because as was mentioned before, such a model implies a continuous decrease of h with magnitude. Also, it is important to note that the condition of large magnitude is not confined to basins with large areas because magnitude depends on the map scale [Mesa and Gupta, 1987].

In Chapter 4, Hack's relationship is explained as the result of basin elongation, where the main factor controlling the bulk of its shape is the organization of the drainage network. The organization of the network embedded in the basin is postulated to occur based on the principles of optimum energy expenditure. Obviously, this explanation will not hold for basins beyond a certain size, above which tectonic processes will control the shape of watersheds.

The results from the random topological model may be interpreted in two different ways:

(1) Natural river networks are topologically random. Their most important dependence is by far on the laws of chance.

(2) Horton's laws, defining R_B, R_L, and R_A, as well as many other well-documented empirical relationships such as Hack's law, Melton's law, etc., exist in most possible networks and thus their observance does not say much about the processes that control network growth and development.

A different perspective from the random topological model maintains that any statement relative to the genesis and development of drainage networks should be fundamentally based on the third dimension – elevation – which is conspicuously missing in any discussion regarding positions (1) and (2) stated earlier.

It seems obvious that there are physical laws that govern and describe the carving and maintenance of individual channels. It is equally obvious that chance plays an essential role even at the level of individual channels, say, through random topographic conditions or through a random field of soil properties. One could foresee that at the level of a whole network, when each channel or reach affects many others and in a sense the full network, both chance and necessity should operate as complementary principles, that is to say, as integral aspects of the same process. Thus from this point of view, to which we subscribe in this book, the crucially important issue is the dynamics through which networks evolve in time and space and from which the various features one observes in natural river basins arise. Thus our point of view is qualitatively different from that expressed in (1) before. Along these lines a good example is given by Smart [1972], who asks the reader to assume the existence of a hypothetical physicist who is an excellent experimentalist and who knows Newton's laws of motion but is completely ignorant about statistical mechanics. Clearly, this physicist will not be able to calculate the behavior of perfect gases, but through repeated experiments of the position and velocity of individual particles he will conclude that positions appear to come from a uniform distribution and velocities appear to have been drawn at random from a Maxwell distribution. In a similar manner, the geomorphologist who finds that channel networks appear to be drawn (in their planar structure) from a topologically random population should not infer that there does not exist a fundamental dynamic behavior behind such an appearance.

Statistical mechanics demonstrates that it is possible to build theories about complex systems that are insensitive to the details, especially concerning simple random systems. In more complicated dissipative systems characterized by a flux of matter and energy, one also finds the emergence of robust generic properties revealing the presence of an order that results from the interaction of the exceedingly many individual components that make up the system.

Position (2) described before is also simplistic. To state, for example, that the most common values of R_L in random topological networks with link lengths drawn at random from a common distribution are near 2 does not say much about natural networks. In effect, one may assume that the initial network controlled by major tectonic conditions is of a primitive character that does not follow Horton's laws. Under this assumption the important question is then: What are the most fundamental characteristics of the dynamics that, under random fluctuations, will produce the evolution of the primitive network toward one that fulfills the empirical facts and laws observed in natural drainage networks? The evolution from the initial conditions and the role of chance in such an evolution will by necessity be inserted in a three-dimensional landscape.

Howard [1971b] points out that all random-walk models give roughly equivalent results, suggesting that the relation between random-walk networks and natural systems may be rather superficial. It would appear that these models are successful because they produce topologically random networks and common link length distributions. We indeed agree that the above relation is rather superficial. To start with, all these models include a dynamics in two dimensions full of ad-hoc assumptions and constraints. These models represented the first steps in the study and modeling of drainage networks and as such they were very valuable, but obviously, these models cannot be considered a realistic attack on the main problems and processes that control network growth and development.

Many others studied the random model. In particular, the expected characters of asymptotic ($N(1) \to \infty$) topologically random networks were studied in detail. Moon [see Waymire, 1989] derived the expected diameter (the maximum length from source to outlet measured in topological units) of random channel networks. Troutman and Karlinger [1984] studied the expected width function of topologically random channel networks. In Chapter 7 we will outline their results in connection with fractal theories of the hydrologic response. Waymire [1992] studied some mathematical problems involving the asymptotic analysis of rooted random tree graphs and branching patterns. Finally, Wang and Waymire [1991] investigated large deviation rates for Horton's ratios for random topology models.

Central to topologically random networks is the assumption that all possible configurations of the drainage systems are equally likely. We will argue in this book that assuming that some spanning and loopless network configurations are more likely than others is both suggestive and of important practical implications. We argue this by associating a probability to configurations, which in an important class of network models will be related to minimum energy dissipation of the entire structure. This element will introduce *necessity* as a factor in network growth. With respect to the role of necessity and its interaction with *chance*, it is important to distinguish between two possible viewpoints in the establishment of the network structure. One viewpoint would impose a dynamics toward some optimal state, whatever the definition of optimality happens to be. In this case, the goal of evolution is established a priori from outside the system itself. One could also envision that optimality in complex systems with many interacting elements is not a program of action imposed from the outside, but rather something that arises from the coevolution and interaction of the many constitutive elements of the system. Thus in one case, the system evolves and rearranges itself as a whole according to what is best for a preestablished goal. In the other case, it moves spontaneously toward a most convenient state. In both cases, the driving force is eminently random in the form of water and energy input from storms that precipitate into the basin, landslides and sediment movement of different causes, tectonic forces and changes in the landscape of the watershed, etc. These random inputs produce fluctuations in one element of the system, which then induces changes in his neighbors and so on. Avalanches of changes of all different sizes may thus be observed throughout time.

In the context of drainage networks, the dynamics toward a preestablished optimal state in terms of energy expenditure leads to what we call optimal channel networks (OCNs), which are the subject of Chapter 4. When the dynamics is such that the networks spontaneously move toward a critical state, we have what we call self-organized critical (SOC) channel networks, studied in Chapter 5. As we will show, it is highly suggestive and of major consequence that OCNs and SOC networks are statistically indistinguishable from one another in their planar forms.

The fundamental character of the stochastic noise, as the driving force of OCNs and SOC-type models, is different from the role that the random component plays in the subset of models that we study in the following sections of this chapter. In these models the driving mechanisms are of deterministic character, and the aspect of chance in the models lies in the conditions over which those mechanisms operate. Thus one could operate those models under completely deterministic conditions. In which case, results will be dramatically different than when the models act upon random fields that describe the spatial structure of one or more of the variables embedded in the deterministic mechanism. Nevertheless, although these differences make clear how important the stochastic nature of those variables is in the representation of the appropriate structure of the drainage network, the role of chance in those types of models and its interplay with necessity is not as fundamental as in the case of OCNs and SOC networks.

1.3.4 Limitations of Statistical Models

Random-walk and random-topology models are statistical in nature and produce fractal structures (as will be seen in Chapter 2) closely resembling many planar features of real river networks. Through such models, network evolution is viewed as a growth process without an explicit linkage with the dynamics of flow and transport processes. Thus the foregoing statistical models do not really link dynamical processes and fractal structures, nor do they provide insight into the geomorphic process responsible for network generation.

In essence, purely statistical models of drainage networks exhibit one of the key ingredients of nature's design, chance, but lack the other, necessity. Elsewhere in this book we will explore new tools, blending statistical and deterministic rules, in which both ingredients will be at work. In our view it is only through the interplay of chance and necessity that different types of fractal growth and their characteristic dynamic processes can possibly be linked.

1.4 Deterministic Models of Drainage Network Development

1.4.1 Introduction

This section does not attempt an exhaustive review of models of deterministic character. Rather, its purpose is to present a description of the main schemes that have been used in the studies of river basin development. As explained before, the driving mechanisms are deterministic although they operate under random conditions, either for some of the variables or in the initial conditions.

The models studied in this section incorporate the altitude dimension $z(\mathbf{x})$ (where $\mathbf{x}$ is the arbitrary vector coordinate) as a fundamental part of their structure.

1.4.2 Models Based on Junction Angle Adjustments

Howard [1971a,b,c; 1990] introduced simulation models of stream networks by headward growth and branching. The models start from an arbitrary initial stream network developed on a square matrix, those produced by random headward growth and described earlier in this chapter. Each stream head is assumed to receive an equal discharge Q_0 at its head. Furthermore, each internal and external segment is assumed to receive lateral drainage at a rate proportional to its length, where the constant of proportionality, α, is set to unity in the simulations. Thus the discharge at a point along a segment is equal to the weighted sum of the lengths of all the segments upstream from the point plus Q_0 times the numbers of sources. Howard [1990] assumes a value for Q_0 of 0.3α. The channel gradients ∇z are assumed to be related to discharges through the relationship

$$\nabla z = KQ^{\theta} \tag{1.34}$$

where the exponent θ is taken as a constant for each simulation, with values fixed at the start between -0.1 and -0.75. The initial network and Eq. (1.34) allow computation of the vertical drop of any segment of the initial network, which is then subjected to successive captures.

In the capture model any given segment may be rearranged from its present course to a different course if the downstream gradient along the alternate path is greater than the existing gradient, subject to limitations on the number of streams (two) allowed to join at a given site.

The phenomenon of stream capture changes the angle of junction, not by gradual shifting of the point of junction because of differential sedimentation and erosion, but by discrete shifting or capture of one stream segment at the junction by another [Howard, 1971a]. For example, for small values of the angle A (Figure 1.48) the gradient from point 1 to point 2 may be greater than the gradient from point 1 to the junction if $\nabla z_1 > \nabla z_2$ is assumed (here for brevity ∇z_1 is the slope of link 1). In such cases, advantageous capture [Howard, 1971b] might occur along the path described in Figure 1.48, resulting in changes in the angles of junction. If ΔH is the vertical drop and L is the distance between points 1 and 2, then

$$\Delta H = \nabla z_1 l_1 - \nabla z_2 l_2 \tag{1.35}$$

and

$$L^2 = l_1^2 + l_2^2 - 2 l_1 l_2 \cos A \tag{1.36}$$

For a fixed point 1 on channel 1, the point 2 on channel 2 at which the gradient is a maximum is found by taking the partial derivative of $\Delta H/L$ with respect to l_2 [Howard, 1971a]:

$$l_2 = l_1 \left(\frac{\nabla z_1 \cos A - \nabla z_2}{\nabla z_1 - \nabla z_2 \cos A} \right) \tag{1.37}$$

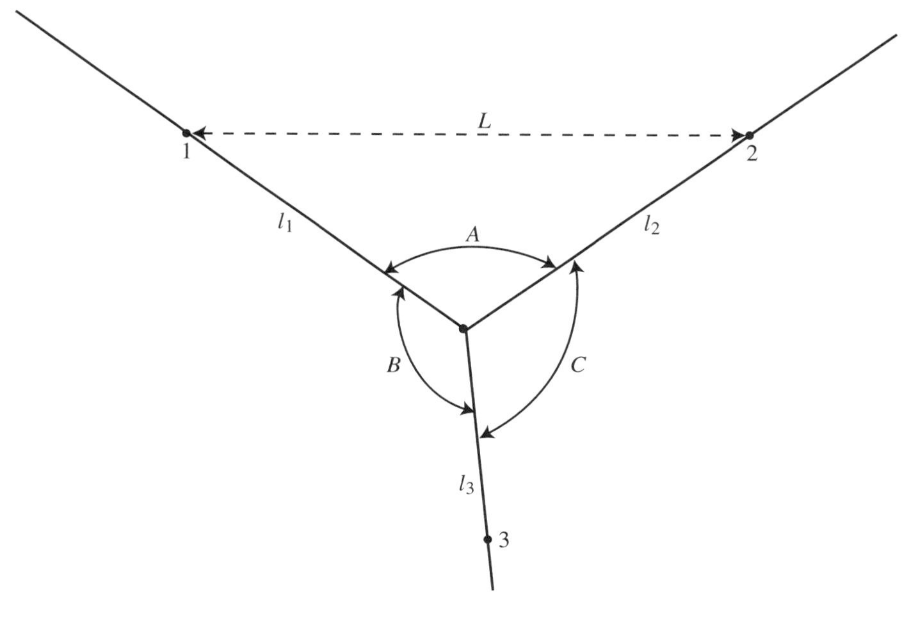

Figure 1.48. Plan view of stream junction illustrating terminology [from Howard, 1971].

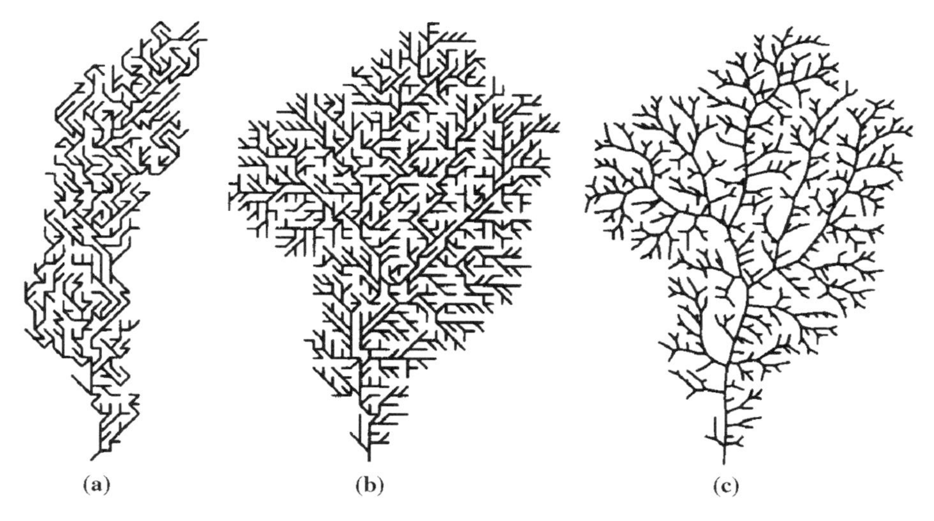

Figure 1.49. Evolution of a representative network developed in a 100 × 100 matrix in capture and minimum power models [from Howard, 1990]: (a) headward growth; (b) headward growth after modification by systematic capture; (c) minimum power optimization of systematic capture [from Howard, 1990].

If l_2 is zero or negative (i.e., along the imaginary downstream projection of stream 2 beyond its junction with stream 1), then ∇z_1 must be equal to or greater than the gradient at any point upstream from the junction, and advantageous capture is impossible. Thus capture will not occur if

$$\nabla z_1 \cos A - \nabla z_2 \leq 0 \tag{1.38}$$

Figure 1.49 from Howard [1990] shows the initial network (Figure 1.49(a)) and modifications it suffers by systematic capture (Figure 1.49(b)). Not only the network is modified but the drainage area increases considerably with significant changes in the basins boundaries. Stream profiles are continuously updated using Eq. (1.34) during the capture process.

The capture model described before is an approximation to an optimization model which operates on the network obtained through successive captures. The simulation adjusts the junction angles with the criteria of minimizing the power or rate of work, P_Ω, occurring at the junction:

$$P_\Omega = \sum_{i=1}^{3} C_i L_i \tag{1.39}$$

with

$$C_i = \rho g Q_i \nabla z_i \tag{1.40}$$

where Q_i is the discharge flowing through the channel segment with gradient ∇z_i, ρ is the fluid density, and g is the gravitational constant.

The power expended along the ith segment of length L_i is formed by integrating the product $\rho g Q \nabla z$ along the segment:

$$(P_\Omega)_i = \rho g \int_0^{L_i} (Q_i + \alpha\sigma)\nabla z(\sigma)d\sigma \tag{1.41}$$

where Q_i is now the discharge entering the segment from upstream, σ is the position along the segment measured from its upstream end, and $\nabla z(\sigma)$ is the gradient of the stream at position σ. The simulations assume the validity of Eq. (1.34) and the invariance of α.

The power expended along the segment is

$$\begin{aligned}(P_\Omega)_i &= \rho g K \int_0^{L_i} (Q_i + \alpha\sigma)^{1+\theta} d\sigma \\ &= \rho g K \left(\frac{(Q_i + \alpha L_i)^{2+\theta} - Q_i^{2+\theta}}{\alpha(2+\theta)} \right)\end{aligned} \tag{1.42}$$

The total power at a junction is the sum of the power expended along segments entering and leaving the node. The gradient of power in the x direction is calculated as $\partial P_\Omega/\partial x$ and similarly in the y direction as $\partial P_\Omega/\partial y$. Thus the x-direction gradient for a junction at (x, y) is [Howard, 1990]

$$\begin{aligned}\frac{\partial P_\Omega}{\partial x} &= \rho g K \sum_{i=1}^{3} \left\{ (Q_i + \alpha L_i)^{1+\theta}(x - x_i)/L_i \right\} \\ &\quad + \rho g K \sum_{i=1}^{2} \left\{ \left[(Q_3 + \alpha L_3)^{1+\theta} - Q_3^{1+\theta} \right](x - x_i)/L_i \right\}\end{aligned} \tag{1.43}$$

which makes use of the relationships

$$Q_3 = Q_1 + Q_2 + \alpha L_1 + \alpha L_2 \tag{1.44}$$

$$L_i = \sqrt{(x - x_i)^2 + (y - y_i)^2} \tag{1.45}$$

Howard's [1990] simulation model proceeds then as follows. Prior to each iteration, the lengths and contributing discharges for individual segments are calculated. Then the gradients of power or rate of work in the x and y directions are computed using the above approach. The individual nodes of the network resulting from the capture model are then moved in the direction opposite to the gradient times the factor $\mathcal{K}_f \bar{L}/\nabla(P_\Omega)_{\max}$, where $\nabla(P_\Omega)_{\max}$ is the maximum gradient of power occurring within the network, $\bar{L}$ is the average link length, and $\mathcal{K}_f$ is a relaxation parameter set to a small

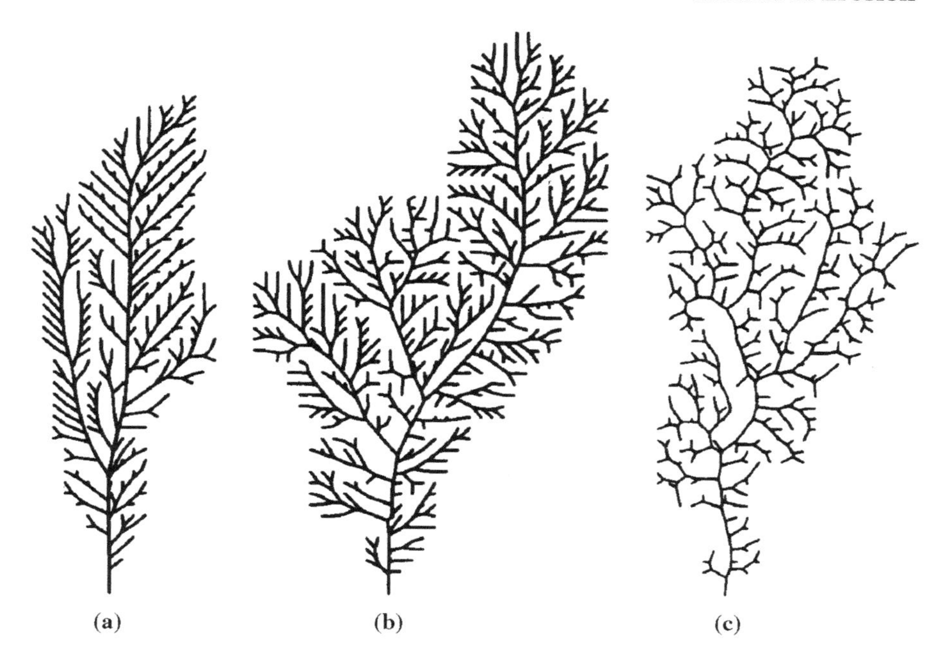

Figure 1.50. Comparison of stream networks generated from the initial network shown in Figure 1.48: by systematic capture and minimum power optimization with (a) $\theta = -0.1$, (b) -0.25, and (c) -0.75, respectively [from Howard, 1990].

enough value that the results are independent of its value (Howard's [1990] simulation uses $\mathcal{K}_f = 0.005$). Only interior nodes are allowed to move during the simulation; the sources and outlet remain fixed. This constraint is necessary because otherwise the solution will tend to a network whose segments all will have zero length.

Howard also describes conditions that, once reached, end the successive recalculations. The most important and controlling one is that the simulation ends when the length of any individual segment reaches a critical lower limit (which Howard sets at 0.05 times the average segment length). Figure 1.49(c) shows the resulting network from this procedure after it has acted on the network obtained from systematic capture (Figure 1.49(b)).

The value of the exponent θ in Eq. (1.34) is very important in the network pattern that is obtained as the final result. Figure 1.50 shows different networks generated using different values of θ from the same initial condition.

Howard's schemes have interesting features that incorporate the altitude dimension of the network structure. Nevertheless, these schemes rely on a number of assumptions and tuning parameters that markedly contrast with the very robust and general features of the self-organized systems that appear to be pervasive in natural structures. Also, it is important to notice that the above scheme is applied locally at each junction and starts from networks such as those produced by random headward growth. This is markedly different from the OCNs we study in Chapter 4, where the evolution of the network takes place from arbitrary initial conditions explicitly following a global optimum search.

1.4.3 Models of Erosion and the Evolution of River Networks

Important processes affect the planimetric and elevation structure of drainage basins. Chief among these are the interplay of the prevailing erosional processes of dispersive or concentrative nature; the spatial and temporal development of channel links, progressing mostly by headward growth and branching and possibly affected by climatic changes; the large-scale migration of valleys and divides and the related capture processes; and the progressive adjustments of junction angles of confluent streams. Many models have been developed in the past to provide realistic simulations of the temporal evolution of landforms based on a deterministic description of the effects of the chief geomorphic agents. However, some general questions [e.g., Howard, 1994] are relevant to the basic motivation behind this book. What is the simplest mathematical model that simulates morphologically realistic landscapes? What should be the effects of initial conditions and inheritance on basin form and evolution? What are the relative roles of deterministic and random processes in basin evolution? Do processes and forms in the drainage basin embody principles of optimization and, if so, why? Is there some characteristic river basin form that is invariant in time even under a change in the relative role of the chief landforming processes? Throughout this book we will address the above questions. In this section we will outline a mathematical description of the chief geomorphic agents active within a river basin.

The development of drainage basins requires at least two superimposed processes [Howard, 1994; Rigon, Rinaldo, and Rodriguez-Iturbe, 1994]. One must be a *diffusional* or dispersive, creep-like mass-wasting process capable of eroding the land surface (by weathering, rainsplash, or other means) with finite gradients ∇z even for vanishingly small contributing areas A. Such a process must be characterized by a progressive loss of efficiency as A increases so that in the average the gradient of land surface increases downslope if rates of surface lowering are essentially uniform in space. The other is an *advective*, or concentrative, fluvial process that increases its efficiency with contributing area, that is, with flow rates, but requires large gradients for small values of A. The resulting combination of processes and the embedded spatial transition from slope-dependent mass-wasting to concentrative runoff processes in channels justifies the essentials of the morphology of the river basin, as recognized by the terse statement by Gilbert [1909], recently quoted by Howard [1994]:

"... On the upper slopes, where water currents are weak, soil creep dominates and the profiles are convex. On lower slopes water flow dominates and profiles are concave."

One example of the interplay of convex and concave landforms is shown in Figure 1.51.

The theoretical treatment of drainage network morphology has been dominated for years by two approaches of different nature. One is based only on deterministic rules, thus recognizing consistent nonrandom patterns in nature. The second approach, discussed in Section 1.3, relies on the fundamental stochastic postulates of random topology and its corollary regarding the existence of recurrent laws in nature as the outcome of most probable states. Nevertheless, both random and nonrandom effects are likely to be simultaneously operating in the development of a drainage network. Scheidegger's [1979] principle of antagonism in geomorphological evolution is a significant example of recognition of such concurrent

Figure 1.51. Aerial view of hillslope and channel patterns. Here the interplay of convex and concave topographies is particularly clear [courtesy of W. E. Dietrich].

characters where forces of endogenic (i.e., tectonic) character (systematic, nonrandom at basin scale) and forces of exogenic or weather-related nature (substantially random) are seen to shape the morphology of river courses and, more generally, the physiography of the Earth's crust.

Drainage processes and forms at the basin-level scale are defined by the evolution of the river basin elevation field. The general form of the governing equation for elevation changes in an open dissipative fluvial system is then basically

$$\frac{\partial z(\mathbf{x}, t)}{\partial t} = \nabla \cdot \mathbf{F} + U + \eta(\mathbf{x}, t) \tag{1.46}$$

where z is the landscape elevation at the arbitrary site of coordinates $(x, y) = \mathbf{x}$ belonging to the catchment; t is time; and $\mathbf{F}$ defines the eroded material flux of sediments per unit bulk density of surface material. Such flux is defined by the modes of transport and its sign is suitably defined; U is an arbitrary uplift rate; and $\eta(\mathbf{x}, t)$ is an additive noise corresponding to exogenic agents. In this context we maintain a distinction between the additive factors U and η in Eq. (1.46) because of the differing physical meaning and mathematical structure. Indeed, for the scope of this book it makes sense to assume $U(\mathbf{x}, t) = U$, that is, a constant uplift rate; whereas, for example, in the context of slope evolution models a random forcing η uncorrelated with U will play an important role. In fact, it will be seen (Chapter 4) that important consequences to landscape evolution may be induced by the interplay of random exogenous fluctuations with deterministic, mass-conserving relaxations.

Eq. (1.46) represents the continuity equation for sediment mass and elevation. In the case of nonactive tectonics and simple, deterministic relaxations we simply have

$$\frac{\partial z(\mathbf{x}, t)}{\partial t} = \nabla \cdot \mathbf{F} \tag{1.47}$$

With the reasonable assumption that the flux $\mathbf{F}$ is oriented in the direction of the gradient (the steepest topographic descent), we have

$$\mathbf{F} = -F \frac{\nabla z}{|\nabla z|} \tag{1.48}$$

where the ratio on the right-hand side is the unit downslope vector.

Smith and Bretherton [1972] (and later Lowenhertz [1991]) enforced continuity of flow as well as continuity of sediment in a one-dimensional setup to yield the following system of coupled equations:

$$\frac{\partial z(\mathbf{x}, t)}{\partial t} = \frac{\partial F(Q, |\nabla z|)}{\partial x} \tag{1.49}$$

$$\frac{\partial Q\,(\mathbf{x}, t)}{\partial x} = p \tag{1.50}$$

where Q is the instantaneous flow rate and p is some constant rainfall rate. Eqs. (1.49) and (1.50) constitute a coupled system of partial differential equations where the rate of change of elevation is dependent on the net flux of sediments, which in this one-dimensional system is forced by a linear increase in discharge. Although clearly removing the limitation of a one-dimensional coordinate system dramatically increases the complexity of the description of the system, the key factor in Eq. (1.50) is the coupling of the developing landscape $z(\mathbf{x}, t)$ with the flow rate Q, a fundamental step in the understanding of landscape self-organization.

The above system of Eqs. (1.49) and (1.50) was initially used to understand channel initiation mechanisms through linear stability analysis of the response of a hypothetical base state to small perturbations in the

Figure 1.52. Gully erosion in a trench displaying quasi-parallel erosion patterns. Small-scale heterogeneities force local aggregation at random locations. The related increase in flow rates and potential erosion stabilizes the aggregation patterns. When the average lengths of the erosion paths become comparable with the scale of heterogeneity, a fully aggregated structure is expected [courtesy of W. E. Dietrich].

landforms of the surface z. Smith and Bretherton [1972] used Eq. (1.49) and (1.50) without specifying the detailed functional relationship $F(Q, |\nabla z|)$ relating the slope $|\nabla z|$ and the water flow Q, assumed to be a function of the area draining through unit contour width A (i.e., $Q \approx A$). They studied the behavior of an initial smooth surface subject to steady and uniform external forcing U and p. This base state appears to be unstable, and small perturbations grow indefinitely when

$$F - Q\frac{\partial F}{\partial Q} < 0 \tag{1.51}$$

A one-dimensional equilibrium form is concave when Eq. (1.51) is satisfied and convex otherwise. A basic equivalence is thus established between concavity of a one-dimensional profile and instability in the two-dimensional landscape. Smith and Bretherton [1972] assumed that instability as characterized by the growth of perturbations leads to rilling and ultimately to channel growth. Figure 1.52 shows an example of growth of perturbations for gully erosion in a trench.

Lowenherz [1991] reevaluated the stability criterion (1.51). She suggested a short wavelength cutoff to represent the lower scale limit of validity of the continuum equations for surface evolution, resulting in a modified stability threshold. Other refinements of the Smith and Bretherton approach have dealt with characterizations of the transport capacity F of the surface flow [e.g., Luke, 1974; Ahnert, 1987; Tarboton, Bras, and

Rodriguez-Iturbe, 1992], and, significantly, it has been established that the overall concavity of the stable profile is related to the dominant mode of transport, whether it be weathering- or transport-limited.

As noted before, the transition from the hillslope system (roughly, characterized by convex topography) to the channel system embedded in concave valleys implies a change in the dominant mode of sediment transport. As an example, in weathering-limited hillslope transport, here indicated by a subscript 1, the flux of sediments can be assumed to be slope-limited; that is, $F_1 = F_1(\nabla z)$. In alluvial channels the rate of sediment removal is generally governed by both slope and water discharge; that is, $F_2 = F_2(Q, \nabla z)$. The landscape evolution equation can therefore be expressed as

$$\frac{\partial z(\mathbf{x}, t)}{\partial t} = \nabla \cdot \big(\mathbf{F}_1(\nabla z) + \mathbf{F}_2(Q, \nabla z)\big) + U \tag{1.52}$$

where regime (1) is identified by the condition $F_1 \gg F_2$ and regime (2) is identified by $F_2 \gg F_1$. In the transition regions we find that F_1 and F_2 are of the same order of magnitude, and thus couplings of the two mechanisms are expected.

In this section we will examine two approaches. In the first approach (Section 1.4.4) both F_1 and F_2 are operating and a model of their interplay is employed. In the second we consider the impact of detachment-limited transport, that is, the common case in which the actual transport rates are in some places considerably less than would be predicted for transport-limited processes. Other approaches [e.g., Ahnert, 1976, 1984; Cordova, Rodriguez-Iturbe, and Vaca, 1982; Roth, Siccardi, and Rosso, 1989] have dealt with the particular case where $F_2 \gg F_1$ and we will not review them in detail here.

Whatever the mathematical specification of diffusive or advective processes, we notice that timescales of the processes resulting in sediment fluxes F_1 and F_2 are likely to be quite different, and thus it is reasonable to assume that the evolution of the land surface is given by

$$\frac{\partial z(\mathbf{x}, t)}{\partial t} = \nabla \cdot \mathbf{F}_2(Q, \nabla z) \tag{1.53}$$

whenever fluvial erosion processes are active in sediment production. This is the case during most rainfall events that trigger sediment production both in the hillslopes and channels. This production lasts much longer than the rainfall itself, but in geological timescales one may think that most of the time rainfall is not active on the hillslopes, and moreover sediment transport by river flow has reached a dynamic equilibrium of production and transport. Thus over long, geological timescales one has

$$\frac{\partial z(\mathbf{x}, t)}{\partial t} = \nabla \cdot \mathbf{F}_1(\nabla z) + U \tag{1.54}$$

Notice that a random component is not explicitly assumed in Eq. (1.54) but may enter the system – and deeply affect its evolution – through initial conditions.

Let us assume for simplicity that diffusive mass wasting, F_1, can be described by the simplest mathematical form:

$$\mathbf{F}_1(\nabla z) = D \nabla z \tag{1.55}$$

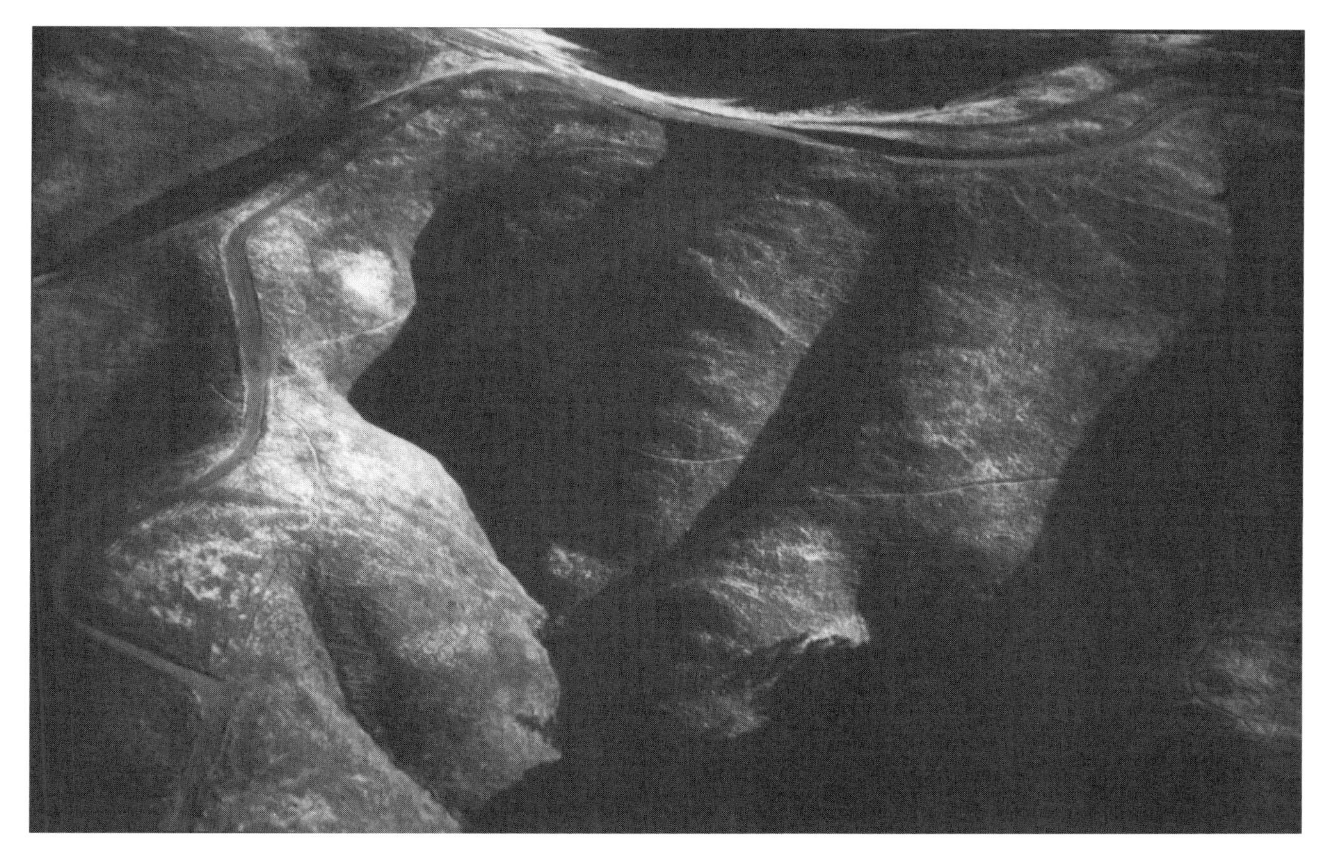

Figure 1.53. Hillslope patterns in vegetated areas with clearly diffusional features [courtesy of W. E. Dietrich].

where D takes on the role of a diffusion coefficient specifying the relevant time and spatial scales of hillslope evolution processes. Figure 1.53 shows an example of a landscape with predominantly diffusional mechanisms.

A complete statement of the mathematical problem requires the coupled determination of Q and ∇z. The mathematical problem implies a system of equations composed of the continuity equations for sediment and flow plus the momentum equation of the carrier flow. Under the reasonable assumption that the flow is assumed locally uniform, that is, the component of weight in the flow direction locally balances resistances, one has [e.g., Kramer and Marder, 1992]

$$\frac{\partial z}{\partial t} = \nabla \cdot F_2(Q, \nabla z)\nabla z + D\nabla^2 z + U \tag{1.56}$$

which defines the mass balance equation for sediments. For the water flow one has

$$\frac{\partial d}{\partial t} = \nabla \cdot Q\nabla z + p \tag{1.57}$$

where $\nabla z \approx \nabla z/\sqrt{1 + |\nabla z|^2}$, which assumes that the flow is locally directed along the steepest descent. Usually one assumes

$$Q\nabla z \propto \frac{d\nabla(d + z)}{\sqrt{1 + |\nabla(d + z)|^2}} \sim \frac{d\nabla z}{\sqrt{1 + \nabla z^2}} \tag{1.58}$$

In the above equations, $z(\mathbf{x}, t)$ is the evolving landscape elevation in the arbitrary site of coordinates $\mathbf{x}$ belonging to the catchment; $d(\mathbf{x}, t)$ is the flow depth; t is time; Q is the flow rate; $F_2(Q, |\nabla z|)$ is any suitable fluvial transport formula; U is the rate of uplift; p is the input of effective rainfall forcing the system, possibly distributed in space and time; geology is specified by U, and the nature of the exposed lithology (affecting surface erodibility) is defined by D, the diffusion term in the sediment continuity equation.

Eq. (1.56) states that the net rate of change of landscape elevation is balanced by the net sediment transport by flowing water, geologic uplift and diffusion, or other landscape-forming processes.

Eq. (1.57) defines the rate of change of flow depth as a result of the accumulation of discharge by flow convergence induced by the forming surface and the injection of flow from precipitation. Notice that flow rates are defined by a scalar intensity, Q, whose direction is the unit downslope gradient ∇z.

Eq. (1.58) is a simplified equation that states that the flow (locally uniform) is directed along the steepest descent.

One important issue concerns the determination of the flow rates, Q, that determine significant mass erosion. From Eq. (1.57), one observes that, given any fixed point i within a basin with total contributing area A_i, integration of both sides over A_i has the interesting corollary that, by Green's theorem, the flux term $\int_{A_i} \nabla \cdot Q\nabla z dA = -Q_i$, where Q_i is the flow rate at the chosen point because fluxes are by definition null along the boundaries of the basin defining A_i. Thus

$$Q_i = \int_{A_i} p\, dA - \int_{A_i} \frac{\partial d}{\partial t}\, dA \tag{1.59}$$

In steady–state conditions ($\partial d/\partial t \sim 0$) one has

$$Q_i \sim \int_{A_i} p\, dA \propto A_i \tag{1.60}$$

the last proportionality is meaningful whenever landscape-forming events are characterized by substantial spatial homogeneity of p. The importance of Eq. (1.60) lies in the fact that determination of the key ingredient of the variation in space of landscape-forming flow rates can be done on a purely topographic basis, that is, via Eq. (1.14).

Many studies have assumed that the transport capacity $\mathbf{F}$ of many geomorphic processes can be approximated as a power function of contributing area, A, and slopes, ∇z, that is [e.g., Armanini and Di Silvio, 1992]

$$\nabla \cdot \mathbf{F} \propto A^m \nabla z^n \tag{1.61}$$

where for most processes, $m, n \geq 0$. In Eq. (1.61), total contributing area clearly surrogates flow rates. Howard [1994] noted that the crossovers between diffusional and concentrative processes may be illustrated by a steady-state landscape in which the erosion rate is areally uniform and balances the rate of tectonic uplift U. Then the transport through any location on the landscape must equal, over the long run, the product of uplift rate and contributing drainage area:

$$\nabla \cdot \mathbf{F} = UA \tag{1.62}$$

If the land surface is everywhere *transport-limited*, so that actual transport rates equal those given in Eq. (1.61), the following relationship is obtained [Willgoose, Bras, and Rodriguez-Iturbe, 1991a]:

$$\nabla z \propto U^{1/n} A^{(1-m)/n} \tag{1.63}$$

Processes with $m < 1$ will be diffusional, and those with $m > 1$ will be concentrative. As argued before, mass-wasting processes and creep are commonly modeled as diffusional with $m = 0$ and $n = 1$. For example, transport of sand in alluvial channels is usually assumed as concentrative with $m \sim 1.5$ and $n \sim 2$.

Diffusional processes are commonly equated with mass wasting and concentrative processes are identified with wash processes, but exceptions occur. For instance, in circumstances where stream discharge decreases in the downstream direction or where transport of coarse bed load in stream networks occurs without any form of downstream fining, a diffusional regime may be appropriate for modeling landforms [e.g., Paola, 1989; Paola, Heller, and Angevine, 1992]. The other extreme case is that of snow and rock avalanches on steep, bedrock slopes that may produce concentrative erosion by eroding chutes and creating spurs and gully terrains [e.g., Howard and Selby, 1994]. Debris flows, altering low-order channels through their powerful action, frequently originate from the accumulation of colluvial deposits in high-relief slopes [e.g., Dietrich et al., 1993] and often act on diverging topographies. Even from this superficial overview the complexity of a detailed specification of all possible processes is manifest, and thus we will concentrate here only on the main modeling structures.

In many landscapes the actual transport rates are in some places considerably less than would be predicted from transport-limited processes. Howard [1994] noted that one case occurs when bedrock is exposed on slopes and the rate of local erosion is determined by weathering rates, that is, in *weathering-limited* conditions. The volume of bed-sediment transport by wash processes on hillslopes and low-order rills and channels is limited by the ability of flow to entrain residual soil or colluvium or bedrock, giving rise to *detachment-limited* conditions.

The transient response of detachment-limited landscapes to variations in rate of uplift (or, equivalently, in base-level lowering) is considerably different than for transport-limited conditions, with important implications for landscape evolutions. Indeed, many models have distinguished suspended-loads (detachment-limited) from local, point-to-point runoff erosion (transport-limited), as we will show later on in this section.

The system whose evolution is described by Eq. (1.52) is in equilibrium when $\partial z(\mathbf{x}, t)/\partial t \approx 0$, which in turn implies that the net rate of sediment transport is balanced by the uplift rate:

$$\nabla \cdot \left(\mathbf{F}_1(\nabla z) + \mathbf{F}_2(Q, \nabla z)\right) = -U \tag{1.64}$$

In the case of nonactive tectonics ($U = 0$) and for any case where sediment transport may be described by Eq. (1.61), equilibrium will only be reached for a peneplain, that is

$$\nabla z(\mathbf{x}, t) = 0 \tag{1.65}$$

However, fluvial processes may exhibit a different behavior. In particular, the exceedence of a critical shearing stress, say τ_c, may be required to trigger erosion. As such fluvial erosion is characterized by a threshold-dependent net flux

$$\nabla \cdot F_2(Q, \nabla z) = \alpha f\left(\tau(Q, \nabla z) - \tau_c\right) \tag{1.66}$$

where α characterizes the timescale of fluvial erosion processes, and $f(\xi) = 0$ for $\xi \leq 0$. The shear stress in dynamic equilibrium is given by

$$\tau = \rho g\, d\, \nabla z \tag{1.67}$$

for a sheet flow of depth d over a slope ∇z. The fluid density is ρ, and g is the acceleration of gravity. The general formulation of the sediment continuity equation can therefore be written in the form

$$\frac{\partial z(\mathbf{x}, t)}{\partial t} = \alpha f(\tau - \tau_c) + \beta F(\nabla z) + U \tag{1.68}$$

where $z(\mathbf{x}, t)$ is, as usual, the elevation field; $f(\cdot)$ is the threshold-dependent erosion function appearing in Eq. (1.66); τ is the local shear stress; τ_c is the critical threshold for erosion; α is the coupling constant with dimension of the inverse of time, defining the characteristic timescale for threshold-dependent erosion activities; and $\beta F(\nabla z) = \nabla \cdot \mathbf{F}_1(\nabla z)$ is the net flux of threshold-independent transport, not necessarily in the diffusive form $D\nabla^2 z$. Hence here $F(\nabla z)$ is the transport law made slope-limited; β is the inverse of the timescale characteristic of slope-limited sediment transport.

An interesting case is characterized by $\alpha \gg \beta$, implying that the two processes, which mimic fluvial and hillslope evolution processes, have very different timescales and thus can be decoupled. Thus the evolution equations are respectively

$$\frac{\partial z(\mathbf{x}, t)}{\partial t} = \alpha f(\tau - \tau_c) \tag{1.69}$$

whenever threshold-dependent processes are acting, in analogy to Eq. (1.53), and

$$\frac{\partial z(\mathbf{x}, t)}{\partial t} = \beta F(Q, \nabla z) + U \tag{1.70}$$

otherwise (as in Eq. (1.54)). Randomness, again not explicitly accounted for, is added to the system through initial conditions.

Whatever the specification of the dynamic parameters, Eq. (1.68) defines a complex nonlinear coupling because of the threshold term. We note that equilibrium surfaces $z(\mathbf{x}, \infty)$ (i.e., for which $\partial z/\partial t = 0$) of fluvial landscapes ($\beta = 0$) do not need to be flat in the absence of uplift ($U = 0$). In fact, once the slopes progress toward a state generating shear stresses lower than the threshold τ_c, the flattening action of erosion is stopped.

In the above framework, heterogeneities in the exposed lithology, in the geologic structure, or in the vegetational cover affecting the shear resistance of the surface layer can be described by arbitrarily complex structures of the field parameters $\tau_c(\mathbf{x})$ and $\beta(\mathbf{x})$. The surface resistance can also be made arbitrarily dependent of the degree of erosion achieved (i.e., $\tau_c(\mathbf{x}, z)$, $\beta(\mathbf{x}, z)$) or on climatic fluctuations (i.e., $\tau_c(\mathbf{x}, t)$).

The crux of the matter is whether the above threshold-dependent dynamic specification yields erosion activities that self-organize themselves into a process reproducing the observational morphology of river networks.

1.4.4 A Process-Response Model of Catchment and Network Development

Willgoose, Bras, and Rodriguez-Iturbe [1989; 1990; 1991a–d; 1992] present the development and applications of a process-response model for the erosional development of catchments and their channel networks. This section is taken from Willgoose et al. [1991a].

The elevations within the catchment – both hillslope and channel – are simulated by a mass transport continuity equation applied over geologic time. Mass transport processes considered include fluvial sediment transport, such as that modeled by the Einstein–Brown equation, and mass movement mechanisms such as creep, rainsplash, or landslide. A crucial component of this model is that it explicitly incorporates the interaction between the hillslope and the growing channel network based on physically observable mechanisms. An important and explicit differentiation between the processes that act on the hillslope and in the channels is made. A point is defined to be in a channel when selected flow and transport processes exceed a threshold value. If a function (called the channel initiation function) is greater than some predetermined threshold at a point, then the channel head advances to that point. The channel initiation function is primarily dependent on the discharge and slope at a point, and the channel initiation threshold is dependent on the resistance of the catchment to channelization. Channel growth is thus governed by the hillslope form and processes that occur upstream of the channel head. The channel initiation function concept is independent of Smith and Bretherton's [1972] definition of channels as points of instability in the flow equations. Nevertheless, the concept is not necessarily contradictory. This is particularly true given the recent realization that Smith and Bretherton's [1972] analysis would lead to a system of rills spaced at an infinitesimal distance apart unless a basic scale is built into the equations [Lowenherz, 1991]. Introducing this realistic scale of separation is conceptually consistent with the threshold analogy. The elevations on the hillslope and the growing channel interact through the different transport process in each regime and the preferred drainage to the channels that results. The interaction of these processes produces the *long-term form of the catchment.* The preferential erosion in the channels results in the familiar pattern of hills and valleys with hillslope flow toward the channel network in the bottoms of the valleys.

This conceptual model of catchment evolution has been used by Willgoose et al. [1991a] to study a number of crucial issues in river basin geomorphology. These issues include sensitivity to changes in exogenous forces such as variations in tectonic uplift, attainment of dynamic equilibrium and the implications involved in this process, a new formulation of landscape evolution, and a process-oriented classification of landscapes. We will not deal with these issues here, limiting the presentation to a brief description of the model as published in Willgoose et al. [1991a].

The structure of the model is as follows. The model describes the long-term changes in elevation with time that occur in a drainage basin as a result of large-scale mass transport processes. The mass transport processes modeled are tectonic uplift, fluvial erosion, creep, rainsplash, and landsliding. Some of these processes are modeled with their own physics, others are aggregated into a diffusive term, as explained below. For long-term elevation changes it is necessary to model the average effect of these processes with time, so events like individual landslides are not modeled; rather, the aggregate effect of many landslide events is modeled. Consequently, the model describes how the catchment is expected to look, on average, at a given time.

A crucial component of the model, which distinguishes it from previous geomorphological process models, is that it explicitly differentiates between the part of the catchment that is channel and the part that is hillslope. Using this distinction, different transport processes are modeled in each regime. Thus channels will typically be dominated by fluvial erosion whereas, on the hillslope, diffusive processes will be more apparent and may even dominate in some circumstances. A channel is assumed to form when a function, called the channel initiation function, exceeds a threshold, hereafter called the channel initiation threshold. The channel initiation function is nonlinearly dependent on the hillslope discharge and slope. The adopted form of the channel initiation function is one that is capable of modeling overland flow velocity, bottom shear stress, and groundwater stream sapping with modification of only two parameters. The general concept of the channel initiation function and its threshold behavior in channel growth may also be used to model channel advance from localized landsliding [Dietrich et al., 1987, 1988]. A restriction of the adopted channel model is that once a channel is formed, by the channel initiation function exceeding the channel initiation threshold, it exists forever. Thus channel heads may only advance, not retreat, and the planar position of the channel may not change with time. The latter restriction on planar position is not believed to be critical because, although channels meander about their floodplain, the general position of the valley floor is more or less fixed. The former restriction related to channel advance is also not considered to be critical because the associated elevation equation is a mean equation, and for compatibility the mean channel network should be used; fluctuations in the channel head position back and forth over short timescales are ignored. It is argued that while channel heads advance and retreat over short timescales, these fluctuations do not dominate the process of fixing the mean position of the channel head and the mean elevations.

The mass transport continuity equation used for describing the changes in elevation is

$$\frac{\partial z}{\partial t} = U(x, y, t) + \frac{1}{\rho_s(1-n)} \nabla \cdot \mathbf{F} + D \nabla^2 z \tag{1.71}$$

where (let [L], [M], [T] represent units of length, mass, and time, respectively): z = landscape elevation [L]; t = time [T]; $U(x, y, t)$ = rate of tectonic uplift [LT^{-1}]; x, y = the horizontal directions [L]; $\mathbf{F} = (F_x, F_y)$ = rates of fluvial sediment transport per unit width in the x, y directions, respectively [$\mathrm{M\ T^{-1}\ L^{-1}}$]; D = elevation diffusivity [$\mathrm{L^2\ T^{-1}}$]; ρ_s = density of sediment [$\mathrm{M\ L^{-3}}$]; and n = porosity of sediment.

In principle, the total fluvial sediment transport rate, $F = |\mathbf{F}|$, can be formulated in a general fashion so that it reflects the fluvial sediment transport processes observed in the field. One of the adopted forms for this equation is

$$F = f(Y)q^{m_1}\nabla z^{n_1} \tag{1.72}$$

and

$$Y = \begin{cases} 1, \text{ for a channel, } f(Y) = \beta_1 \\ 0, \text{ for hillslopes, } f(Y) = \beta_1 O_t \end{cases}$$

where q = mean peak discharge per unit width or $q = Q/w$ for channels; Y = indicator variable that takes the unit value at points where a channel exists and takes 0 on hillslopes; Q = mean peak discharge in channels; w = characteristic width of channels; ∇z = slope in the steepest direction at the point of interest; O_t = ratio of hillslope erosion rate to channel erosion rate; and β_1 = sediment transport rate coefficient.

It has been shown that the Einstein–Brown total-load sediment transport equation is well approximated by the above equation over a range of channel and rill geometries [Willgoose et al., 1989], and it is also believed to approximately model the process of soil wash [Kirkby, 1971]. The factor O_t is the ratio of the rate of overland erosion to channel erosion, and it is believed to be less than 1, though this has not been confirmed in the field. The parameters β_1, n, m_1, and n_1 can vary in space to represent heterogeneity of the soil system.

The discharge and the area contributing to a point in the catchment are related by

$$Q = \beta_2 A^{m_2} \tag{1.73}$$

where A is the area contributing runoff to that point, β_2 is a runoff coefficient, and m_2 is another parameter. Willgoose et al. [1989] propose that in the field the appropriate discharge to use in determining the long-term elevation changes due to fluvial transport is the mean peak discharge determined from a flood frequency analysis of a partial flood series of peak discharges [Langbein, 1947; Langbein and Schumm, 1958]. The lower cutoff for the series is that discharge below which insignificant sediment transport occurs. Experience with frequency analysis of annual flood series suggests that m_2 should be in the range of 0.5 to 1.0, and β_2 follows from the frequency analysis being somewhat analogous to a runoff rate [Strahler, 1964]. In experimental studies of catchment evolution, where the rainfall rate is constant for long periods of time [e.g., Parker, 1977], the value of m_2 is unity and β_2 is the runoff rate.

The diffusion term in Eq. (1.71) results from a mass transport of the form of:

$$\text{mass flux} = D\nabla z \tag{1.74}$$

which is mathematically equivalent to the fluvial sediment transport in Eq. (1.72) if $m_1 = 0$ and $n_1 = 1$. It has been proposed that this linear diffusion transport can model a number of physical processes including creep [Culling, 1963], rainsplash [Dunne, 1980], and landsliding for low slopes [Andrews and Bucknam, 1987].

The process whereby the catchment develops is one in which the channels extend, creating lines of preferred drainage and erosion and producing valleys. These valleys cause convergence of flow around the channel head (be it surface water or groundwater), where as a result the channel initiation function is high. If the channel initiation function exceeds the threshold, then channel extension occurs. Lateral growth off the sides of other channels occurs when the lateral inflow to some part of the existing channel network is high enough that the channel initiation function is greater than the threshold. Channel growth, and thus valley growth, stops when competition between advancing channel heads decreases the contributing area to the channel heads to such an extent that the channel initiation function is below the channelization threshold and the hillslopes are stable against channelization. If hillslope slopes are very small, the channel initiation function will also be small. This process is analogous to one proposed by Dunne [1980], where he expressed channel initiation in terms of groundwater sapping to channel heads.

The channel initiation function adopted, say a, is one with the general form

$$a = \beta_3 q^{m_3} \nabla z^{n_3} \qquad m_3, n_3 > 0 \tag{1.75}$$

so that channel growth is encouraged if discharges or slopes at the channel heads are increased. Discharges may, in turn, be increased by increased runoff or increased source areas to the channel head. The form of Eq. (1.75) is fairly generic. Criteria like surface velocities, shear stress, and groundwater seepage can all be cast in that form [see Willgoose et al., 1989]. The trends with discharge and slope are qualitatively consistent with field observations of natural catchments [Hadley and Schumm, 1961; Patton and Schumm, 1975; Dietrich et al., 1986, 1993; Montgomery and Dietrich, 1988, 1992] and disturbed catchments [Toy and Hadley, 1987]. However, at the regional level these trends may be confused by correlations between vegetation cover and rainfall, and thus runoff [Langbein and Schumm, 1958]. Coefficients β_3, n_3, and m_3 can vary in space.

The channel initiation function (1.75) feeds back as input into the channelization equation, which is of the type

$$\frac{\partial Y}{\partial t} = f\left(\frac{a}{a_t}\right) \tag{1.76}$$

The exact structure of Eq. (1.76) is irrelevant; the important fact is that the form of the function $f(\cdot)$ is such that when a is less than a threshold value a_t, $a/a_t < 1$, $Y = 0$ is the solution of Eq. (1.76), but once a/a_t exceeds unity, $Y = 1$ becomes the stable solution to the equation for all subsequent times regardless of the future value of a/a_t. The rate at which Y approaches 1 is controlled by a parameter; during this transition the affected regions behave with weighted properties of hillslopes and channels. A particular form of Eq. (1.76) is discussed in Willgoose et al. [1991c]. Through the value of Y (0 or 1) that controls Eq. (1.72), the channel initiation function also feeds back into Eq. (1.71), which describes the changes in elevation. Notice that channels are thus permanent once they are formed. Basically, the model involves two differential equations for elevations and channelization, Eqs. (1.71) and (1.76), which describe the two states of the system. It also involves two constitutive equations for sediment transport and channel initiation function, Eqs. (1.72) and (1.75), which feed back as inputs into the two state equations. Thus there is a nonlinear interaction between elevation and channelization and the channel initiation function and sediment transport in space. This interaction is the central feature of the model that drives the network growth as described in detail in Willgoose et al. [1991a–c]. The role of the channel initiation function is to trigger the channelization process when the threshold is exceeded at any point, but the network structure, as well as its growth and development, is controlled by the highly nonlinear system described by the two state equations interacting in space among themselves and with the constitutive equations. This is a hydrological adaption of the well-known conceptual models developed by Meinhardt [1976, 1982], which have been used in a variety of fields.

If one wishes to study the growth and development of the whole network structure, it is imperative to characterize those points of the basin where channels have developed as well as the connectivity between those points. This is the crucial role of the channelization Eq. (1.76) and its interaction with the elevation Eq. (1.71). An issue is whether channel and hillslopes grade continuously into one another in the neighborhood of channel margins or if channels show distinct behavior as soon as they are formed. In the proposed model there is a continuous gradation in space of the channel initiation function (1.75); once it exceeds a threshold, there is a transition from hillslope transport processes to channel sediment transport process. We believe this is a realistic representation of the natural process.

As previously noted, channel development is considered to be a one-way process, from hillslope to channel, ultimately with a deterministically known channel head position. This may be considered a reasonable assumption for the mean channel head position, though it may not adequately describe the channel head at any given time [Calver, 1978]. It must be noted, however, that the channel/hillslope differentiation is used in the model only to distinguish between mean transport rates in different parts of the catchment. Thus the channel head used in the model only indicates that the transport rate changes at the channel/hillslope interface. The model does not indicate whether the actual channel head in the field (if it can be identified) is abrupt or gradual [Montgomery and Dietrich, 1988, 1992]. Here, for convenience, the distinctions in sediment transport rates between the channel and the hillslope and the actual channel head are considered synonymous.

The contributing area, A, and slope, ∇z, are determined from the land surface elevations, z, by analyzing the directions of steepest slope. In the case of surface runoff, discharge can be directly related to the area determined in this way. If, however, Eq. (1.75) is a function of groundwater discharge, then the discharge is related to the elevations and slopes of the phreatic surface. For an aquifer with planar isotropic conductivities, the groundwater discharge can be determined from surface elevations if the surface elevations and phreatic surface elevations are linearly related [e.g., O'Loughlin, 1986].

Figures 1.54 and 1.55 show the evolution of elevations and channel networks with time for a computer simulation using the model described above. In this case, the initial condition was an unchannelized flat surface with a notch in one corner; this might model, for instance, regraded spoil heap. Hence the tectonic effect is an instantaneous and uniform uplift of the bulk of the area relative to the outlet. With time the channel

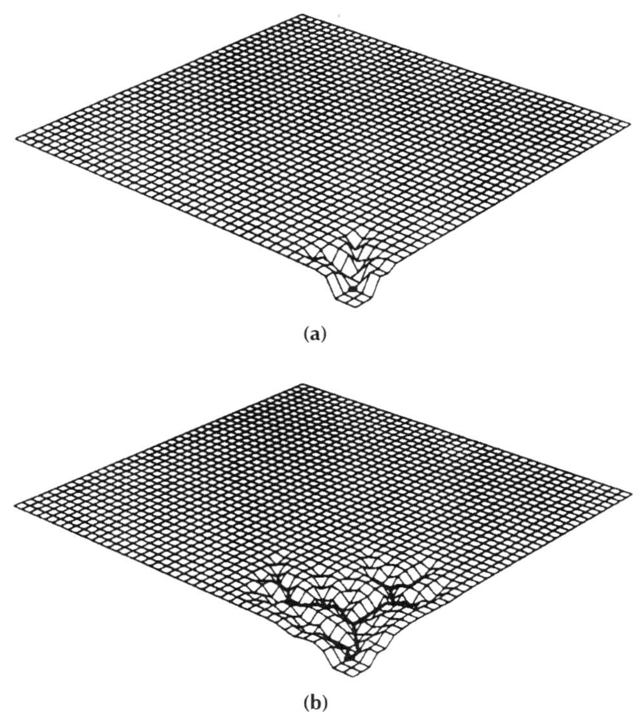

Figure 1.54. Sample simulation of the evolution of a catchment and its channel network with time [after Willgoose et al., 1991a] (b) follows (a) in evolutionary time.

network grows; the preferential erosion in the channels creates 'canyons' around the channels. The high slopes at the heads of the canyon result in high values for the channel initiation function on the hillslopes, causing both headward growth of the existing channel heads and creation of new channel heads from lateral branching. Once the channel network has stopped growing, the hillslopes and channel erosion interact to smooth the landscape into rolling hills and valleys.

Contours in Figure 1.56 of the same simulation demonstrate that qualitatively all aspects of landscape are formed, including isolated hilltops. It is important to note that the only randomness in the simulation of Figures 1.54 and 1.55 is that applied to the initial elevations (a fraction of a percent of the notch drop is randomly allocated over the surface); in this simulation erodibility (β_1 and D), runoff (β_2), resistance to gully erosion (channel initiation function, Eq. (1.75), threshold), and propensity to gully erosion (β_3) were all considered to be spatially uniform. No thorough study of the effects of randomness on these properties has been performed, but the results of Willgoose et al. [1991a,b,c] suggest that for equal coefficients of variation, and flat initial conditions, randomness in the initial elevations is the most important variable.

To demonstrate this effect, the final networks generated for four sets of random initial conditions are shown in Figure 1.57; the standard deviation of the initial elevation perturbations is about 0.2 percent of the height of the initial notch.

It is important to understand what would happen if the simulations in Figure 1.55 were allowed to proceed for a much longer time. Because erosion would continue, eventually the catchment would become a flat plain with the elevation of the outlet notch. This is what is called the peneplain [Chorley, Schumm, and Sugden, 1985]. This results from the catchment being a dissipative system [Huggett, 1988] with no process in the modeled system to oppose erosion. In energy terms, the potential energy of the catchment (i.e., elevation) is being dissipated by the transport of energy out of the catchment by erosion. In the geomorphologic sense there can be no dynamic equilibrium (i.e., $\partial z/\partial t = 0$ everywhere) because the catchment elevations are always declining, even though the channel network is unchanging and thus, by itself, in equilibrium.

However, dynamic equilibrium may be attained if there is a tectonic uplift (U in Eq. (1.71)) opposing the erosion of the catchment at all times. In energy terms, the tectonic uplift is a source of potential energy for the system and opposes the dissipation of energy. Because at dynamic equilibrium the elevation at every point in the catchment is constant with time, from Eq. (1.71) we may write

$$U = \frac{1}{\rho_s(1-n)} \nabla \cdot \mathbf{F} + D \nabla^2 z \tag{1.77}$$

Thus at dynamic equilibrium the mean temporal erosion rate is equal to the mean temporal tectonic uplift irrespective of whether the tectonic uplift is spatially uniform or not. Thus catchments where net deposition or erosion is taking place, for instance, at the toe of a hillslope [e.g., Toy and Hadley, 1987, p. 231], cannot be in dynamic equilibrium. In catchments where tectonic uplift is spatially variable, the mass wastage must also be spatially variable in the same general pattern for equilibrium to occur. In small catchments, where the tectonic uplift can be considered spatially uniform, at dynamic equilibrium mass wastage will, in the mean sense, occur at equal rates through the catchment. This is the classical conceptualization of dynamic equilibrium [Gilbert, 1909; Ahnert, 1976] and this conclusion is independent of the form of the transport processes exemplified in Eq. (1.72).

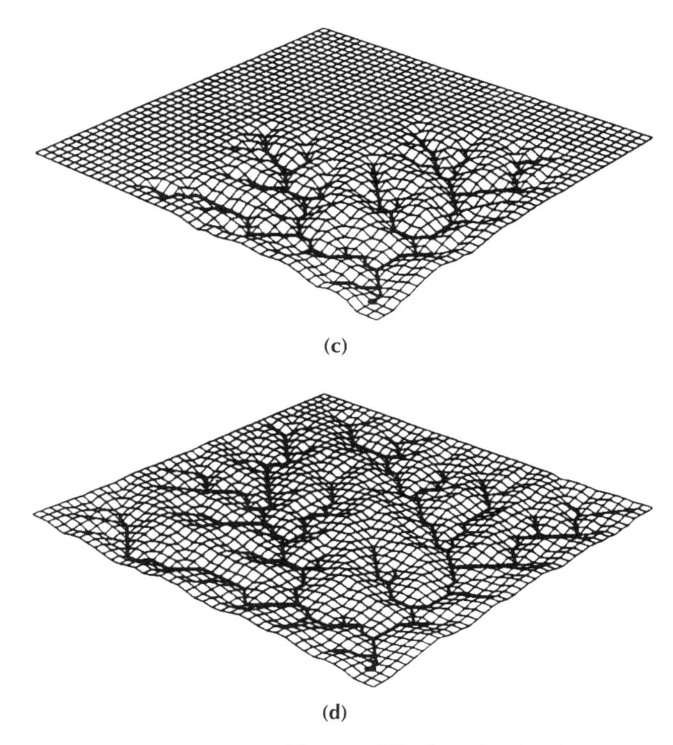

Figure 1.55. Sample simulation of the evolution of a catchment and its channel network with time [after Willgoose et al., 1991a].

Transient landscape effects have also been investigated. If a catchment is not in dynamic equilibrium, then either the mean channel network or the mean elevations must change with time; these will be called transient conditions. The use of mean properties is conceptually important because even if channel heads are advancing and retreating over short timescales, say as a result of landslide events, but the mean position of the channel head is constant, then these fluctuations alone do not negate conditions for dynamic equilibrium. Likewise, short-term elevation fluctuations accompanying pulses of sediment transport [e.g., Parker, 1977] would not necessarily invalidate the mean conditions for dynamic equilibrium.

Of interest here is the transient period of catchment development, the period when the channel network is initially growing or when a previously stable channel network has been destabilized as a result of some environmental disturbance. A convenient means of describing this regime is by plotting, with time, the total length of channel network upstream of some fixed point in the catchment. If the area of the catchment drained through that point does not change, for instance, by stream capture, then this increase in network length is equivalent to an increase in drainage density.

1.4.5 Detachment-Limited Basin Evolution

Howard [1994] proposed a rather complete simulation model incorporating creep and threshold erosion as well as detachment- and transport-limited fluvial processes. It is of interest to review here the main structure and results of his approach.

The model is arranged over a square matrix of cells, within which both slope and channel processes occur. Whatever the embedded width of active channel possibly existing in the cell, it is assumed to be smaller than the cell size; then both mass wasting and fluvial transport/erosion – whatever their relative importance – coexist in each cell. Such an assumption, assumed valid even for alluvial pediments and fans, corresponds to the

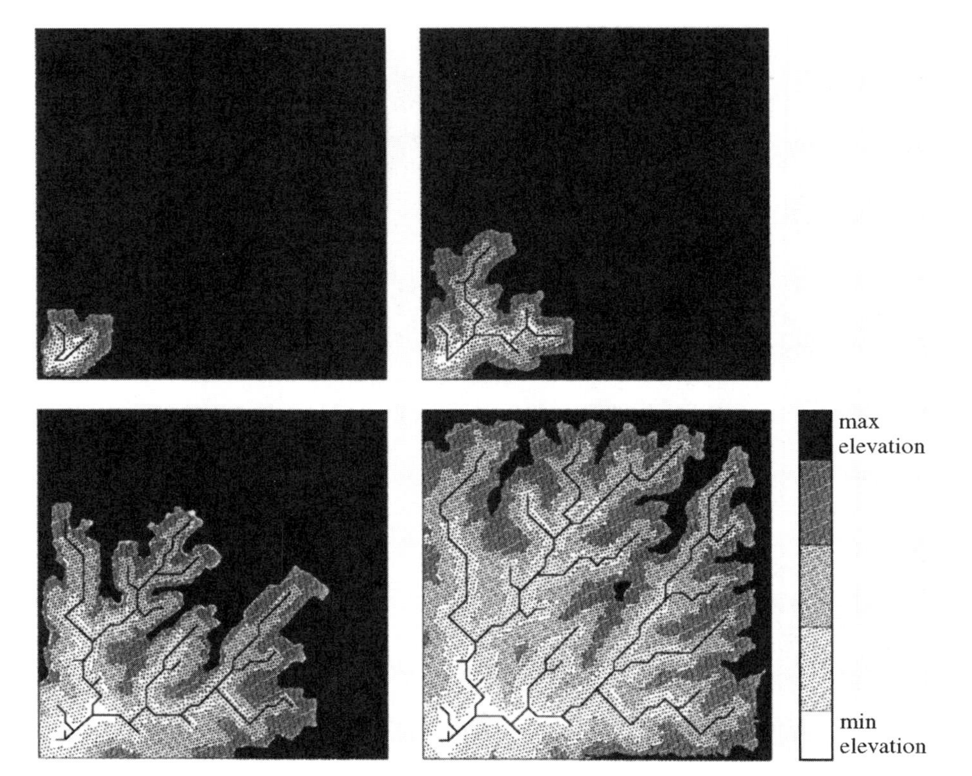

Figure 1.56. Sample simulation showing the evolution of catchment elevation contours and channel network with time [after Willgoose et al., 1991a].

observation that flow – even on such surfaces – is generally channelized and confined to few active channels at any time.

The basic landform evolution equation employed by Howard is Eq. (1.46) with $\eta = 0$. Randomness enters the system only through initial conditions, in the form of small disturbances superimposed to an otherwise flat surface. In this model the net erosion rate $\partial z/\partial t - U$ is relative to a fixed reference frame, and the volumetric transport can result from a variety of processes, excluding only soil production processes (i.e., the conversion of rock volumes to colluvium). Process rate laws are considered in terms of potential mass-wasting (i.e., diffusional), $\partial z/\partial t|_m$, and fluvial (i.e., advective), $\partial z/\partial t|_c$, erosion rates. Given that each simulation cell contains variable proportions of channel and slope components, the actual erosion rate is computed as a weighted sum of the potential rates.

Potential erosion or deposition due to diffusive agents, chiefly weathering, rainsplash, or mass movements, is still mathematically described as the spatial divergence of a vector rate, that is

$$\left.\frac{\partial z}{\partial t}\right|_m = \nabla \cdot \mathbf{F}_m \tag{1.78}$$

The rate of movement is expressed by two additive terms, one for creep-like and rainsplash diffusion and one for near-failure conditions (Figure 1.58):

$$\mathbf{F}_m = \left(K_s F(|\nabla z|) + \frac{K_f}{1 - K_x |\nabla z|^a} \right) \frac{\nabla z}{|\nabla z|} \tag{1.79}$$

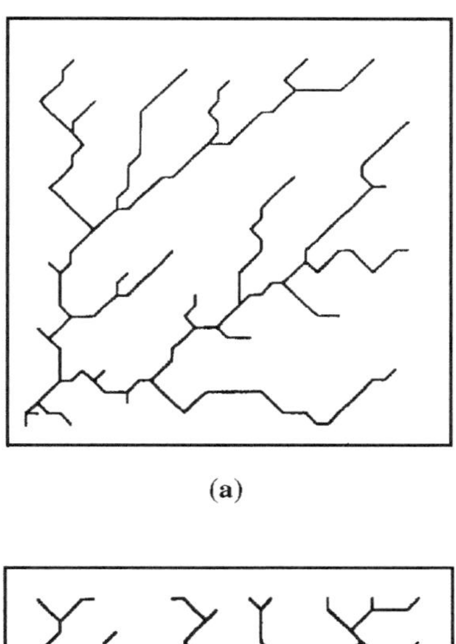

(a)

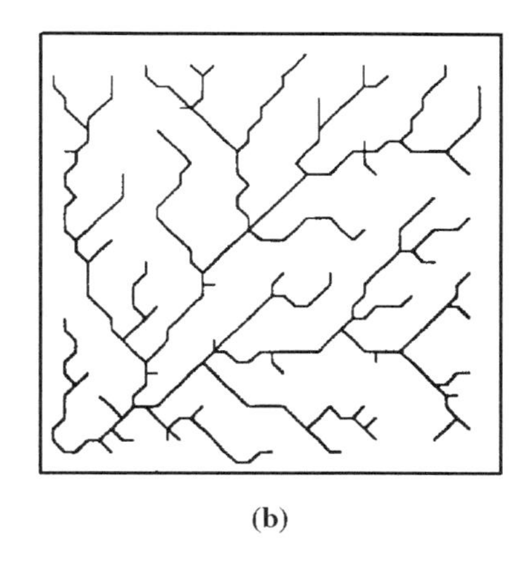

(b)

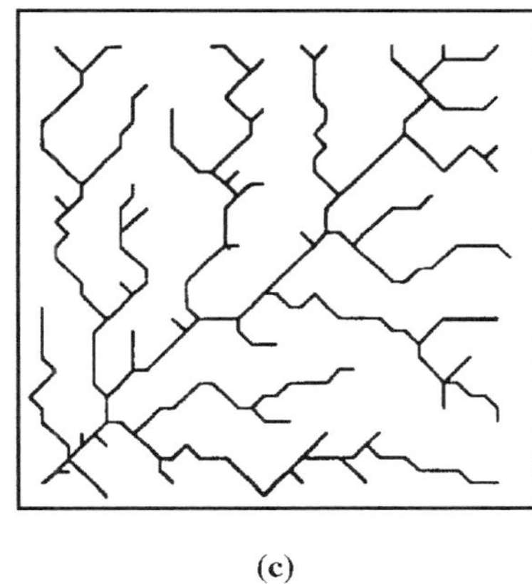

(c)

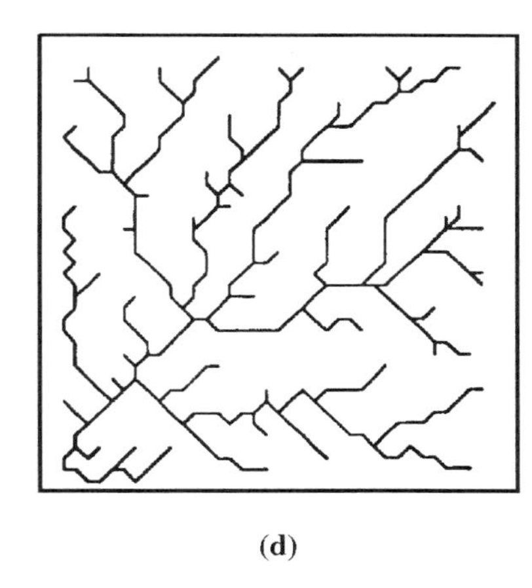

(d)

Figure 1.57. Four random networks generated by small perturbations on the initial conditions describing catchment elevation [after Willgoose et al., 1991a].

where $F(|\nabla z|)$ is an increasing function of the absolute value of slope. The parameters (K_s, K_f, K_x, and a) are assumed constants in space and time. Notice that the ordinary diffusive case dealt with, for instance, in Section 1.4.4 assumes that $F(|\nabla z|) \sim |\nabla z|$ and $K_f \sim 0$ yield

$$\left.\frac{\partial z}{\partial t}\right|_m = K_s \nabla^2 z \tag{1.80}$$

The second term in Eq. (1.79) models near-failure conditions on slopes such that mass movement rates increase without limit as gradients approach a threshold value equal to $1/K_x^{1/a}$. In Howard's [1994] model the spatial divergence in Eq. (1.78) is evaluated in eight directions through a finite-difference algorithm.

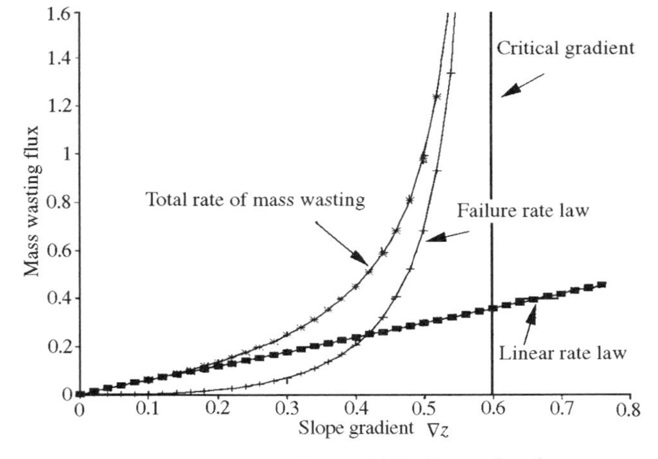

Figure 1.58. Example of mass-wasting flux with two coexisting mechanisms [after Howard, 1994].

The model incorporates both detachment-limited erosion on slopes and steep headwater channels as well as sediment transport in alluvial channels. Potential channel deposition/erosion, $\partial z/\partial t|_c$, is assumed to be dependent on the nature of the channel, that is, whether or not it is *alluvial*.

Mass balance assumes the common one-dimensional form:

$$\left.\frac{\partial z}{\partial t}\right|_{c} = \frac{\partial F_2}{\partial x} \tag{1.81}$$

where F_2 is the (scalar) volumetric transport rate per unit channel width and x is the downstream direction for the flow. Sediment discharge, F_2, is broken down into wash load, which is assumed to never be redeposited except in depressions, and bed sediment (bed load and suspended load), which is carried in capacity amounts if an alluvial bed is present. If the channel is alluvial, then bed sediment is predicted from sediment transport relationships. For nonalluvial channels the transport divergence $\partial F_2/\partial x$ is given by an intrinsic detachment capacity, as in the case of channels flowing on bedrock or on colluvium in which the bed load is less than a capacity load. For example, erosion in bedrock channels depends on the ability to scour or pluck bed material. Exposure of bedrock beds, due to downcutting of resistant rock, slows the erosion response of headwater catchments to baselevel changes [Howard, Dietrich, and Seidl, 1994].

Howard [1994] noticed that most deterministic modeling approaches have oversimplified erosion, transport, and depositional processes, particularly in the fluvial system. Several channel types can occur in a large river basin, and accordingly a single transport or erosion law cannot suffice for a detailed description of the system. We now recall a few examples discussed by Howard [1994].

Fine-bed alluvial channels occur primarily in lowlands and are favored by low-relief, abundant supply of fine sediment and absence of coarse detritus, either because of little production in headwater areas or abrasion or sorting out of coarse debris during transport. Deposits from fine-bed alluvial channels are those most commonly represented in fluvial depositional environments. Interestingly, the characteristic timescale of the response of fine-bed alluvial channels to changes in sediment supply, discharge, base level, or tectonic deformation is the shortest of all bed types [Howard et al., 1994].

Coarse-bed rivers are common in runoff-generating, mountainous regions, where physical weathering processes produce coarse detritus. Such streams range from those carrying an abundant sediment load (i.e., live bed channels) to those with very low transport rates of bed sediment and slopes just steep enough to allow some transport at very high flow stages (i.e., threshold channels). Of course, channels may alternate between live bed and threshold conditions if sediment supply rates are episodic. Live bed gravel channels are similar to sand bed channels in that the gradients are affected by both sediment size and sediment supply rates.

Live bed gravel may occur in mountainous, alpine, arid, and arctic areas where sediment yields relative to discharge are high and physical weathering predominates over chemical actions. Such channels convey a wide range of grain sizes on the bed, and most grain sizes are mobilized at about the same flow stage (Parker's [1978, 1979] concept of equal mobility). The major difference from fine-bed channels is that downstream sorting and abrasion play an important role so that grain sizes decrease rapidly in the downstream direction. In channels where the gravel thinly mantles the bedrock and thus provides a small storage of sediments, the primary cause of downstream fining is downstream reduction in size of locally contributed sediments, although temporary sorting effects may be induced if sediment supply from slopes is episodic.

Threshold gravel channels generally occur in areas where physical weathering has supplied coarse gravel but overall sediment yields are low. In these conditions, reworking of the gravel by floods maintains gradients close to threshold conditions, but the supply rates and sizes of gravel and boulders often vary spatially in a complex way so that there is no unique pattern of downstream change in grain size or channel gradient.

Channel incision into bedrock occurs when the supply of sediment to the channel cannot keep the channel continuously mantled with an alluvial cover, usually because of either a steep gradient or scanty sediment supply. Thus bedrock channels are favored by one or more of the following factors: high relief, high uplift rates, local faulting, resistant bedrock, and possibly a dominance of debris flow transport. The headwater tributaries of many river systems draining mountainous areas are bedrock floored. Because scouring and plucking occur during high-flow stages, channels with a thin alluvial cover can slowly erode the underlying bedrock while maintaining an alluvial cover during low flow conditions. Yet, the erosional capacity of alluvial channels is limited so that if the downstream erosion rates exceed this capacity, local gradients steepen and bedrock becomes exposed, thus substantially altering the overall resistance to erosion.

Finally, channels in which bedrock exposures alternate with short alluvial sections are common under a number of different scenarios.

Mathematically, for *nonalluvial channels*, in which bed load sediment fluxes are less than a capacity load, the detachment capacity is assumed proportional to a shear stress exceedence $\tau - \tau_c$ as in Eq. (1.68). Clearly, the shear stress is thought to be exerted on the bed and banks of the embedded channelization by a dominant discharge, that is, leading to landforming events for the fluvial structure. As seen in Section 1.2.4, Howard [1994] assumes a bankfull discharge type relationship for the dominant discharge, that is, $Q \propto A^e$ and a hydraulic relationship of the same type, that is, $W \propto Q^b \sim A^{be}$. Notice that the use of such relationships for runoff on slopes as well as in channels implies that little regolith (both colluvium material or residual soils) or depression storage of precipitation occurs during erosion events, as runoff production is assumed areally uniform. For landscapes with appreciable infiltration capacity, particularly where vegetated, it is assumed that most runoff erosion on slopes occurs during infrequent, very intense rainfall events owing to saturation overland flow or shallow interflow through large macropores. Combining the geometric relationship with the usual hydraulic relationship of uniform flow (Eqs. (1.13) and (1.12)), one has an evolution equation of the following type:

$$\left.\frac{\partial z}{\partial t}\right|_c = K_t \left(K_z A^c \nabla z^d - \tau_c\right) \tag{1.82}$$

where τ_c is a critical shear stress depending essentially on substrate type, K_t is a constant that includes effects of substrate erodibility as well as magnitude of the dominant discharge; and K_z, $c \propto e(1 - b)$, d are suitable constants whose definition is unnecessary in this context.

The actual detachment rate thus decreases from the intrinsic detachment capacity (for zero sediment load) to zero as the actual bed sediment transport rate approaches the flow transport capacity.

Fluvial processes may be eroding colluvium material (i.e., the regolith) that is delivered into the channel by mass wasting, or emerging rock. Regolith material is assumed to be more erodible than bedrock by a factor F, where $F \geq 1$. The fluvial erosion rate for bedrock is given by Eq. (1.82) so that the equivalent rate for the regolith is $F\partial z/\partial t|_c$. In landscapes characterized by a shallow weathering regolith, such as in steep mountain slopes, bedrock is commonly exposed in headwater rills and channels, and a large ratio of bedrock to regolith erodibility ($F \gg 1$) is appropriate. However, in landscapes in deeply altered bedrock, till, or uncemented alluvium, the parent material and the surface soil involved in mass wasting may have nearly equivalent erodibility ($F \sim 1$).

When calculating net erosion in matrix cells containing nonalluvial channels, it is necessary to take into account that erosion is occurring in a channel of width W and that the channel may be eroding both regolith and bedrock during a prefixed time step. The eroded regolith comprises both that delivered by mass wasting into the cell from the adjacent cells and that which is locally derived. During each time step the channel is assumed to first erode regolith delivered to the channel by slope erosion and then, if it is capable of eroding all the regolith, bedrock. Within each time step, the fraction of time during which the channel is eroding regolith is termed Δt ($0 \leq \Delta t \leq 1$). Three cases are possible:

- The simplest case occurs when a sufficiently large amount of regolith is delivered to the cell and the channel never erodes bedrock ($\Delta t = 1$). The volume of regolith delivered to the cell (of, say, size δ) per unit time is $\partial z/\partial t|_c \delta^2$, and the volume of channel erosion is $\partial z/\partial t|_c F\delta W$. The net elevation change, $\partial z/\partial t$, is the sum of the volume of regolith mass movement and of volumetric channel erosion divided by the proper area of the cell:

$$\frac{\partial z}{\partial t} = \left.\frac{\partial z}{\partial t}\right|_m + F\frac{W}{\delta}\left.\frac{\partial z}{\partial t}\right|_c \tag{1.83}$$

  which is valid for $\partial z/\partial t \geq 0$.
- If the channel is capable of eroding all the regolith delivered to it, then the value of Δt must be determined. Weathering is assumed to be capable of keeping pace with the overall net rate of lowering. Thus the net volumetric rate of regolith production and delivery to the channel by mass wasting for erosion is determined from the balance of regolith volume imported from adjacent cells, the within-cell contribution from weathering adjacent to the cell, and the maximum volume of regolith that can be eroded per unit time by the channel, that is,

$$\Delta t = -\frac{\left.\frac{\partial z}{\partial t}\right|_m \delta^2 - \frac{\partial z}{\partial t}\delta(\delta - W)}{FW\delta \left.\frac{\partial z}{\partial t}\right|_c} \tag{1.84}$$

  and the net erosion into bedrock is given by

$$\frac{\partial z}{\partial t} = (1 - \Delta t)\left.\frac{\partial z}{\partial t}\right|_c \tag{1.85}$$

  where the above relationships hold for $\partial z/\partial t|_m \geq 0$ and $\partial z/\partial t < 0$.
- The final possible case occurs when $\partial z/\partial t|_m < 0$ and is similar to the previous case except that it assumes that regolith removal by channel erosion

pertains only to the volume surrounding the stream and below the level of removal by slope erosion.

For *alluvial channels* the potential rate of erosion is determined in a different way, and the spatial divergence of the volumetric unit bed sediment transport rate is specified by bed load and sediment load transport equations for sediment discharge. Sediment transport equations require the thorough specification of bed material size distribution, including mean grain size, alluvium porosity, specific weight of sediment grains, and fall velocity of grains in turbulent flows, among other things. Transport equations are then recast into a relationship involving, as usual, total contributing area and slope, whose final mathematical structure does not differ dramatically from Eq. (1.82).

The resulting set of differential equations is solved via a straightforward finite-difference approach, whose description is not of interest here.

Interestingly, transitions in the main modes of transport within the same cell may occur in time according to the evolution of the landscape. In fact, although all cells in the simulation lattice contribute to erosion or deposition from either an alluvial or a nonalluvial channel, the two types of channel are mutually exclusive. Thus a model of this type must provide criteria for temporal and spatial transitions.

If the channel in a cell is nonalluvial, the sediment volume arriving from upstream and that locally eroded by the channel are considered to be transported without deposition and to be routed instantaneously in the simulation timescale through the nonalluvial network. During each iteration the total potential bedload transport capacity of the channel in each cell is calculated. If the actual transport rate exceeds the potential rate, then the cell is converted to an alluvial channel. When such a transition occurs, the base of the alluvial deposit is set equal to the local elevation.

During an iteration, as long as erosion due to sediment divergence in an alluvial channel is less than the thickness of the alluvium, the channel remains alluvial. However, alluvial channels in nature and in the model may also occur under conditions of slow downcutting (such that the base of the alluvial deposit is constantly reset to the local elevation during each iteration). The maximum amount of erosion that an alluvial channel on shallow alluvium over bedrock can accomplish is considered to be equal to the rate for erosion of nonalluvial channels into bedrock. If the erosion due to sediment divergence or the routing procedure employed by Howard [1994] exceeds this value within a channel on shallow alluvium, then the channel is converted to a nonalluvial channel. It is assumed that $\partial z/\partial t = -1$ at the lower matrix edge, simulating the effects of uplift. When the rate of decrease is stabilized everywhere, a steady state is defined.

Illustrating some specific examples discussed by Howard [1994] is of interest. The simplest case is obtained when (i) all channels are nonalluvial, (ii) the critical shear stress is set to zero, (iii) no near-failure contribution to slope erosion is assumed, and (iv) regolith and bedrock are equally erodible ($F = 1$). In this case, the basic evolution equation reduces to a structure in which the erosion rate is a linear combination of a diffusive mass-wasting term in Laplacian form tending to smooth the landscape and a channel erosion term that becomes more important – for a given gradient – as drainage area increases. All cells have some contribution from

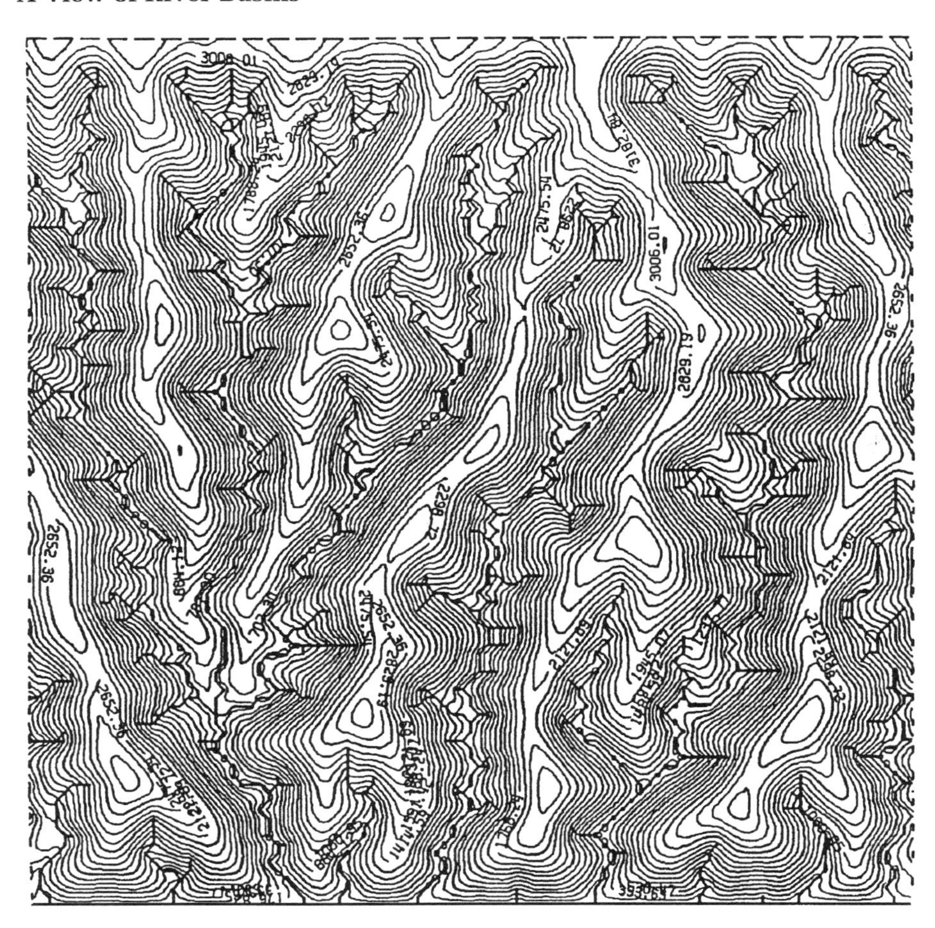

Figure 1.59. Steady state produced in the case $F = 1$ and no critical shear stress [after Howard, 1994].

fluvial erosion and mass wasting. An example of the results is shown in Figure 1.59. The particular choice of parameters chosen to produce Figure 1.59 yields

$$\frac{\partial z}{\partial t} = -0.04\,\nabla^2 z - 0.05\,A^{0.5}\nabla z^{0.7} \tag{1.86}$$

The initial condition is a randomly perturbed planar surface that provides the necessary randomness for the evolution of a realistic landscape. Howard [1994] noticed that the landscape in Figure 1.59 is partitioned into a majority of cells where diffusive terms dominate and a dendritic valley network (which is not fixed by the approach) is dominated by fluvial erosion. It is worth noting that first-order statistics of the three-dimensional structure ($\langle \nabla z(A) \rangle$ vs A) of the landscape in Figure 1.59 show the trend observed in nature (Section 1.2.11).

Figure 1.60 shows another significant case, where the regolith is much more erodible than the bedrock ($F = 500$), all other parameters and initial conditions being the same as in the case shown in Figure 1.59. The topographies produced with $F = 1$ (Figure 1.59) and $F = 500$ (Figure 1.60 are similar in broad patterns and drainage density but differ in details of morphology and pattern of branching. In the case $F = 1$ (Figure 1.59), low-order tributaries are relatively short compared with higher-order

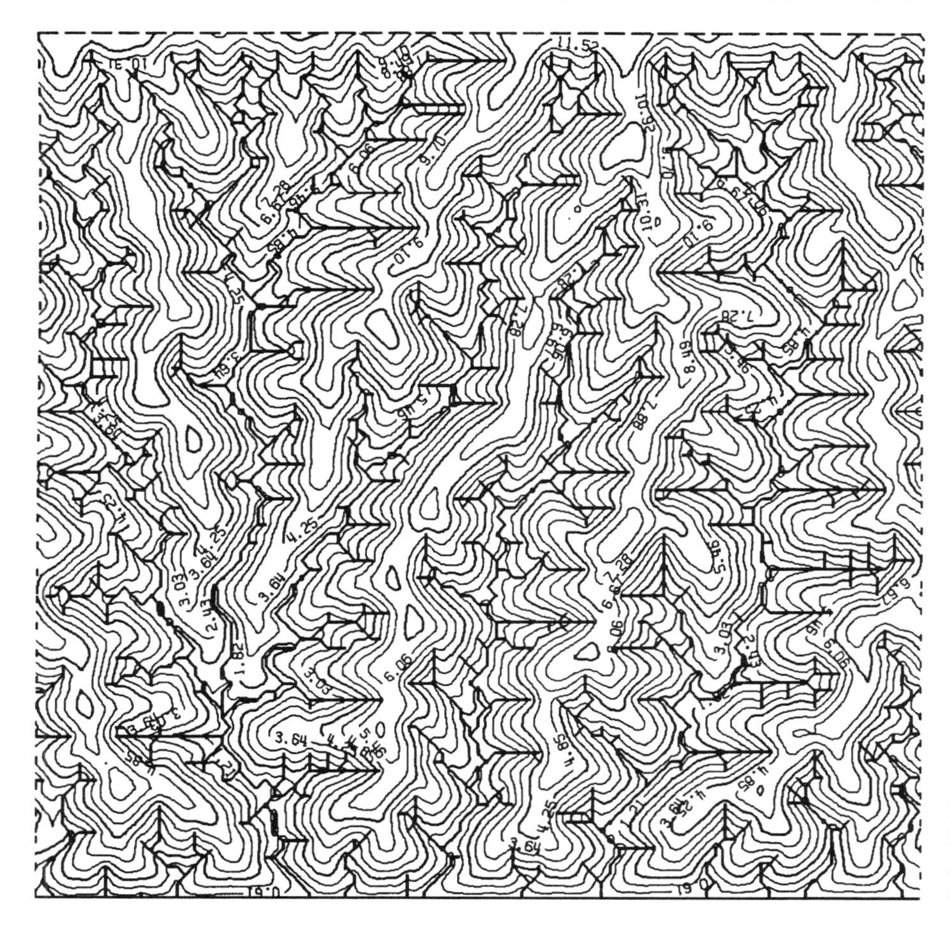

Figure 1.60. Steady-state topography produced with $F = 500$ and no critical shear stress [after Howard, 1994].

streams (low exterior–interior link length ratio), whereas the $F = 500$ channel system has nearly equal interior and exterior link lengths. In the latter case, division of the landscape into low-order networks is more complete, and the direct character of the drainage pattern (the mean value of the width function, i.e., the average pathlength measured along the branching structure) is higher than for the $F = 1$ case. In this case, shown in Figure 1.60, transitions from convex divides to channeled low-order streams are abrupt, whereas for the case $F = 1$, broad hollows occupy a comparatively larger proportion of the landscape. Howard [1994] notes that the above morphological differences largely result from the more concave channel profile for the case $F = 1$ than for the case $F = 500$, as one would anticipate.

Including a critical shear stress for fluvial erosion produces sharp transitions between slopes and divides (Figure 1.61). The imposition of finite τ_c lowers the drainage density and produces strongly convex slope profiles. Owing to the absence of fluvial erosion, the actual divergence of fluxes on slopes near the divide equals the contribution of mass wasting. Although a critical shear stress to represent the flow condition necessary to induce entrainment is included in Howard's scheme, it might also be used to represent a variety of processes, as also acknowledged by Howard[1994].

Finally, Figure 1.62 illustrates the case where a critical threshold slope is assumed in mass wasting, as described by a finite value of $K_f > 0$ in

Figure 1.61. Steady-state topography with $F = 500$ and $\tau_c = 5$ (pixel units) [after Howard, 1994].

Eq. (1.79). To emphasize its effects, a small value of the diffusive component (small K_s in Eq. (1.79)) is adopted. The rapid increase in mass-wasting rates with slope gradient near the threshold means that the downslope increase in volume of mass-wasting debris in steady-state topography can be accommodated through slight changes in gradient. Slope profiles are therefore nearly linear, and divides are narrow and sharp (Figure 1.62). A nice example of a similar situation is described by badlands, which have nearly linear profiles and very narrow divides because of mass-wasting control by sagging and small slumps under near-threshold conditions. Figure 1.63, also taken from Howard [1994], provides convincing evidence for this explanation of badland morphology.

Howard's approach is very accurate in the description of different processes affecting fluvial erosion, and the complexity of the possible scenarios for landscape evolution is manifest and likely to be realistic. However, the indefinite persistence of initial conditions, common to all deterministic models, and the need to tune the large number of parameters required to describe the many processes introduce important elements of arbitrariness. Why do natural networks show recurrently common features at large scale, both in plan and in the altitude dimension, if so many different processes are quite likely at work? If one is capable of dissecting and understanding the processes in their minute details, is the ability to explain the general

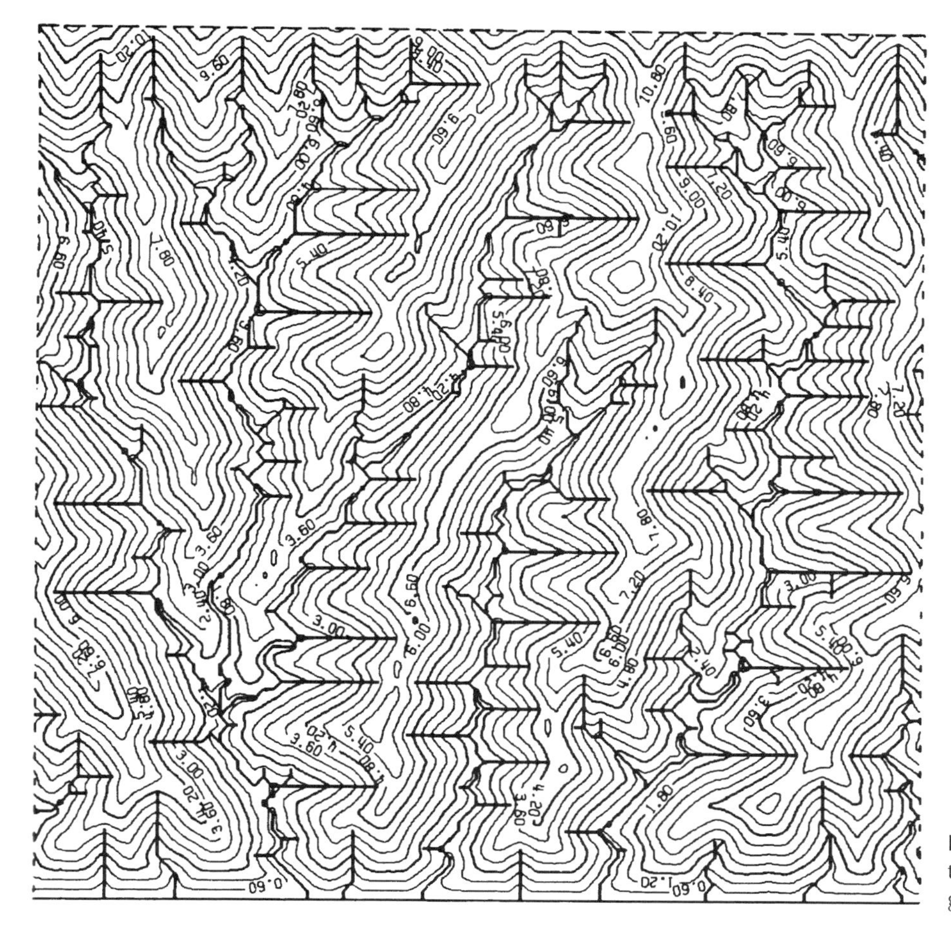

Figure 1.62. Steady-state topography with a critical slope gradient ($K_f > 0$) [Howard, 1994].

picture – including the full complexity of such large dissipative systems with many degrees of freedom – consequently granted? Or is the origin of recurrence somewhat imprinted in the dynamics itself, regardless of the detailed dynamic specification of the processes?

It is clear that Howard's views agree with the reductionist, process-oriented tenet to which most geomorphology is committed. As such it is only through specifying in detail the many processes active in nature and the parameters of their mathematical formalization that landscape evolution can be understood. Although we agree with Howard that specific processes need to be modeled so that specific landforms may be described, we place our main interest in something related but distinct: the dynamic origins of the broad, ubiquitous, self-similar characters shown by the geometry of natural river basins, to which the following chapters of this book are devoted.

1.4.6 Limitations of Deterministic Models

All the approaches described so far in this section allow for growth and development of meaningful three-dimensional river basin structures. Nevertheless, they do not clarify the linkages between fundamental aspects of the dynamics and the existence of general types of scaling relationships in

Figure 1.63. Natural badland slopes in the Mancos shale badlands near Caineville, Utah [courtesy of A. Howard, 1994].

the elements of the network and the landscape itself. This does not mean that these approaches are not valuable in the reproduction of specific natural landforms. On the contrary, the simulation of particular landforms relies on the detailed specification of the dominant dynamics and on the calibration of the relative importance of many processes. As valuable as it is for some purposes, the reductionist perspective is probably not the most adequate in the search for an explanation for the dynamic reason behind the extraordinary diversity and yet the deep symmetries and recurrences of the morphology of river basins. In this book we do not aim at capturing all the features of the phenomena, just those that are *fundamentally* relevant.

The lack of a clear framework within which to study the dynamic origin for fractal river basin structures initiated the work leading to this book. We believe this lack of clear cause–effect links is a limitation embedded in many approaches taken to the problem of growth and differentiation of natural patterns. Other types of models, such as those ones in this section, do not have the same limitations but often suffer from strong dependence and indefinite persistence of the effects of initial conditions. They also operate under the stringent need to tune many parameters in order to obtain observational features.

The requirement that simulated landscapes reproduce general observational scaling characteristics – especially in the altitude structures – is a demanding one, especially if coupled with the requirement that models must be *robust* with respect to quenched disorder and parameter values in order to acceptably explain the recurrence of statistical patterns in natural forms.

Natural structures are produced by highly nonlinear processes, and most of the time, the representation of such nonlinearities in deterministic models is either simplistic, in order to allow for an analytic or numerical solution to the problem, or arbitrarily represented, in the sense that it does not incorporate the various feedbacks and threshold-types of phenomena that appear to be central to geomorphology. To explain the ubiquity of recurrent forms in nature something else is needed. This will be the central theme of Chapters 4, 5, and 6 of this book.

1.5 Lattice Models

In the continuum approach described in the previous sections, nonlinear partial differential equations for mass and momentum balance were employed. The solutions almost inevitably rely on numerical methods to describe complex landforms. Lattice models represent a different approach in which both deterministic and statistical rules are embedded in the dynamics. The dynamics is specified through a set of discrete rules operating on a lattice of sites. The local rules incorporate transport of matter and flow and may be similar in structure to the probabilistic connectivity rules employed in random-walk types of growth.

In principle, lattice models include cellular automata [e.g., Wolfram, 1988]. In cellular automata, a discrete analog to a partial differential equation is set up, and an entire grid (or *lattice*) of computed values is changed according to a set of rules that update the system to the next time step in a simultaneous manner. A viable alternative to this procedure is a Monte Carlo approach, where random sites are selected and the basic set of rules is locally enforced. Random search procedures are interesting per se because they embed rules of chance into the model rather than into initial or boundary conditions.

Lattice models have generated great interest with respect to studies of the dynamics behind fractal growth [e.g., Bak, Tang, and Weisenfeld, 1987, 1988]. Therefore we will review in Chapters 5 and 6 many approaches connected to the framework of critical self-organization, some of which deal explicitly with river networks and basins. However, there exist a number of approaches [e.g., Chase, 1992; Kramer and Marder, 1992; Marder, 1993; Leheny and Nagel, 1993] that deal with lattice growth of networks without reference to the framework of critical self-organization. We thus believe it is appropriate to review some that are of particular interest for the subject at hand, mainly to examine the ingredients and limitations of these approaches.

Chase [1992] proposed a quantitative three-dimensional lattice model of landform development through fluvial erosion and deposition for scales from hill size to mountain range. The model incorporates the evolution of regional topography and the effects of tectonic motions and climate on landscapes. Fractal geometries arising in that context were also studied. In our view the most interesting feature of Chase's [1992] model is the study of the effects of randomly seeded storms (*precipitons*) that cause diffusional smoothing on the topographic lattice. Portions of elevation differences between adjacent sites are eroded by slope-limited transport, and alluvium deposition is allowed when sediment carrying capacities are exceeded. The lattice model works at grid resolutions up to 1 km, and thus it does not attempt to distinguish between hillslopes and channels. The rules for erosion are spatially uniform for all grid cells. The grid of digital topography, with cell dimension L, is characterized by numbered cells labeled by their elevation and material properties, especially erodibility. Into such a grid a precipiton is dropped. This causes diffusive action on the four nearest cells. The precipiton then moves to the lowest adjoining cell, where it erodes an amount of material proportional to the elevation difference multiplied by a tunable empirical coefficient termed the erodibility factor. Deposition takes place when the precipiton's sediment-carrying capacity, assumed proportional to the slope, is exceeded. The rules for diffusion are linear through slope-proportional rules. The model concept is indeed appealing, but, unfortunately, no attempt was made to investigate the resilience of the landform statistical characteristics to the tunable parameters and/or quenched randomness. Moreover, even the search for fractal features in the landscape structure and particularly the drainage network appears doubtful because there is no distinction between hillslopes and channels.

Kramer and Marder [1992] and Marder [1993] proposed nonconservative and conservative lattice models of coupled flow and sediment transport. In the nonconservative scheme, any erosion action removes an amount of height Δz fixed by a rule and does not redistribute it among neighbors. Erosion is assumed to take place where the amount of water being transported is sufficiently large. The set of rules for the Kramer and Marder [1992] and Marder [1993] model is the following:

- At each site i of a lattice (i stands for two integers i, j) two integers are specified, z_i and d_i, corresponding to the height of the landscape surface and the depth of water. The total surface height is given by $s_i = \mathrm{k}z_i + d_i$, where k is a scaling factor indicating the relative heights of water and landscape and giving an effective erodibility.
- A lattice site is chosen at random. If the surface height is lower than that of any of the nearest neighbors, water units are moved to bring the surfaces as close to even as possible.
- For each water unit transported out, a unit of land is dissolved, but only if the land is lower at the destination site.
- Additional water falls on a random site at a specified interval T_0.

Networks grown according to the above rules are shown in Figure 1.64. The scheme develops in a 200×500 rectangular lattice over a period of 2×10^9 iterations with $\mathrm{k} = 1$. The initial conditions are an initially flat plane without water ($d_i = 0$ for every i). The boundary conditions are periodic on the short left and right edges, a no-flux barrier at the top edge, and an outflow boundary at the bottom edge.

In the conservative lattice model the matter removed at a site is redistributed among neighbors. This requires some technicalities in dealing with the relative elevations of source and destination for both landscape and water surface elevations, accurately described in Marder [1993]. An

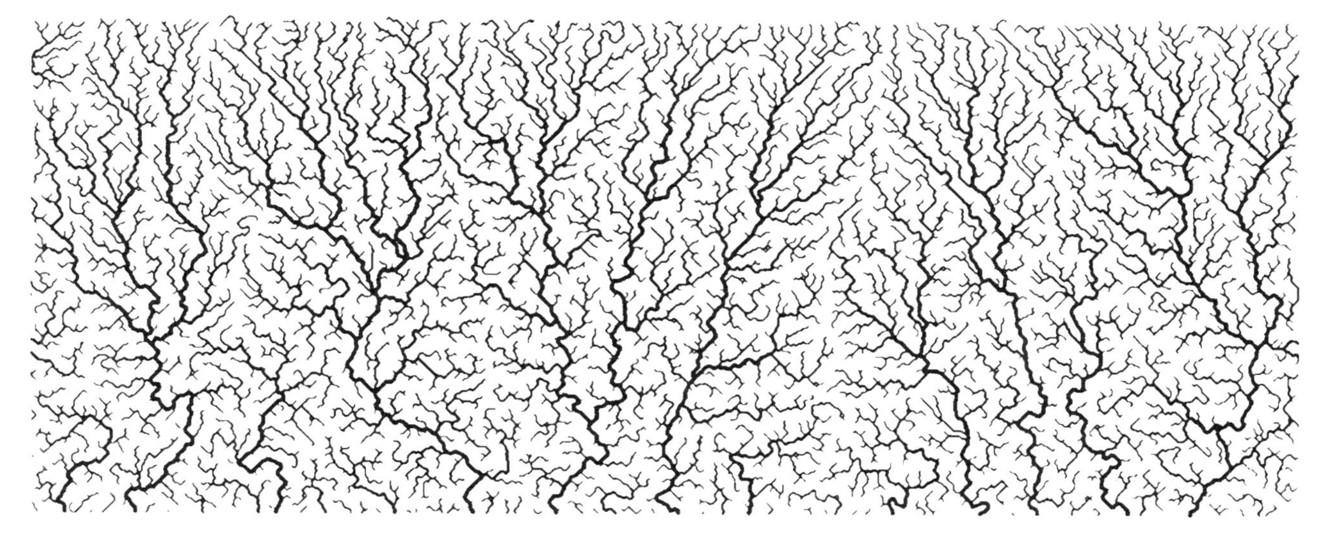

Figure 1.64. Channel networks produced by nonconservative lattice models in a 200×500 rectangular lattice after 2×10^9 iterations [after Marder 1993].

example of a channel network produced by the soil-conserving lattice model, starting from an initially flat, dry 200×500 lattice over a time period of 1.5×10^{10} iterations is shown in Figure 1.65.

The results look realistic. Some scaling relationships investigated match relatively well those found in natural drainage networks, including probability distributions of aggregated flow and the (assumed single) fractal dimension of the resulting landscape. However, the existence of parameters that need to be tuned to obtain correct scaling characteristics, the unnecessary details in the lattice scheme, as well as the lack of a critical examination of the robustness of the procedure, and the resulting three-dimensional properties of the network and its accompanying landscape make it doubtful that this model will provide a general framework for studying river basin landscapes.

The model of Leheny and Nagel [1993] suffers, in our view, from the same limitations. In their model a rectangular lattice of points simulates the terrain, and the height of the land $z(x, y)$ is specified at each lattice point. The lattice is assumed to have periodic boundary conditions in one direction, say x. Simulations start from a featureless surface. Eroding

Figure 1.65. Channel networks produced by conservative lattice models in a 200×500 rectangular lattice after 1.5×10^{10} iterations [after Marder 1993].

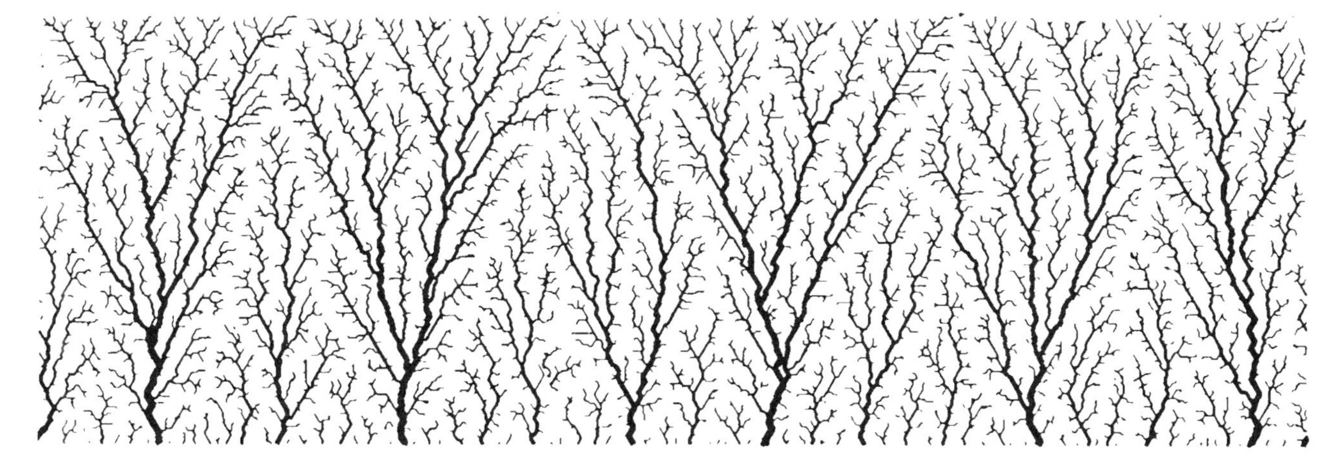

water enters the simulation as precipitation landing at a random site on the lattice. The water flows across the landscape by moving from the site it occupies to one of the (four) nearest neighbors, according to a probability distribution in which a threshold is imposed:

$$P(\Delta h_i) = \frac{e^{E\Delta h_i}}{\sum_{j=1,\Delta h_j \geq 0}^{j=4} e^{E\Delta h_j}} \qquad \Delta h_i \geq 0 \tag{1.87}$$

($P(\Delta_i) = 0$ if $\Delta h_i < 0$) where Δh_i is the height difference between the site the water occupies and the neighbor i and E is a tunable parameter in the model. Other details complete the needed dynamical specification [Leheny and Nagel, 1993].

The model is interesting for its simplicity and for its random character. However, the resulting landscapes were only tested on planar statistics, and not the most stringent ones. Also, after inspecting the resulting landscapes we feel that important structural properties related to scaling of the elevation field with drainage area are not respected, thus limiting the geomorphological interest of this approach. Furthermore, the need for tuning external parameters of unclear physical meaning reduces the interest of this approach.

In conclusion, we believe that the lattice models examined somewhat lack generality, in particular because tuning of parameters is required to reach states that are characterized by the absence of preferential scales and the robust appearance of meaningful statistics. It is doubtful whether tuned models are indeed capable of explaining nature's resiliency in the formation of its patterns of self-organization, regardless of heterogeneity and randomness. This important point is central to the framework of critical self-organization, which we review in the introductory sections of Chapter 5.

CHAPTER 2

Fractal Characteristics of River Basins

This chapter studies the fractal characteristics of river basins through analysis of the experimental data discussed in Chapter 1. Scale invariance, self-similarity, and self-affinity applied to river basin forms are studied. Here we also outline important methodological aspects of fractal geometry that will be used throughout the book.

2.1 Introduction

2.1.1 Fractals and Fractal Dimensions

Fractals provide a mathematical framework for treatment of irregular, ostensibly complex shapes that display similar patterns or geometric characteristics over a range of scales. The concept of fractals was introduced by Mandelbrot [1967, 1977, 1983] and characterizes those objects in which properly scaled portions are identical (in a deterministic or statistical sense) to the original object. In river basins we will always deal with the statistical description of components and thus the 'fractal' scaling property refers to the invariance of the probability distributions describing the object's composition under geometric transformations or changes of scale.

Common examples of fractals include the shore of continents, the shape of clouds, the profile of mountains, and river systems. In river systems fractal scalings can be observed at two different levels, either in the organization of the river network structure at different scales or in the individual wandering watercourse. Fractal properties of river systems at both levels have been analyzed by Tarboton et al. [1988], La Barbera and Rosso [1987, 1989], Nikora [1991, 1994], and Nikora and Sapozhnikov [1993a,b] among many others. These studies have described the scaling properties of the geometry of river systems and have calculated the corresponding fractal dimensions which, as will be seen in this chapter, are convenient descriptors of the scaling behavior.

Self-similarity is a concept associated with fractal geometry that refers to invariance, not with respect to additive translations but rather to multiplicative changes. A self-similar object appears unchanged after increasing or decreasing its size, and thus under closer and closer scrutiny it reveals more and more fine-grained structures. The smaller structures are identical to the large-scale ones when they undergo an expansion in their scale. This similarity of the parts to the whole is called self-similarity, and it is described by the lack of any characteristic scale, a property called scaling. Self-similarity may be strict or statistical. In the case of strict self-similarity the magnification of any of the small parts of the system brings in a richness of detail that perfectly matches the larger picture. This is not the case in most natural systems such as river basins. Thus when one focuses on the denditric structure of the drainage network of a small subbasin (defining more and more of smaller channels through a finer degree of resolution), it immediately resembles the network structure of the whole basin. The two objects look the same in a statistical sense when the small subbasin is viewed under magnification. In other words, looking at the magnified version of the small subbasin we cannot say it is not indeed the network of the larger basin. Statistical self-similarity may be also defined in terms of certain invariances of the probability distributions describing the objects. For instance, the average properties of a part of the whole are the same as for the whole itself except for a similarity transformation on the metric properties that underlie the property analyzed.

Examples of self-similarity abound in nature. Figure 2.1 shows that each frond in a thicket of ferns sprouts miniature copies of itself. Those copies, in turn, sprout smaller versions of the same basic shape and structure. This repeated duplication, remarkably regular in Figure 2.1, yields the characteristic feature of natural fractals. Notice that a few possible mathematical constructions of a fern are reproduced [Neill and Murphy, 1993; Peitgen et al., 1992], where the basic regularity embedded in the recurrent shapes of ferns is captured.

A strict self-similar object of special interest for hydrologists and geomorphologists is the Koch curve, introduced in 1904 by the Swedish mathematician Holze Van Koch. Take a segment of a straight line (the initiator) and replace its middle third by an equilateral triangle as shown in Figure 2.2. The result is called the generator, which in this case has length 4/3 if the length of the initiator is unit. We now repeat the procedure on every straight segment of the object and the length of the fractured line becomes $(4/3)^2$. Repeating the procedure ad infinitum results in a *curve* of infinite length, which, although continuous everywhere, is nowhere differentiable. These properties are characteristics of the 'curves' now called *fractals*.

An example from the earth sciences is shown in Figure 2.3, where the Ria coastline in northeastern Japan is illustrated. It is fascinating to see the resemblance to the larger picture contained in an apparently smooth and featureless segment when it is examined through a magnifying lens [Nakano, 1983].

The length of a 'smooth' curve can be measured by stepping along it with dividers, or a ruler, of length r. The length is then approximately $L(r) = Nr$, where N is the number of divider steps. When the sinuosity of the curve increases, one needs to resort to smaller and smaller rulers, r, in order to measure the length correctly. In fact, as the step size r goes to zero, one converges to the exact length L:

$$L = \lim_{r \to 0} L(r) = \lim_{r \to 0} Nr \tag{2.1}$$

In his classic study of the length of coastlines, Richardson [1961] found that the above limit often does not exist. As r goes to zero the product Nr diverges to infinity because of the infinite amount of finer structures contained in the fractal. The important fact, however, is that for a particular exponent $D > 1$ the product Nr^D stays finite. For exponents smaller than D the product tends to infinity, whereas for exponents larger than D the product goes to zero. It is at a particular critical value D that the limit

Figure 2.1. Self-similarity in natural and mathematical ferns [after Neill and Murphy, 1993; Peitgen et al., 1992].

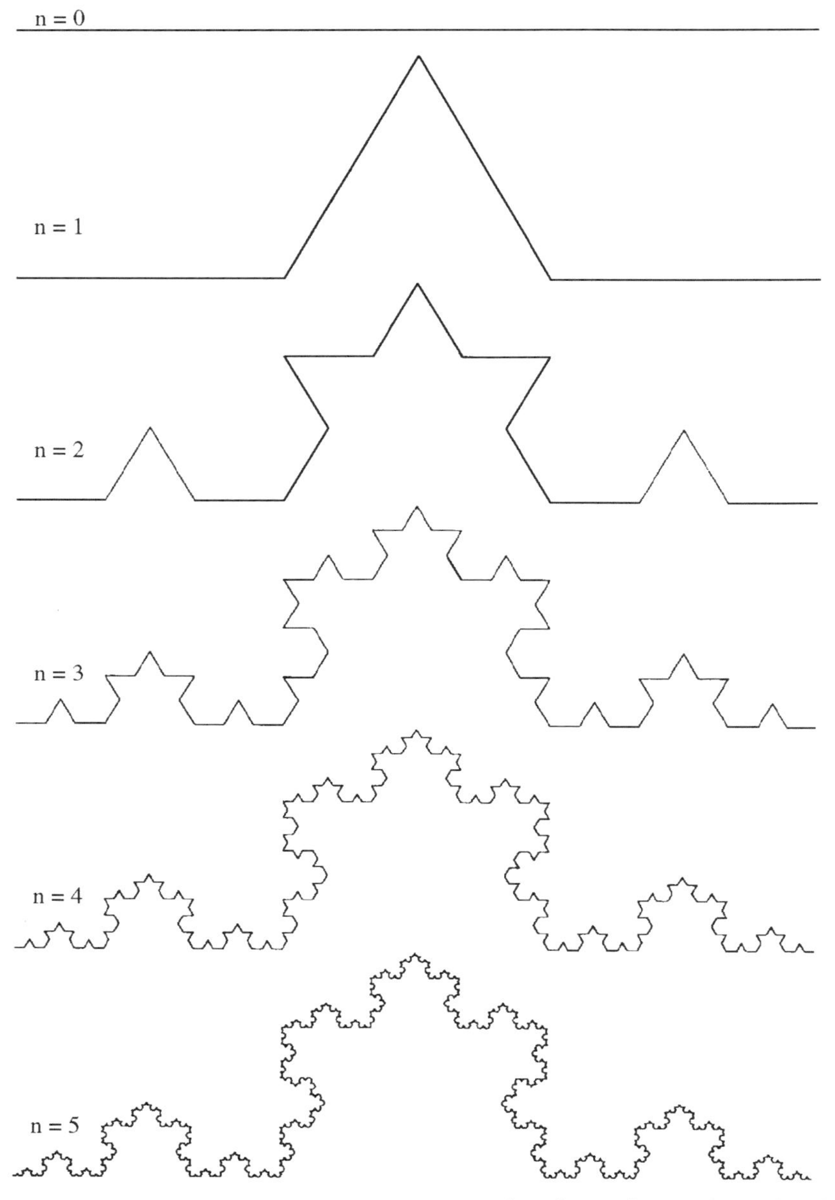

Figure 2.2. From top to bottom, Initiator, generator, the next stage in the construction, and high-order approximation for the Koch curve [from Schroeder, 1991].

procedure yields a finite and nonzero value:

$$\lim_{r \to 0} N(r)\, r^D = \text{constant} \qquad (2.2)$$

where the notation $N(r)$ has been used to emphasize the dependence of the number of divider steps on the size of the step. Eq. (2.2) implies that at small values of r one has

$$N(r) \propto r^{-D} \quad \text{or} \quad L(r) \propto r^{1-D} \qquad (2.3)$$

The critical exponent, D, is the most common definition of *fractal dimension*. The critical exponent D is sometimes assumed equal to the so-called *Hausdorff dimension*, D_H, defined rigorously by a limit covering of a set by suitable subsets of decreasing 'diameter' [e.g., Falconer, 1985]. Here we will use the names Hausdorff and fractal dimensions interchangeably although some theoretical problems arise towards a general definition. These aspects are beyond the scopes of this book but it seemed nevertheless appropriate to briefly discuss this issue given the common use of D_H [e.g., Schroeder, 1991]. Here it suffices to recall that the correct identification of D_H is not practical and that in all cases of geomorphological interest the estimations of D and D_H coincide. Equation (2.3) indicates that on a log-log plot of length versus ruler size for small rs the fractal (Hausdorff) dimension is one minus the slope.

Equivalently we have, from the definition of D

$$D = \lim_{r \to 0} \frac{\log N(r)}{\log(1/r)} \qquad (2.4)$$

where $N(r)$ is the number of *disks* of diameter r needed to cover the fractal.

In the case of the Koch curve we have $r = (1/3)^n$ and the number of pieces is $N = 4^n$, thus the fractal dimension is

$$D = \frac{\log 4}{\log 3} = 1.2619 \qquad (2.5)$$

Similar to what happens in the Koch curve, Richardson [1961] found that typical coastlines do not have a meaningful length. As will be seen

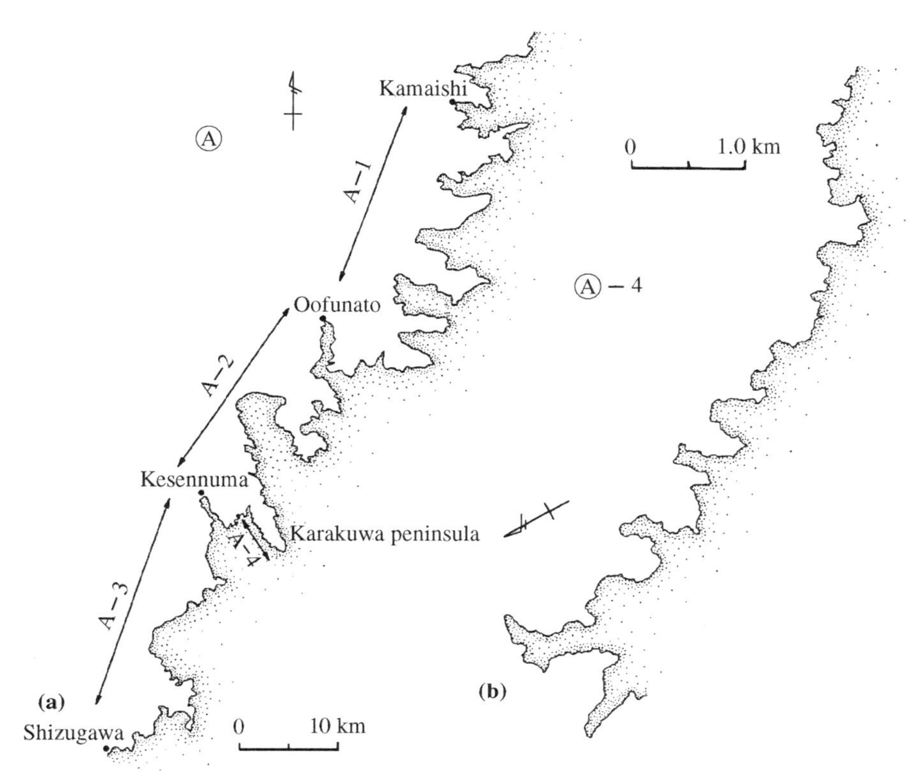

Figure 2.3. (a) Ria coastline in Northeastern Japan. (b) Magnified version of its smaller part (A-4) [from Nakano, 1983].

later in this chapter this is also the case for river courses. As one increases the resolution, more and more details of the interface between water and land become apparent, and the length diverges to infinity.

The fact that typically D is not an integer gives this parameter the peculiar character of a *fractal* dimension. The term fractal – first coined by Benoit Mandelbrot – also reflects the geometric structure of the object, which is characterized by a fragmentation of any part into many self-similar smaller constituents when observed at higher and higher resolutions.

For a smooth curve one sees from Eq. (2.2) that $D = 1$. Similarly, for a smooth surface the number, N, of area elements, for example, squares or circles, necessary to cover the surface increases as $1/r^2$, and therefore $D = 2$. Likewise, for a smooth three-dimensional object, $D = 3$. This demonstrates that not all self-similar objects are fractals. Thus a line segment, a square, or a cube can be broken into small components obtained by similarity transformation. The relation between the number of pieces N and the reduction factor c, $c = 1/r$, is given by

$$N = \frac{1}{c^{D_S}} \tag{2.6}$$

where $D_S = 1, 2, 3$ for the line, the square, and the cube, respectively.

In the case of the Koch curve we have $N = 4, 16 \cdots$ with $c = 3, 9 \cdots$. Thus Eq. (2.6) is valid with

$$D_S = \frac{\log 4^k}{\log 3^k} = 1.2619 \tag{2.7}$$

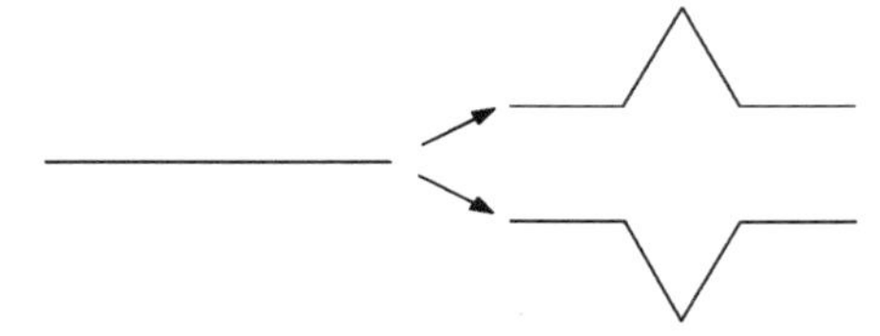

Figure 2.4. Two possible replacement steps in the Koch construction [from Peitgen, Jurgens, and Saupe, 1992].

When D_S is computed through Eq. (2.6) for a strictly self-similar object it is frequently called the self-similarity dimension, and in this case it coincides with the fractal and the Hausdorff dimensions defined in Eq. (2.4).

When the fractal dimension D is determined via Eq. (2.4) through divider-type length measurements, it is also called the divider or ruler dimension.

Deterministic fractals may be randomized to serve as more realistic models of natural forms. The related procedures may vary from highly sophisticated to very simple ones. An example of the latter is the randomization of the Koch curve on the basis of the position of the generator. Figure 2.4 shows the alternative positions and Figure 2.5 shows one realization of the Koch curve, whose generator is now oriented randomly in each step.

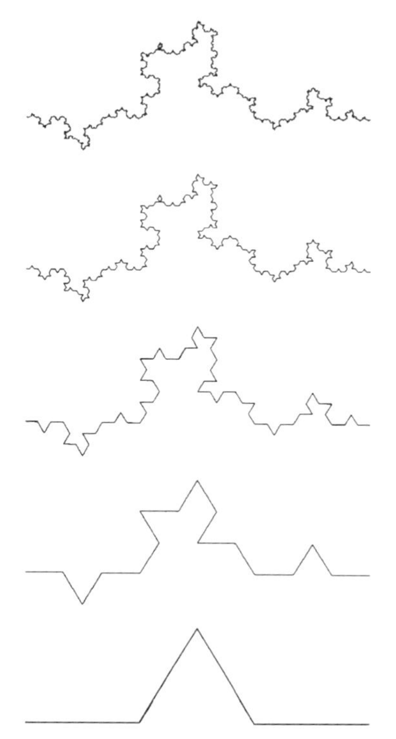

Figure 2.5. One realization of the random Koch curve. The replacement steps are the same as in the original Koch curve, with the exception that the orientation of the generator is chosen randomly in each step [from Peitgen et al., 1992].

Fractals come in all kind of structures and in very different disguises. A fractal structure very different from the Koch curve is the Cantor set, which appears in many places throughout this book under a wide variety of settings. It is a fractal whose structure is made up by a "dust" or set of isolated points. One example of this type of object is obtained as follows. One starts with the closed interval [0, 1] and removes the open interval [1/3, 2/3] (i.e., the points 1/3 and 2/3 are not removed). The process is then repeated on the remaining two segments of length 1/3 as shown in Figure 2.6.

When the procedure is repeated ad infinitum one is left with a "dust" made up of an infinite – and uncountable – number of points. There is not a single connected line segment, and therefore the total length measure is zero. However, we have an infinite number of points scattered over the interval. We observe that when changing the length scale by a reduction factor $c = 1/3$, we need $N = 2$ such pieces to cover the original set. After n removal steps we have $N = 2^n$ segments, each of length $r = (1/3)^n$. Thus the similarity dimension – and also the Hausdorff dimension – is in this case equal to

$$D_S = D = \frac{\log 2}{\log 3} = 0.6309 \qquad (2.8)$$

Figure 2.7 from Feder [1988] shows the construction of two 'Cantor-like' sets that look different despite the fact that they have the same similarity dimension $D_S = 1/2$.

In most practical applications the Hausdorff approach is not the most appropriate for quantifying the fractal structure of an object. Depending on

Figure 2.6. Construction of the "middle-third-erasing" Cantor set. Its fractal dimension equals $\log 2/\log 3 = 0.63$ [from Schroeder, 1991].

Figure 2.7. Two constructions of the Cantor set with $D = 1/2$. Top: $N = 2$ and $r = 1/4$. Bottom: $N = 3$ and $r = 1/9$ [from Feder, 1988].

the method used, one emphasizes different aspects of the fractal structure of the object, which are in turn characterized by different versions of the fractal dimension. These different dimensions may coincide numerically for some fractal objects, but quite frequently they yield different numbers. We will now undertake a rapid overview of some of the fractal dimensions that will be used throughout this book.

2.1.2 The Box-Counting Dimension

In many cases of interest the geometric structure of an object is such that traditional measures prove unsuitable; that is, it cannot simply be measured by a ruler. An example of this is the branching structure of a drainage network, whose fractal structure is reflected in the fact that smaller areas, when studied at finer resolution, display drainage flowpaths that are similar to those of much larger areas. In other words, we cannot attach a characteristic scale to the structure without having analyzed beforehand the size of the region being studied.

To estimate the box-counting dimension, one covers the object with a regular grid of size r and simply counts the number of grid boxes, $N(r)$, that contain some part of the object. Figure 2.8 illustrates the procedure. The value of r is progressively reduced to obtain a series of smaller and smaller sizes and the corresponding numbers $N(r)$. As $r \to 0$ we find that $\log N(r)/\log(1/r)$ converges to a finite value defined as the *box-counting* dimension.

An example of box counting applied to the determination of the fractal dimension of a contour line of elevation in a natural basin is shown in Figure 2.9. In (a) the set of points defining the contour line is isolated; (b) shows the box-counting procedure when a grid of size $r = 1$ is drawn, resulting in a number of covering boxes $N(r) = N(1) = 11$ (notice that here the 'covering' boxes are crossed for clarity); (c) shows the covering with a grid of size 1/2: $N(1/2) = 32$; (d) shows the grid of size 1/8, resulting in $N(1/8) = 100$.

The box-counting dimension is most frequently used in the natural sciences. It is easy to implement in a computer code even in the common case of objects embedded in higher dimensional spaces (e.g., in a

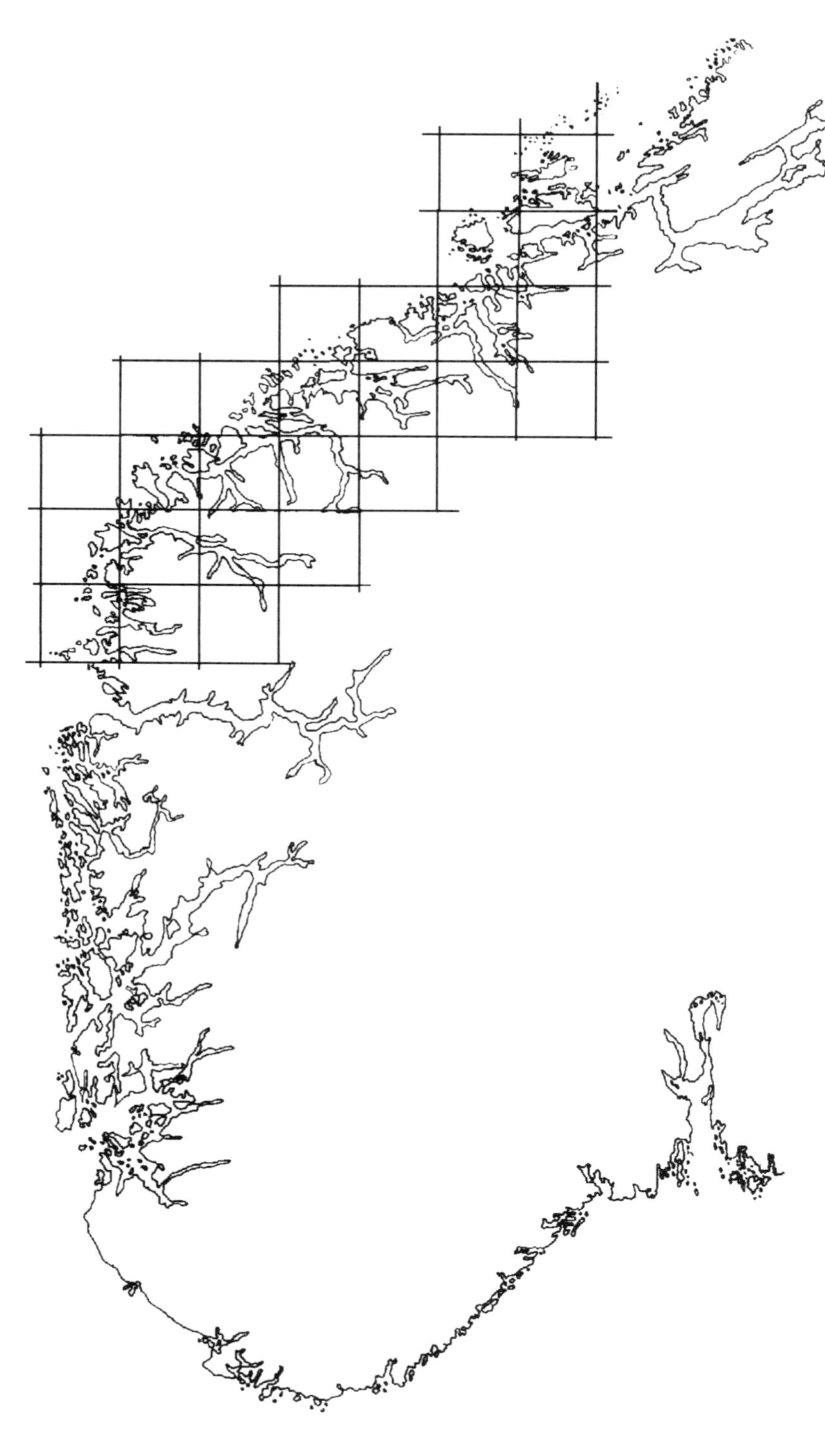

Figure 2.8. Determining the fractal dimension of the coast of Norway by counting how many boxes the outline of the coast penetrates [from Feder, 1988].

three-dimensional case the boxes are cubes with side size r). Moreover, we will assume that for all practical purposes fractal, Hausdorff, and box-counting dimensions coincide in the applications pursued here.

2.1.3 The Cluster Dimension or Mass Dimension

The cluster dimension or mass dimension is a measure of how the structure fills its embedding space. One example is the dendritic structure produced by the viscous fingering of a fluid injected into a porous medium and displacing another immiscible fluid. If one measures the mass, M, of the invading fluid as a function of the radius R from the injection point, one finds it increases according to a power law:

$$M \propto R^{D_c} \tag{2.9}$$

where D_c is a fractional exponent depending on the viscosity of the liquid and the porosity of the medium. The exponent D_c is called the cluster or mass fractal dimension and contrasts with the integer values (of, say, 2 or 3) that one would obtain for a similar experiment with a smooth figure in the plane (e.g., a circle) or in three-dimensional space (e.g., a sphere).

In the case of the Koch curve, we can consider the generator itself as the smallest cluster. The first generation contains the mass of four elements and has a radius of $R = 3$. The second generation has a mass of $N = 16$ and a radius $R = 9$. In the nth generation we have $N = 4^n$ and $R = 3^n$. Thus the clustering structure of the Koch curve follows Eq. (2.9) with $D_c = \log 4/\log 3$. Figure 2.10 from Feder [1988] shows the iterative construct and the scheme for the calculation.

A dendritic structure for which the cluster dimension has been analyzed in detail is that corresponding to diffusion-limited aggregation (DLA),

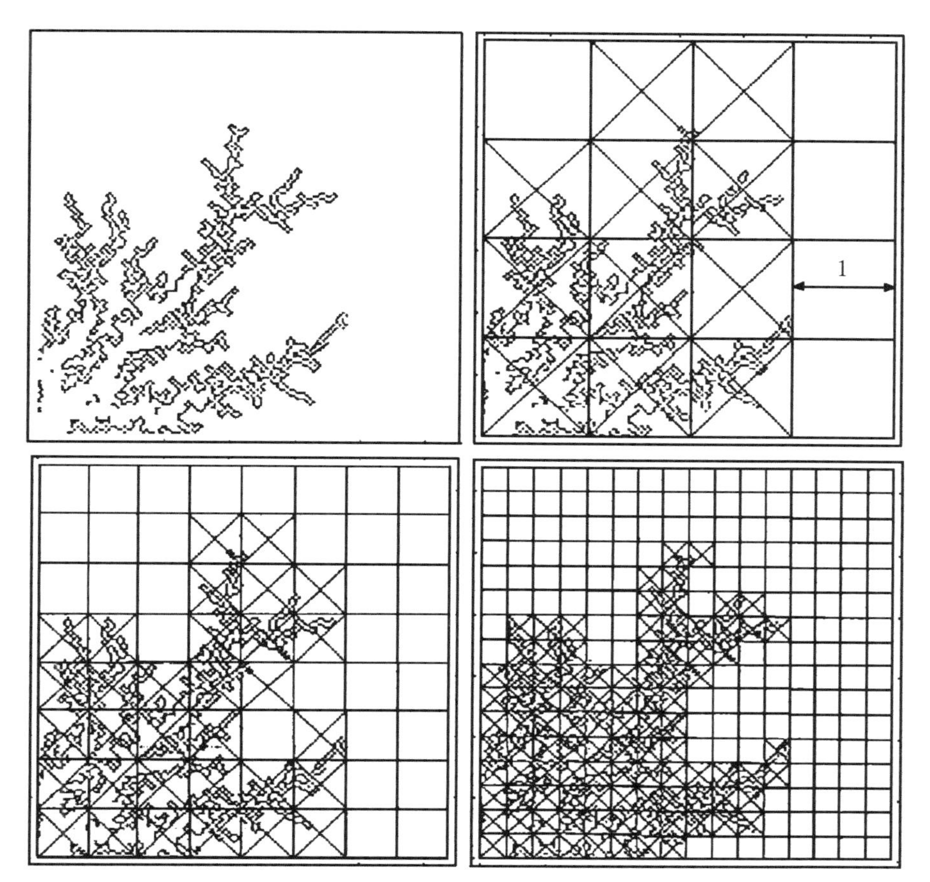

Figure 2.9. Box-counting procedure applied to a contour line of elevation of a subset of a real map [after Rigon, 1994].

which is a dynamical description of aggregate growth [Witten and Sander, 1981; Meakin, 1987a,b]. DLA starts with a seed particle, say, at the origin of a two-dimensional coordinate system. Next a region is chosen around the seed particle (e.g., a circular area of 500 particle diameters). Particles are then released one at a time from the boundary of the region and allowed to diffuse randomly until they either attach themselves to the aggregate or leave the region. Figure 2.11 shows the procedure. The aggregate behaves like a sticky material that grows through the attachment of particles that hit any part of its structure. Figure 2.12 shows an example of the resulting cluster of connected particles with the typical dendritic structure of DLA. There are many different extensions of DLA, including those where many particles move simultaneously or those where the dendritic cluster is allowed to move to pick up nearby particles.

A conceptually different model of the same type of phenomena is one that, instead of tracing individual particles, considers a continuous density function. This function is the Laplace partial differential equation that governs the electrostatic potential. Aggregation occurs where the potential is greatest, which is in turn along the boundaries of the dendritic structure. The fractal dimension of the aggregate depends on the applied voltage, and it remains approximately constant until a critical voltage is reached, after which it increases rapidly.

Figure 2.10. Triadic Koch clusters [from Feder, 1988].

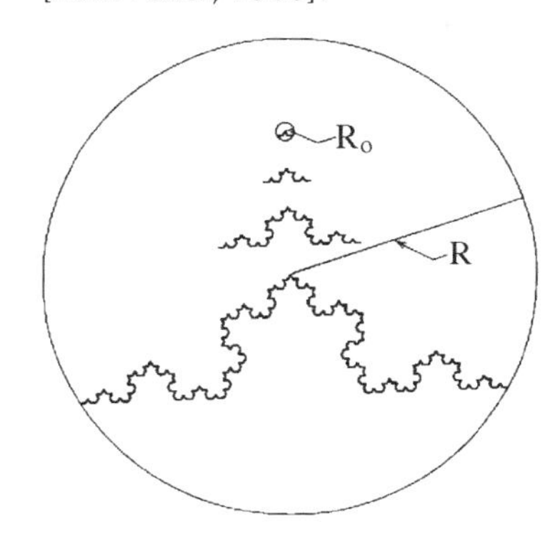

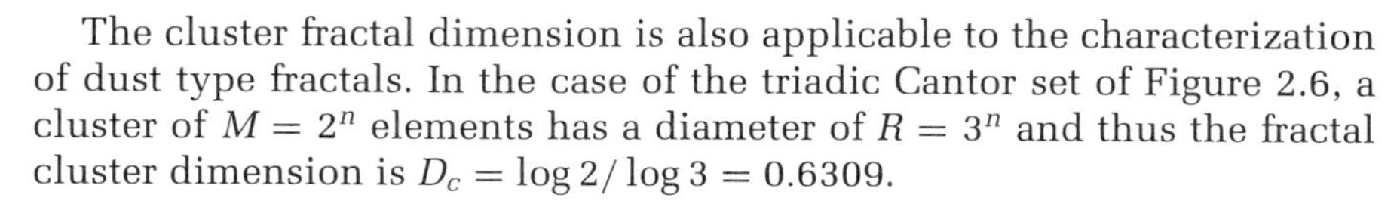

The cluster fractal dimension is also applicable to the characterization of dust type fractals. In the case of the triadic Cantor set of Figure 2.6, a cluster of $M = 2^n$ elements has a diameter of $R = 3^n$ and thus the fractal cluster dimension is $D_c = \log 2/\log 3 = 0.6309$.

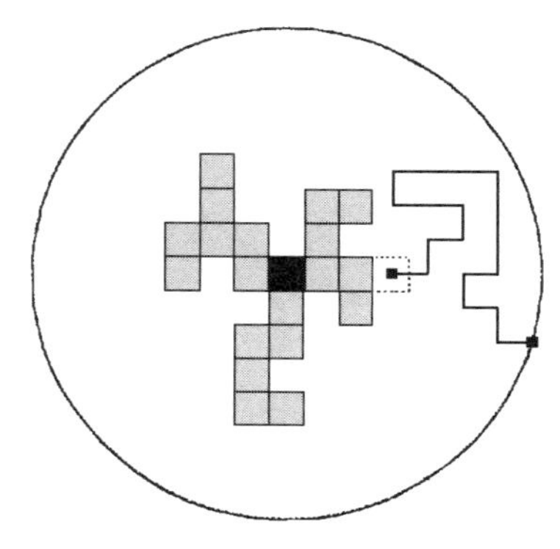

Figure 2.11. Diffusion-limited aggregation (DLA) simulation in two dimensions. Particles move from a pixel until they 'attach' to the existing dendrite [from Peitgen et al., 1992].

2.1.4 The Correlation Dimension

In many practical cases one of the fractal dimensions that better characterizes the spatial aggregation of the object is the so-called *correlation dimension*. It frequently is one of the easiest dimensions to estimate and it is most useful if the fractal is a dust, for example, a set of points distributed over a certain region. The closeness of the points in the structure is determined by how many of them have pairwise distances less than a distance r. Normalizing the count by the square number of points yields the value of $C(r)$, and the correlation dimension is then defined as

$$D_2 = \lim_{r \to 0} \frac{\log\ C(r)}{\log\ r} \tag{2.10}$$

Figure 2.12. Result of the numerical simulation of DLA based on Brownian motion of single particles [from Peitgen et al., 1992].

The subindex 2 in the nomenclature of Eq. (2.10) reflects the fact that the correlation dimension is a particular case of a whole spectrum of fractal dimensions called Rényi dimensions. The Rényi dimension of the qth order is defined as

$$D_q = \lim_{r \to 0} \frac{1}{q-1} \frac{\log \sum_k p_k^q}{\log r} \qquad -\infty \le q \le \infty \tag{2.11}$$

The term p_k^q is the probability that q points drawn randomly from the object fall in the same box k of size r. The sum $\sum_k p_k^q$ stands for the probability that the q random points fall together into any one of the boxes of size k that cover the whole object. In the case $q = 2$ we deal with pairs of points and $\sum_k p_k^2$ is equivalent to $C(r)$, which in turn represents the relative number of 2-tuples where all pairwise distances are less than r. Thus when $q = 2$, Eq. (2.11) gives the correlation dimension of Eq. (2.10).

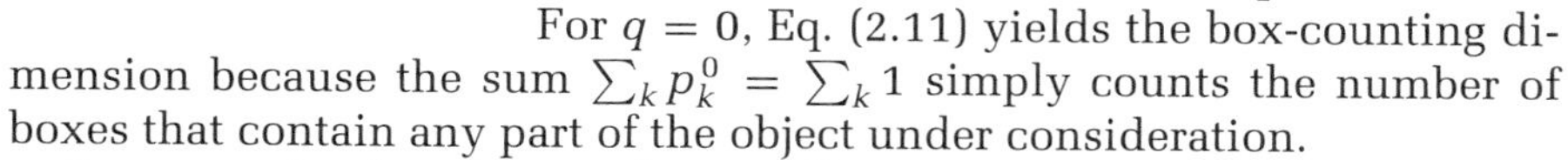

For $q = 0$, Eq. (2.11) yields the box-counting dimension because the sum $\sum_k p_k^0 = \sum_k 1$ simply counts the number of boxes that contain any part of the object under consideration.

For $q = 1$, Eq. (2.11) gives the so-called *information dimension*:

$$D_1 = \lim_{r \to 0} \frac{-\sum_k p_k \log p_k}{\log(1/r)} \tag{2.12}$$

The numerator of Eq. (2.12) is Shannon's entropy, the cornerstone of information theory and the reason for the term information dimension.

The information dimension is frequently used in the analysis of time series data when investigating whether the process represents random noise or is generated by some characteristic or chaotic mechanism.

2.1.5 Self-Similarity and Power Laws

Implicit in the definition given in the introduction to this chapter, a fractal is a set (an object, a geometrical construct) whose parts are somewhat *congruent* to the set itself under affine transformations. These transformations are defined as geometric operations through which the set is transformed into a copy of itself possibly translated, rotated, and zoomed. Choosing an arbitrary point of vector coordinates $\mathbf{x}$ belonging to the set, an affine point $\mathbf{y}$ is defined by the relationship $\mathbf{y} = \mathbf{A}(\lambda, \theta) + \mathbf{T}$. The matrix $\mathbf{A}$ defines a magnification of parameter λ and a rotation of parameter θ; $\mathbf{T}$ is a translation operator. Although we will return in Section 3.6.1 to some particular properties of the distortion θ, here it suffices to underline the properties of the parameter λ that identifies the characteristic length of the transformation.

In particular, a fractal set may be identified by an average property, say S, of any subset G' of the whole, say G. We address the average properties of any subset in relation to those of the whole. An important property of fractal object G is the validity of a functional relationship of its average property S of the type

$$S(G) = f(\lambda)S(\lambda G') \qquad \forall G' \subset G \tag{2.13}$$

where f is a suitable function. Importantly, if the property S is dependent on a single variable, say q, describing the set (e.g. its volume, its area, some intensive properties), that is, $S(G) = S(q(G))$, then the validity of Eq. (2.13) requires that the relationship $S(q)$ be a power law, that is

$$S(q) = cq^{\alpha} \tag{2.14}$$

where c, α are suitable coefficients [e.g., Korvin, 1992]. As seen before, in this case, the set S is said to *scale* and the property of obeying a power law is referred to as *scaling*.

The topology of river basins as well as the hydraulic geometry of the drainage network and even the hydrologic response of the basin to different kinds of precipitation input are characterized by power law relationships between the variables involved in their description. Some of these were mentioned in Chapter 1. These power law relationships may or may not be of a statistical nature but are in many cases the reflection of self-similar types of phenomena.

Simple power functions like

$$g(x) = ax^{\alpha} \tag{2.15}$$

where a and α are constants, are abundant in the hydrology and geomorphology of river basins. These power laws are said to be *scale free* or to possess *scaling invariance*. When x is rescaled through multiplication by a constant, $g(x)$ is still proportional to x^{α}.

Power laws are an abundant source of self-similarity because they represent phenomena that have the same characteristics or features at very small and at very large scales. An example is shown in Figure 2.13, which shows the power law character of fragmentation. Because fragments have a variety of shapes, the cube root of volume is an objective measure of size. The number N of fragments with cube root of volume greater than r

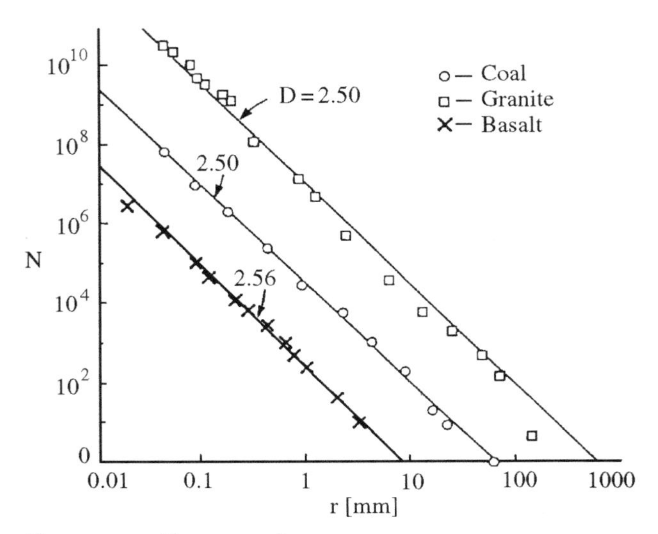

Figure 2.13. The power law character of fragmentation. The best-fit distribution is shown for each data set [reproduced from Turcotte, 1992].

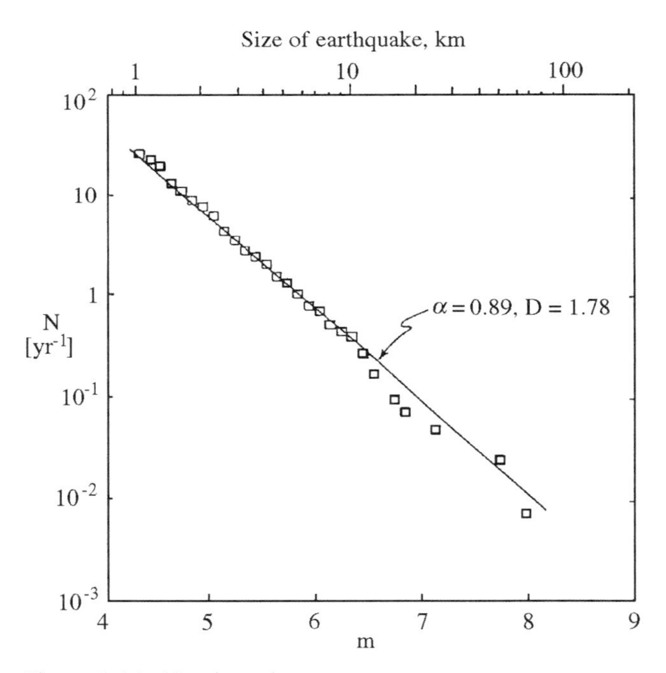

Figure 2.14. Number of earthquakes per year N with magnitudes greater than m as a function of m. The line represents Eq. (2.16) with $\alpha = 0.89$ and $a = 1.4 \times 10^5$ [reproduced from Turcotte, 1992].

is given as a function of r for broken coal [Bennett, 1936], broken granite from a 61 kt underground nuclear detonation [Schoutens, 1979], and impact ejecta due to a 2.6 kms^{-1} polycarbonate projectile impacting on basalt [Fujiwara, Kamimoto, and Tsukamoto, 1977]. For a given material (e.g., granite) the number of fragments in a fixed span of sizes remains constant and independent of the absolute sizes. Thus the number only depends on the length of the span. The phenomenon is the same at all scales and thus it does not involve a characteristic unit (e.g., unit length, unit mass, or unit time) that would make it scale dependent and *nonscaling*. Obviously, it is expected that there will exist upper and lower limits to the scaling behavior. These limits depend on the physical characteristics of the processes involved. In the case of fragmentation the upper limit is controlled by the size of the object being fragmented, whereas the lower limit most likely depends on the grain size.

A different example of a well-known power law is shown in Figure 2.14 which gives the number of earthquakes occurring per year in southern California with magnitudes greater than m as a function of m. The horizontal axis is also logarithmic since m measures the energy released by the earthquake. The solid circle is the expected rate of occurrence of great earthquakes in southern California. At magnitudes greater than 6.5 the data deviate a little from the power law line, and this may be due either to the small number of large earthquakes that make the statistics unreliable or to physical reasons peculiar to large earthquakes.

The examples given above refer to statistical scaling where the power law of Eq. (2.15) takes a probabilistic character of the kind

$$P[X \geq x] \propto x^{-\alpha} \tag{2.16}$$

(or, conversely, the probability density, say $f(x)$, is $f(x) \propto x^{-(\alpha+1)}$) that describes the scaling or self-similar character of the relative occurrences or of the fluctuations of the process being analyzed. The fractal structure of river basins is reflected in the many power law relationships that appear to be valid independently of climate, soil, or geological conditions. Some of these will be studied in detail later in this chapter.

2.2 Self-Similarity in River Basins

The recent availability of digital elevation models (DEMs) of river basins and the continuously increasing power of computers have allowed careful study of the self-similarity characteristics of the drainage network and its accompanying landscapes. Section 1.2.9 described in some detail the determination of drainage networks from DEMs. One of the techniques studied there is based on work of O'Callaghan and Mark [1984] as described by Tarboton et al. [1988]. DEM data are generally supplied on a rectangular grid with each point representing the elevation of a pixel or unit area. In the data of Tarboton et al. [1988], the pixels have an area of 30 m by 30 m, corresponding to the grid size commonly used in the U.S. Geological Survey DEMs. A drainage direction is assigned, linking each pixel with one of its eight neighbors, based on the steepest slope. This effectively defines a drainage path or flow field. The number of pixels draining through each pixel is then counted to give the accumulated area that drains into each pixel. Channel networks may then be defined in many ways (see Section 1.2.9), the easiest of which is to define channelized pixels as those that have total contributing drainage area greater than a threshold support area. Decreasing the support area results in a finer network of channels and is thus tantamount to increasing the resolution used to study the basin. The limit to this refinement, in early studies of this issue [Tarboton et al. 1988], was the 30 m by 30 m pixel size of the U.S. Geological Survey's DEMs. Figure 1.15 [after Tarboton et al., 1988] shows networks in a river basin (i.e., Walnut Gulch, Arizona) defined with varying support areas. Also, Figure 1.16, from the same source, shows an example of the extraction of drainage networks of subbasins of different area within the same river basin (St. Joe River). Figure 2.15 shows an image of a large river network taken from space, where the richness of the planar patterns found in nature is evident. Figures 2.16 and 2.17 show the complexity of intertwined patterns arising from the interplay of geologic and fluvial phenomena. The figures underline the need for a detailed analysis of the individual roles and mutual interrelations of both planar landforms and structures in the third dimension.

From observing such figures one can pose two questions:

- Is the structure fractal? (implying that finer structures are statistically indistinguishable from grosser representations, and hence there is no dominating scale)
- If so, what is the fractal dimension?

The fact that by definition a river network drains its entire basin suggests the hypothesis that the fractal dimension of river networks (if it exists) should approach 2, that is, the network is space filling. Tarboton et al. [1988] applied Richardson's divider method to different drainage networks throughout the United States. This method was described in Section 2.1 of this chapter. In applying the divider procedure to river networks, rules are required to deal with bifurcations. Tarboton et al. [1988] measured the length of each Strahler stream (defined according to Strahler's [1952] network-ordering convention, described in Section 1.2.1) separately. At the end of streams there is, generally, a leftover piece of stream shorter than r. If the distance from the last stepping point to the end was greater than $0.5r$, it was counted in N in Eq. (2.4); otherwise, it was not included in the length. Figure 2.18 gives results for several different networks. The Souhegan is a 440-km^2 basin in southern New Hampshire that was digitized from

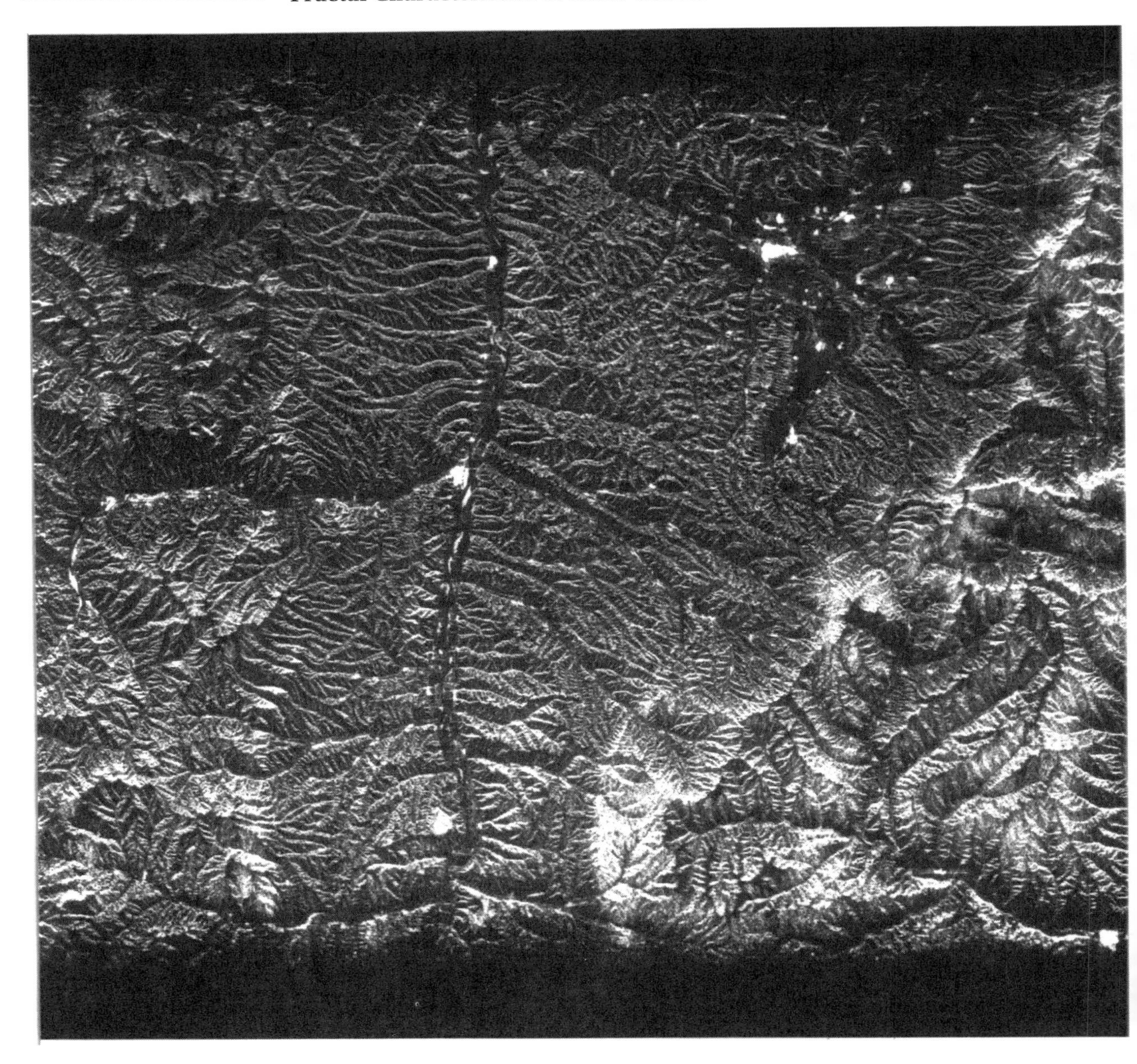

1:24,000 U.S. Geological Survey maps. The Hubbard (area 75 km^2) in Connecticut, and W15 (area 23 km^2) in Walnut Gulch, Arizona, were extracted from DEMs. The eight networks consisted of two hand-digitized networks in New Hampshire and six networks obtained from three DEM basins with support areas of 50 and 20 pixels. The DEM basins used were the Hubbard, W15, and W7. The pattern for all the networks was the same, a gently sloping line with slope about 0.05 for small ruler lengths, followed by an abrupt change to slope of about one for large ruler lengths. This clearly indicates two distinct regions of scaling. The first, with fractal dimension $D \approx 1.05$ is due to the sinuosity of individual rivers and corresponds to the self-similarity formed in coastlines, like the one shown in Figure 2.3.

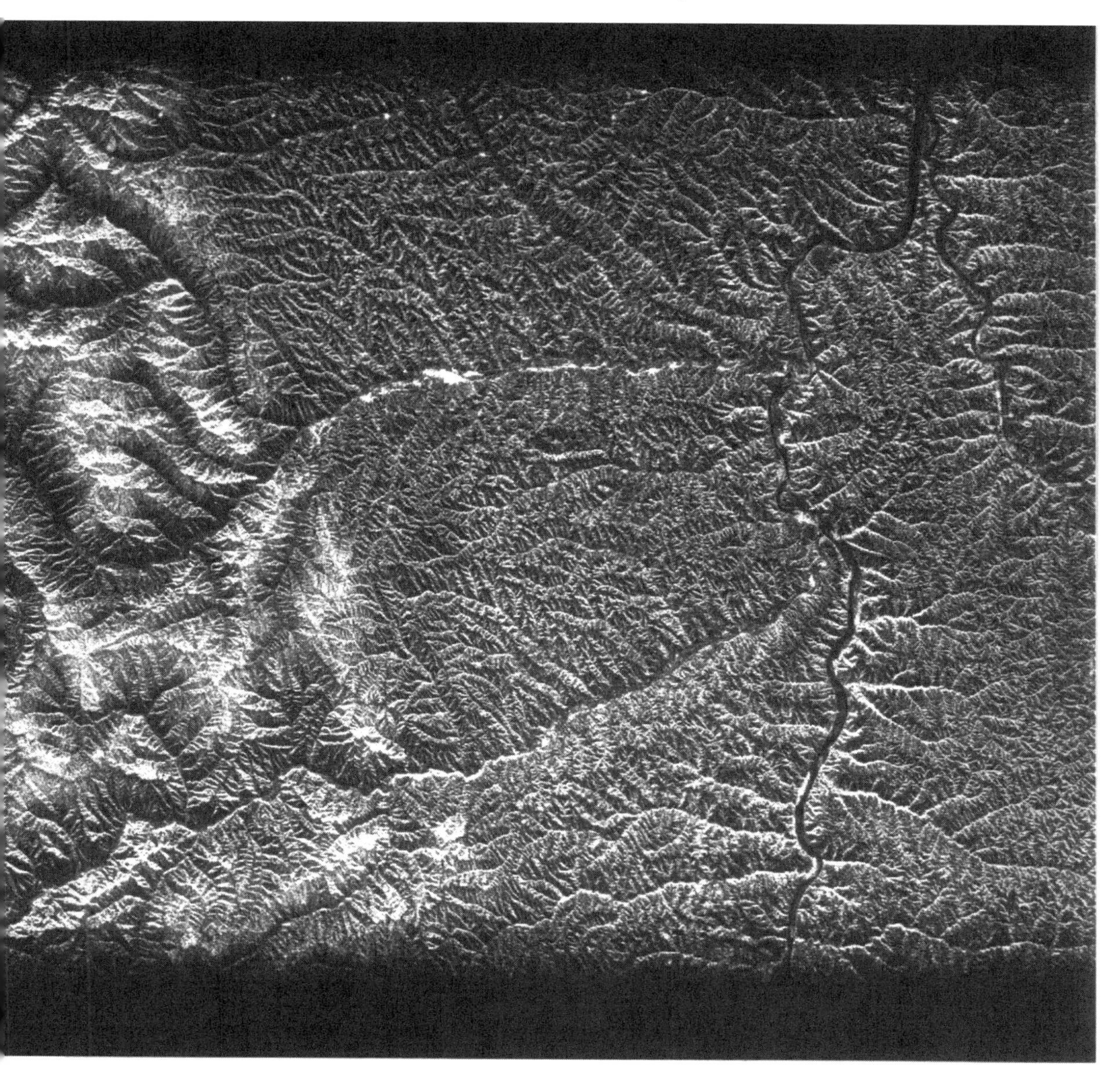

Figure 2.15. In this spectacular Landsat image, in the visible-red wavelength, extensive dendritic patterns in a portion of the river basin of the Yellow River (Shanxi Province, China) are shown. Beautiful fractal patterns are evident over a large range of scales [courtesy of NASA].

The second region, with D near 2, is due to the branching characteristic of networks. More precisely, the value $D < 2$ may be attributed to streams shorter than $0.5r$ not being counted at all, reflecting the fact that at coarse resolution we see fewer streams.

Tarboton et al. [1988] also applied the box-counting technique described in Section 2.1.2 to estimate the fractal dimension of the drainage network. Some of their results are shown in Figure 2.19. We see that there are basically two asymptotic slopes. The first, a slope close to -1 for a small box size, implies that at scales that are small relative to the resolution of the map, channels have dimensions close to that of the line. The second, at the large box-size end of the scale where the slope is -2, indicates

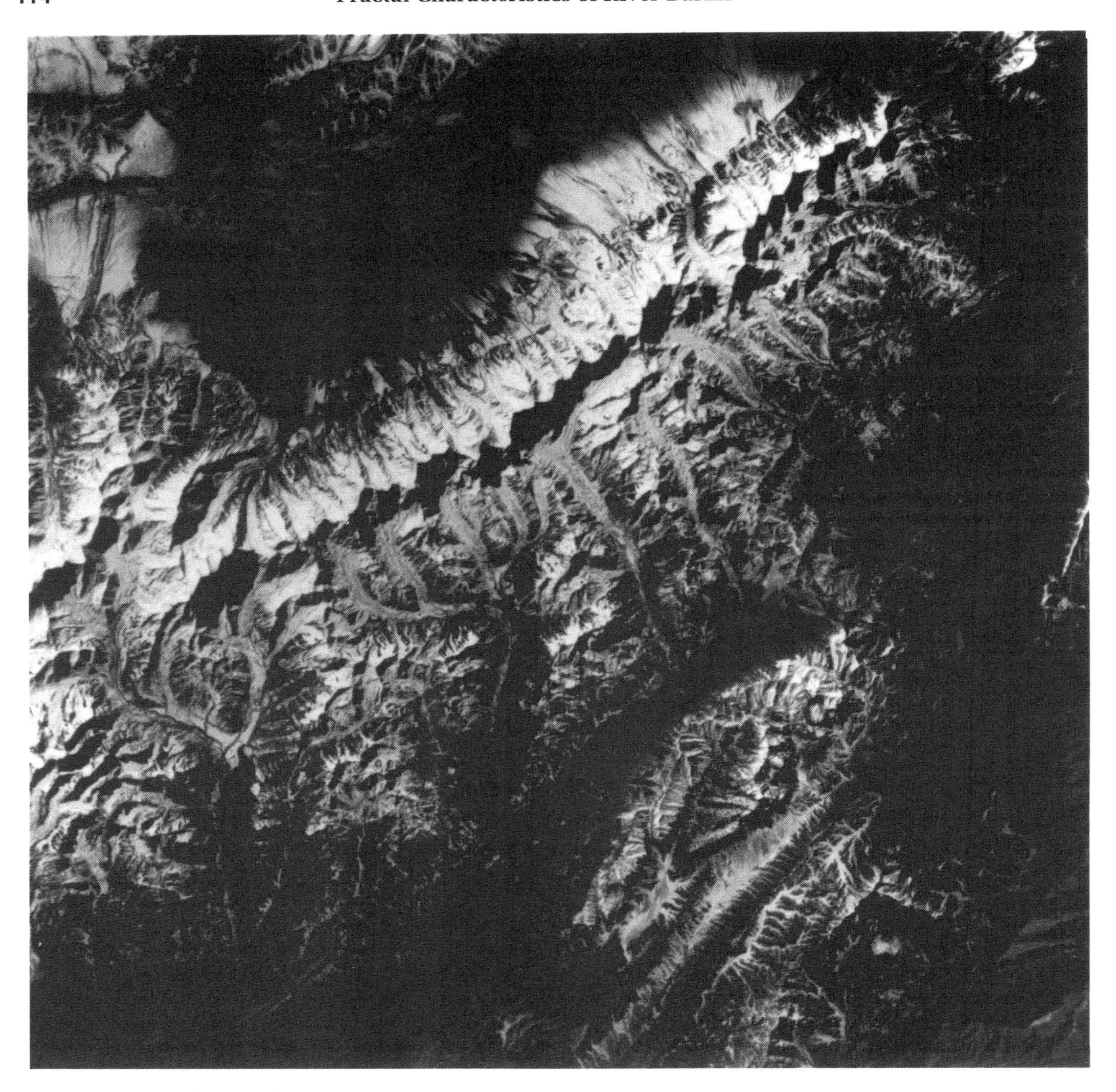

Figure 2.16. An earth view of King Ata Tag mountain range, western China. The landscape shows a powerful interplay of geologic controls and fluvial patterns. Here fractal landforms appear both in the planar view and in the elevation structure, as seen throughout this chapter [courtesy of NASA].

that practically all boxes are intersected by a channel. At this scale the network is space filling with $D = 2$. Note that the more detailed the network (smaller support area), the smaller the scale above which the network is space filling. It is important to point out that a fractal dimension equal to 2 does not necessarily mean the structure is space filling. This is only so if the object does not allow self-overlapping.

A classical example of a fractal process that is not space filling but that has fractal dimension equal to 2 is Brownian motion in two dimensions. Figure 2.20(a) shows the motion of a very fine particle in a liquid as

Figure 2.17. Earth view of Mount Everest and of the "roof of the world". Abundant details of glacier surfaces including moraines, crevasse fields, and ice falls are shown. Remarkably, a dendritic structure of the valleys with a definite hierarchy is shown, yielding another type of fractal structure both in plan and in the third dimension [courtesy of NASA].

observed under a microscope. When the shutter speed of the experiment is 100 times faster, the particle is observed 100 times between the points A and B, which were connected by a straight line in the original observation. Figure 2.20(b) shows the motion at the 100-times-faster shutter speed when the displacements are magnified 10 times with respect to the original scale. One observes that the two pictures are statistically indistinguishable, and thus the process is statistically self-similar when the spatial dimension is changed by a factor r and the time dimension by a factor r^2. Objects or processes that are indistinguishable (strictly or statistically speaking) at different scales of observation but that need to be scaled in different directions with different factors are called *self-affine*. These processes are very abundant in river basins, as will be seen later in this chapter. The fractal dimension of two-dimensional Brownian motion can be obtained from Eq. (2.4), observing that when the spatial resolution is increased by a factor of $1/r$, we obtain $N(r) \approx 1/r^2$ more pieces to cover [Schroeder, 1991]. Thus

$$D = \frac{\log N(r)}{\log(1/r)} = 2 \tag{2.17}$$

Although $D = 2$, Brownian motion is not plane-filling but instead possesses a very large degree of overlapping.

In the case of the drainage network the experiments of Tarboton et al. [1988] suggest a plane-filling structure with fractal dimension equal or very

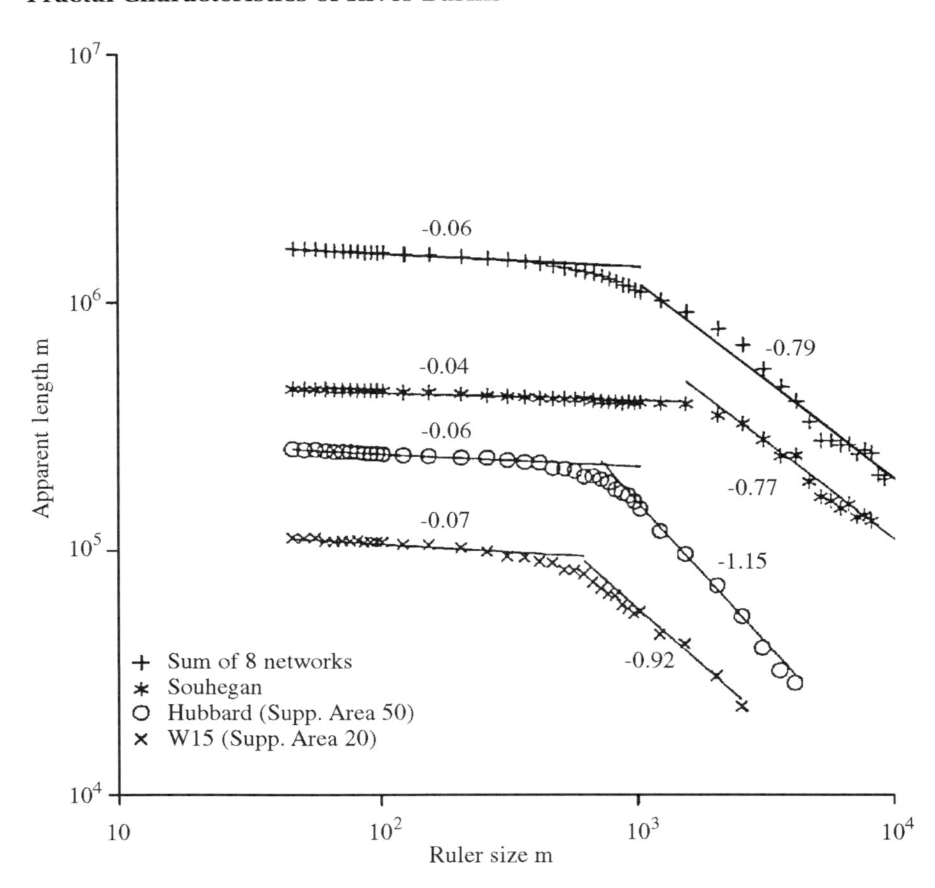

Figure 2.18. Divider method results for typical river networks. The numbers give slope of the fitted straight lines [after Tarboton et al., 1988].

close to 2. This value was also suggested by Mandelbrot [1967, 1983] in analogy with the fractal dimension of the Peano plane-filling curve. The Peano structure displays useful analogies with the drainage network, as we will see in detail later in this chapter and throughout this book. Space-filling structures play a fundamental role in the architecture and maintenance of many natural systems. Thus, for example, the kidney houses three interwoven tree-like vessel systems, the arterial, the venous, and the urinary systems and each one of them has to access every part of the kidney [Peitgen et al., 1992]. Fractals are very useful for the analysis and description of these complicated space-filling structures. In the case of the drainage network one may think that the network must penetrate everywhere to drain an area, and thus in the limit would be a plane-filling tree.

As previously mentioned, the region with slopes near -1 in Figure 2.19 is due to short streams being excluded as r increases. In this region, Eq. (2.3), $N(r) \propto r^{-D}$, can be interpreted as saying that $N(r)$ is proportional to the number of streams with length greater than r. Mandelbrot [1983] notes that the probabilistic counterpart to this is a power law or hyperbolic distribution:

$$P\,[\text{length} > l] \propto l^{-D} \tag{2.18}$$

where D is again the fractal dimension and l refers to stream length.

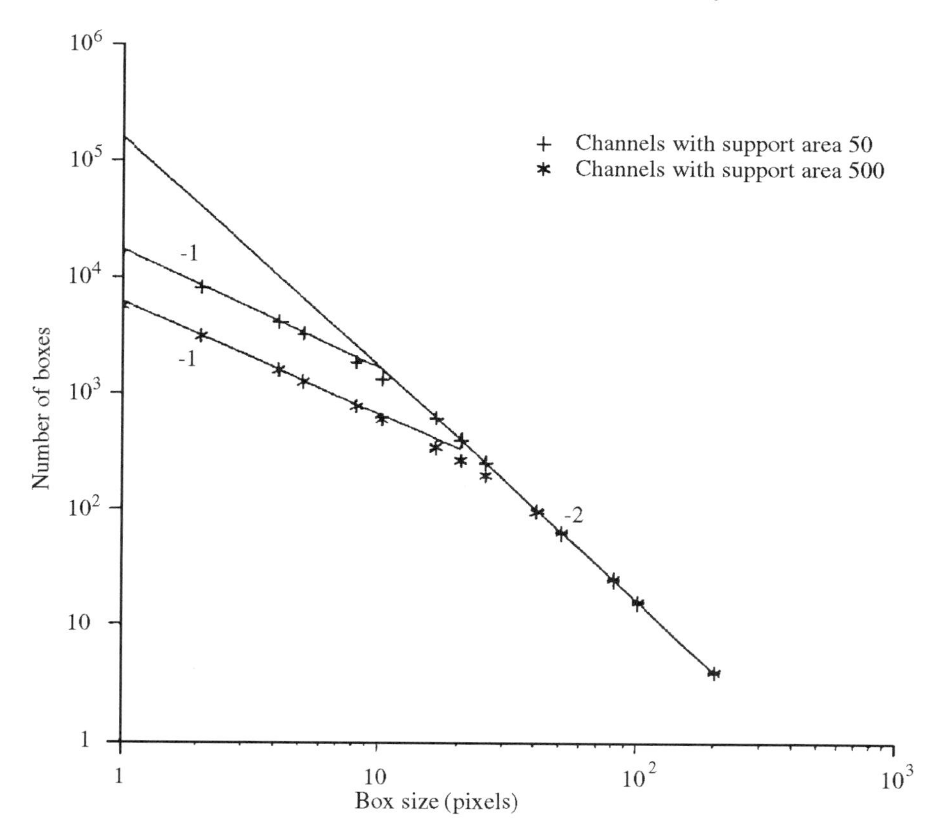

Figure 2.19. Box-counting results from Walnut Gulch, Arizona. The numbers give slope of the fitted straight lines [after Tarboton et al., 1988].

Figures 2.21 and 2.22 [from Tarboton et al., 1988] give the exceedence probability of stream length aggregated from several river basins. Points were plotted using

$$P = \frac{m}{n+1} \tag{2.19}$$

where m is the ranking from longest to shortest stream length and n is the number of streams in the sample. The figures indicate a hyperbolic tail with $D \approx 2$. Figure 2.21 uses geometric length, defined as the straight-line distance between endpoints of a stream. Figure 2.22 uses length measured along the stream, naturally limited by the resolution of the map or DEM from which the network is obtained. The slight difference in slope between these figures may be due to length along the stream itself being a fractal measure with dimension D slightly in excess of 1. As an example, suppose we have, from Eq. (2.18), fitted to Figure 2.22

$$P\,[\text{length} > l] \propto l^{-\lambda} \tag{2.20}$$

Now if l is itself a fractal, with dimension D_l, one has

$$l \propto r^{D_l} \tag{2.21}$$

Combining, we get

$$P\,[\text{length} > l] \propto r^{-\lambda D_l} \tag{2.22}$$

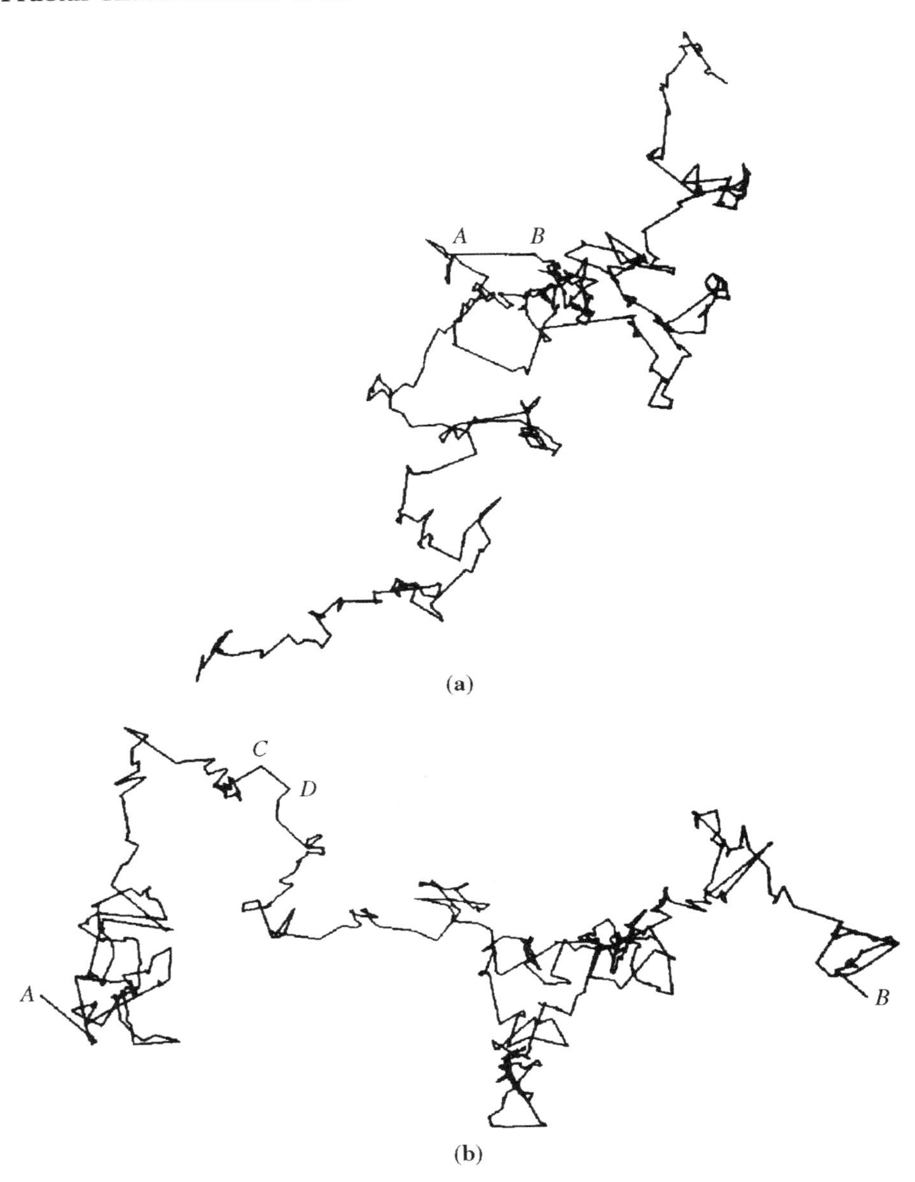

Figure 2.20. (a) Brownian motion; (b) the segment AB of the Brownian motion in (a) sampled 100 times more frequently and magnified 10 times [from Schroeder, 1991].

Thus, as pointed out by Tarboton et al. [1988], the fractal dimension of the whole network is $D = \lambda D_l$. The result $D = 2$ is therefore consistent with slope $\lambda = 1.8$, seen in Figure 2.22, and $D_l = 1.1$ as suggested by the flatter slopes of Figure 2.18. Tarboton et al. [1988] interpret these figures as evidence that the network is space filling with $D = 2$.

The existence of two different fractal dimensions in the same object is not unusual in the earth sciences. In the case of the drainage network the value of approximately 1.1 reflects the large degree of sinuosity of the river channels. Measurements carried to the limit along the interface between flow and banks make the concept of stream length meaningless. The fractal dimension of 2 refers to a different characteristic of the network, namely, its bifurcation in smaller and smaller subbasins that are statistically similar at different scales, which when carried to the limit can be thought of as invading all the region that needs to be drained.

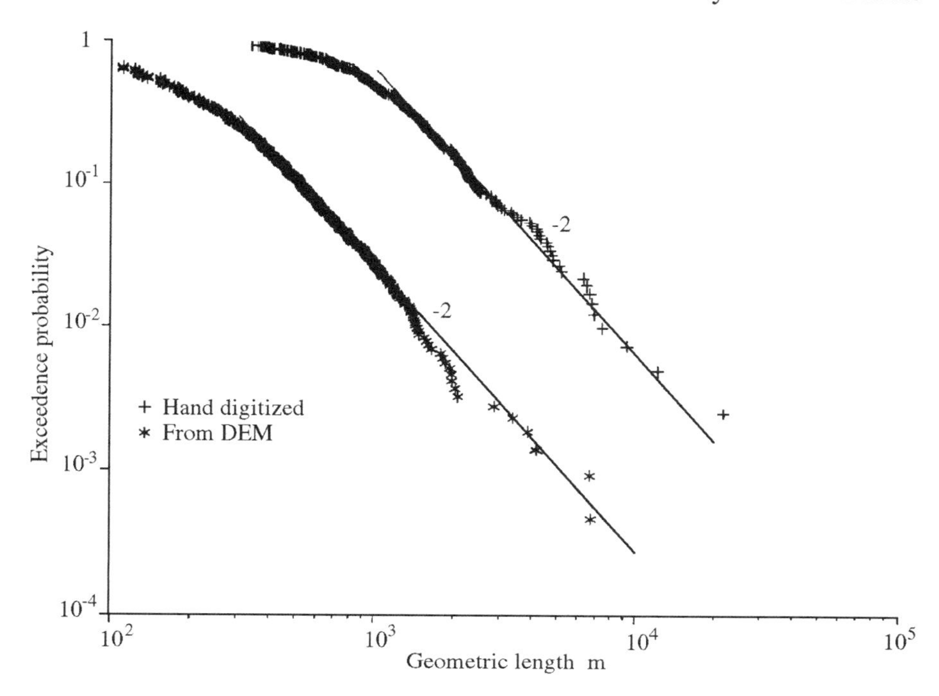

Figure 2.21. Geometric stream length exceedence probability. The DEM data are based on 2,178 streams from three networks with support area of twenty pixels [after Tarboton et al., 1988].

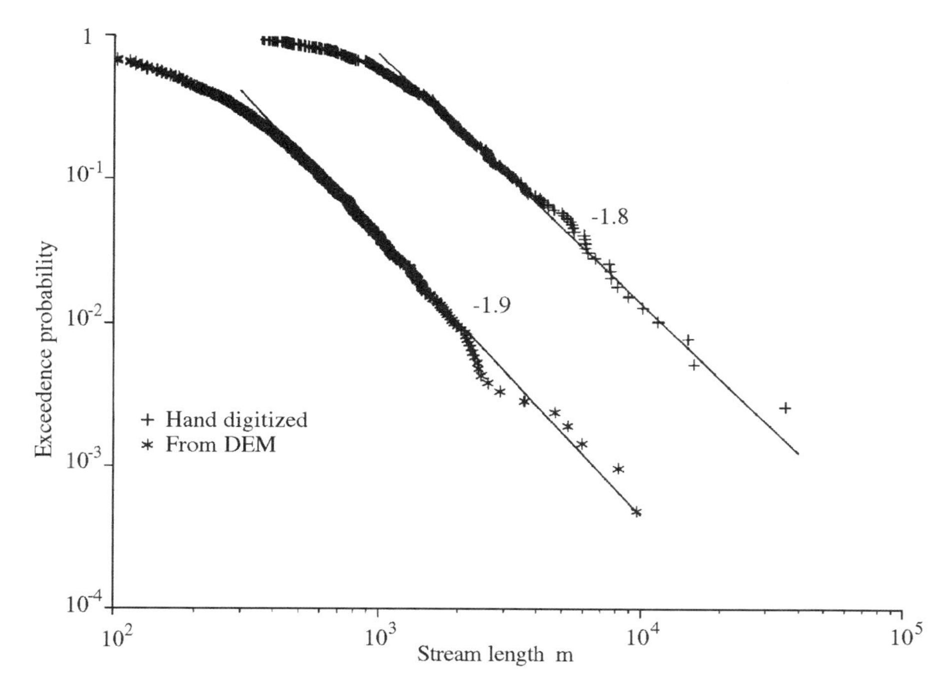

Figure 2.22. Stream length (along stream) exceedence probability [after Tarboton et al., 1988].

Figure 2.23 shows the analysis by Richardson's divider method of the shoreline of Gull Lake in Ontario, Canada, carried out by Kent and Wong [1982]. They attribute the *bifractal* nature of the graph to the presence of at last two different geomorphological fractal processes. Each process modifies the shoreline but acts at different scales. This is not uncommon in river basins where hillslopes and valleys are under the effect of much different influences such as tectonic forces and diffusion processes. The

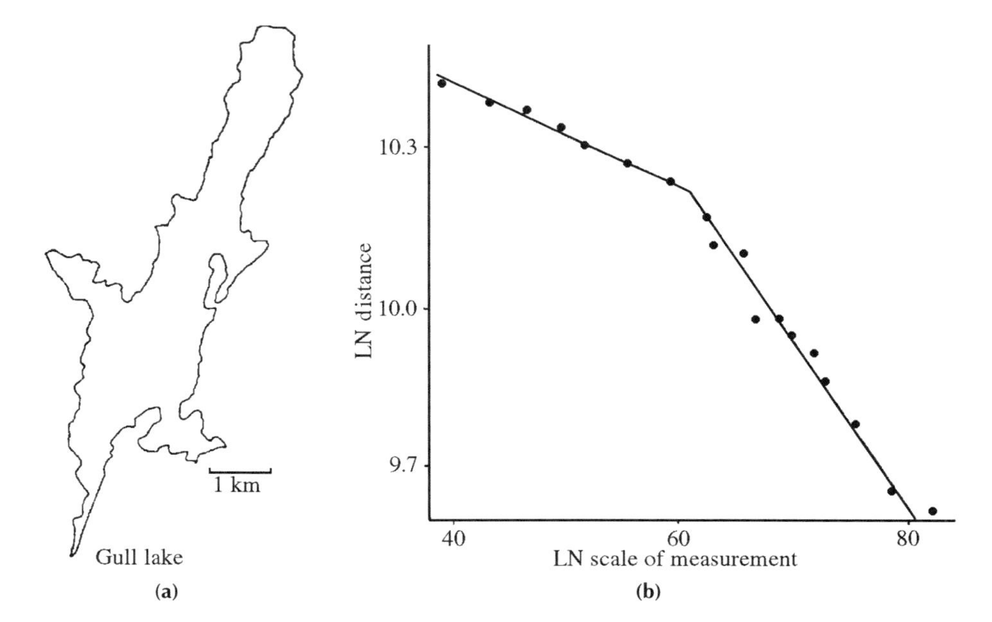

Figure 2.23. (a) Gull Lake, Ontario, Canada, and (b) the fractal analysis of its shoreline. [from Kent and Wong, 1982 (reproduced from Korvin, 1992)].

relation of the tree-like fractal structure of the river basin to Horton's laws will be studied in the next section.

2.3 Horton's Laws and the Fractal Structure of Drainage Networks

This section, based on Tarboton et al. [1988], relates Horton's [1945] laws of network composition to the fractal structure of drainage networks. As described in detail in Chapter 1, Horton's laws, and in particular the length and bifurcation laws, are usually stated in terms of Strahler's [1952] ordering scheme. According to such a scheme, source streams are of order one. When two first-order streams join, they become second order and, in general, when two streams of equal order merge, a stream of higher order is formed. When low- and high-order streams join, the continuing stream retains the order of the higher-order stream.

As seen in Chapter 1, the set of empirical laws collectively referred to as Horton's laws include the bifurcation law

$$R_B = \frac{N(\omega - 1)}{N(\omega)} \tag{2.23}$$

and the length law

$$R_L = \frac{\bar{L}(\omega)}{\bar{L}(\omega - 1)} \tag{2.24}$$

where $N(\omega)$ is the number of streams of order ω and $\bar{L}(\omega)$ is the average length of streams of order ω.

The above are geometric-scaling relationships, because they hold no matter the order or resolution at which we view the network. If we regard a channel network as the collection of paths where water flows, it is possible to imagine, getting lower and lower orders of streams with higher and higher resolution. Viewed this way, the limiting channel network is a fractal, with properties related to R_B and R_L.

Based on Horton's laws, La Barbera and Rosso [1987] reported that the fractal dimension of river networks should be

$$D = \max\left(\frac{\log R_B}{\log R_L}, 1\right) \tag{2.25}$$

A derivation of the above equation requires that Horton's bifurcation and length ratios hold exactly at all scales in the network. Then the total length of streams in the network is the sum of a geometric series with multiplier R_B/R_L. Let a network of order Ω have main stream length $L(\Omega)$. Then, using Horton's length ratio, the mean length of a stream of order ω (with $\omega \leq \Omega$) is $L(\Omega)/R_L^{\Omega-\omega}$. By Horton's bifurcation law there are $R_B^{\Omega-\omega}$ of these streams, so that the total length of stream of order ω is $L(\Omega)(R_B/R_L)^{\Omega-\omega}$. Adding over all ω to get the length of the whole network, L_T, we get the geometric series:

$$L_T = \sum_{\omega=1}^{\Omega} L(\Omega)(R_B/R_L)^{\Omega-\omega} = L(\Omega)\frac{1-(R_B/R_L)^{\Omega}}{1-(R_B/R_L)} \tag{2.26}$$

Strahler [1964] gives this result. If $(R_B/R_L) \leq 1$, the series converges to a finite L as Ω approaches infinity and we have $D = 1$. Remember that this is a limit process where $L(\Omega)$ is held constant and Ω increases as the resolution is refined. However, if $(R_B/R_L) \geq 1$, as is most often the case in river channel networks, the series diverges and for large Ω we get

$$L_T \propto (R_B/R_L)^{\Omega-1} \tag{2.27}$$

Now, for convenience setting $L(\Omega) = 1$, first-order streams have average length

$$L(1) = (1/R_L)^{\Omega-1} \tag{2.28}$$

which we will use as the ruler or resolution to measure the length of the network because $L(1) \to 0$ as $\Omega \to \infty$. From Eq. (2.28) one has

$$\Omega - 1 = -\frac{\log L(1)}{\log R_L} \tag{2.29}$$

which in Eq. (2.27) gives

$$L_T \propto L(1)^{1-(\log R_B/\log R_L)} \tag{2.30}$$

By analogy to Eq. (2.3) we find that the fractal dimension of a drainage network is given by

$$D = \frac{\log R_B}{\log R_L} \tag{2.31}$$

Tarboton et al. [1990] observed that if the individual streams that make up the entire network are themselves fractal, say characterized by a dimension D_l, then Eq. (2.31) needs a correction. The equivalent count of the number N of fractal measures yields $N = L_T/L(1) \sim L(1)^{-\log R_B/\log R_L}$ and, because the fractal measure $L(1)$ has dimension D_l, it is related to a linear measure r by

$$L(1) \propto r^{D_l} \tag{2.32}$$

Substitution yields

$$N \sim r^{-D_l(\log R_B/\log R_L)} \tag{2.33}$$

which, compared with Eq. (2.3), yields the fractal dimension of the whole network as

$$D = D_l \frac{\log R_B}{\log R_L} \tag{2.34}$$

This extension of the original result was invoked by Tarboton et al., [1990] and La Barbera and Rosso [1990] to explain the observation that generally D is less than 2. In fact, Mandelbrot [1977] and Hjelmfelt [1988], with different motivations and rationale, suggested that individual streams have fractal dimension $D_l \approx 1.1$. We will not pursue this matter here, because we have not yet introduced important tools related to the self-affinity of watercourses. This will be pursued in Section 2.7.7. We also note that the issue of the signatures of nonchannelized parts of the landscape (discussed in Section 1.2.12) was overlooked in the early studies of the fractal nature of river basins.

Horton's laws can also be used to derive the fractal dimension of the network from the stream length exceedence probability distribution. Following Tarboton et al. [1988], one reasons that in a river network with Horton's bifurcation and length ratio laws holding exactly, there are $R_B^{\Omega-\omega}$ streams of order ω with length $L(\Omega)/R_L^{\Omega-\omega}$. So the total number of streams exceeding a length $l = L(\Omega)/R_L^k$ is

$$\sum_{i=0}^{k} R_B^i = \frac{R_B^{k+1} - 1}{R_B - 1} \tag{2.35}$$

where

$$k = \frac{\log\left(L(\Omega)/l\right)}{\log R_L} \tag{2.36}$$

If the total number of streams is N_T, we can write

$$P[\text{length} \geq l] = \frac{(R_B^{k+1} - 1)(R_B - 1)}{N_T} \propto R_B^{k+1} \quad \text{for} \quad k \gg 1 \tag{2.37}$$

For k large (so that R_B^{k+1} dominates 1 and $e^{\log R_B^{k+1}} \propto l^{\log R_B/\log R_L}$) this becomes

$$P[\text{length} \geq l] \propto l^{\log R_B/\log R_L} \tag{2.38}$$

Comparing Eq. (2.38) with Eq. (2.22), we again obtain the fractal dimension given by Eq. (2.31).

It is worthwhile to emphasize that the above derivations assume Horton's laws hold exactly at all scales in the network. In real basins it is frequently found that there are significant variations of R_B and R_L when they are estimated from different spatial resolutions. It is also important that there always exists a degree of scattering along the regression lines used in the estimation. This makes Eq. (2.31) much less attractive for estimating the fractal dimension than other direct methods such as box-counting already described in this chapter. Table 2.1 from Tarboton et al. [1988] gives

Table 2.1. *Network geometry data for several river networks*

Basin	Magnitude	R_B	R_L	$\log R_B / \log R_L$
Souhegan, N.H.	177	3.5	2.0	1.8
Squanacook, N.H.	133	3.5	1.7	2.5
W 15 Support A 20, Arizona	329	4.2	2.1	1.9
W 15 Support A 50, Arizona	107	3.3	1.6	2.4
Hubbard Support A 20, Connecticut	1217	4.1	2.1	1.9
Hubbard Support A 50, Connecticut	486	4.7	2.3	1.8
Youghiogheny River, Maryland	1798	4.6	2.2	1.9
Daddys Creek, Tenn.	1181	4.1	2.2	1.8
Allegheny River, Pa.	5966	4.5	2.4	1.7

Source: Tarboton et al., 1988.

Horton ratios for the river basins used in the box-counting methodology. The resulting fractal dimensions from Eq. (2.31) are scattered around 2, which tends to confirm the earlier results that river networks are fractal objects with dimension near 2.

If the result $D = 2$ is accepted, it implies $R_B = R_L^2$, thus providing a fundamental link between Horton's ratios. This appears to be borne out in practice within the scatter in estimation of R_B and R_L. Both R_B and R_L appear to describe the same scaling property evident in river basins. It is also worthwhile to recall from Chapter 1 that the random-topology model of Shreve [1967] gives average values $R_B = 4$ and $R_L = 2$. The random-topology model is therefore consistent with $D = 2$.

2.4 Peano's River Basin

Among the many types of Peano [1890] constructions, the so-called Peano basin is of special usefulness to hydrologic studies. It appears under different contexts throughout this book, both as a tool in the study of the fractal and multifractal characteristics of river networks and as an idealized condition under which one may pursue the derivation of analytical results related to the hydrologic response.

The construction of the Peano basin is shown in Figure 2.24. At the prefractal stage when the Horton order of the basin is $\Omega = 1$, the generator is simply a segment of unit length cutting the unit area diagonally. The generator is not shown in Figure 2.24. In the case of the Peano basin and in other studies of channel networks, it is frequently convenient to work with a sequence of rulers, that is

$$\delta = \delta(\Omega) \qquad \Omega = 1, 2, \ldots, \infty \tag{2.39}$$

which let the Strahler order, Ω, increase by one unit on changing $\delta(\Omega - 1)$ to $\delta(\Omega)$. Thus the value of Ω also identifies the Ωth stage of the iterative process. This is shown in Figure 2.24.

A sequence of areal rulers, $\alpha(\Omega)$, corresponding to $\delta(\Omega)$ ($\Omega = 1, 2, \ldots, \infty$) is also chosen. The rulers are then $\delta(\Omega) = L(1, \Omega)$, the length of the shortest streams embedded in a network of order Ω, and $\alpha(\Omega)$ is defined as the drainage area of each link of length $L(1, \Omega)$. We will often use the notation $L(j, \Omega)$ to stress the dependence on the scale of the resolution measured

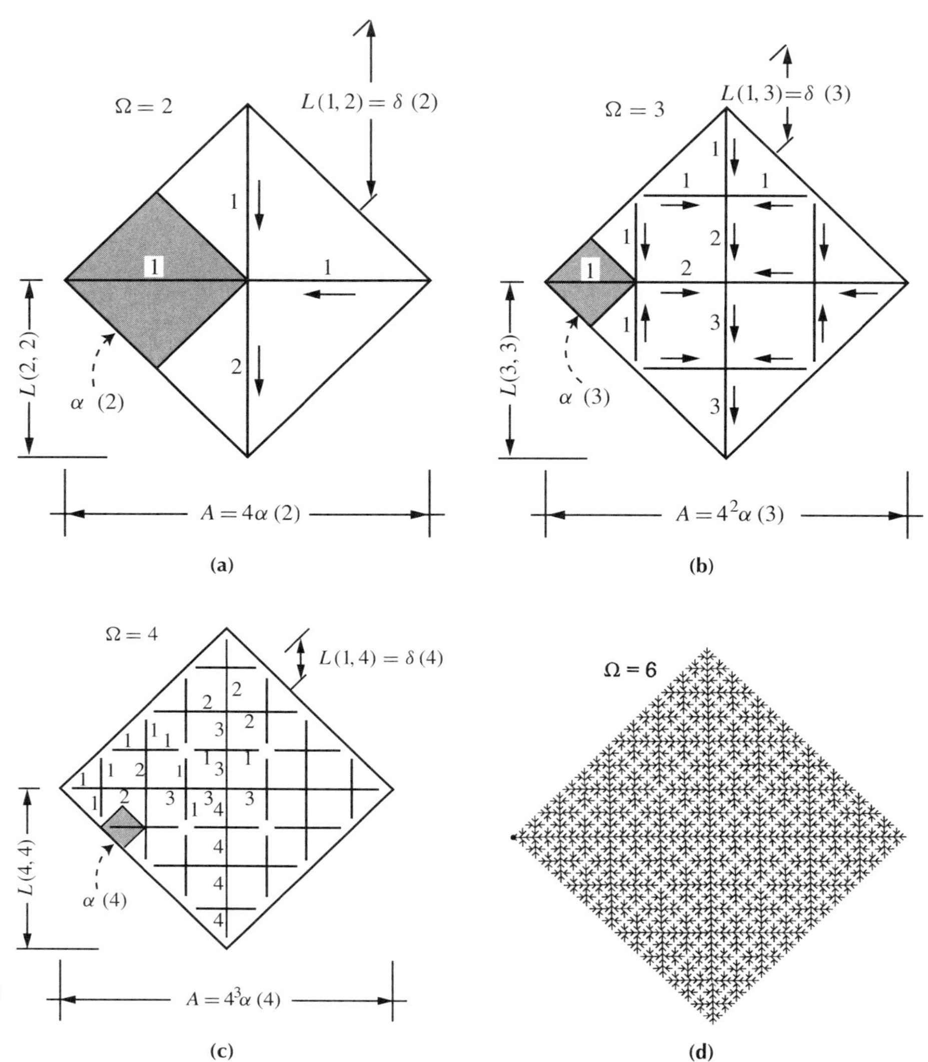

Figure 2.24. Peano's basin at the stages of generation $\Omega = 2, 3, 4, 6$ and geomorphological details [after Marani et al., 1991].

by the order Ω. This notation is an obvious extension of the notation $L(j)$ used in Chapter 1.

When $\Omega = 1$ we simply have a segment of unit length as the diagonal of the area ($L(1, 1) = 1$). At the second stage of generation ($\Omega = 2$, Figure 2.24) four links are generated with length $L(1, 2)$, where $L(1, 2)$ is the length of Strahler streams of order 1 embedded in a network of order 2. As explained before we take

$$\delta(\Omega) = L(1, \Omega) \tag{2.40}$$

where $L(1, \Omega)$ is the length of first-order streams embedded in a network of order Ω. Note that in the case of exact geometric constructs like that in Figure 2.24 all first-order streams have the same length, and therefore there is no need to assume an average for such a length. Note also that at $\Omega = 2$, $\delta(2) = L(1, 2) = 1/2$, and the length of the longest stream is $L(2, 2) = 1/2$ (it is easily seen by inspection that the length of the longest stream remains unaffected by the stage of generation). For $\Omega = 2$ the number of links in the longest path from source to outlet is 2.

At the stage $\Omega = 3$, one obtains $16 = 4^2$ links of length $\delta(3) = L(1, 3) = 1/4$, whereas the length of the longest stream is $L(3, 3) = 1/2$. The number of links in the longest path from source to outlet is $4 = 2^2$. A the stage $\Omega = 4$, similarly, we have $4^3 = 64$ links of length $\delta(4) = L(1, 4) = 1/2^3 = 1/8$. The number of links in the longest path from source to outlet is $8 = 2^3$.

At the arbitrary stage of generation, Ω, we have a total number of $4^{(\Omega-1)}$ links, each of length $\delta(\Omega) = L(1, \Omega) = 1/2^{\Omega}$. The maximum path from source to outlet is made up by $2^{\Omega-1}$ links.

From Figure 2.24 it is readily observed that Peano's basin has Horton's bifurcation ratio R_B exactly equal to 4 (for $\Omega > 1$) and length ratio R_L exactly equal to 2. From Eq. (2.31) we find that the fractal dimension of Peano's deterministically self-similar basin is $D = 2$. The Peano basin also has other important similarities and differences with the planar structure of river networks.

We recall a few other analytical results of Marani et al., [1991], which are useful when using Peano's basin as a research tool for investigating fractal properties in river basins. Let

$$A = \alpha(1) = 4\alpha(2) = \ldots = 4^{\Omega-1}\alpha(\Omega) \tag{2.41}$$

$$L(1, 1) = \delta(1) = L(2, 2) = 2\delta(2) = \ldots = L(\Omega, \Omega) = 2^{\Omega-1}\delta(\Omega) = \ldots \tag{2.42}$$

where A is the drainage area. Notice that $\alpha(i)/\alpha(i-1) = 4$ and $\delta(i)/\delta(i-1) = 2$. Notice also that $L(1, \Omega) = L(1, 1)/2^{\Omega-1} = 2^{1-\Omega}$ is the length of the shortest Strahler stream embedded in a network of order Ω. It is also of interest to observe that the maximum length from source to outlet, $l_{\max}$, is

$$l_{\max} = \delta(1) = 2\delta(2) = \cdots = 2^{\Omega-1}\delta(\Omega) \tag{2.43}$$

Let $A = A(\Omega)$ and $L_T = L_T(\Omega)$ be the finite measures of the basin area and of the total basin length (i.e., scale independent), it follows that

$$L_T = L_T(\Omega) = N_\delta(\Omega)\delta(\Omega)^D \tag{2.44}$$

and

$$A = A(\Omega) = N_\alpha(\Omega)\alpha(\Omega)^{D_A} \tag{2.45}$$

where D, D_A are the fractal dimensions of length and area. Applying the definitions we obtain

$$\frac{L_T(\Omega)}{A(\Omega)} = \frac{\delta(\Omega)^D}{\alpha(\Omega)^{D_A}} = \frac{1}{\mathcal{D}} \tag{2.46}$$

(i.e., $A(\Omega) \propto L_T(\Omega)$) where $\mathcal{D}$ is a finite constant analogous to the drainage density (Section 1.2.2). The conceptual suggestion of the previous equation is that all the hydrological quantities dependent on ratios of drainage areas can be equally calculated by the ratio of the corresponding total basin lengths.

We will now explore some of the analytical results that can be obtained for Peano's network. Let $a(\omega, \Omega)$ be the contributing area drained by a stream of order ω embedded in a network of order Ω referred to the areal ruler $\alpha(\Omega)$; let $A(\omega, \Omega)$ be the total contributing area drained by subbasins of order ω; let $L_T(\Omega, \Omega)$ be the total length of the network measured by the ruler $\delta(\Omega)$; and let $N_\delta(\omega, \Omega)$, $N_\alpha(\omega, \Omega)$ be the number of rulers δ, α needed to cover the area A and total length L_T, respectively, of the basin of order Ω up to its order ω. The Euclidean measures are then

$$A(\omega, \Omega) = N_\alpha(\omega, \Omega)\alpha(\Omega), \qquad L_T(\omega, \Omega) = N_\delta(\omega, \Omega)\delta(\Omega) \tag{2.47}$$

and they diverge as $\Omega \to \infty$. The corresponding fractal measures are $A(\omega) = N_\alpha(\omega, \Omega)\alpha(\Omega)^{D_A}$ and $L_T(\omega) = N_\delta(\omega, \Omega)\delta(\Omega)^D$, where D_A and D may be called area and bifurcation fractal dimensions, respectively. Using the relationships

$$N_\alpha(\omega, \Omega) = \frac{1}{\alpha(\Omega)} \sum_{i=1}^{\omega} N(i, \omega)a(i, \omega), \qquad N_\delta(\omega, \Omega) = \frac{1}{\delta(\Omega)} \sum_{i=1}^{\omega} N(i, \omega)L(i, \omega)$$

and recalling that $\sum_{i=j}^{k} x^k = (x^j - x^{k+1})/(1-x)$, the following properties may be derived [Marani et al., 1991]:

- The total Euclidean basin length is:

$$L_T(\Omega, \Omega) = \delta(\Omega)\frac{R_B^\Omega - R_L^\Omega}{R_B - R_L} = L(\Omega, \Omega)(2^\Omega - 1)$$

  and clearly diverges as $\Omega \to \infty$; the number of rulers, $\delta(\Omega)$, needed to cover the network is

$$N_\delta(\Omega, \Omega) = \frac{R_B^\Omega - R_L^\Omega}{R_B - R_L} \tag{2.48}$$

- The fractal measure $L_T(\Omega)$ is

$$L_T(\Omega) = \frac{R_L^D}{R_B - R_L} L(\Omega, \Omega)^D \tag{2.49}$$

- The contributing area ratios [Smart, 1972] are defined as

$$\begin{aligned} R_a(\omega, \Omega) &= \frac{a(\omega, \Omega)}{a(\omega - 1, \Omega)} = \frac{N_\alpha(\omega, \Omega)\alpha(\Omega)}{N_\alpha(\omega - 1, \Omega)\alpha(\Omega)} \\ &= \frac{N_\delta(\omega, \Omega)\delta(\Omega)}{N_\delta(\omega - 1, \Omega)\delta(\Omega)} = \frac{L(\omega, \Omega)}{L(\omega - 1, \Omega)} = 2 \end{aligned} \tag{2.50}$$

  and

$$\begin{aligned} R_A(\omega, \Omega) &= A(\omega, \Omega)/A(\omega - 1, \Omega) \\ &= \frac{N(\omega - 1, \Omega)}{N(\omega, \Omega)} \frac{\sum_{j=1}^{\omega} N(j, \Omega)a(j, \Omega)}{\sum_{j=1}^{\omega-1} N(j, \Omega)a(j, \Omega)} \\ &= 4\frac{1 - (R_L/R_B)^\omega}{1 - (R_L/R_B)^{\omega-1}} \end{aligned} \tag{2.51}$$

  when $\omega \to \infty$, Eq. (2.51) yields

$$R_A = R_B = 4 \tag{2.52}$$

  This agrees with the result of Rosso, Bacchi, and La Barbera [1991], who found that for Hortonian networks $R_B = R_A^{D/2}$ for $R_A \geq R_B \geq 2$ when $D = 2$.

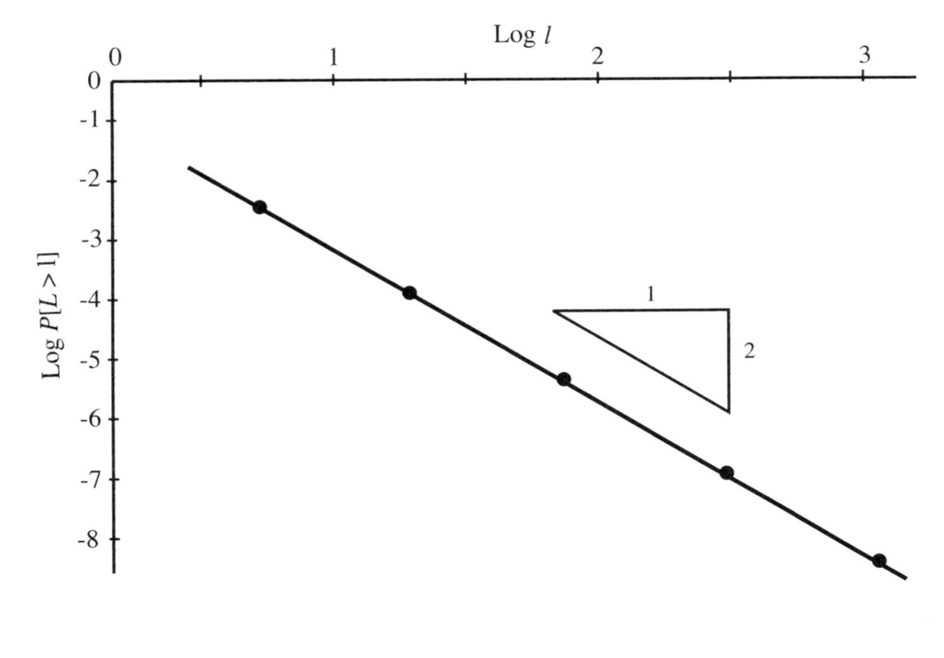

Figure 2.25. Exceedence probability of Strahler's stream lengths for Peano's basin. The slope is equal to 2.

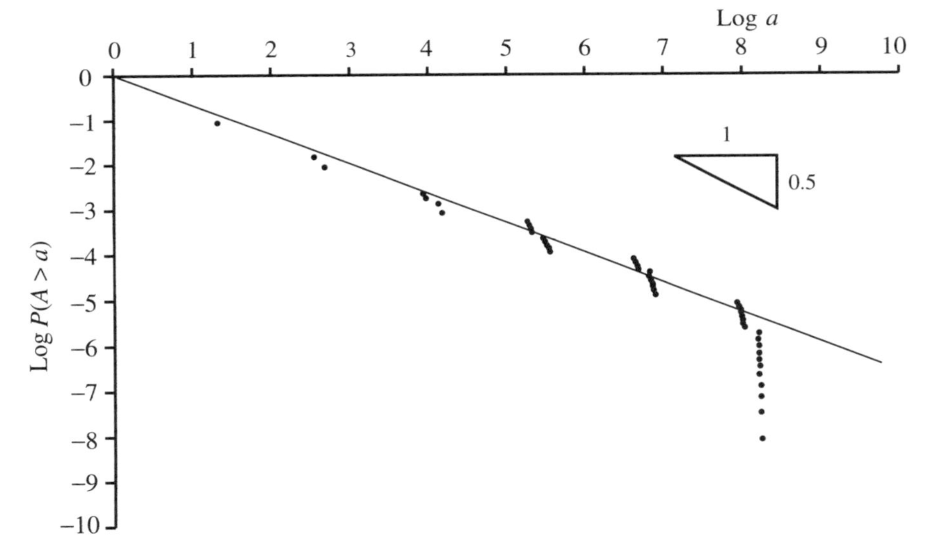

Figure 2.26. Exceedence probability of cumulative areas for Peano's basin. The mean slope is $\beta = 0.5$ but a more complex structure emerges.

Before leaving Peano's basin, it is instructive to look at the probability distributions of Strahler stream lengths and contributing drainage area at any point in the basin. These are shown in Figures 2.25 and 2.26. In contrast to natural river basins (see Section 2.5.2), in Peano's deterministic construct neither area nor lengths are aggregated over all scales, certain structures being preferred by the exactly recursive scheme of Peano's construction.

We observe that the slope of Strahler length distribution is 2, as one would also infer from the Hortonian characters (with $R_B = 4$ and $R_L = 2$) of Peano's network [Marani et al., 1991].

One can also prove that the area distribution can be approximated by a power law (i.e., $P[A \geq a] \propto a^{-\beta}$) with scaling exponent $\beta = 0.5$ [Colaiori et al., 1996]. In fact, one can associate to each site, say i, a total contributing area, say, $A_i(N)$ at the nth stage of generation of Peano's construct. Let ν_N denote the set of distinct values assumed by A_i at stage N. As seen previously in this section, ν_N contains 2^N distinct values obtained by adding 2^{N-1} new values to ν_{N-1} previously available ones. One can show that the following iterative rule holds: $A_i = 3(\sum_k c_k(i)4^k) + 1$ (where $i = 0, 1, \ldots$ and $c_k(i)$ are the coefficients of the binary expansion of i, that is, $i = \sum_k c_k(i)2^k$).

Let also M_i^N be the number of sites i with area A_i at the Nth stage of generation. One can define: (1) $t(A_i)$ as the first 'time', or step, where an area with value A_i appears in the hierarchical order $t = 0$ to $t = 2^N$, that is, from source to outlet. This implies that $t(A_i) = 0$ if $i = 0$ and $t(A_i) = 1 + I[\log_2(i)]$ for $i > 0$ (where $I[.]$ means integer part of); (2) $M_i^N = 0$ for $N < t(A_i)$; (3) $M_i^N = 1$ for $N = t(A_i)$; and (4) $M_i^N = 4M_{i-1}^{N-1}$ for $N > t(A_i)$.

Solving for M_i^N yields [Colaiori et al., 1996] the relationship $M_i^N = \frac{2}{4}4^{N-t(A_i)} + \frac{1}{3}$ for $N \geq t(A_i)$.

By noticing that all A_is born at the same 'time' step have the same probability $p_N(A_i) = M_i^N/4^N$, one observes that for $A_i = 4^t$ the exact result is obtained: $P[A_i \geq a = 4^t, N \to \infty] \propto a^{-\beta}F(a/L^{1+H})$ with $\beta = 0.5$ and $H = 1$. Also, $F(x) = (1 - x)/3, 0 \leq x \leq 1$, with $F(x) = 0$ when $x \geq 1$. This follows from the direct computation of $P[A_i > a = 4^t] = \sum_{s=t+1}^{N}(\frac{2}{3}4^{N-s} + \frac{1}{3})(2^{s-1}/4^N) = \frac{1}{3}2^{-t}(1 - 2^{2(t-N)}) = \frac{1}{3}a^{-1/2}(1 - a/L^2)$.

The above exact results will be useful when comparing different network structures in light of optimality principles.

2.5 Power Law Scaling in River Basins

As discussed in Section 2.2, the structure of river basins is characterized by a great number of power law relationships, which reflect the scaling invariance of the processes involved. An example of these power laws is the probability distribution describing stream length exceedence probability shown in Figures 2.21 and 2.22. This section studies the power law structure of elevations, aggregated areas, and power expenditure in river basin networks. Most of these power laws are of a probabilistic character, implying the statistical self-similarity of the variable under study. Especially interesting is the linkage between the three-dimensional structure of the basin and its planar description in terms of power laws.

One of the main obstacles in understanding surficial hydrologic processes is the high spatial variability of surface features in river basins to which those processes are intimately linked. River runoff is a key flux in climate systems. It occurs over a wide range of spatial scales, from the microscale of the individual channel link in a drainage network, through the mesoscale of drainage basins, to the macroscale of continents. The search for invariance properties across scales as a basic hidden order in hydrologic phenomena is one of the main themes of hydrologic science [National Research Council, 1991]. This search is crucial for the development of new models and measurement efforts.

The objective of this section is to present some invariance properties for river basins. These properties refer to the invariance of the probability distributions under wide changes of spatial scales.

2.5.1 Scaling of Slopes

As studied in detail in Section 1.2.11, there is a structural linkage between the architecture of slopes in a river basin and the corresponding drainage area at the different links. The slopes of the different links with a common drainage area in a river basin are realizations of a random variable, which we characterize through its mean and variance. Both mean and variance scale with contributing area. In the case of the mean we have

$$E\left[\nabla z(A)\right] \propto A^{-\theta} \tag{2.53}$$

where $\theta \approx 0.5$. Eq. (2.53) is frequently written using the *magnitude* of the link, n, as a surrogate variable for the contributing area. In this case one has

$$E\left[\nabla z(n)\right] \propto n^{-0.5} \tag{2.54}$$

Examples of the results analyzed here can be found in Section 1.2.11.

Eqs. (2.53) and (2.54) were postulated by Tarboton et al. [1989a], who continued along Gupta and Waymire's [1989] suggestion of a model of link drops with a self-similar probability distribution function and area or magnitude taking the role of scaling parameter. The equivalence, for these purposes, of area and magnitude is clear from the discussion in Section 1.2.5. Magnitude, n, is the number of exterior links of the network or, equivalently, the number of first-order streams. The total number of links is $2n - 1$, and this quantity exhibits an excellent linear relation with the area of the basin (section 1.2.4). Thus in this section, area and magnitude will be used interchangeably in all of our discussions.

Gupta and Waymire's [1989] model for link drops is given by

$$\frac{H(\alpha n)}{\mu(\alpha)} \stackrel{d}{=} H(n) \tag{2.55}$$

where $H(n)$ is the random link drop in altitude for a link magnitude n and α is a scaling factor.

The term $\mu(\alpha)$ is a normalization function, and $\stackrel{d}{=}$ denotes equality of probability distribution. Eq. (2.55) is the definition of self-similarity, or scaling invariance, of a random variable $H(n)$, dependent on – and therefore indexed by – the scale parameter n. Gupta and Waymire [1989] suggest that $\mu(\cdot)$ is of the form

$$\mu(n) = n^{-\theta} \tag{2.56}$$

They also suggest that link *slopes* are independent of link lengths, an assumption later justified by the analysis of Tarboton et al. [1989a], who found these variables to indeed be uncorrelated in real basins. The implication is that link drop is dependent on link length, which is also verified in the data (see Section 1.2.11). Thus Tarboton et al. [1989a] start with a self-similar assumption, analogous to Eq. (2.55), for slopes:

$$\nabla z(\alpha n) \stackrel{d}{=} \mu(\alpha)\nabla z(n) \tag{2.57}$$

where $\nabla z(n)$ is the link slope, dependent on magnitude, with $\mu(\alpha)$ given by Eq. (2.56). The justification for this assumption lies in the empirical observations [e.g., Flint, 1974] relating slope and contributing area. Let L

denote the random link length assumed independent of magnitude. Tarboton et al. [1989a] did not in fact detect significant trends of L with magnitude. The link drop is

$$H(n) = \nabla z(n) L \tag{2.58}$$

so

$$H(\alpha n) = \nabla z(\alpha n) L \stackrel{d}{=} \mu(\alpha) \nabla z(n) L = \mu(\alpha) H(n) \tag{2.59}$$

equivalent to Eq. (2.57).

Gupta and Waymire [1989] started with Eq. (2.57) and used the inverse argument to derive Eq. (2.59). Eq. (2.59) can alternatively be stated as

$$Z = \frac{\nabla z(n)}{\mu(n)} \stackrel{d}{=} \nabla z(1) \tag{2.60}$$

that is, the variable $\nabla z(n)/\mu(n)$, which we call Z, is independent and identically distributed (iid) as $\nabla z(1)$. The self-similar model is simply

$$\nabla z(n) = \mu(n) Z = n^{-\theta} Z \tag{2.61}$$

This clearly shows the nature of the self-similar model, which is characterized by power law scaling, the scale parameter n being raised to the power $-\theta$. Moments of the self-similar model scale proportionally to $(n^{-\theta})^k$, that is

$$\begin{aligned} E\left[\nabla z(n)\right] &= n^{-\theta} E(Z) \\ Var\left[\nabla z(n)\right] &= n^{-2\theta} Var(Z) \\ M_k\left[\nabla z(n)\right] &= n^{-k\theta} M_k(Z) \end{aligned} \tag{2.62}$$

where M_k denotes the kth moment.

The quantiles of the self-similar model all scale proportionally to $n^{-\theta}$, that is

$$Q\nabla z_\gamma(n) = QZ_\gamma n^{-\theta} \tag{2.63}$$

where $Q\nabla z_\gamma(n)$ is a quantile (nonrandom value) of the slope corresponding to a probability of exceedence γ, that is

$$P\left[\nabla z(n) > Q\nabla z_\gamma(n)\right] = \gamma \tag{2.64}$$

and, similarly, QZ_γ is a quantile of Z:

$$P[Z > QZ_\gamma] = \gamma \tag{2.65}$$

Gupta and Waymire [1989] incorporated this scaling in terms of link drops into the random topology model.

Tarboton et al. [1989a] carried out an extensive empirical study of the scaling of slopes in real basins through DEMs. Part of those results was shown in Section 1.2.11. Of particular interest are Figures like 1.24 and 1.25, which show the scaling of the mean and variance of local slope with contributing area. As explained in Section 1.2.11, the links in Figures 1.24 and 1.25 were grouped into bins containing at least twenty links that cover a narrow range of areas and magnitudes. The group sample means are plotted as circles and show power law scaling approximately proportional to $n^{-0.5}$. An example of the group sample variances is plotted in Figure 1.25

and shows power law scaling approximately proportional also to $n^{-0.5}$. If the exponent θ is estimated from the mean as 0.5, the self-similar model (Eq. (2.62)) would predict that slope variances should scale proportionally to n^{-1}. With θ estimated as 0.6 the slope variances should scale proportionally to $n^{-1.2}$. This is not the case in Figure 1.25 and many others reported in Chapter 1. This is the first indication of failure of the self-similar model.

Further evidence of such failure can be obtained by looking at the distribution of the normalized variable $Z = \nabla z(n)/\mu(n)$ (Eq. (2.60)). With θ estimated from the mean values, we can compute Z for each link. This gives $N = 2n - 1$ realizations of Z, so for N large we can get an idea of the probability distribution of Z by ranking and plotting (according to the plotting position $P = i/(N+1)$, where i is the rank from largest (1) to smallest (N)) of the N realizations of Z. Tarboton et al. [1989a] performed this analysis using $\theta = 0.6$ for Big Creek and the St. Joe River, whose link slope means and variances scale with area as shown in Figures 1.23 to 1.25. The results are shown in Figures 2.27 and 2.28. Also shown are the probability distributions estimated from subsets consisting of all links with a given Strahler order. The 95% confidence limit error bars in the plotting position probability computed from the incomplete beta distribution are shown. The incomplete beta distribution is used because it is the theoretical distribution of nonparametric estimates of probability [Tarboton et al., 1989a]. The Z are supposed to be iid, that is, independent of Strahler order, ω, so sets of links from different Strahler orders should have the same probability distribution. This is, however, not the case and is another indication of the failure of the self-similar model of slopes. Tables 2.2 and 2.3 from Tarboton et al. [1989a] give statistics of link properties for Big Creek and the St. Joe River. The tables show how the mean and variance of slope and drop both decrease with order. The normalization accounts for the trends in the mean but makes the normalized variance, and hence the coefficient of variation, increase with order, counter to what the self-similar

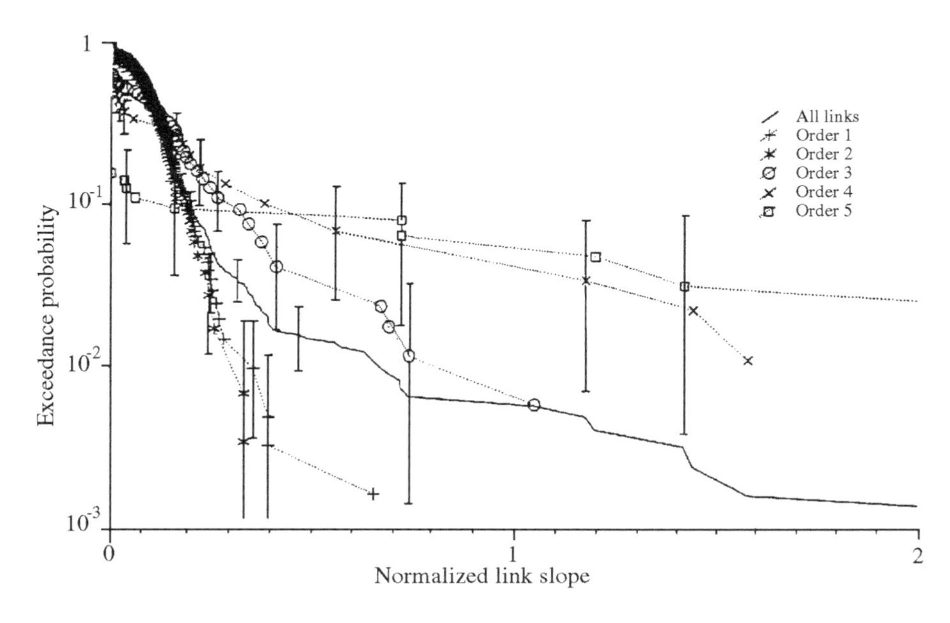

Figure 2.27. St. Joe River link slopes normalized with $n^{-0.6}$ [after Tarboton et al., 1989a].

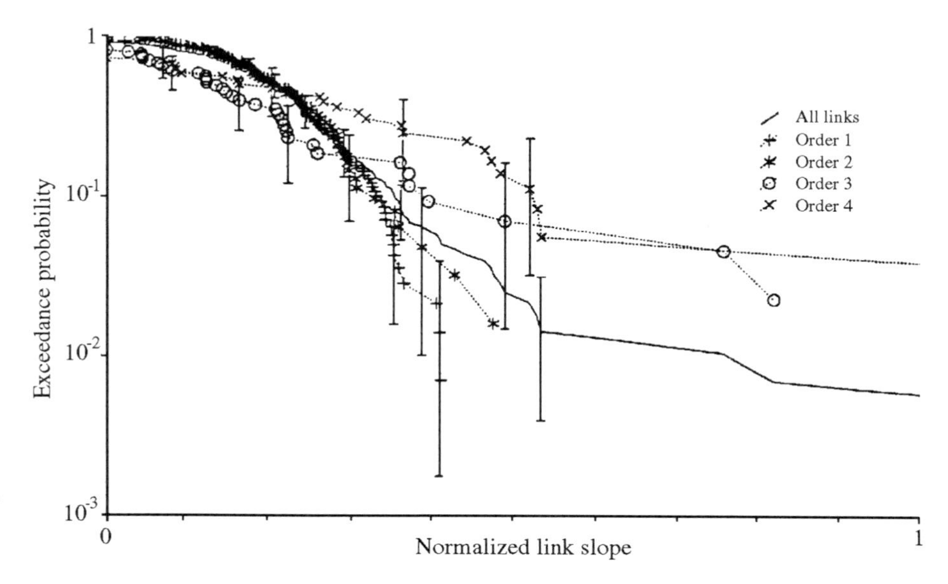

Figure 2.28. Big Creek link slopes normalized with $n^{-0.6}$ [after Tarboton et al., 1989a].

Table 2.2. *Big Creek link statistics (magnitude 139, order 4)*

	All Order	$\omega = 1$	$\omega = 2$	$\omega = 3$	$\Omega = 4$
$E[H]$ [m]	82	121	79	21	9.4
$E[\nabla z]$	0.136	0.203	0.114	0.039	0.020
$E[L]$ [m]	614	591	735	612	497
$Var[H]$ [$m^2 10^3$]	7.36	9.0	3.6	0.54	0.12
$Var[\nabla z]$ 10^{-3}	11.1	9.2	2.9	1.9	0.55
$Var[L]$ 10^3 m^2	240	221	228	384	141
$CV[H]$	1.04	0.79	0.77	1.10	1.16
$CV[\nabla z]$	0.78	0.47	0.48	1.12	1.20
$CV[L]$	0.80	0.79	0.65	1.01	0.76
$E[H/\mu(n)]$	123	121	145	102	117
$E[\nabla z/\mu(n)]$	0.204	0.203	0.208	0.173	0.238
$Var[H/\mu(n)]$	11.9	9.00	12.6	16.2	16.7
$Var[\nabla z/\mu(n)]$	22	9.2	9.0	34.7	78.1
$CV[H/\mu(n)]$	0.89	0.79	0.77	1.24	1.10
$CV[\nabla z/\mu(n)]$	0.72	0.47	0.46	1.07	1.17
$R(H, L)$	0.61	0.84	0.66	0.77	0.55
$R(\nabla z, L)$	−0.02	0.009	−0.20	−0.09	−0.02
$R(H/\mu(n), L)$	0.77	0.84	0.67	0.85	0.59
$R(\nabla z/\mu(n), L)$	−0.04	0.009	−0.17	−0.03	−0.01

Source: Tarboton et al., 1989a.

model for full distribution would predict. Also, note that correlation coefficients between slope and length are negligible, whereas the correlation between drop and length is not. This gives credibility to the assumption of independence between slopes and lengths and thus on this basis Tarboton et al. [1989a] consider slope as the most fundamental variable rather than link drop. Significance tests on the difference between the mean (t test) and variance (F test) of the normalized slopes given in Tables 2.2 and 2.3 were carried out by Tarboton et al. [1989a]. These indicate no significant differences between the mean normalized slope

Table 2.3. *St. Joe River link statistics (magnitude 621, order 5)*

	All Orders	$\omega = 1$	$\omega = 2$	$\omega = 3$	$\Omega = 4$	$\Omega = 5$
$E[H]$ [m]	90.6	135	72.8	30.5	14.9	4.8
$E[\nabla z]$	0.066	0.100	0.047	0.026	0.013	0.0038
$E[L]$ [m]	1356	1383	1465	1233	1180	1176
$Var[H]$ [m$^2 10^3$]	11.2	14.7	5.1	1.27	0.53	0.26
$Var[\nabla z]$ 10^{-3}	4.70	5.6	1.4	1.30	0.78	0.20
$Var[L]$ 10^3 m^2	1209	1527	1054	785	616	675
$CV[H]$	1.17	0.90	0.98	1.17	1.55	3.33
$CV[\nabla z]$	1.04	0.75	0.82	1.41	2.08	3.66
$CV[L]$	0.81	0.89	0.70	0.72	0.66	0.70
$E[H/\mu(n)]$	136	135	135	131	145	149
$E[\nabla z/\mu(n)]$	0.102	0.100	0.088	0.109	0.13	0.126
$Var[H/\mu(n)]$	32.6	14.8	16.9	24.8	65.1	262
$Var[\nabla z/\mu(n)]$	26.3	5.60	5.1	24.8	78.5	264
$CV[H/\mu(n)]$	1.33	0.90	0.96	1.20	1.75	3.43
$CV[\nabla z/\mu(n)]$	1.59	0.74	0.81	1.44	2.16	4.08
$R(H, L)$	0.67	0.79	0.73	0.61	0.32	0.16
$R(\nabla z, L)$	0.01	−0.03	0.09	−0.03	−0.04	0.02
$R(H/\mu(n), L)$	0.51	0.78	0.74	0.57	0.28	0.14
$R(\nabla z/\mu(n), L)$	−0.01	−0.09	0.09	−0.03	−0.04	0.004

Source: Tarboton et al., 1989a.

of links of different orders, indicating that normalization works for the mean. However, for the variances the hypothesis that the different order links are from the same population is rejected at the 0.05 level for the great majority of cases. This is a clear failure of the self-similar model to characterize the link slope distribution.

In Table 2.2, $E[x]$, $Var[x]$, and $CV[x]$ are the expected value, the variance, and the coefficient of variation of x, respectively. Also, as in the notation employed before, H is the link drop, L is the link length, and ∇z is the link slope defined as H/L. The number of links for each order (from 1 to 4) is 139, 61, 42, and 35 (total 277). Finally, $R(x, y)$ is the correlation coefficient of x and y.

In Table 2.3 the notation is analogous to that of Table 2.2. The number of links for each order (from 1 to 5) is 621, 295, 173, 89, and 16 (total 1,241).

The above analysis showed in a conclusive manner that link slopes are not self-similar with respect to area as a scale index. Rather than simple or single scaling, where the same scaling exponent is capable of describing all the moments of the probability distribution, one finds a *multiscaling* behavior, where different moments of the link slopes distribution scale with area according to different scaling exponents. The richness and implications of a multiscaling structure will be first studied in Chapter 3 and will then become an integral part of our description of river basins throughout the reminder of this book.

2.5.2 Scaling of Contributing Areas, Discharge, and Energy

Surficial processes, and more specifically the organization of river runoff around the arborescent geometry of drainage networks, involve mass transport and energy dissipation phenomena. We will now study the scaling features of total contributing area, surrogating a meaningful discharge or aggregated mass, and energy dissipation.

Our starting point is the study of total contributing area at a site. In a river basin the total contributing area of a site i, that is, A_i, is identified through drainage directions and is measured (in DEMs) by the number of sites upstream of the site *connected* by the network (Section 1.2.9). From a computational viewpoint, it can also be regarded as a measure of the flow rate if a unit, constant rate of rainfall is applied uniformly over the basin. Thus throughout this book we will assume for convenience that a significant measure of discharge, say Q, be it the mean annual value or some landscape-forming value, can be surrogated by total contributing area A, that is, $Q \sim A$. This holds as long as the assumption of spatially uniform rainfall rate is meaningful. In Chapter 4 we will investigate the inference of spatial heterogeneity in the input rate on the relationship between Q and A.

As we have seen in detail in Section 1.2.9, in natural basins total contributing areas are amenable to detailed experimental investigation through the analysis of DEMs. For completeness we will repeat the basic equation that describes total contributing area at the arbitrary site i of a connected structure.

Let A_i be the area at a given site i. If $nn(i)$ are the nearest neighbors to i and constant unit injection is assumed, the equation for total contributing area is

$$A_i = \sum_{j \in nn(i)} A_j \, W_{i,j} + 1 \qquad (2.66)$$

where $W_{i,j}$ is a matrix that has value 1 if i collects flow from its neighbor j through a drainage direction and 0 otherwise.

This section specifically addresses the issue of scaling in total contributing areas, or their surrogated measures, in dendritic structures and, chiefly, river networks. As seen in Section 2.1.5, a scaling structure postulates that the distribution function obeys a power law, that is

$$P[A \geq a] \propto a^{-\beta} \qquad (2.67)$$

where a is the arbitrary value of the random variable A. The coefficient β characterizes the entire aggregation pattern.

Analogously, in this section we investigate through experiment the scaling properties of energy expenditure $E \propto Q\nabla z \sim A\nabla z(A)$. As we observed in Section 2.5.1, scaling of slopes ensures that $\langle \nabla z(A) \rangle \propto A^{-0.5}$. Therefore $E \propto A^{1/2}$ and one obtains the result

$$P[E \geq e] \propto P[A^{1/2} \geq a] \propto \int_{a^2}^{\infty} x^{-(\beta+1)} \, dx \propto a^{-2\beta} \qquad (2.68)$$

from which one expects that the slope of the scaling relationship for energy dissipation is roughly twice that of total contributing area.

2.5.2.1 Observational Evidence A key feature is the power law character exhibited by the observational evidence for total contributing area at an arbitrary site of a real river basin. Later on (Section 2.10) we will focus

Table 2.4. *Characteristics of the basins analyzed by Rodriguez-Iturbe et al. [1992a]*

Name	Location	Area (km^2)	DEM m × m	A_t	Number of Links	Slope $-\beta$ $P[A \geq a]$	Slope $P[E \geq e]$
Brushy	Alabama	322	30×30	25	7251	−0.43	−0.91
Caldwell	Idaho	147	30×30	40	2187	−0.41	−0.93
Racoon	Pennsylvania	448	30×30	50	5033	−0.42	−0.92
Schoharie	New York	2408	68.3×92.7	70	3089	−0.43	−0.90
St. Joe	Idaho	2834	62.2×92.7	50	4997	−0.44	−0.92

Note: A_t is the support area in pixels.

on a more comprehensive view of the scaling characters of this important quantity, that is, of finite-size scaling.

Rodriguez-Iturbe et al., [1992a] used five basins of different characteristics throughout North America for study of the scaling properties of discharge and energy in river basins. All of the basins have areas above 100 km^2 and drainage networks that, when defined through DEMs, have more than 2, 000 links, allowing a reliable probability distribution analysis of discharge and energy in the population of links. The basins are described in Table 2.4. Any link of a drainage network carries a wide range of discharges resulting from a variety of rainfall events and antecedent conditions of soil moisture. It is assumed that the mean annual flow is a representative discharge of the conditions at every link. Flow measurements at every link of a network have not been collected, and because the mean annual flow has been observed to be proportional to the drainage area in many regions of the world, the total cumulative area draining into a link will then be used as a surrogate variable for discharge.

As discussed in Chapter 1, channelized pixels may be defined in a number of ways. The easiest way, which was implemented in early studies on this subject, defines channels by their contributing area automatically from the DEM. Streams are then identified as those pixels with cumulative contributing area greater than a support threshold (Section 1.2.9).

Figure 2.29 shows the probability distributions $P[A \geq a]$ for the basins analyzed through DEMs by Rodriguez-Iturbe et al. [1992a]. The distributions show good agreement with a power law form for nearly three logarithmic scales. Rodriguez-Iturbe et al.'s [1992a] result is

$$P[A \geq a] \propto a^{-0.43 \pm 0.02} \tag{2.69}$$

which is an important character of seemingly general nature, that is, regardless of, say, size, vegetation, geology, climate, or orientation of the basin. The deviation that we observe (Figure 2.29) at very large values of areas is a finite-size effect (analyzed in Section 2.10), the number of links with such a large area is very small, and the statistics lose significance. These deviations are thus expected to start earlier for those basins with fewer total links. The value of the slope β is unaffected by the size of the support threshold used to identify the network. Figure 2.30 shows the power law distributions of mass (i.e., contributing drainage area at any point) as a function of the support threshold for one of the basins analyzed. The reduction of the support threshold with a corresponding increase in

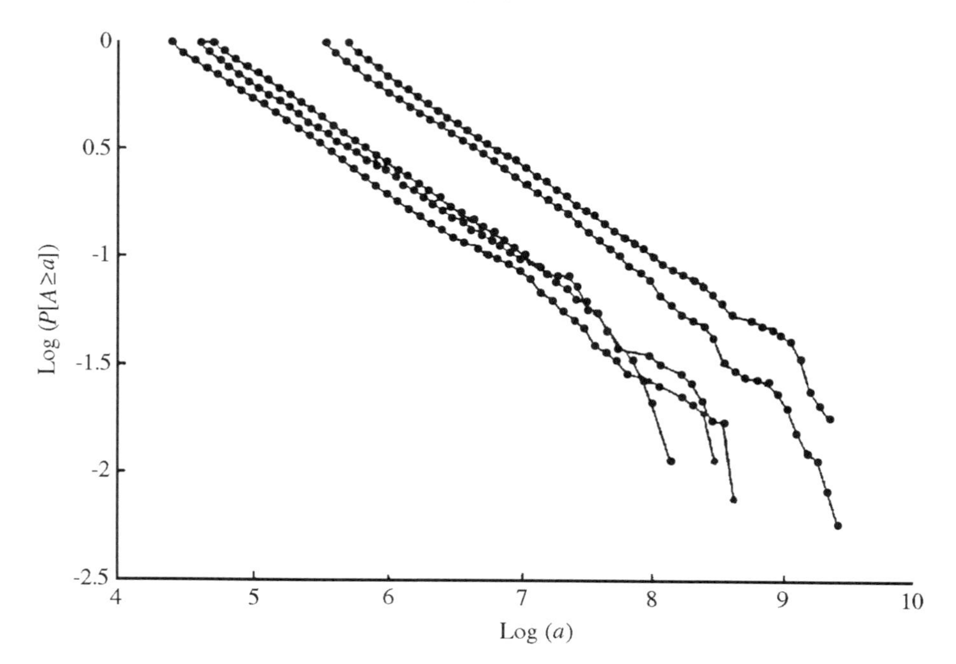

Figure 2.29. Cumulative distribution of contributing drainage area at any point in five different basins. Notice that area is measured in pixel units [after Rodriguez-Iturbe et al., 1992a].

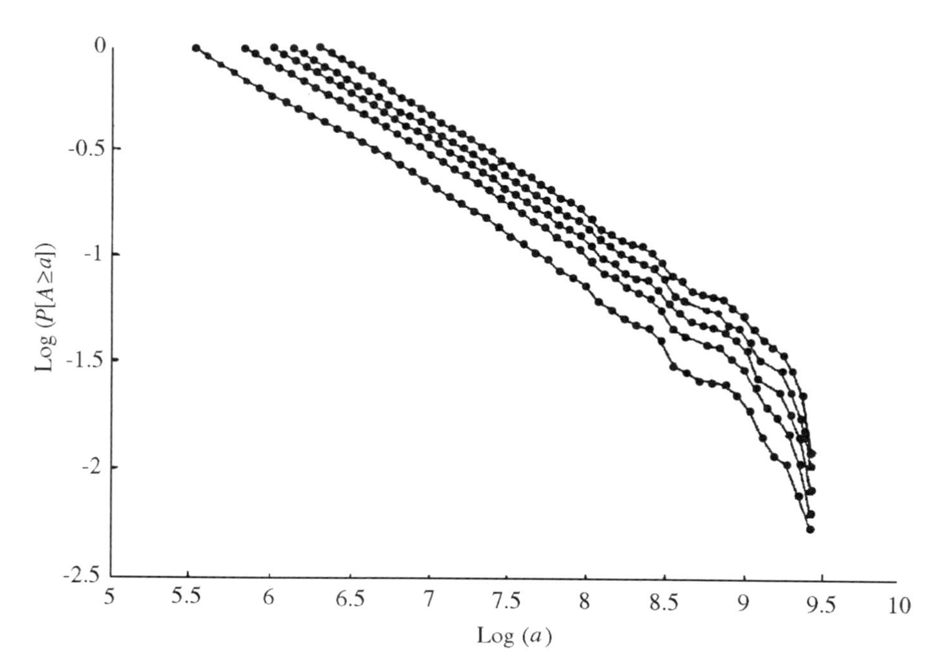

Figure 2.30. Cumulative distribution of contributing drainage area at any point for the same basin as a function of the support threshold [after Rodriguez-Iturbe et al., 1992a].

the number of links does not change the slope but only tends to define it better over a longer range. Obviously, there will be a limit after which one would not be dealing with the drainage network, but rather with the contributing hillslopes. It is important to notice that often the exponent β is statistically indistinguishable among different basins and is approximately equal to 0.43. The recurrence of similar values for the exponent β suggests some resemblance to a self-organized critical phenomenon, as described by Bak et al. [1987, 1988], where a spatially extended dissipative dynamical system naturally evolves into states with no characteristic time or length scales. The spatial signature of critical self-organization is the emergence of a scale-invariant (fractal) structure that becomes apparent through a power law distribution. The dynamic background of this phenomenon will be discussed in Chapter 5. An explanation of the value $\beta = 0.43$ based on the fractal nature of rivers will be presented in this chapter after we introduce the concept of self-affine fractal objects. Some analytical results, available for particular structures reminiscent of river networks, are shown in Section 2.5.6.

The scaling features of energy expenditure in river basins are of great practical and theoretical interest. From the practical point of view such features could provide new avenues for investigating the possibility of energy sources in largely ungauged river basins. From the theoretical point of view it is through energy expenditure that river flow carves the networks that display so many features of such general character throughout different climates, soils, and geological conditions.

Mandelbrot [1974] has shown that uniform energy input in extended dissipative systems frequently results in a power law spatial distribution of energy storage and a fractal energy dissipation. Bak and coworkers [Bak et al., 1987, 1988; Bak and Chen, 1989, 1991] have interpreted this as a manifestation of a critical state, which is an attractor for the dynamics being robust with respect to variations in the parameters, the initial conditions, and the presence of randomness. Thus many dynamic systems with extended spatial degrees of freedom in two or three dimensions naturally evolve in self-organized states in which the energy is dissipated at all length scales. It is essential that the systems are dissipative (energy is released), that they are spatially extended with many degrees of freedom, and that energy is fed into the system.

River networks are precisely this kind of system; the different channel links and their contributing areas support each other in a way that cannot be understood by studying the individual components in isolation, and the holistic view offered by self-organized critically is of great use [Rinaldo et al., 1993; see also Chapter 5]. Thus one would expect river networks to exhibit a power law spatial distribution of energy dissipation similar to the one found in other dissipative phenomena like earthquakes. For example, the well-known Gutenberg–Richter law that gives the number of earthquakes with magnitude greater than a certain value has been shown to be equivalent to a power law of the energy released by the events [Turcotte, 1989; Bak and Tang, 1989]. The rate of energy expenditure per unit length of channel at any link of a river network is proportional to the product of discharge and slope, $Q\nabla z$. Figure 2.31 shows the cumulative distributions of energy released through the links of the different networks given in Table 2.4, where again area A has been used as a surrogate variable of discharge. The oscillations for large values of energy expenditure are likely to be due to the small number of links with large energies, which makes the statistics unreliable in the upper part of the distribution. The flattening of the distribution for small energy releases is similar to that observed in earthquakes [Bak and Tang, 1989] and may be due either to problems of resolution in the case of small slopes or to the fact that power law behavior takes place after a certain threshold. This is not crucial because the region of interest is the right-hand side of the distribution. The power

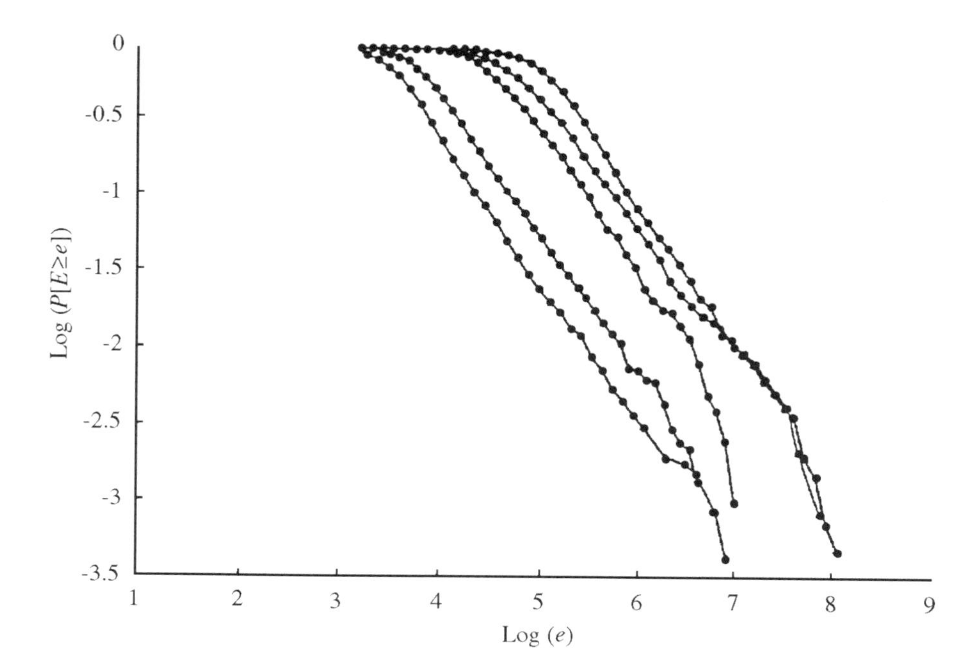

Figure 2.31. Cumulative distribution of energy release in five different basins. Energy $E_i = A_i \nabla z_i$ at the ith site is measured in pixel units [after Rodriguez-Iturbe et al., 1992a].

law exponent is in all cases close to 0.90, which now will be related to the exponent found for the mass distribution, which was close to 0.45.

Rewriting Eq.(2.68) with $\beta = 0.45$ yields

$$P[E \geq e] \propto P[A^{1/2} \geq a] \propto \int_{a^2}^{\infty} x^{-1.45} dx \propto a^{-0.90} \tag{2.70}$$

which is close to the observational results. We will also see in Section 2.10 that finite-size scaling slightly affects the scaling in Eq. (2.70), yielding a closer match of empirical and theoretical results.

In conclusion, discharge and energy, two highly important variables from the theoretical and practical points of view, have been shown to follow power law distributions in their spatial variations in river networks. The apparent universality of the power law exponents suggests that river basins are dynamic systems that naturally evolve toward a state that is an attractor for a very complex dynamics.

2.5.2.2 Exact Results Takayasu et al. [1988] have shown that aggregation systems with constant injection of mass follow a power law mass distribution. Their model is described as follows: Particles with integer units of mass are placed on each site of a geometric lattice and are allowed to aggregate by random-walk processes with discrete time steps. Besides this aggregation process there is an injection effect: A particle with unit mass is added at every site at every time step. As a result, there is no site without particles at any time step. An example of dendritic structures formed by this model is shown in Figure 2.32. Although in Takayasu et al.'s interpretation, Figure 2.32 has time as the vertical axis, the network can also be seen as a two-dimensional object; it is equivalent to the stochastic model of river networks developed by Scheidegger [1967] described in

Section 1.3.2. The model operates in a one-dimensional lattice where each point has two neighbors (the so-called directed lattices). At any moment in time one finds that the connectivity relates each point to those in the one-dimensional lattice, which, up that that moment, are linked to the point under construction. As pointed out by Takayasu et al. [1988], in this model the particle masses are exactly equivalent to the size of Scheidegger's rivers (the size of its total contributing basin) obtained for drainage network patterns that operate in two-dimensional space lattices rather than in time coupled with a one-dimensional lattice.

Figure 2.32. Dendritic structure formed by the aggregation model of Takayasu et al. [1988] with $p = 1/2$.

It was seen numerically that after a long time the size distribution of aggregated areas, say $p(a) = p(a, t \to \infty)$, converges to a power law size distribution independent of the initial conditions, that is

$$p(a) \propto a^{-1.331 \pm 0.006} \tag{2.71}$$

The cumulative probability distribution of a site having mass, that is, area, larger than a satisfies Eq. (2.67). Thus the value of the exponent β, which depends on the characteristics of the random walk, is $\beta = 1/3$ when at each time step particles will either jump to the nearest neighbor site or stay at the same site with probability $q = 1/2$.

The above result was obtained analytically. It is instructive to analyze in some detail a few of the analytical results obtained for Scheidegger's trees [Takayasu et al., 1988; Huber, 1991].

Let $A_i(t)$ be the total mass, that is, contributing area, at time t. The general evolution equation is a modification of Eq. (2.66) as follows:

$$A_i(t+1) = \sum_j W_{i,j}(t) A_j(t) + \eta_i(t) \tag{2.72}$$

where j spans all sites (long-range interactions are not excluded a priori, see below); $W_{x,y}(t) = W_{y \to x}(t)$ has the form

$$W_{x,y}(t) = \begin{cases} 1 \text{ with probability} & q(x)|x-y| \\ 0 \text{ with probability} & 1 - q(x)|x-y| \end{cases} \tag{2.73}$$

subject to the constraint of mass conservation, that is, $\sum_j q(j) = 1$ and $\sum_j W_{i,j} = 1 \forall i$. Finally, $\eta_i(t)$ is the distributed injection term.

Although there exist solutions to Eq. (2.72) of more general nature, we will only describe the solution for Scheidegger's short-range interaction model, for which only nearest-neighbor connections are possible and probabilities are uniform in space, i.e. $q(x) = 1/2 \; \forall x$, that is

$$W_{x,y}(t) = \begin{cases} 1 \text{ with probability } 1/2 \\ 0 \text{ with probability } 1/2 \end{cases} \tag{2.74}$$

if $x = y$ or $|x - y| = 1$, and $W_{x,y} = 0$ otherwise. Also, the injection term is simply $\eta_i(t) = 1$ for all times and sites. Let us define for convenience the

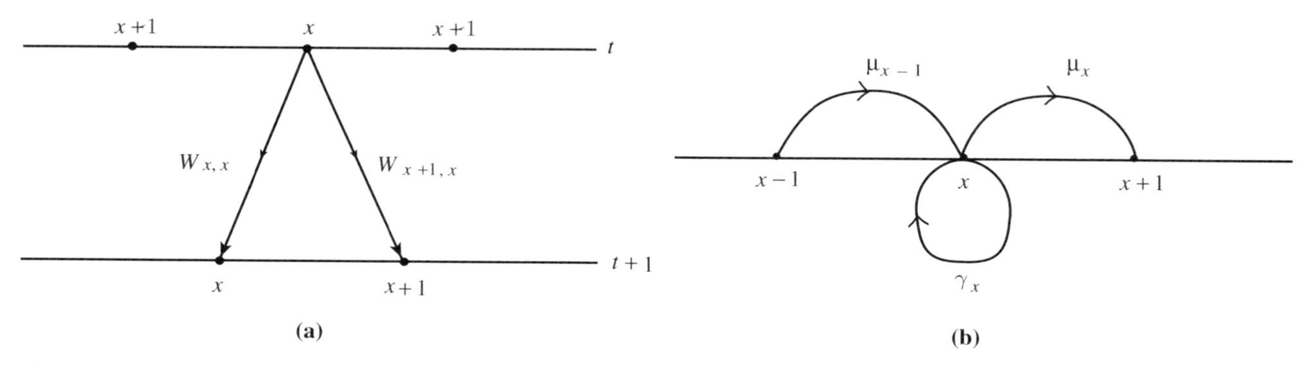

Figure 2.33. (a) Scheidegger's scheme from time t to $t+1$; (b) the random-walk analog of (a).

auxiliary variables γ_x and μ_x as follows:

$$\gamma_x = W_{x,x} = \begin{cases} 1 \text{ with probability } 1/2 \\ 0 \text{ with probability } 1/2 \end{cases} \tag{2.75}$$

and

$$\mu_x = W_{x+1,x} = \begin{cases} 1 \text{ with probability } 1/2 \\ 0 \text{ with probability } 1/2 \end{cases} \tag{2.76}$$

The auxiliary variables are related by mass conservation, that is, $\gamma_x + \mu_x = 1$. The formal equivalence with a one-dimensional random walk is established [Huber, 1991] (Figure 2.33) and the evolution equation for the arbitrary mass $A_x(t+1)$ is

$$A_x(t+1) = \gamma_x A_x(t) + \mu_{x-1} A_{x-1}(t) + 1 \tag{2.77}$$

The meaning of Eq. (2.77) is that at every time step the mass stays at the site or moves to the next site with probabilities γ_x and μ_x, respectively, while it may only receive from the neighboring site mass $x-1$ with probability μ_{x-1} (Figure 2.33). From the condition $\gamma_{x-1} = 1 - \mu_{x-1}$, one obtains the evolution equation in the form

$$A_x(t+1) = \gamma_x A_x(t) + \big(1 - \gamma_{x-1}(t)\big) A_{x-1}(t) + 1 \tag{2.78}$$

Thus the set $\{\gamma\}$ is a random variable and the individual values γ_x are statistically independent by construction. Hence $P(\{\gamma\}) = \prod_x p(\gamma_x)$. To get mean values, one has to average over the values of $\{\gamma\}$ and obtain the limit as $t \to \infty$. This is a subtle point that clouds the main points of the derivation.

Huber [1991] introduces a suitable r-point characteristic function

$$\mathcal{Z}_r(s,t) = \left\langle e^{-s\sum_{x=1}^{r} A_x(t)} \right\rangle_\gamma \tag{2.79}$$

where the average $\langle\rangle_\gamma$ is taken with respect to the values of γ. Notice that for $r = 1$ we obtain the characteristic function

$$\left\langle e^{-sa(t)} \right\rangle = \int ds\, p(a,t) e^{-sa(t)} \tag{2.80}$$

(where $p(a,\infty)$ is the probability density in Eq. (2.71) whose exceedence probability distribution is $P[A \geq a]$ in Eq. (2.67)) which is a regular Laplace transform. In particular, we will study whether the limiting form of $p(a,t)$

is $\propto a^{\beta+1}$ by determining its characteristic function $\mathcal{Z}_1(s, \infty)$ to which it is related by inverse Laplace transformation.

Expansion of the sum in Eq. (2.79) yields

$$\mathcal{Z}_r(s, t+1) = \left\langle e^{-s(A_1(t+1)+\cdots+A_r(t+1))} \right\rangle$$
$$= e^{-sr} \left\langle e^{-s\sum_{x=1}^{r-1} A_x(t)} e^{-s\gamma_0(t)A_0(t)} e^{-s\gamma_r(t)A_r(t)} \right\rangle_\gamma \tag{2.81}$$

because of the substitution of Eq. (2.78). One therefore has to carry out the averages on all values of $\gamma_0(t)$ and $\gamma_r(t)$ in the four possible cases, that is, $\gamma_0 = \gamma_r = 0$; $\gamma_0 = 1, \gamma_r = 0$; $\gamma_0 = 0, \gamma_r = 1$; and $\gamma_0 = 1, \gamma_r = 1$. The result is the characteristic equation:

$$\mathcal{Z}_r(s, t+1) = \frac{e^{-sr}}{4} \left[2\mathcal{Z}_r(s, t) + \mathcal{Z}_{r-1}(s, t) + \mathcal{Z}_{r+1}(s, t)\right] \tag{2.82}$$

The solution of the above difference equation for our particular case of interest, that is, $\mathcal{Z}_1(s, t)$ in the limit $t \to \infty$, will now be obtained. Let $\hat{\mathcal{Z}}_r(s) = \lim_{t\to\infty} \mathcal{Z}_r(s, t)$. From Eq. (2.82) one gets

$$\hat{\mathcal{Z}}_{r+1}(s) + (2 - 4e^{sr})\hat{\mathcal{Z}}_r(s) + \hat{\mathcal{Z}}_{r-1}(s) = \phi \tag{2.83}$$

with boundary condition

$$\hat{\mathcal{Z}}_0(s) = 1 \tag{2.84}$$

Defining $R_r(s) = \hat{\mathcal{Z}}_r(s)/\hat{\mathcal{Z}}_{r+1}(s)$ and $R_0(s) = 1/\hat{\mathcal{Z}}_1(s)$, one obtains

$$R_{r-1}(s) = (4e^{sr} - 2) - \frac{1}{R_r(s)} \tag{2.85}$$

and using the boundary condition Eq. (2.84), one gets

$$\frac{1}{\hat{\mathcal{Z}}_1(s)} = (4e^s - 2) - \frac{1}{R_1(s)} \tag{2.86}$$

Thus substituting, one gets the continued fraction

$$\hat{\mathcal{Z}}_1(s) = \frac{1}{4e^s - 2 - \frac{1}{4e^{2s} - 2 - \frac{1}{4e^{3s} - 2 - \cdots}}} \tag{2.87}$$

By using suitable series expansions of the continued fraction, Huber [1991] obtained a general solution. He linearized the individual terms in Eq. (2.87) around $s = 0$, corresponding to the large area limit $a \to \infty$ in Eq. (2.67),

$$\hat{\mathcal{Z}}_1(s \to 0) \approx \frac{1}{2 + 4s - \frac{1}{2 + 8s - \frac{1}{2 + 12s - \cdots}}} \tag{2.88}$$

Exact recurrence relations [e.g., Churchill and Brown, 1979] of Bessel's functions allow the limit relation

$$\hat{\mathcal{Z}}_1(s \to 0) \approx 1 - c_1 s^{1/3} - \frac{2}{5}s + \mathcal{O}(s^{5/3}) \tag{2.89}$$

where $0.4 < c_1 < 1$ [Huber, 1991]. By working back through the Laplace transform, one sees that the leading exponent 1/3 is exactly equal to the exponent β in Eq. (2.67), that is

$$\beta = 1/3 \tag{2.90}$$

We will use the above result for comparison in different places throughout this book, mainly to show that the characteristics of directed trees, like those of Scheidegger's rivers, are not attained by real rivers in nature.

Also of direct physical interest is a continuum version of the river model (hence the notation employing x for i even in the integer case), through which a steady-state recursion relation of the same type is obtained. We note that a geometric explanation of $\beta = 1/3$ is based on the fact that the mass distribution is the area surrounded by two random-walk trajectories, which are the ridges in the boundary of the basin [Takayasu et al., 1988].

Finally, we show that invariance and scaling properties in the distributions of total contributing area in drainage basins can be derived, for an exactly ordered drainage system, from Horton's laws of network composition. This was studied in detail by La Barbera and Roth [1994] and by De Vries, Becker, and Eckhardt [1994], whose exposition we closely follow.

Recall the definition of link as the reach of stream between two neighboring junctions or between a source and the next junction. An area that directly drains into the link is associated with each link. We assume an ideal network that is characterized by (i) the length of the links of the network and their associated areas, which are described by random variables with distributions that are independent of location within the network; and (ii) the network exactly follows Horton's laws with $R_B > R_L$.

The assumption is made that mean annual flow, say Q, is assumed proportional to total contributing area A at any point. The value $Q(\omega) \sim A(\omega)$ for the mean (annual) discharge in a link of order ω is then proportional to the number of links, say $N_{\text{links}}(\omega)$, draining through this link. Here the above, somewhat involved, notation is required in order to avoid confusion with $N(\omega)$, the number of Strahler's streams of order ω.

One can compute N_{links} directly from Horton's laws. Measuring the length of the streams in units of the mean link length, the mean number of links in a stream of order ω is $R_L^{\omega-1}$. Notice that in a subnetwork of order i there are, on the average, $R_B^{\omega-i}$ such streams. Thus the total mean number of links in a subnetwork of order ω is the sum over all terms $R_B^{\omega-i}\, R_L^{i-1}$ for $i \leq \omega$, that is

$$\bar{N}_{\text{links}}(\omega) \sim \sum_{i=1}^{\omega} R_B^{\omega-i}\, R_L^{i-1} = \frac{R_B^{\omega} - R_L^{\omega}}{R_B - R_L} \tag{2.91}$$

Thus one has (here we use the same symbol $A(\omega)$ for the mean and the actual area, that is, discharge, in a link of order ω at no risk of confusion)

$$A(\omega) \propto R_B^{\omega} - R_L^{\omega} \tag{2.92}$$

We now consider the entire network up to the order Ω. The mean number of links of order ω in the network, say $N_{\text{links}}^{\Omega}(\omega)$, is

$$N_{\text{links}}^{\Omega}(\omega) = R_L^{\omega-1}\, R_B^{\Omega-1} \tag{2.93}$$

The mean total number $N_{\text{links}}^{\Omega}(\Omega)$ of links in a network of order Ω is thus

$$N_{\text{links}}^{\Omega}(\Omega) = \sum_{i=1}^{\Omega} R_B^{\Omega-i}\, R_L^{i-1} = \frac{R_B^{\Omega} - R_L^{\Omega}}{R_B - R_L} \tag{2.94}$$

So the probability $f(a) = f(A(\omega))$ (notice the use of the usual lowercase symbol a for the arbitrary value of the random variable) of drawing a link of order ω with mean flow a becomes

$$f(a) = \frac{N^{\omega}_{\text{links}}(\Omega)}{N^{\Omega}_{\text{links}}(\Omega)} \sim \left(\frac{R_L}{R_B}\right)^{\omega} \tag{2.95}$$

The distribution function for the discharge (i.e., total contributing area) $P[A \geq a]$ is then derived by De Vries et al. [1994] as

$$P[A \geq a] \sim \sum_{i=\omega}^{\Omega} \left(\frac{R_L}{R_B}\right)^{\omega}$$

$$= \frac{(R_L/R_B)^i - (R_L/R_B)^{\Omega+1}}{1-(R_L/R_B)} \tag{2.96}$$

an equation also obtained by La Barbera and Roth [1994].

Under the assumption of large orders Ω, it is possible to find the order ω such that both $\Omega \gg \omega$ and $\omega \gg 1$ are satisfied. Because $R_B > R_L$, one can therefore approximate the preceding equation as

$$P[A \geq a] \sim \left(\frac{R_L}{R_B}\right)^{\omega} \tag{2.97}$$

Under the same conditions, Eq. (2.92) gives

$$a = A(\omega) \sim R_B^{\omega} \tag{2.98}$$

De Vries et al. [1994] used then the result of La Barbera and Rosso [1989] for the topological dimension of the network $D = \log R_B / \log R_L$ (Section 2.3) to express the mutual relationship of Horton's ratios, that is

$$R_L = R_B^{1/D} \tag{2.99}$$

Substituting Eqs. (2.98) and (2.99) into Eq. (2.97), one obtains

$$P[A \geq a] \propto a^{(1/D)-1} \tag{2.100}$$

from which one concludes that the exponent of the power law is (Eq. 2.67)

$$\beta = 1 - \frac{1}{D} \tag{2.101}$$

or, explicitly accounting for Horton's parameters

$$\beta = 1 - \frac{\log R_L}{\log R_B} \tag{2.102}$$

Substituting the measured value of 1.8 for D [Tarboton et al., 1988] (see Section 2.2) in Eq. (2.101), one finds $\beta \approx 0.45$, not far from the measured values of Rodriguez-Iturbe et al. [1992a]. Table 2.5, from De Vries et al. [1994], lists the values of R_B and R_L together with the calculated values for β from Eq. (2.102) for several large drainage basins. All exponents are relatively close to the values derived in Table 2.4 from direct computation.

The above analysis is not entirely satisfactory for it does not embed important geomorphological quantities related, for instance, to the extent of unchannelized portions of the river basin. However, as we will analyze in detail later, the power law part of the probability of total contributing area

Table 2.5. *Exponents β from Horton's parameters*

River Basin	R_B	R_L	$\beta = 1 - \frac{\log R_L}{\log R_B}$
Daddys Creek, Tenn.	4.1	2.2	0.44
Allegheny River, Penn.	4.5	2.4	0.42
Youghiogheny River, M.	4.6	2.2	0.47
Hubbard A50	4.7	2.3	0.46

Source: De Vries et al., 1994.

relates directly to fluvial characters that are well defined by the Hortonian rules.

Independently of De Vries et al. [1994], La Barbera and Roth [1994] carried out roughly the same analysis leading to a more general expression. In fact, introducing the fractal dimension of river courses $D_l \sim 1.1$ (Section 2.3), they find that the coefficient of the power law of total contributing area is

$$\beta = \frac{1}{D} - \frac{1}{D_l} \tag{2.103}$$

which, for the value $D = 1.8$ used before, yields an expected value of the slope $\beta = 0.432$, which compares even better with the experimental figures in Table 2.4.

De Vries et al. [1994] also compute the distribution function for the discharge (i.e., total contributing area) for the infinite topologically random channel networks in the random model [Shreve, 1967, 1969] described in Section 1.3.3. Because these networks follow Horton's laws with $R_B = 4$ and $R_L = 2$, one gets for the exponent β

$$\beta = 1/2 \tag{2.104}$$

This result could also be obtained in another analytical way. Recall the definition of the magnitude of a link is the number of sources draining through the link (see Section 1.3.3). The probability of drawing a link of magnitude n out of an infinite topologically random channel network, say $f(n)$, is [Shreve, 1967]

$$f(n) = \frac{2^{-(2n-1)}}{2n-1} \binom{2n-1}{n} \tag{2.105}$$

Again, we assume the mean discharge $Q(n) \sim A(n)$ in a link of magnitude n is proportional to the number of links draining through this link. This number equals $2n - 1$ and for $n \gg 1$ one has

$$a = A(n) \approx 2n \tag{2.106}$$

De Vries et al. [1994] proceed to find that the distribution function is given by (recall that we adopt the notation a for the arbitrary value of total contributing area $A(n)$ in a link of magnitude n)

$$P[A \geq a] = \sum_{k=n}^{\infty} \frac{2^{-(2k-1)}}{2k-1} \binom{2k-1}{k} = 2^{-2n} \frac{(2n)!}{(n!)^2} \tag{2.107}$$

Using Stirling's approximation of the factorial, that is,

$$n! \approx e^{n \log n - n} (2\pi n)^{1/2} \qquad n \gg 1 \tag{2.108}$$

and Eq. (2.106), one obtains

$$P[A \geq a] \approx \frac{1}{(\pi n)^{1/2}} \propto a^{-1/2} \qquad (2.109)$$

in agreement with the result of Eq. (2.104)

De Vries et al. [1994] point out that the difference between the calculated exponent in Shreve's model and the observed exponents for real river networks provides further evidence that river networks are not purely topologically random.

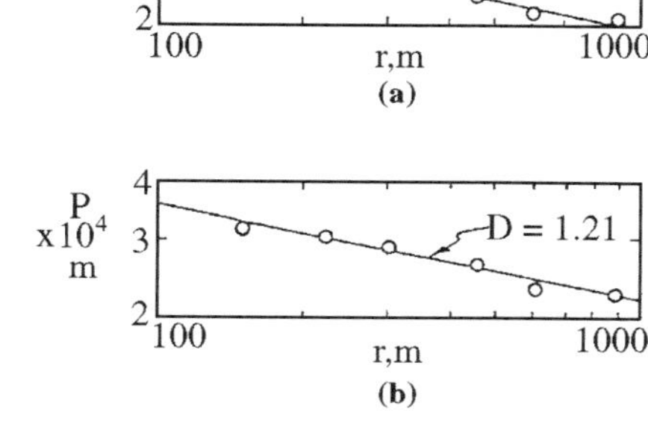

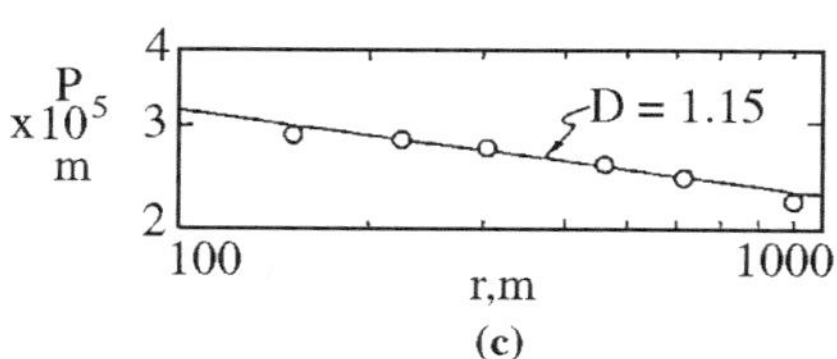

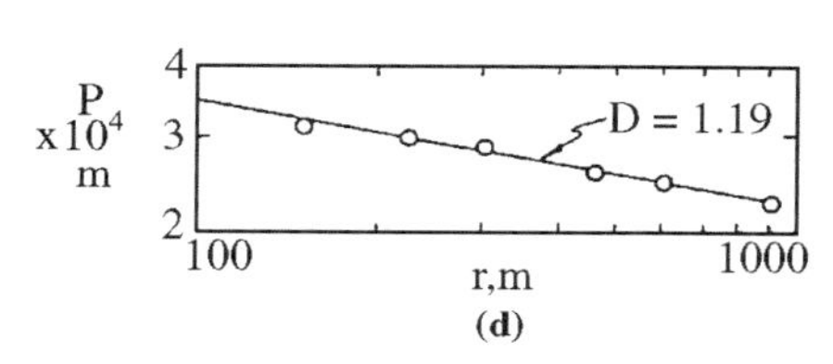

Figure 2.34. The lengths P of specified contours in several mountain belts as a function of the length r of the measuring rod: (a) is a 3, 000-ft contour of the Cobblestone Mountain quadrangle, transverse ranges, California ($D = 1.21$), (b) a 5, 400-ft contour of the Tatooh Buttes quadrangle, Cascade Mountains, Washington ($D = 1.21$), (c) a 10, 000-ft contour of the Byers Peak quadrangle, Rocky Mountains, Colorado ($D = 1.15$), and (d) a 1, 000-ft contour of the Silver Bay quadrangle, Adirondack Mountains, New York ($D = 1.19$). [after Turcotte, 1992].

2.6 Self-Similarity of Topographic Contours

The contour lines on a topographic map of a river basin are analogous to the coastlines studied earlier in this chapter. In fact, one might think they are coastlines produced by seas filling the basin up to a certain elevation. Because of this similitude, it is reasonable to expect that contour lines are self-similar objects that can be characterized in terms of their fractal dimension. This is indeed the case, as can be seen in Figure 2.34 from Turcotte [1992], where the divider or ruler method as been applied to estimate the fractal dimension of different contour lines across the United States.

It is reasonable to hypothesize that the geometric properties of natural landscapes in general may not be described by specifying a single fractal dimension. This is likely to be reflected in the characteristics of the contour lines, where fractal dimensions will arise from a large number of spatially variable field properties, reflecting natural heterogeneity in the interplay of exposed lithology of the soil mantle, geology, climate, etc. Many other descriptions of the landscape, such as the so-called exceedence sets, $S_{\geq T}$, defined as the set of points whose elevation is at or above a certain height T; or transects, defined as the set of elevation values sampled along one-dimensional tracks, also reflect the impact of natural heterogeneities in their fractal properties. They are usually not characterized by a single fractal dimension, but rather by a spectrum of fractal dimensions; that is, they are (somewhat loosely defined as) *multifractal* objects. The study of multifractality in river basins starts in Chapter 3 and continues through the rest of the book. The study of landscape transects as well as further properties of the topographic contours of river basins requires the concept of self-affinity, which is the topic of the next section.

2.7 Self-Affinity in River Basins

As emphasized throughout this chapter, fractals are objects in which properly scaled portions are identical (in a deterministic or statistical sense) to the original object. A description of this scaling behavior is the fractal dimension.

There are two types of fractal objects: self-similar and *self-affine* sets. The difference between them resides in whether the appropriate rescaling of the parts to obtain the original object is isotropic or anisotropic (i.e., the amplification scales may not be the same in different directions). Two classical examples of self-similar and self-affine fractals are coastlines and mountains, respectively. As seen previously in this chapter, coastlines are made up of small peninsulas and bays that at the beach scale are indistinguishable from the peninsulas and gulfs at the continental scale when amplified isotropically. Mountains, on the other hand, look very flat when viewed at a planetary scale but very rough when viewed at a human scale. However, if different scales are used for the horizontal and vertical amplifications, the rescaled mountain profiles look the same. Thus in order to characterize the scaling invariance of self-similar fractals, one needs a single number, the fractal dimension, while for self-affine fractals, the anisotropic scaling requires two or more numbers [e.g., Matsushita and Ouchi, 1989].

The purpose of this section is to investigate the scaling properties of individual watercourses and basin topography and to study whether these objects are self-affine or self-similar fractals. Similar structures in other fields have shown self-affine scaling, for example, directed polymers [Kardar and Zhang, 1987] and the boundary of growing interfaces [Kardar, Parisi, and Zhang, 1986, Meakin et al., 1985, 1986].

2.7.1 Brownian Motion and Fractional Brownian Motion

Brownian motion is the classical self-affine process and a brief review of some of its major features in one dimension is useful introduction to self-affinity.

Consider a random walk in one dimension in which, at intervals τ, a random displacement or step of length ξ is taken orthogonal to the time axis. The distribution p of the random variable ξ is taken as Gaussian and given by

$$p(\xi, \tau) = \frac{1}{\sqrt{4\pi\sigma\tau}} e^{-\xi^2/(4\sigma\tau)} \tag{2.110}$$

The variance of the process of steps, or mean square jump distance, is

$$E[\xi^2] = \int_{-\infty}^{\infty} \xi^2 p(\xi, \tau) d\xi = 2\sigma\tau \tag{2.111}$$

where σ is called the diffusion coefficient, usually defined through the Einstein relation

$$\sigma = \frac{1}{2\tau} E[\xi^2] \tag{2.112}$$

The sequence of $\xi's$ is the sequence of steps of the random walk, where the position X of the walker on the x axis perpendicular to the time axis is given by

$$X_{-i}(t = n\tau) = \sum_{i=1}^{n} \xi \tag{2.113}$$

When τ tends to zero, the position becomes a random function $X(t)$, which Mandelbrot calls a "Brown" function denoted by $B(t)$. In practice, the position of the walker (or particle) is observed at intervals $b\tau$. Suppose we observe the process only at every second step, i.e. $b = 2$. The increment ξ in the particle's position is now the sum of two independent increments ξ' and ξ'' [e.g., Feder, 1988], and thus the joint probability distribution is equal to the product of the two densities because of the statistical independence of ξ' and ξ'':

$$p(\xi', \xi'', \tau) = p(\xi', \tau)p(\xi'', \tau) \tag{2.114}$$

The two increments must add to the total value of ξ and thus we have

$$\begin{aligned} p(\xi, 2\tau) &= \int_{-\infty}^{\infty} p(\xi - \xi', \tau)p(\xi', \tau)d\xi' \\ &= \frac{1}{\sqrt{4\pi\sigma 2\tau}} e^{-\xi^2/(4\sigma 2\tau)} \end{aligned} \tag{2.115}$$

Thus the increments of the process, when observed with half the original resolution, are still Gaussian with zero mean but with a variance of twice the original one, $E[\xi^2] = 4\sigma\tau$.

In general, for observations at intervals $b\tau$ one has a Gaussian distribution with zero mean and variance:

$$E[\xi^2] = 2\sigma t \qquad \text{where} \qquad t = b\tau \tag{2.116}$$

This can be interpreted as the fact that Brownian motion looks the same, that is, is statistically unchanged, when the timescale is changed by a factor of b and at the same time the length scale is changed by a factor $b^{1/2}$. More formally

$$p(\xi^* = b^{1/2}\xi, \tau^* = b\tau) = b^{-1/2}p(\xi, \tau) \tag{2.117}$$

Processes like Brownian motion that remain statistically (or deterministically) unchanged under transformations that scale different axes, for example, time and distance, by different factors are called self-affine. A fundamental equation describing this property may be written as

$$X(t + \lambda\Delta t) - X(t) \stackrel{d}{=} \lambda^H \left[X(t + \Delta t) - X(t)\right] \tag{2.118}$$

where $\stackrel{d}{=}$ means equality in probability distribution and H – to be discussed later – is called the Hurst exponent.

In the case of Brownian motion, the probability distribution of the particle position $X(t)$ may also be found as [e.g., Feder, 1988]

$$P\left(X(t) - X(t_0)\right) = \frac{1}{\sqrt{4\pi\sigma|t - t_0|}} e^{-(X(t)-X(t_0))^2/(4\sigma|t-t_0|)} \tag{2.119}$$

which satisfies the scaling relation:

$$P\left(b^{1/2}\left(X(bt) - X(bt_0)\right)\right) = b^{-1/2}P\left(X(t) - X(t_0)\right) \tag{2.120}$$

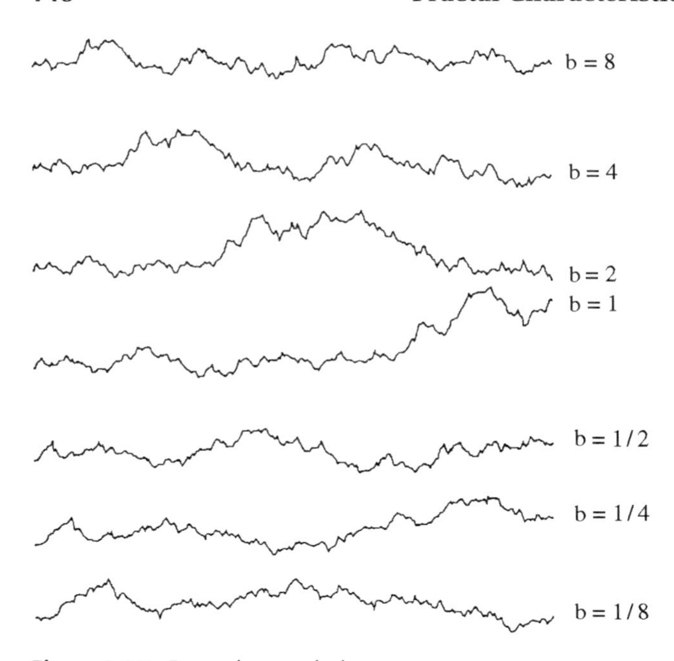

Figure 2.35. Properly rescaled Brownian motion [after Peitgen and Saupe, 1988].

The mean and variance of the particle position are

$$E\left[X(t) - X(t_0)\right] = 0$$
$$E\left[\left(X(t) - X(t_0)\right)^2\right] = 2\sigma|t - t_0| \tag{2.121}$$

where $X(t_0)$ is the particle position at a reference time t_0. Figure 2.35 shows an example of Brownian motion and its proper scaling under different factors. In the figure the graph in the center shows a small section of Brownian motion $X(t)$. In the other graphs the properly rescaled random functions of the form $b^{-1/2}X(bt)$ are displayed. The scaling factor b ranges from $b = 1/8$ to $b = 8$, corresponding to expanding and contracting the original function in the time direction. Note that the visual appearance of all the samples is the same.

Mandelbrot introduced the concept of *fractional Brownian motion*, $B_H(t)$, as a generalization of the function $X(t)$ such that its increments have zero mean but their variance is proportional to $|t - t_0|^{2H}$, with $0 < H < 1$, rather, than to $|t - t_0|$. Thus Eq. (2.121) is generalized for the case of fractional Brownian motion to become

$$E\left[B_H(t) - B_H(t_0)\right] = 0$$
$$E\left[\left(B_H(t) - B_H(t_0)\right)^2\right] = 2\sigma\tau\left(\left|\frac{t - t_0}{\tau}\right|\right)^{2H} \propto |t - t_0|^{2H} \tag{2.122}$$

Fractional Brownian motion is self-affine with parameter H. In other words, $B_H(t_0 + t) - B_H(t_0)$ and $(1/b^H)(B_H(t_0 + bt) - B_H(t_0))$ are statistically identical for any t_0 and $b > 0$. With $t_0 = 0$ and $B_H(t_0) = 0$ we find that $B_H(t)$ and $(1/b^H)B_H(bt)$ are statistically indistinguishable.

The generalization of fractional Brownian motion to multiple dimensions is straightforward, that is, a random field $B_H(t_1, t_2, \ldots, t_n)$ is defined where the main properties are the following:

- The increments $B_H(t_1, \ldots, t_n) - B_H(s_1, \ldots, s_n)$ are Gaussian with zero mean.
- The variance of the increments $B_H(t_1, \ldots, t_n) - B_H(s_1, \ldots, s_n)$ depends only on the distance $(\sum_{i=1}^{n}(t_i - s_i)^2)^{1/2}$ and it is given by

$$E\left[\left(B_H(t_1, \ldots, t_n) - B_H(s_1, \ldots, s_n)\right)^2\right] \propto \left(\sum_{i=1}^{n}(t_i - s_i)^2\right)^H \tag{2.123}$$

Specific information about simulation algorithms in one and multiple dimensions may be found, for instance, in the book by Peitgen and Saupe [1988].

The variance of both ordinary and fractional Brownian motion diverges with time. Mandelbrot baptized the coefficient H as the Hurst coefficient, to honor Harold Edwin Hurst (1900–1978), a hydrologist who carried out pioneering work in the persistence of hydrologic time series, more specifically in the apparently abnormal length of time the Nile River spends in excursions above or below its average flow discharge. The special value of $H = 1/2$ gives the familiar Brownian motion.

Persistence or very long (e.g. infinitely long) correlations are most important characteristics of fractional Brownian motion. The study of persistence is intimately related to the correlation and spectral structure of fractional Brownian motion, which are the subjects of Section 2.7.2.

2.7.2 Power Spectrum and Correlation Structure of Fractional Brownian Motion

Random functions $X(t)$ are often characterized by their power spectrum or power spectral density, say $S_X(f)$. Because the Fourier transform of X is not easily defined in the general case, we first restrict $X(t)$ to a finite sample interval $0 \leq t \leq T$. Thus the sample, say $X(t, T)$, coincides with X within $0 \leq t \leq T$ and is zero elsewhere. Defining, as $F_X(f, T)$, the Fourier transform of the specific sample of $X(t)$ for $0 < t < T$ so that

$$F_X(f, T) = \int_0^T X(t) e^{-2\pi i f t} dt \tag{2.124}$$

then the power spectrum is given by

$$S_X(f, T) = \frac{1}{T} \left| F_X(f, T) \right|^2 \tag{2.125}$$

where $|\cdot|$ indicates absolute value of a complex number and the notation is consistent with our definition of the sample. The associated spectral density of X, $S_X(f)$, is then obtained in the limit $T \to \infty$. Technically, $S_X(f)$ is a nonnegative and even function. In most applications relevant to this book we will assume that T is large enough, and therefore $S_X(f, T) \simeq S_X(f)$ without the need for a precise distinction. The power spectrum of a stationary process is related to the two-point covariance function of the process given by

$$C_X(\tau) = E\left[X(t) \cdot X(t+\tau)\right] - E^2\left[X(t)\right] \tag{2.126}$$

$C_X(\tau)$ provides a measure of how the fluctuations of a process separated by a distance τ are related. For stationary processes, $S_X(f)$ and $C_X(\tau)$ depend on each other through the Wiener–Khintchine relation:

$$C_X(\tau) = \int_0^\infty S_X(f) \cos(2\pi f \tau) df \tag{2.127}$$

The normalized version of $C_X(\tau)$ is the correlation function $\rho(\tau)$

$$\rho(\tau) = \frac{C_X(\tau)}{C_X(0)} \tag{2.128}$$

and its Fourier transform is the normalized spectral density function that integrates to one. The terms power spectrum and spectral density will be used interchangeably throughout this book.

We will now show, following Saupe [1988], that self-affine processes like fractional Brownian motion have spectral densities in the form of a power law, that is, $1/f^{\zeta}$

$$S_{B_H}(f) \propto \frac{1}{f^{\zeta}} \qquad \text{with} \quad \zeta = 2H + 1 \qquad \text{and} \quad 0 < H < 1 \tag{2.129}$$

Let $X(t)$ be a self-affine fractional Brownian motion, which by definition scales according to

$$X(t) = \frac{1}{b^H} X(bt) \tag{2.130}$$

where the equality should be interpreted in a statistical sense. A sample of $X(t)$ is designated by $Y(t, T)$ where

$$Y(t, T) = \frac{1}{b^H} X(bt) \qquad \text{if} \quad 0 < t < T \tag{2.131}$$

and 0 otherwise. The Fourier transform of $Y(t, T)$ is then

$$F_Y(f, T) = \int_0^T Y(t, T) e^{-2\pi i f t} dt = \frac{1}{b^H} \int_0^{bT} X(s) e^{-2\pi i \frac{f}{b} s} \frac{ds}{b} \tag{2.132}$$

where we have used s/b for t and ds/b for dt in the second integral. Thus

$$F_Y(f, T) = \frac{1}{b^{H+1}} F_X\left(\frac{f}{b}, bT\right) \tag{2.133}$$

and from Eq. (2.133) we find that the spectral density of $Y(t, T)$ is given by

$$S_Y(f, T) = \frac{1}{b^{2H+1}} \frac{1}{bT} \left| F_X\left(\frac{f}{b}, bT\right) \right|^2 \tag{2.134}$$

and in the limit when $T \to \infty$ or equivalently as $bT \to \infty$ we obtain

$$S_Y(f) = \frac{1}{b^{2H+1}} S_X\left(\frac{f}{b}\right) \tag{2.135}$$

Because $Y(t)$ and $X(t)$ are statistically identical, their spectral densities must coincide and thus

$$S_X(f) = \frac{1}{b^{2H+1}} S_X\left(\frac{f}{b}\right) \tag{2.136}$$

Finally, we set $f = 1$ and replace $1/b$ by f to obtain the desired result:

$$S_X(f) \propto \frac{1}{f^{2H+1}} = \frac{1}{f^{\zeta}} \tag{2.137}$$

Eq. (2.137) constitutes a fundamental characteristic of self-affine records. The spectral density is a power law that implies that fluctuations in the process appear on all scales and that the structure of such fluctuations is statistically the same at all frequencies.

Thus, roughly speaking, $S_x(f) \propto f^{-\zeta}$ corresponds to a fractional Brownian motion with $H = (\zeta - 1)/2$. When $0 < \zeta < 1$, the correlation function is given by

$$C_X(\tau) \propto \tau^{\zeta - 1} \tag{2.138}$$

For $\zeta > 1$, Eq. (2.138) is not valid and the fluctuations are nonstationary [e.g., Takayasu, 1990]. Many geophysical fields indeed have $\zeta > 1$ and hence cannot be considered stationary processes. When this is the case, one should not use autocorrelation functions but rather variograms [Lavallée et al., 1993].

For ordinary Brownian motion, $H = 1/2$ and $\zeta = 2$. The effect of H in the persistence characteristics of the process can be well appreciated in Figure 2.36, which shows $B_H(t)$ as a function of time using $B_H(0) = 0$ in all cases. The simulation of $B_H(t)$ is carried out from the previously obtained increments of the process. Details of this are given, for example, in Feder [1988] and Peitgen and Saupe [1988]. Figure 2.36 shows that as the exponent H is increased the record of the particle position undergoing a fractional Brownian motion increases in amplitude, and the noise is proportionally reduced. As compared with ordinary Brownian motion, fractional Brownian motion with $H > 1/2$ has excursions above and below different thresholds that last must longer and also moves anomalously large distances from the origin. We say that fractional Brownian motion with $H > 1/2$ exhibits *persistence*. The case $H < 1/2$ behaves in the opposite way, showing what is called *antipersistence*.

The large departures from the origin undergone by fractional Brownian motion lead to the definition of an anomalous diffusivity for fractal diffusion [e.g., Feder, 1988]:

$$\sigma_H = \sigma |t|^{2H-1} \tag{2.139}$$

which obviously coincides with the classical one when $H = 1/2$. The anomalous character of the diffusion coefficient of Eq. (2.139) is due to the fractal nature of the process itself taking place in Euclidean space. If the process takes place in a fractal set that is in turn embedded in Euclidean space, we still find an anomalous diffusivity but with a changed exponent in Eq. (2.139) [Feder, 1988; Aharony, 1986]. This distinction is important in the context of surface hydrologic processes because if we talk, say, about the ridges or boundaries of a basin, we have a fractal process evolving in an Euclidean space, but if we talk about the diffusion of a particle of water or sediment in the river network, we have a process evolving in a fractal set.

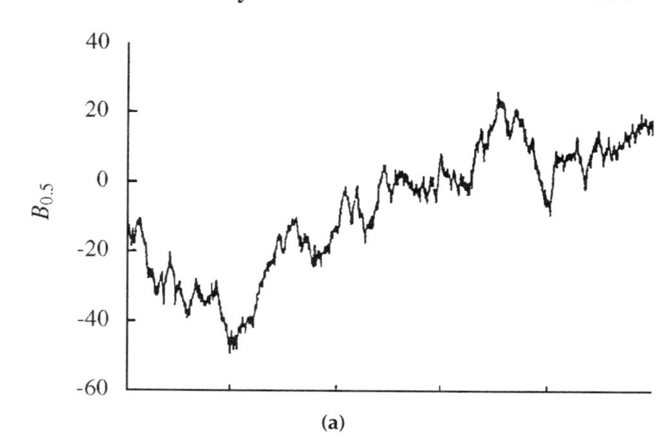

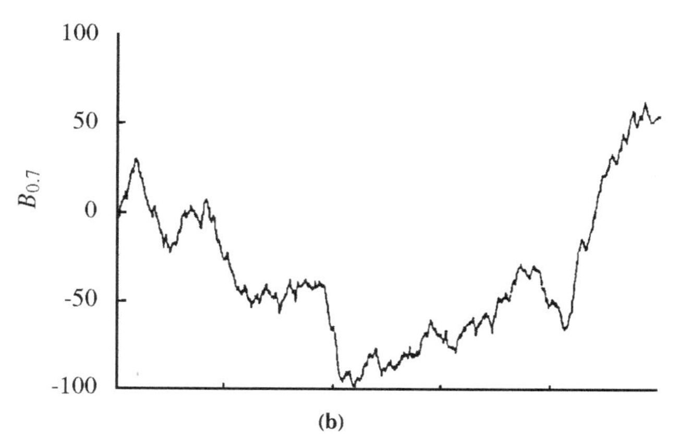

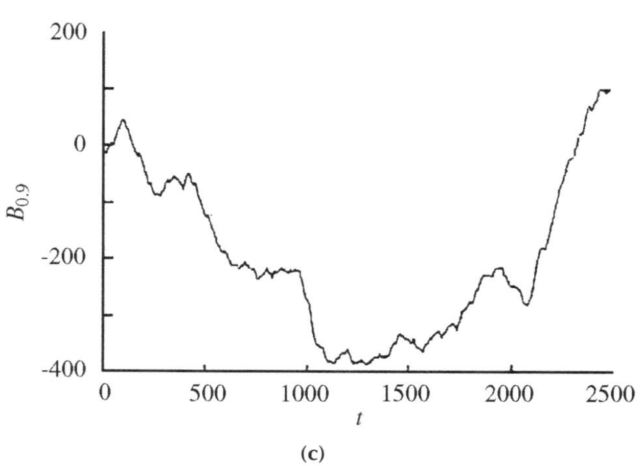

Figure 2.36. Examples of Fractional Brownian functions $B_H(t)$; (a) $H = 1/2$, (b) $H = 0.7$, (c) $H = 0.9$ [from Feder, 1988].

The persistence characteristics of fractional Brownian motion can be best observed through the correlation structure between future and past increments of $B_H(t)$. This can be computed by first deriving the covariance between the increments from $-t$ to 0 and the increments between 0 and t. Because the mean is zero, the covariance of the increments $\Delta B_H(t)$ is

$$C_{\Delta B_H}(t) = E\Big[\big(B_H(0) - B_H(-t)\big)\big(B_H(t) - B_H(0)\big)\Big] \tag{2.140}$$

and, for convenience, set $B_H(0) = 0$ so that the correlation function is given by

$$\rho_{\Delta B_H}(t) = \frac{E\big[-B_H(-t)B_H(t)\big]}{E\big[B_H^2(t)\big]} \tag{2.141}$$

Using Eq. (2.122) with $t_0 = -t$, one obtains

$$2E\big[B_H^2(t)\big] - 2E\big[B_H(-t)B_H(t)\big] = 2\sigma\tau\left(\frac{2t}{\tau}\right)^{2H} \tag{2.142}$$

Choosing the units so that $\tau = 1$ and $2\sigma\tau = 1$ and comparing with Eq. (2.141), it is found that

$$\rho_{\Delta B_H}(t) = 2^{2H-1} - 1 \tag{2.143}$$

We observe from Eq. (2.143) that for $H = 1/2$ the correlation is always zero for the past and future increments of the process. Nevertheless, and most importantly, for $H \neq 1/2$ the correlation is different from zero regardless of how large the separation in time is between the increments. For $H > 1/2$ (persistence) a positive fluctuation in the process will lead, on the average, to a positive fluctuation in the future, no matter how far into the future. On the other hand, for $H < 1/2$ (antipersistence) the opposite effect takes place; that is, a decreasing trend in the past implies a continued decrease in the future, and furthermore this applies for arbitrarily large t [Feder, 1988].

One last methodological observation concerns the fact that regular, deterministic signals with suitable properties may have power law spectra. Thus, in general, a power law spectrum does not warrant scale-invariant properties for the underlying signal. Recall that usually Fourier power spectra are evaluated by discrete Fourier transforms (DFT), possibly via FFT algorithms, which operate on the generalized function obtained by sampling X_j discrete values of X. The basic relationship is $X_j = \sum_{k=-N/2}^{N/2} \hat{X}_k \exp(2i\pi j k/N) = \sum_{k=0}^{N/2} \mathcal{X}_k \sin(2\pi j k/N + \theta_k)$ where $\hat{X}_k$ is the complex amplitude of the kth component of the spectrum, $\mathcal{X}_k$ is a spectral coefficient related to $\hat{X}_k$, N is a suitable cutoff, and θ_k is the kth phase. It can be shown that an algebraic decay of the spectrum becomes a necessary and sufficient condition for a fractal signal if the phases θ_k are randomly distributed [e.g., Hough, 1989; Russ, 1994].

2.7.3 Characterization of Self-Affine Records

We have seen that the scaling properties of self-affine processes may be characterized equivalently by either the spectral exponent ζ or the parameter H, which are related through Eq. (2.129), $\zeta = 2H + 1$. Yet another characterization is provided by the fractal dimension D, but this deserves careful attention because its assignment to a self-affine fractal may lead to ambiguous results [Voss, 1986]. Here we will follow Feder's [1988] exposition regarding the ambiguity and scale dependence of the box-counting dimension and Voss's [1988] exposition for the divider dimension applied to self-affine fractals.

In the case of the box-counting dimension, let the time span of the record be T and let us cover the record with boxes of width $b\tau$ in time and ba in position. In each segment of length $b\tau$ the range of the record is $\Delta B_H(b\tau) = b^H \Delta B_H(\tau)$. Thus we need $T/(b\tau)$ boxes to cover the time span of the record and $\Delta B_H(b\tau)/(ba)$ boxes to cover its range in position. Therefore we need

$$N(b; a, \tau) = \frac{b^H \Delta B_H(\tau)}{ba} \frac{T}{b\tau} \approx b^{H-2} \tag{2.144}$$

boxes to cover the record. Eq. (2.144) yields a box-counting dimension of

$$D = 2 - H \tag{2.145}$$

for self-affine records.

The derivation given above is not valid when the boxes are not small with respect to the range of the record. Thus if a is chosen to have a size of the same order of the typical step length $a = \sqrt{E[\xi^2]} = 1$, then in each time segment $b\tau$ we need on the order of only one box to cover the span $\Delta B_H(b\tau)$. Therefore

$$N(b; a, \tau) \approx 1 \cdot \frac{T}{b\tau} \approx b^{-1} \tag{2.146}$$

Thus in this case, the box-counting dimension is $D = 1$. When a is small the analysis has high resolution and $D = 2 - H$ is called the *local* box-counting fractal dimension of the self-affine process. When a is large the analysis is one of global character and thus it is said that self-affine records have a *global* box-counting fractal dimension equal to 1. Globally, when looking from far apart a self-affine record is not a fractal. Because ordinary Brownian motion has $H = 1/2$, it is said that its fractal dimension is $D = 2 - 0.5 = 1.5$.

In the case of the divider dimension, one can divide the self-affine curve into N segments by fitting a ruler of size ℓ along the curve. The length along each segment is

$$\begin{aligned} \ell &= [\Delta t^2 + \Delta B_H^2]^{1/2} = \Delta t \left[1 + \frac{\Delta B_H^2}{\Delta t^2}\right]^{1/2} \\ &\propto \Delta t \left[1 + \frac{1}{\Delta t^{2-2H}}\right]^{1/2} \end{aligned} \tag{2.147}$$

where we have used the fact that the typical $B_H(t)$ variation scales as $\Delta B_H = \Delta t^H$. On small scales with $\Delta t \ll 1$, the second term dominates and $\ell \propto \Delta t^H$ so that the number of rulers of size ℓ, $N(\ell)$, is proportional to

$$N(\ell) \propto \Delta t^H \cdot N \propto \Delta t^H \frac{1}{\Delta t} = \Delta t^{H-1} \tag{2.148}$$

or

$$N(\ell) \propto [\ell^{1/H}]^{H-1} \propto \ell^{1-1/H} \tag{2.149}$$

and thus the *local* divider dimension of a self-affine record is $D = 1/H$. On the other hand, at large scales, with $\Delta t \gg 1$, in Eq. (2.147) ℓ varies linearly with Δt and the *global* divider dimension of a self-affine record is $D = 1$.

The above analysis shows that the same self-affine record can have an apparent *self-similar* dimension D of either 1, $1/H$, or $2 - H$ depending on the technique used and the choice of scale. Thus care should be exercised when a fractal dimension is assigned to a self-affine record. Notice that the derivations leading to $D = 2 - H$ and $D = 1/H$ as the local box-counting and local divider dimension, respectively, are valid for a self-affine *record*. Later in this chapter we will study the characterization of self-affine curves that are not necessarily univalued functions.

A different methodology, especially attractive in the case of nonstationary processes, is the *method of variations* proposed by Dubuc et al. [1989a,b]. This technique proves more precise for avoiding effects due to the finite size of the data set.

We will now describe the basic elements of the method of variations. Let $G = (x, f(x))$ be a set defined in the plane where $f(x)$ is the record. We define the ν-oscillation as

$$\nu(x, \delta) = \sup_{y \in U_\delta(x)} f(y) - \inf_{y \in U_\delta(x)} f(y) \quad U_\delta(x) = \{s : |x - s| < \delta\} \tag{2.150}$$

where $\delta > 0$. Therefore the variation $\nu(x, \delta)$ is defined for the neighborhood of size δ of points around x. The *variation* V of f is defined by

$$V(\delta, f) = \int_0^1 \nu(x, \delta) dx \tag{2.151}$$

where it should be noted that the sets on which the ν-oscillation are computed refer to different points x and that they do not need to be disjoint. We also assume that the distances x in the set have been normalized by the maximum length such that $x \in (0, 1)$.

The key ingredient of the variation method is the theorem by Tricot et al. [1989], which relates the process of covering the surface of a fractal object and the process of estimating the variations exhibited by the surface.

Let us introduce here an arbitrary covering function $\Omega(G, \delta)$ of G made up by disjoint elements of *diameter* δ. A fractal dimension D of G (of the box-counting type) is defined by

$$D = \lim_{\delta \to 0} \frac{\log \Omega(G, \delta)}{\log \delta} \geq D_H \tag{2.152}$$

[Falconer, 1990], where D_H is defined as in Section 2.1.1.

Tricot's theorem [1995] states the equivalence between the above dimension calculated through the *volume* $\Omega(G, \delta)$ and that evaluated through the variation V of diameter δ. Thus

$$D = \max\left[1, \lim_{\delta \to 0}\left[2 - \frac{\log V(\delta, f(x))}{\log \delta}\right]\right] \tag{2.153}$$

To estimate fractal dimensions through the variation method, the following steps are taken:

- The function $(x, f(x))$ is discretized into N elements $(f_1, \ldots, f_N)$ (Figure 2.37).

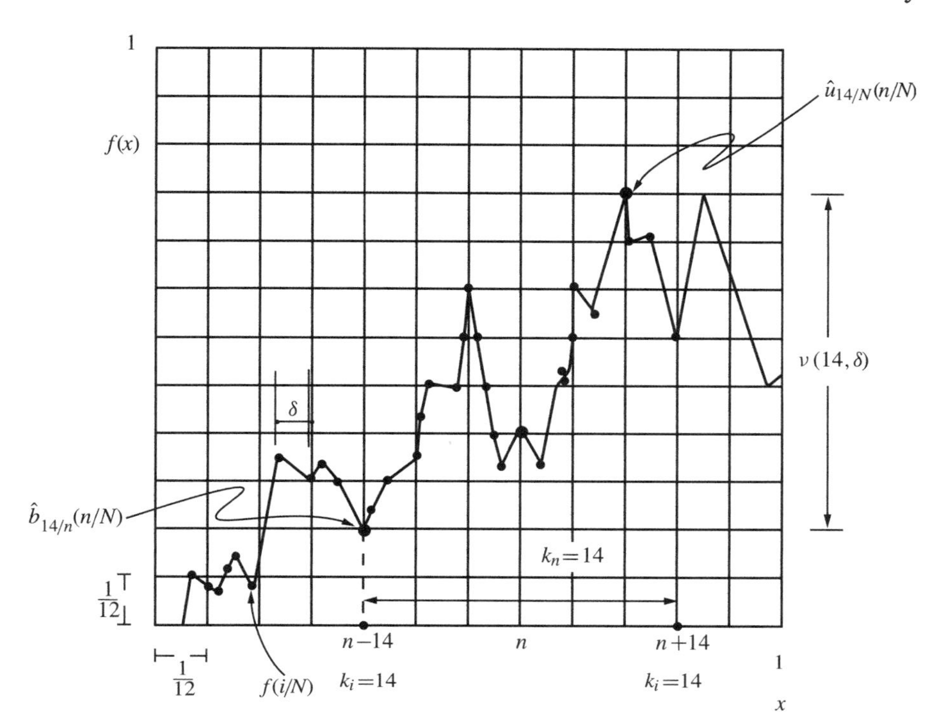

Figure 2.37. Discretization of a curve and the method of variation.

- A set of integers $k_1 < \ldots < k_i \ldots < k_{i_{\max}}$ is chosen so as to define the size of the set of diameters surrounding each point where the ν-oscillation is computed. Instead of using the diameter δ, the set where variations are computed is defined by the integer number k_i corresponding to the number of elements of the discretization included in the set.
- Because the maximum and minimum of the function in each set are to be computed, we treat them separately. The maxima for every choice of k_i are

$$u_{k_i/N}\left(\frac{n}{N}\right) = \max_{(n-k_i)/N \leq j/N \leq (n+k_i)/N} f\left(\frac{j}{N}\right) \tag{2.154}$$

where n is the integer labeling the arbitrary site x around which one computes the variations. Eq. (2.154) may be rewritten as [Rigon, 1994]

$$\hat{u}_i(n) = \max_{(n-k_i) \leq j \leq (n+k_i)} f\left(\frac{j}{N}\right) \tag{2.155}$$

where $j = 1, \ldots, N$ and $k_i = 1, \ldots, k_{i_{\max}}$. An analogous construction is performed for the set of minima, which we will denote as $\hat{b}_i$.
- The ν-oscillation in the arbitrary point n is given by $\hat{u}_i(n) - \hat{b}_i(n)$, where i indicates the width of the interval. The variation is computed by

$$\hat{V}_i(f) = \frac{1}{N}\sum_{n=1}^{N} \hat{u}_i(n) - \hat{b}_i(n) \tag{2.156}$$

- From the computational viewpoint an important feature is that if the sequence of diameters k_i is chosen so as to yield $k_i < 2k_{i-1}$, then to compute the variation one may use the following recursive relations:

$$\begin{aligned} \hat{u}_{i+1}(n) &= \max[\hat{u}_i(n-\delta_i), \hat{u}_i(n+\delta_i)] \\ \hat{b}_{i+1}(n) &= \max[\hat{b}_i(n-\delta_i), \hat{b}_i(n+\delta_i)] \\ \delta_i &= k_{i+1} - k_i \end{aligned} \tag{2.157}$$

- The fractal dimension D is obtained from the interpolation of the log-log plot of the variation V versus the diameter δ as in Eq. (2.153).

A few computational steps should be taken to yield to more reliable results. Because for $k_1 = 1$ the variation is computed only on three points (the current point and the nearest neighbors), only a subset R of the N points of discretization is used. The above operations are repeated by varying the surrounding sets k_i considering progressively more of the N points in the computation of the ν-oscillation. Thus one varies R to get a more precise estimate of the fractal dimension. Boundary effects are also likely to arise [Rigon, 1994]. The ν-oscillation measured for the extreme intervals of the domain of f is smaller than the ν-oscillation that the function would have for an indefinitely extended domain. This introduces a systematic convexity in the log-log plots through which the dimension is computed. Furthermore, the error induced increases as the size of the diameter increases. A possible procedure consists in eliminating from the estimate of the variations the elements placed at distances smaller than k_i from the extremes of the domain. This obviously reduces the number of points on which the variation is computed, depending on the distance k_i. This, and a few other more technical requirements (aimed at a *renormalized* variation method [Dubuc et al., 1989a]), renders the method quite efficient in estimating the fractal dimension even with relatively little data. It is also easily generalized to d dimensions.

Figure 2.38 shows an example of the estimation of the fractal dimension of a single vertical transect of the surface of the Fella River landscape in northern Italy. The agreement of the log-log plot to a straight line is excellent, and the resulting estimate of the fractal dimension is $D = 1.5 \pm 0.1$.

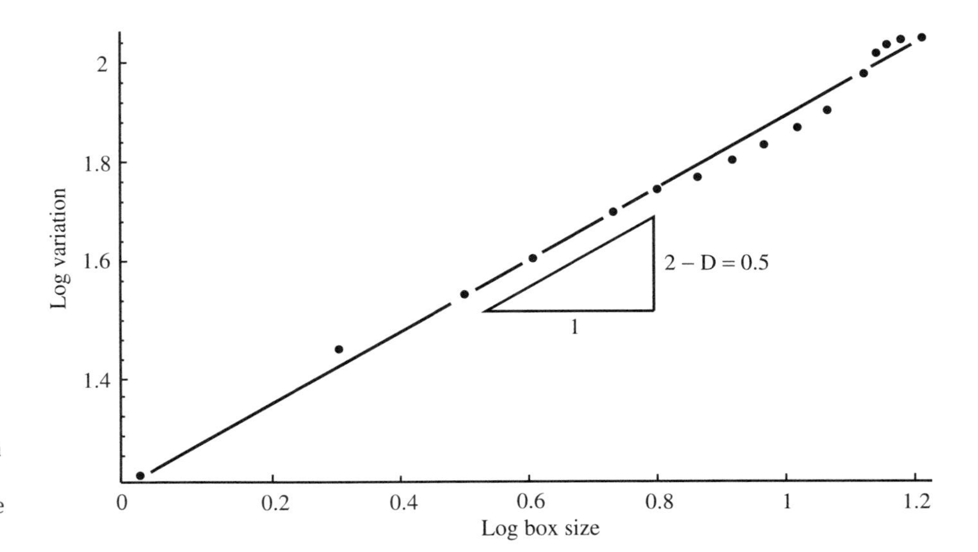

Figure 2.38. Results of the method of variations applied to a transect of the Fella River landscape [after Rodriguez-Iturbe et al., 1994].

Finally, we note that once the ensemble average of all resulting fractal dimensions is obtained for repeated longitudinal and latitudinal transects within a large portion of the Fella River Basin, one gets an average fractal dimension of $D \approx 1.6$ and quite a scatter in the results. This result will be discussed and compared with other methods, for example, Matsushita and Ouchi [1989], in Section 2.8 and further rediscussed in the framework of multifractal analysis in Chapter 3.

2.7.4 Self-Affine Characteristics of Topographic Transects

A characteristic and typical record of river basin topography is the transect of elevations defined as an equally spaced series of terrain heights taken along a straight line. Obviously, such records cannot be self-similar and investigating whether they are self-affine and, if so, determining their characteristics is of great interest.

Turcotte [1992] carried out spectral analyses of topographic transects from three different parts of Oregon. The Willamette lowland is characterized by sedimentary processes, the Wallowa Mountains are characterized by a major tectonic uplift, and the Klamath Falls area belongs to the basin and range tectonic regime. The topography was digitized along latitude and longitude lines at seven points per kilometer. For each region, 20 equally spaced transects of 512-point length were analyzed in both the latitudinal and longitudinal directions. Figure 2.39 from Turcotte [1992] shows eight typical examples from the above basins.

Through the slope of the spectral density, ζ, and the relations $H = (\zeta - 1)/2$ and $D = 2 - H$, Turcotte provides the average values for the local box-counting fractal dimensions of the Oregon data. These are given in Table 2.6.

The first thing to notice in Turcotte's analysis is the consistent approximately linear character of the individual spectrum in double logarithmic graphs. This shows that the transects may be considered self-affine records, at least at the scale of resolution used by Turcotte [1992]. Nevertheless, it is important to emphasize the regional character of the values given in Table 2.6. The spectral slope and thus the fractal dimension vary from one individual transect to the other, although the variations are not of a major

Table 2.6. *Regional average of the local box-counting fractal dimension for one-dimensional vertical profiles of Oregon topography*

Region	Fractal Dimension
Willamette lowland	
Latitude	1.436
Longitude	1.507
Wallowa Mountains	
Latitude	1.499
Longitude	1.485
Klamath Falls	
Latitude	1.492
Longitude	1.500

Source: Turcotte, 1992.

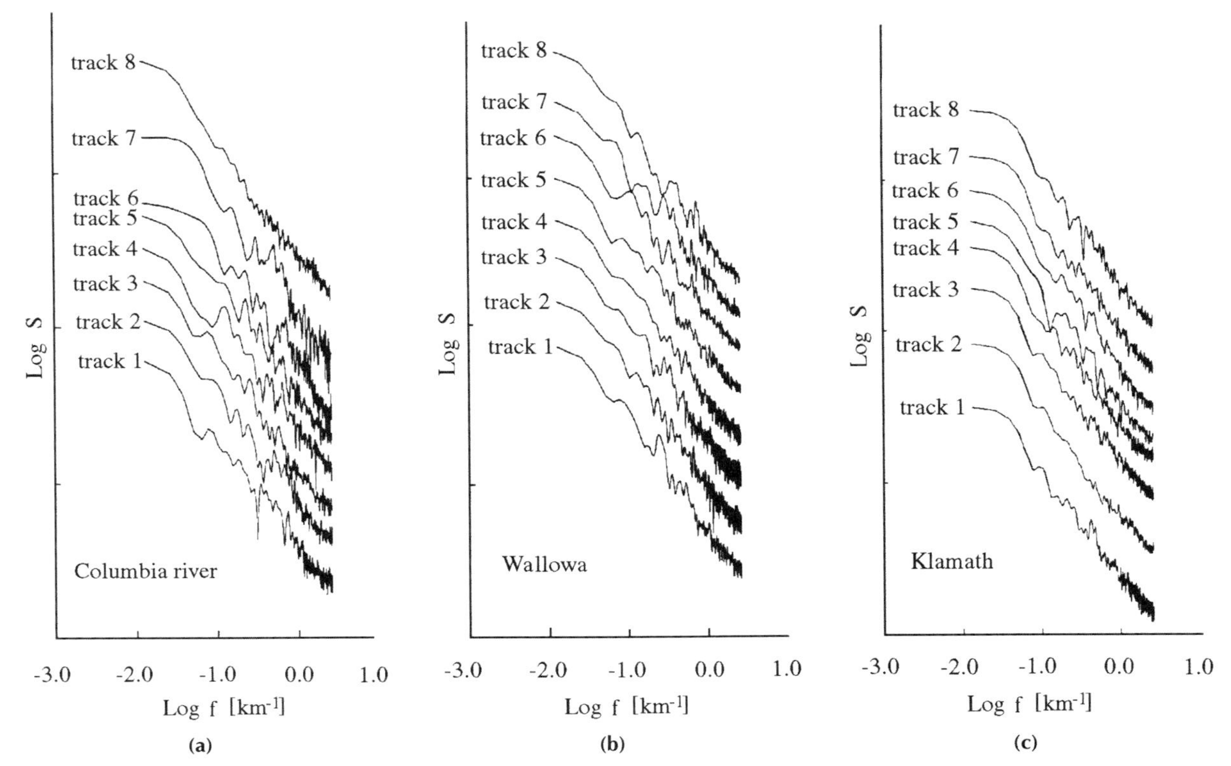

Figure 2.39. Power spectral density versus wave number for transects of elevation fields in three basins in Oregon. The profiles are offset vertically so as not to overlap [after Turcotte, 1992].

character. This multiplicity of fractal dimensions points to the complexity of real topographies from which the transects result when the surface intersects with vertical planes. The characteristics of topographic relief are easy to visualize in river basins, where different types of erosional processes act at different spatial locations depending on the local geology, climate, vegetation, and soil. The interplay of these erosional processes may lead to landscapes of different degrees of roughness, where the changes of slope in the hillslopes may be quite abrupt or, in contrast, very smooth. Chapter 3 will present further analyses in which the multifractal aspects of transects will be evident.

Because in Figure 2.39 the digitization of the topography was carried out every 150 meters, it may be too coarse to contain the finer details of erosional characteristics. The drainage network itself plays an important role in this characterization, and its linkage with the hillslope process leads to the landscape of the river basin. Chapter 6 will place a special emphasis on the basin landscape when both network and hillslope processes are taken into account.

Rodriguez-Iturbe et al. [1994] compared different estimators of fractal dimensions on the transects obtained from the real landscape of the Fella River basin in northern Italy (see Figure 1.14). Figure 2.40 shows the comparison of the Matsushita-Ouchi algorithm (described in Section 2.8.6) and the power spectrum analysis, both applied to a 64×64 subgrid of

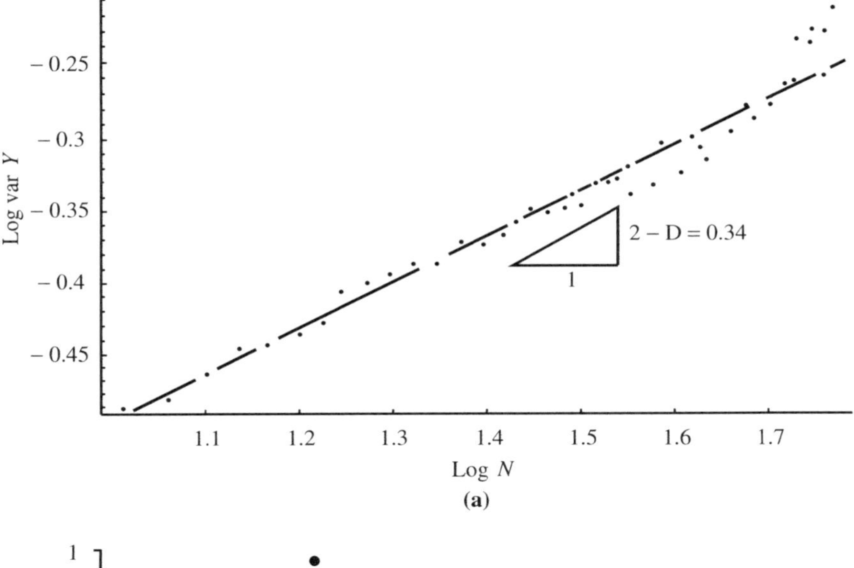

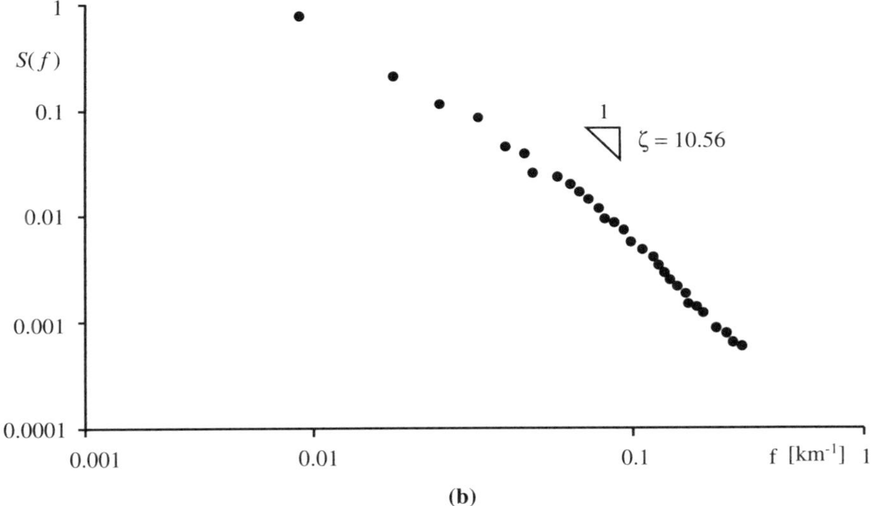

Figure 2.40. Landscape fractal dimension (the Fella River landscape, northern Italy) (a) via Matsushita and Ouchi and (b) via spectral analysis [after Rodriguez-Iturbe et al. 1994].

elevations extracted from a high-altitude (and hence rough) portion of the Fella River landscape. The results were obtained by averaging over the 64 transects extracted from the DEM. We note that the value of fractal dimension arising from both analyses is consistently close to $D \approx 1.6$. The Matsushita-Ouchi analysis yields $D = 1.66$ and the power spectra yield $D \approx 1.6$. However, when we apply the method to a larger portion of the DEM we observe a large variation in the resulting values of fractal dimensions. Figure 2.41 illustrates this point. When applied to a significant number of transects without averaging, the estimation of D by Matsushita and Ouchi's method (as well as all other methods tested, including the spectral method and the variation method described in this chapter) varies in the wide range $1.32 \leq D \leq 1.66$.

Figure 2.41 shows the fractal dimensions plotted as a function of the average elevation of the transect. We observe on one side the variability of the fractal dimension D and on the other the lack of clear correlation with

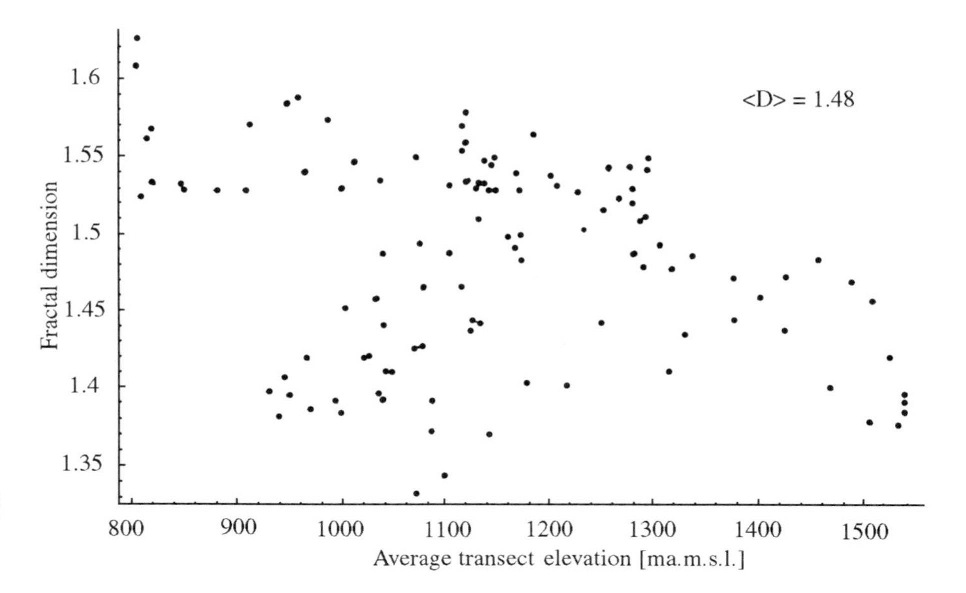

Figure 2.41. Fractal dimensions of profiles of elevations of the Fella River Basin, as a function of the average transect elevation [after Rigon, 1994].

the average elevation. This again suggests a more complex picture than that described by the scaling assumption postulated by a single fractal dimension.

2.7.5 Self-Affine Characteristics of Width Functions

The width function, $W(x)$, of a river basin gives the number of links in the network at a flow distance x from the outlet. This concept, of fundamental hydrologic importance, was discussed in Section 1.2.7. It is important to emphasize the fact that the distance is measured along the network rather than, say, radially from the outlet. Figure 1.11 gives examples of the width function of different basins, where a great variety of fluctuations and shapes are observed.

Figure 2.42 shows the spectra of different width functions (apart from vertical shifting), allowing for comparison of the slopes of the log-log fits [after Marani et al., 1994]. Figure 2.42 suggests that width functions are indeed approximately described by a single, average fractal dimension. The exponents ζ vary in the rather narrow range 1.7–1.9 with very good fit in a linear plot on log-log paper. This consistency translates into a similar conclusion for local box-counting fractal dimensions that are thus found to be in the range $1.55 \leq D \leq 1.65$.

It should be noted that although a linear approximation of the power spectra in the log-log plane represents the data quite well at high frequencies, departures from it are observed at the lowest frequencies. This can be interpreted by assuming the existence of a common mechanism regulating the organization of natural drainage network at smaller scales. At the larger scales the structure is, as expected, controlled by the overall shape of the basin [Marani et al., 1994].

It is interesting to observe (Figure 2.43) the power spectrum of the width function of Peano's basin (produced by a binomial multiplicative process, as described in Section 3.2). The wider range of fractal dimensions produced by the multiplicative process results in a spectrum without power-law behavior. It thus follows that it would be artificial to analyze this process under the framework of a single fractal dimension and that a meaningful interpretation may be given only by assuming an entire hierarchy of fractal dimensions.

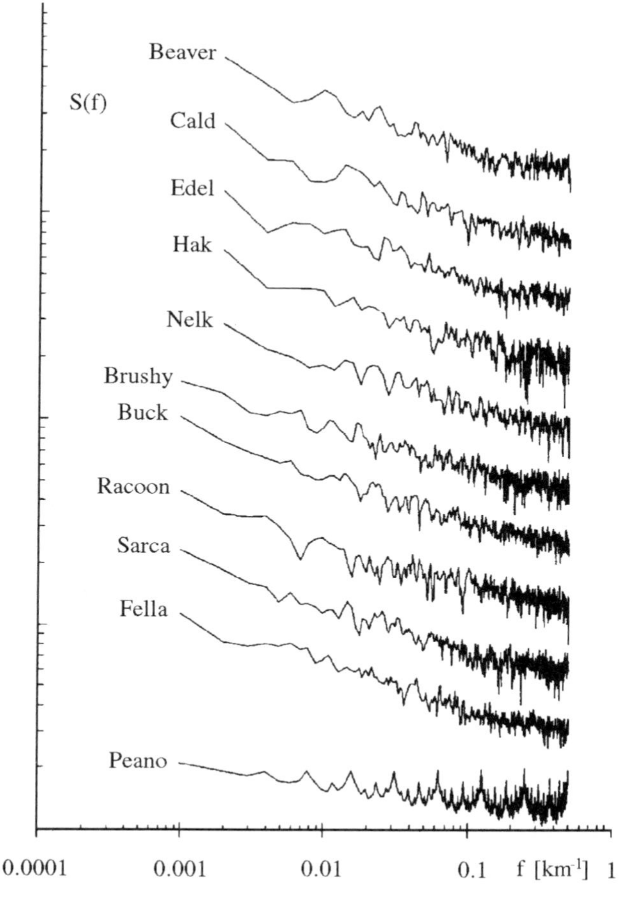

Figure 2.42. Power spectra of real width functions have a power law form with good agreement of the exponents ($1.7 \leq \zeta \leq 1.9$) at high frequencies [after Marani et al., 1994].

2.7.6 Other Self-Affine Characterizations

Vertical profiles of topography (transects) and width functions constitute extremely important characteristics of river basins and are expressed in the form of records of an univalued function sampled at regularly spaced intervals. The horizontal projections of river channels and basin boundaries are fundamentally different from transects in the sense that they are curves, which are not necessarily univalued functions. Matsushita and Ouchi [1989] developed an algorithm to study this type of curve, which we now present and then apply to the study of river channels and basin boundaries.

Let us consider a curve in which the smallest available resolution is l (defined as the unit length). In the case of a digital elevation map, l is the pixel size. Consider two arbitrary points P_i and P_{i+N} separated by N units (i.e., by a distance Nl) along the curve, and calculate the variance of the x and y coordinates of the points between P_i and P_{i+N}:

$$X^2 = \frac{1}{N}\sum_{j=i}^{i+N}\left(x_j - \bar{x}_{i,i+N}\right)^2 \qquad Y^2 = \frac{1}{N}\sum_{j=i}^{i+N}\left(y_j - \bar{y}_{i,i+N}\right)^2 \tag{2.158}$$

where

$$\bar{x}_{i,i+N} = \frac{1}{N}\sum_{j=i}^{i+N} x_j \quad \text{and} \quad \bar{y}_{i,i+N} = \frac{1}{N}\sum_{j=i}^{i+N} y_j \tag{2.159}$$

Iterations of the calculation for many pairs of points at different distances N, yield, in the case of some curves, a scaling behavior:

$$X^2 \propto N^{2\nu_x} \qquad Y^2 \propto N^{2\nu_y} \tag{2.160}$$

If the scaling exponents ν_x and ν_y in Eq. (2.160) are the same, we have a self-similar fractal with fractal dimension $D = 1/\nu_x = 1/\nu_y$. If ν_x and ν_y are different, then we have a self-affine fractal because the scaling behavior is anisotropic [Matsushita and Ouchi, 1989]. The Hurst coefficient H is given by the ratio

$$H = \frac{\nu_y}{\nu_x} \tag{2.161}$$

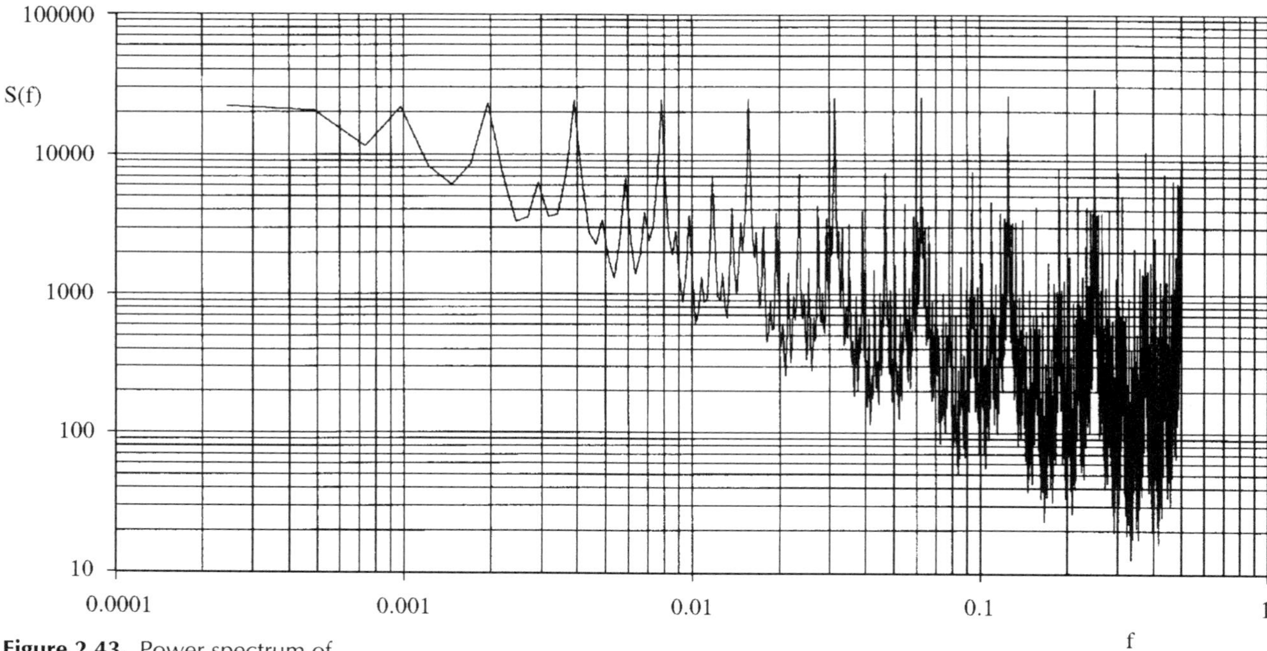

Figure 2.43. Power spectrum of the width function of Peano's basin (i.e., of a binomial multiplicative process of parameter $p = 0.25$. See Section 3.2 for details).

Methods commonly used to measure fractal dimensions, like box-counting, are not able to identify this kind of anisotropy.

The method may be applied to a discretization of a curve obtained by sampling its values in N points P_i of coordinates (x_i, y_i) (Figure 2.44) via the following sequential procedure:

- An (arbitrary) pair of points A and B, at distance ℓ measured along the curve, is chosen.
- One computes

$$X_\ell^2 = \frac{1}{N_\ell} \sum_{i=1}^{N_\ell} (x_i - \bar{x})^2 \tag{2.162}$$

$$Y_\ell^2 = \frac{1}{N_\ell} \sum_{i=1}^{N_\ell} (y_i - \bar{y})^2 \tag{2.163}$$

  where $\bar{x}$, $\bar{y}$ are the averages of the values within the span ℓ and N_ℓ is the number of elements contained in the span of the curve of length ℓ from A to B (Figure 2.44).
- The above operation is repeated for a suitably large number of intervals obtained by varying the endpoints $A = P_0$ and $B = P_{N_\ell}$. We observe that, in principle, $N(N-1)/2$ partitions are possible.
- The variances X_ℓ^2, Y_ℓ^2 are ranked according to their ℓ value. Then a value of $\Delta\ell$ is chosen, and an average is taken on the variances obtained for intervals of length ℓ in the range $(n-1)\Delta\ell < \ell < n\Delta\ell$.
- The functions $X_\ell^2 \approx \ell^{\nu_x}$ and $Y_\ell^2 \approx \ell^{\nu_y}$ are plotted in a log-log diagram. The exponents ν_x, ν_y are then estimated.

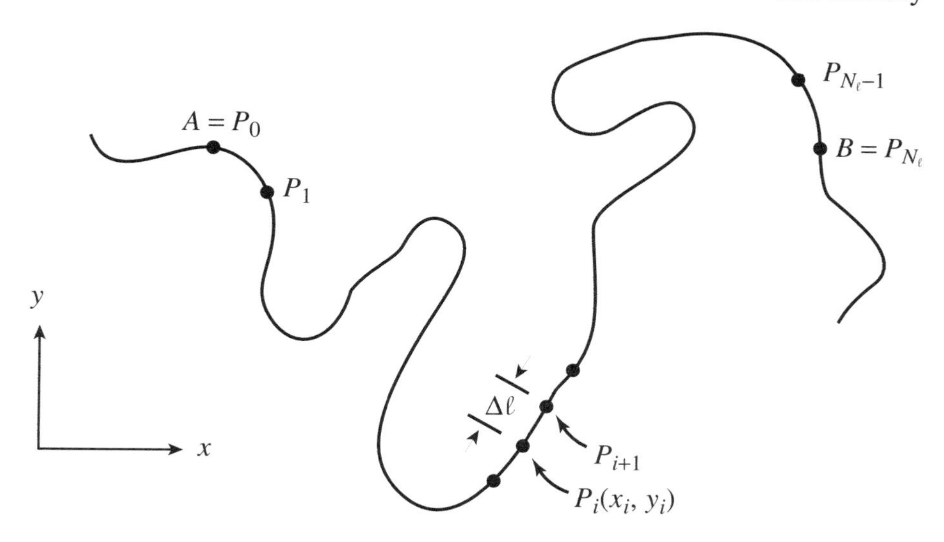

Figure 2.44. Fractal measures with Matsushita and Ouchi's method.

- If $\nu_x = 1$, then $\nu_y = H$. If instead $\nu_x = \nu_y$, the set is isotropic and self-similar and the box-counting dimension D – see the discussion below – is $D = 1/\nu_x = 1/\nu_y$. Otherwise, the ratio of the two exponents can be computed although it will not strictly represent a fractal dimension.

The following reasoning provides the basis for the estimation procedure. Let $Y_\ell^2 = \frac{1}{\ell}\sum_{i=1}^{\ell}(y_i - \bar{y}_\ell)^2$ be the estimation of Y^2 based on a particular span of size ℓ. The notation $\bar{y}_\ell$ emphasizes the fact that the average of y is taken within a span of curve of size ℓ. Consider the M intervals delimited by endpoints separated by the same number, ℓ, of points. Then one may average the resulting M values of Y_ℓ^2, obtaining $\langle Y_\ell^2 \rangle$ as

$$\begin{aligned}
\langle Y_\ell^2 \rangle &= \frac{1}{M\ell}\sum_{j=1}^{M}\sum_{i=j}^{j+\ell}(y_i - \bar{y}_\ell)^2 \\
&= \frac{1}{\ell}\sum_{i=1}^{\ell}\left(\frac{1}{M}\sum_{j=1}^{M}(y_j - \bar{y}_\ell)^2\right)_{i=\text{const}} \\
&= \frac{1}{\ell}2\sum_{i=1}^{\ell}C_y(i) \approx \frac{1}{\ell}\sum_{i=1}^{\ell} i^{2H} \approx \frac{\ell^{2H+1}}{\ell} = \ell^{2H}
\end{aligned} \tag{2.164}$$

(where i indexes the arbitrary point P_i (Figure 2.44) and $C_y(i)$ is the co-variance of lag i of y). We have employed the fact that the covariance of a fractal object scales as ℓ^{2H} (see Eq. (2.122), where H is, as usual, Hurst's exponent $(0 < H < 1)$).

Given a distance ℓ and N points in the discretization, there are N_ℓ sample points within such distance. Therefore the number of intervals where the evaluation of the mean is performed depends on the distance ℓ. Clearly, at distances close to N, the number of intervals on which the global variances (Eq. (2.164)) are computed is significantly reduced and the estimation loses statistical significance. On the other hand, intervals separated by too small a distance may not yield a significant contribution to the variance because the stability of the method depends also on the number of points M on which every individual variance is computed. Following Matsushita and Ouchi [1989], we deem that a reliable criterion is to restrict the range of feasible distances ℓ, for a given size N of the sample data, to a meaningful subset, like, for example, $10 < \ell < N/2$.

Finally, the accuracy of the method is a function of the number of sampled points. A test of the goodness of estimation may be obtained by using a window of N_0, say $N_0 < 10$, adjacent points and by letting such a window move, sampling the entire record. By applying the above procedure, one obtains a local estimate of the dimension that should not vary at all if the points indeed lie on a perfect straight line in the log-log plot. The variations in the dimensions resulting from the application of the constrained procedure described above are interpreted as an index of the goodness of the estimation [Rigon, 1994].

Notice that the total number of operations required to estimate the fractal dimension with this method is $O(N^3)$. Such a number may be reduced by sampling the intervals at random, although it is advisable to check that the samples remain statistically meaningful.

A major advantage of the method proposed by Matsushita and Ouchi [1989] is that it allows the analysis of closed or unconnected curves.

To illustrate the procedure described above, let us use the trace of a simple Brownian motion, which was seen before to be a fractional Brownian motion with Hurst coefficient $H = 0.5$. In this illustration, the particle moves one unit up or down with equal probability at each time step. Figure 2.45 shows a realization of such motion. Figure 2.46 shows the scaling of X^2 and Y^2 versus N. The slopes of the lines give, from Eq. (2.160), $\nu_x = 1.0$ and $\nu_y = 0.5$, as predicted by the theory of fractional Brownian motion [Feder, 1988]. These exponents indicate that a rescaling of the horizontal axis by a factor $1/b$ needs a rescaling of the vertical axis by a factor $1/\sqrt{b}$ in order to leave the distribution invariant in a statistical sense.

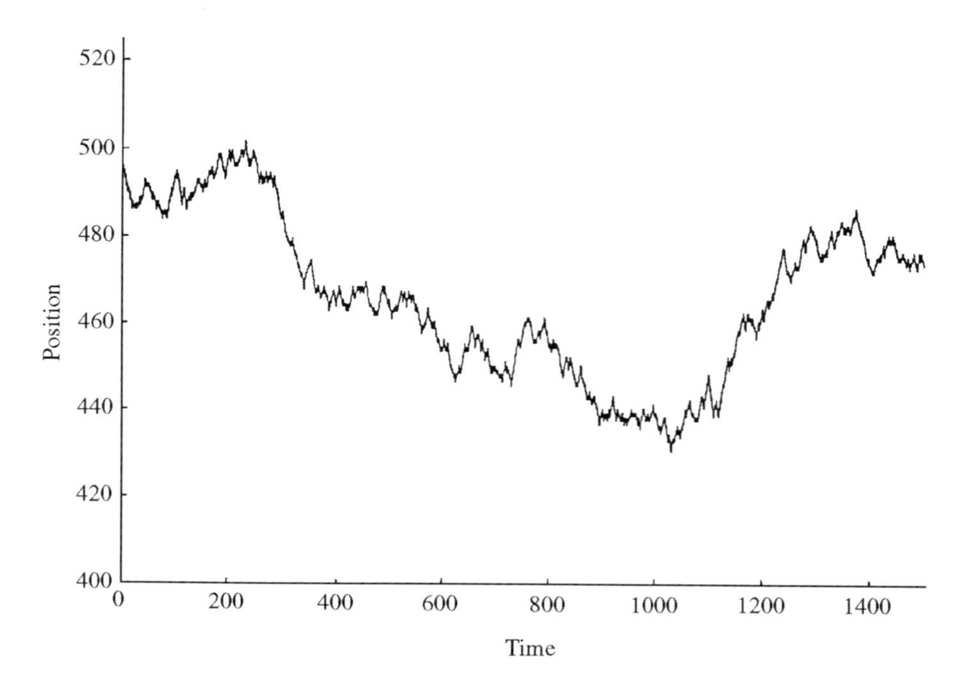

Figure 2.45. Trace of a Brownian function with unit jumps. The horizontal axis represents time and the vertical axis represents position [from Ijjasz-Vasquez, Bras, and Rodriguez-Iturbe, 1994].

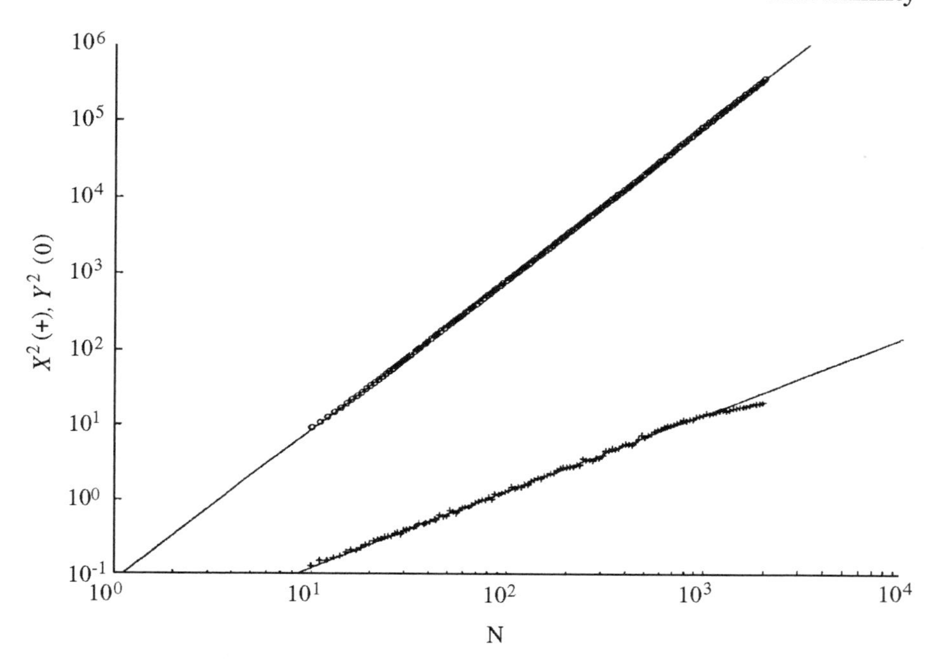

Figure 2.46. Self-affine scaling of the trace of a Brownian motion: (o) scaling of X^2; (+) scaling of Y^2 [from Ijjasz-Vasquez et al., 1994].

Although it is possible to define a fractal dimension $D = 2 - H$, this dimension is not the local box-counting dimension studied in Section 2.7.3.

2.7.7 Self-Affine Scaling of Watercourses

Ijjasz-Vasquez, Bras, and Rodriguez-Iturbe [1994] studied the scaling properties of the main channel in a river basin using the previously described procedure of Matsushita and Ouchi [1989]. From DEMs, streams were identified as those pixels with contributing area greater than a threshold value (see Chapter 1). The main channel is identified by traveling upward from the outlet. At each bifurcation the path with the largest contributing area was followed. As an example, Figure 2.47 shows the trace of the main watercourse for the Fella River Basin (northern Italy) described in Chapter 1 (see also Rodriguez-Iturbe et al. [1994]).

Two points should be noted when analyzing the scaling properties of channels. First, the overall flow direction of rivers is usually not oriented along the horizontal axis, as was the case in the Brownian motion simulation in Figure 2.45. Therefore it is necessary to repeat the analysis using different orientations to find the principal axis of anisotropy. Second, in order not to overestimate the scaling and roughness of self-affine curves, linear trends have to be removed [Malinverno, 1989]. The method of Matsushita and Ouchi [1989] handles linear trends well as long as this trend is not significantly altered. It is not difficult to observe an overall trend in actual rivers, but sometimes such trends shift near the top of the basin as tributaries of similar size merge to form a larger channel. In Section 2.7.8 we will describe the procedure of Ijjasz-Vasquez et al. [1994] for detrending the whole river. Different methodologies for this purpose do not appear to have any significant effect on the estimation of the self-affine properties of watercourses.

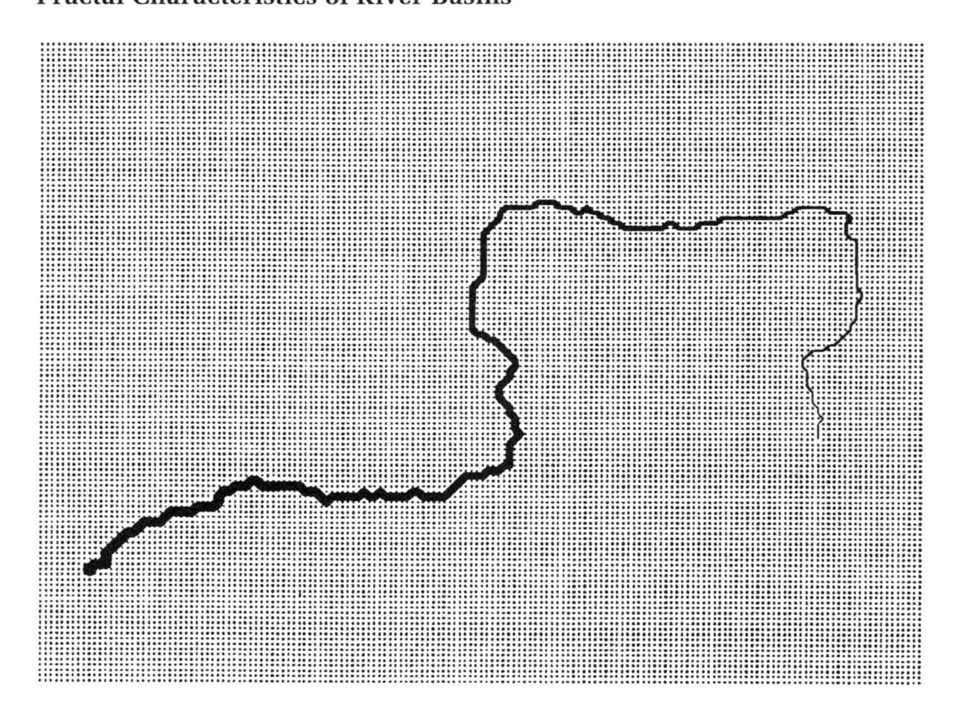

Figure 2.47. Plot of the main stream of the Fella River.

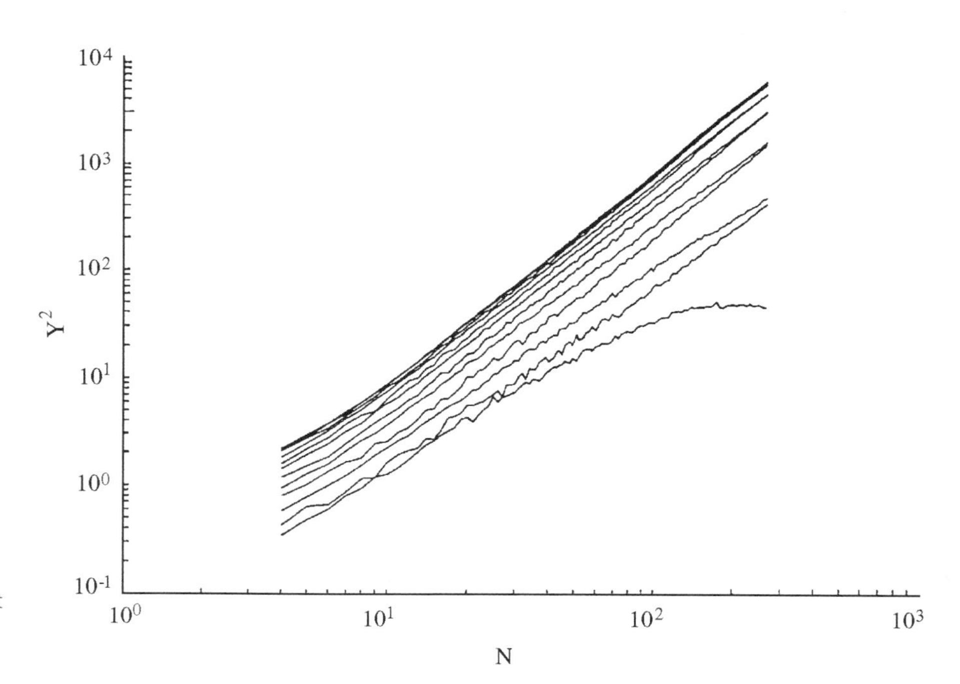

Figure 2.48. Scaling of Y^2 of main channel of the East Delaware River Basin for different axis orientations [from Ijjasz-Vasquez et al., 1994].

Ijjasz-Vasquez et al. [1994] first estimated the scaling of X^2 and Y^2 for different axis orientations. An example is shown in Figure 2.48 for the case of the main channel of the East Delaware River Basin. It gives the log-log scaling of Y^2 versus N for different axis oriented at every 15°. If the watercourse was a self-similar object, the slopes would be the same for different orientations. However, this is not the case and Figure 2.49

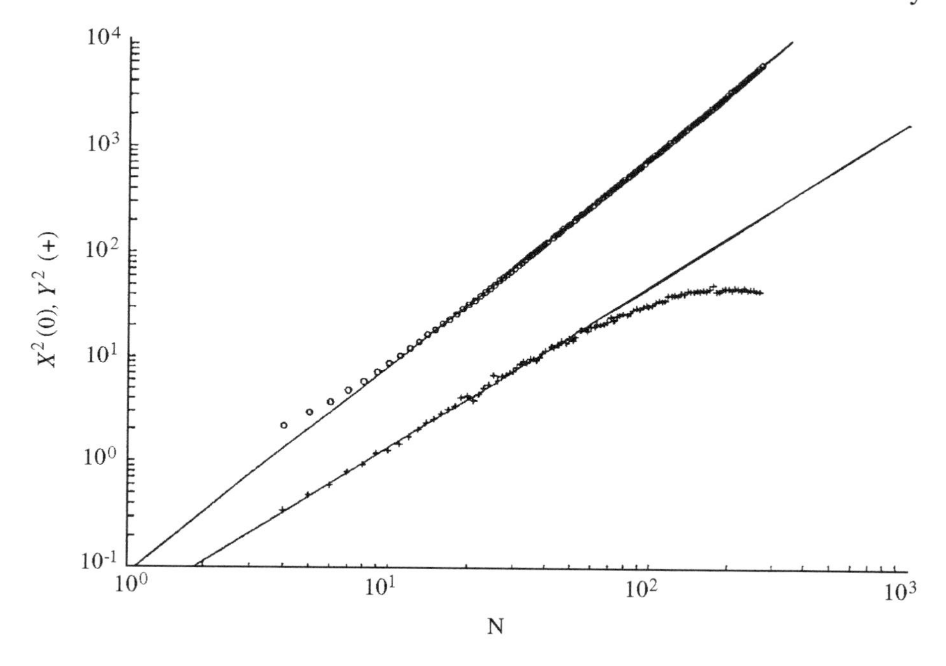

Figure 2.49. Scaling of X^2 and Y^2 with the appropriate anisotropy axis for main channel of the East Delaware River Basin (New York) [from Ijjasz-Vasquez et al., 1994].

Table 2.7. *Characteristics and self-affine scaling of main channels from the eight basins analyzed by Ijjasz-Vasquez et al. 1994*

Basin	Location	Area (km^2)	ν_x	ν_y
Buck Creek	CA	606	0.98	0.73
East Delaware River	NY	933	0.99	0.79
Schoharie Creek Headwaters	NY	98	0.97	0.75
Big Creek	ID	147	0.99	0.71
Racoon Creek	PA	448	1.00	0.76
St. Joe River	ID,MO	2834	1.00	0.76
Schoharie Creek	NY	2408	1.00	0.76
Beaver Creek	PA, OH, MN	1223	1.00	0.75

presents the scaling of X^2 and Y^2 for the principal anisotropy axis in the case of the main channel of the East Delaware River Basin. The slopes give $\nu_x = 1.0$ and $\nu_y = 0.75$, showing that we do indeed have a self-affine object. The principal anisotropy axes are located, not surprisingly, along the overall direction of the channel and perpendicular to it.

Table 2.7 shows the scaling exponents of the main channels of eight different basins located throughout the United States. In all cases they were found to exhibit self-affine scaling with approximately $\nu_x = 1.0$ and $\nu_y = 0.75$. Given that these rivers are located in regions where the relief can be appropriately measured with DEMs, it is possible that in regions where meandering is a dominant feature of the river the scaling parameters will change. Related work on fractal dimensions of meandering rivers was presented by Nikora [1994]. Different tools to identify the river courses were used in such cases.

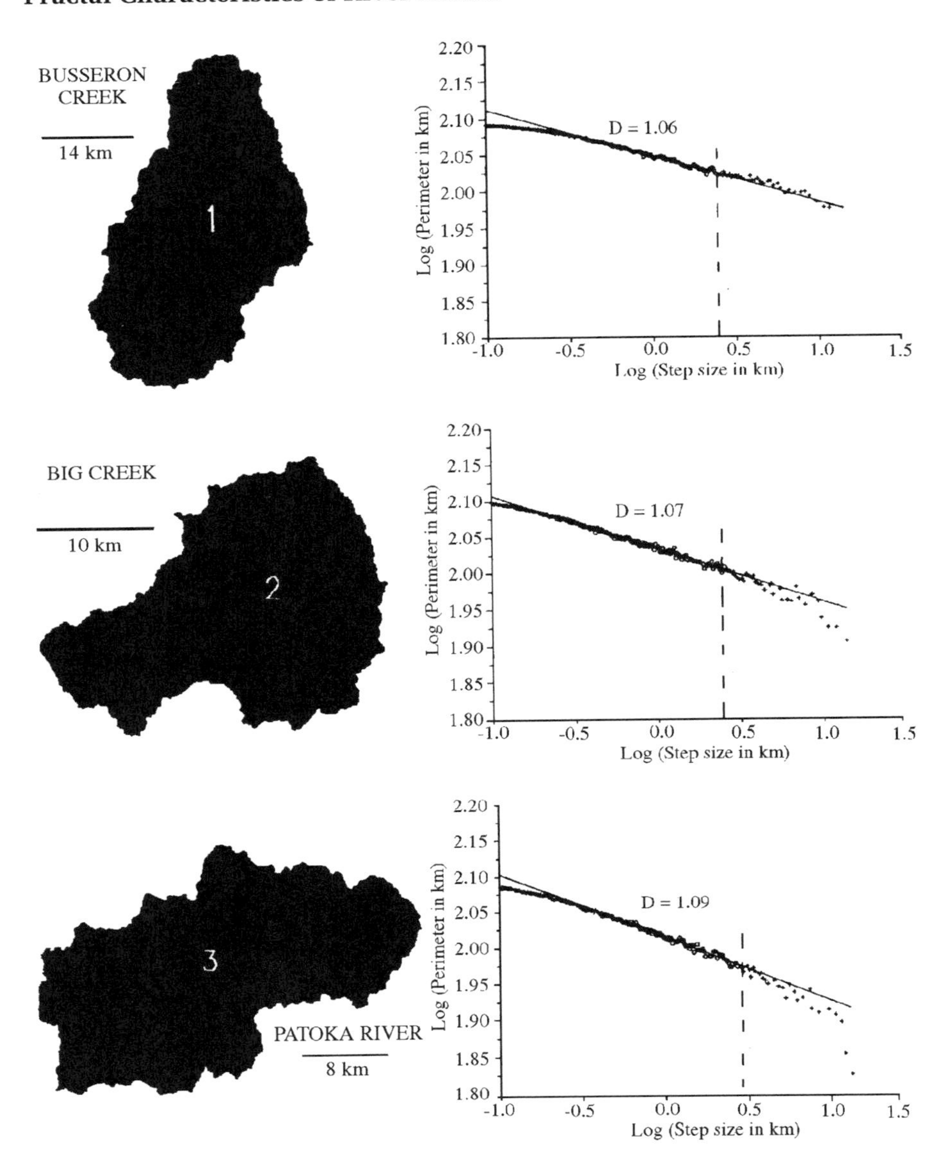

2.7.8 Self-Affine Scaling of Basin Boundaries

It has been suggested that basin boundaries and river courses are in essence mirror images of each other. Also, it has been shown that, under certain general assumptions, the topological characteristics of the channel and ridge networks are identical [Werner, 1991]. It is of interest then, to compare the scaling characteristics of basin boundaries and river courses.

Breyer and Snow [1992] carried out the estimation of the fractal divider dimension of several drainage basin perimeters in southern Indiana and other regions of the United States. The values they obtained were in the range 1.06–1.12. An example of the results is shown in Figure 2.50.

Numerous models, which produce boundaries similar to those of basins, have been shown to have self-affine behavior. Examples include the Eden model [Eden, 1961; Family and Vicsek, 1991] and ballistic deposition

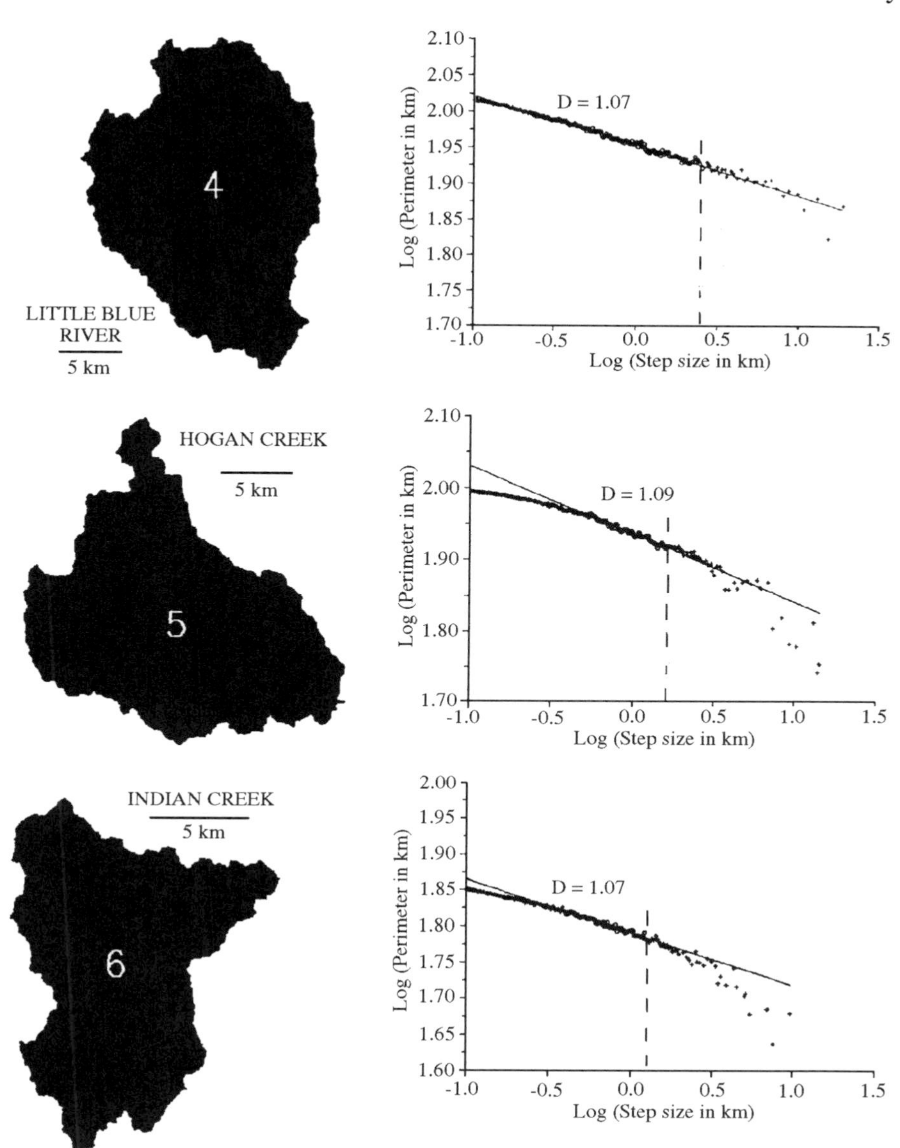

Figure 2.50. Examples of the estimation of fractal divider dimension for drainage basin perimeters [after Breyer and Snow, 1992].

models [Meakin, 1991]. In all these experiments the boundary has a clear anisotropy axis.

Ijjasz-Vasquez et al. [1994] studied the scaling behavior of the tortuosity of basin boundaries normal to the basin domain. In previous studies of interfaces, the direction normal to the boundary is clearly defined. In basin boundaries, on the other hand, in order to perform the analysis the boundary was divided into sections; the linear trend was taken out from these sections and then the sections were put back together as a single detrended curve to which the self-affinity analysis can be applied. In Figure 2.51 the circles show the scaling behavior of the oscillations normal to the basin domain taken along the boundary for the case of the East Delaware River Basin. The slope gives a value of $\nu_y = 0.75$. The individual boundary

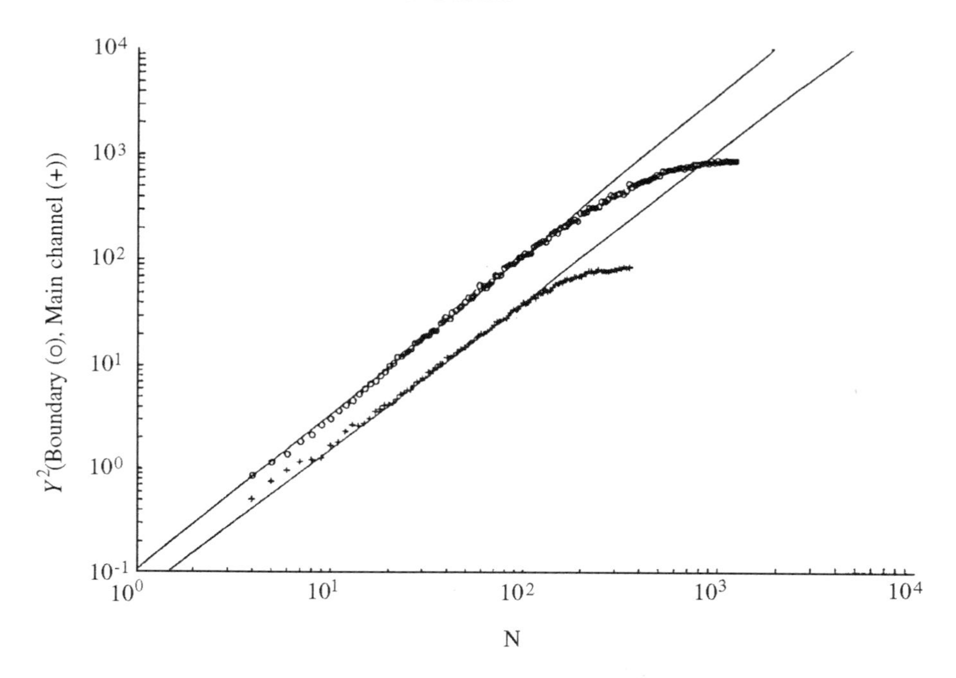

Figure 2.51. Y^2 scaling (o) of detrended boundaries of the East Delaware River Basin (New York) ($\nu_y = 0.75$); (+) detrended main channel ($\nu_y = 0.74$) [after Ijjasz-Vasquez et al., 1994].

Table 2.8. *Self-affine scaling of detrended main channels and boundaries from eight different basins*

Basin	ν_y Boundary	ν_y Main Channel
Buck Creek	0.74	0.77
East Delaware River	0.75	0.74
Schoharie Creek Headwaters	0.79	0.72
Big Creek	0.79	0.74
Racoon Creek	0.74	0.77
St. Joe River	0.75	0.79
Schoharie Creek	0.76	0.75
Beaver Creek	0.75	0.77

Source: Ijjasz-Vasquez et al., 1994.

sections detrended for this analysis were 200 pixels in length. Similar results were obtained with sections of different lengths.

Figure 2.51 also presents the results of the analysis for the entire main channel of the East Delaware River Basin after detrending. Table 2.8 presents the values of ν_y for the main channels and the boundaries of the nine basins analyzed.

The results in Table 2.8 indicate that the self-affine scaling of rivers and boundaries is in fact very similar. However, at a given scale boundaries visually seem to have larger oscillations than channels, and one would tend to say that they look rougher. One can appreciate how this effect manifests itself in Figure 2.51. Although the slopes are similar for the boundary and the main channel, notice that the scaling of the latter breaks at a smaller distance N between points along the curve. This behavior is observed for all the basins analyzed. What this means is that even if we take points farther and farther apart along the channel, the variance does not increase because oscillations of larger magnitude do not appear. On the other hand, such oscillations do appear on the boundary of the basin and the linear portion of the scaling is more extended. Thus basin boundaries have similar self-affine characteristics to watercourses, but the linear log-log scaling is more extended in boundaries as a result of self-affine oscillations of larger magnitude. So, although the scaling properties are similar, boundaries and rivers cannot be considered mirror images of each other in their planar configurations.

2.8 Transects, Contours, Watercourses, and Mountain Ridges as Parts of the Basin Landscape

This section will briefly review the fractal characteristics of different elements of the landscape under the perspective that they should all fit together in a coherent framework. Absent among the landscape components we have studied this far are the hillslopes of a basin, and their study and modeling are postponed until Chapter 6 of this book. At this point, our objective is to connect the characteristics of the different elements we have studied in this chapter and to provide a natural motivation for the analysis of the multifractal characteristics of river basins, which is the subject of Chapter 3.

The landscape of a river basin is a complex rough surface whose characterization may be attempted in several different ways. It is convenient to consider a field of elevations $z(\mathbf{x})$ defined at every point $\mathbf{x}$ in a subset of R^2. If $z(\mathbf{x})$ has the property defined by Eq. (2.118), namely

$$z(\mathbf{x}_1 + \lambda \mathbf{e}) - z(\mathbf{x}_1) \stackrel{d}{=} \lambda^H [z(\mathbf{x}_2 + \mathbf{e}) - z(\mathbf{x}_2)] \tag{2.165}$$

(where $\mathbf{e}$ is a unit vector, λ is a separation distance, $\mathbf{x}_{1,2}$ are two arbitrary points, and the exponent H is the Hurst exponent), then it is said to be simple scaling. There is a unique local, box-counting, fractal dimension D throughout the entire field that characterizes its scaling characteristics. The unique dimension D is related to the Hurst exponent via the relationship $D = 3 - H$. It is important to note that a simple scaling field is postulated to have constant statistical characteristics of its increments with respect to position and orientation. Eq. (2.165) holds, in fact, for any choice of $\mathbf{x}_{1,2}$ and $\mathbf{e}$.

The transects of the simple scaling landscape defined by Eq. (2.165) are records of the intersections of the landscape with vertical planes. Thus they represent the altitude variations of a hiker following any straight-line path at a constant speed in the xy plane and constitute a fractional Brownian motion [Voss, 1986]. If the hiker travels a distance Δr in the xy plane ($\Delta r^2 = \Delta x^2 + \Delta y^2$), the typical altitude variation Δz is given by

$$\Delta z \propto \Delta r^H \tag{2.166}$$

The intersection of a vertical plane with the simply scaling surface $z_H(\mathbf{x})$, where the subscript has been added to emphasize its self-affine character, is thus a self-affine fBm trace characterized by H and has a box-counting fractal dimension

$$D = 2 - H \qquad \text{for transects} \tag{2.167}$$

Another manner in which to characterize the basin landscape is through its intersection with a horizontal plane. These intersections will produce the topographic contours or isolines, where again in the case of a monofractal, simple-scaling surface $z_H(\mathbf{x})$, the resulting fractal dimension is one less than the surface itself:

$$D = 2 - H \qquad \text{for contours} \tag{2.168}$$

However, in the case of contour lines and coastlines of the landscape, the coordinates x and y are equivalent and the family of curves (possibly disconnected) we obtain from the intersection with a horizontal plane are self-similar [Voss, 1988]. Thus in Eq. (2.165), H has no structural significance with respect to the contour lines themselves but is instead related to the structure of the simple-scaling, self-affine landscape from which the contour lines are derived.

All the equations of this section are only valid for the case of a simple-scaling or monofractal landscape characterized by the exponent H in Eq. (2.165).

The analysis of Turcotte [1992] for transect records in different parts of Oregon (Table 2.6) yielded values of the local box-counting dimension between 1.4 and 1.5, estimated on the basis of the double logarithmic slope of the power spectra. On the other hand, Turcotte [1992] also reports results for topographic contour in different parts of the United States. The contours are self-similar, with fractal dimensions between 1.1 and 1.2 estimated by the divider method. One should remember that the two analyses mentioned above are for different regions, but it is interesting that in both cases there is little spread in the particular values of D among transects and in the values of D among contours. Nevertheless, the transects suggest a value of H for a self-affine simple-scaling elevation field of around 0.55, whereas the contours suggest a value of H near 0.85. This discrepancy casts doubts on the assumption that generally a fluvial landscape is a self-affine monofractal surface and suggests that it may indeed be an object to which Eqs. (2.165) through (2.168) are not applicable.

Matsushita and Ouchi [1989] carried out analyses on the Mount Yamizo area in Fukushima Prefecture, Japan. Working on 1:25,000 scale maps of the area, they studied one transect and one contour line (700 m) using the method described in Section 2.7.6. Figures 2.52 and 2.53 show the related analyses.

The transect record for the dependence of the standard deviations of horizontal and vertical coordinates x and z on the curve length N between many pairs of points yields the self-affine exponents $\nu_x = 1$ and, on the average, $\nu_z = 0.59$ in Eq. (2.160). This implies a self-affine process with $H = 0.59$, similar to the values obtained by Turcotte [1992] for different regions in the United States.

The 700-m contour line of the Mount Yamizo area (Figure 2.53) proved to be self-similar with $\nu_x = \nu_y = 0.73$, $D \approx 1.37$. Notice that the dotted line indicates the position of the transect in Figure 2.52. Using $H = 0.59$ from the transects in Eq. (2.167), we get $D = 1.41$, which compares well with the value obtained from the transects. Thus the hypothesis of a self-affine single-scaling landscape cannot be rejected for this region on the basis of the preliminary analyses of Matsushita and Ouchi [1989].

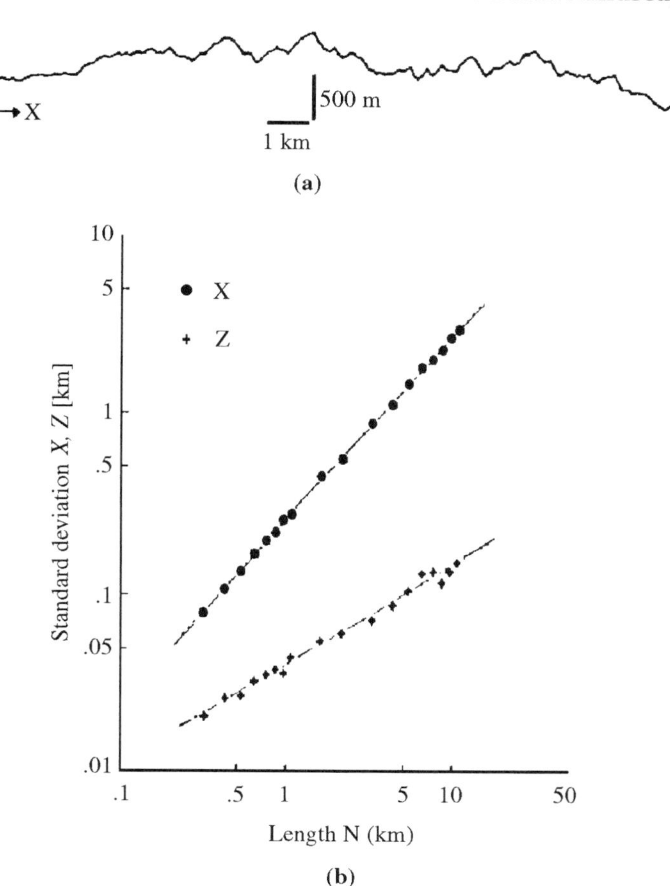

Figure 2.52. (a) Transect profile near Mt. Yamizo in Fukushima Prefecture, Japan, and (b) dependence of the standard deviations of x and z on the curve length of the curve shown in (a) [after Matsushita and Ouchi, 1989].

The purpose of the above examples is mainly to point out the apparent contradictions that may surface if a *multifractal* process is analyzed under a single-scaling monofractal framework. There is some evidence that river basin landscapes are indeed multifractal. This evidence and the appropriate techniques of analysis will be presented in the next chapter. These include the study of the elevation field itself, $z(\mathbf{x})$, as well as the intersections of the landscape with vertical and horizontal planes.

Regarding mountain ridges, basin, and subbasin boundaries, they are not intersections of the landscape with planes, and thus the most interesting part of their study is related to their planar projection. As shown in Section 2.7.8, the planar projections may be considered self-affine curves with an H value of approximately 0.75, although again this value could change for different boundaries in the same basin. The fact that the power law for the contributing area at any randomly chosen point, $P[A \geq a] \propto a^{-\beta}$, holds very well among basins of very different characteristics with a small range of variation around $\beta \approx 0.43$, and furthermore that this law and the value of β can be explained assuming that basin boundaries are self-affine fractional Brownian motion with $H \approx 0.75$, leads one to believe that the range of variation in H for boundaries is indeed small.

With respect to the channel network itself, we have seen that it presents self-similar characteristics on its planar projection when it is seen as a

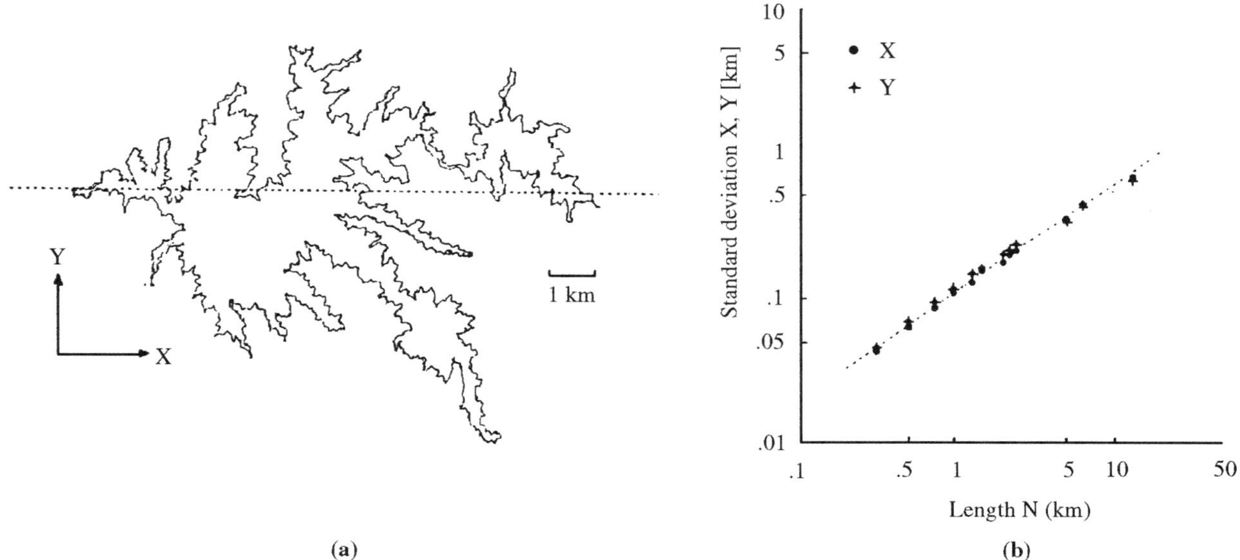

Figure 2.53. (a) Contour line at 700 m taken from the 1:25,000 scale map of Mt. Yamizo area. (b) Dependence of the standard deviations of two horizontal coordinates on N between many pairs of points in (a) [after Matsushita and Ouchi, 1989].

whole. One should distinguish between a fractal dimension related to the sinuosity of individual channels, $D_l \approx 1.1$ or 1.2, and a fractal dimension related to the fractal branching characteristics of the network, $D \approx 2$. In the case of a specific watercourse, and especially for the main channel in the basin, the scaling characteristics tend to correspond to a self-affine process with $H \approx 0.75$, revealing a preferred planar direction. This would correspond to a local box-counting dimension of approximately 1.25. It is appropriate to point out that the analysis by Tarboton et al. [1989] reported in Figures 2.21 and 2.22 involves the whole drainage network. In the case of the main course, the object is self-affine and thus, for instance, the explanations of Hack's law based on a sinuosity or divider's fractal dimension of approximately 1.15 are to be taken with caution. The three-dimensional structure of the network is specially interesting; we have seen that there are clear signals of multiscaling behavior in the scaling of link slopes with contributing areas.

Chapters 3 and 6 will study how we can generally expect that different transects will correspond to self-affine records with different values of H. Similarly, we will see that we can expect some variation in the fractal dimension of different self-similar contour lines. The dynamics that leads to the fractal characteristics of the landscape and its embedded network, both in its planar projection and its three-dimensional structure, is a fundamental part of this book. Its study is covered in Chapters 4, 5, and 6.

2.9 Hack's Law, the Self-Affinity of Basin Boundaries, and the Power Law of Contributing Areas

2.9.1 Does Hack's Law Imply Elongation?

A river basin is an anisotropic system defined by a longitudinal typical length $L_\parallel$ (which we will identify with the linear size of the system) and a typical transverse length $L_\perp \leq L_\parallel$ (see Figure 2.54). We will call $L_\parallel$ the diameter of the basin and $L_\perp$ its width.

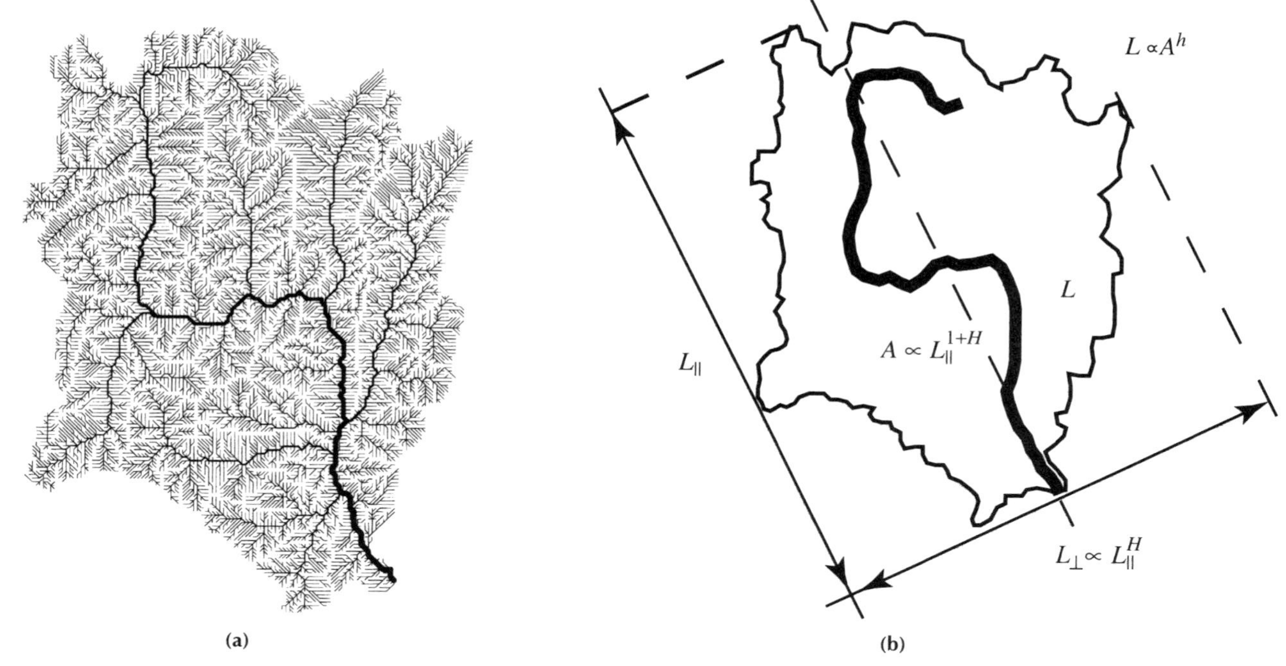

Figure 2.54. A sketch of a river basin. $L_{\|}$, $L_{\perp}$, and L are the longitudinal, transverse, and mainstream lengths, respectively. [after Maritan et al., 1996a].

We shall assume that the width scales as

$$L_{\perp} \propto L_{\|}^{H} \qquad 0 \leq H \leq 1 \tag{2.169}$$

and we will call basins self-affine if $H < 1$ and self-similar if $H = 1$. Here H is still called the Hurst exponent because of the similarity with its analog in the fractional Brownian motion context.

Eq. (2.169) postulates that basin boundaries are self-affine *curves* for which $L_{\|}$, $L_{\perp}$ can be seen as diameter and width, respectively [Feder, 1988]. It is a fairly general property of self-affine boundaries that their embedded area, say A, is related to the diameter and width via

$$A \propto L_{\|}^{1+H} \tag{2.170}$$

valid whatever the self-affine scalings of the boundaries [Tricot, 1995].

A summary of the experimental values obtained for L and $L_{\|}$ for a sample of DEMs is shown in Table 2.9.

Hack's [1957] law was introduced in Section 1.2.3 and further discussed in Section 1.3.3 with reference to its implications for the random topological model. We briefly recall here some of the main aspects of these discussions.

Hack's law relates the length of the longest stream L in the drainage region, measured to the divide, with the drainage area of the basin, say A, as $L \sim A^h$, where we have seen that the commonly accepted values for the exponent are in the range $h = 0.56 - 0.6$.

Hack's equation can be rewritten as

$$\frac{A}{L^2} \propto A^{-0.15 \pm 0.05} \tag{2.171}$$

thus implying that catchments of all sizes are not entirely similar in shape.

Table 2.9. *A summary table of relevant features of the DEMs used in the analysis of Hack's Law*

Name	Location	Area [km^2]	L [km]	$L_{\parallel}$ [km]
Salt Creek	Idaho	2091	97.8	66.8
Guyandotte	W. Va.	2088	145.1	75.8
Tug Fork	W. Va.	1442	97.8	66.9
Salmon River	Calif.	1936	94.9	64.5
Boise River	Idaho	1847	94.8	63.3
Little Coal	W. Va.	984	90.5	57.5
North Fork River	Va.	825	70.1	37.0
South Fork Smith River	Calif.	606	52.3	28.3
Dry Fork	W. Va.	586	63.7	41.6
Johns Creek	Ky.	484	68.8	45.7
Big Coal	W. Va.	449	56.2	40.7
Racoon Creek	Pa.	448	53.6	35.2
Pingeon Creek	W. Va.	405	49.1	35.3
Moshannon Creek	Pa.	393	49.7	33.8
Wooley Creek	Calif.	384	38.75	30.41
Clear Fork	W. Va.	333	59.0	34.9
Brushy Creek	Ala.	322	52.4	29.9
Rockcastle Creek	Ky.	310	45.9	33.5
Sturgeon Creek	Ky.	295	46.3	27.1
Edinburgh Creek	Va.	294	43.8	27.9
Reynolds	Idaho	266	36.0	30.4
Island Creek	W. Va.	260	30.5	23.0
Wolf Creek	Ky.	212	30.3	21.7
Sexton Creek	Ky.	186	33.3	23.4
Elkhorn Creek	W. Va.	190	36.5	24.6
Big Creek	Idaho	147	25.6	16.9
Indian Creek	W. Va.	148	41.0	28.3
Big Creek (Cald)	Idaho	146	21.5	16.9
Muddy Run	Pa.	141	28.8	21.0
Schoharie Creek Headwater	N.Y.	108	18.5	15.8
Pound Creek	Ky.	105	23.5	17.8
Tipton River	Pa.	64	18.8	12.5
Blackberry Fork	Ky.	53	18.0	13.3
Catskill Creek	N.Y.	33	15.4	10.9

Rather, as the area increases, A/L^2 decreases, leading to the conclusion that there is a tendency toward elongation of the larger catchments.

Mandelbrot [1983] suggested that an exponent larger than 0.5 in $L \propto A^h$ could arise from the fractal character of river channels, which causes measured length to vary with the spatial scale of the object. The argument is that the length of a stream channel is expected to increase as the power $D/2$ of its drainage area, where D is the divider fractal dimension of the stream. Thus Eq. (2.171) would be a reflection of a fractal dimension of river channels close to $D = 2 \times 0.6 = 1.2$. However, as explained in detail by Feder [1988], to derive the fractal dimension of a river from the previous argument, we need to make specific invariance assumptions about the shape of the basins.

A different interpretation of Hack's law was provided by Ijjasz-Vasquez, Bras, and Rodriguez-Iturbe [1993b] under the framework of optimal channel networks (OCNs), which are the result of searching the fluvial systems for a drainage configuration whose total energy expenditure is minimized

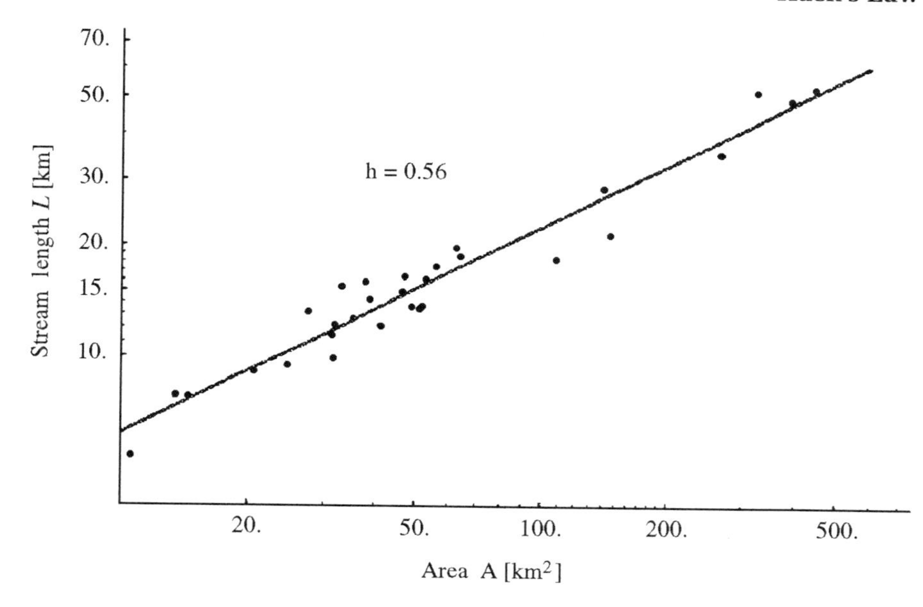

Figure 2.55. Hack's relationship L versus A for the basins in West Virginia described in Table 2.9 [after Rigon et al., 1996].

[Rodriguez-Iturbe et al., 1992b; Rinaldo et al., 1992]. This interpretation is explained in detail in Chapter 4. Thus Hack's relationship may not necessarily be a direct consequence of the fractal sinuosity of river courses but could instead result as a consequence of competition for drainage and global minimization of energy dissipation in river basins. Ijjasz-Vasquez et al. [1993b] showed that OCNs lead to relations of the type $L \propto A^h$, both when L is the straight-line distance between outlet and furthest point in the divide, say $L_{\parallel}$, and when L is the length measured along the stream channel. In the first case, h is equal to 0.57 and in the second case it is equal 0.58 (Section 4.17); both values are significantly higher than 0.5 and very close to those obtained through the mainstream length by Hack [1957] and Gray [1961]. These results imply that in the case of OCN's basins do indeed elongate with size.

Figure 2.55 shows the result of Hack's analysis carried out for the DEMs of river basins from a reasonably homogeneous region taken from the basins described in Table 2.9. Although the best estimate of the slope is close to the commonly found value of 0.56, we notice that the closer look at the morphology of the fluvial basins allowed by DEMs suggests statistical fluctuations that were not considered in the original work by Hack or by subsequent experimental work. These fluctuations are enhanced if we consider all the basins presented in Table 2.9.

Figure 2.56 shows the result obtained by plotting a modified Hack's relationship, as obtained by plotting $L_{\parallel}$ versus A for the basins analyzed in Figure 2.55. We obtain a relationship of the type $L_{\parallel} \propto A^{h'}$ with $h' \sim 0.52$. This suggests a statistical relationship of the type $L \propto L_{\parallel}^{h/h'}$ with $h/h' \sim 1.15$, which we will exploit later.

It is also clearly observed that the scatter that arises in the above relationships, neglected in earlier experimental observations, calls for some statistical interpretation, which we will provide in the framework of scaling structures.

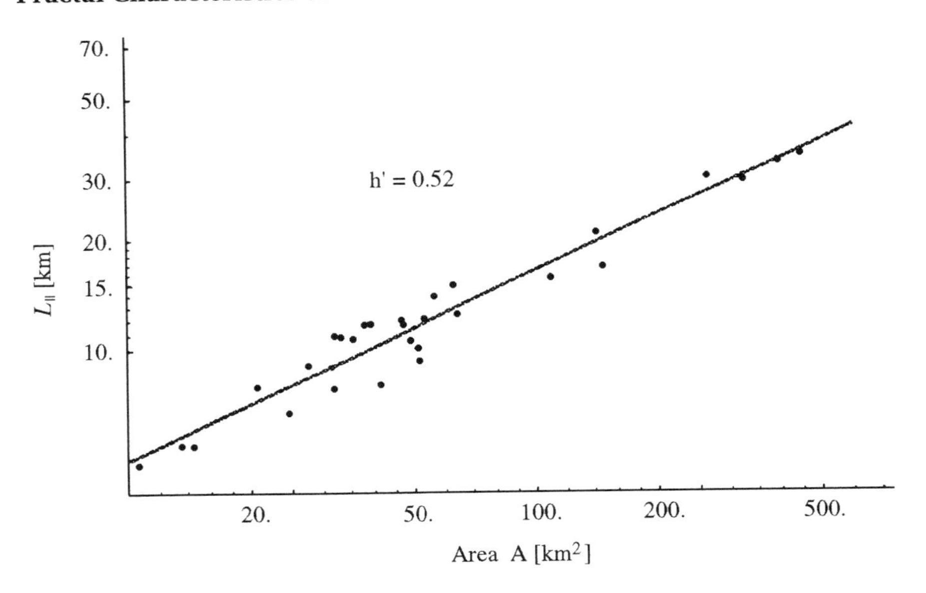

Figure 2.56. Modified Hack's relationship $L_\parallel$ versus A for the basins described in Table 2.9 [after Rigon et al., 1996].

Given the importance of Hack's law and its connection to different aspects of drainage basin structure, it is relevant to attempt a clarification in general terms of its relation to the fractal sinuosity of stream channels and to the elongation of river basins.

In any given simply connected planar figure we have $A \propto L_\parallel^2$, where A is the area and $L_\parallel$ is a characteristic length. The constant of proportionality depends on the shape of the figure. Define elongation as a property related to the magnification of the basin that describes the effect that when $L_\parallel$ increases, $A/L_\parallel^2$ decreases. The shape remains geometrically similar when $A/L_\parallel^2$ is constant, regardless of changes in $L_\parallel$.

Consider now a set of river basins and express the area of the ith basin as a function of the characteristic distance by

$$A\,(L_\parallel^i) = a(L_\parallel^i)\;(L_\parallel^i)^2 \tag{2.172}$$

where $A\,(\cdot)$ and the proportionality constant, $a(\cdot)$, have been explicitly written as functions of $L_\parallel$. If one orders the basins such that $L_\parallel^1 < L_\parallel^2 < .... < L_\parallel^i < ...$, then elongation will occur if $a(L_\parallel^1), a(L_\parallel^2), ...a(L_\parallel^i), ...$ constitutes a decreasing sequence. Thus, assuming a power law dependence of $a(L_\parallel)$ on $L_\parallel$, we will say that the series of river basins shows statistical elongation when

$$a(L_\parallel) \propto L_\parallel^{-q} \qquad q > 0 \tag{2.173}$$

and define q as the elongation exponent. We will now explicitly relate Hack's law to the elongation exponent, q, and a scaling exponent, ϕ_L, related to the divider fractal dimension of stream channels, through the following relationships:

$$L \propto L_\parallel^{\phi_L} \tag{2.174}$$

$$A = a(L_\parallel)\,L_\parallel^2 \tag{2.175}$$

where Eqs. (2.174) and (2.175) express the fractal character of mainstream

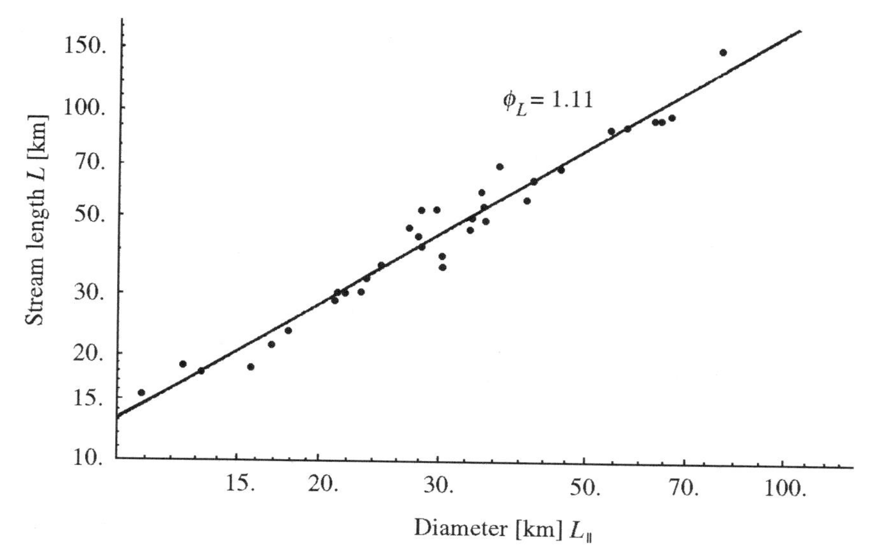

Figure 2.57. An experimental assessment of the scaling law $L \propto L_\parallel^{\phi_L}$ from the DEMs described in Table 2.9.

lengths and the general dependence of area on the square value of a characteristic length, which in this case is the longest straight line distance from the outlet to the divide, respectively.

Figure 2.57 shows an experimental validation of Eq. (2.174) for the basins given in Table 2.9. Here $\phi_L \sim 1.1$, thus matching the ratio h/h' derived from Figures 2.55 and 2.56.

Also, from Eq. (2.175) one obtains

$$a(L_\parallel) \propto L_\parallel^{\phi_L/h \, - \, 2} \tag{2.176}$$

Thus Hack's law is consistent with fractality and elongation as long as $q = 2 - \phi_L/h > 0$. If $q < 0$, the basins experience contraction and if $q = 0$, they retain similar shapes in the statistical sense. Hack's exponent, the scaling exponent of lengths, and the elongation exponent are thus related through

$$q = 2 - \frac{\phi_L}{h} \tag{2.177}$$

One observes that Eq. (2.177) imposes constraints between ϕ_L and h. Thus Hack's exponents between 0.57 and 0.6 require values of ϕ_L smaller than 1.14 and 1.2, respectively, for elongation to occur. Increasing values of ϕ_L imply contraction of river basins when increasing with size. Our experimental evidence regarding values of ϕ_L and h fulfills the condition that $q > 0$ without exceptions, implying that indeed there seems to be an elongation of river basins with size.

2.9.2 Power Law of Contributing Areas, Hack's Relationship, and the Self-Affinity of Basin Boundaries

As discussed in detail in Section 2.5.2, Rodriguez-Iturbe et al. [1992a] found that the probability distribution for the contributing area to a randomly chosen point in a drainage network follows the power law

$P[A > a] \propto a^{-\beta}$, where β has been found to be approximately in the range between 0.41 and 0.45 in basins of very different geologic conditions, climate, vegetation, and soils.

We will now present an interpretation for the values of β found in nature. The analysis is an adaptation that follows the reasoning proposed by Takayasu et al. [1988] for Scheidegger's [1967] basin model, and it is a revised version of the analysis presented in Rodriguez-Iturbe et al. [1992a].

The straight-line distance $L_\parallel$ from the outlet to the most distant point in the boundary can be regarded as the first collision time of two self-affine fractals starting from a common origin. The two colliding self-affine fractals represent the boundaries of the basin. As pointed out by Takayasu et al. [1988], in the case of Brownian motion the distribution of $L_\parallel$ is given by the well-known recurrence time distribution of Brownian motion in one dimension. This density function $f(L_\parallel)$ was derived by Feller [1971], who made use of the fact that $L_\parallel$ is, in this case, a random vector in $\boldsymbol{R}^2$ whose probabilistic properties can be completely determined by those of its projection on the x axis. This enables us to reduce the original random-walk problem to a simpler problem in $\boldsymbol{R}^1$ corresponding to the recurrence time distribution. Thus, using the notation $L_\parallel$ for the recurrence time t, one obtains [Feller, 1971; Takayasu et al., 1988]

$$f(L_\parallel) = \frac{1}{\sqrt{2\pi\sigma L_\parallel^3}} \exp^{-1/(4\sigma L_\parallel)} \propto L_\parallel^{-3/2} \tag{2.178}$$

When compared with Brownian motion, a fractional Brownian motion type of trail undergoes much larger departures from the origin, which can be accounted for through a fractal diffusivity defined as [Feder, 1988]

$$\sigma_H = \sigma \ |t|^{2H-1} \tag{2.179}$$

Thus in this case (recall that $L_\parallel$ surrogates t) we obtain

$$f(L_\parallel) \propto (L_\parallel^3 L_\parallel^{2H-1})^{-1/2} = L_\parallel^{-(H+1)} \tag{2.180}$$

which reduces to Eq. (2.178) when $H = 1/2$.

As discussed in Section 2.7.8, the results of Ijjasz-Vasquez et al. [1994] support the assumption of the self-affine fractional Brownian motion character of basin boundaries.

The area of the basin is approximately proportional to the product of its length, characterized by $L_\parallel$, and its width. For the case of a basin whose boundaries are made up by the intersection of two fractional Brownian motions starting from a common origin, the width is expected to be proportional to the value of $\sqrt{2\sigma_H t} \propto t^H$, the quantity inside the square root representing the mean square displacement. We then find that the area, A, is expected to be proportional to

$$A \propto L_\parallel \, L_\parallel^H = L_\parallel^{H+1} \tag{2.181}$$

It also implies a relation between the self-affine character of the basin boundaries and Hack's exponent. In fact, by employing Eq. (2.174) and

Hack's law ($L \propto A^h$) together with Eq. (2.181), one obtains

$$h = \frac{\phi_L}{H+1} \tag{2.182}$$

with the usual notation. In the case of Scheidegger's [1967] networks (where $\phi_L = 1$ and $H = 0.5$) we obtain the perfect result $h = 2/3$, also found by Huber [1991]. From Eqs. (2.181) and (2.182) we also obtain $H = 1 - q$.

The probability distribution $P[A > a]$ is obtained as

$$P[A > a] \propto P[L_\parallel > a^{1/(H+1)}] \tag{2.183}$$

and using Eq. (2.180), we finally get

$$P[A > a] \propto \int_{a^{1/(H+1)}}^{\infty} L_\parallel^{-(H+1)} dL_\parallel \propto a^{-\frac{H}{H+1}} \tag{2.184}$$

and therefore, from Eq. (2.67), $\beta = H/(H+1)$. One interesting example is Scheidegger's [1967] model for which it is known (Section 2.5.2) that $P[A > a] \propto a^{-1/3}$ [Huber, 1991]. In that model the boundary is a Brownian motion for which $H = 0.5$. Our result, Eq. (2.184) gives $\beta = 1/3$, in perfect agreement with the model. Thus we conclude that the self-affine characteristics (e.g., the Hurst exponent) of the basin boundaries are intimately related to Hack's exponent and the distribution law of contributing areas.

Ijjasz-Vasquez et al. [1994] studied the scaling properties of the main channel and the boundaries of river basins (see Sections 2.7.7 and 2.7.8). They concluded that both are self-affine objects. In particular, the basin boundaries they studied had a well-defined Hurst exponent between 0.74 and 0.80. For those values of H we obtain β between 0.425 and 0.45. This shows an excellent agreement between the self-affine characteristics of the basin boundaries and the power law of aggregated areas.

Eq. (2.181) contains Hack's law, because by using $H = 0.75$, one gets Hack's exponent of 0.57. The above arguments show again that the self-affine fractal nature of the basin boundaries leads to an elongation of basins with size.

One word of caution is in order at this point. The above derivation relies on the assumption of self-affinity for basin boundaries. It should be noted, however, that the inference of a meaningful fractal dimension from the value of H poses serious theoretical problems. In fact, the common relationship $D_H = 1/H$ [e.g., Feder, 1988], where D_H is the divider fractal dimension (Section 2.1.1) of the boundaries, is valid only for self-affine records like those resulting from the intersection of a landscape with a vertical plane, with altitude measurements at equal intervals along the line. Also, the relationship of H with the local box-counting fractal dimension (Section 2.1.2), $D_B = 2 - H$, is valid only in the one-dimensional fractional Brownian motion like that in the above case [Tricot, 1995]. In general, the description of a self-affine curve needs two anisotropic scaling parameters, say η and ξ. In the case of a record, one of them, say η, is equal to one. When $\eta = \xi$, one encounters self-similarity.

As shown by Ijjasz-Vasquez et al. [1994], basin boundaries are not self-similar, and hence the divider fractal dimension cannot be functionally related to the H value appearing in this section. Because direct measurements of H are possible and the data support Eq. (2.181), we will use H without the need for further linkages to fractal dimensions.

Finally, Peckam [1995] has shown a relationship between Hack's exponent h and a fractal dimension, say D_{SST}, of a self-similar tree as $h = 1/D_{SST}$. In the case of a space-filling self-similar tree, one has $D_{SST} = 2$ and h would be 0.5, whereas only for $D_{SST} < 2$ can Hack's exponent be higher than 0.5. We observe that for real basins, strict self-similarity is untenable and, furthermore, if the network is not truly space filling, not much can be said about elongation from this result from a theoretical viewpoint.

2.9.3 Hack's Law and the Probability Distribution of Stream Lengths to the Divide

Several characterizations of the stream lengths that make up a drainage network have been proposed, as seen in Section 2.2. A common characterization is the probability distribution of Horton stream lengths. Thus a population of stream lengths may be defined by using the Horton–Strahler stream-ordering procedure. The lengths may be defined as straight-line distances between the endpoints of a stream or as distances obtained by following the stream course. In both cases, the stream itself is defined by the ordering procedure. As discussed in Section 2.2, in both cases the exceedence probability distributions of Horton–Strahler stream lengths follow a power law distribution with exponents of 2 and 1.9, respectively.

It is appealing to define stream length properties without linking the analysis to any particular ordering scheme. One way to do this is to analyze the random variable defined as the longest distance, $\mathcal{L}$, measured through the network from a randomly chosen point to the boundary of the basin. The probability distribution of this distance may be derived by making an assumption about Hack's law and the distribution of contributing areas. We will assume that Hack's law is valid when considering not only different basins but also subbasins defined by the choice of any point inside a basin. This is not the same type of analysis on which Hack's law is based, and therefore one should not necessarily expect either the same value of the exponent or the same excellent fit observed when dealing with different basins. Nevertheless, one would hope for the exponent to be larger than 0.5 with a good fit in the relationship. In Section 2.10 we will also provide an interpretation based on scaling properties.

Calling $\mathcal{L}$ the random stream length to the divide, we have

$$P[\mathcal{L} > L] \propto P[A^h > L] = P[A > L^{1/h}] \propto L^{-\beta/h} \tag{2.185}$$

where β is defined in Eq. (2.67) and h is Hack's exponent. The power law of lengths defined in Eq. (2.185) has been verified in numerous basins with an excellent fit throughout several logarithmic scales. It links the aggregation pattern of areas with the stream length structure of the network and Hack's law. The exponent of Eq. (2.185) would fall between 0.70 and 0.80 for values of β between 0.43 to 0.45 and h between 0.57 and 0.60.

Figures 2.58 to 2.60 show experimental values of $P[\mathcal{L} \geq L]$ for river basins whose DEM features were described in Table 1.1 (Section 1.2.9). In particular, Figure 2.58 shows the detailed results obtained for the Brushy River Basin where we used a support area of fifty pixels and the lengths were measured in kilometers. We note that the choice of a suitable support

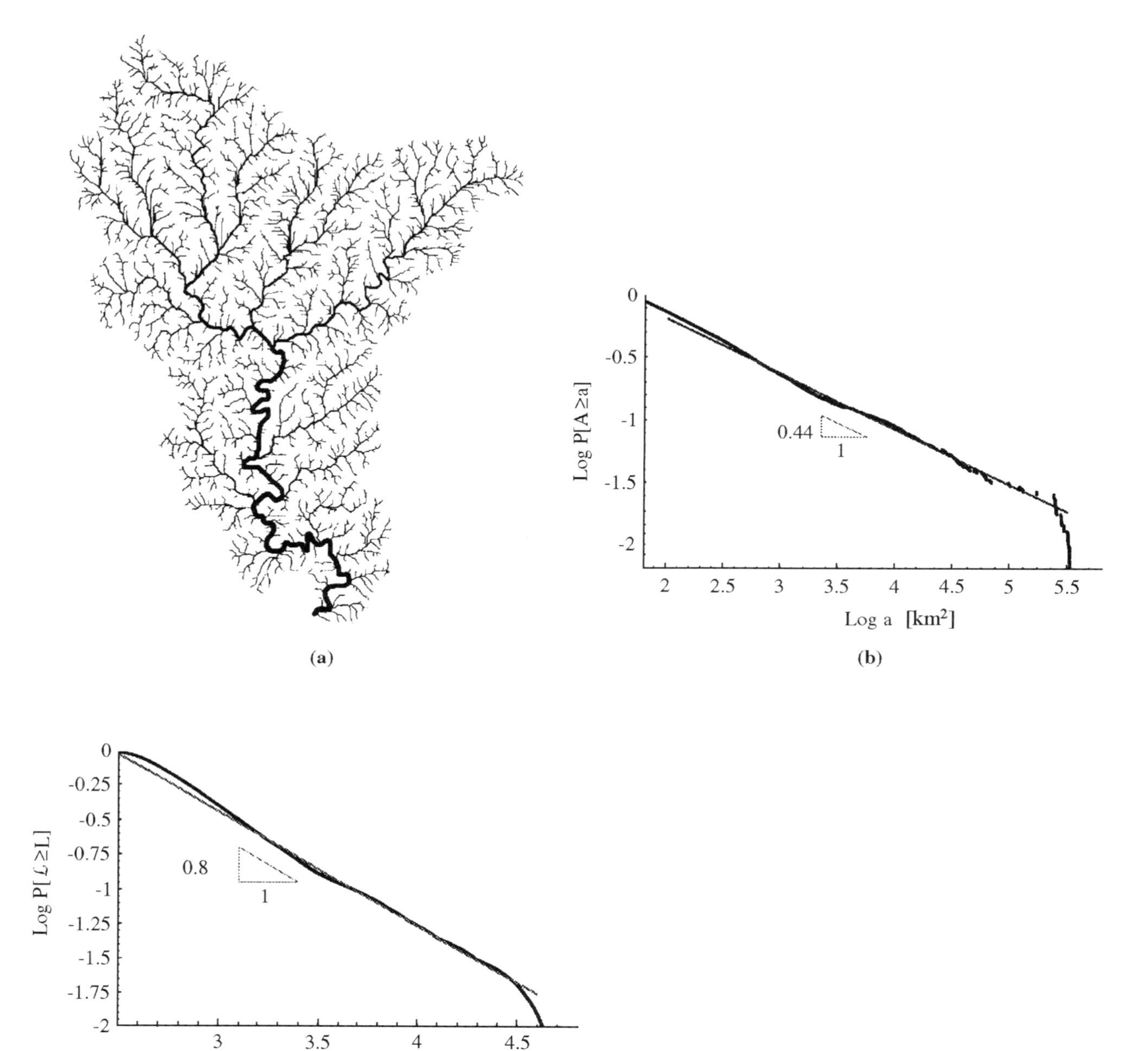

Figure 2.58. (a) A view of the Brushy River Basin with support area of 50 pixels; (b) $P[A > a]$ versus a ($\beta \sim 0.45$); (c) $P[\mathcal{L} \geq L]$ versus L (slope is ~ 0.8).

area for extracting the network slightly affects the estimates of the exponents. The exponent of the power law of lengths, in this case, is 0.8, the value of β is 0.45, and the value of h resulting from Eq. (2.185) is ≈ 0.56.

Figure 2.59 shows the same results for the Racoon River Basin. In this case, we observe a slightly higher value for the slope of the exceedence probability of lengths and a worse adaptation of the theoretical prediction in Eq. (2.185) to the common values of Hack's exponent. Nevertheless, as

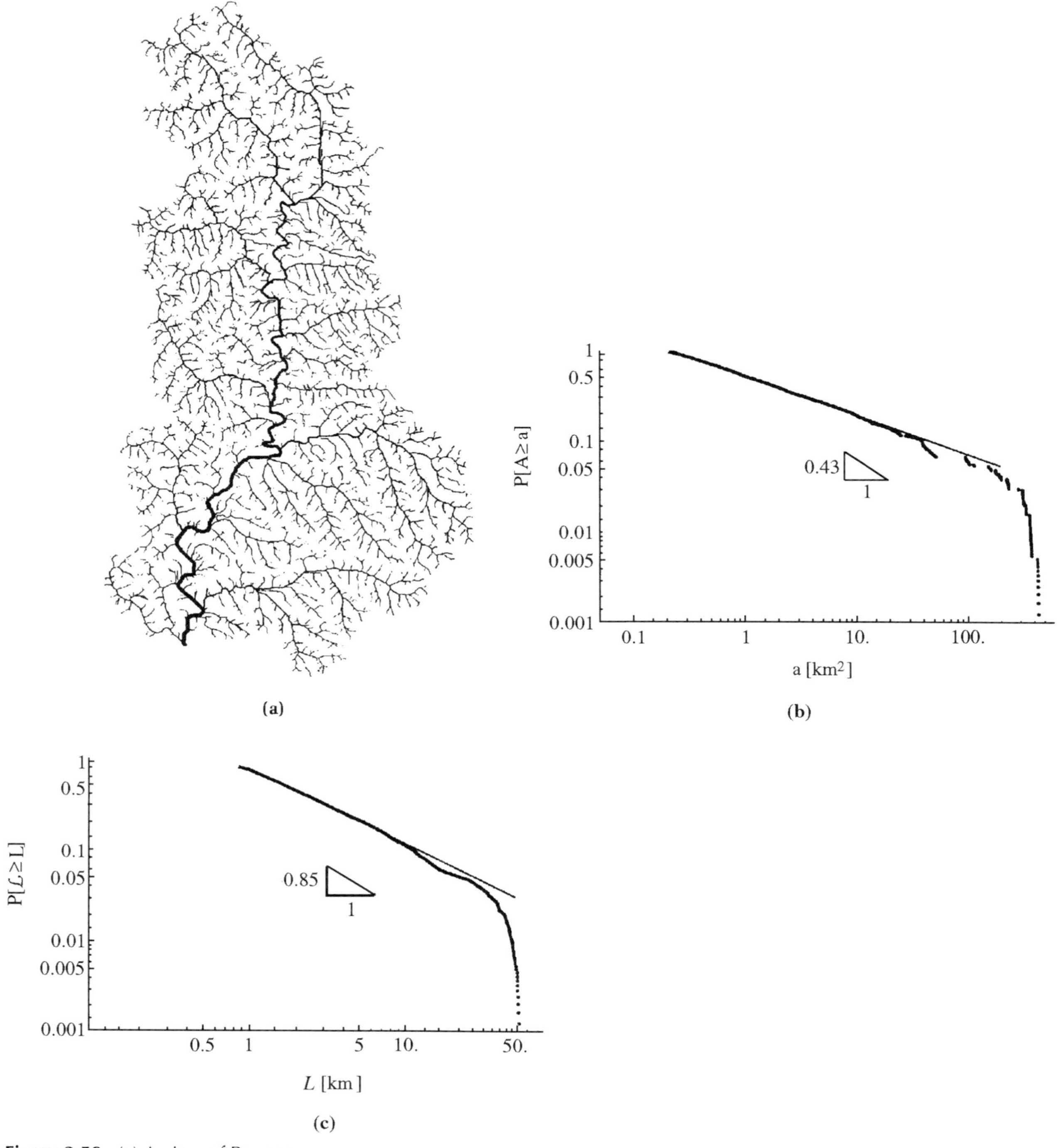

Figure 2.59. (a) A view of Racoon River Basin with support area of 50 pixels; (b) $P[A > a]$ versus a ($\beta \sim 0.43$); (c) $P[\mathcal{L} \geq L]$ versus L (slope ~ 0.85).

we will see in Section 2.10.2, we expect statistical departures from Hack's values.

Figure 2.60 shows a synthesis of the experimental analyses conducted on the DEMs of the basins in Table 1.1, all of them analyzed – somewhat arbitrarily – with the same support area of fifty pixels. The average

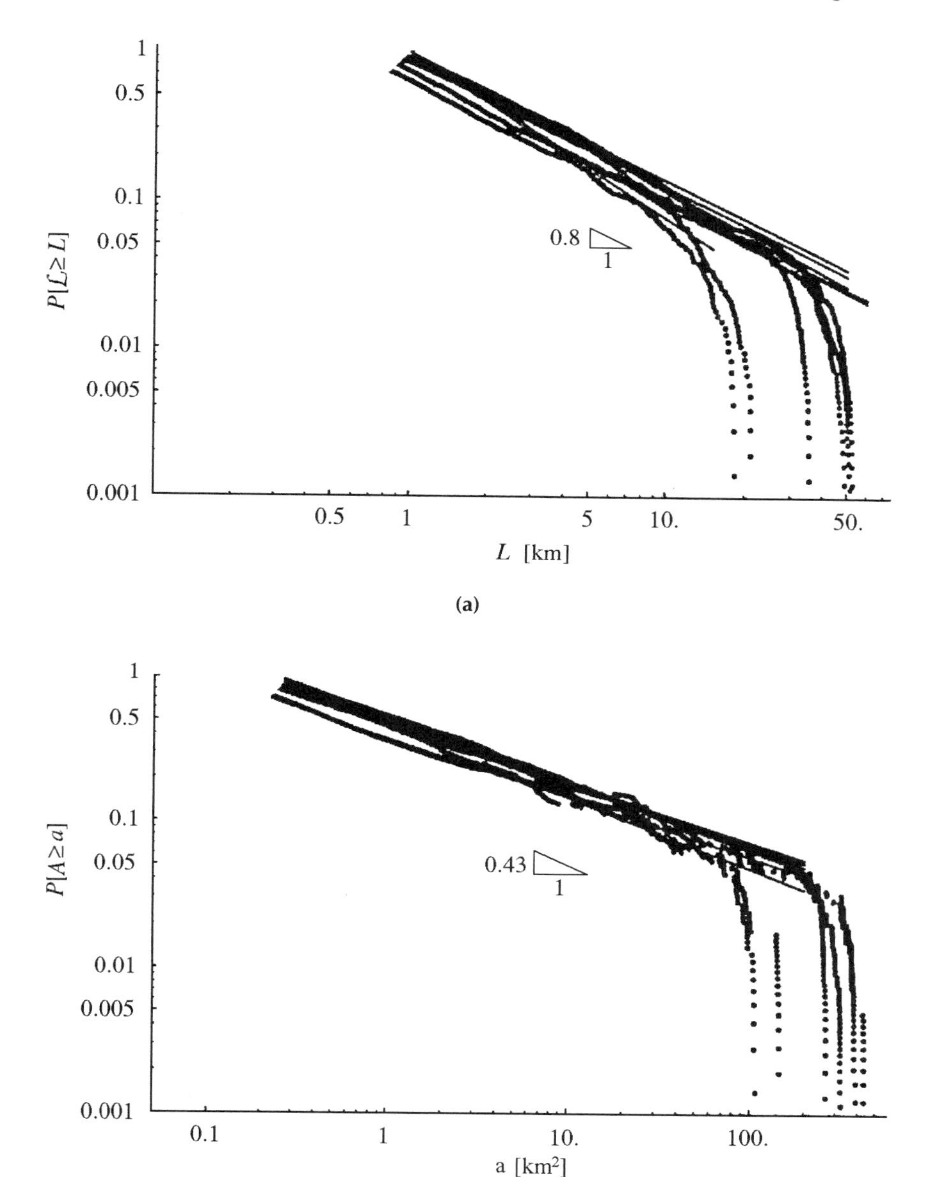

Figure 2.60. (a) $P[\mathcal{L} \geq L]$ versus L for the basins in Table 1.1; (b) $P[A \geq a]$ versus a for the same basins.

exponents of the probability distributions are 0.80 for the lengths and 0.43 for total contributing areas.

2.10 Generalized Scaling Laws for River Networks

In this section we shall show how a simple finite scaling ansatz leads to a natural explanation for many empirical and experimental facts as well as for the scaling properties exhibited by real natural basins and their suitable numerical simulations. Our exposition closely follows that of Maritan et al. [1996a].

2.10.1 Scaling of Areas

As discussed earlier in this chapter, many observations of real basins reveal that the length L is related to the linear size of the system through

$$L \sim L_{\parallel}^{\phi_L} \qquad 1 \leq \phi_L \leq 1 + H \tag{2.186}$$

where $\phi_L \sim 1.1$. Here the bounds mean that the main stream ranges from straight ($\phi_L = 1$) to space filling ($\phi_L = 1 + H$).

In addition, the fact that Hack's exponent is $h > 0.5$ is an indication of anisotropy in the basin shape. Here we recall that the *elongation exponent* q defined in Eq. (2.173) is

$$\frac{A}{L_{\parallel}^2} \sim L_{\parallel}^{-q} \quad q = 1 - H > 0 \tag{2.187}$$

In terms of the elongation exponent this means that when $q > 0$, basins elongate, whereas if $q < 0$, the basins contract. As seen in Section 2.6.2, real river basins exhibit a power law probability of exceedence of total contributing area, say $P[A \geq a] = P(a)$ (notice the slight change of notation, chosen for convenience), that scales like $P(a) \propto a^{-\beta}$ with $\beta = 0.43 \pm 0.02$ (Section 2.5.2). This was shown for several log scales, and the finite-scale effect shown by all probability plots was not considered therein. We will now explore the finite-size effect that yields the functional dependence on size of the probability distribution $P(a, L_{\parallel})$, where $L_{\parallel}$ is a *given* linear size of the system. Figure 2.61 shows $P(a, L_{\parallel})$ in four subbasins of different $L_{\parallel}$ size within the Fella River region [Maritan et al., 1996a]. In the figure the dashed line indicates a reasonable threshold for channelized areas (i.e., the suitable minimum area supporting a channel head). Notice also that the largest basin covers four log (ten) scales of a.

Let us define the probability density distribution, $p(a, L_{\parallel}) = -\partial P(a, L_{\parallel}) / \partial a$, for a basin as having area a for a given linear size $L_{\parallel}$. As in many other scaling systems [e.g., Fisher, 1971; Meakin, Feder, and Jossang, 1991], one might expect that the finite-size distribution $p(a, L_{\parallel})$ obeys a scaling form of the type

$$p(a, L_{\parallel}) = a^{-\tau} f\left(\frac{a}{a_c(L_{\parallel})}\right) \qquad 1 \leq \tau \leq 2 \tag{2.188}$$

where $a_c(L_{\parallel})$ is a characteristic area and $f(x)$ is a scaling function satisfying the following properties:

$$\lim_{x \to \infty} f(x) = 0 \quad \text{sufficiently fast} \tag{2.189}$$

$$\lim_{x \to 0} f(x) = c \tag{2.190}$$

where c is a suitable constant [e.g., Meakin et al., 1991; Nagatani, 1993; Sun, Meakin, and Jossang, 1994a; Maritan et al., 1996a]. The first equation ensures the correct behavior at infinity, and the second equation gives a power law behavior in the infinite size limit, that is, $p(a, \infty) = c\ a^{-\tau}$. Owing to Eq. (2.170), we find that the characteristic area $a_c(L_{\parallel})$ obeys

$$a_c(L_{\parallel}) \sim L_{\parallel}^{\phi} \tag{2.191}$$

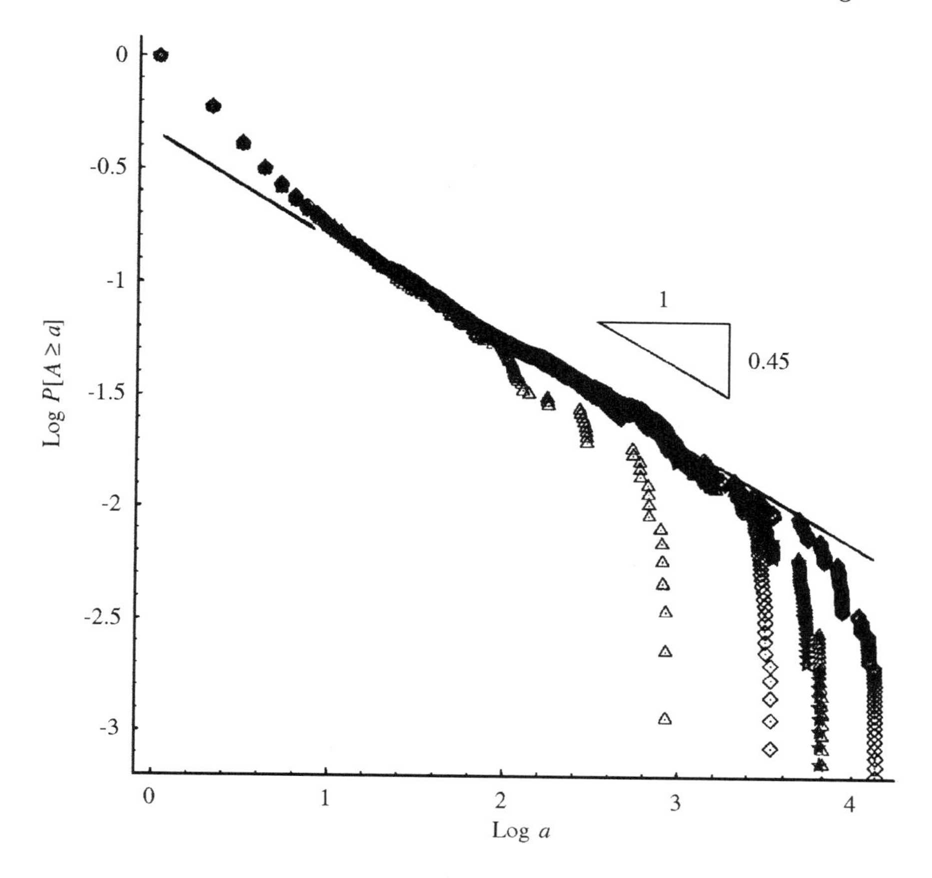

Figure 2.61. $P(a, L_\parallel)$ for four subbasins of the Fella River of different $L_\parallel$ size [after Maritan et al., 1996]. Here we employ pixel units for which a source has $a = 1$. Logarithms are in base 10.

with $\phi = 1 + H$. Because we require normalization

$$1 = \int_1^\infty da\, a^{-\tau} f\left(\frac{a}{L_\parallel^\phi}\right) = L_\parallel^{\phi(1-\tau)} \int_{L_\parallel^{-\phi}}^\infty dx\, x^{-\tau} f(x) = \frac{c}{\tau - 1} \qquad (2.192)$$

The last equality follows by noting that, because $1 \leq \tau \leq 2$, we cannot allow the lower cutoff to go to zero. Therefore we find

$$p(1, L_\parallel \to \infty) = c = \tau - 1 \qquad (2.193)$$

that is, the proportion of the number of sources is independent of the size of the system.

The probability of exceedence of total contributing area is then

$$P(a, L_\parallel) = \int_a^{+\infty} dA \quad p(A, L_\parallel) \qquad (2.194)$$

Because $\phi = 1 + H$, and using Eqs. (2.188) and (2.194) jointly with the fact that $a_c(L_\parallel) \sim L_\parallel^\phi$, one gets

$$P(a, L_\parallel) = a^{-\beta} \mathcal{F}\left(\frac{a}{L_\parallel^\phi}\right) \qquad (2.195)$$

with

$$\beta = \tau - 1 \qquad (2.196)$$

where we have defined

$$\mathcal{F}(x) = x^{\beta} \int_{x}^{+\infty} d\chi\ \chi^{-(1+\beta)} f(\chi) \tag{2.197}$$

Many network models that have been studied in different contexts have a *directed* character due to the fact that they are typically grown on a slope that gives a preferred direction to the flow. Under this condition one expects [Takayasu, 1990] that

$$\langle a \rangle = \int_{1}^{+\infty} da\ a\ p(a, L_{\parallel}) \sim L_{\parallel} \tag{2.198}$$

Eq. (2.198) can be obtained assuming that the spanning tree defining the river network is strictly direct, as, for example, in the Scheidegger [1967] model. This means that along the path from any site to the outlet the tangent always has a positive projection along a fixed direction. If the river network is not direct, a correction is needed, as will be seen in the following. In this case the lower cutoff can be allowed to go to zero, because the integral is convergent for $L_{\parallel} \to \infty$, and thus one easily finds

$$\phi = \frac{1}{2 - \tau} \tag{2.199}$$

which means that there is only one independent exponent in Eq. (2.195). This result had already been observed by Meakin et al. [1991] and Maritan et al. [1996a]. Notice, however, that Eq. (2.198) holds only if statistically relevant river configurations are directed.

Using Eqs. (2.191), (2.196), and (2.199), we get

$$\beta = \frac{H}{H+1} \tag{2.200}$$

in agreement with the theoretical results derived in Section 2.10.2 and with experimental data from which we expect H to be in the range 0.75–0.8 and $\beta = 0.43 \pm 0.02$.

We will now show that a nondirect character yields a minor modification to Eq. (2.200). In the general case (nondirected networks), one obtains [Maritan et al., 1996a]

$$\langle a \rangle = L_{\parallel}^{\phi(2-\tau)} \int_{1/L_{\parallel}^{\phi}}^{\infty} dx\ x^{1-\tau}\ f(x) \tag{2.201}$$

where the lower cutoff is irrelevant because the integral converges for $L_{\parallel} \to \infty$. It thus follows that

$$\langle a \rangle \propto L_{\parallel}^{\phi(2-\tau)} \tag{2.202}$$

The above results also allow the relation of exponents τ and ϕ thus linking the planar aggregation structure to the elongation because

$$\langle a \rangle \propto \ell \propto L_{\parallel}^{\varphi} \tag{2.203}$$

where φ is a new scaling exponent, derived as follows. If we compute the sum of the lengths to the outlet $\sum_i x_i$ (where i spans all sites from 1 to N, N is the total number of sites and x_i is the length to the outlet from site i measured along the network) we have $\sum_i x_i = \sum_i \sum_{j \in \gamma(i)} \Delta x_j \propto \sum_i a_i$, where j indexes the intermediate steps in the path $\gamma(i)$ from i to the outlet and Δx_j is the spatial step, that is, the lattice size, unity in isotropic lattices.

The last proportionality follows from the observation that a_i measures the number of times every Δx_i appears in the sum. Thus we obtain the equivalence $\sum_i x_i \propto \sum_i a_i \Delta x_i = \sum_i a_i$. The mean distance to the outlet, ℓ, is $\ell \propto \sum_i x_i$ and the mean total contributing area is $\langle a \rangle \propto \sum_i a_i$. Thus the length ℓ, which is characteristic of the entire aggregation pattern, and the mean area $\langle a \rangle$ are proportional.

The relationship of ℓ with an Euclidean length $L_\parallel$ has been established as $\ell \propto L_\parallel^\varphi$ with $\varphi = 1.05$. Figure 2.62 shows an experimental evaluation of φ. Notice that φ does not coincide, in principle, with ϕ_L (see Eq. (2.186)) because ℓ is the mean distance to the outlet computed along the network of all sites (i.e., the centroid of the width function $W(x)$), whereas L is just the *mainstream* length.

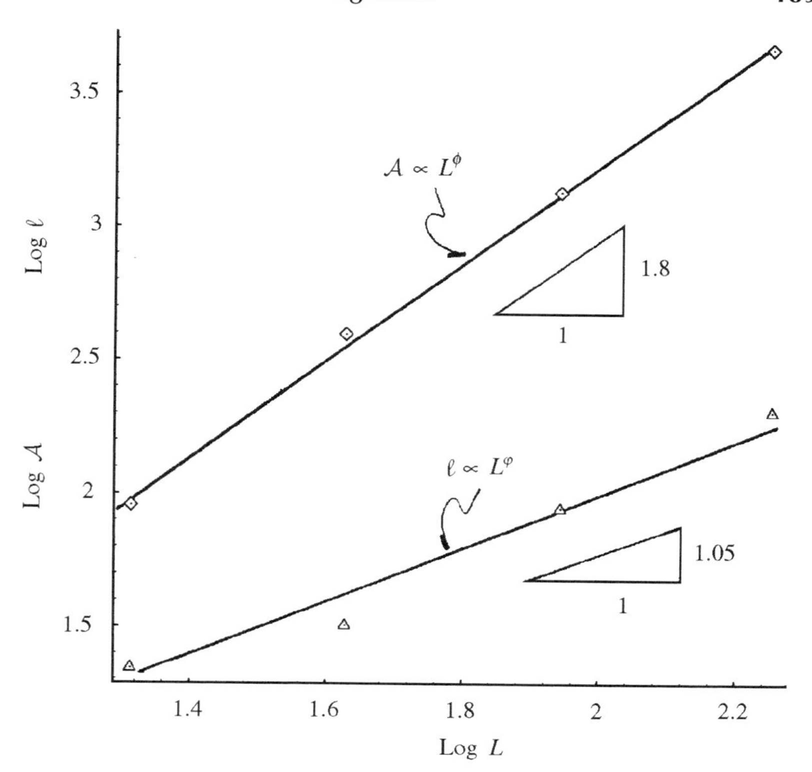

Figure 2.62. Experimental scaling relationships for the Fella River Basin: an independent verification of the relationship $\mathcal{A} \propto L^\phi$ and $\ell \propto L^\varphi$ with $\phi = 1.8 \pm 0.01$ and $\varphi = 1.05 \pm 0.02$. Here we employ pixel units, i.e., $\mathcal{A}$ is obtained from the second moment of the total contributing area whose units are discussed in Figure 2.61. Lengths, L, ℓ, are multiples of the unit ruler identified by the linear size of the pixel. Logarithms are in base 10. [after Maritan et al., 1996a].

Thus we find the relationship linking the scaling coefficients:

$$\phi = \frac{\varphi}{2 - \tau} \qquad (2.204)$$

where $\varphi \approx 1.05$. Notice that in the case $\varphi = 1$, one recovers Eq. (2.198). The above relationship is analytically verified, for instance, in the simple Scheidegger model of network development described in Sections 1.3.2 and 2.5.2, where $\tau = 4/3$ and $\varphi = 1$ yield exactly $\phi = 3/2$. We also note that in the general case one obtains

$$\beta = \frac{1 + H - \varphi}{H + 1} \qquad (2.205)$$

We finally note that, from Eqs. (2.188) and (2.191), one obtains

$$\langle a^n \rangle \propto L_\parallel^{(n-\tau-1)\phi} \qquad (2.206)$$

and thus we can provide an alternative definition of the characteristic size of contributing area, $\mathcal{A}$, as

$$\mathcal{A} = \frac{\langle a^2 \rangle}{\langle a \rangle} \propto L_\parallel^\phi \qquad (2.207)$$

which, once substituted in Eq. (2.188), yields the final scaling relationship

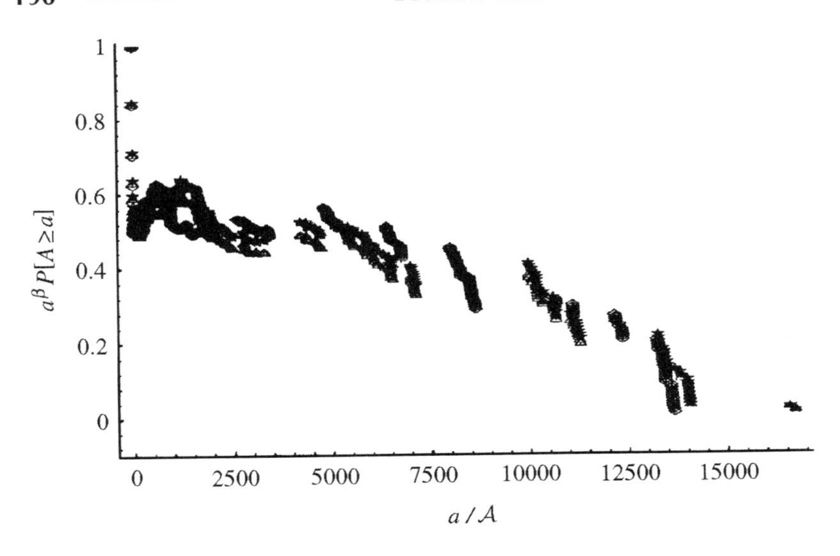

Figure 2.63. Collapse of the scaling curves $a^\beta P(a, \mathcal{A})$ versus $a/\mathcal{A}$ for the four subbasins in Figure 2.61. The collapse is obtained for $\beta = 0.45$ [after Maritan et al., 1996a].

in the form

$$p(a, \mathcal{A}) = a^{-\tau} f\left(\frac{a}{\mathcal{A}}\right) \tag{2.208}$$

From the data shown in Figure 2.61 one obtains the collapse of the various curves shown in Figure 2.63 for the experimental value $\tau = 1.45$ obtained for the Fella River Basin, in domains ranging from 2200 km^2 to 140 km^2. The collapse of the various curves is good, and thus we may conclude that the generalized scaling law (2.188) is supported by experimental and theoretical data.

2.10.2 Scaling of Lengths

The stream length at any point, say L, is defined, as usual, as the main distance measured through the network from the point to the boundary of the basin. Technically, one defines the mainstream pattern upstream of any junction following the site as that having maximum area (in case of equal contributions one chooses at random) until a source is reached. In keeping with the notation adopted in this chapter, we note that for subbasins of area a we indicate the mainstream length as l, and obviously at the closure of the basin, where $a = A$, one has $l = L$. Let us now define the probability distribution $\pi(l, a)$ of the lengths defined in this way *for points with a given area a*. The constraint of choosing sites with a given area a plays an important role in a revised interpretation of Hack's law (see below). The length distribution is then given by

$$\hat{\pi}(l, L_\parallel) = \int_1^{+\infty} da\, \pi(l, a)\, p(a, L_\parallel) \tag{2.209}$$

We shall further denote by $\Pi(l, a)$ and $\hat{\Pi}(l, L_\parallel)$ the corresponding probabilities of exceedence, for example, $\Pi(l, a) = \int_l^\infty \pi(x, a)\, dx$. In general, nothing can be said about the distribution $\hat{\pi}(l, L_\parallel)$ or $\hat{\Pi}(l, L_\parallel)$ unless the distribution $\pi(l, a)$ is known. However, it is generally accepted that the function $\pi(l, a)$ is a relatively sharply peaked function of one variable with respect to the other, thus leading to an effective constraint between areas and lengths. Thus one may assume that

$$\pi(l, a) = \delta(l - ca^h) \tag{2.210}$$

where $\delta(\cdot)$ stands for Dirac's delta function and c is a suitable constant, see Eq. (1.8). Equation (2.210) is a mathematical statement of Hack's law when the law is assumed to hold without dispersion and, moreover, when it is assumed to hold for all subbasins embedded into a basin. In this case, it is easily derived that in the $L_\parallel \to \infty$ limit

$$\hat{\Pi}(l, L_\parallel \to \infty) \propto l^{-\beta/h} \tag{2.211}$$

Table 2.10. *Summary table of scaling coefficients. Column 3 refers to results that will be described in Chapter 4.*

Exponent	Scheidegger Trees	Optimal Channel Networks	Real Basins
$\tau = 1 + \beta$	4/3	1.43 ± 0.02	1.43 ± 0.02
$\phi = 1 + H$	3/2	$1.8 - 1.9$	1.8 ± 0.1
H	0.5	–	$0.75 - 0.80$
h	2/3	$0.57 - 0.58$	$0.57 - 0.60$
ϕ_L	1	1.1	1.1 ± 0.01
φ	1	1.05	1.05 ± 0.01
$\gamma = 1 + \beta/h$	1.5	1.8 ± 0.05	$1.8 - 1.9$

Source: Maritan et al., 1996a.

and $\hat{\pi}(l, L_{\parallel}) \propto l^{-(1+\beta/h)}$, where Eq. (2.188) has been used. This result is equal to that of Eq. (2.185). The same result (Eq. (2.211)) can be derived if we assume for $\pi(l, a)$ a scaling form like

$$\pi(l, a) = \frac{1}{l}\, g\left(\frac{l}{a^h}\right) \tag{2.212}$$

which is a generalization of Eq. (2.210). We notice that Eq. (2.212) does not strictly presume Hack's law, because the original assumptions do not consider statistical fluctuations about the mean value.

A summary of scaling exponents in real basins, optimal channel networks, and Scheidegger's trees (Sections 1.3.2 and 2.5.2) is shown in Table 2.10. Notice that only results of direct measurements are shown in the table. For example, the value of $\phi = 1 + H$ is measured through the scaling relationship of $\mathcal{A}$ versus $L_{\parallel}$ (Eq. (2.207)), whereas, H has been measured from the self-affine properties of the basin boundaries [Ijjasz-Vasquez et al., 1994]. The matching of corresponding (and independent) values of ϕ and H can be appreciated, as well as other related quantities.

We will now show how the above scaling properties naturally lead to Hack's law. It is important to stress the fact that a priori the areas a and the lengths l are dependent random variables. As we said above this means that the lengths of the streams should be measured with respect to a given value of the area a. Let us define

$$\overline{l_a} \equiv \langle l \rangle_a = \int_1^{+\infty} dl\, \pi(l, a)\, l \tag{2.213}$$

where the distribution $\pi(l, a)$ has been assumed to be normalized to unity and is given by either Eq. (2.210) or Eq. (2.212).

Use of Eqs. (2.169) and (2.186), identifying $\overline{l_a}$ with the length l and $a_c(L_{\parallel})$ with the area a, leads to

$$h = \frac{\phi_L}{1 + H} \tag{2.214}$$

It should be noted that this equation was already derived in Section 2.9.2 by a different argument. Notice also that it would be inconsistent to infer $\phi_L \neq 1$ without assuming $\varphi \neq 1$ in the determination of the proper value of H (Eqs. (2.200) and (2.205)).

In the important case where one assumes $\varphi \sim \phi_L$, we obtain the result

$$\beta = 1 - h \tag{2.215}$$

which is another intriguing result. Notice that one cannot theoretically justify the equality of φ and ϕ_L because the former is related to the scaling of the mean distance to the outlet and the latter to the scaling exponent of the mainstream L with the diameter $L_{\parallel}$. Nevertheless, the observational coincidence of $\beta \sim 0.43$ with $h \sim 0.57$ is noteworthy. Using the inequality $1 \leq \phi_L \leq 1 + H$ it is straightforward to show that the elongation exponent q in Eq. (2.187) is

$$q = 1 - H = 2 - \frac{\phi_L}{h} > 0 \tag{2.216}$$

which implies again that basins elongate with size.

Finally, we observe that a powerful scaling argument is obtained when describing the statistical variability of the mainstream lengths at any point in the basin, l, for a given area a through a finite-size probability distribution. Following Rigon et al. [1996], we generalize Eq. (2.212) as

$$\pi(l, a) = l^{-\xi}\, g\left(\frac{l}{a^h}\right) \tag{2.217}$$

where, as usual, $\pi(l, a)$ is the probability density of length l given a drainage area a, and ξ is a scaling exponent. Eq. (2.217) is another mathematical statement of Hack's law. It postulates fractality because it embeds a basic similarity in the distribution of lengths when rescaled by a factor a^h. Notice that, as shown in Eq. (2.206), one has

$$\langle l^n \rangle \propto a^{h(n-\xi-1)} \tag{2.218}$$

where $\langle l^n \rangle \equiv \langle l^n \rangle_a$; that is, it is the nth moment of the distribution of mainstream lengths for a given area a. Straightforward manipulation yields

$$\frac{\langle l^n \rangle}{\langle l^{n-1} \rangle} \propto a^h \tag{2.219}$$

Rigon et al. [1996] have tested the assumption behind Eqs. (2.217) and (2.219) through the analysis of double logarithmic plots of the ratio of consecutive moments versus area from large DEMs of different basins.

Figure 2.64 shows the ratios of five consecutive moments for the Guyandotte River Basin (Table 2.9), defining a different subbasin with its contributing area at every link of the network. We observe that a clear common set of straight line emerges, whose slopes are 0.56 ± 0.01 for at least three decades. The case $n = 1$ also suggests that $\xi \sim 1$; that is, Eq. (2.217) implies the validity of Hack's law for the mean as seen in Eqs. (2.210) and (2.212). The deviations appearing at small contributing areas reflect the presence of unchannelized areas (Section 1.2.12). For large areas (say, $a > 500$ km^2) the statistics progressively lose significance because of the scarcity of data available.

Although technicalities arise (in particular related to the required binning of lengths in adequate intervals of area Δa, as described in Rigon et al. [1996]), it is clearly seen that data support the validity of Eq. (2.217). We emphasize that this consistency in what concerns the internal structure of a basin suggests that Hack's relationship is to be viewed within a statistical framework and not necessarily in connection with arbitrary definitions of suitable basins or subbasins, for example, at predefined outlets.

Scaling arguments provide an appealing framework because they bypass the somewhat artificial techniques for estimating Hack's law from

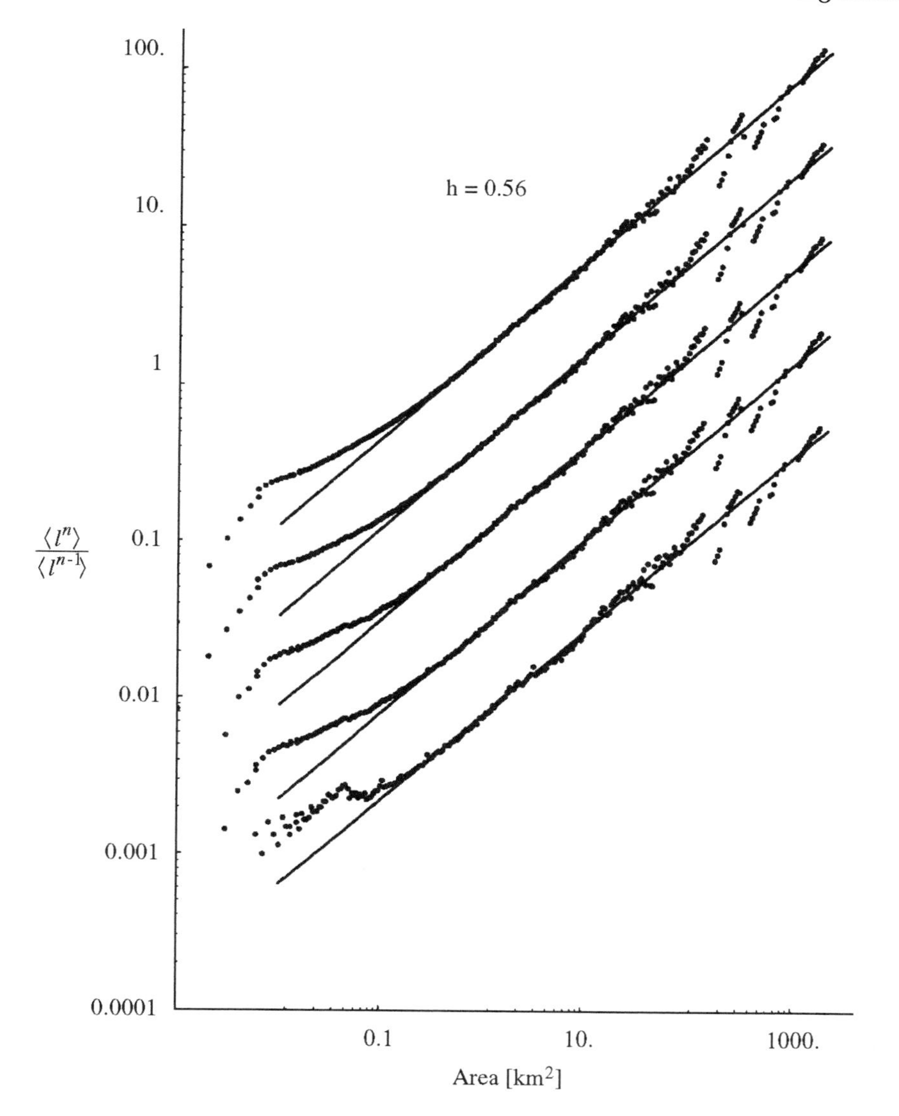

Figure 2.64. A double logarithmic plot of the scaling ratio $\langle l^n \rangle / \langle l^{n-1} \rangle$ versus a for the Guyandotte River Basin, where, from top to bottom, $n = 1, 2, 3, 4$, and 5. The first curve for $y = x$, is simply the plot of $\langle \ell \rangle$ versus area. The vertical scales are offset vertically to avoid overlapping of the curves. Notice that the slope, according to Eq. (2.219), is the exponent h [after Rigon et al., 1996].

regression analysis. Our assessment of a value of Hack's exponent defined through scaling arguments proves to be close to the original one in which suitable nonoverlapping basins were plotted. In our framework, such an occurrence is viewed as likely whenever each realization is close to the mean value of the distribution.

It is of interest to analyze results analogous to those in Figure 2.64 for two basins developed in different geologic contexts. Figures 2.65 and 2.66 illustrate the finite-size scaling assumption, Eq. (2.217), for the Tug Fork and Boise River Basins described in Table 2.9. The quality of the fit is also excellent over at least three decades, and the slope is $h = 0.54 \pm 0.01$ and $h = 0.53 \pm 0.01$, respectively.

From the above results we suggest that Hack's law is indeed an outgrowth of fractality. The aggregation characteristics of river basins cannot be isolated from their geologic context because the features of the boundaries are related to the internal organization of the basin. As such, Hack's exponent cannot be universal. Nevertheless, because a free competition

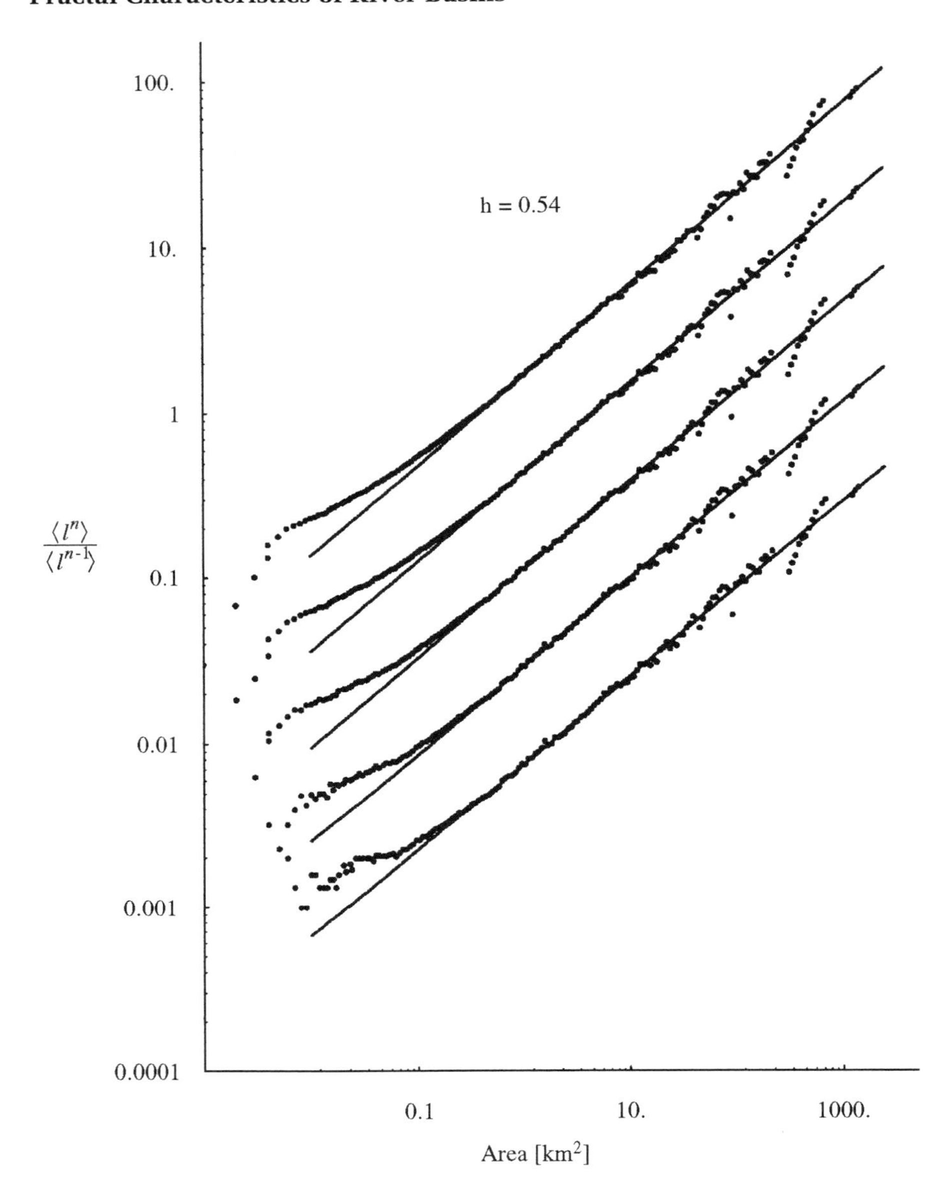

Figure 2.65. A double logarithmic plot of the scaling ratio $\langle l^n \rangle / \langle l^{n-1} \rangle$ versus a for the Tug Fork River Basin, where, from top to bottom, $n = 1, 2, 3, 4$, and 5. The vertical scales are offset vertically to avoid overlapping of the curves.

for drainage commonly occurs for the migration of divides, 'average' characters appear consistent – and seemingly universal – for certain choices of basins (e.g., nonoverlapping, closed at a confluence, control-free, etc.).

The above scaling analysis essentially shows that Hack's law is a reflection of the basic similarity of the part and the whole in river basin landforms. Mandelbrot's idea (Section 2.9) of the inference of the fractal characters of the mainstream from Hack's relationship was therefore incomplete but visionary because many fractal forms in the river basin are related. The structural similarity in shape carries throughout the length scales of the fluvial basin, as clearly illustrated by the scaling plots of the ratios of moments of mainstream lengths (Figures 2.64, 2.65, and 2.66). This similarity is conditional on the constraints (or the absence of them)

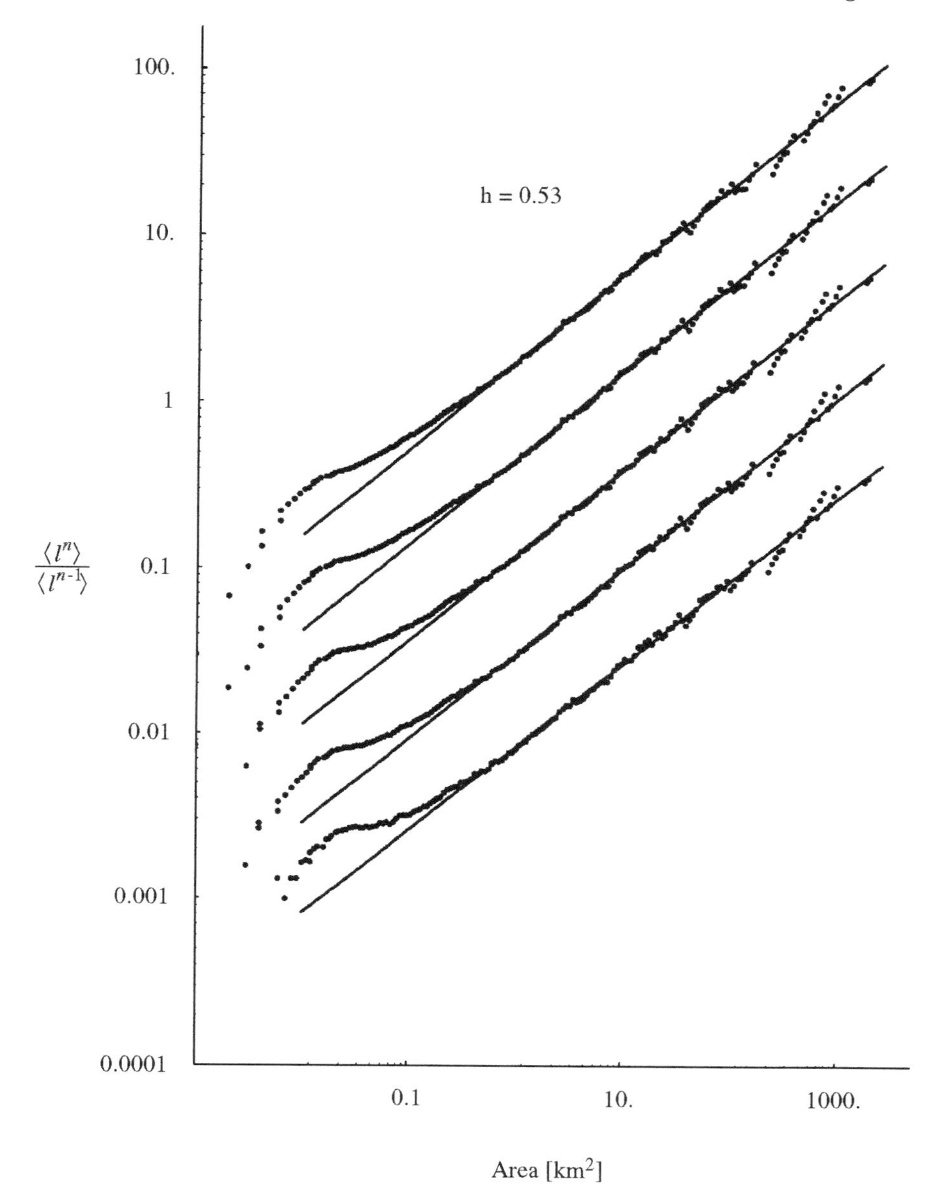

Figure 2.66. A double logarithmic plot of the scaling ratio $\langle l^n \rangle / \langle l^{n-1} \rangle$ versus a for the Boise River Basin, where, from top to bottom, $n = 1, 2, 3, 4$, and 5. The vertical scales are offset vertically to avoid overlapping of the curves [after Rigon et al., 1996].

imposed on the developing basin. The consistency of the scaling structure with the results that we found from our data is remarkable.

We also suggest that the extensions of Hack's original result to basins of completely different regions may be misleading. Although the 'average' characters point toward a similar behavior, the claimed universality is untenable unless clearly referred to a common dynamic and geologic framework. This will be addressed in Section 4.17, where we show that optimal characters of the network organization indeed tend to an elongation close to that of the average experimental values of Hack's exponents.

CHAPTER 3

Multifractal Characteristics of River Basins

Multifractal measures concern the study of the distribution of a physical quantity on a geometric support, and multifractals have been defined as geometric objects that exhibit different local fractal dimensions in different regions within the support. Multifractals require that the fractal concept is generalized to include intricate structures with more than one scaling exponent, that is, a spectrum of exponents. We show that multifractal concepts enlarge our ability to describe geomorphic structures. We start our study from exact multifractal measures, defined precisely for idealized structures that nevertheless bear a resemblance to river networks. Similarities and departures between idealized and real structures prove instructive and suggestive. Later we focus on the analysis of data from real basins. This chapter conveys important methodological aspects that will be used later in the analysis of both simulated and real landscapes.

3.1 Introduction

Our analysis starts by considering a population whose members are distributed over, say, a volume of linear size L. Whatever the physical nature of the population (from the spatial distribution of ore in the bulk to the distribution of the human population over the Earth), we observe that the distribution may fluctuate remarkably from point to point in space. Gold, for instance, is found in high concentrations only at a few places, in lower concentrations at many places, and in very low concentrations almost everywhere [Feder, 1988; Schroeder, 1991]. The important point underlying scaling, or fractal, processes is that this description holds remarkably well whatever the scale of observation. Here we extend the concept of scale invariance by allowing the description of a different type of complex, yet ordered, spatial fluctuations.

Multifractal measures concern the study of distributions of physical quantities on a geometric *support*. The support may be a line segment, a plane, the surface of a sphere, or a volume. In particular cases, the support may be a fractal.

The concepts underlying the development of multifractals were originally introduced by Mandelbrot [1972, 1974] in his discussion of turbulence and then expanded [Mandelbrot, 1983] to many other contexts. The term multifractal can be traced to Frisch and Parisi [1985] and Benzi et al. [1984], who developed applications of multifractals to the study of turbulence. The analysis of experimental results through the introduction of the so-called $f(\alpha)$ spectrum, defined later in this chapter, was proposed by Frisch and Parisi [1985], Halsey et al. [1986a,b], and Jensen et al. [1985], who demonstrated a remarkable agreement between observations and a theoretical model. It is particularly relevant in our context that they demonstrated the usefulness of multifractals for describing an underlying structure to different types of data.

The idea that a fractal measure may be represented in terms of intertwined fractal subsets that have different scaling exponents has opened entire new fields because of the capabilities of this type of analysis for describing the geometry of physical systems. To avoid an exhaustive literature review on this issue, we limit our attention to work of direct interest in the context of river basins. Such work concerns the distribution of currents in fractal resistor networks; see, for example, De Arcangelis, Redner, and Coniglio [1985], Rammal et al. [1985], Aharony [1989], and Blumenfeld et al. [1987]. Also of interest are multifractals in network-like structures that arise in the context of diffusion-limited aggregation (and the dynamics of the related growth processes), which are discussed by Meakin et al. [1985, 1986], Meakin [1987a,b], and Halsey, Meakin, and Procaccia [1986a].

In the geophysical literature regarding river basins, multifractal analyses have been applied to investigate the nature of spatial fluctuations of the distribution of aggregated areas and related quantities surrogating energy dissipation. The analyses have been carried out both using DEMs and models of aggregation with injection [Ijjasz-Vasquez, Rodriguez-Iturbe, and Bras, 1992; Takayasu and Takayasu, 1989]. The Scheidegger [1967] model of river networks discussed in Chapter 1 has also been thoroughly examined in light of the multifractal formalism. Some of the theoretical arguments presented in the foregoing references may need to be revised in light of subsequent theoretical results from the so-called random cascade theory [Holley and Waymire, 1992; Peckam and Waymire, 1992; Gupta and Waymire, 1993].

A hydrologically interesting measure is the width function $W(x)$ (see Section 1.2.7) of a river network. We recall here that width functions are operators that map the area of the basin, a two-dimensional object, onto a one-dimensional support. In fact, given the DEM of an arbitrary basin and its outlet, a subdivision into elementary areas is performed and the length of the path from each elementary area to the outlet is computed. One then adds the elementary areas sharing the same value of the length computed in such a manner. The width function contains important information about the mechanisms of development of the drainage network and incorporates the essential characters of the hydrologic response through its linkage with residence time distributions (see Chapter 7). It has been suggested by Marani et al. [1991] that in idealized basins, deterministically self-similar and Hortonian with $R_B = 4$ and $R_L = 2$ like, for example, the so-called Peano's basins (cf. Section 2.4), the spatial evolution of contributing areas epitomized by $W(x)$ has a multifractal structure generated by a very specific deterministic multiplicative process, described in detail in Section 3.2. We will later analyze this result and its implications for the width functions of real basins in detail, pointing out similarities and departures from Peano's exact results.

A new set of analyses of topographic fields such as the landscape of river basins was provided by Lavallée et al. [1993], exploiting the concepts of generalized scale invariance developed by Schertzer and Lovejoy [1989a,b] in the context of rain and cloud fields. Schertzer and Lovejoy argued that geographic and geophysical fields are generally multifractal, that is, characterized by an infinite hierarchy of fractal dimensions, and that inconsistencies are often inevitable when the study of natural fields is forced into geometric frameworks involving single fractal dimensions. We will examine these concepts (which require specific analytical techniques) with respect to real topographic data collected in river basins in northern Italy.

This chapter is thus centered on the study of multifractal characteristics in real river basins. We will also show that common characters and implications emerge from real data regardless of basin size, orientation, geology, and climate. We will also show that the recurrence of certain multifractal characters has important implications for comparison of basin structures; that is, not all tree-like, loopless, plane-filling random structures reproduce such characters. On the contrary, we will see that the study of drainage area distribution along a support, naturally defined as the length from the farthest source to the outlet measured along the network, has important distinctive characters.

The results we obtain will be used throughout the rest of the book for comparing structures generated by theoretical models. The reproduction of such features constitutes a more stringent test than the simple affinity of a few planar statistics.

3.2 Peano's Basin and the Binomial Multiplicative Process

Consider Peano's basin (Chapter 2) at the stages of generation identified by $\Omega = 2, 3, 4$, and 6, respectively (Figure 2.24), where Ω is Strahler's order of the resulting networks. Recall, from Section 2.4, that Strahler's maximum order is directly related to the stage of generation of an iterative process leading to the construction of Peano's basin. In fact, for each change of scale, every link generates four links, two resulting from the subdivision at the half length of the previous link and two new links. The sequence of linear rulers, $\delta(\Omega)$, capable of describing the geometry of the construct is chosen in such a way that δ is the length of the shortest link embedded in a network of order Ω, that is, $L(1, \Omega)$, where, as in Chapter 2, $L(1, \Omega)$ is the length of first-order streams embedded in a network of order Ω. We will often use the notation $L(j, \Omega)$ to stress that the dependence on the scale of resolution defined by Ω is an obvious extension of the notation $L(j)$ of Chapter 1. The related *area* ruler (Figure 2.21) is the square whose diagonal is δ.

Recall from Section 2.4 that in the formalism of Strahler's numbering we assume, with Marani et al. [1991], that

$$\delta(\Omega) = L(1, \Omega) \tag{3.1}$$

where $L(1, \Omega)$ is the length of first-order streams embedded in a network of order Ω. Also, at the arbitrary stage of generation, Ω, we have a total number of $4^{\Omega-1}$ links, each of length $\delta(\Omega) = L(1, \Omega) = 1/2^{\Omega}$. The maximum path from source to outlet is made up of $2^{\Omega-1}$ links.

In Chapter 2 we noted that Peano's basin has Horton's bifurcation ratio R_B exactly equal to 4 and length ratio R_L exactly equal to 2. It has other important similarities with the planar structure of river networks. Also, Peano's network total length is characterized by a fractal dimension $D = \log R_B / \log R_L = 2$, and the plots of the probability of exceedence of total contributing areas and Strahler's lengths follow the tendency of a power law, although this happens only through the discrete range of scales occupied by the recursive construct. We will now examine in detail the width function of Peano's basin, to show that it has peculiar features that can be investigated analytically.

In the case of Peano's construct, the plane-filling structure allows a one-to-one correspondence of the number of links and the number of elementary areas covering the basin [Marani et al., 1991]. Hence the relative proportion of links at distance x from the outlet is equal to the relative proportion of drainage area at the same distance. All connected links of size δ in Figure 2.21 are grouped in distinct classes defined by their distance x to the outlet. In practice we choose the distance to the outlet from the link's *endpoint*, the difference from choosing, say, the midpoint being immaterial as $\Omega \to \infty$. Choosing fractions x of the maximum (unit) distance $1 = \delta(\Omega)2^{\Omega-1}$ from source to outlet, a partition is defined by the subintervals of length $\delta(\Omega)$, that is, $0, 1/2^{\Omega-1}, 2/2^{\Omega-1}, 3/2^{\Omega-1}, \ldots, x, \ldots, 1$. We also may substitute x by the integer multiple, say i, of the ruler $\delta(\Omega) = 2^{1-\Omega}$, that is, $x = i\delta(\Omega)$.

The geometry of the connections then defines the required partition of the drainage area into subregions at the same distance x. Let $\mathcal{N}(i, \Omega)$, $i = 1, \ldots, i_{\max}$ be the number of links at distance i, that is, in the ith partition of the maximum distance from source to outlet. Also, let $W(x, \Omega)$ be the width function measuring such a partition. Here the change of notation with respect to that used in Chapter 1 and later in this book (i.e., $W(x)$) stresses the fact that the paths to the outlet are embedded into a stage Ω of the fractal structure generation. The two notations will be used interchangeably in the text, hopefully at no risk of confusion. If $\mathcal{N}$ is the total number of links in the network (as shown below by induction, $\mathcal{N} = 4^{\Omega-1}$), the connection of the two quantities is simply $i \to x = i\delta(\Omega)$ and $\mathcal{N}(i, \Omega) \to W(x, \Omega) = \mathcal{N}(i, \Omega)/\mathcal{N}$.

Let us review in detail the quantities used in the computation. We notice from Figure 2.21 that at the stage $\Omega = 2$ of generation, one has 1/4 of the area concentrated in the first half of the maximum distance, the remaining 3/4 being in the second half. For $\Omega = 3$, we need to partition sixteen links: 1/16 in the first fourth of the unit length, 3/16 in the second, 3/16 in the third, and 9/16 in the fourth. We thus obtain a multiplicative process – to be defined in a more general context later – through which we split the total number of links with a recursive rule.

We first note the recursive character of the $\mathcal{N}(i, \Omega)$ throughout the same generation stage, that is, $\mathcal{N}(2i, \Omega) = 3\mathcal{N}(i, \Omega)$ for $i \leq i_{\max}/2$. This suffices for recursively constructing the function, because the first half of the function is exactly the full set of the values at the previous stage of generation, that is, $\mathcal{N}(i, \Omega) = \mathcal{N}(i, \Omega - 1)$ for $i \leq i_{\max}/2$.

By induction, we observe (Table 3.1) that the general term of the width function $W(x, \Omega)$ can be given by [Marani et al., 1991]

$$W(x, \Omega) = \begin{cases} 1/4\ W(x, \Omega - 1) & x \leq 1/2 \\ 3/4\ W(2x - 1, \Omega - 1) & 1/2 < x < 1 \end{cases} \tag{3.2}$$

because the ratio of the total number of links in any two successive generations ($\Omega \to \Omega + 1$) is $4^{\Omega}/4^{\Omega+1} = 1/4$.

The plot of the width function $W(x, 11)$ of Peano's basin is shown in Figure 3.1 for the 11th stage of generation.

Table 3.1. *Dimensional width functions $\mathcal{N}(i, \Omega)$ of Peano's basins of order Ω*

Order, Ω	Total number of links, $\mathcal{N}$	Links in the longest path, $i_{\max}$	$\mathcal{N}(i, \Omega)$ $i = 1, i_{\max}$	
1	4^0	2^0	$\mathcal{N}(1, 1) = 1$	
2	4^1	2^1	$\mathcal{N}(1, 2) = 1$	$\mathcal{N}(2, 2) = 3$
3	4^2	2^2	$\mathcal{N}(1, 3) = 1$	$\mathcal{N}(2, 3) = 3$
			$\mathcal{N}(3, 3) = 3$	$\mathcal{N}(4, 3) = 9$
4	4^3	2^8	$\mathcal{N}(1, 4) = 1$	$\mathcal{N}(2, 4) = 3$
			$\mathcal{N}(3, 4) = 3$	$\mathcal{N}(4, 4) = 9$
			$\mathcal{N}(5, 4) = 3$	$\mathcal{N}(6, 4) = 9$
			$\mathcal{N}(7, 4) = 9$	$\mathcal{N}(8, 4) = 27$
5	4^4	2^4	$\mathcal{N}(1, 5) = 1$	$\mathcal{N}(2, 5) = 3$
			$\mathcal{N}(3, 5) = 3$	$\mathcal{N}(4, 5) = 9$
			$\mathcal{N}(5, 5) = 3$	$\mathcal{N}(6, 5) = 9$
			$\mathcal{N}(7, 5) = 9$	$\mathcal{N}(8, 5) = 27$
			$\mathcal{N}(9, 5) = 3$	$\mathcal{N}(10, 5) = 9$
			$\mathcal{N}(11, 5) = 9$	$\mathcal{N}(12, 5) = 27$
			$\mathcal{N}(13, 5) = 9$	$\mathcal{N}(14, 5) = 27$
			$\mathcal{N}(15, 5) = 27$	$\mathcal{N}(16, 5) = 81$
...	...	...	...	
Ω	$4^{\Omega-1}$	$2^{\Omega-1}$	$\mathcal{N}(i, \Omega) = \begin{cases} \mathcal{N}(i, \Omega-1) & i \le 2^{\Omega-2} \\ 3\mathcal{N}(i - 2^{\Omega-2}, \Omega-1) & 2^{\Omega-2} < i \le 2^{\Omega-1} \end{cases}$	

Source: Marani, Rigon, and Rinaldo, 1991.

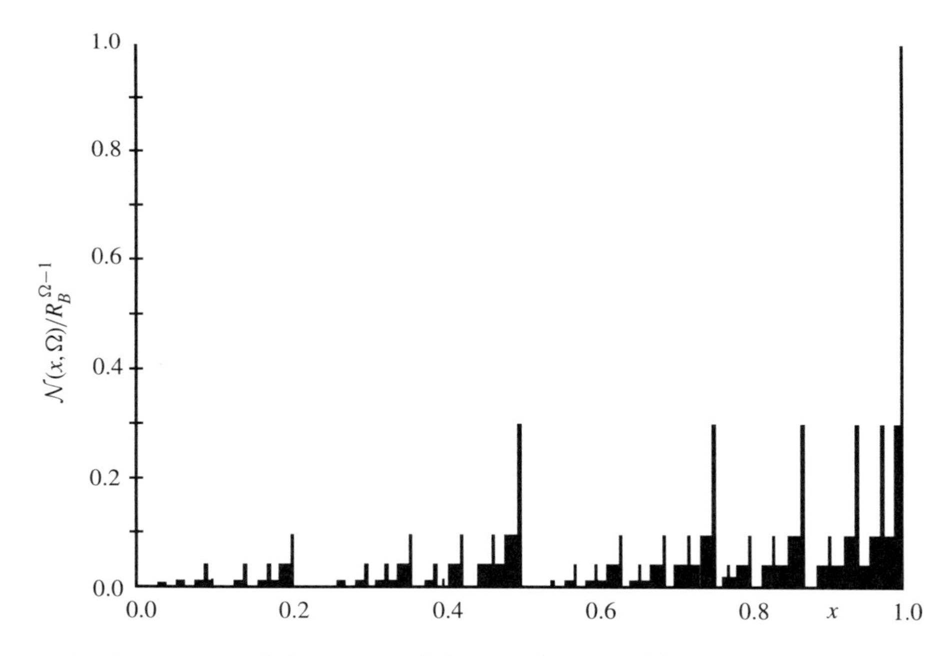

Figure 3.1. The width function $W(x, 11)$ of Peano's basin of order 11.

An interesting elaboration of the results in Table 3.1 and Figure 3.1 is illustrated in Figure 3.2, which shows the cumulated width function $M(x, \Omega)$ defined as

$$M(x, \Omega) = \sum_{j=1}^{x} W(j, \Omega) \tag{3.3}$$

The values of the resulting quantities are shown in Table 3.2.

Table 3.2. *The measure $M(x, \Omega)$ of Peano's basis or order Ω ($x = i\delta(\Omega) = i2^{1-\Omega}$)*

Order, Ω	$M(x, \Omega)$ $0 \le x \le 1$	
1	$M(1, 1) = 1$	
2	$M(0.5, 2) = 1/4$	$M(1.0, 2) = 1$
3	$M(0.25, 3) = 1/16$	$M(0.50, 3) = 4/16$
	$M(0.75, 3) = 7/16$	$M(1.00, 3) = 1$
4	$M(0.125, 4) = 1/64$	$M(0.250, 4) = 4/64$
	$M(0.375, 4) = 7/64$	$M(0.500, 4) = 16/64$
	$M(0.625, 4) = 19/64$	$M(0.750, 4) = 28/64$
	$M(0.875, 4) = 37/64$	$M(1.000, 4) = 1$
...	...	
Ω	$M(x, \Omega) = \begin{cases} 1/4\ M(2x, \Omega) & 0 \le x \le 1/2 \\ 1/4 + 3/4\ M(2x-1, \Omega) & 1/2 < x \le 1 \end{cases}$	

Source: Marani, Rigon, and Rinaldo, 1991.

By induction, it can be easily seen that the measure $M(x, \Omega)$ obeys the recursive rule:

$$M(x, \Omega) = \begin{cases} 1/4\ M(2x, \Omega) & 0 \le x \le 1/2 \\ 1/4 + 3/4\ M(2x-1, \Omega) & 1/2 < x < 1 \end{cases} \tag{3.4}$$

The 'curve' shown in Figure 3.2 is $M(x, 11)$. It is a famous construct, the so-called *devil's staircase* [Mandelbrot, 1983; Feder, 1988]. The curve remains constant on the intervals that correspond to the length of the ruler δ, and its limiting properties for $\Omega \to \infty$ constitute a deep mathematical subject, of which we will give a somewhat detailed account in the following.

We begin our mathematical analysis of the width function of Peano's basin by observing the local character of the fractal dimension of the set $W(x, \Omega)$. In fact, to do so, we examine the divergence of the quantity $\mathcal{N}(x, \Omega)$ (where $i \to x$ from our previous notation)

$$\lim_{\Omega \to \infty} \mathcal{N}(x, \Omega)\delta^d = \begin{cases} 0 & d < D(x) \\ \infty & d > D(x) \end{cases} \tag{3.5}$$

The value of the fractal dimension D, which controls the divergence of the width function at x, is therefore

$$D(x) = \frac{\log \mathcal{N}(x, \Omega)}{\log \delta} = \frac{\log \mathcal{N}(x, \Omega)}{(\Omega - 1)\log 2} \quad \text{for } \Omega \to \infty \tag{3.6}$$

Hence in this case the value of $D(x)$ can be obtained through straightforward computations by evaluating the slope of semilog plots of $\mathcal{N}(x, \Omega)$ versus Ω for fixed positions x. The results are shown in Figure 3.3. Notice that the minimum fractal dimension is $D(0) = 0$, because $\mathcal{N}(0, \Omega) = 1$ for every Ω. Hence $D(0) = \log 1/((\Omega - 1)\log \delta) = 0$. Also notice that the maximum fractal dimension is obtained at

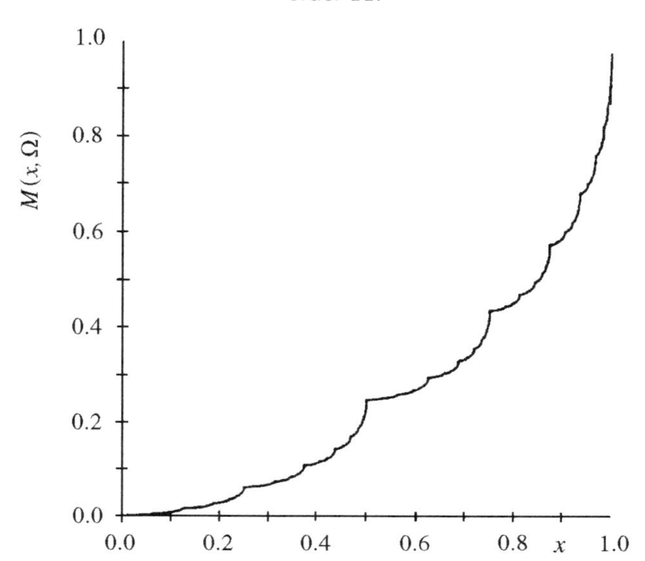

Figure 3.2. The cumulated width function (the devil's staircase) $M(x, 11)$ of Peano's basin of order 11.

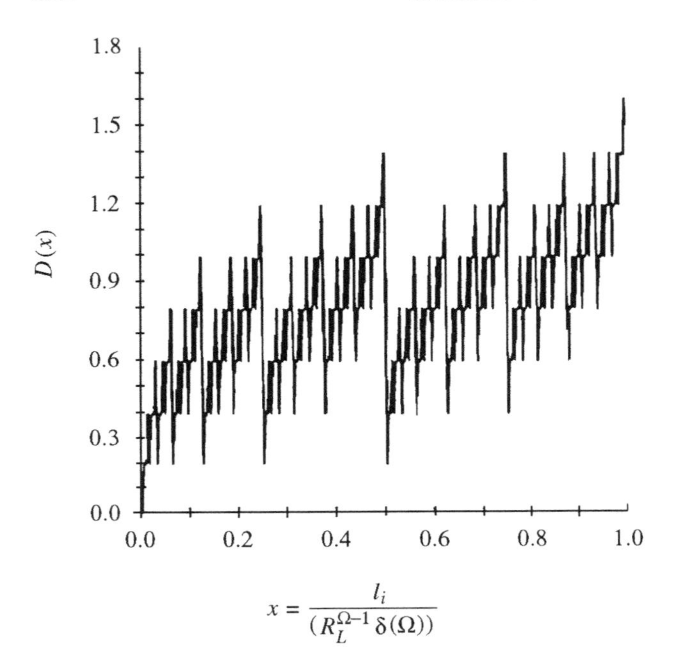

Figure 3.3. Fractal dimensions $D(x)$ of the width function of Peano's basin. The order of the computations is $\Omega = 11$ [after Marani et al., 1991].

$x = 1$, where we observe by inspection that as many as $3^{\Omega-1}$ links (out of a total number of $4^{\Omega-1}$) are located. Therefore one has $D(1) = \log 3^{\Omega-1}/\log 2^{\Omega-1} = \log 3/\log 2 = 1.5354$.

It is very interesting that the fractal dimension $D(x)$ crucially depends on x, that is, the coordinate along the one-dimensional support defined by the maximum length from source to outlet. This defines, still in an informal manner, a *multifractal* character in the sense that the set of widths observed with various rulers has fractal characters that depend on location within the network. The presence of different scaling laws in different spatial positions within a support is a basic ingredient of multifractality.

We now discuss the following mathematical exercise, taken from Feder [1988]. Let a statistical sample of a population consist of $\mathcal{N}$ members distributed over the line segment $\mathcal{S} = [0, 1]$. We consider $\mathcal{N}$ to be a large sample of an underlying distribution when $\mathcal{N}$ is finite. In order to characterize this distribution, we divide the line segment into partitions of length $\delta = 2^{-n}$ so that $N = 2^n$ partitions are needed to cover $\mathcal{S}$. Here n is the *number of generations* in the binary subdivision, of the line segment, as shown in the following. Notice that with the convention adopted for Peano's basin (based on Strahler's ordering and thus meaningless until a tree is generated) it would be $n = \Omega - 1$. Here we will not use Ω to avoid a more involved notation.

We label the partitions by the index $i = 0, 1, 2, \ldots, N - 1$. The distribution of the population over the line is specified at the resolution δ and by the number, $\mathcal{N}_i$, of members of the population in the ith partition. The fraction of the total population $\mu_i = \mathcal{N}_i/\mathcal{N}$ is a convenient measure for the actual content in the ith partition. The set $\mathcal{M}$, given by

$$\mathcal{M} = \mu_i\Big|_{i=0}^{N-1} \tag{3.7}$$

gives a complete description of the distribution of the population at the resolution δ. The measure $M(\mathcal{L})$ of a subregion $\mathcal{L}$ of the line segment $\mathcal{S}$ is

$$M(\mathcal{L}) = \sum_{i \in \mathcal{L}} \mu_i \tag{3.8}$$

We will show, following Feder [1988], that the set $\mathcal{M}$ has scaling properties and that a particular choice of μ_i can be cast in the form of Peano's width function.

Consider the following Besicovitch *multiplicative process* (popularized by Mandelbrot [1983]), which generates a measure on the unit interval $\mathcal{S} = [0, 1]$. First divide $\mathcal{S}$ into two parts of equal length $\delta = 2^{-1}$. The left half is assigned a fraction p of the population and therefore the left segment has measure $\mu_0 = p$. The right-hand segment is given the remaining fraction $1 - p$ and therefore has the measure $\mu_1 = 1 - p$. Thus $\mathcal{M}_1 = \mu_0, \mu_1$.

We choose now to increase the resolution to $\delta = 2^{-2}$. We thus partition the total population in each part in the same way as before. We find four pieces with the fractions of the population in the cells given by

$$\mathcal{M}_2 = \mu_i\Big|_{i=0}^{2^2-1} \tag{3.9}$$

that is, $\mathcal{M}_2 = \mu_0\mu_0, \mu_0\mu_1, \mu_1\mu_0, \mu_1\mu_1$.

The next generation, that is, $n = 3$, is obtained by dividing each partition of the unit interval into two new cells. The ith cell with content μ_i is separated into a left-hand cell with measure $\mu_j = \mu_i\mu_0$ and a right-hand cell with measure $\mu_{j+1} = \mu_i\mu_1$. As a result, the whole line segment [0, 1] is divided into cells of length $\delta = 2^{-3}$ and the set $\mathcal{M}$ at the third generation is given by the list of measures

$$\mathcal{M}_3 = \mu_i\Big|_{i=0}^{2^3-1} \tag{3.10}$$

and the measures are $\mu_0 = \mu_0\mu_0\mu_0$, $\mu_1 = \mu_0\mu_0\mu_1$, $\mu_2 = \mu_0\mu_1\mu_0$, $\mu_3 = \mu_0\mu_0\mu_1$, $\mu_4 = \mu_1\mu_0\mu_0$, $\mu_5 = \mu_1\mu_0\mu_1$, $\mu_6 = \mu_1\mu_1\mu_0$, and $\mu_7 = \mu_1\mu_1\mu_1$.

As this process is iterated, it produces shorter and shorter segments that contain less and less of the total measure. It is of interest that in the case $p = 0.25$ the measure $\mu(x)$ of the cell located at x and the measure $M(x, n) = \sum_{i=0}^{x\cdot 2^n} \mu_i$ (here the index x specifies the cell index $x = i \cdot 2^{-n}$) for the subregion of $\mathcal{S}$ defined by $[0, x]$ (and for the case $n = 11$) are the curves shown in Figures 3.1 and 3.2 [Marani et al., 1991]. Thus Peano's width function is precisely obtained by the Besicovitch multiplicative process with parameter $p = 0.25$. The parameter p describes the bifurcation structure as it defines the number of elements generated by each element of the lower hierarchical order in the ordering scheme, that is, $R_B = 1/p$. In particular, Figure 3.2 also shows the cumulated measure $M(x, 11)$ for the region $[0, x]$, after eleven generations of the multiplicative process of the parameter $p = 0.25$ described above.

We note that the measure $M(x, n)$ scales in the sense that the left and right halves of Figure 3.2 are obtained from the whole by the following relations:

$$M(x, n) = \begin{cases} p\,M(2x, n) & 0 \le x \le 1/2 \\ p + (1-p)M(2x-1, \Omega) & 1/2 < x < 1 \end{cases} \tag{3.11}$$

which are identical to those obtained by induction for Peano's basin (Eq. (3.4)).

Note the embedded self-affinity (see Chapter 2) in the recursive construction. In fact, the right half of each distribution (see Table 3.1) equals the left half times $p/(1-p)$, and the entire distribution is invariant as the left half is stretched by a factor of 2 in the horizontal direction and a factor of $1/(1-p)$ in the vertical direction [e.g., Schroeder, 1991]. In the limit as $n \to \infty$, the distribution $M(x, \infty) = M(x)$ over the entire unit interval equals the distribution over the left-half interval stretched horizontally by 2 and vertically by $1/(1-p)$:

$$M(x) = \frac{1}{1-p} M(x/2) \tag{3.12}$$

where the factors 2 and $1/(1-p)$ are the two scaling factors of this self-affine fractal.

After n generations there are $N = 2^n$ cells labeled sequentially by the index $i = 0, 1, \ldots, N-1$. The length of the ith cell is $\delta(n) = 1/2^n$, and the measure μ_i, or fraction of the population, in the cell can be written [Feder, 1988] as

$$\mu_i = \mu_0^k \mu_1^{n-k} \tag{3.13}$$

where k is the number of zeroes in the binary fraction representation of the number

$$x = i/2^n = \sum_{j=1}^{n} 2^{-j}\epsilon_j \tag{3.14}$$

where the digits ϵ_j have only two possible values, 0 or 1. For example, for $n = 3$ the first cell, $j = 0$, is represented by 0.000, the $j = 1$ cell by 0.001, the next cell by 0.010, and so on, until finally the last cell, $j = 7$, is represented by 0.111. Let us spell out the above in detail for the reader less familiar with binary notation. A binary representation is made up by 0s and 1s, say $x = 0.001$ means $x = 0 \cdot 2^0 + 0 \cdot 2^{-1} + 0 \cdot 2^{-2} + 1 \cdot 2^{-3} = 1/2^3 = 1/8$. We may prove by induction that indeed binary fractions having n digits are needed to represent all the cells in the nth generation. For $n = 3$ we have partitioned (Figure 2.24) the maximum distance from source to outlet in eight links. Thus the distances to the outlet of the farthest point of the link are in decimal units $x = 0, 1/8, 1/4, 3/8, 1/2, 5/8, 3/4, 7/8, 1$, respectively. In binary representation they are $x = 0.000, 0.001, 0.010, 0.100, 0.101, 0.110, 0.111$, respectively. We therefore have a one-to-one relation among n digits $\epsilon_1, \epsilon_2, \ldots, \epsilon_n$ and a real number x, that is, $x = (\epsilon_1, \epsilon_2, \ldots, \epsilon_n)$. Again, the key point is that binary fractions having n digits are needed to represent all the cells in the nth generation. Note that the reason for the binary notation of x to describe the above distribution is that each 0 in the binary expansion of x corresponds to a left-half interval, and each 1 to a right-half interval.

As a note, we observe that the notation μ_i or $\mu(x)$ for the distribution replaces the correct notation $\mu_i(B_\delta)$ or $\mu B_\delta(\mathbf{x})$, where B_δ, $B_\delta(\mathbf{x})$ denote the hypercubes defined by the mesh of side δ. In simpler words, B_δ denotes a closed ball of radius $\delta > 0$ located at $\mathbf{x}$ [e.g., Holley and Waymire, 1992].

Using the above binary representation, an alternative formulation is the following [Holley and Waymire, 1992]. Let $p + q = 1$, $0 < p \neq q < 1$, and μ is the measure on $[0, 1]$ defined by the prescription:

$$\mu_i = p^{\sum_{j=1}^{n} \epsilon_j} q^{n - \sum_{j=1}^{n} \epsilon_j} \tag{3.15}$$

For a suitable notion of dimension (see Chapter 2), the structure of the measure μ on a d-dimensional real space, R^d, may be defined by the scaling relationship

$$\mu \propto \delta^\alpha \quad \text{as } \delta \to 0 \tag{3.16}$$

where the *scaling exponent* α is the so-called Lipschitz–Holder exponent, which we will discuss in detail in Section 3.3. The richness of the scaling structure is enhanced if we observe a different behavior in different spatial positions, that is

$$\mu(x) \propto \delta^{\alpha(x)} \quad \text{as } \delta \to 0 \tag{3.17}$$

The set of values of α covers a spectrum whose width is characteristic of the process, as we will see in the following. From Eq. (3.15) we thus have at the nth generation, where $\delta = 2^{-n}$

$$p^{\sum_{j=1}^{n} \epsilon_j} q^{n-\sum_{j=1}^{n} \epsilon_j} = (2^{-n})^{\alpha} \tag{3.18}$$

and, on taking the logarithm in base 2 of both sides [Holley and Waymire, 1992] we obtain

$$\alpha = \alpha(x) = -\frac{1}{n}\sum_{j=1}^{n} \epsilon_j \log_2 p - \left(1 - \frac{1}{n}\sum_{j=1}^{n} \epsilon_j\right) \log_2 q \tag{3.19}$$

which defines a spectrum of values of α that correspond to different spatial positions x. The dependence on x is implicit because different points are represented by different sequences of zeros and ones described by the digits ϵ_j. We will show that a reasonable way of ordering such a set is to correlate it with a particular fractal dimension associated with the multiplicative process. In fact, every measure μ_i may be characterized by one exponent α such that $\mu_i \propto \delta^{\alpha}$, where, as in the above, δ is the ruler of the geometric support of μ_i. Hence the singularity exponent α is a function of the position i, but many sites i may share the same exponent when a regular covering of size δ is chosen. Therefore let

$$N_{\delta}(\alpha) = \#\{i : \mu_i \geq \delta^{\alpha}\} \quad \delta \to 0 \tag{3.20}$$

be the number of sites i that share the same measure μ_i (here the symbol # means "the number of"). We then define in a rough manner (a more precise definition based on limit properties follows in the section devoted to the computation of the singularity spectrum) the scaling properties of the set $N_{\delta}(\alpha)$ as

$$N_{\delta}(\alpha) \propto \delta^{-f(\alpha)} \tag{3.21}$$

where therefore

$$f(\alpha) = \lim_{\epsilon \to 0} \lim_{\delta \to 0} -\frac{\log(N_{\delta}(\alpha + \epsilon) - N_{\delta}(\alpha))}{\log \delta} \tag{3.22}$$

The preceding point is now explained with a precise example. Consider a width function, say the set of data obtained from a DEM, as shown in Section 1.2.9. Let $\mathcal{N}_i$ be the number of pixels, that is, the drainage area, placed at the ith distance (cell) from the outlet. Let $\mathcal{N}$ be the total number of pixels in the DEM of the river basin considered. Thus $\sum_{1=1}^{i_{\max}} \mathcal{N}_i = \mathcal{N}$. We define the quantity

$$\mu_i = \frac{\mathcal{N}_i}{\mathcal{N}} \tag{3.23}$$

which represents, for an established resolution, the normalized value of the width function, which is a probability measure ($\sum_i \mu_i = 1$). A plot of the distribution of the measure is shown in Figure 3.4. Let us choose, as in

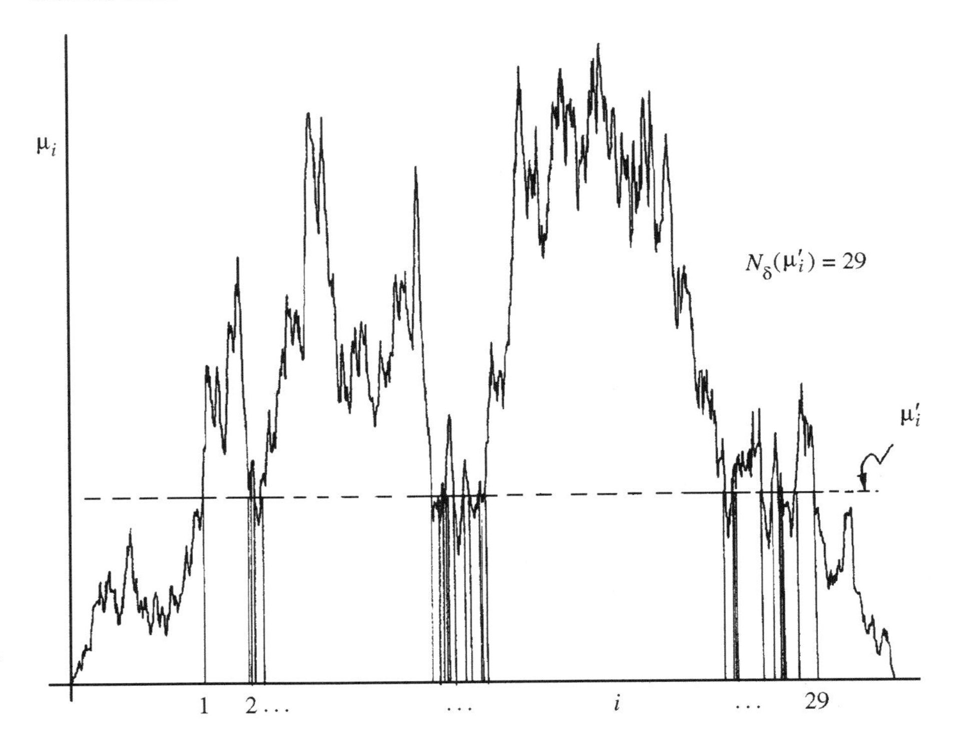

Figure 3.4. A partition of the set of points of the normalized width function obtained by isolating the distances i characterized by the same probability μ_i.

Figure 3.4, an arbitrary value μ_i and let $\mathcal{S}_i$ be the set of points i characterized by the same value of the measure μ_i. The sets $\mathcal{S}_i$, $i = 1, \ldots, i_{\max}$ are a partition of the set $\mathcal{S}$ of all points belonging to the width function. To compute the fractal dimension of the arbitrary set $\mathcal{S}_i$, we cover it by rulers of size δ and determine the exponent D that yields a finite (nonzero) value of the sum:

$$\sum_{\mathcal{S}_i} \delta^D = N_\delta(\mu_i)\, \delta^D \tag{3.24}$$

where, as in Figure 3.4, $N_\delta(\mu_i)$ is the number of μ_i-crossings as $\delta \to 0$. In general, we will have, as suggested by Figure 3.4

$$D = f\,(\mu_i) \tag{3.25}$$

and because we have argued that $\mu \propto \delta^\alpha$, we observe that

$$D = f\,(\alpha) \tag{3.26}$$

From the standpoint of the correct notation, we observe (for the last time, because in the following we will follow the less rigorous simplified notation) that we should have written $\mu B_\delta(\mathbf{x}) \propto \delta^\alpha$ and $\alpha = \lim_{\delta \to 0} \log \mu B_\delta(\mathbf{x}) / \log \delta$.

From all the above we therefore conclude that Figure 3.1 also shows, in addition to the width function of an $\Omega = 11$ Peano's basin, the (probability) measure $\mu(x)$ of the partitions of the total area generated in the 11th stage of generation of the multiplicative process with the parameter $p = 0.25$. As shown by Eq. (3.13), one cell has the highest measure $(1 - p)^{11}$. There are eleven cells with measure $(1 - p)^{10} p^1$ and so on. One can show –

by complete induction – that at the level n of generation we have, with $\xi = k/n$ and $k = 0, 1, \ldots n$,

$$N_n(\xi) = \binom{n}{\xi n} = \frac{n!}{(\xi n)!((1-\xi)n)!} \tag{3.27}$$

intervals with measure

$$\mu_\xi = \mu_0^{n\xi} \mu_1^{n(1-\xi)} = p^{n\xi}(1-p)^{n(1-\xi)} \tag{3.28}$$

The total measure of the segments representing the population is

$$M(x=1,n) = \sum_{i=1}^{2^n} \mu_i = \sum_{\xi=0}^{1} N_n(\xi)\mu(\xi) = \big(p + (1-p)\big)^n = 1 \tag{3.29}$$

because of the binomial expansion theorem. The cells, describing the distribution of the population, cover the line completely and contain all of the measure, that is, every member in the population.

In the nth generation, $N_n(\xi)$ line segments are generated with length $\delta_n = 2^{-n}$, all having, by construction, the same measure $\mu(\xi)$. These segments form a subset, $\mathcal{S}_n(\xi)$, of the unit interval $\mathcal{S} = [0, 1]$. The points in the set have the same number k, $k = \xi n$, of zeroes among the n first decimal places in the binary expansion of the x coordinate of the points. One finds [Feder, 1988], in the limit $n \to \infty$, that ξ is the fraction of zeros in the infinite binary fraction representation of the points in the set $\mathcal{S}_\xi$. We now show, following Feder [1988], that this set is a *fractal* set of points. To see this, we cover the set with intervals of length δ and form the d-dimensional measure $M_d(\mathcal{S}_\xi)$. We then compute the fractal dimension $D(\xi)$ of this set by studying how M_d behaves as $\delta \to 0$, that is, $n \to \infty$:

$$M_d(\mathcal{S}_\xi) = \sum_{\mathcal{S}_\xi} \delta^n = N_n(\xi)\,\delta^d \tag{3.30}$$

which, in the fractal case, should yield $N_n(\xi)\delta^d \to 0$ for $d > D(\xi)$ and $N_n(\xi)\delta^d \to \infty$ for $d > D(\xi)$, $D(\xi)$ being the fractal dimension of the set.

We see that this is indeed the case for Peano's basin and for its underlying deterministic multiplicative process. If we make use of Stirling's formula for the factorial $n!$

$$n! = \sqrt{2\pi}\, n^{n+1/2} e^{-n} \tag{3.31}$$

we find an approximate expression for $N_n(\xi)$ in Eq. (3.27) as

$$N_n(\xi) \approx \frac{1}{\sqrt{2\pi n\xi(1-\xi)}} e^{-n(\xi\ \log\xi + (1-\xi)\log(1-\xi))} \tag{3.32}$$

Noting that $n = -\log\delta/\log 2$, we find that the measure M_d in Eq. (3.30) may be written, after neglecting the term $n^{-1/2}$ [Feder, 1988], as

$$M_d(\mathcal{S}_\xi) \propto \delta^{-f(\xi)}\delta^d \tag{3.33}$$

where the new exponent $f(\xi)$ is defined by the following relationship:

$$f(\xi) = -\frac{\xi\log\xi + (1-\xi)\log(1-\xi)}{\log 2} \tag{3.34}$$

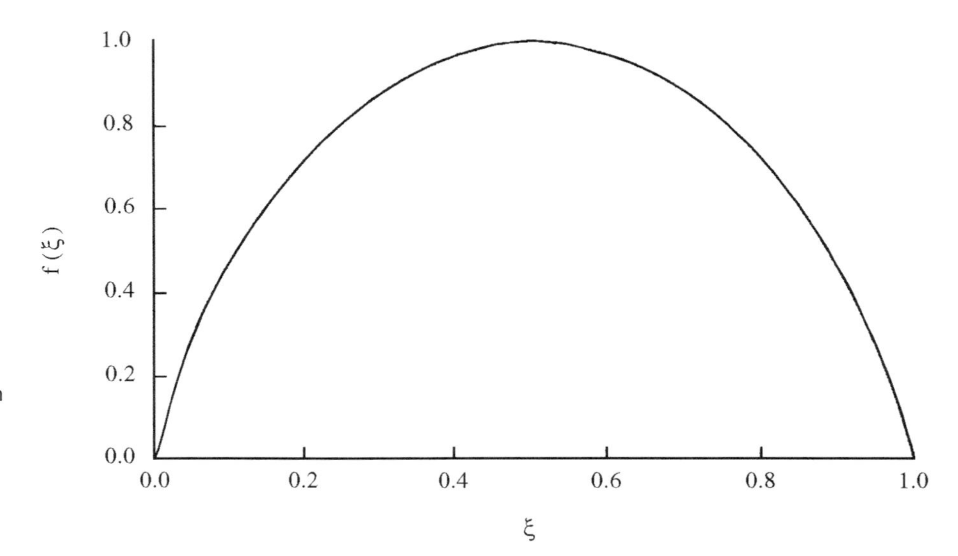

Figure 3.5. The fractal dimension $f(\xi)$ of subsets $\mathcal{S}_\xi$ of the interval that contains points x having a fraction ξ of zeroes in the binary expansion of x as a function of ξ [after Feder, 1988].

It then follows that the d-measure M_d remains finite as $\delta \to 0$ only for $d = f(\xi)$, and therefore the fractal dimension, $D(\xi)$, of the set $\mathcal{S}_\xi$ is precisely $f(\xi)$.

It also follows that the population generated by the multiplicative process is spread over a set of points in the unit interval $\mathcal{S} = [0, 1]$. This set is a union of subsets $\mathcal{S}_\xi$ that are fractal, with fractal dimensions $f(\xi)$ given by Eq. (3.34). The fractal dimension depends on the position parameter ξ as shown in Figure 3.5 and thus we have created a *multifractal* object as already suggested by Figure 3.3. From Figure 3.5 we note that the maximum fractal dimension $f(\xi)$ is unity, that is, the fractal dimension of the support, and that the minimum $f(\xi)$ is zero, owing to the fact that the number N_δ of sites that share the value $\mu = \mu(0) = 1/2^n$ is one (for $x = 0$) at every stage n of generation, that is, $N_\delta \propto \delta^0 = 1$.

The measure $M(x)$ of the population distributed over the unit interval $\mathcal{S} = [0, 1]$ is completely characterized by the union of the fractal sets $\mathcal{S}_\xi$. Note that each set in the union is fractal with its own fractal dimension. This is one reason for the term *multifractal*. We will show in Section 3.3 that the position parameter ξ defined from the binary representation of x is not very handy in practice and that another parametrization is more commonly used, that is, the multifractal spectrum $f(\alpha)$ versus α. In fact, organizing the data according to the position $0 \le x \le 1$ as in Figure 3.2 focuses on a very specific partition process, although organizing the data by counting the proportion of mass μ and by the number of its occurrences is more general.

3.3 Multifractal Spectra

The Lipschitz–Holder exponent α is useful for characterizing the complex scaling structure of a multifractal measure $M(x)$ [e.g., Mandelbrot, 1983]. Our exposition again closely follows that of Feder [1988]. Consider, for instance, the measure $M(x) = M(x, n \to \infty)$ discussed in Section 3.2 that was generated by a multiplicative process at the nth level of generation. This measure is a nondecreasing function of x, with *increments* $\mu(\xi) = \Delta^n(\xi)$. As seen above, in all x that have $\xi \cdot n$ zeroes among the first n digits of the corresponding binary fraction of x, one has

$$x = \sum_{j=1}^{n} 2^{-j}\epsilon_j \tag{3.35}$$

with the same notation used in Eq. (3.34). We choose a value $x(\xi)$ corresponding to a given value of ξ, which belongs to the set $\mathcal{S}_\xi$. The measure $M(x)$ is also given at a point $x(\xi) + \delta$ with, as usual, $\delta = 2^n$. The increment in $M(x)$ between the points $x(\xi)$ and $x(\xi) + \delta$ is μ_ξ, which is defined by the following equation:

$$\mu(\xi) = M\big(x(\xi) + \delta\big) - M\big(x(\xi)\big) \propto \delta^\alpha \tag{3.36}$$

where we have defined the Holder–Lipschitz exponent α as in Eq. (3.16), that is, $\mu(\xi) \propto \delta^\alpha$.

In any of the subsequent generations more and more points in the set $\mathcal{S}_\xi$ are obtained although Eq. (3.36) is valid throughout, that is, even in the limit $n \to \infty$.

It follows from Eqs. (3.27) and (3.35), and from the relationship $\delta = 2^{-n}$, that the measure for a multiplicative population has Lipschitz–Holder exponents α given by

$$\alpha(\xi) = \frac{\log \mu(\xi)}{\log \delta} = -\frac{\xi\ \log p + (1-\xi)\ \log(1-p)}{\log 2} \tag{3.37}$$

We observe that α is a linear function of ξ and is a function of the weight p defining the subdivision of the intervals of the multiplicative process. We find for the multiplicative measure with $p \le 0.5$ that $\alpha_{\min} \le \alpha \le \alpha_{\max}$ with

$$\alpha_{\min} = -\frac{\log(1-p)}{\log 2} \quad \text{for } \xi = 0 \tag{3.38}$$

and

$$\alpha_{\max} = -\frac{\log p}{\log 2} \quad \text{for } \xi = 1 \tag{3.39}$$

In the case of Peano's basin, where $p = 1/4 = 0.25$, then $\alpha_{\min} = 0.41$ and $\alpha_{\max} = 2$.

It is useful to note that a one-to-one correspondence between the parameters α and ξ is established. Thus the subsets $\mathcal{S}_\xi$ may also be written as $\mathcal{S}_\alpha$. The measure $M(x)$ is characterized by the sets $\mathcal{S}_\alpha$ whose union set is the unit interval $\mathcal{S} = [0, 1]$.

The measure has *singularities* with Lipschitz–Holder exponent α on the fractal sets $\mathcal{S}_\alpha$, and it is crucial to realize that they have fractal dimension $f(\alpha) = f(\xi(\alpha))$ [Feder, 1988].

Thus technically, we can generate the $f(\alpha)$ curve for the measure of the population generated by the multiplicative process by generating a sequence of values of ξ, $0 \le \xi \le 1$. For each ξ we then obtain a value of α,

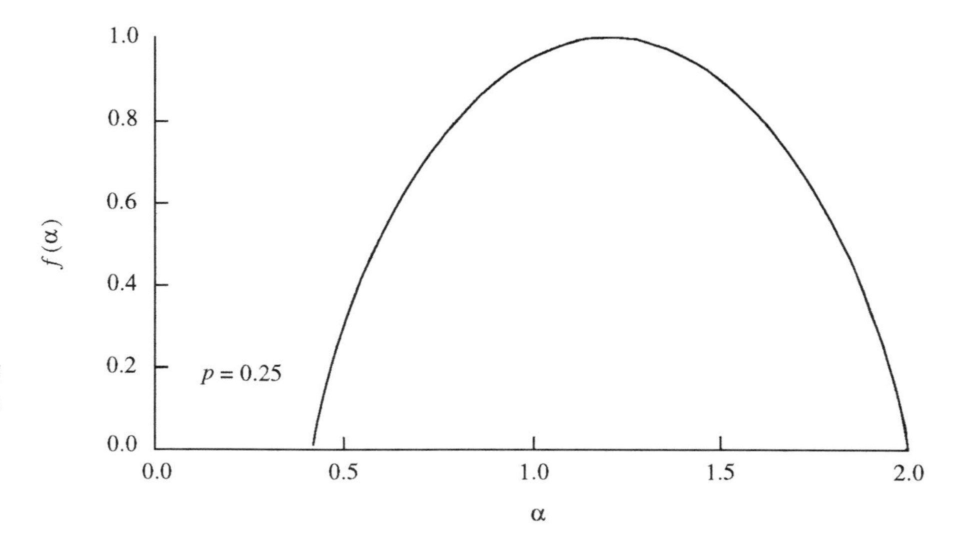

Figure 3.6. The multifractal spectrum of the width function of Peano's basin and of the measure constructed by the binomial multiplicative process of parameter $p = 0.25$ [after Feder, 1988].

$\alpha_{\min} \leq \alpha \leq \alpha_{\max}$, via Eq. (3.37) and the corresponding value of $f(\alpha(\xi))$ via Eq. (3.34).

Figure 3.6 shows the resulting curve $f(\alpha)$ versus α, called the *multifractal spectrum*, for the binomial multiplicative process with $p = 0.25$, corresponding to the normalized width function of Peano's basin.

The multifractal spectrum in Figure 3.6 has a few general features. The derivative of $f(\alpha)$ is

$$\frac{df(\alpha)}{d\alpha} = \frac{df}{d\xi}\frac{d\xi}{d\alpha} = -\frac{\log\xi - \log(1-\xi)}{\log p - \log(1-p)} \tag{3.40}$$

having used via Eq. (3.34), $df/d\xi = -(\log\xi - \log(1-\xi))/\log 2$ and, via Eq. (3.37), $d\alpha/d\xi = (\log p - \log(1-p))/\log 2$. The maximum is $f(\alpha_0) = 1$, with $\xi = 1/2$ and $\alpha_0 = -(\log p + \log(1-p))/(2\log 2)$.

Feder [1988] notes that as a general result the maximum value of f equals the fractal dimension of the *geometric support* of the measure μ, which in this case is one because the measure is defined over the length of the unit interval. For measures defined on fractal geometric supports, that is, with fractal dimension D, one finds $f_{\max}(\alpha) = D$.

In the case of Peano's basin (i.e., of the multiplicative process of parameter $p = 0.25$), the maximum occurs at $\alpha_0 = 1.207$.

The discussion of the $M(x)$ function properties is somewhat delicate and involves questions such as whether or not the limit points of the sequence of points generated by the multiplicative process are included [see Mandelbrot, 1983; Feder, 1988]. We just comment here that the curve $M(x)$ has zero derivative almost everywhere, that is, at all points except on a set of points with measure zero. Nevertheless, it increases from 0 to 1 as x increases from 0 to 1. It is the devil's staircase shown in Figure 3.2. We will not pursue mathematical issues of the nature of the devil's staircase any further because our focus is on the description of natural river basins. Rather, we will concentrate on the computation of the $f(\alpha)$ curve for measures obtained from real data.

Another special point on the $f(\alpha)$ curve occurs where $\xi = p$, that is

$$\frac{df(\alpha_S)}{d\alpha} = 1 \qquad f(\alpha_S) = \alpha_S = -\frac{p\ \log p + (1-p)\log(1-p)}{\log 2} \tag{3.41}$$

where a line through the origin is tangent to $f(\alpha)$.

The multifractal spectrum, as we have introduced it, partitions the support of a measure into subsets that contain points characterized by the same exponent α in the scaling behavior of local integrals (i.e., the 'mass' contained in an interval of length δ). A more general definition of multifractal spectrum partitions the support of a measure by use of a generic local property P_α [Veneziano, Moglen, and Bras, 1995].

Let $y(\mathbf{x})$ be a measure, defined in a region Ω of the n-dimensional space $\boldsymbol{R}^n$ and P_α be a local property of y, where α is a parameter. Denote by Ω_α a subset of Ω for which $\alpha(\mathbf{x}) = \alpha$. The multifractal spectrum of y relative to property P_α may be defined as $(\alpha, f(\alpha))$, where $f(\alpha)$ is a fractal dimension of Ω_α.

One may now introduce different multifractal spectra making use of relevant properties of $y(\mathbf{x})$. For example, one may divide Ω into sets Ω_α for which $y(\mathbf{x}) > \alpha$. This characterization is equivalent to a cluster analysis that studies the fractal character of 'islands' for which $y(\mathbf{x}) > \alpha$.

The usual definition of a multifractal spectrum is retrieved if one considers the sets for which the integral of $y(\mathbf{x})$ over a 'box' of size δ centered in $\mathbf{x}$ scales as δ^α.

We now turn to the problem of estimating the sequence of exponents of singularities and of detecting a possibly multifractal structure for data observed experimentally. This estimation was done, for example, for coastlines and real topographies [Mandelbrot, 1983, 1990; Lavallée et al., 1993], for cloud and rain fields [Lovejoy, Schertzer, and Tsonis, 1987; Gupta and Waymire, 1993], or for viscous fingering patterns [e.g., Feder, 1988]. In the case of the geometry of real river basins, a few studies [e.g., Rodriguez-Iturbe et al., 1992c; Rinaldo et al., 1992] have focused on the scaling structures of measures constructed through the width function.

The evaluation of the spectrum of singularities is not an easy task, but it is essential when complex intertwined processes shape the geometry under study. In fact, experimental observations and the results of numerical simulations of such processes basically provide sets of points $\mathcal{S}$ that need to be analyzed. Perhaps the most widely used method for the analysis of the structure of such sets is the box-counting methodology described in Chapter 2. In this method the E-dimensional space of the observations is partitioned into (hyper-)cubes with side δ, and one counts the number $N(\delta)$ of cubes that contain at least one point of the set $\mathcal{S}$. Clearly, as noted by Feder [1988], this is the crudest form of measure for the set, and it gives limited information on the structure of the set. For instance, if a coastline folds back and forth so that it crosses a given 'box' a number of times, that box still contributes only once to the number of boxes needed to cover the set. In the same vein, the usual fractal dimension D, for the multiplicative process based on the limit as $\delta \to 0$ of $\log N/\log(1/\delta)$, is of no help because after n iterations the number of pieces N equals 2^n and the length δ of each piece equals 2^{-n}.

The singularity spectrum partially addresses this problem. It allows, through the definition of the fractal dimensions $f(\alpha)$, the study of the scaling structure of the number of sites with the same measure. Also, the measure has unequal scaling depending on the spatial position within the geometric support on which the measure is distributed.

Let us review the necessary mathematical tools for understanding the procedure for the extraction of the singularity spectrum from data sets. A set $\mathcal{S}$ consisting of $\mathcal{N}$ members will have $\mathcal{N}_i$ members in the ith cell in which the geometric support is partitioned. In the case of the width function, the cell is naturally the pixel size of the DEM as seen in Chapter 1. The points $\mathcal{N}_i$ are assumed to be sample points of an underlying measure. Let us use the *mass* (i.e., *probability*) $\mu_i = \mathcal{N}_i/\mathcal{N}$ in the ith cell to construct the measure, which may be written in the following manner [see, e.g., Feder, 1988, Section 6.7]:

$$M_d(q,\delta) = \sum_{i=1}^{N} \mu_i^q \delta^d = N(q,\delta)\delta^d \tag{3.42}$$

This measure depends on the scale δ and on the exponent q, which play a crucial role that is discussed throughout this section. Also, $M_d(q,\delta)$ being a fractal object, one argues that

$$N(q,\delta)\delta^d \to \begin{cases} 0 & \text{if } d > \tau(q) \\ \infty & \text{if } d < \tau(q) \end{cases} \tag{3.43}$$

as $\delta \to 0$. This measure has therefore a *mass* exponent $d = \tau(q)$ for which the measure neither vanishes nor diverges as $\delta \to 0$. The mass exponent $\tau(q)$ is sometimes called the Rényi exponent [Rényi, 1970; Holley and Waymire, 1992]. It depends on the moment order q chosen. The measure is then characterized by a whole sequence of exponents $\tau(q)$ that controls how the moments of the probabilities μ_i scale with δ. It follows from Eq. (3.43) that the weighted number of boxes $N(q,\delta)$ has the form

$$N(q,\delta) = \sum_{i=1}^{N} \mu_i^q \propto \delta^{-\tau(q)} \tag{3.44}$$

and the mass exponent is therefore given by the following equation:

$$\tau(q) = -\lim_{\delta\to 0} \frac{\log N(q,\delta)}{\log \delta} \tag{3.45}$$

Operationally, we need to relate $\tau(q)$ in Eq. (3.45) to the singularity spectrum. To do so, we first note that if we choose $q = 0$ for the moment order q, then we have $\mu_i^0 = 1$. Therefore we find that $N(0,\delta) = N(\delta)$ is simply the number of boxes needed to cover the set, and $\tau(0) = D$ therefore equals the fractal dimension of the support. Also, the probabilities are normalized, that is, $\sum_i \mu_i = 1$, and it follows from Eq. (3.45) that $\tau(1) = 0$.

As Feder [1988] points out, choosing large values of q, say 10 or 100, in Eq. (3.45) favors contributions from cells with relatively high values of μ_i because $\mu_i^q >> \mu_j^q$ if $\mu_i > \mu_j$ with $q >> 1$.

Conversely, an exponent $q \ll -1$ highly favors cells i with relatively low values of the measure μ_i. These limits are best discussed by considering the derivative $d\tau(q)/dq$ of the mass exponents, given by the following equation:

$$\frac{d\tau(q)}{dq} = -\lim_{\delta \to 0} \frac{\sum_i \mu_i^q \log \mu_i}{\left(\sum_i \mu_i^q\right) \log \delta} \tag{3.46}$$

because $d(\log \sum_i \mu_i^q)/dq = (d \sum_i \mu_i^q / dq)/(\sum_i \mu_i^q) = (\sum_i de^{q \log \mu_i}/dq)/(\sum_i \mu_i^q) = (\log \mu_i e^{q \log \mu_i})/(\sum_i \mu_i^q) = (\sum_i \mu_i^q \log \mu_i)/(\sum_i \mu_i^q)$.

Following Feder [1988], we will now examine some properties that link the mass exponents $\tau(q)$ to the Holder–Lipschitz exponents α. Let μ_- be the minimum value of μ_i in the sum. The limit for $q \to -\infty$ of Eq. (3.46) is (recall the definition of the Holder–Lipschitz exponent via Eq. (3.37))

$$\frac{d\tau(q)}{dq}|_{q \to -\infty} = -\lim_{\delta \to 0} \frac{\left(\sum_i' \mu_-^q\right) \log \mu_-}{\left(\sum_i' \mu_-^q\right) \log \delta} = -\lim_{\delta \to 0} \frac{\log \mu_-}{\log \delta} = -\alpha_{\max} \tag{3.47}$$

where the prime on the sum indicates that only a fraction of the $\mathcal{N}$ cells contribute, that is, only the cells with $\mu_i = \mu_-$. An analogous argument holds in the limit $q \to \infty$, that is

$$\frac{d\tau(q)}{dq}|_{q \to \infty} = -\lim_{\delta \to 0} \frac{\log \mu_+}{\log \delta} = -\alpha_{\min} \tag{3.48}$$

where μ_+ is the largest value of μ_i that leads to the smallest value of α. We will show later that indeed in general $\alpha = -d\tau/dq$.

Before we proceed any further into the analysis of the mathematical properties of the mass, or Rényi, exponents $\tau(q)$, it is both expedient and interesting to compute them for the process that describes the width function of Peano's basin. The general behavior of the sequence of mass exponents $\tau(q)$ is illustrated by the measure on the interval generated by the multiplicative binomial process in which n partitions of the unit interval are generated. For this process we find that (recall Eq. (3.27))

$$N(q, \delta) = \sum_{k=0}^{n} \binom{n}{k} p^{qk}(1-p)^{q(n-k)} = \left(p^q + (1-p)^q\right)^n \tag{3.49}$$

because of the binomial expansion formula. Substituting the generation number given by $n = -\log \delta / \log 2$ in the above developments, and using Eq. (3.45), one obtains

$$\tau(q) = \frac{\log(p^q + (1-p)^q)}{\log 2} \tag{3.50}$$

The resulting sequence of mass exponents is shown in Figure 3.7. For $q = 0$, we find $\tau(0) = 1$, which is the dimension of the support, that is, the unit interval.

Feder [1988] shows that the sequence of mass exponents is also related to the $f(\alpha)$ curve in a general way that is useful for applications of the analysis of real data sets. A *multifractal measure* is supported by a set $\mathcal{S}$, which is the union of fractal subsets $\mathcal{S}_\alpha$ with α chosen in the allowed spectrum of values. Assuming that the complete set $\mathcal{S}$ is fractal, with a fractal dimension D, the subsets have fractal dimensions $f(\alpha) \leq D$. For fractal subsets, with a fractal dimension $f(\alpha)$, the number $N(\alpha, \delta)$ of segments of

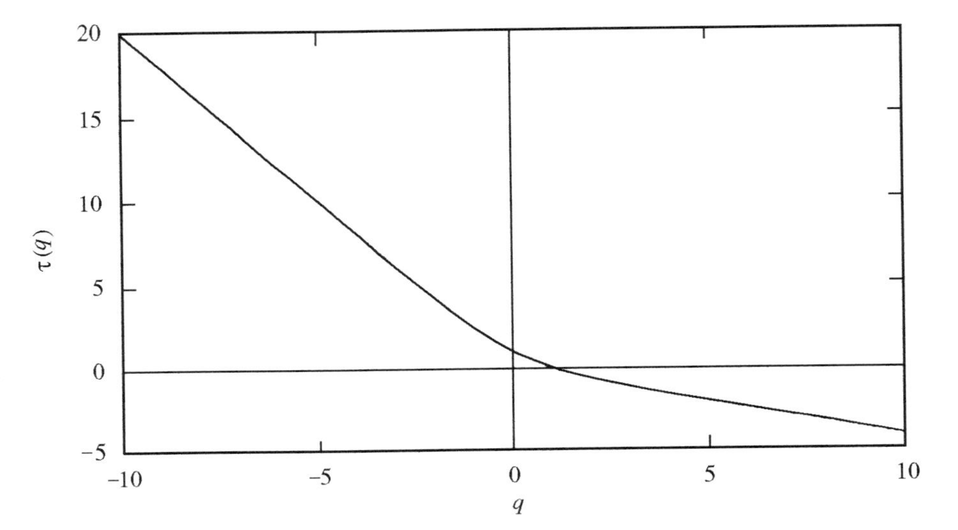

Figure 3.7. The sequence of mass exponents $\tau(q)$ as a function of the moment order q for the width function of Peano's basin, that is, the binomial multiplicative process [after Feder, 1988].

length δ needed to cover the sets $\mathcal{S}$ with α in the range $(\alpha, \alpha + d\alpha)$ is

$$N(\alpha, \delta) = \rho(\alpha)d\alpha \quad \propto \quad \delta^{-f(\alpha)}d\alpha \tag{3.51}$$

where $\rho(\alpha)$ is the density of segments from $\mathcal{S}_\alpha$ to $\mathcal{S}_{\alpha+d\alpha}$, that is, $\rho(\alpha) \approx \delta^{-f(\alpha)}$. For such sets the measure μ_α in a cell of size δ has a power law dependence (i.e., $\mu_\alpha \propto \delta^\alpha$) on the length scale δ. Arguing that $\sum_i \mu_i^q \approx \int_{\alpha_{\min}}^{\alpha_{\max}} d\alpha \; N(\alpha, \delta)\mu_\alpha^q$, the measure M given in Eq. (3.42) may be written as follows as $\delta \to 0$:

$$M_d(q, \delta) = \int_{\alpha_{\min}}^{\alpha_{\max}} d\alpha \quad \delta^{-f(\alpha)}\delta^{\alpha q}\delta^d = \int_{\alpha_{\min}}^{\alpha_{\max}} d\alpha \quad \delta^{-f(\alpha)+\alpha q+d} \tag{3.52}$$

(a formally rigorous proof can be found, for instance, in Falconer [1990]). Feder [1988] notes that the integral in Eq. (3.52) is dominated by the terms where the integrand has its maximum value because $\delta \to 0$. Thus

$$\int_{\alpha_{\min}}^{\alpha_{\max}} d\alpha \quad \delta^{\alpha q - f(\alpha)} \approx \delta^{\sup_{\alpha_{\min} \le \alpha \le \alpha_{\max}}(\alpha q - f(\alpha))} \tag{3.53}$$

The maximum value of the exponent is achieved for the value of $\alpha(q)$ such that

$$\left.\frac{d(q\alpha - f(\alpha))}{d\alpha}\right|_{\alpha=\alpha(q)} = 0 \tag{3.54}$$

and therefore

$$\frac{d\,f(\alpha(q))}{d\alpha} = q \tag{3.55}$$

Thus the integral in Eq. (3.52) is asymptotically given by

$$M_d(q, \delta) \approx \delta^{q\alpha(q)-f(\alpha(q))+d}(\alpha_{\max} - \alpha_{\min}) \propto \delta^{q\alpha(q)-f(\alpha(q))+d} \tag{3.56}$$

Finally, we note that M_d remains finite in the limit $\delta \to 0$ if d equals the mass exponent $\tau(q)$ given by

$$\tau(q) = f(\alpha(q)) - q\alpha(q) \tag{3.57}$$

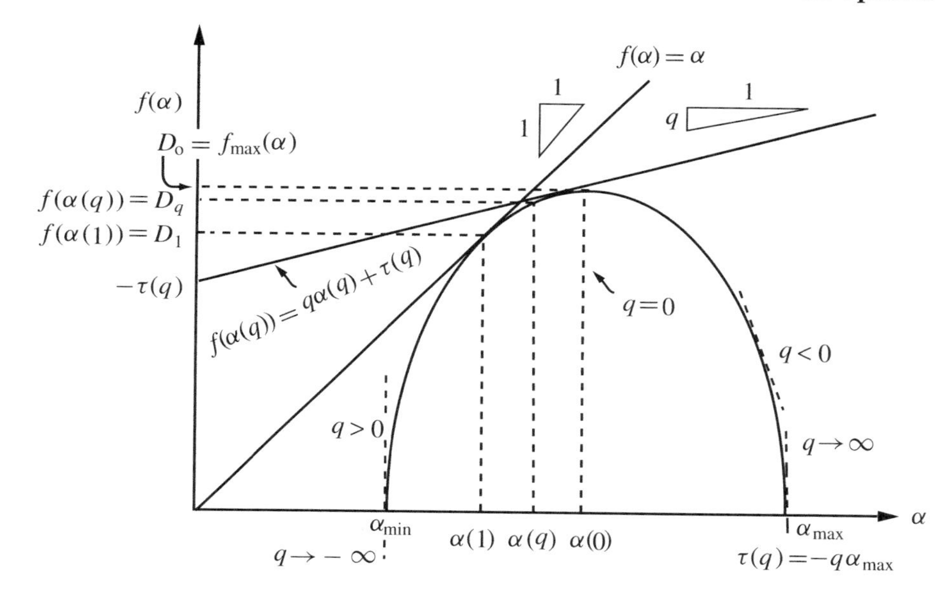

Figure 3.8. A multifractal spectrum, where the most significant quantities relating f, α with q, $\tau(q)$ are indicated.

where $\alpha(q)$ is the solution to Eq. (3.55). It is thus shown that the mass exponent $\tau(q)$ is given in terms of the Lipschitz–Holder exponent $\alpha(q)$ and the fractal dimension $f(\alpha(q))$ of the support.

Conversely, once the mass exponents $\tau(q)$ are known, one may determine the spectrum of singularities using Eqs. (3.55) and (3.57), yielding the so-called Legendre transformation [Feder, 1988] of the variables τ and q to the variables f and α:

$$\alpha(q) = -\frac{d}{dq}\tau(q)$$

$$f(\alpha(q)) = q\alpha(q) + \tau(q) \tag{3.58}$$

where the first relation follows from taking the derivative of Eq. (3.57) because $d\tau(q)/dq = \alpha + q d\alpha/dq - df/dq = \alpha + df/d\alpha\, d\alpha/dq - df/dq = \alpha$ and making use of Eq. (3.55). We thus conclude that the slope of the singularity spectrum is the value of q. These two equations give a parametric representation of the singularity spectrum, and they constitute a Legendre transformation of the independent variables τ and q to the independent variables α and f.

The maximum of the $f(\alpha)$ curve occurs for $df/d\alpha = 0$. From the above mathematical description, it follows that the maximum is obtained for $q = 0$, and we conclude from Eq. (3.57) that $f(\alpha)_{\max} = D$, because $\tau(0) = D$, where D is the fractal dimension of the support of the measure. Thus $N(0, \delta) = N_\delta$ is the number of hypercubes required to cover the support of the measure. Also, a generalized dimension is defined as $D_q = \tau(q)/(q-1)$ [e.g., Schroeder, 1991].

The various relations relating $\tau(q), q, \alpha,$ and $f(\alpha)$ are synthesized in Figure 3.8.

Using Eq. (3.58) for the simple example of the binomial multiplicative process with $p = 0.25$ given by Eq. (3.34), we recover the curve shown in Figure 3.6.

The estimation of $f(\alpha)$ versus α from data is difficult to perform. In fact, the direct determination of $\tau(q)$ followed by the application of the Legendre transform, Eq. (3.58), may induce a noteworthy artificial smoothing via the required intermediate step caused by the determination of the Rényi coefficient. As a result, one may hide discontinuities of $f(\alpha)$ corresponding to physical phenomena. Also, it has been noted [e.g., Chhabra and Jensen, 1989] that the application of the above method suffers from mathematical ambiguities, related, for example, to whether f is a Hausdorff dimension rather than a box dimension and to large computational errors due to logarithmic corrections [e.g., Meneveau and Sreenivasan, 1987] that arise from scale-dependent prefactors in Eq. (3.21). However, despite these inaccuracies in the quantitative assessment of the spectrum, the methods provide important qualitative information about the statistical properties of the measure [Mandelbrot, 1990]. Chhabra and Jensen [1989] proposed the following method, which circumvents the above difficulties and proves superior to other (known to date) techniques for estimation of the singularity spectrum.

From Eqs. (3.46) and (3.58) we have

$$\alpha(q) = \lim_{\delta \to 0} \frac{\sum_i \mu_i^q \log \mu_i}{\log \delta \sum_i \mu_i^q} \tag{3.59}$$

and then

$$\begin{aligned} f(\alpha(q)) &= \tau(q) + q\,\alpha(q) \\ &= \lim_{\delta \to 0} \frac{1}{\log \delta} \left(-\log \sum_i \mu_i^q + q \frac{\sum_i \mu_i^q \log \mu_i}{\sum_i \mu_i^q} \right) \\ &= \lim_{\delta \to 0} \frac{1}{\log \delta} \left(\frac{\log(\sum_i \mu_i^q \log \mu_i^q / \sum_i \mu_i^q)}{\sum_i \mu_i^q} \right) \end{aligned} \tag{3.60}$$

The above set of equations directly provides the values of f and α parametrized by q. Rigon [1994] pointed out an important computational problem related to the way the limit in Eq. (3.60) is actually performed. The problem is succinctly described as follows. Let $f(\alpha)_\delta$ be the value of f at the resolution δ. On changing δ, the quantity $\log(\sum_i \mu_i^q \log \mu_i^q / \sum_i \mu_i^q)/\sum_i \mu_i^q$ should vary linearly with $\log \delta$, as required by Eq. (3.60). Therefore $f(\alpha)$ may be obtained as the slope of the straight line found by linear interpolation of the pairs of points obtained at different values of δ.

Technically, one has to establish a sequence j of rulers δ_j of decreasing length with which to cover the geometric support. With each choice of $\delta = \delta_j$ one needs to evaluate the terms appearing in Eq. (3.60). To change scales, one observes that because of the additive property of the measure (let $\mu_i(\delta)$ be the measure at the ith site with ruler δ), one has

$$\mu_j(\delta_{n-1}) = \sum_{i \in j} \mu_i(\delta_n) \tag{3.61}$$

where δ_{n-1} is an integer multiple of δ_n and $i \in j$ denotes the set of intervals of diameter δ_n included into the j th interval of diameter δ_{n-1}. Various strategies [Rigon, 1994] might be adopted for the choice of the sequence of intervals, and all of them must obey the condition in Eq. (3.61). Notice that the procedures embedded in Eq. (3.61) postulate concepts of *coarse graining* (cf. Section 4.18) and their related suitable aggregation laws whenever the measure has dimensions $D > 1$. In simpler cases one has $D \leq 1$, as in the noteworthy case of the width function.

In summary, the procedure leading to the determination of the singularity spectrum is the following:

- The discretized data $(i\delta_n, \mu_i(\delta_n))$, $i = 1, \ldots, N$, are collected and an ordered sequence of rulers of diameters $\delta_1, \ldots, \delta_n$ is chosen.
- For every value of q, the moment exponent, one constructs the sequence $(\alpha_q, f(\alpha(q)))_{\delta_j}$ for $j = 1, \ldots, n$ via the coarse-graining procedure in Eq. (3.61). The value of $f(\alpha)$ is obtained by the slope of the graph $\log(\sum_i \mu_i^q(\delta_j) \log \mu_i^q(\delta_j))/\sum_i \mu_i^q(\delta_j)$ versus $\log \delta_j$. Similarly, α is obtained by the slope of the plot $\sum_i \mu_i^q(\delta_j) \log \mu_i(\delta_j)/\sum_i \mu_i^q(\delta_j)$ versus $\log \delta_j$.
- The entire operation is repeated for another value of q.

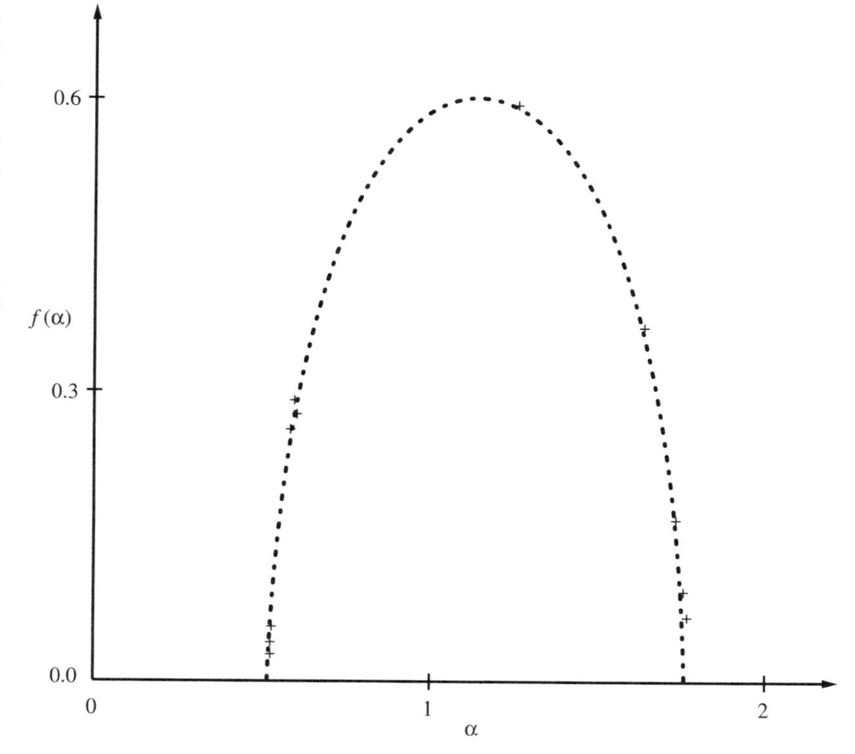

Figure 3.9. Comparison of the analytical and the computed spectra of the binomial multiplicative process with $p = 0.7$ [after Chhabra and Jensen, 1989].

The result is a set of pairs of values (α and $f(\alpha)$), each corresponding to a different value of q, which acts as a parameter. It is important to note that the error of estimation may be simply evaluated by computing the variance of the estimation of the slope in the linear interpolation contained in the second step of the above procedure.

The results of the above method applied to the computation of the binomial multiplicative process spectrum with parameter $p = 0.7$ are shown in Figure 3.9, where the analytical solution obtained via Eqs. (3.34) and (3.37) is plotted as a dashed line.

Other methods suitable for determining the singularity spectrum, for example, the scaling histograms method [Meneveau and Sreenivasan, 1991], seem to yield less reliable results than those obtained with the methodology just described.

One important feature of the singularity spectrum concerns its possible symmetric or asymmetric shape. In fact, turbulence data from Meneveau and Sreenivasan [1987] have provided a comparison of experimental curves with those predicted by multiplicative processes of the type described by Eq. (3.13). Observations of fully developed turbulence are very well described by the binomial process discussed earlier in this section, leading to a symmetric spectrum. In fact, the binomial multiplicative process, with $p = 0.7$, leads to an $f(\alpha)$ curve that accurately describes the observed multifractal spectrum of the dissipation field.

The symmetry of the singularity spectrum has been addressed by Peckham and Waymire [1992]. We will exploit their results when dealing with the so-called random multiplicative cascade, a concept that complements the multiplicative processes discussed in this Section.

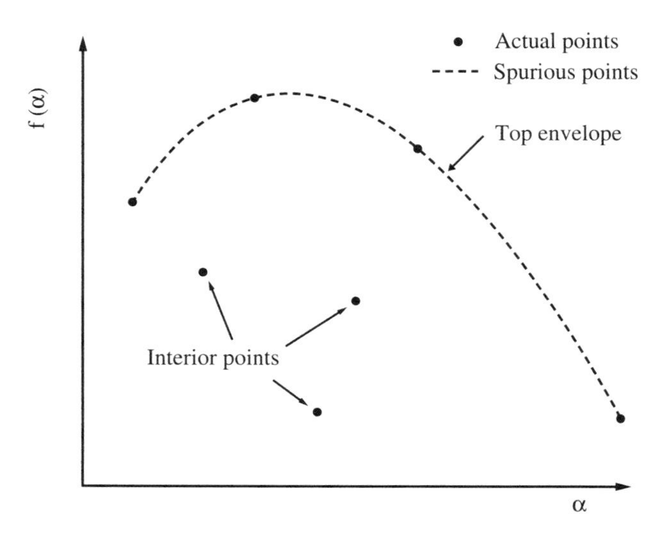

Figure 3.10. Illustration of a discrete multifractal spectrum. The interior points remain invisible to the algorithms that generate top envelope estimates of the actual spectrum [after Veneziano et al., 1995].

We deem it important to stress that, as noted by Veneziano et al. [1995], the numerical procedures for extracting the singularity spectrum suffer from severe estimation errors and that much attention should be devoted to the interpretation of numerically obtained spectra. In essence, such limitations stem from the assumption that $f(\alpha)$ is a continuous and convex function [Mandelbrot, 1990]. When this is not the case, the numerical procedures will yield an envelope, or convex hull, of the true $f(\alpha)$ curve. That is, the product of these algorithms is the smallest convex set containing $(\alpha, f(\alpha))$. This has two serious consequences: (i) when the true spectrum is defined over a discrete set of points, spurious points may be introduced, and (ii) 'interior' points, that is, points not belonging to the envelope, will not be captured.

A schematic representation of the possible situation is represented in Figure 3.10.

We will now consider an example following Veneziano et al. [1995]. Let us consider the parabola $y(x) = 2x - x^2$ in the interval [0, 2]. Local integrals P_δ define the 'mass' of $y(x)$ within an interval $(x - \delta/2, x + \delta/2)$. They have the form $P_\delta(x) = (2x - x^2)\delta - \delta^3/12$ (that scale like δ) for $0 < x < 2$, whereas at $x = 0$ and $x = 2$ one finds $P_\delta(0) = P_\delta(2) = \delta^2 - \delta^3/3$ (i.e., $\propto \delta^2$). The theoretical singularity spectrum is thus composed of just two points: (1, 1) and (2, 0). Figure 3.11 reproduces the $\tau(q)$ versus q curve for the parabola considered. Notice that, owing to discretization errors, the numerical values do not reproduce the correct values in the neighborhood of the 'transition' from one scaling exponent to the other. Instead of just two values of α (corresponding to the two values of the slope of $\tau(q)$ versus q), one obtains a number of spurious points between the true values (which are correctly identified). This is illustrated in Figure 3.12, where the numerical spectra of the parabola, obtained both by the Chhabra–Jensen method and by the qth moments, are represented. Therefore we concur with Veneziano et al. [1995] that caution should be used when interpreting numerical estimates of the multifractal spectra.

Figure 3.11. Function $\tau(q)$ for the parabola $y(x) = 2x - x^2$ in the range [0, 2]. Theoretical function (solid line) and numerical estimate (circles) [after Veneziano et al., [1995].

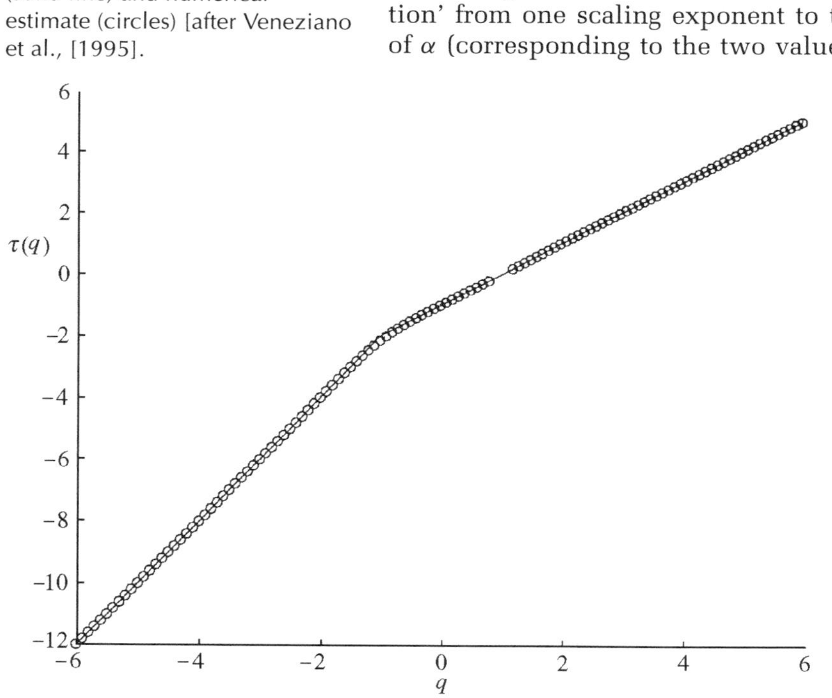

Some generalizations of the multifractal formalism may be pursued. Typically, one extension is the study of the distribution of multifractal measures on *fractal* supports. One such example is obtained by studying a multiplicative process on fractions of the unit intervals, as in the case of the binomial multiplicative process described in Section 3.2, except that a number of open intervals are removed, leaving individual line segments separated by empty holes. Other length scales thus enter the picture because the interplay of empty and full line segments, in the partition of a unit measure, yields a different kind of multiplicative process.

It is worth observing that the generalization of the multiplicative process to cases in which more than one spatial scale enters the partition may explain the fact that the singularity spectrum may not be continuous [Halsey et al., 1986b]. Consider, for example, the two-scale Cantor set shown in Figure 3.13. Two rescaling parameters ℓ_1 and ℓ_2 are used, as well as two measures p_1 and p_2 through which the self-similar process shown in Figure 3.13 is derived.

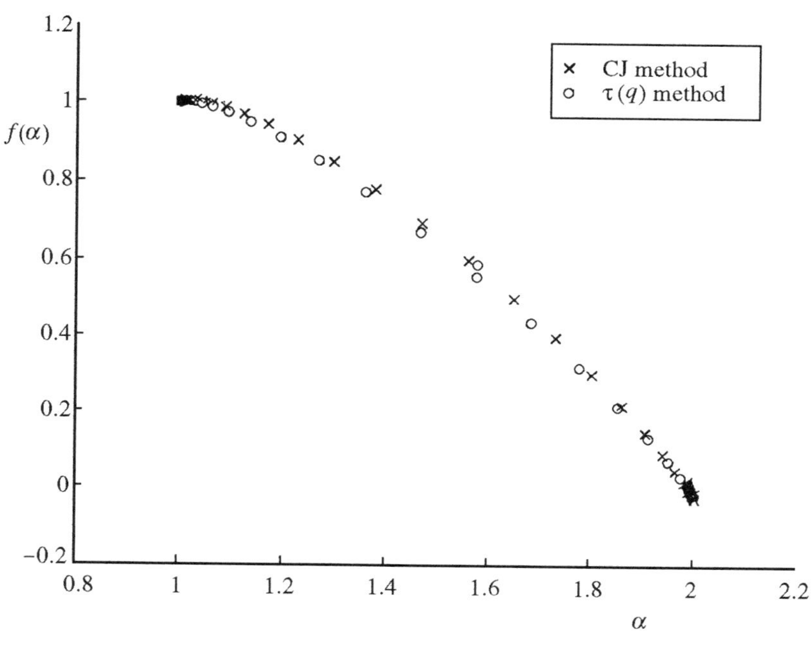

Figure 3.12. Multifractal spectra obtained by applying a Legendre transform method and the Chhabra–Jensen method to the parabola $y(x) = 2x - x^2$ in the range [0,2] [after Veneziano et al. [1995].

The resulting multifractal spectrum may be derived analytically and is shown in Figure 3.13(b). In the figure we notice that the fractal dimension of the support is $D_0 \approx 0.61$.

It is interesting to note the effects of a modification of the above set (Figure 3.14). In this figure a unit interval is subdivided into three segments, two of length ℓ_2 and one of length ℓ_1. The two former intervals each receive a proportion of the total measure given by p_2, and the latter interval receives a proportion given by p_1. We have $\ell_1 + 2\ell_2 = 1$ and $p_1 + 2p_2 = 1$. Each of these three intervals is then subdivided in the same manner, and so forth.

Although the measure on the line segment is rearranged at each step of the recursive process, at each step the support for the measure remains the original line segment. Thus we expect that D_0 for this measure will be 1. Furthermore, the densest intervals on the line segment contract not to one point (as was the case in the two-scale Cantor set) but to a set of points of finite dimension. Thus the lowest value of α and hence the value D_∞ corresponds to *a nonzero value* of f. On the other hand, there is always only one segment at the lowest value of the density so that we still expect $D_{-\infty}$ to correspond to a value of $f = 0$. Thus the singularity spectrum (Figure 3.15) is hook-like.

Finally, there exist partitioning processes that allow only *discrete*, rather than continuous, $f(\alpha)$ curves. Here it suffices to notice that sometimes we might observe hook-like shapes or discrete values of $f(\alpha)$ from data analyses. It is not trivial to ascertain whether the characters of the computed singularity spectra are due to numerical errors or to characters embedded in the processes. Nevertheless, a certain tendency for the densest sets to contract on a set of finite dimension is expected because we organize the area by integer multiples of pixels.

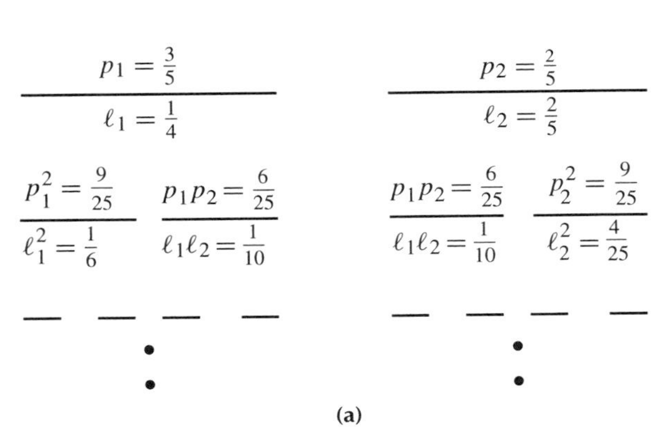

Figure 3.13. (a) A Cantor set construction with two rescalings: $\ell_1 = 0.25$ and $\ell_2 = 0.4$ and $p_1 = 0.6$ and $p_2 = 0.4$; (b) its multifractal spectrum [after Halsey et al., 1986].

Figure 3.14. A particular partitioning process [after Halsey et al., 1986b].

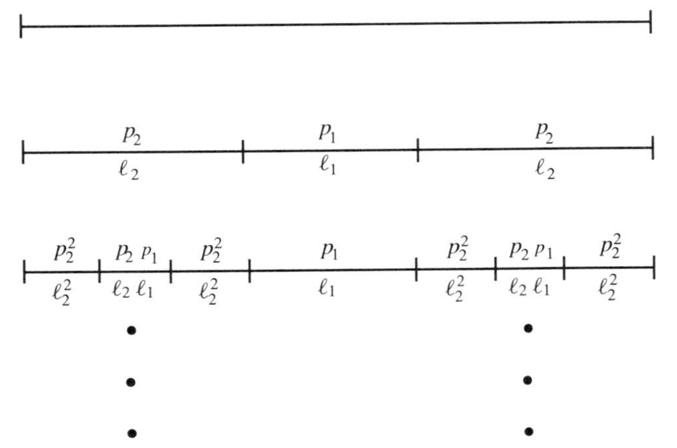

3.4 Multifractal Spectra of Width Functions

We now turn to the analysis of the width functions of several natural river basins characterized by different geology, size, location, climate, vegetation cover, and exposed lithology. The characteristics of the basins and of their DEMs have been described in Chapter 1 (Section 1.2.7 and Table 1.1). One important consideration is that the width function determined as in Section 1.2.7 (i.e., by subdividing a basin into elementary areas and adding those areas sharing the same value of the distance, measured along the network from the outlet) carries information on all geometric scales of the basin pertaining to both channels and hillslopes. However, the content of detailed information on the hillslope scales is often hampered by the cutoff imposed by the pixel size (i.e., usually of the order of 30 m) and therefore reliable information on the fine occupation of areas by the basin structure may not be granted (see Section 1.2.12). Figure 2.39 shows some of the raw data that will be analyzed through multifractal formalisms.

Figure 3.16 shows the multifractal spectra $f(\alpha)$ versus α for various width functions of real basins. All spectra have been computed via the Chhabra–Jensen [1989] method, where 100 values of the moment order q ranging from -20 to $+20$, including noninteger orders, have been imposed.

As noted in Section 3.3, much care must be exerted in interpreting the numerical multifractal spectra. It is particularly important to determine whether spurious points are introduced by the estimation algorithm. To do so, following Veneziano et al. [1995], one may exclude from the analysis the two end portions of the width function, which are expected (recall the example of the parabola in the previous section) to exhibit a different scaling behavior with respect to interior points. The result of this analysis is presented in Figure 3.17. This result suggests that large portions of the multifractal spectra of Figure 3.16 are indeed spurious. Three important features of such spectra should however be considered genuine: the minimum and maximum values of α and the maximum value of $f(\alpha)$. These points must in fact belong both to the true spectrum and to its envelope. They carry important information on the scaling structure of the width functions and show consistent values in all cases.

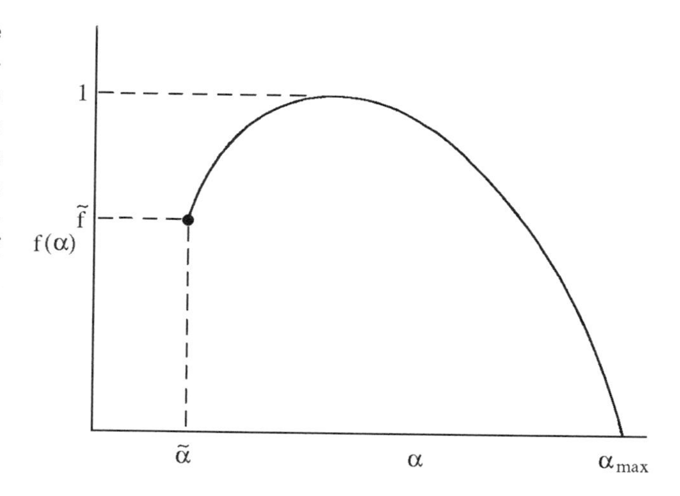

Figure 3.15. The $f(\alpha)$ function for a particular measure defined by the partitioning process of Figure 3.14 [after Halsey et al., 1986b].

We observe, in fact, that the maximum value of the singularity spectra is correctly, $f_{max}(\alpha) = 1$, reproducing for $q = 0$ the fractal dimension of the support. Notice that the value α_0 yielding $f(\alpha_0) = 1$ is larger than unity, as in Peano's basin where $\alpha_0 = 1.207$, and is very close for all basins analyzed. Also notice the very consistent reproduction of the maximum and minimum values of α.

Some variation of the value of α_{max} is observed, as opposed to a very robust identification of α_{min}. Because α_{max} is determined by the measures μ_- with the minimum content of area (see Eq. (3.47)) its fluctuations from basin to basin probably reflect local changes of threshold for channelizations, which we expect to vary from basin to basin because of a number of geomorphological factors. The right-hand side of the spectrum is also known to carry most uncertainties for the numerical determination [e.g., Chhabra and Jensen, 1989].

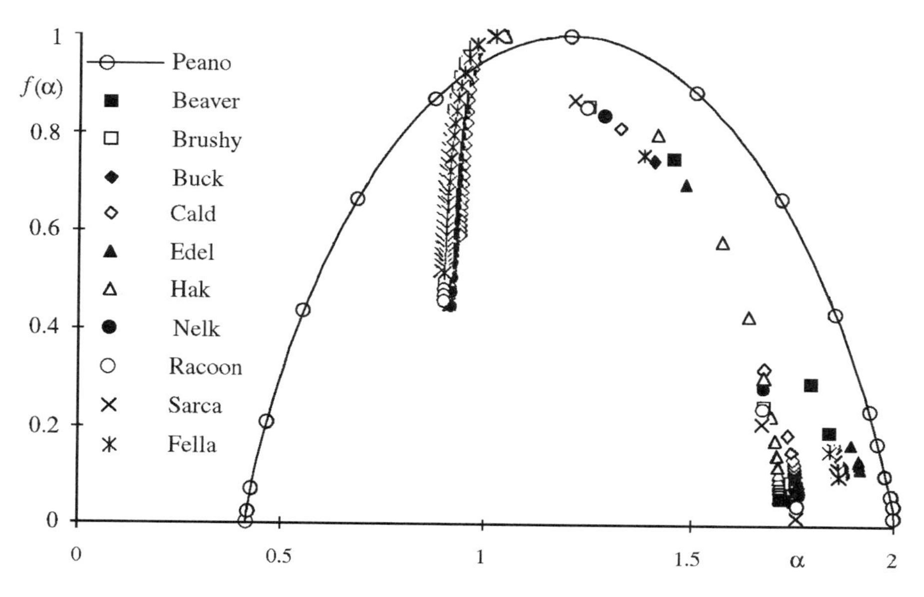

Figure 3.16. A comparison of the singularity spectra of different width functions [after Rinaldo et al., 1993]. The spectrum of Peano's basin is also shown.

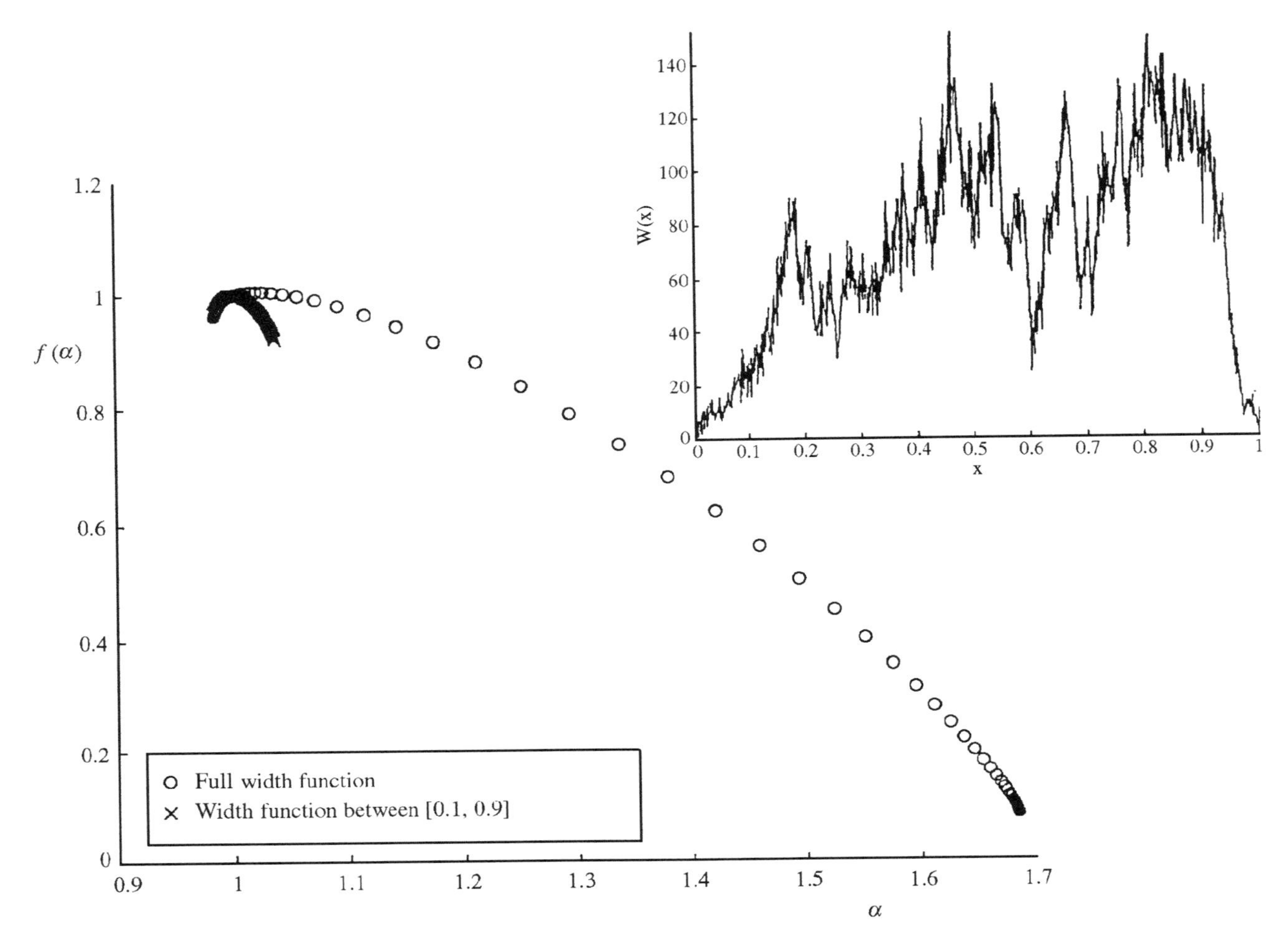

Figure 3.17. Singularity spectrum estimates for the North Fork width function generated by the Chhabra–Jensen algorithm [after Veneziano et al., 1995].

Nevertheless, in spite of the estimation problems dealt with above, the main (and genuine) characters of the spectra are remarkably similar regardless of diversities in the morphology of the basins. These characters can be well reproduced, as suggested by the power spectra in Section 2.7.5 and by an ad hoc analysis of Veneziano et al. [1995], by a single scaling exponent.

Thus there appears to exist an underlying unity behind the variety in the structure of the width functions that naural basins exhibit. This has important consequences. Such a unity postulates that the mechanisms of occupation of drainage area by a river network as described by the width function are consistently the same. These characters withstand the test of size from a few hundred square kilometers to 10^5 km^2 and of quite different geologic and lithologic conditions. This should not lead us to conclude that the width function carries no distinctive information about plane-filling aggregation patterns. In fact, if we compare the spectra with that of Peano's basin – shown in Figure 3.16 – we observe a substantial difference in the range of αs covered. Thus the left side of the spectrum is especially significant and is related to the positive moment order qs and to measures with larger values. In the left side, Peano's values are double the values of the Holder–Lipschitz' exponents observed in nature, resulting in a major change of fractal dimensions. In fact, the maximum fractal dimension of Peano's width function is observed at $x = 1$ (where as much as $(R_B - 1)^{\Omega-1}$ pixels out of $R_B^{\Omega-1}$ are located) as $D(1) = \log(R_B - 1)/\log R_L = 1.585$. The maximum values $D(x) = 2-\alpha(x)$ observed in nature never exceed 1.1. This is indeed a major difference. Indeed, real-life basins do not exhibit the systematic structure of Peano basins in their construction, yielding a width function of more regular features.

Figure 3.18. An example of off-lattice diffusion-limited aggregates. The cluster contains $N = 50,000$ particles [after Meakin, 1991].

3.5 Multiscaling and Multifractality

We will now consider a phenomenon called *multiscaling* that is often associated with multifractality. Multiscaling is a general concept of scale invariance introduced [see, e.g., Coniglio and Zannetti, 1989] to explain the anomalies present in many complex structures, such as those arising in critical phenomena. The invariance refers to the probability distribution function of a certain variable of interest.

Multiscaling has been observed in the density profile $g(r, R)$ of diffusion-limited aggregates (DLAs) [Coniglio, 1986; Coniglio, Amitrano, and Di Liberto, 1986; Amitrano et al., 1991], defined as

$$g(r, R)d^d r = dN \tag{3.62}$$

where dN is the number of particles in the infinitesimal d-dimensional volume $d^d r$ at distance r from the origin, and the dependence on the total number of particles N is expressed via the radius of gyration $R = R(N)$. The reader is referred to Feder [1988] or to the review by Meakin [1991] for a detailed explanation of the rules for the growth of DLA. Here it suffices to recall that the original model of DLA, described in Section 2.1.3, employs Brownian particles released from random positions chosen within a predefined outer circle. When particles collide, they stick together and stop. When particles exit the circle, they are abandoned. Tree-like structures are grown in this way (or in many alternative ways that basically employ similar ideas). The radius of gyration, R, is a basic measure of the extent of the coral-like structure in all d Euclidean dimensions of the embedding space. One example of DLA is shown in Figure 3.18.

The theory of DLA usually assumes, in analogy with critical phenomena, that $g(r, R)$ satisfies the standard scaling form:

$$g(r, R) \propto r^{-(d-D)} = C(r/R)r^{-(d-D)} \tag{3.63}$$

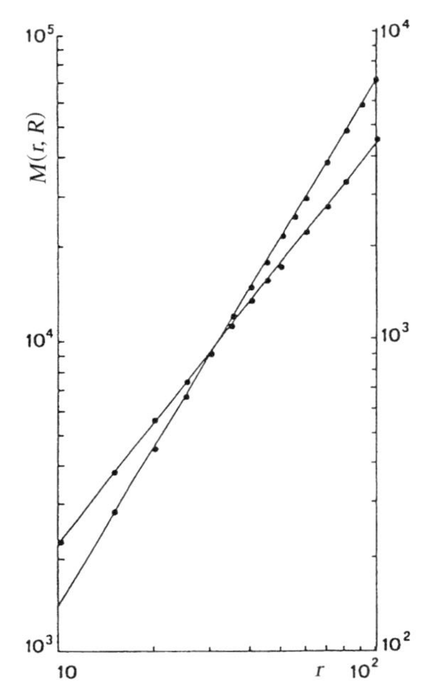

Figure 3.19. Log-log plot of the mass $M(r, R)$ versus r for $x = 1.5$ and $x = 1.9$ [after Amitrano et al., 1991].

where $C(x)$ is a scaling function and D is the fractal dimension of the aggregate. However Amitrano et al. [1991] suggest that the scaling structure of DLA is much richer than that allowed by standard scaling and that the above relation ought to be replaced by the *multiscaling* form:

$$g(r, R) = r^{-d+D(r/R)} C(r/R) \tag{3.64}$$

where $D(r, R)$ is a function of r and R instead of being chosen a priori to be a constant, as assumed in standard scaling. Eq. (3.64) is derived from the consideration that the density profile, apart from a constant factor, is invariant under rescaling all lengths by a factor ℓ:

$$\tilde{r} = \frac{r}{\ell} \qquad \widetilde{R} = \frac{R}{\ell} \qquad g(\tilde{r}, \widetilde{R}) = \ell^{d-D(r/R)} g(r, R) \tag{3.65}$$

As explained by Coniglio and Zannetti [1989], one then requires that the transformation obeys group properties. Namely, if rescaling by a factor ℓ_1 transforms $g(r, R)$ into $g(\tilde{r}_1, \widetilde{R}_1)$ and successively rescaling by another factor ℓ_2, $g(\tilde{r}_1, \widetilde{R}_1)$ transforms into $g(\tilde{r}_2, \widetilde{R}_2)$, then $g(\tilde{r}, \widetilde{R})$ transforms into $g(\tilde{r}_2, \widetilde{R}_2)$ when rescaling by a factor $\ell_1\ell_2$. It is easy to verify that this requirement implies that $D(r, R) = D(\tilde{r}, \widetilde{R})$, which in turn implies that $D(r, R)$ is a homogeneous function of r and R, that is, $D(r, R) = D(r/R)$. Therefore, from Eq. (3.65), choosing $\ell = r$ one finds the multiscaling form (3.64) [Coniglio and Zannetti, 1989].

Apparently the two forms of Eqs. (3.63) and (3.64) may seem to coincide in the limit $R \to \infty$. This is true if r is finite. However, if both r and R diverge with their ratio $x = r/R$ fixed, the two forms differ considerably [Amitrano et al., 1991]. The differences were shown by performing two sets of calculations, called A and B. In A, Amitrano et al. [1991] grew 2,000 DLA clusters on the square lattice of $N = 10{,}000$ particles. In B, they grew 200 off-lattice DLA clusters of $N = 10{,}000$ particles. The results were studied by dividing the growth stages corresponding to twenty different values of the radius of gyration R. At each stage of growth corresponding to a radius R, the cluster was divided into twenty shells of radius $r_n = (n - 0.5)R/10$, with n being an integer larger than 2. The authors then calculated the mass $M(r_n, R)$ in the shell between r_{n-1} and r_{n+1} corresponding to an average radius r_n and to a value $x_n = r_n/R = (n - 0.5)/10$. The relation of the mass $M(r, R)$ – where the index n has been suppressed – with the density profile is given by $M(r, R) = 2\pi r g(r, R)\Delta r$, where $\Delta r = R/5$ is the width of the shell. From the preceding equations we have

$$M(r, R) = \frac{2\pi C(x)}{5x} r^{D(x)} \tag{3.66}$$

For a fixed value of x the slope of log M versus log r gives $D(x)$ (Figure 3.19): The overall effects of multiscaling, described by the value of the resulting fractal dimensions $D(x)$, are shown in Figure 3.20, where the uncertainty in the estimation ranges from 0.01 to 0.03.

Coniglio and Zannetti [1989] addressed the question of whether there is a connection between the multiscaling and the multifractal behaviors obeyed by the growth probability distribution of DLAs. This is the case if the Holder–Lipschitz exponent α obeys

$$\alpha = \alpha(r/R)$$

Thus the set characterized by the smallest values of α is localized at the tips of the aggregate on the external part of the structure, whereas larger values of α correspond to regions closer to the origin. This could also be the case for the width functions of drainage basins.

An important theoretical point was made by Jensen, Paladin, and Vulpiani [1991]. They showed that multiscaling may be a direct consequence of an imperfect representation of a truly multifractal object whenever a lower cutoff impairs the complete description of the probability measure.

The proof goes as follows. A cutoff ϵ is imposed on the weights of the elements (say, boxes Λ_i of size δ) making up the partition of a multifractal set. Technically, we disregard a box Λ_i in the multifractal set if

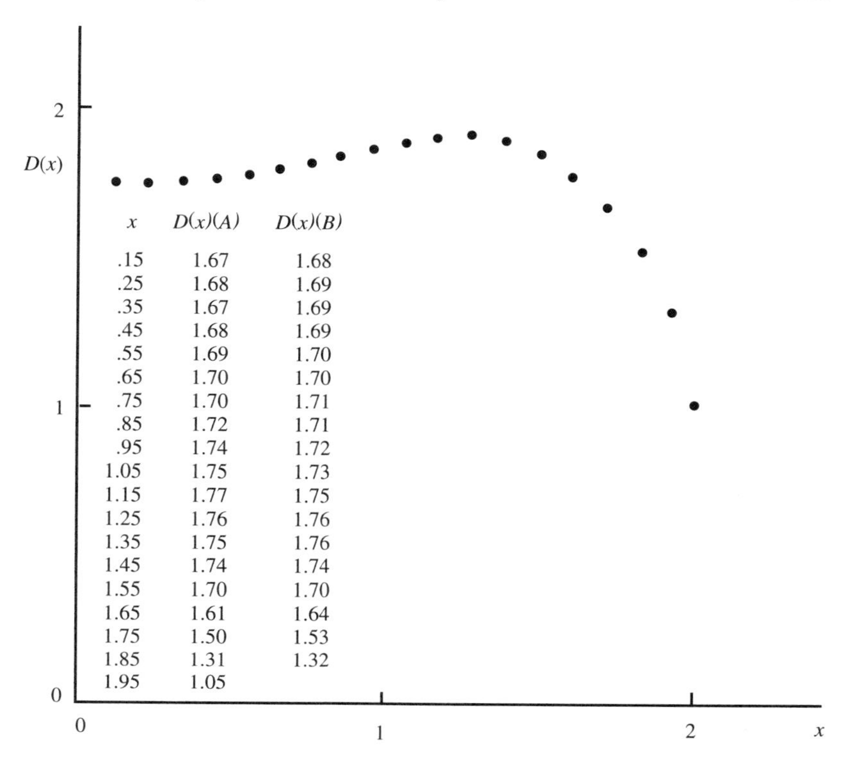

x	D(x)(A)	D(x)(B)
.15	1.67	1.68
.25	1.68	1.69
.35	1.67	1.69
.45	1.68	1.69
.55	1.69	1.70
.65	1.70	1.70
.75	1.70	1.71
.85	1.72	1.71
.95	1.74	1.72
1.05	1.75	1.73
1.15	1.77	1.75
1.25	1.76	1.76
1.35	1.75	1.76
1.45	1.74	1.74
1.55	1.70	1.70
1.65	1.61	1.64
1.75	1.50	1.53
1.85	1.31	1.32
1.95	1.05	

Figure 3.20. Plot of $D(x)$ for the sets of configurations A and B [after Amitrano et al., 1991].

$$C_q(\delta, \epsilon) = \sum_{i \in \Lambda_i} \mu_i^q(\delta) < \epsilon \tag{3.67}$$

Varying ϵ, one therefore computes a series of curves $C_q(\delta, \epsilon)$, which Jensen et al. [1991] relate to the $f(\alpha)$ curve. This relation is established by noticing that the probability measure, $\mu_i(\delta)$, of a box Λ_i of a self-similar set scales as $\mu_i(\delta) \propto \delta^{\alpha_i}$ (see Section 3.3). In the multifractal framework, for example, as in Eq. (3.51), the number of boxes indexed by $\alpha_i \in [\alpha, \alpha + d\alpha]$ is $\delta^{-f(\alpha)}\rho(\alpha)d\alpha$ (where $\rho(\alpha)$ is a smooth function independent of δ) and the sum defining the generalized (continuous) correlation C_q is written as

$$C_q(\delta) = \int_{\alpha_{\min}}^{\alpha_{\max}} \delta^{\alpha q - f(\alpha)} \rho(\alpha) d\alpha \propto \delta^{\tau(q)} \quad \text{as } \delta \to 0 \tag{3.68}$$

where $\tau(q) = q\alpha(q) - f(\alpha(q))$ is estimated at the value $\alpha(q)$ when $\tau(q)$ reaches its minimum.

In the presence of a cutoff ϵ, an empty box Λ_i cannot be distinguished from a box with probability less than ϵ. A cutoff, therefore, makes it impossible to determine large α values corresponding to the less probable regions. If

$$\mu_i \propto \epsilon \approx \delta^{\bar{\alpha}} \tag{3.69}$$

then the boxes Λ_i indexed by

$$\alpha > \bar{\alpha} = \frac{\log \epsilon}{\log \delta} \tag{3.70}$$

do not appear in the sum of Eq. (3.67) because they are assumed to be empty. The upper integration limit in Eq. (3.68) then becomes

$$C_q(\delta, \epsilon) = \int_{\alpha_{\min}}^{\bar{\alpha}} \delta^{\alpha q - f(\alpha)} \rho(\alpha) d\alpha \tag{3.71}$$

This will not modify the estimation of the minimum of $\alpha q - f(\alpha)$, that is, $\alpha(q)$, if $\alpha(q)$ still falls inside the integration interval. However, when $\alpha(q) > \bar{\alpha}$, the minimum is reached at the upper limit $\bar{\alpha}$. We therefore obtain [Jensen et al., 1991]

$$C_q(\delta, \epsilon) = \begin{cases} \delta^{\tau(q)} & \text{if } \bar{\alpha} > \alpha(q) \\ \delta^{q\bar{\alpha}(\epsilon,\delta) - f(\bar{\alpha}(\epsilon,\delta))} & \text{if } \bar{\alpha} \leq \alpha(q) \end{cases} \tag{3.72}$$

The *multiscaling* for $\bar{\alpha} \leq \alpha(q)$ is a power law with a slowly varying exponent proportional to $\log \delta$. Jensen et al. [1991] proposed a rescaling procedure by considering the function $F_q = \log C_q / \log \epsilon$ as a function of the parameter $\theta = \log \delta / \log \epsilon = 1/\bar{\alpha}$, that is

$$F_q(\delta) = \begin{cases} \theta \tau(q) & \text{if } \theta < 1/\alpha(q) \\ q - \theta f(1/\theta) + \cdots & \text{for } \theta \geq 1/\alpha(q) \end{cases} \tag{3.73}$$

The first regime is the single-scaling regime, where the function $F_q(\theta)$ is a straight line with slope $\tau(q)$. In the second regime, F_q has a multiscaling behavior conditioned by the $f(\alpha)$ spectrum. Thus for $\alpha < \alpha(q)$ one has [Jensen et al., 1991]

$$C_q = \delta^{q\bar{\alpha} - f(\bar{\alpha})} I_q(\alpha_{\min}, \bar{\alpha}) \tag{3.74}$$

where, to first order in the expansion of $(\alpha - \bar{\alpha})$, the second term in the right-hand side is given by

$$I_q(\alpha_{\min}, \bar{\alpha}) = \int_{\alpha_{\min}}^{\bar{\alpha}} \delta^{[q - f'(\alpha)](\alpha - \bar{\alpha})} \rho(\alpha) d\alpha \tag{3.75}$$

where use has been made of the relationship:

$$\frac{\alpha q - f(\alpha)}{d\alpha} = \begin{cases} q - f'(\alpha) < 0 & \text{for } \alpha < \alpha(q) \\ 0 & \text{for } \alpha = \alpha(q) \end{cases} \tag{3.76}$$

Using the relation $\delta = \epsilon^{1/\bar{\alpha}}$, we obtain the following relationship:

$$I_q(\alpha_{\min}, \bar{\alpha}) = \frac{1 - \epsilon^{(f'(\bar{\alpha}) - q)(\bar{\alpha} - \alpha_{\min})/\bar{\alpha}}}{|\log \epsilon| (f'(\bar{\alpha}) - q)/\bar{\alpha}} R(\bar{\alpha}) \tag{3.77}$$

where $R(\bar{\alpha}) \approx 1/(\bar{\alpha} - \alpha_{\min})$ [Jensen et al., 1991]. Notice that in the limit when $\bar{\alpha} \to \alpha_{\min}$, I_q must approach unity because when only boxes with $\alpha = \alpha_{\min}$ survive the cutoff procedure, a homogeneous fractal is obtained where

$$C_q \propto \delta^{\alpha_{\min} q - f(\alpha_{\min})} \tag{3.78}$$

Thus the additive correction needed in Eq. (3.73) for $\theta \geq 1/\bar{\alpha}$ is

$$F_q(\delta) = q - \theta f(1/\theta) + \frac{\log I_q(\alpha_{\min}, 1/\theta)}{\log \epsilon} \tag{3.79}$$

Following Jensen et al. [1991], we see how the above information works with reference to the binomial multiplicative process with, for example, parameter $p = 0.6$ and length scale 1/3. At the nth stage of construction ($\delta = 3^{-n}$) the correlation sum $C_q(\delta)$ takes on the form

$$C_q(\delta) = \sum_{m=1}^{n} \binom{n}{m} p^{mq}(1-p)^{(n-m)q} \tag{3.80}$$

and the introduction of a cutoff ϵ results into a correlation sum $C_q(\delta, \epsilon)$ obtained by omitting terms in Eq. (3.80) whose probabilities $p^m(1-p)^{n-m}$ are less than ϵ. Figure 3.21 shows plots of $\log C_2(\delta, \epsilon)$ versus $\log \delta$ at various cutoffs ϵ. Figure 3.22 shows the rescaled data according

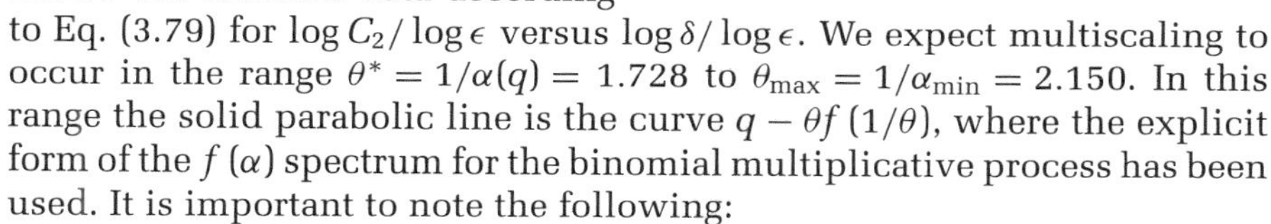

to Eq. (3.79) for $\log C_2/\log\epsilon$ versus $\log\delta/\log\epsilon$. We expect multiscaling to occur in the range $\theta^* = 1/\alpha(q) = 1.728$ to $\theta_{\max} = 1/\alpha_{\min} = 2.150$. In this range the solid parabolic line is the curve $q - \theta f(1/\theta)$, where the explicit form of the $f(\alpha)$ spectrum for the binomial multiplicative process has been used. It is important to note the following:

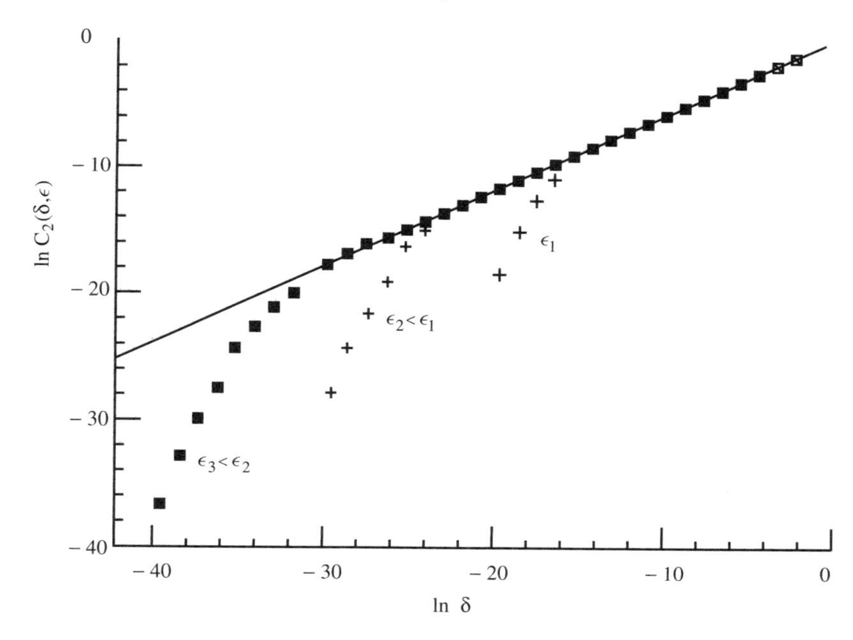

Figure 3.21. Scaling of $C_2(\delta, \epsilon)$ with δ at different cutoffs ϵ for the binomial multiplicative process, with $\delta = 3^{-n}$ and $p = 0.6$ [after Jensen et al., 1991].

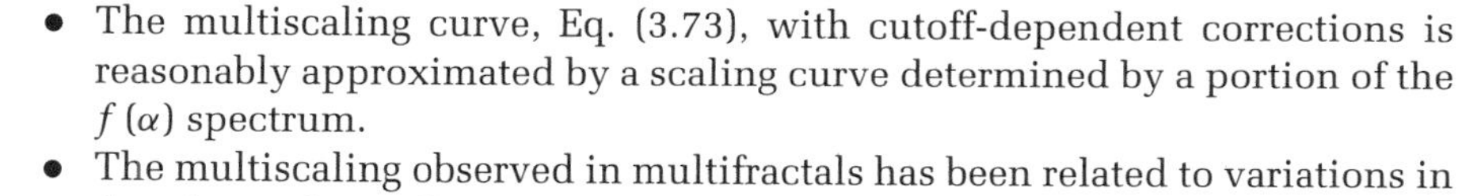

- The multiscaling curve, Eq. (3.73), with cutoff-dependent corrections is reasonably approximated by a scaling curve determined by a portion of the $f(\alpha)$ spectrum.
- The multiscaling observed in multifractals has been related to variations in the physical cutoff parameter. This suggests the possibility of determining the underlying multifractal structure (if any) when multiscaling is empirically observed in a physical system.
- The cutoff directly affects the right-hand side of the singularity spectrum of a multifractal measure, moving it to the left proportionally to the cutoff ϵ, depending on the nature of the underlying $f(\alpha)$ curve.

In Chapter 4 we will examine peculiar artificial multiscaling introduced by a cutoff in the description of the width function of simulated networks where we could identify different stages of growth. This is not possible in DEMs. However, we can experimentally verify

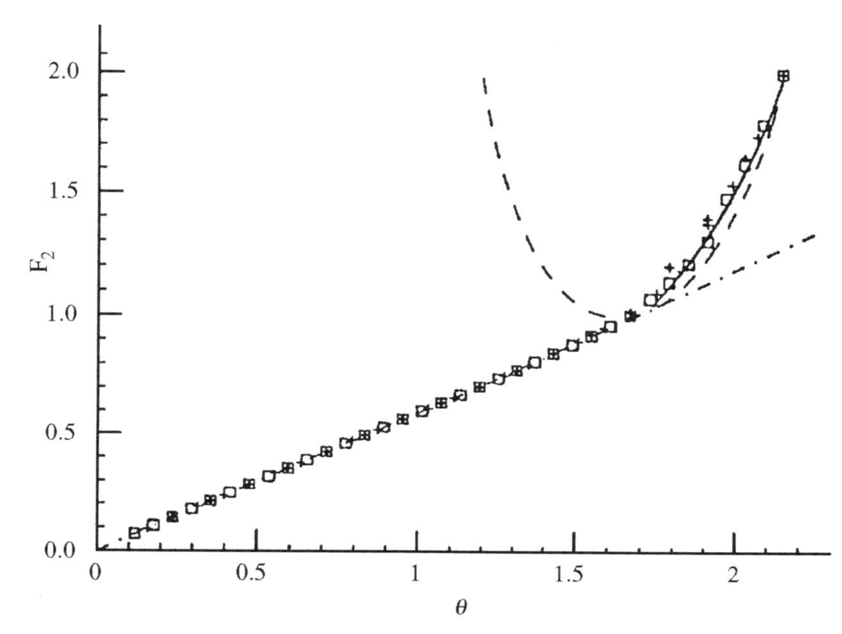

Figure 3.22. Multiscaling transformation for the data in Figure 3.21 [after Jensen et al., 1991].

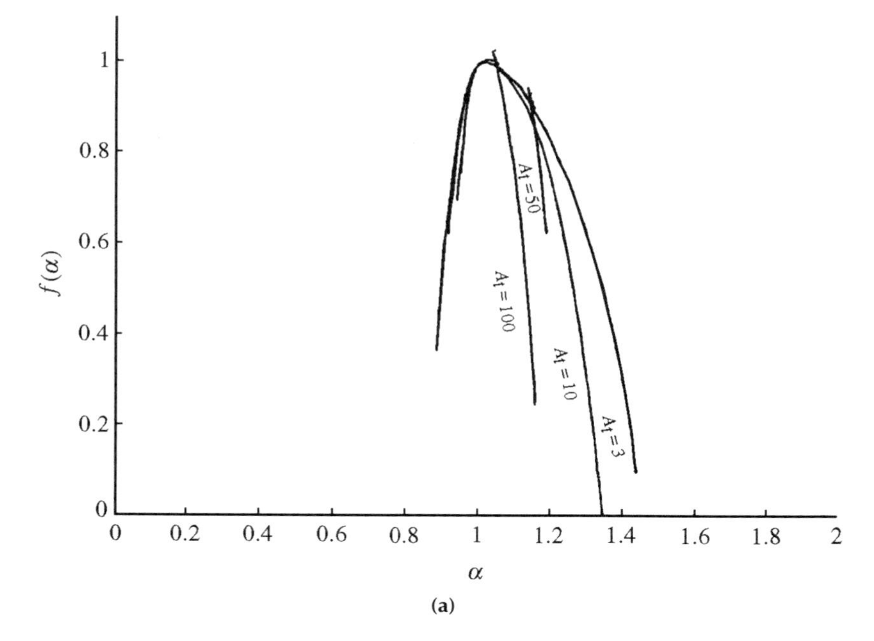

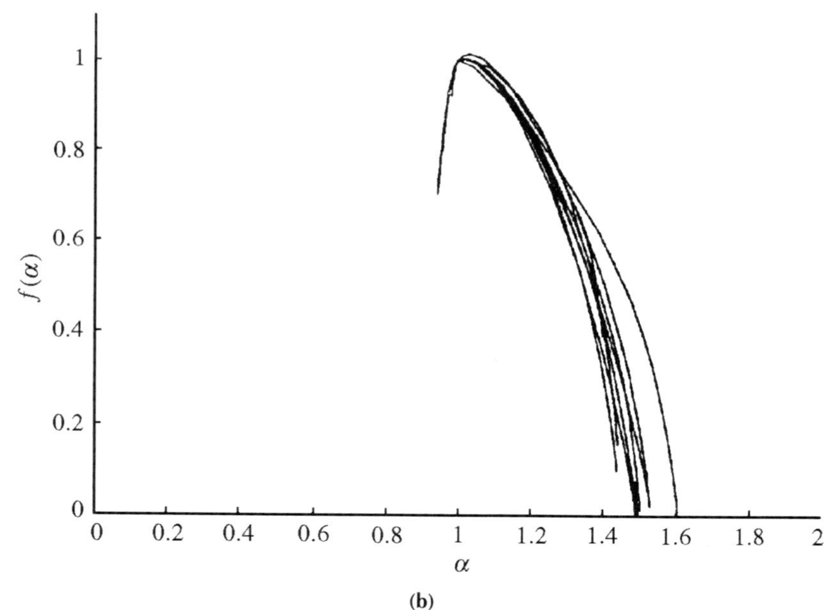

Figure 3.23. Multifractal spectra of the width function of natural basins; (a) the effect of different threshold areas for Nelk; (b) spectra for nine basins with support area $A_t = 3$ [after Rinaldo et al., 1992].

the effects of different cutoffs that were used in the description of the width functions of natural basins considered in Chapter 2.

A reasonable cutoff parameter is the support (or threshold) area A_t (see Section 1.2.9), originally used to identify the channelized part of a DEM. On changing the support area, the resulting aggregation process reflects a cutoff in the quantity of area that can be assigned to each link and thus implies a cutoff in the estimation of the width function. Figures 3.23(a) and 3.23(b) show the multiscaling effect of the support area on the resulting singularity spectrum of real basins. Notice that the spectra with full support ($A_t = 1$) are shown in Figure 3.16. It is clear that the effect of the cutoff results in a contraction of the spectrum. In fact, the right-hand side of the $f(\alpha)$ curve relies on the moment orders q dominated by the measures, μ_-, with the smallest values.

Later in this book (Chapter 4) we will return to the effects of multiscaling detection. This will be possible only when analyzing growth processes related to river basin evolution through controlled numerical experiments.

3.5.1 Other Multifractal Descriptors

Let us consider the graph of a given width function $W(x)$, made up of the points $(x, W(x))$, $x \in [0, 1]$. An alternative notation is (i, W_i), where $i = 1, M$ indexes the locations $x = i\Delta x$ (where Δx denotes the arbitrary spatial ruler) along the maximum path (of length $M\Delta x$) from source to outlet. As stated before, W_i is positive and normalized.

One feature of a width function that we want to study in more depth is the presence of wide fluctuations over all the observable scales. We saw in Section 2.7.5 that the power spectrum has the power law form $S(f) \propto f^{-\zeta}$, thus supporting a self-affine interpretation of the width function. This, in turn, is consistent with a scaling relationship of the type

$$W(x + \lambda\Delta x) - W(x) \stackrel{d}{=} \lambda^H\left[W(x + \Delta x) - W(x)\right] \tag{3.81}$$

(where $\stackrel{d}{=}$ means equality in distribution), where $H = (\zeta - 1)/2$. The underlying assumption is stationarity in the increments of $W(x)$.

Let us now consider a way to analyze the properties of a multifractal set $(x, W(x))$ (or alternatively (i, W_i)) deliberately restricting ourselves to the one-dimensional case. A function of the measure W_i is defined as

$$z(r) = \sum_{i=0}^{r} W_i \tag{3.82}$$

The analysis presented here is based on the study of generalized variograms or qth moments $C_q(\lambda)$ [Lavallée et al., 1993]:

$$C_q(\lambda) = \left\langle \left| z(r+\lambda) - z(r) \right|^q \right\rangle \tag{3.83}$$

where $\langle\,\rangle$ is the ensemble averaging operator. The qth variogram is said to be scaling if

$$C_q(\lambda) = \delta^{K(q)} C_q(1) \propto \lambda^{K(q)} \tag{3.84}$$

When Eq. (3.81) holds, it can be shown [Lavallée et al., 1993] that the exponent $K(q)$ depends linearly on q, that is

$$K(q) = qH \tag{3.85}$$

(see also the detailed discussion in the context of topographical fields in Section 3.6). In such a case, $W(x)$ is said to be *simple scaling*. A more general form for $K(q)$ could be observed whenever the theoretical tools do not assume the existence of simple scaling from the onset. This is the case when using the generalized covariance analysis based on the study of Eq. (3.84), where the departures from a straight line in the $K(q)$ versus q diagram detect *multiple* scaling features.

It is also worthwhile to recall that the power spectrum $S(f) \propto f^{-\zeta}$ is the Fourier transform of the second moment $C_2(\lambda)$. Because $C_2(\lambda) \propto \lambda^{K(2)}$, a relation between the exponents ζ and $K(2)$ can be established as [Lavallée et al., 1993]

$$\zeta = 1 + K(2) \tag{3.86}$$

Notice that in the case of simple scaling, one has $K(2) = 2H$ and $\zeta = 1+2H$.

The computations of the generalized covariances in Eq. (3.84) have been performed with the techniques described in Bellin, Salandin, and Rinaldo [1992] in the context of heterogeneous transport in groundwater. Figure 3.24 shows the $K(q)$ versus q curve for the width function of one of the basins described in Table 2.4.

Figure 3.24 also shows an intermediate step involving the computation of the generalized covariances $C_q(\lambda)$, which are shown in a log-log plot versus λ to underline their scaling structure. In Figure 3.24 the reference results obtained through the same procedure for Peano's basin are also illustrated.

The graph shows that the width of the singularity spectrum of Peano's basin yields a significant departure from simple scaling. Departures from simple scaling are much less evident for the width function of real basins. This seems consistent with the analyses previously performed (see, e.g., Figure 3.16), as a relatively narrow multifractal spectrum renders reasonable a description based on simple-scaling assumptions. On the contrary, a

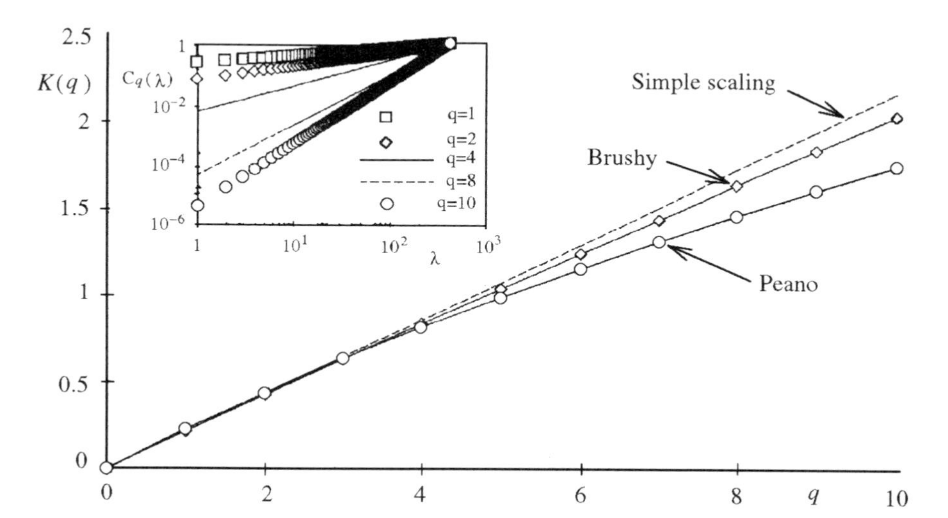

Figure 3.24. $K(q)$ versus q and (insert) generalized covariances of the width functions of the Brushy River Basin and Peano's curve. The straight line ($K(q) = qH$) is simple scaling [after Marani et al., 1994].

wide range of αs makes such an approximation untenable. Hence, although spurious points may be artificially induced by the estimation procedure of the singularity spectrum (Section 3.4), the variogram analysis suggests that multiscaling may indeed be present in the width function of real basins.

We now show that the above observation is substantiated theoretically by inferring the relationship between the multifractal spectrum and the generalized variogram in Eq. (3.84). Specifically, we show that the qth moment analysis is equivalent to that based on the singularity spectrum and the results of the two approaches are connected by the relation [Lavallée et al., 1993]

$$K(q) = \inf_{\alpha}\left[q(E - \alpha) - E + f(\alpha)\right] \tag{3.87}$$

where E is the dimension in which the set is embedded ($E = 1$ in the case of $W(x)$).

In what follows we will refer, without loss of generality, to the study of a spatial distribution in E dimensions of an arbitrary measure μ. We make use of the following notation:

- λ is the resolution scale, that is, the dimension of the ruler at the resolution adopted;
- L is a linear dimension characterizing the span of the sample data; and
- δ is the reduction coefficient ($\delta > 1$) that identifies the resolution, that is, $\lambda = L/\delta$; Thus μ_δ indicates the measure at the resolution δ.

Following Lavalleé et al. [1993], we make a distinction between the scaling properties of the measures of mass and of mass density (defined as $\epsilon_\delta = \mu_\delta/\lambda^E$ at the resolution λ). We recall, from the definition of the qth moment, that the following relationship holds:

$$C_q(\lambda) = \left\langle \epsilon_\delta^q \right\rangle \propto \lambda^{K(q)} \tag{3.88}$$

for the density and, analogously, for the mass one has

$$C'_q(\lambda) = \left\langle \mu_\delta^q \right\rangle \propto \lambda^{K'(q)} \tag{3.89}$$

because the difference in the cumulated measures z at lag λ yields precisely the mass μ_δ as defined above. Given that $\epsilon_\delta = \mu_\delta/\lambda^E$, we have the relation

$$K'(q) = K(q) + qE \tag{3.90}$$

Let us consider the probability of exceedence for the continuous density ϵ, which, in the case of a multifractal object, by definition has the form

$$P\left[\epsilon \geq \lambda^{-\gamma}\right] \propto \lambda^{c(\gamma)} \tag{3.91}$$

where $c(\gamma)$ is called the *codimension*. The related probability density $p\left(\lambda^{-\gamma}\right)$ is given by (recall that $d(\lambda^{-\gamma}) = d(\exp^{-\gamma \log \lambda}) \propto \lambda^{-\gamma}\, d\gamma$)

$$p\left(\lambda^{-\gamma}\right) = \frac{dP}{d(\lambda^{-\gamma})} \propto \lambda^{c(\gamma)+\gamma} \tag{3.92}$$

Now consider (Section 3.3) the basic relationships defining the multifractal spectrum. The mass μ_δ is given by $\mu_\delta \propto \lambda^\alpha$, and from Eq. (3.91) $\mu_\delta = \epsilon\lambda^E$. Moreover, ϵ takes on the values $\lambda^{-\gamma}$, and thus $\mu_\delta \propto \lambda^{E-\gamma}$. Therefore we have

$$\alpha = E - \gamma \tag{3.93}$$

Following the proof given in Section 3.3, we also note that the number, say N_i, of sites i where $\mu_i = \lambda^\alpha$ (given as $N_i \propto \lambda^{-f(\alpha)}$) is equal to the number of sites where $\epsilon_i = \mu_i/\lambda^E = \lambda^{-\gamma}$ (with $\gamma = E - \alpha$). N_i is proportional to the probability of ϵ_i at a site times the total number of sites ($L^E/\lambda^E \propto \lambda^{-E}$). Thus we may write

$$N_i \propto p\left(\lambda^{-\gamma}\right) d\left(\lambda^{-\gamma}\right)\lambda^{-E} \propto \lambda^{\gamma + c(\gamma)}\,\lambda^{-\gamma}\,\lambda^{-E} \propto \lambda^{c(\gamma)-E}, \tag{3.94}$$

from which, in turn, we find

$$f(\alpha) = E - c(\gamma) \tag{3.95}$$

Furthermore, it follows by the above definitions that

$$C'_q(\lambda) = \left\langle \mu_\delta^q \right\rangle = \int_{-\infty}^{+\infty} \mu_\delta^q\, dp\,(\mu) \tag{3.96}$$

$$\propto \int_{-\infty}^{+\infty} \left(\lambda^{E-\gamma}\right)^q \lambda^{c(\gamma)}\, d\gamma \propto \lambda^{\max_\gamma [c(\gamma) - q\gamma + qE]}$$

The last statement results as $\lambda \to 0$ (i.e., large reduction scale), as seen in Eq. (3.52). Because, from Eqs. (3.97) and (3.89)

$$K'(q) = \max_\gamma \left[c(\gamma) - q\gamma + qE\right] \tag{3.97}$$

one finally gets, making use of Eqs. (3.93) and (3.95)

$$K(q) = \max_\gamma \left[c(\gamma) - q\gamma\right] \tag{3.98}$$

from which Eq. (3.87) is derived, noting that $\gamma = E - \alpha$ (Eq. (3.93)) and $c(\gamma) = E - f(\alpha)$ (Eq. (3.95)), and that $\max_\gamma$ is substituted by $\inf_\alpha$ because of the minus sign.

The Rényi exponent $\tau(q)$ is defined by $\tau(q) = \min_\alpha[f(\alpha) - q\alpha]$ (see Section 3.3), and thus by substituting Eqs. (3.93), (3.95), and (3.98) one gets

$$\tau(q) = (1 - q)E - K(q) \tag{3.99}$$

Therefore we infer that the width of the singularity spectrum is indeed linked to departures from the simple-scaling behavior, as suggested by Figure 3.24. We thus conclude that based on generalized covariance analysis the simple-scaling assumption for the fluctuations of $W(x)$ could be accepted as reasonable. Nevertheless, the slight departure from simple scaling observed in Figure 3.24 – together with the results of Figure 3.18, notwithstanding the effect of spurious points – suggest, that a final word on the true fractal nature of width functions may be yet to come.

3.6 Multifractal Topographies

3.6.1 Fractal versus Multifractal Descriptors

In Chapter 2 we studied the impact of fractal geometry on our ability to describe nature following the ground-breaking work of Benoit Mandelbrot. In this section we will examine data from the elevation field of real river basins, mainly to conclude (in agreement with the suggestion of Lavallée et al. [1993]) that the variability of landscape topography yields to multifractal analyses and simulations. The data to which we specifically refer were collected in northern Italy, but analogous conclusions were drawn from data collected under completely different climates, geologies, etc. (Figure 3.25).

If geographic and geophysical fields are generally multifractal, that is, characterized by a hierarchy of fractal dimensions, then inconsistencies are inevitable when the fields are forced into geometric frameworks involving single fractal dimensions. Here we will examine the following:

- multifractal descriptors enlarge our ability to describe nature; and
- the commonly used monofractal relationships (seen in Chapter 2) relating fractal dimensions to spectral and variogram exponents give inaccurate or misleading results when applied to multifractal objects.

In Chapters 5 and 6 we will examine some dynamic processes that are possibly responsible for multifractal growth processes.

Our analysis starts from the observation that geophysical systems like landscape topography are characterized by extreme spatial variability, which has traditionally been associated with scaling properties. In Chapter 2 we analyzed in detail self-similar and self-affine transformations, where the scaling exponents were unique with respect to the relative spatial position within the field.

Up to this point two classes of scale-changing operations have been studied, one where small scales are reduced copies of the large scales (self-similarity) and the other where compression along certain coordinate axes is permitted (self-affinity). An interesting concept that generalizes the above has been called generalized scale invariance [Schertzer and Lovejoy, 1989b] and allows for the analysis of broader types of scaling. Scale-invariant systems include fields (and thus measures, possibly multifractal) and may allow types of scale transformations such as rotations and anisotropic compression that varies from point to point. The basic idea can be simplified as follows. A simple change of scale in a one-dimensional system can be written as $\ell_1 = \lambda^{-1}\ell_0$, $\lambda > 0$ being the scaling

Figure 3.25. An earth view of Nepal (China). Two major physiographic features are in evidence in this photograph. First, a small segment of the rugged Himalayan mountains where local relief is large. Several glaciers radiate from the northernmost regions. An extensive fluvial valley is also evident in the central part of the photograph. We expect that an analysis of the roughness of this landscape would exhibit signs of multifractality as a result of the many intertwined processes shaping the landscape's forms [courtesy of NASA].

parameter and ℓ_1, ℓ_0 the spatial scales related by the transformation. In a three-dimensional system the change of scale will be performed through a matrix, say $\mathbf{T}_\lambda$, where

$$\mathbf{B}_\lambda = \mathbf{T}_\lambda \mathbf{B}_1 \tag{3.100}$$

where $\mathbf{B}_1$ is a ball of unit radius. Schertzer and Lovejoy [1989b] have observed that the scale-changing matrix $\mathbf{T}_\lambda$ may be generally written as

$$\mathbf{T}_\lambda = \lambda^{-\mathbf{G}} = \exp(-\mathbf{G}\log\lambda) \tag{3.101}$$

where $\mathbf{G}$ is a suitable matrix. We observe that if $\mathbf{G}$ is not the identity matrix, $\mathbf{T}_\lambda$ is no longer a mere contraction. For example, in the isotropic case leading to a self-similar transformation we have

$$\mathbf{G} = \begin{pmatrix} 1 & 0 & 0 \\ 0 & 1 & 0 \\ 0 & 0 & 1 \end{pmatrix} \tag{3.102}$$

In the case of self-affinity we have

$$\mathbf{G} = \begin{pmatrix} g_{11} & 0 & 0 \\ 0 & g_{22} & 0 \\ 0 & 0 & g_{33} \end{pmatrix} \tag{3.103}$$

and in the general case where rotational differences may appear we have $\mathbf{G} = g_{ij} \neq 0 \; \forall i, j$. We will not detail the property that the generalised 'ellipsoid' defined by $\mathbf{G}$ has to have in order to define a meaningful transformation.

The above scale-changing transformation defines new ways of covering objects to determine anisotropic fractal dimensions [e.g., Schertzer and Lovejoy, 1989b]. As a result, some generalizations of the concept of Hausdorff dimension have been proposed that are potentially capable of describing complex objects characterized by features like marked anisotropy.

The topography of a river basin is indeed a complex object. Let us review the formalism needed to describe it without assuming a monofractal framework.

Let us consider a topographic field of elevations $z(\mathbf{x})$, defined at every point $\mathbf{x}$ in a subset of $\boldsymbol{R}^2$. As described in Section 2.8, the field $z(\mathbf{x})$ is defined to be simple scaling if its increments obey the following relationship:

$$z(\mathbf{x}_1 + \lambda\mathbf{e}) - z(\mathbf{x}_1) \stackrel{d}{=} \lambda^H [z(\mathbf{x}_2 + \mathbf{e}) - z(\mathbf{x}_2)], \tag{3.104}$$

where $\mathbf{e}$ is a unit vector, λ is a separation distance, $\mathbf{x}_1, \mathbf{x}_2$ are two arbitrary points, and $\stackrel{d}{=}$ indicates equality in distribution, that is, $x \stackrel{d}{=} y$ iff $P(x > q) = P(y > q) \; \forall q$; the exponent H is called the Hurst exponent, as seen in Chapter 2 [Mandelbrot, 1983]. The unique fractal dimension D characterizing the simple-scaling field is related to the Hurst exponent H via the relationship $D = 3 - H$ (see Section 2.8). It is important to note that a simple-scaling field is postulated to have constant statistical characteristics with respect to position and orientation. Eq. (3.104) holds, in fact, for any choice of $\mathbf{x}_{1,2}$ and $\mathbf{e}$.

A scaling function in the sense of Eq. (3.104) may, for instance, be produced by summing correlated or independent random noises and is characterized by a single scaling exponent. A special case in which the probability distribution of the altitude increments $\Delta z = z(\mathbf{x}_1 + \lambda\mathbf{e}) - z(\mathbf{x}_1)$ is Gaussian is Mandelbrot's fractional Brownian motion [Mandelbrot and Van Ness, 1968].

As discussed in Chapter 2, many examples of fractal analyses of topography and bathymetry measured along a linear track can be found in the literature. Most of these studies are based on monofractal (or, simple scaling) models that do not seem entirely consistent with the properties of measured field data. Recent contributions [Schertzer and Lovejoy, 1989b; Lavallée et al., 1993] give a different interpretation of the fractal characters observed in real topographies, arguing that geographic fields are generally multifractal and that inconsistencies seem inevitable when these entities are forced into narrower geometric frameworks involving a single fractal dimension.

There exist many ways to characterize rough surfaces. One way to study the fractal properties of a field $z(\mathbf{x})$ is to define, through the introduction of a threshold value of elevation T, sets that have theoretically determined characteristics with respect to the fractal geometry of the field.

To obtain *sets*, we typically establish thresholds and define the sets of interest. One such set is the exceedence set, $S_{\geq T}$, introduced as the set of all the points $\mathbf{x}$ satisfying $z(\mathbf{x}) \geq T$. Likewise, one can define the perimeter set P_T as the subset of the points of $S_{\geq T}$ having at least one nearest neighbor belonging to the set $S_{<T}$. In general, because P_T is a subset of $S_{\geq T}$, we have $D(P_T) \leq D(S_{\geq T})$. In the case of a simple-scaling structure the sets $S_{\geq T}$ are characterized by a single fractal dimension $D(S_{\geq T}) = D$, whereas $D(P_T) = 2 - H$, both values being independent of T. These properties provide a means of checking whether simple scaling arises for $z(\mathbf{x})$. One such example is shown in Figure 3.26, taken from Lavallée et al. [1993]. The lines (bottom to top) are the box-counting results for altitude thresholds that decrease by a factor of 2 from 3, 600 m. The corresponding dimensions are in the range 0.84–1.92.

From Figure 3.26 it is clear that the above topographic data from France do not meet the simple-scaling assumption, because in such cases the functional box-counting dimensions at various thresholds should be identical.

A few notes of clarification are needed at this point. An exceedence set comprises all the points above a threshold T and is closed by isolines. This, in turn, relies on an intrinsic spatial resolution typically much larger than the smallest scale of the phenomenon we seek to describe [Lavallée et al., 1993]. As one imposes an inner resolution scale, a finite resolution field is obtained through sampling and coarse graining under a monofractal framework. Because the initial averaging may not smooth out the fluctuations resulting in complex scaling structures, from the outset it seems logical to avoid constraining the analysis of the process through a monofractal framework.

Let us consider the volume n of a set S at a scale δ:

$$n(\delta) \propto \delta^{D(S)} \tag{3.105}$$

where, as seen in Chapter 2, $D(S)$ is the dimension (i.e., the length is the 'volume' of the line $\propto \delta$; the area is the volume of a plane $\propto \delta^2$, and so on). Volume is therefore the *measure* of the set, and a simple-scaling (power law) function. The dimension $D(S)$ is, of course, scale-invariant. Consider $N(\delta)$; that is, the number of boxes of size δ needed to cover the set S. Because the number of points of S covered by each box is $n(\delta)$, we have

$$N(\delta)\ n(\delta) = \text{constant}$$

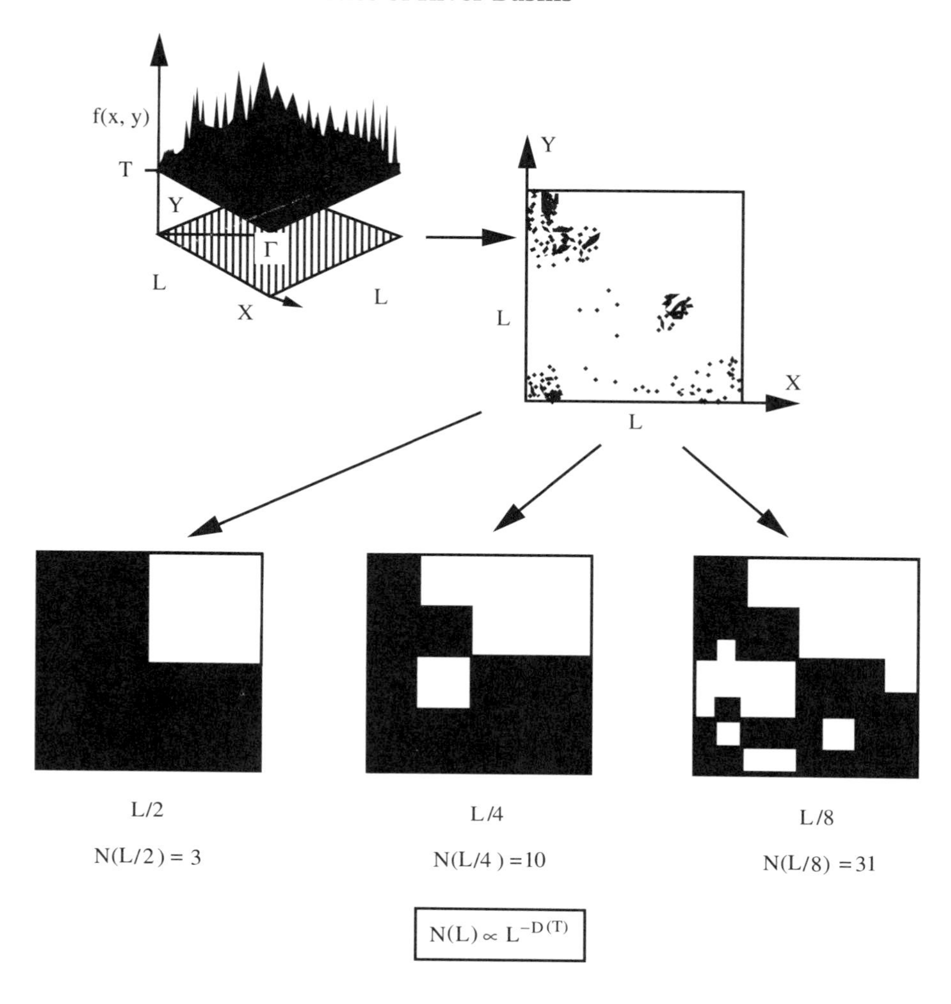

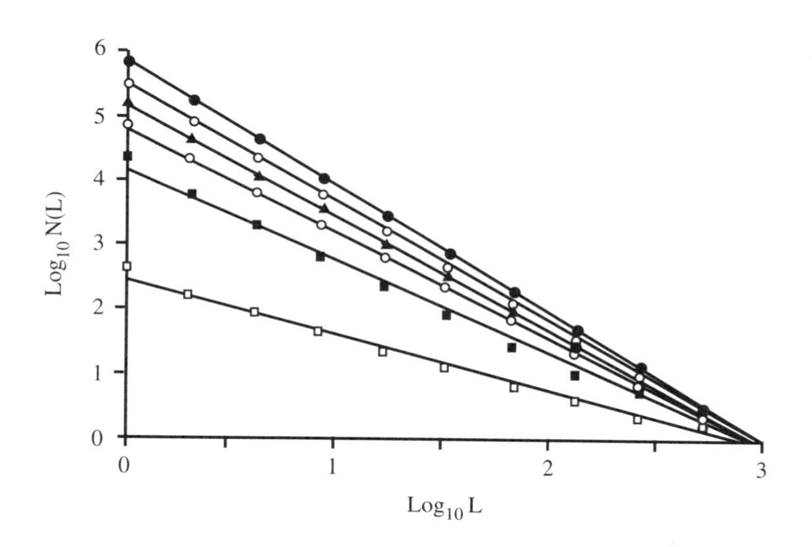

Figure 3.26. Functional box counting applied to a 1024×1024 km^2 map of France at 1-km resolution [after Lavallée et al., 1993].

and therefore $N(\delta) \propto \delta^{-D(S)}$. The value of $D(S)$ obtained by box counting is an estimate of the fractal dimension of the set.

In dealing with fields we are usually interested in distributions. Distributions involve scaling exponents called codimensions. The codimension $C(S)$ of a set S is the scaling exponent of the fraction of space occupied by S, that is, of the probability $P(\delta)$ of finding S in a given box of size δ. Thus

$$P(\delta) \propto \delta^{C(S)} \tag{3.106}$$

where the exponent $C(S)$ determines the scaling of the probability, and it is an intrinsic property of the set regardless of the space on which it is observed. On the other hand, the fractal dimension one measures depends on the intersection of the process with an observational space of dimension, say D, because the number of box-counting covering elements $N(\delta)$ is given by

$$N(\delta) = P(\delta)\, \delta^{-D} \tag{3.107}$$

where δ^{-D} is the total number of boxes in the observation set. If the observing set has large enough dimension, that is

$$D > C(S) \tag{3.108}$$

then the observation is possible and we obtain

$$D(S) = D - C(S) \tag{3.109}$$

Lovejoy and Schertzer [1990] and Lavallée et al. [1993] have observed that natural processes can often be modeled by stochastic processes defined on probability spaces with infinite dimensions. This would postulate an observational space with $D \to \infty$. In this case, the use of codimensions to characterize the sparseness of the probability space is necessary, as it is in all cases when a multifractal process is studied on low-dimensional *cuts* with $D < C(S)$.

The above considerations underline the importance of the dimension of the observation space when analyzing a field. As an example, a topographic field $z(\mathbf{x})$ may be analyzed by taking transects, or one-dimensional cuts ($D = 1$), or by cutting the field with threshold planes where the set S is defined by the condition $z(\mathbf{x}) \geq T$ ($D = 2$). The differences are noteworthy as, for example, a one-dimensional transect is a self-affine curve, whereas the border set curve (defined by the definition of isolines $z(\mathbf{x}) = T$) may be self-similar.

Hence the perimeter set P_T associated with $S_{\geq T}$ is defined as the border set of $S_{\geq T}$ resulting from the T-crossing set of the topographic field, that is, P_T is defined by the set of points $\mathbf{x}$ for which $z(\mathbf{x}) = T$. It is that set of points where arbitrarily small neighborhoods of $\mathbf{x}$ contain at least one point with $z(\mathbf{x}) < T$ and $z(\mathbf{x}) \geq T$. We have already observed that P_T is a subset of $S_{\geq T}$, and hence $D(P_T) \leq D(S_{\geq T})$. In general, the change in dimension of $D(S)$ reflects the connectivity properties of $S_{\geq T}$. It may increase or decrease with the threshold elevation T and is bounded above by the decreasing function $D(S_{\geq T})$.

If we finally define $S_{=T}$ as the set of points $\mathbf{x}$ for which we have $z(\mathbf{x}) = T$, the condition

$$D(P_T) \geq D(S_{=T}) \tag{3.110}$$

is simply a corollary of the fact that $S_{=T}$ is a subset of P_T. In simple-scaling fields sampled by two-dimensional cuts of the field $z(\mathbf{x})$ we have

$$C(P_T) = C(S_{=T}) = H \tag{3.111}$$

and hence isolines in simple-scaling topographies have dimensions

$$D(P_T) = D(S_{=T}) = 2 - H \tag{3.112}$$

whatever the value of the elevation T.

3.6.2 Generalized Variogram Analysis

There are other ways to determine the scaling nature of a field, namely, those related to the definition of generalized variograms [Lavallèe et al., 1993], which are essentially a generalization of height–height correlation analyses. The qth moment of a range, or generalized variogram, $C_q(\lambda)$, is defined as (Eq. (3.83))

$$C_q(\lambda) = \left\langle \left| z(\mathbf{x}) - z(\mathbf{x}+\mathbf{r}) \right|^q \right\rangle_{|\mathbf{r}|=\lambda} \tag{3.113}$$

where $\langle\,\rangle$ is the ensemble averaging operator and λ is the separation distance. The qth moment is said to be scaling if the following relation holds:

$$C_q(\lambda) = \lambda^{K(q)} C_q(1) \quad \propto \lambda^{K(q)}. \tag{3.114}$$

It is worth noting that the definition of $C_q(\lambda)$ given above implies that the random field of the elevation differences is stationary, because the qth moment is supposed to depend only on the separation distance, or lag, λ.

As seen before, in the case of simple scaling, the exponent $K(q)$ is linear; that is

$$K(q) = qH \tag{3.115}$$

as can be shown by taking qth powers of the moduli of both sides of Eq. (3.104) and taking ensemble averages. In multifractals a more general form for $K(q)$ has to be assumed. We refer to this case as multiple scaling.

Restricting ourselves, without loss of generality, to the case of values sampled along a one-dimensional track, we consider the relation between the form of the power spectrum obtained from the Fourier transform of the data and the exponent $K(q)$ governing the scaling of the moments. It is known that for a series whose graph is self-affine the power spectrum has the form of a power law, that is, $S(f) \propto f^{-\zeta}$, except for cutoff effects. The energy spectrum is the Fourier transform of a second-order moment ($q = 2$) and so it is easy to show that $\zeta = 1 + K(2)$ (i.e., Eq. (3.86)). In fact, for stationary random processes the power spectrum is the Fourier transform of the autocorrelation function $R(\lambda) = \langle z(x+\lambda) z(x) \rangle$ (a function of the lag only), which is a power-law as $R(\lambda) \propto \lambda^{\zeta-1} (0 < \zeta < 1)$ for a self-affine graph $z(x)$, as seen in Section 2.7.2.

The autocorrelation function can, in turn, be related to the second-order moment $C_2(\lambda)$ via

$$\left\langle \left| z(x+\lambda) - z(x) \right|^2 \right\rangle = 2\left\langle z(x)^2 \right\rangle - 2R(\lambda) \tag{3.116}$$

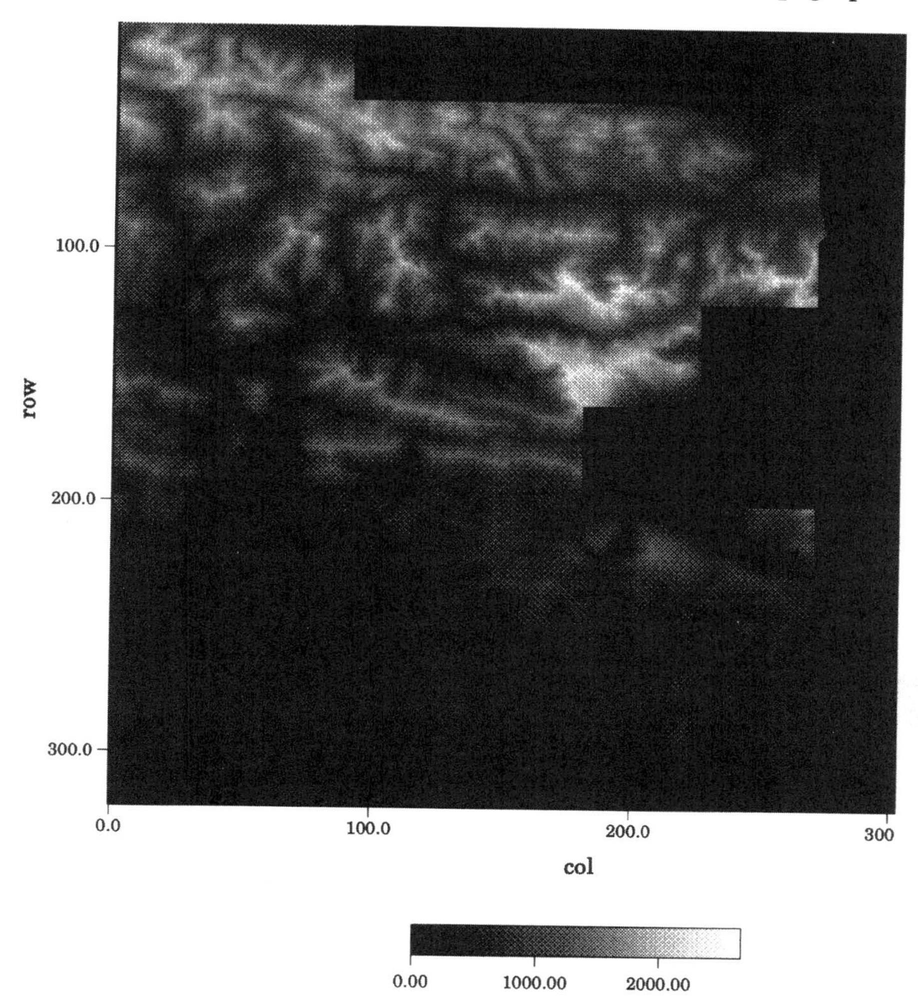

Figure 3.27. Density plot of the DEM of the Fella River basin in northern Italy where generalized covariances were estimated.

giving $\zeta = 1 + K(2)$. As seen already, in the case of simple scaling $K(2) = 2H$ and $\zeta = 1 + 2H$. Because the fractal dimension of the self-affine profile, $z(x)$, is $D = 2 - H$, we have $D = (5 - \zeta)/2$.

The above relations are commonly used in evaluating the fractal dimension of a two-dimensional profile and can be useful in determining the fractal or multifractal nature of such a profile. Lavallèe et al. [1993] contended that because for multifractals $K(2) \neq 2K(1) = 2H$, the above relation between D and ζ will not hold in most applications to real geophysical profiles. In fact, to justify $D = 2 - H$, we only need $C_1(\lambda) = \lambda^H C_1(1)$, whereas $D = (5 - \zeta)/2$ also requires that $C_2(\lambda) = \lambda^{2H} C_2(1)$.

We studied a 2,200 km^2 portion of the digital terrain map (DTM) of northern Italy produced by the Italian Servizio Geologico Nazionale (pixel size: 250 × 310 m, Figure 3.27) to test the scaling properties of real topographies. The results of Lavallée et al. [1993] and of our computations (Figure 3.28) show that natural topographies carry the signatures of multiscaling, that is, a nonlinear relation between $K(q)$ and q. The generalized covariances have been computed with computational techniques

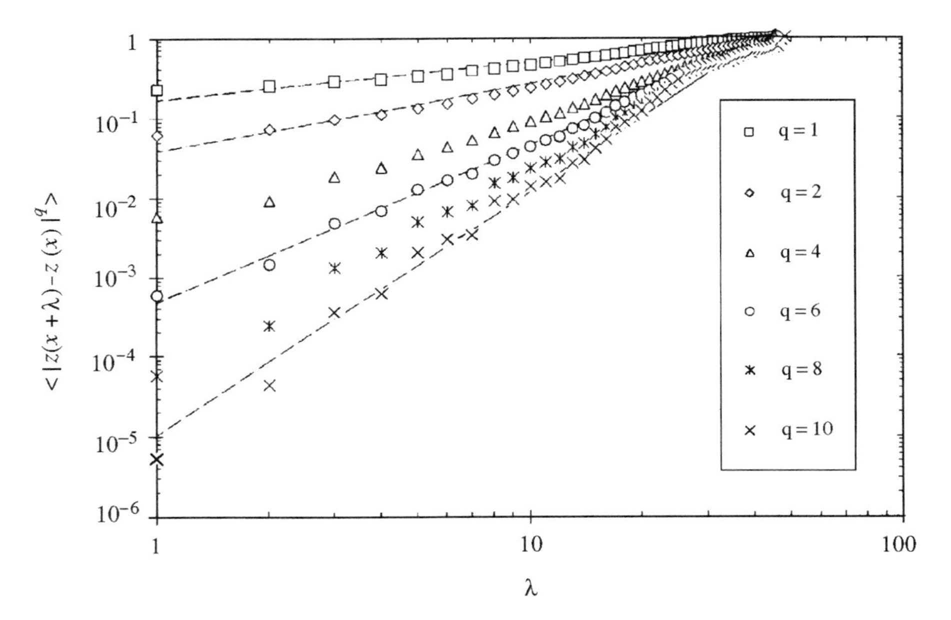

Figure 3.28. Generalized covariances of the field $z(\mathbf{x})$ shown in Figure 3.27.

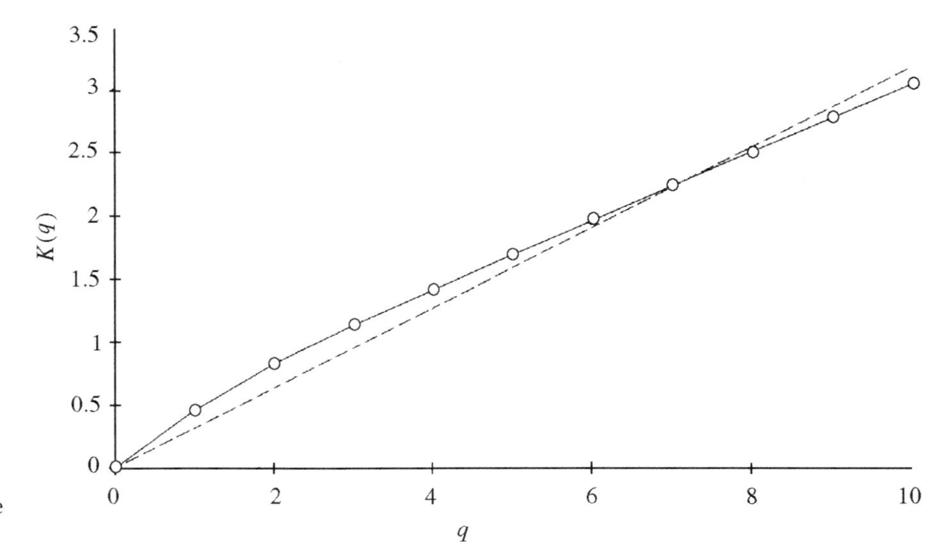

Figure 3.29. The nonlinear relationship of $K(q)$ versus q for the generalized covariances of the landscape in Figure 3.27.

described in Bellin et al. [1992]. The nonlinear relationship between $K(q)$ and q implies that the geometric properties of natural landscapes in general may not be described by specifying a single fractal dimension. It is important to observe the straight generalized variogram $C_q(\lambda)$ for varying moment orders q shown in Figure 3.28. This implies that the field is actually scaling, although differently from order to order. Without a clear scaling relationship we may easily misinterpret a nonscaling field as being a multifractal.

One important operational issue is, in fact, to avoid confusing multiple scaling exponents with fields that show no scaling in their variograms. If the variograms do not plot as straight lines in a double logarithmic plot of $C_q(\lambda)$, it can hardly be inferred that the field is multifractal. In Chapter 6 we will examine thoroughly the impact of landform evolution processes that destroy the scaling structure of the variograms. A best fit of this slope, $K(q)$, may yield a nonlinear relationship with q, but the poor fit of the linear curves, $\log C_q(\lambda)$ versus $\log \lambda$, suggests that the conclusion of multifractality is untenable.

Also in Chapter 6 we will suggest that multiscaling in river basin topography may be a by-product of heterogeneity in the erodibility of the soil mantle because in nature this property varies strongly, not only with position on the surface but also with depth, that is, with the degree of erosion achieved.

From the above analyses a few observations may be drawn:

- Because in many fractal analyses of topographic fields, $z(\mathbf{x})$, simple scaling is assumed from the onset, the estimations of $H = K(2)/2$ tend to be inaccurate (because $K(2) \neq 2K(1)$) whenever the field $z(\mathbf{x})$ shows nonlinear scaling properties.
- General variogram analyses provide an accurate tool for analyzing field properties, as the analyses emphasize large fluctuations that may be obscured by analyses centered on low-order moments. However, variogram analyses postulate the stationarity of the increments $\Delta z(\lambda)$ because the statistical properties of $\Delta z(\lambda) = |z(\mathbf{x} + \lambda) - z(\mathbf{x})|_{|\lambda|=\lambda}$ are assumed to depend on λ and not on $\mathbf{x}$. This may be quite a restrictive assumption, as is documented, for instance, in the literature on turbulence [e.g., Batchelor, 1953] and heterogeneous transport [see, e.g., Dagan, 1989].

The stationarity issue is important. In fact we know (see Section 1.2.11) that landscape topography in the channelized regime of fluvial landscapes shows a scaling relationship for local slopes, with total drainage area at a point. Such a relationship implies nonstationarity in the increments in elevation when sampled along flow directions identified by the channels because the increments Δz will largely depend on the spatial position, that is, $\Delta z(\mathbf{x}, \lambda)$.

Rodriguez-Iturbe et al. [1994] argued that a sampling procedure of Δz along coordinate axes (say, along E–W and S–N tracks) on a large enough landscape does not detect the nonstationarity induced by the network structure, and a global stationarity of increments then results. At any rate, care should be exerted in evaluating the presence of multiscaling when using tools that postulate stationarity.

A basic question is whether the nonlinear variability we have been discussing is embedded in the dynamics of the landscape development or whether it arises because of the heterogeneity of field properties. Leaning toward the second hypothesis, we will investigate this issue in detail in Chapter 6. We will deal with the identification of dynamic processes of landscape evolution that reproduce the experimental features discussed in Chapters 1 through 3 and possibly allow for multifractal variability.

3.7 Random Cascades

In this section we examine the geomorphological width functions of several river basins through the formalism of conservative random cascades and we compare the results with those of more conventional fractal and multifractal analyses. We suggest that the spatial patterns of aggregation described by the width functions analyzed through this procedure also show common recurrent characters and that some implications of the previously described multifractal analyses are also supported by the random cascade rationale.

3.7.1 Canonical Random Cascades

Random cascades are models useful for the analysis of fields exhibiting large fluctuations in space and/or time. They play an important role in current theories of rainfall distribution in space and time and in statistical theories of turbulence [e.g., Gupta and Waymire, 1993]. The construction of such models will be outlined here for the one-dimensional case with a view toward their application to the study of width functions, but the results apply whatever the dimensionality of the problem.

The procedure starts with the quantity considered initially having a uniform value W_0 over the unit interval $J_0 = [0, 1]$. This interval is then divided into b subintervals of length $1/b$. Let $J_1(\sigma)$, $\sigma = 1, 2, \ldots, b$, denote this partition of J_0 into b subintervals. To each interval $J_1(\sigma)$ is assigned a value $W(\sigma_1) = W_0\, p_1(\sigma_1)\, b^{-1}$, $\sigma_1 = 1, 2, \ldots b$, where $p_1(\sigma_1)$ is a (mean one) random variable. Each interval is then further subdivided into b subintervals, and the above operation is iteratively repeated n times. At the nth generation the interval identified by the sequence $\sigma_1, \ldots, \sigma_n$ (which can be thought of as specifying a path through a b-ary tree) is therefore assigned the value

$$W_n(\sigma_1, \ldots, \sigma_n) = W_0\, p_1(\sigma_1)\, p_2(\sigma_2) \ldots p_n(\sigma_n)\, b^{-n} \qquad \sigma_i = 1, 2, \ldots b \tag{3.117}$$

where the $p_i(\sigma_i)$s are iid as a random variable p, called the generator. It is stipulated that $\langle p \rangle = 1$, which means that the ensemble average of the quantity W_0 is conserved after the redistribution. As $n \to \infty$ a continuous distribution $W_\infty(x)$ over the unit interval is obtained.

We note that the total mass W_n in the arbitrary segment, say J_n, of the nth stage of generation of a one-dimensional cascade may be given by the relationship

$$W_n(J_n) = W_0 \int_{J_n} \varphi(x)\, dx \tag{3.118}$$

where the equivalent of a mass density $\varphi(x)$ is given by [Gupta and Waymire, 1993]

$$\varphi(x) = \prod_{j \in path\ to\ J_n} p_j \qquad p_j \geq 0 \tag{3.119}$$

In the two-dimensional case (or in higher dimensions), x becomes a vector, $\mathbf{x}$. Notice that one can associate the interval J_n to a position x on the unit interval, and hence one may write, for convenience, $J_n(x)$.

Random cascades study the limit behavior of the mass distribution $W_n \to W_\infty$ as $n \to \infty$, which may or may not be degenerate, that is, almost surely zero throughout the paths to any J. It is of interest that the nondegenerate structure of W_∞ is a consequence of anomalously large deviations from the mean because with probability 1, the product $W_0 p_1 p_2 \ldots p_n \to 0$

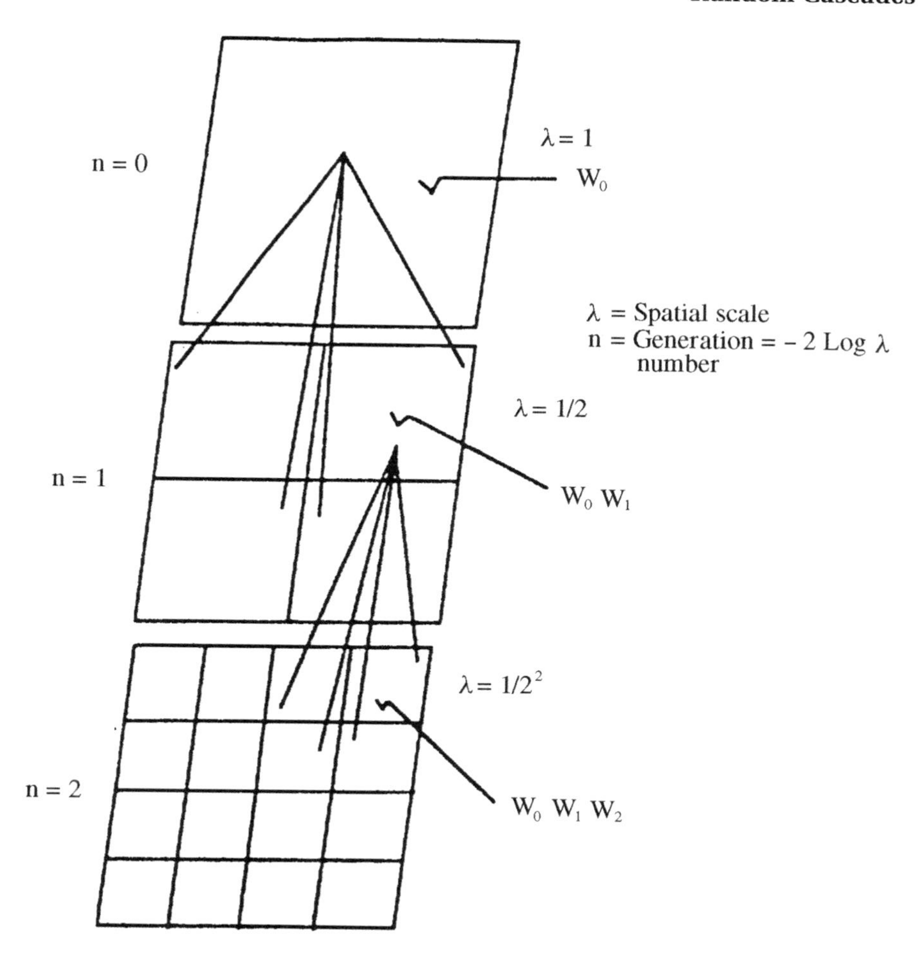

Figure 3.30. A schematic plot of random cascade geometry [after Gupta and Waymire, 1993].

as $n \to \infty$ along any individual path of the tree defined by the cascade [Gupta and Waymire, 1993]. A schematic plot of the random cascade geometry is shown, for the two-dimensional case, in Figure 3.30. The definition of the properties of the random cascade rests on the specification of the parameter b, called the branching number, and the probability distribution of the generator p.

Following Mandelbrot [1974] and Kahane and Peyrière [1976], Gupta and Waymire [1993] show that the main characters of the limit distribution are controlled by a function (usually termed the modified cumulant generating, or Mandelbrot–Kahane–Peyrière (MKP), function) defined as

$$\chi_b(h) = \log_b \left[\langle p^{\,h} \rangle\right] - (h - 1) \tag{3.120}$$

where p is the random variable whose realizations are the values p_i assigned to each subinterval into which the original unit interval has been partitioned.

The MKP function $\chi_b(h)$ controls the nondegeneracy of the limit distribution W_∞ and the finiteness of its moments. We are not interested in such properties, but rather in the link between the MKP function and the singularity spectrum $f(\alpha)$. A theorem by Kahane and Peyriere [1976] yields insight into the nondegeneracy of W_∞, the divergence of its moments, and

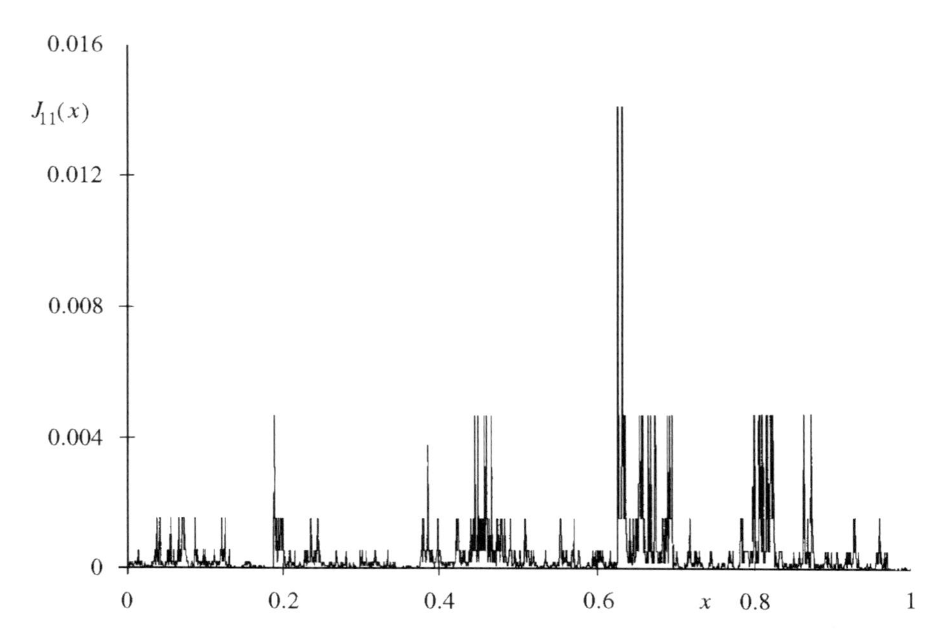

Figure 3.31. A plot of a random binomial cascade with $q = 0.25$. The notation $J_{11}(x)$ denotes the path to the x location of the cascade tree at the 11th stage of generation.

the dimension of its support, where the key condition for nondegeneracy is given by

$$\left.\frac{d\,\chi_b}{dh}\right|_{h=1} < 0 \tag{3.121}$$

Before pursuing the interpretation of width functions through random cascades, it is instructive to describe a few processes produced by prescribed generators p.

One interesting example is the *random binomial* cascade. This example generalizes the deterministic binomial multiplicative process studied in Section 3.2. Specifically, let $b = 2$ and $0 < q < q' < 1$, $q + q' = 1$. The generator is fixed by assuming

$$p = \begin{cases} 2q & \text{with probability } P = 0.5 \\ 2q' & \text{with probability } P = 0.5 \end{cases} \tag{3.122}$$

In the above case, the MKP function is

$$\chi_b(h) = \log_2\left(q^h + (1-q)^h\right) \tag{3.123}$$

because $\sum_i P_i p_i^h = \langle p^h \rangle = 0.5(2q)^h + 0.5(2(1-q))^h = 2^{(h-1)}(q^h + (1-q)^h)$. It is easy to show that the condition in Eq. (3.121) is met.

The gross features of the resulting distribution at the $n = 11$ stage of generation are shown in Figure 3.31. We notice that the random binomial

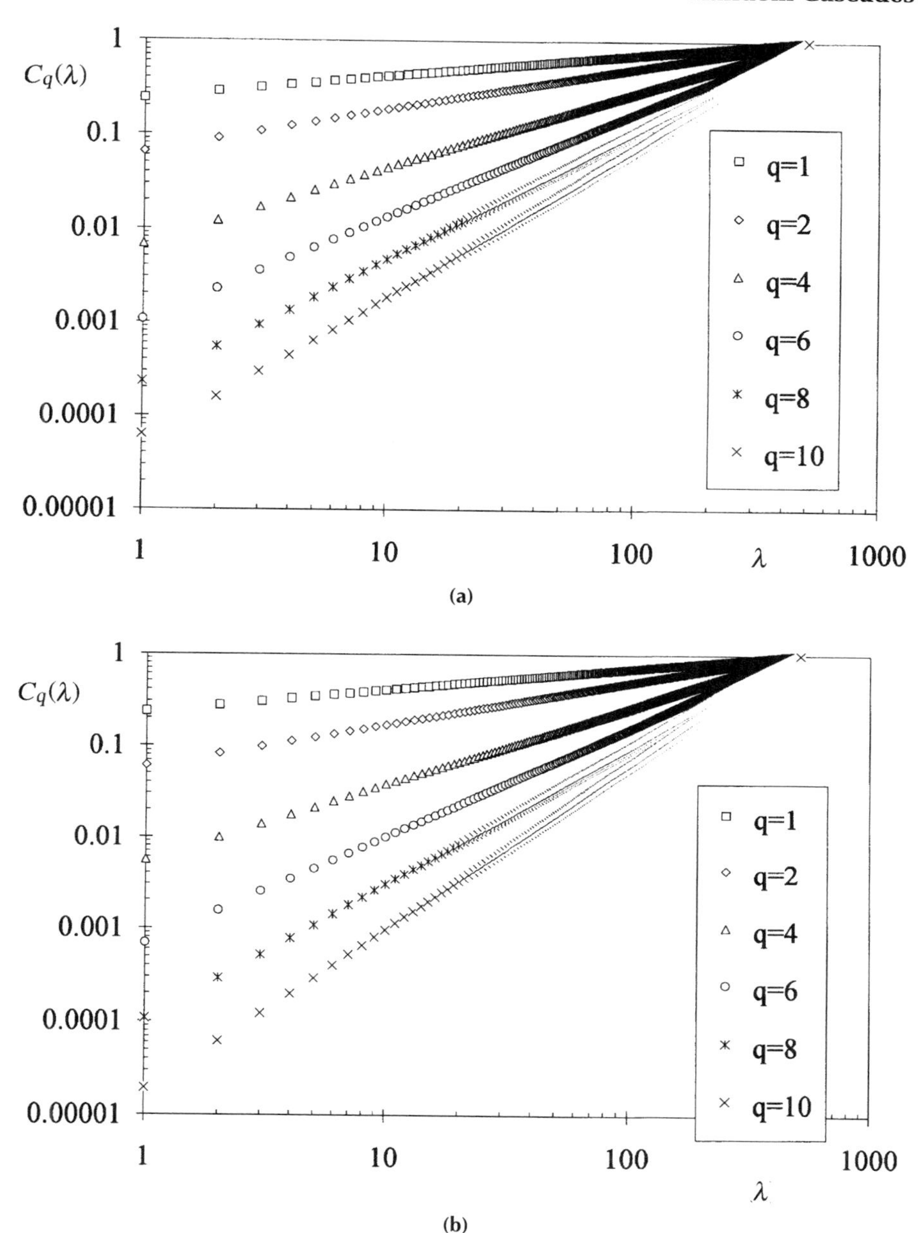

Figure 3.32. (a) Generalized covariance analysis for a deterministic cascade with $q = 0.25$; (b) same as in (a) but for a random cascade with $q = 0.25$.

cascade differs noticeably from the deterministic cascade with the same parameters shown in Figure 3.1. Nevertheless, their singularity spectra coincide [Holley and Waymire, 1992]. This fact can be explained heuristically by observing that the scaling of the number of sites with the same amount of mass is not tied to the particular position of the sites. Thus the $f(\alpha)$ spectrum is insensitive to the spatial organization that the randomization basically provides. Figures 3.32(a) and 3.32(b) show the scaling of the generalized covariances (see Section 3.5.1) for the cascades shown in

Figures 3.1 and 3.31. Figure 3.33 shows their respective $K(q)$ versus q diagrams. The departures from a straight line clearly underline the presence of the multiple scaling structure induced by construction.

Another model of interest to our study is the so-called α model [Schertzer and Lovejoy, 1987; Gupta and Waymire, 1993]. This model is defined as follows. Assume $P\,(p = a_1) = q$ and $P\,(p = a_2) = 1 - q$ in Eq. (3.122), such that $0 \leq a_1 < 1 < a_2$ with the constraint that $\langle p \rangle = 1$. In this case, the MKP function

$$\chi_b(h) = \log_b \left[a_1^h q + a_2^h (1-q)\right] - (h-1) \tag{3.124}$$

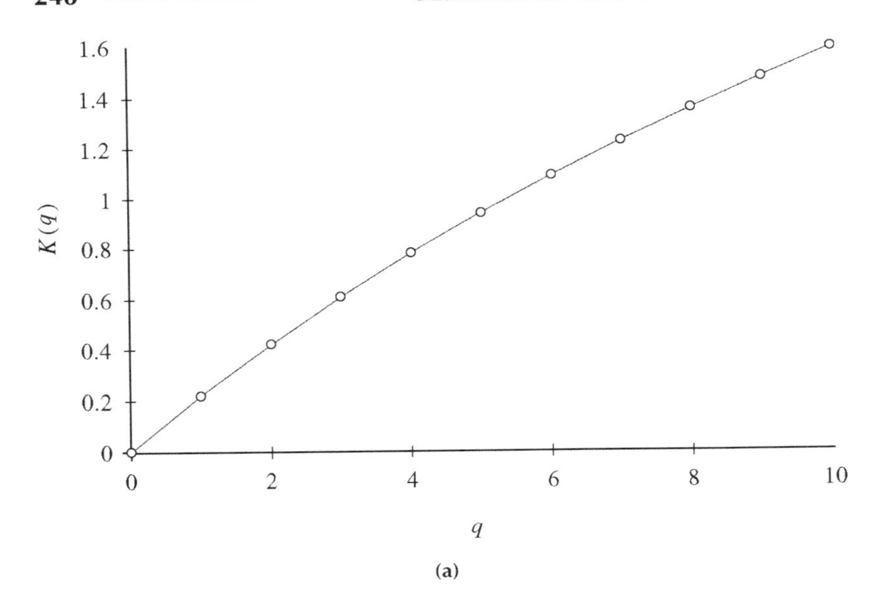

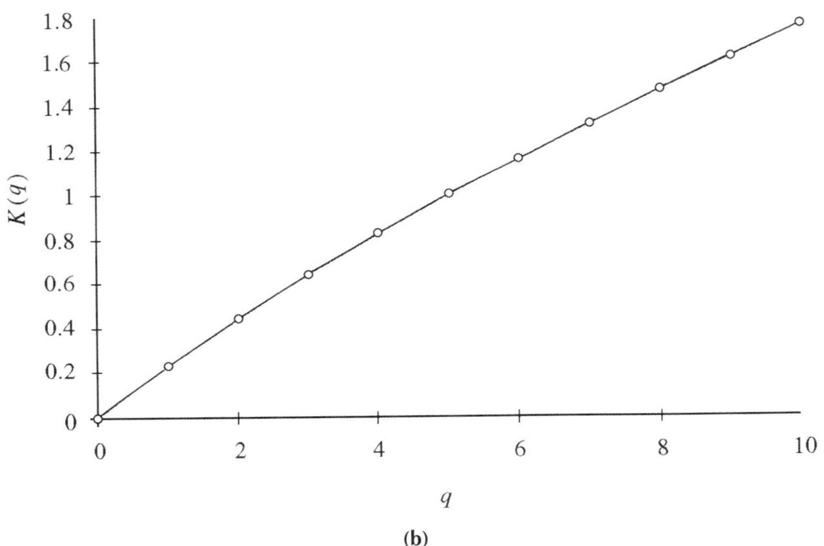

Figure 3.33. Scaling coefficients $K\,(q)$ versus q for (a) deterministic (Figure 3.1) and (b) random binomial cascades (Figure 3.31) with $q = 0.25$.

satisfies the conditions (3.121) for nondegeneracy. Figure 3.34 shows two cases of α processes obtained for different values of the parameters. The model allows the simulation of different degrees of intermittency and different ranges of values covered by the process. Figure 3.35 shows the perfect scaling characters shown by the α model.

A theorem by Holley and Waymire [1992] is useful in the analysis of random cascades. If the generator p of a random cascade satisfies the conditions

$$P\,(p > a) = 1 \qquad (a > 0) \tag{3.125}$$

the process is called strongly bounded below. It is said to be strongly bounded above if

$$P\,(p < b) = 1 \tag{3.126}$$

where b is the branching number. If $\langle p^{\,2h} \rangle / (\langle p^{\,h} \rangle)^2 < b$, then one has

$$\bar{f}\,(\alpha) = E \inf_h \left[h\,\frac{\alpha}{E} + \chi_b(h)\right] \tag{3.127}$$

where, as seen in Section 3.5.1, E is the dimension of the support ($E = 1$ for $W(x)$) and $\bar{f}\,(\alpha)$ denotes the *convex hull* of $f\,(\alpha)$. The convex hull of a set is the smallest convex set containing it (e.g., for three points it is a triangle). In the case of a continuous and convex function, its convex hull coincides with the function itself. Holley and Waymire [1992] noticed that

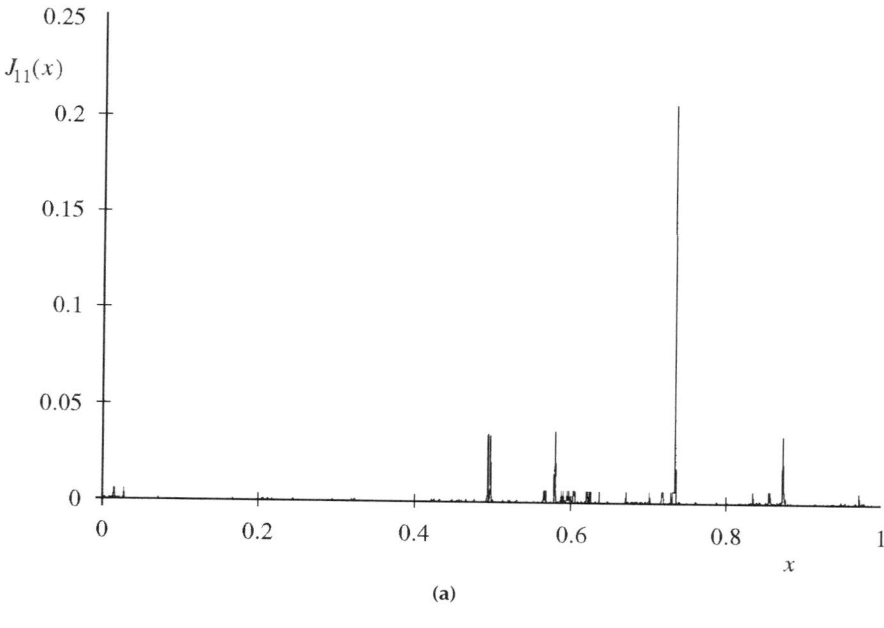

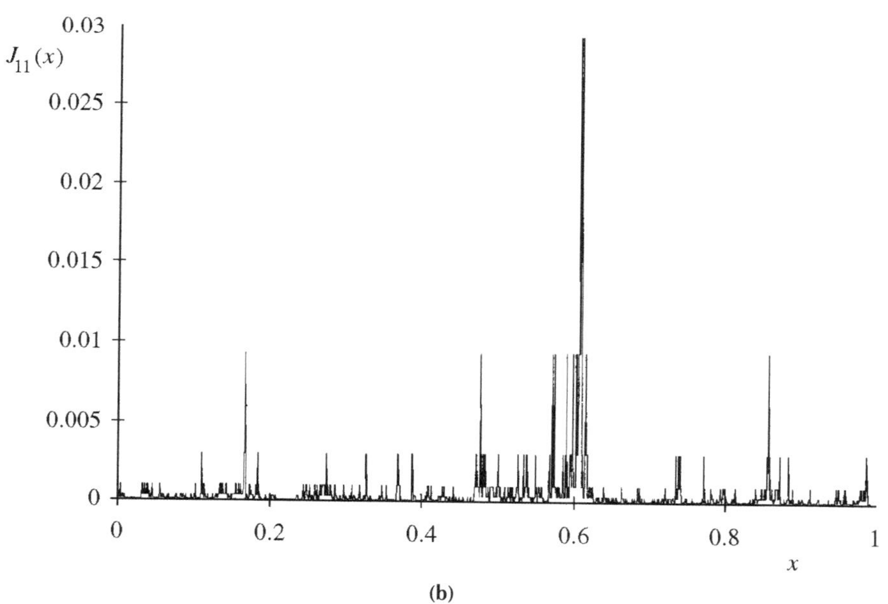

Figure 3.34. Random cascade at the 11-th stage of generation through the α model with $b = 4$; (a) $P\ (p = 2.4) = 0.3$ and $P\ (p = 0.4) = 0.7$; (b) $P\ (p = 2.2) = 0.2$ and $P\ (p = 0.7) = 0.8$.

the singularity spectrum defined through Legendre transformation (as in Eq. (3.58)) is always convex, whereas in the general case (Eq. (3.127)) it may not be so.

The above results establish a connection between the properties of the random cascade and its corresponding singularity spectrum $f(\alpha)$.

3.7.2 Conservative Random Cascades and Width Functions

The application of random cascade models to the study of width functions can be physically justified only in the context of conservative

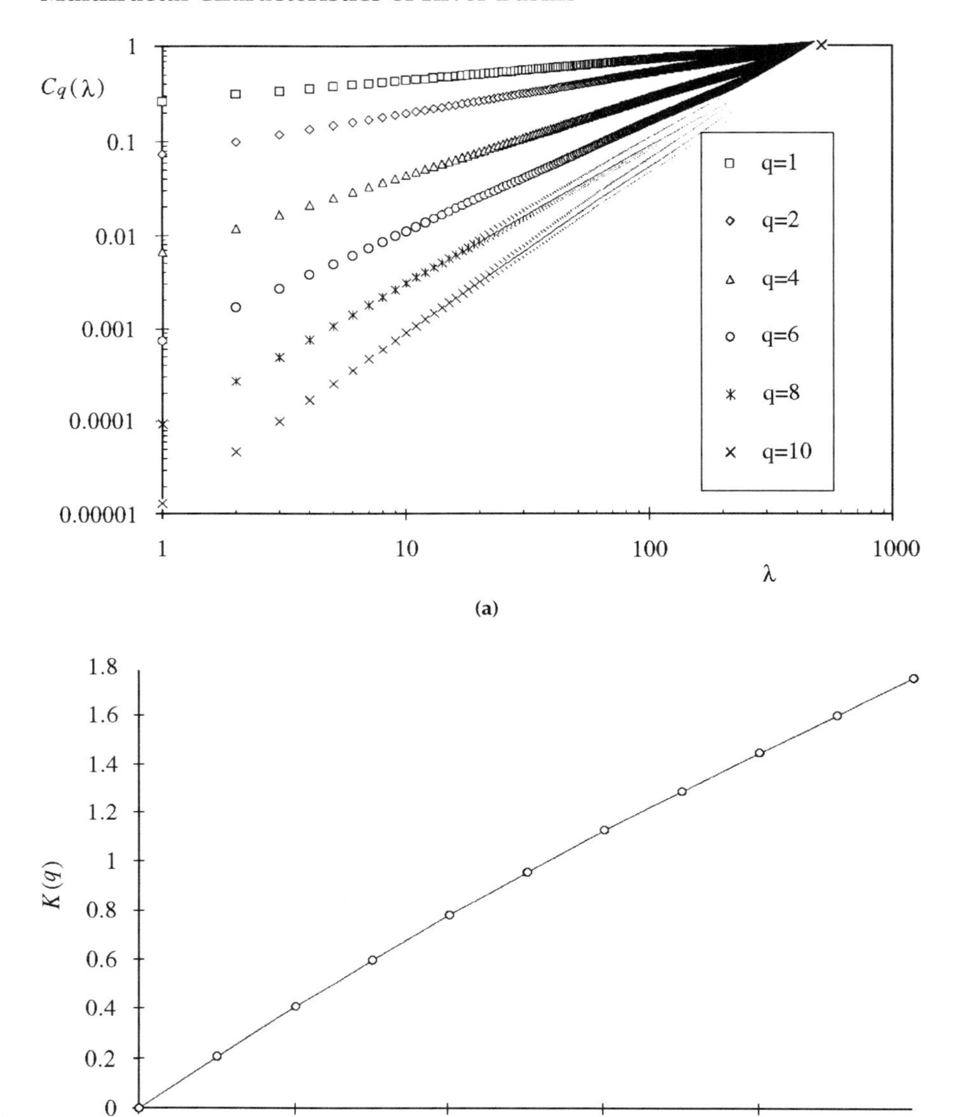

Figure 3.35. (a) Generalized covariances and (b) the $K(q)$ versus q curve for the α model of a random cascade ($P(p = 1.4) = 0.5$ and $P(p = 0.6) = 0.5$).

cascades for which mass is strictly conserved in any realization, that is, $W_n(\sigma_1, \ldots, \sigma_n) = W_0 b^{-n}$ in Eq. (3.117). In fact, a width function may be viewed as obtained by progressively dissecting the total drained area (the strictly conserved quantity) along the backbone of the network in the same way as a multiplicative process is obtained by dissecting a uniform measure along the unit interval. Furthermore, an analytically known deterministic binomial cascade is the one corresponding to the width function of Peano's basin.

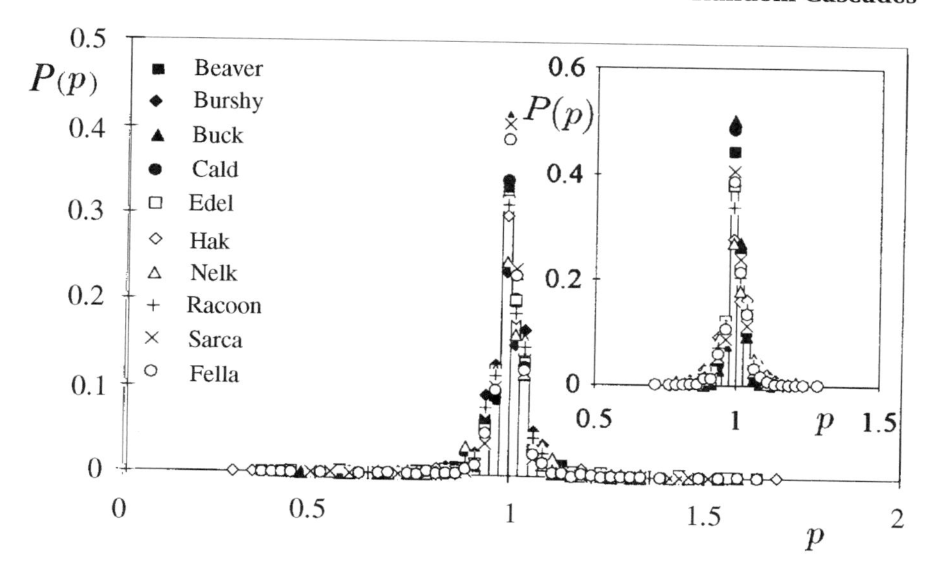

Figure 3.36. Frequency classes computed for the generator p of the width functions considered and (smaller graph) of the filtered signals [after Marani et al., 1994].

In the case of width functions of real basins, it is interesting to study the properties of the frequencies of a conservative generator p, both to determine the main features of its distribution and to test its possible recurrence in different basins.

The statistical properties of the generator p of $W(x)$ are estimated through a coarse-graining technique. Starting at the maximum resolution, coarse graining is accomplished by adding the values of W on pairs ($b = 2$) of neighboring intervals. The corresponding p_is are computed as the ratios between this sum and the original two values of W. This procedure was recursively repeated and stopped when the unit value of W was obtained [Marani et al., 1994].

This estimation procedure has been tested by applying it to the random binomial cascade, in a strictly conservative scheme, where in each subdivision the mass is partitioned by randomly drawing one number p from the generator and assigning fractions p and $(1-p)$ of the initial mass to the resulting halves. The bimodal distribution $P(p)$ is perfectly reproduced in any realization by coarse graining with $b = 2$. Because $b = 2$ a basic symmetry of the conservative cascade is forced. This is not the case for the odd partitions that were also tested by Marani et al. [1994].

It is important to notice the robustness shown by the resulting frequency distributions of the p_is. The histogram of the values of p obtained for $b = 2$ is shown in Figure 3.36. The distributions computed are always symmetric – even in the case of odd partitions – and show consistent values of their variance (the coefficient of variation ranges from 0.06 to 0.08).

An even better consistency is shown when the original width function is filtered by removing the spectral components of the twenty lowest frequencies of the discrete Fourier transform (see the insert in Figure 3.36).

Marani et al. [1994] have shown that the values of p_i obtained by the coarse-graining procedure satisfy the condition $\langle p^{2h}\rangle/(\langle p^{h}\rangle)^2 < b$. Nevertheless, the amount of data does not allow any conclusions beyond the third moment. Notice that the generator p also obeys the boundedness requirements $P(p > 0) = 1$ and $P(p < b) = 1$, as seen in Figure 3.36.

We conclude that the application of random cascade arguments suggests once again the recurrence of many common features in the spatial organization of real basins.

CHAPTER 4

Optimal Channel Networks : Minimum Energy and Fractal Structures

Optimal channel network (OCN) configurations are obtained by minimizing the total rate of energy expenditure in the river system as a whole and in its parts. OCNs are obtained by robust, parameter-free random search procedures. Striking similarities are observed for natural and optimal networks in their fractal aggregation structures and in their morphology. Specific comparisons among fractal constructions, DEMs of natural basins, and OCNs suggest that nature seems to reject the type of exact self-similarity exhibited by certain (e.g., Peano's) constructs in favor of different shapes, implying statistical self-similarity not only because of chance acting through random conditions but also because of necessity as reflected by least total energy expenditure. Scaling and thermodynamic properties also yield a comprehensive theoretical framework supporting the likelihood of OCN growth in nature.

4.1 Introduction

As seen in Chapters 2 and 3, theoretical and experimental digital elevation model (DEM) studies suggest that natural channel networks exhibit many fractal characters. Since Mandelbrot's [1983] classical studies, fractal measures have been extensively used in the characterization of natural patterns and physical structures. The general character of some of these measures and their ubiquity suggest the possibility of some general physical principle connected with the intimate structure of natural processes. This chapter pursues the hypothesis that the process of network formation, as well as of a broader class of patterns in nature, is characterized by minimum total energy dissipation of the resulting three-dimensional geometry and topology.

The idea that patterns in nature may be obtained by optimality principles of an energetic character is not new [e.g., Stevens, 1974]. Following Stevens and others [e.g. Murray, 1926; Howard, 1971a, 1990; Leopold and Langbein, 1962], Rodriguez-Iturbe et al. [1992b] have suggested new local and global optimality principles linking energy dissipation and runoff production with the three-dimensional structure of river basins. Their theory explains the most important structural characteristics observed in the geomorphology of drainage systems. This chapter builds on the original idea and further developments to argue that optimality criteria may indeed lead to fractal growth. In the specific case of the physics of fluvial networks we show that, regardless of initial conditions, if the evolution is guided by minimum energy requirements, any network evolves toward one whose fractal properties are essentially indistinguishable from those observed in nature.

The approach presented in this chapter may serve to analyze a broad range of phenomena that lead to the formation of fractal structures through optimality rules in which explicit consideration of the dynamics of the processes involved is embedded. We hope this chapter makes clear that, paraphrasing L. Kadanoff [1995], models are fun and sometimes even instructive.

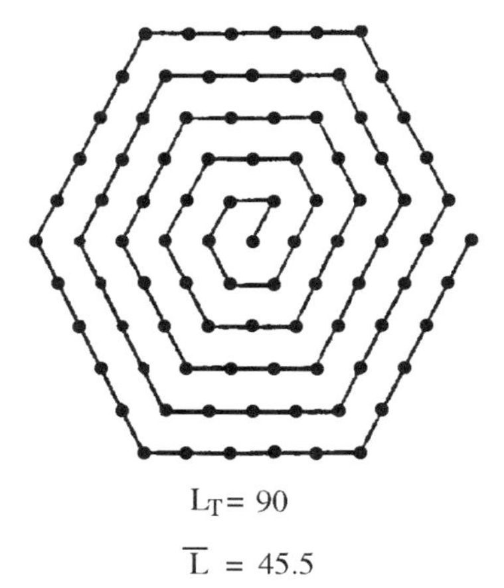

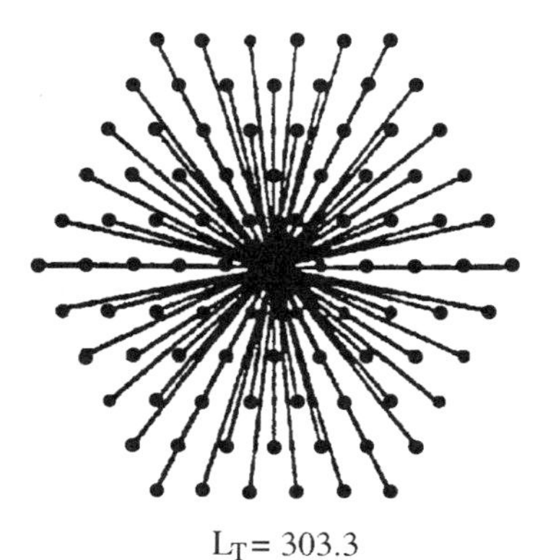

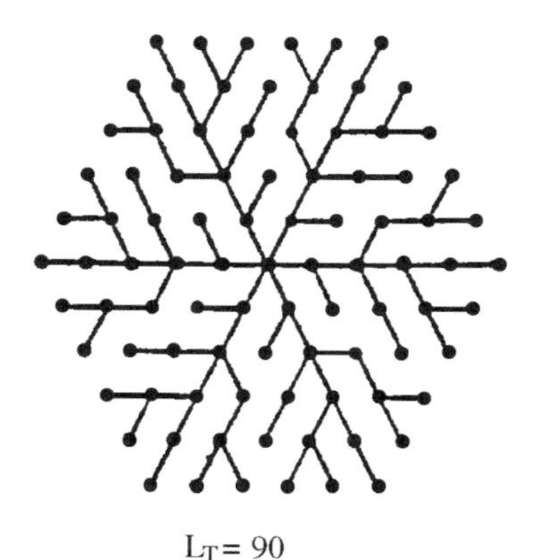

Figure 4.1. Different patterns of connectivity for a set of equally spaced points connected to a common outlet. L_T and $\bar{L}$ are the total and the average lengths of the paths [from Stevens, 1974].

4.2 The Connectivity Issue

Well-developed river basins are composed of two interrelated systems: the channel network and the hillslopes. As described in Chapter 1, the hillslopes control the production of runoff, sediments, and solutes, which in turn are transported through the channel network toward the basin outlet. Every branch of the network is linked to a downstream branch for the transportation of water and sediment, but for its viability it is also linked to every other branch in the basin through the hillslope system. Hillslopes are the runoff-producing elements that the network connects, transforming the spatially distributed potential energy arising from rainfall in the hillslopes to kinetic energy in the flow through the channel reaches. We focus our attention on the drainage network as it is controlled by energy dissipation principles.

It is the need for effective connectivity that leads to the tree-like structure of the drainage network. Figure 4.1, from Stevens [1974], illustrates this point. Assume one wishes to connect a set of points in a plane to a common outlet and, for illustration purposes, assume that every point is equally distant from its nearest neighbors. Two extreme ways to establish the connection would be through spiral- and explosion-type patterns. The explosion pattern has an advantage in that it connects every parcel of the system to the outlet in the most direct manner. Nevertheless, it rejects any kind of interaction between the different parts, and the total path length for the system as a whole is extremely large. Thus, although the pattern gives the minimum average path connecting each parcel to the outlet, it lacks shortness as a whole. The spiral pattern, on the other hand, is quite short for the system as a whole, but it leads to an extremely large average path from a point to the outlet. One is tempted to say that from an organizational point of view the spiral represents pure *socialism* and the explosion pure *capitalism*. In the spiral case, the system is supposed to operate at its best as a whole with a total disregard for the average individual, who finds himself in the worst possible condition. In the explosion case, each individual is supposed to operate at his best when completely oblivious of his neighbors, but the system as a whole cannot survive.

Branching patterns accomplish connectivity by combining the best of the two extremes; they are short as well as direct. The drainage network, as well as many other natural connecting patterns, is basically a transportation system for which the tree-like structure is the most appealing structure for efficiency in the construction, operation, and maintenance of the system.

The drainage network accomplishes connectivity for transportation in three dimensions by working against a resistance force derived from the friction between the flow and the bottom and banks of the channel, the resistance force itself being a function of the flow and channel characteristics. This makes the analysis of optimal connectivity a complex problem that cannot be separated from the individual optimal channel configuration or from the spatial characterization of runoff production inside the basin. The question is whether there are general principles that relate the structure of the network and its individual elements to the rate of energy expenditure that takes place in the system as a whole and in each of its elements.

4.3 Principles of Energy Expenditure in Drainage Networks

A link in a drainage network carries a wide range of discharges resulting from a variety of rainfall events (of different intensities and duration) and antecedent conditions of soil moisture. The individual channel characteristics are commonly assumed to be controlled by the bankfull discharge that the channel is capable of transporting. However, it is also true that most of the work the flow performs throughout time takes place at discharges smaller than the bankfull capacity. From this point of view the mean annual flow may be considered a more representative discharge condition for characterizing the work being done by the flow. Thus it is likely that any principles of optimal energy expenditure that are responsible for the three-dimensional structure of the drainage network will yield the same type of results whether applied to the case of bankfull discharges or when the flow is characterized at every link by the corresponding mean annual value.

Following Rodriguez-Iturbe et al. [1992b], three principles are now postulated:

- the *principle of minimum energy expenditure in any link of the network* for the transportation of a given discharge;
- the *principle of equal energy expenditure per unit area of channel* anywhere in the network; and
- the *principle of minimum energy expenditure* in the network as a whole.

It will be shown that the combination of these principles is a sufficient explanation for the tree-like structure of the drainage network and, moreover, that they generally explain the most important empirical relationships observed in the internal organization of the network and its linkage with the flow characteristics.

The first principle expresses a local optimal condition for any link of the network once its characteristic discharge is defined. The second principle expresses an optimal condition throughout the network regardless of its topological structure (later in this chapter this principle will be interpreted in a probabilistic framework). It postulates that energy expenditure is the same everywhere in the network when normalized by the area of the channel in which it takes place. Thus with the first principle there will be channels that spend much more energy per unit time than others only because of their larger discharge. The second principle makes all channels equally efficient when adjusted for size. The third principle addresses the topological structure of the network and refers to the optimal arrangement of its elements, which in turn is what controls the discharges draining throughout the different links.

The first principle is similar to the principle of minimum work used in the derivation of Murray's law in physiological vascular systems. Murray [1926] derived a relation that states the cube of the radius of the parent vessel should equal the sum of the cubes of the radii of the daughter vessels [see, e.g., Sherman, 1981]. He assumed that two energy terms contribute to the cost of maintaining blood flow in any vessel:

- the energy required to overcome friction as described by Poiseuille's law; and
- the energy metabolically involved in the maintenance of the blood volume and vessel tissue.

Minimization of the cost function leads to the radius of the vessel being proportional to the 1/3 power of the flow. Uylings [1977] has shown that when turbulent flow is assumed in the vessel, rather than laminar flow, the same approach leads to the radius being proportional to the 3/7 power of the flow. The second principle was conceptually suggested by Leopold and Langbein [1962] in their studies of landscape evolution.

It is of interest to add that minimum rate of work principles have been applied in several contexts in geomorphic research, for instance, by Howard [1971a, b], Roy [1983, 1984] and by Woldenberg and Horsfield [1986] among others. Also, the concept of minimum work as a criterion for the development of stream networks has been discussed in different perspectives by authors like Yang [1971a] and Howard [1990]. Of particular interest is the work of Howard, described in Section 1.4.2, who assumed an optimal local rule for the slopes equal to that postulated by Eq. (1.34), that is

$$\nabla z \propto A^{-\theta} \tag{4.1}$$

when $\theta = 0.5$. Capture models embedding randomness in their initial conditions and enforcing Eq. (4.1) are shown to reproduce natural patterns well (see, e.g., Figures 1.49 and 1.50).

However, the key factor missing in Howard's analysis is the *global* principle of minimum total energy dissipation, that is, requiring that the arrangement of the structure with respect to local optimal energy expenditures attains a minimum as a whole. Later in this chapter we will show theoretically that minimum energy dissipation as a whole is attained by three-dimensional structures that minimize the constrained potential energy of the system; that is, they are stable. Thus the outcomes of the approaches based on local optimality rules arise from the fact that whenever randomness is forced into the system through suitable initial conditions, it is likely that capture models will yield structures characterized by low total energy. Thus the apparent success of some simulations obscures the real reason for such success, that is, the low value of total energy dissipation, which – though not required – was nevertheless attained by the simulations.

A counterexample can be easily constructed. A comb-like parallel drainage structure can be deliberately generated (see Figures 5.15 and 5.16) where the local optimal condition, Eq. (4.1), is exactly satisfied everywhere. What distinguishes unrealistic, nonaggregated – though locally optimal – network patterns from real-like network structures is the value of total energy dissipation.

4.4 Energy Expenditure and Optimal Network Configurations

Consider a channel of width w, length L, slope ∇z, and flow depth d. The force responsible for the flow is the downslope component of the weight, $\mathcal{F}_1 = \rho\, g\, d\, L\, w\, \sin\theta \approx \rho\, g\, d\, L\, w\, \nabla z$, where, for small angles θ characteristic of actual slopes, $\sin\theta \approx \tan\theta = \nabla z$. The force resisting the movement is the stress per unit area times the wetted perimeter area, $\mathcal{F}_2 \approx \tau(2d + w)L$ (here τ is the shear stress, assumed constant throughout the cross section), where a rectangular section has been assumed in the channel.

Under the conditions of no acceleration of the flow, $\mathcal{F}_1 = \mathcal{F}_2$; then $\tau = \rho g \nabla z R$, where, R is the hydraulic radius $R = A_w/P_w = wd/(2d + w)$, and A_w, P_w are the cross-sectional flow area and the wetted perimeter, respectively.

In turbulent incompressible flow, the boundary shear stress varies proportionally to the square of the average velocity, $\tau \propto v^2$, such that $\tau = C_f \rho v^2$ where, C_f is a dimensionless resistance coefficient. Thus, equating the two expressions for τ, one obtains the well-known relationship $\nabla z = C_f v^2/(Rg)$, which gives the loss due to friction per unit weight of flow per unit length of channel.

There is also an expenditure of energy related to the maintenance of the channel, which may be represented by a function of soil type and flow conditions, that is, $f(\text{soil, flow})\, P_w L$, where $f(\cdot)$ is a complex function of soil and flow properties representing the work per unit time and unit area of channel involved in the removal and transportation of the sediment that otherwise would accumulate in the channel surface. For instance, from the equations of bed load transport (Section 1.4.5) one may assume that $f \propto \tau^m = K\tau^m$, where K depends only on the soil and fluid properties and m is a constant.

In a channel of length L and flow Q the rate of energy expenditure may then be written as

$$P = C_f \rho \frac{v^2}{R} Q L + K\tau^m P_w L = C_f \rho P_w \frac{Q^3}{A_w^3} L + K C_f^m \rho^m v^{2m} P_w L \qquad (4.2)$$

The coefficient C_f depends mainly on the channel roughness, which tends to decrease only slightly in the downstream direction. On the whole the downstream reduction in roughness resulting from a decrease in particle size is compensated for by other forms of flow resistance like, for example, that offered by bars and channel bends [Leopold et al., 1964].

According to the second principle of energy expenditure, the unit expenditure $P_1 = P/(P_w l)$ is the same anywhere in the network. Substituting P from Eq. (4.2), one obtains

$$P_1 = C_f \rho v^3 + K C_f^m \rho^m v^{2m} = \text{constant} \qquad (4.3)$$

which implies that the velocity tends to be constant throughout the network. As explained in Section 1.2.6, this has been corroborated by the field investigations of Leopold and Maddock [1953], Wolman [1955], and Brush [1961], who obtained values of $z < 0.1$ in the relation between velocity and discharge:

$$v \propto Q^z \qquad (4.4)$$

this being the case for both mean annual flow conditions or bankfull discharges throughout the network. Also, the field experiments of Carlston [1969] and Pilgrim [1977] corroborate this finding, although as pointed out by Howard [1990], this may not be a universal kind of behavior.

Substituting the width $w = Q/(vd)$ in Eq. (4.2) one gets

$$P = \frac{QL}{d}(C_f \rho v^2 + K C_f^m \rho^m v^{2m-1}) + Ld(2C_f \rho v^3 + 2K C_f^m \rho^m v^{2m}) \qquad (4.5)$$

the bracketed terms being constant throughout the network for a given flow condition.

According to the first principle of energy expenditure, P should be a minimum in any link of the network. If the link is transporting a discharge

Q, this postulates a zero derivative of P with respect to depth d, yielding

$$Q \propto d^2 \quad \text{or} \quad d \propto \sqrt{Q} \tag{4.6}$$

Thus in any link of a network the mean annual flow or the bankfull discharge is proportional to the square of their corresponding flow depths, the constant of proportionality being the same everywhere in the network. The above result has been observed by field investigators. As described in Section 1.2.6, Leopold et al. [1964] found

$$d \propto Q^f \tag{4.7}$$

with $f \approx 0.4$ for the dependence of depth on flow in the downstream direction and with the same exponent valid both for bankfull conditions and for mean annual flow conditions. Figure 1.9 shows the dependence on depth in mean annual flow in different sites of a river network. Using Eq. (4.6) in the expression for width, $w = Q/(vd)$, gives

$$w \propto \sqrt{Q} \tag{4.8}$$

Leopold et al. [1964] found a very good relationship between width and the square root of discharge in the downstream direction for both bankfull and mean annual flow conditions. An example of this is also shown in Figure 1.9.

A rectangular cross section assumption was made in the previous derivations to streamline the computations and to avoid clouding the central idea with details. It is not a necessary assumption and, because it greatly restricts the degrees of freedom of the channel section, it leads in the above derivations to an unrealistic and fixed relationship of the ratio d/w.

In fact, the above derivation can be repeated in an analogous manner for a trapezoidal cross section to obtain $d \propto Q^{1/2}$ and $w \propto Q^{1/2}$ without implying the constant ratio d/w of the rectangular section. In fact, if θ' is the internal side angle and l_1 is the length of the base and l_2 of the side (thus $\cos\theta' = d/l_2$), area and wetted perimeter are given by $A_w = (w + l_1)d/2$ and $P_w = l_1 + 2l_2$. The rate of energy expenditure in a link of length L is now

$$P = k_1 v^3 (l_1 + 2l_2)L + k_2 v^{2m}(l_1 + 2l_2)L \tag{4.9}$$

where the unit expenditure, Eq. (4.3), has been multiplied by the total wetted surface. The expressions for l_1 and l_2 are $l_1 = w - 2d\tan\theta'$ and $l_2 = d/\cos\theta'$. The flow width is obtained from $Q = vA_w = vdw - vd^2\tan\theta'$ as

$$w = d\,\tan\theta' + \frac{Q}{vd} \tag{4.10}$$

Substituting Eq. (4.10) in the expression for l_1 and then l_1 and l_2 in Eq. (4.9), one may take the derivative of P with respect to d to minimize the energy expenditure. The relevant results are

$$d = Q^{1/2}\frac{1}{v^{1/2}(2/\cos\theta' - \tan\theta')^{1/2}} \tag{4.11}$$

$$w = Q^{1/2}\left(\frac{\tan\theta'}{v^{1/2}(2/\cos\theta' - \tan\theta')^{1/2}} + \frac{(2/\cos\theta' - \tan\theta')^{1/2}}{v^{1/2}}\right) \tag{4.12}$$

Thus we obtain the same basic scaling relationships as in the case of the rectangular cross section. The important point here is that the rectangular section does not offer any degrees of freedom regarding how to adjust the stream's cross section, reflected in a very restrictive fixed w/d relation, which in the case of the trapezoidal section depends on the angle θ'.

Substituting Eq. (4.6) in Eq. (4.5), we obtain the optimal power expenditure at any link as

$$P = kQ^{0.5}L \tag{4.13}$$

The power expenditure in the form given by Eq. (4.13) can be derived in other manners. The easiest formulation recognizes that the power expenditure in any link can be written as $P \propto Q\,\Delta z$, where Δz is the drop in elevation. Hence one may write, for a link of length L, $P \propto Q\,\nabla z L$. The next step is to employ the scaling relationships observed for $\nabla z(A)$ (see Section 1.2.11), that is, $\langle \nabla z(A)\rangle \propto A^{-0.5}$. If one assumes $A \sim Q$ (Section 1.2.4), and under the assumption that fluctuations from the mean do not matter, one indeed recovers Eq. (4.13).

Adding over all links of the network, we obtain the total rate of energy expenditure under optimal conditions:

$$P = \sum_i P_i = k \sum_i Q_i^{0.5} L_i \tag{4.14}$$

where k varies with the discharge but is constant throughout the network if the same characteristic discharge, such as mean annual flow or bankfull conditions, is operating throughout the basin.

In an explosion pattern like the one in Figure 4.1 the values of Q_i are small because there is no aggregation of flows from tributary links; on the other hand, the sum of the L_i is extremely large and so is P. If each node in Figure 4.1 has a constant discharge, Q, then in the explosion pattern

$$P = kQ^{0.5}L_T \tag{4.15}$$

where L_T is the total path length. In the case of the spiral pattern with a constant discharge at every node, one has

$$P = kLQ^{0.5}(1 + 2^{0.5} + 3^{0.5} + \cdots + N^{0.5}) \tag{4.16}$$

where L is the (constant) distance between neighboring sites (here we are not concerned with the issue of anisotropy of the lattice). Thus, although L_T is small for the spiral, P is again prohibitively large.

The tree-like pattern combines a piecewise aggregation of flows throughout the system at the same time that it keeps the total length of the flowpaths quite short. This yields a much smaller total rate of energy expenditure P. In the case of Figure 4.1, if the input flow at any node is taken as equal to 1, the corresponding values of P are as follows:

- spiral network (Figure 4.1a), $P = 574k$;
- explosion network (Figure 4.1b), $P = 303k$; and
- tree-like network (Figure 4.1c), $P = 151k$.

The explosion pattern is only relatively competitive when most of the points are close to the common outlet. If one keeps adding points further away from the outlet, the total length of the explosion pattern increases dramatically and so does the total energy expenditure P.

Note that the above comparison, although illustrative, is only correct if one assumes that k is the same in all cases, which implies that the flow velocity has remained the same in all the different connectivities, which is not necessarily true in natural networks.

As observed above, Rodriguez-Iturbe et al. [1992b] have postulated principles of optimality of local and global nature. Sinclair and Ball [1996] have relaxed the condition of principle (whose substantiation comes from its consequences) by showing through theory that landscapes and their embedded drainage networks may evolve under erosion rules to obey a variational problem. The essence of their method is to obtain a necessary condition for the minimization of a global quantity, P, as defined below. This condition is then shown to be equivalent to a particular slope–discharge relationship of the type $|\nabla z(\mathbf{x})| \propto Q^{-0.5}(\mathbf{x})$ obtained in the steady state of an evolution equation. As we shall see, a particular explanation is thus provided for the mechanism through which local erosion can minimize a global quantity, in analogy to the principles of least action in mechanics.

Let us examine in detail the mathematical result of Sinclair and Ball [1996]. Specifically, we will show that the vector form of the slope–discharge relationship $\nabla z(\mathbf{x}) \propto \mathbf{Q}^{-0.5}(\mathbf{x})$ (where $\mathbf{Q} = Q\nabla z/|\nabla z|$, i.e., flow directions are in the steepest descent direction, as seen in Section 1.2.4) is the Euler–Lagrange equation corresponding to the condition that the functional

$$P = \int_{\mathbf{x}} Q^{0.5}(\mathbf{x}) d\mathbf{x} \tag{4.17}$$

is made stationary. This is subject to the constraint that

$$\nabla \cdot \mathbf{Q}(\mathbf{x}) = R(\mathbf{x}) \tag{4.18}$$

where $R(\mathbf{x})$ is a precipitation rate specified at each site $\mathbf{x}$. Clearly, Eq. (4.17) is a continuous version of Eq. (4.14). It is also clear that the continuity equation (4.18) yields the usual relationship $Q \propto A$ if R is constant (say, equal to P, as seen in Eq. (1.60)).

After introducing $\lambda(\mathbf{x})$ as a Lagrange multiplier field (recall that the constraint is local), the constrained variational problem is equivalent to the unconstrained variation

$$\frac{\partial}{\partial Q} \int_{\mathbf{x}} \left[Q^{0.5}(\mathbf{x}) + \lambda(\mathbf{x})\big(\nabla \cdot \mathbf{Q}(\mathbf{x}) - R(\mathbf{x})\big)\right] d\mathbf{x} = 0 \tag{4.19}$$

Following standard procedures of calculus of variations, Eq. (4.19) leads to [Sinclair and Ball, 1996] the Euler–Lagrange equation

$$\nabla\lambda(\mathbf{x}) - 0.5\, Q^{-0.5} \frac{\nabla z}{|\nabla z|} = 0 \tag{4.20}$$

under appropriate boundary conditions. Eq. (4.20) enables the identification $\lambda(\mathbf{x}) \equiv z(\mathbf{x})$ to be made [see Sinclair and Ball, 1996], where $z(\mathbf{x})$ is, as usual, the landscape elevation. In fact, Eq. (4.20) requires that there exist at least one scalar field $\lambda(\mathbf{x})$ such that the equation is satisfied. One such field is the landscape elevation under the slope–area law. This implies that the necessary condition for a minimum of P subject to the constraint (4.18) is equivalent to the general relation between height and discharge, directly derived from Eq. (4.20) with $\lambda = z$:

$$\nabla z \propto Q^{-0.5} \sim A^{-0.5} \tag{4.21}$$

which is the local principle adopted above – and a well-established experimental fact (see Section 1.12). Thus Sinclair and Ball [1996] conclude that a state with global minimum P implies a local principle of the type (4.21).

This result requires some discussion. The identification of the scalar field of flow rates $Q(\mathbf{x})$ cannot, in general, be decoupled from the scalar elevation field, $z(\mathbf{x})$, because drainage directions are defined by the local gradients ∇z. The local rule, Eq. (4.21), was analytically derived by Sinclair and Ball [1996] as the steady state of an evolution equation of the type

$$\frac{\partial z(\mathbf{x}, t)}{\partial t} \propto \nabla z(\mathbf{x}, t)^2 Q(\mathbf{x}) \tag{4.22}$$

where a crucial assumption, besides the mechanistic implications for the relevant transport laws, is that no site changes drainage direction after a transient, that is, discharge is only a function of $\mathbf{x}$, $Q(\mathbf{x})$, but elevation depends on both space and time, $z(\mathbf{x}, t)$. This, of course, means that the self-organization process leading to an optimal aggregation pattern has already taken place in the transient and the network is not changing its connectivity anymore. Thus the set of aggregated discharges $Q(\mathbf{x})$ assumed here is a very small subset of all possible configurations of the system, selected by a dynamic process that involves the complex roles of randomness and the interaction between the many parts of an open, dissipative system with many degrees of freedom. We thus interpret the slope–area law (4.21), in a general context, as a necessary, not sufficient, condition for a globally optimal pattern to occur.

Thus one cannot say that any landscape that obeys the slope–area law must have a minimum value of P. Indeed, in the global optimal state the local optimal condition is met, but, in general, not vice versa. Hence, through the results of Sinclair and Ball [1996], we reinforce the diversity of our approach with the network formation capturing mechanisms that impose the local condition and do not specifically require global optimality.

4.5 Stationary Dendritic Patterns in a Potential Force Field

The idea that stable *dendritic* patterns of minimal energy dissipation develop fractal loopless arrangements has been suggested in other fields of science. In this section we will illustrate observational and theoretical evidence supporting OCN concepts.

Ramified, simply connected (i.e., dendritic), fractal patterns have been observed experimentally in electric fields characterized by nonpoint injection of charges [Merte' et al., 1988, 1989, 1990; Hadwich et al., 1990]. One example is shown in the experimental setup illustrated in Figure 4.2. The setup consisted essentially of a cylindrical cell made up by a 3-mm layer of castor oil (because of its high dielectric constant) within an acrylic 114-mm-diameter dish and a 47-mm layer of air above. The inner perimeter of the dish was equipped with a grounded metal ring. The potential between a metallic tip and this ring was fixed at 20 kV.

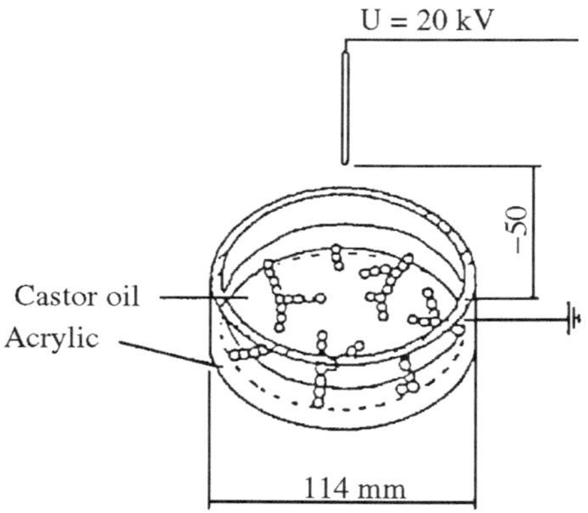

Figure 4.2. The experimental setup of Merte' et al. [1988].

Charges were sprayed quasi-homogeneously upon the oil surface. Inside the dish there were N metallic ball bearings of 2-mm diameter, which undergo a certain arrangement in order to transport charges to the metal ring. The experiment was repeated several times for different numbers of balls ($N = 400, 600, 800, 1000$). At stationary conditions the dish was surveyed and the results were analyzed. Two experimental arrangements are shown in Figure 4.3, where one may observe the resulting dendritic patterns.

The experiments clearly showed that the structure is stable: after displacing some balls slightly from their position they tended to return to their original position.

Further analyses of structures like those in Figure 4.3 have been performed through the accurate digitization of the position of the metal balls. This has allowed the evaluation of the potential field, say $\varphi(\mathbf{x})$, obtained by setting the potential of the ball bearings that were directly or indirectly in contact with the ground (i.e., connected by the structure) equal to zero. The procedure is as follows. For each point the potential is evaluated by numerically solving the Poisson equation $\nabla^2\varphi = 1$ in a mesh of grid points superimposed to the circular domain of the experiment (see the following theoretical developments). Further boundary conditions result from the actual position of the grounded metal balls, that is, zero potential is imposed for all grid points, where a digitized position of a ball has been recorded. The computation of the point values of the potential $\varphi(\mathbf{x})$ allows the determination of the total resistance, say R, defined by the relationship $R \propto \int_A \varphi dA$.

In the different experimental cases the resulting total resistance R has been plotted against the number of particles. Figure 4.4 shows a clear relationship of the type

$$R \propto N^{\xi} \tag{4.23}$$

with $\xi = 0.73 \pm 0.03$. Box-counting and cluster dimension analyses (as discussed in Chapter 2) were applied to the digitized structure, and the results are shown in Figures 4.4(b) and 4.4(c), respectively, where $N(l)$ is the number of boxes of size l making up the structure and $N(L)$ is the number of balls inside a radius L from a suitable seed.

We note that these functions clearly have a power law behavior. The fractal features have cluster and box-counting dimensions approximately equal to 1.6 ± 0.1.

It is of interest to show that the above system obeys minimum total energy dissipation. The proof has been sketched in Hadwich et al. [1990].

Figure 4.3. Dendritic structures for (a) $N = 1000$ and (b) $N = 400$ ball bearings [after Merte' et al. 1988].

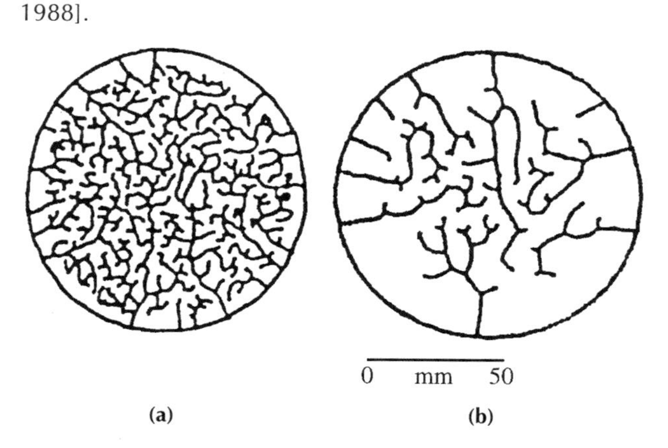

In steady state the continuity equation for an electric charge is

$$\nabla \cdot \mathbf{j} = S_0 \tag{4.24}$$

where $\mathbf{j}$ is the (vectorial function) current density and S_0 is the source term. In the case of Figure 4.2, the source of charges is generated by the 20-kV difference in the applied potential between the suspended charge sprayer and the grounded ring. The proof given here does not assume that the spraying of the charge is uniform and we argue that the randomness embedded in the spraying mechanism is an important component in the development of many stable fractal arrangements found in nature.

Making use of Ohm's law:

$$\mathbf{j} = -\sigma_0 \nabla \varphi \tag{4.25}$$

where φ is the electric potential and σ_0 is the specific conductivity of the castor oil in which the system is allowed to develop. We obtain Poisson's form of the continuity equation:

$$\nabla^2 \varphi(\mathbf{x}) = -S_0/\sigma_0 = \rho(\mathbf{x}) \tag{4.26}$$

where ρ is the charge density.

The total electric potential $\mathcal{W}$ is now derived. In fact, if $\rho(\mathbf{x})$ is the charge density at $\mathbf{x}$, the charge at $\mathbf{x}$, say $q(\mathbf{x})$, is $q(\mathbf{x}) = \rho d\mathbf{x}$. The potential at $\mathbf{x}$ generated by a charge distribution $\rho(\mathbf{x}')$ over a domain A is given by Coulomb's principle, where we recall that the potential at $\mathbf{x}$ generated by a charge at $\mathbf{x}'$ is $q(\mathbf{x}')/|\mathbf{x} - \mathbf{x}'|$, that is, $d\mathcal{W}(\mathbf{x}) \propto \int_A d\mathbf{x}' \rho(\mathbf{x})\rho(\mathbf{x}')/|\mathbf{x}-\mathbf{x}'|$. The total potential $\mathcal{W}$ is obtained through integration over the domain A:

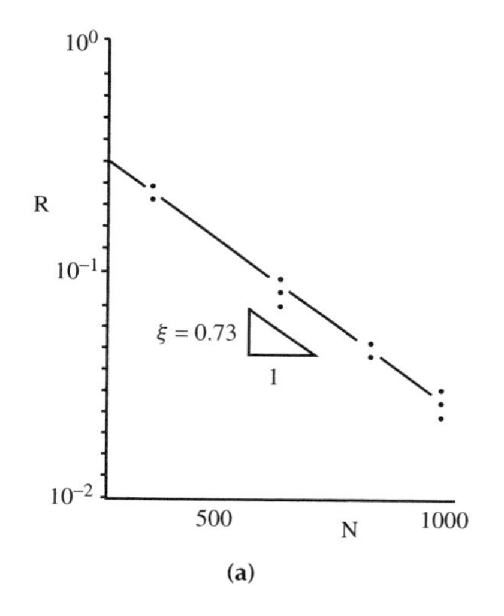

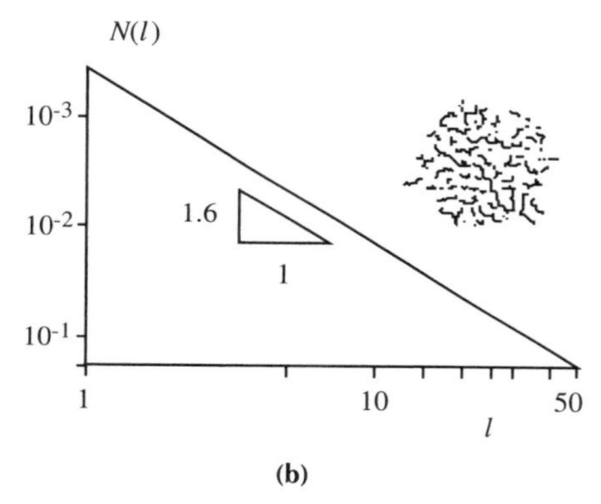

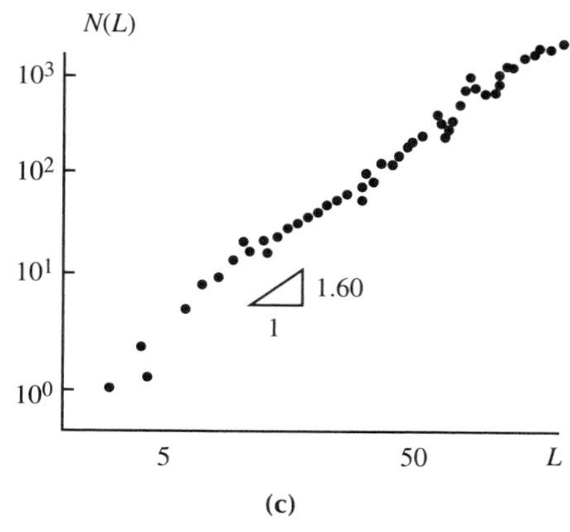

Figure 4.4. (a) Total electric resistance as a function of N; (b) box counting applied to the digitized structure: (c) as in (b) via cluster analysis.

$$\mathcal{W} = \int_A d\mathcal{W} \propto \int_A d\mathbf{x} \int_A d\mathbf{x}' \frac{\rho(\mathbf{x})\rho(\mathbf{x}')}{|\mathbf{x} - \mathbf{x}'|}$$

$$\propto \int_A d\mathbf{x}\rho(\mathbf{x}) \int_A d\mathbf{x}' \frac{\rho(\mathbf{x}')}{|\mathbf{x} - \mathbf{x}'|} \tag{4.27}$$

The second integral may be viewed as the solution of the Poisson equation, Eq. (4.26). In fact, the Green's function $G(\mathbf{x}, \mathbf{x}')$ of the Poisson equation is defined as the solution of the following nonhomogeneous problem:

$$\nabla^2 G(\mathbf{x}, \mathbf{x}') = \delta(\mathbf{x} - \mathbf{x}') \tag{4.28}$$

where δ is the Dirac delta distribution. The general solution to the above problem embeds boundary conditions and is symmetric, i.e., $G(\mathbf{x}, \mathbf{x}') = G(\mathbf{x}', \mathbf{x})$ [Morse and Feshback, 1967, p. 584]. The arbitrary solution to Poisson's Eq. (4.26) is obtained by convolution as

$$\varphi(\mathbf{x}) = \int_A G(\mathbf{x}, \mathbf{x}')\rho(\mathbf{x}')d\mathbf{x}' \tag{4.29}$$

The total potential $\mathcal{W}$ can therefore be written as

$$\mathcal{W} \propto \int_A \varphi(\mathbf{x})\nabla^2\varphi(\mathbf{x})d\mathbf{x} = \int_A dA[\nabla \cdot (\nabla\varphi\varphi) - (\nabla\varphi)^2] \tag{4.30}$$

where the second integral has been derived by integration by parts.

The final result is

$$\mathcal{W} \propto \int_A (\nabla\varphi)^2 d\mathbf{x} \tag{4.31}$$

This follows, observing that if $\varphi = 0$ on the boundary, as is the case for the setup in Figure 4.2 because the metal ring is grounded, the first term in the last integral of Eq. (4.30) is zero (i.e., $\int_A dA\,(\nabla \cdot \nabla\varphi\varphi) = \int_{\partial A} d\sigma\,(\varphi\nabla\varphi) \cdot \mathbf{n}$, where $d\sigma$ is the line element on the boundary ∂A and $\mathbf{n}$ its outer normal vector).

In the case of constant injection, that is, $\rho(\mathbf{x}) \sim \rho$, a convenient rewriting of the potential yields

$$\int_A (\nabla\varphi)^2 d\mathbf{x} = \int_A \varphi(\mathbf{x})\nabla^2\varphi(\mathbf{x})d\mathbf{x} \propto \bar{\varphi} \tag{4.32}$$

where $\bar{\varphi} = \int_A \varphi d\mathbf{x}$ and use has been made of Eq. (4.26).

We now show that in such a system the configuration that minimizes $\mathcal{W}$ also minimizes total energy dissipation. In fact, particle motion for the ith metal ball of position q_i is described by the classic mechanics equation:

$$-\chi'\dot{\mathbf{q}}_i - \nabla_{\mathbf{q}_i}\mathcal{W} = m\ddot{\mathbf{q}}_i \tag{4.33}$$

where $\mathbf{q}_i$ is the position vector of the ith particle, $\dot{\mathbf{q}}_i$ its velocity, and $\ddot{\mathbf{q}}_i$ its acceleration; m is the mass of the ball and χ' is a damping constant, assumed so large that inertial effects become negligible. Notice that the above is correct because one may exchange gradients of the local potential, say $\nabla_{\mathbf{x}}\varphi(\mathbf{x})|_{\mathbf{x}=\mathbf{q}_i}$, with the gradients of $\mathcal{W}$, that is, $\nabla_{\mathbf{q}_i}\mathcal{W}$. This follows from observing that the density field associated with $e_1, e_2, \ldots, e_N$ charges located at positions $\mathbf{q}_1, \mathbf{q}_2, \ldots, \mathbf{q}_N$ is $\rho(\mathbf{x}, \mathbf{q}_1, \ldots, \mathbf{q}_N) = \sum_{i=1}^N e_i\delta(\mathbf{x} - \mathbf{q}_i)$; one has, from Eq. (4.29)

$$\varphi(\mathbf{x}) = \sum_i e_i G(\mathbf{x}, \mathbf{q}_i) \tag{4.34}$$

and

$$\mathcal{W} = \int d\mathbf{x}\varphi(\mathbf{x})\rho(\mathbf{x}) = \sum_i \sum_j e_i e_j\, G(\mathbf{q}_i, \mathbf{q}_j) \tag{4.35}$$

Hence one obtains, through the symmetry of G

$$\nabla_{\mathbf{q}_k}\mathcal{W} = e_k \nabla_{\mathbf{x}}\varphi(\mathbf{x})|_{\mathbf{x}=\mathbf{q}_k} \propto \nabla_{\mathbf{x}}\varphi(\mathbf{x})|_{\mathbf{x}=\mathbf{q}_k} \tag{4.36}$$

because all derivatives other than that for $i = k$ in Eq. (4.34) are null. Thus the equivalence of gradients of global and local potentials is established.

In a stationary state $\ddot{q}_i = \dot{q}_i = 0$ and thus

$$\nabla_{q_i}\mathcal{W} = 0 \tag{4.37}$$

which implies that $\mathcal{W}$ is extremal. Thus the stationary state is stable when $\mathcal{W}$ is minimal.

It also follows from Ohm's law that the total energy dissipation P is defined as

$$P = \int_A (\mathbf{j}\cdot\nabla\varphi)dA \propto \int_A (\nabla\varphi)^2 dA = \mathcal{W} \tag{4.38}$$

Thus a configuration that minimizes P also minimizes $\mathcal{W}$ and, conversely, in stable stationary conditions P is minimal [Hadwich et al., 1990].

Hadwich et al. [1990] therefore concluded that in open potential force fields the minimization of total rates of energy dissipation is theoretically necessary for a stable stationary structure to form. In the above case, it is of interest to note that the authors have observed the fractal shapes and assessed the need for them to obey the principle of minimum total energy dissipation. In OCNs we follow a different pattern, assuming that minimum energy dissipation is the desired state and then deriving the structures that actually minimize energy expenditure by perturbing the shape of the stable configurations.

It is interesting to observe that the above experimental system develops power laws in the description of the structure's geometry as a consequence of the growth of a stationary structure known to obey minimum total dissipation.

Biological applications of the above properties were also reported [Athelogou et al., 1989, 1990a,b] in the study of root systems of plants. Experiments show that the biological potential of root systems tends to be minimal when the roots are in a stationary state. Also of interest is the study of blood vessels (the embryos of Japanese quail incubated in vitro [Athelogou et al., 1990b]), which again reveals optimality of the resulting structures and fractal characters of the growth–stage system. Tsonis [1990] has also suggested that fractal structures of phenomena such as lightning could be a consequence of optimality principles.

4.6 Scaling Implications of Optimal Energy Expenditure

The rate of energy expenditure per unit area of a channel, P_1, may be written as

$$P_1 \sim \frac{\rho g Q \nabla z}{w + 2d} + K C_f^m \rho^m v^{2m} \tag{4.39}$$

Substituting for w and d the expressions obtained from the joint application of the two principles of energy expenditure

$$d \propto Q^{0.5} \quad \text{and} \quad w \propto Q^{0.5} \tag{4.40}$$

one obtains, with constant velocity throughout the network

$$P_1 \sim c_1 Q^{0.5} \nabla z + c_2 \tag{4.41}$$

For a given flow condition, c_1 and c_2 are constant throughout the network, and under the second principle, P_1 is constant in all links. Thus

$$Q^{0.5} \nabla z = \text{constant} \tag{4.42}$$

implying that the left-hand side of Eq. (4.42) is constant throughout the network when mean annual flow conditions exist everywhere in the basin. Under bankfull discharge (or any significant landscape-forming flood) this would still be valid with a different constant on the right-hand side of the equation.

Studies of stable channel sections of gravel rivers by Parker [1978] and Ikeda, Parker, and Kimura [1988] have shown that the product $d\nabla z$ is only a function of sediment characteristics; with $d \propto Q^{0.5}$, this agrees with the previous result. Leopold and Wolman [1957] report a large number of rivers in the United States and India that show on the average a constant value of the product $\nabla z Q^{0.44}$. Leopold et al. [1964] also report the values for the exponent a in the relation $\nabla z \propto Q^a$ for observations in the downstream direction. An average value of $a = -0.49$ was obtained for streams in the midwestern United States for both bankfull and mean annual flow conditions. Nevertheless, for ephemeral streams in semiarid areas they quote an exponent closer to -1. Interesting as those values are, the above field data are probably quite unreliable for studying the relationship $\nabla z \propto Q^a$. In fact, one needs to measure both the discharge and the slope along individual links. As discussed in Chapter 1, the identification of the network itself is not a trivial matter, and it is only recently, through digital elevation maps (DEMs), that a network along with the slopes of its links and their individual contributing areas has been objectively studied.

Discharge measurements in every link of a network are not available, and because the mean annual flow has been observed to be proportional to the drainage area in many regions of the world, area may then be used as a surrogate variable for discharge and $Q^{0.5}\nabla z$ becomes $A^{0.5}\nabla z$, which from Eq. (4.42) is equal to a constant. This relationship can be studied in detail using DEMs. The magnitude of a link, n, is defined as the number of sources upstream of the link. For topological reasons the total number of links draining through the outlet of a link of magnitude n is $2n - 1$. The area draining directly to any link, A^*, varies randomly from link to link but does not depend on the magnitude. Thus the total area, $A(n)$, draining through a link of magnitude n is itself a random variable:

$$A(n) = \sum_{i=1}^{2n-1} A_i^* \tag{4.43}$$

Thus rather than considering the energy expenditure per unit area of channel as a constant anywhere in the network, it is now considered as a random variable, χ, whose expected value is the same throughout the network. This is expressed as

$$\langle Q^{0.5}(n)\nabla z(n)\rangle = \langle \chi(n)\rangle \tag{4.44}$$

where the expected value $\langle \chi(n)\rangle$ is constant. Using $A(n)$ as a surrogate variable of $Q(n)$, Eq. (4.44) yields in first-order analysis

$$\langle \nabla z(n) \rangle \propto (2n-1)^{-0.5} \tag{4.45}$$

Thus the principle of optimal energy expenditure leads to the scaling of the mean link slopes as a function of areas or their surrogate variables, magnitudes. The basic scaling is in terms of discharges, but with discharges proportional to areas, Eq. (4.45) is a natural consequence. The scaling of Eq. (4.45) is precisely the type suggested by theoretical studies (Section 2.5.1) and the value of −0.5 is the one found in the analysis of drainage networks through DEMs (Section 1.2.11). An example was shown in Figure 1.22.

If rainfall is not highly variable over the basin the variance of link slopes may be approximately written as

$$Var[\nabla z(n)] \propto (2n-1)^{-1} Var[\chi(n)] \tag{4.46}$$

Flint [1974] and Tarboton et al. [1989a] found that $Var[\nabla z(n)]$ scales approximately as $(2n-1)^{-0.5}$, which then implies that

$$Var[\chi(n)] \propto (2n-1)^{0.5} \tag{4.47}$$

Figure 4.5 shows an example of the mean and variance of $A^{0.5}\nabla z(n)$ as functions of magnitude for the Racoon River Basin in Pennsylvania. The graph is constructed by grouping links according to magnitude so that there are at least twenty-five links with identifiable slopes in every interval of the histogram. Owing to the 1-m vertical resolution of the DEM there are links whose altitude drop and corresponding slopes cannot be identified. In these cases a random altitude drop between 0 and 1 m is assigned to the link and the corresponding slope is then computed as the ratio of altitude drop and link length. The lines in Figure 4.5 were fitted by least squares through the whole set of points. It is observed that the mean is approximately independent of magnitude, as predicted by the second principle of optimal energy expenditure, and the variance is proportional to $(2n-1)^{0.5}$. This is an indication of multiscaling in $\nabla z(n)$, where there is not just one scaling relation determining all the moments of the process but rather changes of scale affect different moments with different scaling laws.

The multiscaling character of $\nabla z(n)$ points toward the fact that variance of the random variable energy expenditure per unit area of channel, whose expected value is the same throughout the network, increases proportionally to the average travel time for the flow to reach any site in the system. The reason for this lies in the nature of the energy dissipation along any flowpath in the network. Energy is spent along a succession of pools and riffles, analogous to a diffusion of energy along a flowpath or, equivalently, to a random walk in the altitude space through which small drops of random height occur randomly along the flowpath [Tarboton et al., 1989a]. For the optimal operation of the network, it is desirable that the expected value of the energy spent per unit area of channel be the same everywhere in the basin, but the variance of such energy expenditure, similar to that in a diffusion process, will be proportional to the length of the path or equivalent to the average travel time for the flow to reach a particular link from all its tributary links. The average length of flowpath at any point may be considered proportional to the square root of the total area draining at the point, and thus $Var[\chi(n)] \propto (2n-1)^{0.5}$.

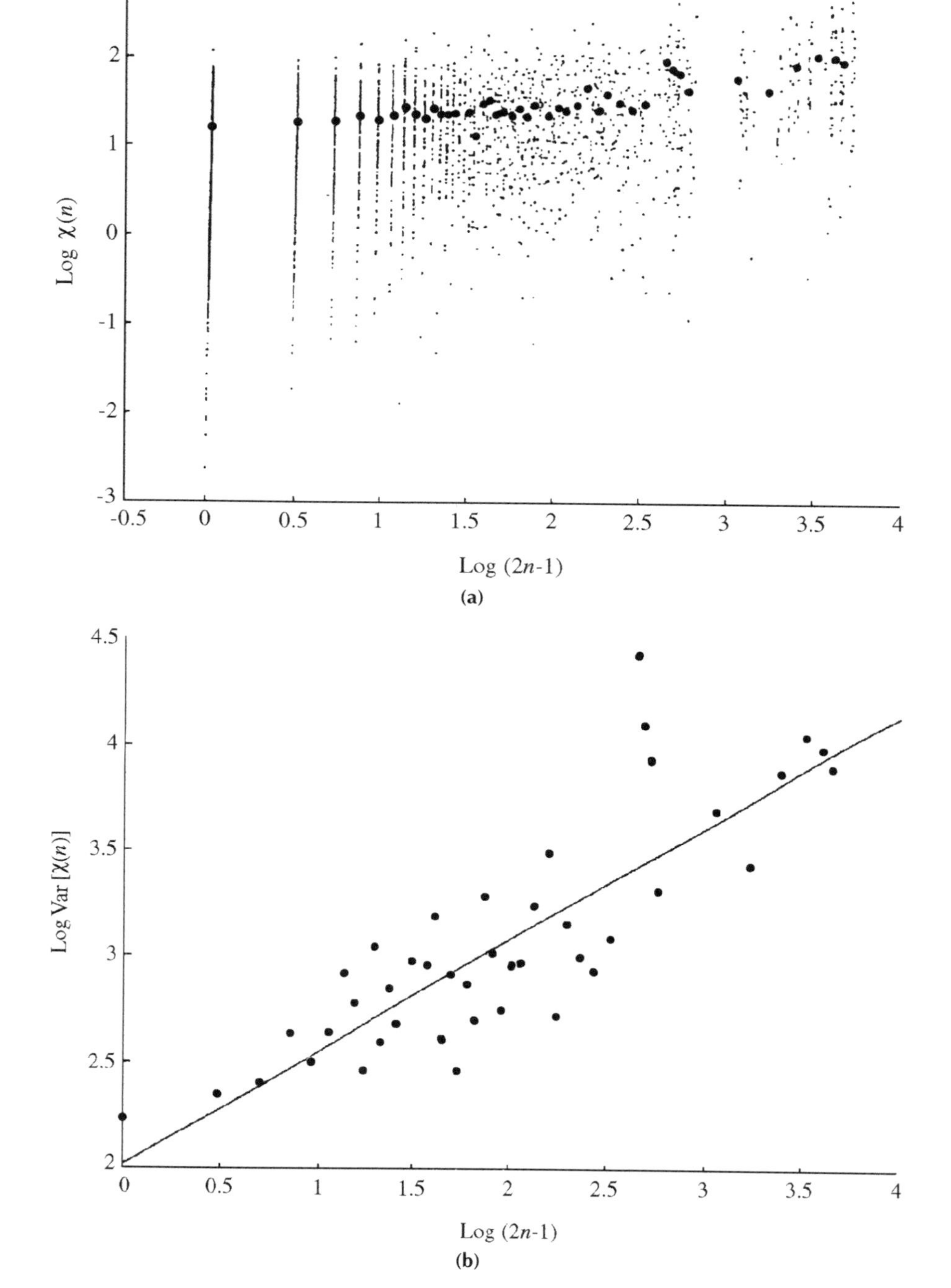

Figure 4.5. (a) Mean and (b) variance of $\chi(n)$ versus $(2n-1)$ for the Racoon River [after Rodriguez-Iturbe et al., 1992b].

The spatial structure of runoff production is intimately linked to the scaling of the drainage network. Equation (4.44) may be written as

$$[i_1 r_1 A_1^* + i_2 r_2 A_2^* + \cdots + i_{(2n-1)} r_{(2n-1)} A_{(2n-1)}^*]^{0.5} \nabla z(n) = \chi(n) \tag{4.48}$$

where i is the mean annual rainfall input and r is the mean annual runoff coefficient of the area draining directly into the individual links contributing to the link of magnitude n. The fact that $E[\nabla z(n)] \propto (2n-1)^{-0.5}$ and $E[\chi(n)] \sim$ constant suggests that, in first-order analysis, basins tend to be organized so that the expected value of annual runoff production per unit area, ir, remains approximately the same throughout the basin.

The above analyses show that the combination of the three principles of energy expenditure results in a unified picture of the most important empirical findings related to the three-dimensional structure of drainage networks. Among them are

- the velocity of flow for a representative discharge tends to be constant throughout the network;
- the depth of flow is proportional to the square root of the discharge, the constant of proportionality being the same everywhere in the network;
- in the downstream direction the channel width varies proportionally to the square root of the discharge;
- the mean value of the slopes of links with magnitude n scale proportionally to $(2n-1)^{-0.5}$.

Further geomorphological consequences of the three principles (including, for instance, Horton's and Hack's laws) will be investigated in Section 4.7. Most important, it will be shown that the fractal characters observed in nature in the aggregation patterns of drainage networks may be seen as a consequence of the principles of minimum energy expenditure.

4.7 Optimal Channel Networks

The above framework linking energy dissipation and runoff production to the three-dimensional structure of river basins hinges on local and global principles of optimal energy expenditure. The two local principles postulate that for prescribed discharges throughout the basin there should exist (i) minimum energy expenditure in any link of the network and (ii) equal energy expenditure per unit area of channel anywhere in the network, yielding the rate P_i of optimal energy expenditure at any (the ith) link of the form

$$P_i = k Q_i^{0.5} L_i \tag{4.49}$$

where k is approximately constant throughout the network, Q_i is the mean annual flow in the ith link (assumed to be representative of the work done by the flow in conveyance and maintenance), and L_i is the link length.

The optimal channel network, OCN, configuration, say s, is derived by (iii) the global principle of minimum energy expenditure in the network as a whole, yielding the condition

$$\sum_i P_i = k \sum_i Q_i^{0.5} L_i = \text{minimum} \tag{4.50}$$

where i spans all the links developing in the s configuration of the network. In unbiased lattices (where $L_i = 1$ for all links), OCN configurations s are obtained by minimizing the functional in Eq. (4.50) under the assumption that network-forming discharges can be surrogated by total contributing area, that is, $Q_i \sim A_i$. Thus the functional determining the total energy expenditure of the system, that is, its Hamiltonian $H(s)$, to be minimized is in general

$$H(s) = \sum_i A_i^{0.5} = \text{minimum} \tag{4.51}$$

which is the basis for the computations described in this chapter. Notice that no parameter is involved in the search for OCNs defined by Eq. (4.51).

Optimal arrangements of network structures and branching patterns result from the direct application of Eq. (4.51).

The basic operational problem for obtaining OCNs for a given area is to find the connected path draining the network that minimizes $H(s)$ without postulating predefined features, for example the number of sources or the link lengths such as assumed in topologically random networks.

One key problem is the assessment of the robustness of the OCN configurations selected by any procedure. This has been studied by Rinaldo et al. [1992] with respect to

- the strategy for minimum search,
- the role of initial conditions,
- the robustness of the functional dependences in $\sum_i P_i$,
- the role of possible lattice anisotropies, and
- the effects of 'quenched' randomness.

The basic optimization strategies implemented are similar to the algorithms developed in the context of the traveling salesman problem [Bounds, 1987; Johnson, 1987]. It has, in fact, been argued that in nonnumerable (NP-complete) problems (in which the exponential growth of possible combinations prevents complete enumeration and screening), approximate solutions based on a random search of improvements provide the best strategies. Different algorithms have been implemented to test the robustness of OCNs with respect to procedure. They are Lin's [1965] approach, Lin and Kernighan's [1973] strategies, and the Metropolis algorithm [Metropolis et al., 1967].

The basic algorithm proceeds as follows. An initial network configuration s is chosen to drain an area made up of pixels (pixels are naturally defined as the area rulers). This defines an orientation and a connection for each pixel that states into which of the eight neighboring pixels its area is draining, neighbors being assumed at unit distance from the centroid. This in turn needs both preliminary and a posteriori speculations on whether a triangular lattice – with six neighboring nodes – or an anisotropic scheme in which diagonal connections were weighted by a $\sqrt{2}$ factor would be a better model of local interactions. This point will be addressed in the following.

A scalar, $A_i(t)$, denotes the total contributing drainage area at a point i, that is, the number of pixels draining – at "time" t of the optimization process – into the ith cell, defined by (i_1, i_2), where in our discrete scheme i_1, i_2 are integers. At the stage t of the optimization process, the total drainage area at i is

$$A_i(t) = \sum_j W_{ij}(t) A_j(t) + 1 \tag{4.52}$$

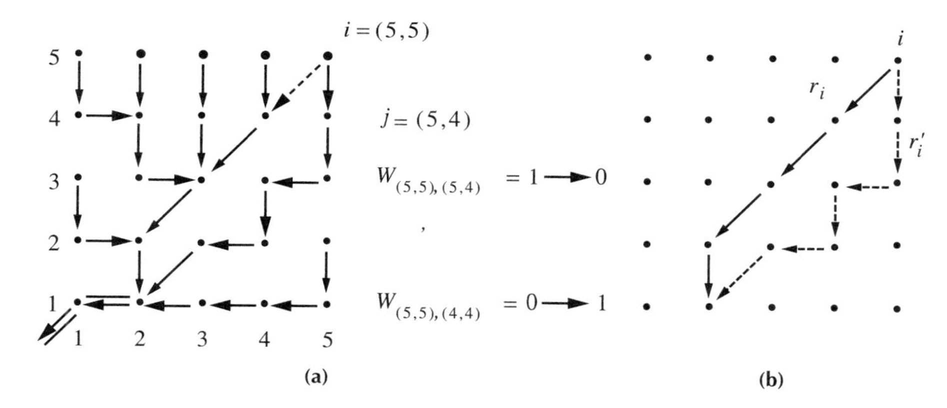

Figure 4.6. Illustration of Lin's basic strategy.

where W_{ij} is a functional operator defined by

$$W_{ij}(t) = \begin{cases} 1, & \text{if } i, j \text{ are connected, i.e. if } i \to j \text{ is a drainage direction} \\ 0, & \text{otherwise} \end{cases}$$

and j spans the eight neighboring pixels of the arbitrary ith site. The unit mass added in Eq. (4.52) refers to the area of the actual site and represents a distributed injection term.

Moving from an initial network configuration (stage $t = 0$), the basic strategy consists of drawing $i = (i_1, i_2)$ at random and perturbing the system (assigning a change δW_{ij}) by modifying at random its connection to the former 'outlet' (or receiving) pixel (Figure 4.6). Hence $W_{ij}(t+1) = W_{ij}(t) + \delta W_{ij}$. This corresponds to perturbing the configuration s (i.e., $s \to s'$).

Adjusting to such a local modification, all aggregated areas A_i are modified in the downstream region until the original and the modified path reconvene (Figure 4.6b). The change is accepted if the modified value of the total rate of energy expenditure $H(s') = \sum_{i \in s'} P_i \propto \sum_{i \in s'} A_i^{0.5} L_i$ is lowered by the random change ($H(s') < H(s)$) and no loops are formed. Loops in fact prevent the definition of a total contributing area and are energetically unfavorable.

As the new configuration is adopted as a base configuration, the process is iterated. Otherwise, the change in $\sum_i P_i$ being positive ($H(s') \geq H(s)$), the t-stage configuration s is perturbed again.

This procedure leads to a network in which no improvement on total energy expenditure appears after a fixed (and large) number of iterations. The whole process is then reset and restarted from the same initial configuration. This is usually done several times to allow the random process a fair chance to capture nonlocal minima. The configuration s with lowest value of $\sum_i P_i$ is then chosen as OCN.

This basic procedure, which we term Lin's approach because of the similarities with the early N-city traveling salesman algorithm [Lin, 1965], respects the rules of a fair search for approximate solutions but is apt to yield trapping in local minimum energy configurations. Variants of the basic procedure, implemented to test the importance of choice of strategy for minimum search, include the Lin–Kernighan strategy and simulated annealing strategies [Metropolis et al., 1953; Bounds, 1987; Wejchert, 1989]. In simulated annealing the main idea is to avoid local minima by accepting perturbations of the current configuration ($s \to s'$) that yield nonimproving

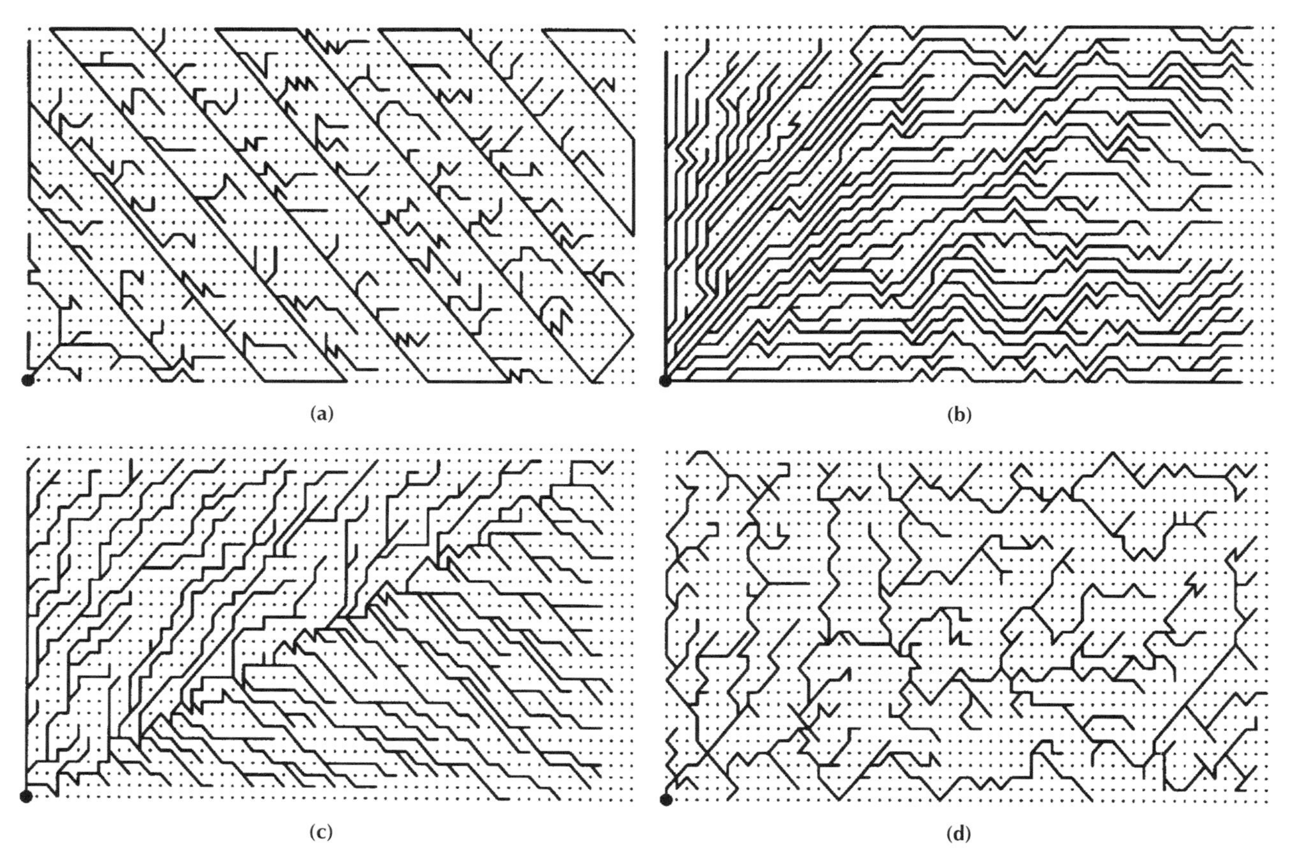

Figure 4.7. Four types of initial conditions for the 30×60 grid optimization [from Rinaldo et al., 1992].

$H(s')$ with a probability depending on some predefined state parameters. In practice, the probability of acceptance of the perturbation is given by the Metropolis rule. The new configuration s' is first checked to see whether it contains any loops. If not, the new energy function $H(s')$ is calculated. If $H(s') \leq H(s)$ the change is accepted. If $H(s') < H(s)$ the new configuration s' is accepted with probability $\exp[-(H(s')-H(s))/T]$, where T is a parameter mimicking temperature of a thermodynamic system. To carry out annealing, one makes changes in the parameter T from relatively high values at the start to low values toward the end of the analysis. Clearly, for high values of T the likelihood of accepting unfavorable changes is high, whereas for $T \to 0$ the rule is equal to that of the basic algorithm. A 'cooling' schedule for decreasing values of T as the procedure evolves is thus required.

The Lin–Kernighan strategy, instead, tries to avoid trapping by perturbing the configuration in more than one location simultaneously. A semideterministic search has also been implemented in which, after a random perturbation proves successful, the process deterministically explores whether connections to the downstream links of larger magnitudes would improve the configurations.

The OCN computations shown in this section illustrate the results of the application of the basic algorithm. Rinaldo et al. [1992] noticed, through a comparison of convergence of OCNs by different algorithms, that the resulting structures show interesting differences in their structural characteristics. This point will be investigated in detail in Section 4.13.

Each run, from initial conditions to convergence with, say, 2^8–2^9 unsuccessful attempts to improve $H(s)$, even for relatively small lattices, requires considerable computer time to succeed.

The effect of different initial conditions has been tested. Figures 4.7(a) to 4.7(d) illustrate examples of initial networks, defined for clarity and in analogy with DEM technology by a support area of five pixels (Section 1.2.9), that is, a link of order and magnitude 1 is born whenever five or more pixels cumulatively drain into it. Figure 4.7(a) is an initial network with a spiral-like pattern and in a sense represents the opposite extreme of Figure 4.7(b), which approaches an explosion-like pattern; both networks are reminiscent of the patterns in Figure 4.1. Figure 4.7(c) represents a spine-like initial configuration and Figure 4.7(d) is a randomly generated starting network.

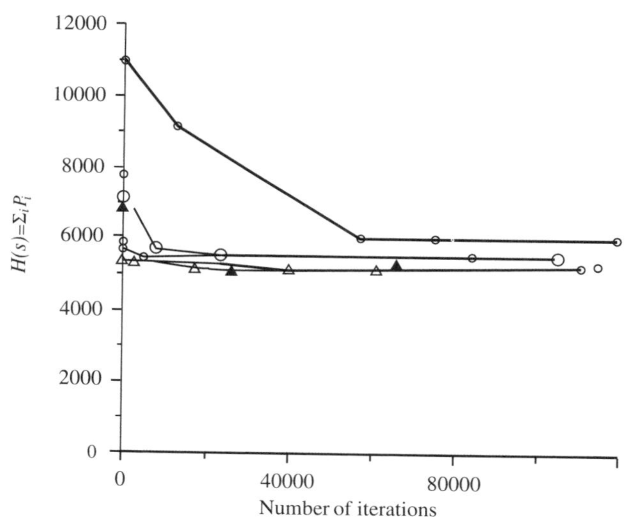

Figure 4.8. The decrease of total rate of energy expenditure along the optimization process for different initial conditions [after Rinaldo et al., 1992].

Based on these configurations, optimization procedures were repeated to find the OCNs for the 30 × 60 grid. Figure 4.8 shows the evolution of the total energy expenditure ($H(s) = \sum_i A_i^{0.5} L_i$, L_i is the assumed isotropic unit) for the different initial conditions in the 30 × 60 grid domain. Although the strict degree of global optimality of the OCNs obtained is unclear, the convergence of total energy is quite robust, rendering a more refined search uninteresting for our purposes. Connected points show the development of the same optimization process, in which computations were stopped to record the features of intermediate networks. We observe that the lines connecting computed points in Figure 4.8 are shown for clarity, although they do not represent the real optimization pattern. A more precise plot would have required the recording of more intermediate configurations, which would have been very time consuming and pointless. Unconnected points in Figure 4.8 represent results of repeated runs from the same initial configuration of which only the final stage was analyzed.

Figures 4.9(a) to 4.9(d) show the OCNs developed from the initial conditions shown in Figures 4.7(a) to 4.7(d). They have nearly the same values of total energy expenditure. The initial conditions distinctly influence the final configuration, but, as will be shown later, this does not substantially affect the statistical and geomorphological properties of the network [Rinaldo et al., 1992]. Such figures show the network with the lowest computed total energy expenditure among all runs performed. The question of whether such patterns are actually of absolute minimum energy not only requires a very complex assessment but is likely to be irrelevant. Figures 4.9(a) to 4.9(d) in fact show examples of OCNs developing from very different initial conditions, whose marks are imprinted in a lasting manner on the final configuration but not in its statistical properties. They exhibit virtually identical statistics and all have nearly the same value of total rate of energy expenditure. A salient result of the OCN analyses is indeed the robustness of the convergence to properties with the same statistics regardless of initial conditions.

Figures 4.10(a) and 4.10(b) illustrate OCNs for a different geometry, that is, a square 60 × 60 lattice. Case (a) is the fully textured optimal connected pattern and case (b) defines a channel with a support area of 5 pixels.

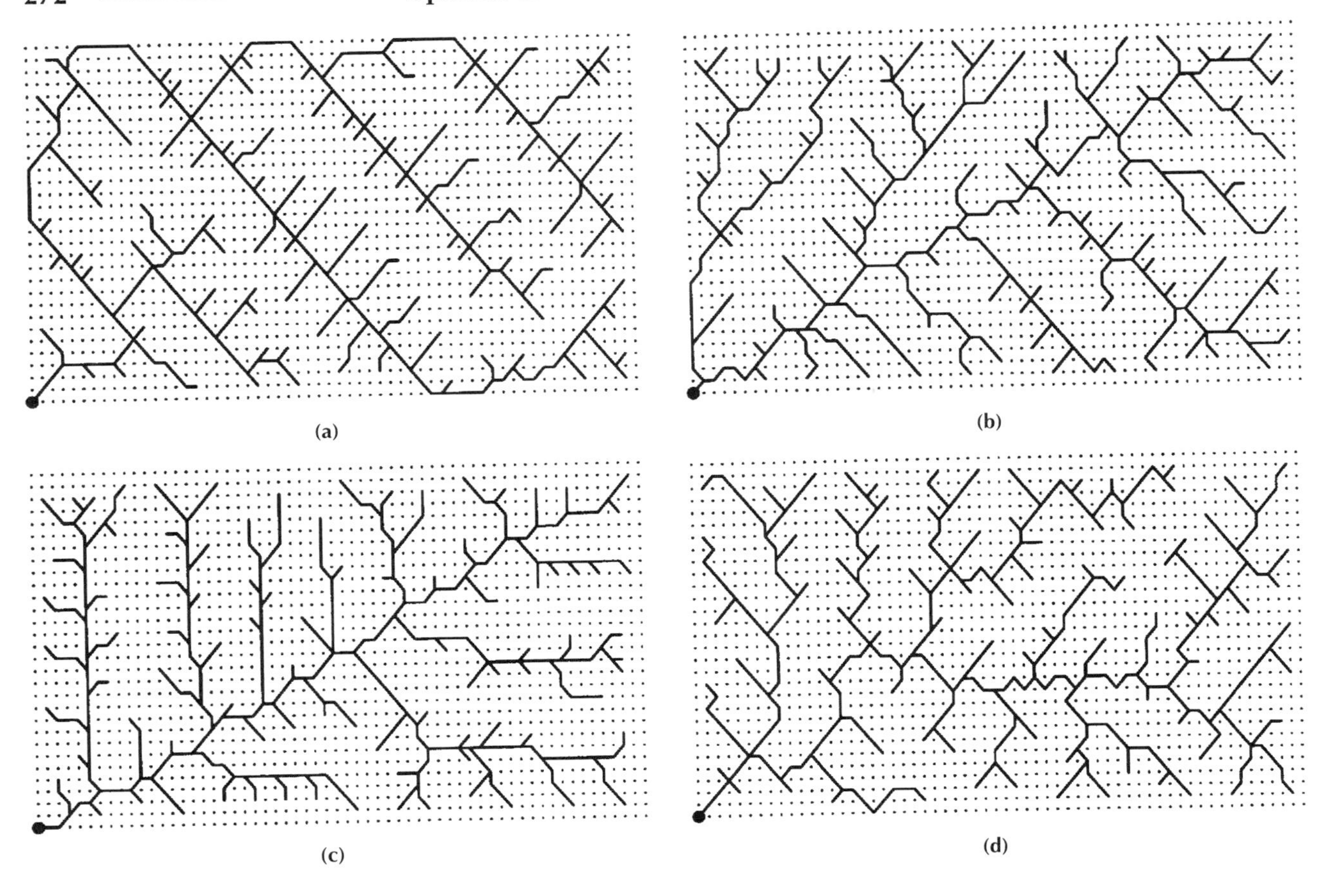

Figure 4.9. OCNs developed from the initial conditions shown in Figures 4.7(a)–4.7(d).

The boundary conditions in the lower right-hand corner are a possible example. Boundary conditions can be arbitrary, both in the position of a single outlet or in the allowance for multiple outlets. Figure 4.11 shows one example of an OCN where the (single) outlet has been randomly placed.

As we will see in the next sections, the statistics of the OCN in Figure 4.11 match perfectly those shown by Figures 4.9 and 4.10 as well as

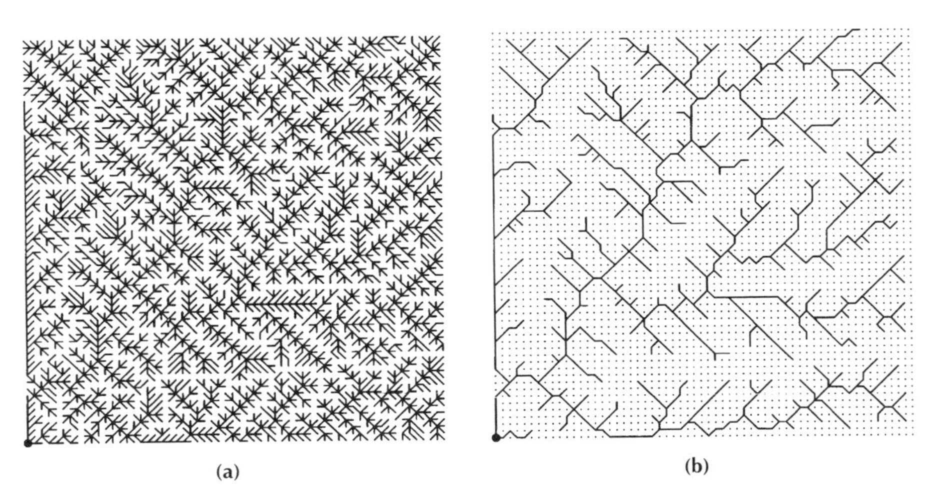

Figure 4.10. 60 × 60 OCN with (a) full texture and (b) support area of five pixels.

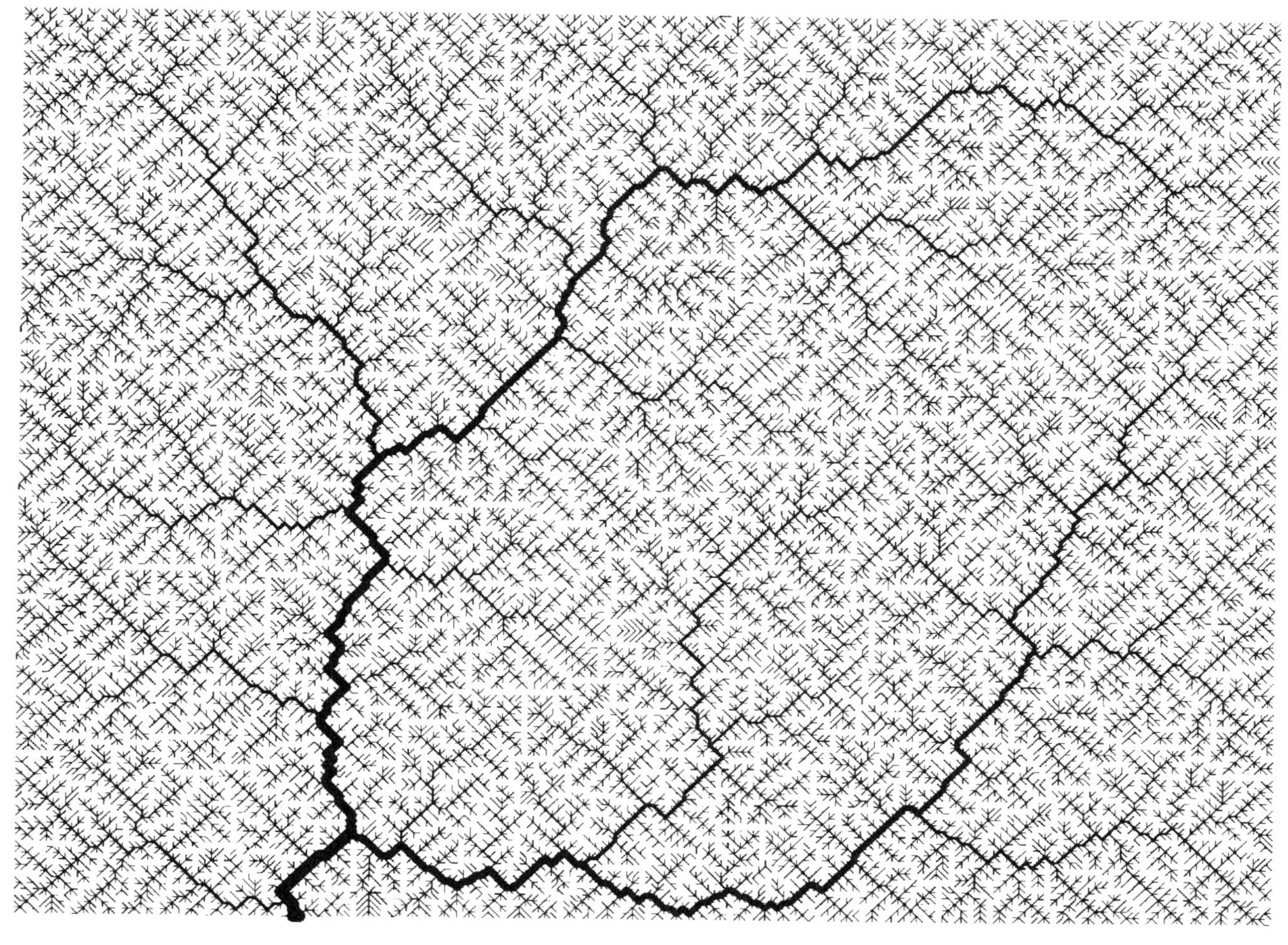

Figure 4.11. OCN with arbitrary single outlet.

those shown in nature. If multiple outlets are allowed, as first suggested by Sun, Meakin, and Jossang [1994a] to simulate competition among drainage networks, OCNs look like the pattern shown in Figure 4.12. Notice that a support area $A_t = 10$ has been used to better capture visually the structure of the nested basins.

Lattice effects may also potentially play a role in the definition of OCNs. The runs shown before had eight neighbors for every pixel, each assumed to be at unit distance from the centroid (e.g., $L_i = 1$ for every i). A triangular lattice, where every pixel has six neighbors, or an anisotropic scheme in which diagonal connections have $L_i = \sqrt{2}$ are two other options. An isotropic hexagonal lattice is widely used in cluster analysis via diffusion-limited aggregates [e.g., Feder, 1988]. Although convenient for the applications, the hexagonal lattice has not been employed because DEMs (see Chapter 1) use eight-node neighbors and thus comparison with real data is much simplified with the present lattice.

Figure 4.13 shows two networks started from identical initial conditions in which the only change in the optimization procedure was to introduce the $\sqrt{2}$ weight for diagonal connections. The differences are marginal in that local preferences for straight connections obviously appear because of

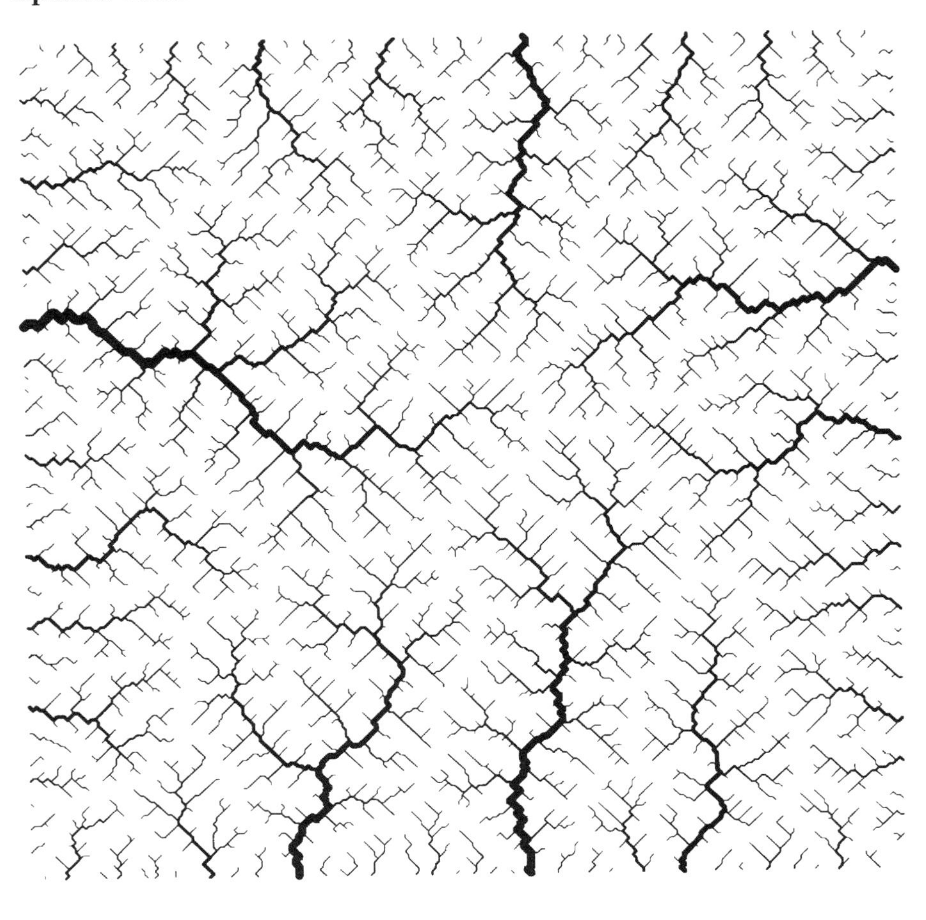

Figure 4.12. Multiple outlet OCN.

the prefactor. However, whenever a support area different from one pixel is defined, the differences tend to disappear and the statistics of the resulting networks (according to the procedures outlined in the next section) are virtually identical. Thus the scheme with eight neighbors at unit distance from the centroid has been adopted throughout our OCN analyses.

A further test imposed on OCN results concerns the effects of perturbations in the form of the energy expenditure function to be minimized. Specifically, Rinaldo et al. [1992] studied whether changes in the energy

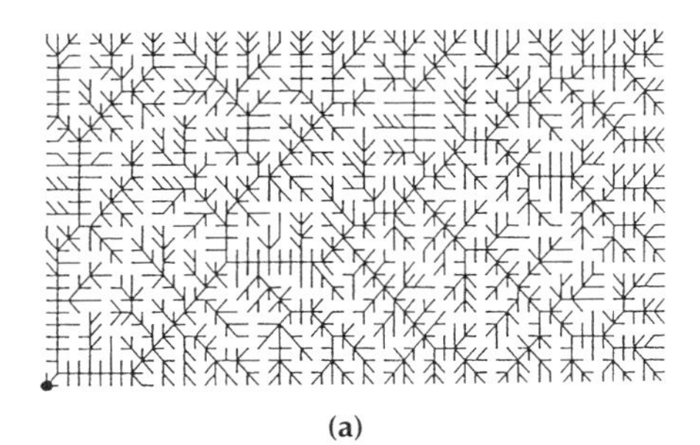

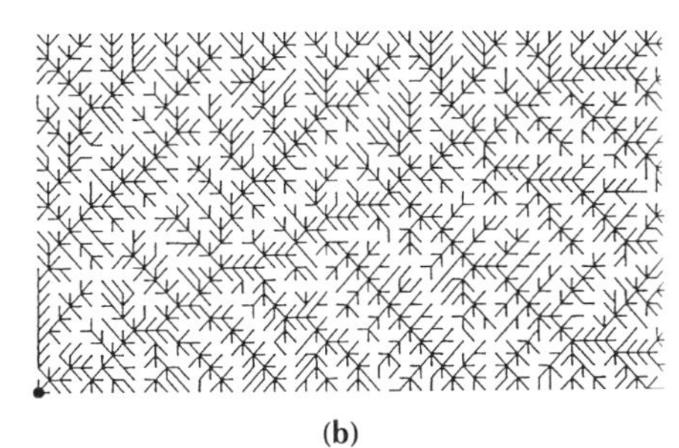

(a) (b)

Figure 4.13. Lattice effects in a 30×60 mesh: (a) OCN with diagonal weight; (b) isotropic OCN.

function Eq. (4.50), that is, in the form

$$H_\gamma(s) = \sum_i A_i^\gamma = \text{minimum} \tag{4.53}$$

with $\gamma \neq 0.5$, lead to dramatically different results. This has conceptual bearings because the principles underlying Eq. (4.50) rely on a set of assumptions whose relaxation might imply modifications in the γ exponent. Moreover, the commonality of network features under extremely variable conditions suggests that nature is resilient regarding the specific value of γ in the characterization of the role of energy expenditure. Thus even if $\gamma = 0.5$ is the theoretically derived result in Rodriguez-Iturbe et al. [1992b] and such a value is the one that matches the most commonly observed empirical facts of drainage networks, one would nevertheless expect that properties related to the topological structure of the network be somewhat resilient to the specific choice of γ.

Rinaldo et al. [1992] and Sun et al. [1994a] have suggested that OCNs with $0 < \gamma < 1$ tend to be characterized by grossly similar planar aggregation patterns. Differences in the most demanding statistics arise and grow with respect to observed natural features as γ departs from 0.5.

For $\gamma \geq 1$ or $\gamma \leq 0$, minimization of the functional in Eq. (4.53) does not yield aggregation patterns at all. A negative exponent γ creates spiral-like, nonaggregated patterns of the type in Figure 4.1 (a) because aggregation is obviously not favored.

The value $\gamma = 1$ is critical in the change of behavior of OCNs. For $\gamma = 1$, Cieplak et al. [1996] have shown that the OCN pattern belongs to the Scheidegger class (Sections 1.3.2 and 2.5.2) of directed river networks. For $\gamma > 1$, explosion-like patterns similar to those in Figure 4.1(c) are achieved by minimization of Eq. (4.53). Directed spanning trees are thus obtained.

Figure 4.14 illustrates the above points by means of five examples of optimization (up to convergence to the first local minimum encountered) of the functional $H_\gamma(s) = \sum_i A_i^\gamma$ with $\gamma = 0.25, 0.5, 0.75, 1$, and 2, respectively. The minimum search was stopped early, but the tendencies described above are clearly shown in the examples.

Randomness in the initial condition prevents the establishment of a perfect explosion pattern by the numerical algorithm. Figure 4.15(a) shows the network obtained for $\gamma = 2$ in a rectangular lattice starting from an Eden growth [1961] random pattern. Once the network in Figure 4.15 (a) is taken as the initial condition for the development of an OCN (e.g., with $\gamma = 0.5$), the resulting network (Figure 4.15(b)) shows the common, recurrent features of OCNs.

It is also of interest to discuss the network obtained by 10^6 iterations from random initial conditions with $\gamma = 0.9$ in a 64×64 lattice (Figure 4.16). While an aggregation pattern is still identifiable, the statistics of the resulting patterns depart visibly from a realistic tree-like structure. If multiple outlets are allowed, a result like in Figure 4.17 is obtained. Besides the statistical and morphological differences to be investigated in the following, notice the unrealistic regularity of the main divides.

Thus, although $\gamma = 0.5$ is not the only exponent that yields planar aggregation patterns close to those observed in nature as a product of energy minimization processes, the fact that only a value $\gamma = 0.5$ implies the

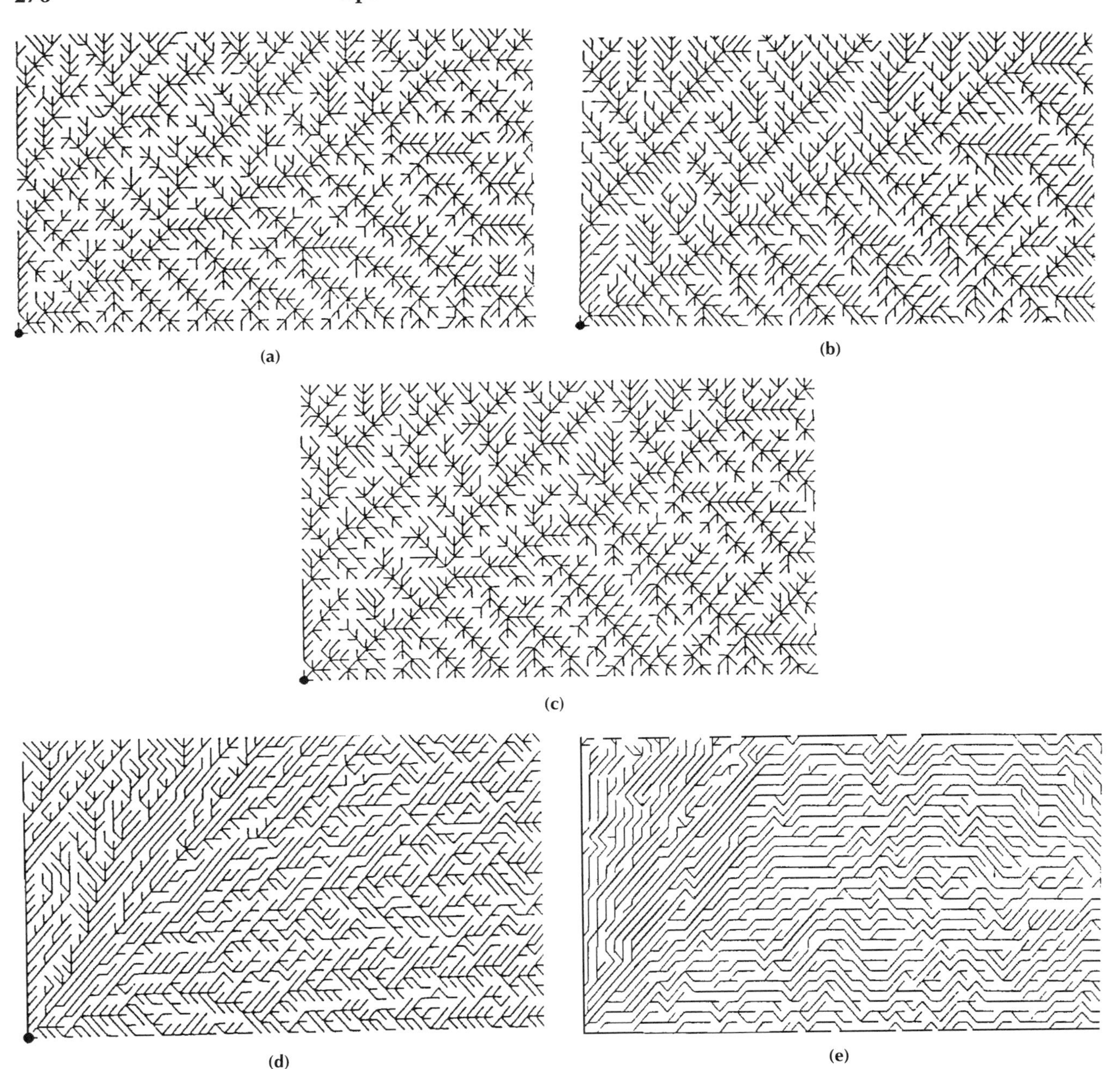

Figure 4.14. Networks obtained minimizing $H_\gamma(s)$ with (a) $\gamma = 0.25$; (b) $\gamma = 0.5$; (c) $\gamma = 0.75$; (d) $\gamma = 1$; and (e) $\gamma = 2$, starting from the same initial conditions.

correct scaling of slopes, along with many of the empirical facts observed in channel networks [Rodriguez-Iturbe et al., 1992b], leads us to conclude that OCNs are robust but also far from independent of the model used.

A final question remains: How robust are these results with reference to the influence of defects in the domain of analysis? It is important that OCN results be insensitive to terrain randomness so as to be useful in the explanation of natural drainage patterns beyond the fulfillment of a few similarities. To test this important aspect, the original model has been modified by the introduction of 'quenched randomness' [Bak et al., 1988].

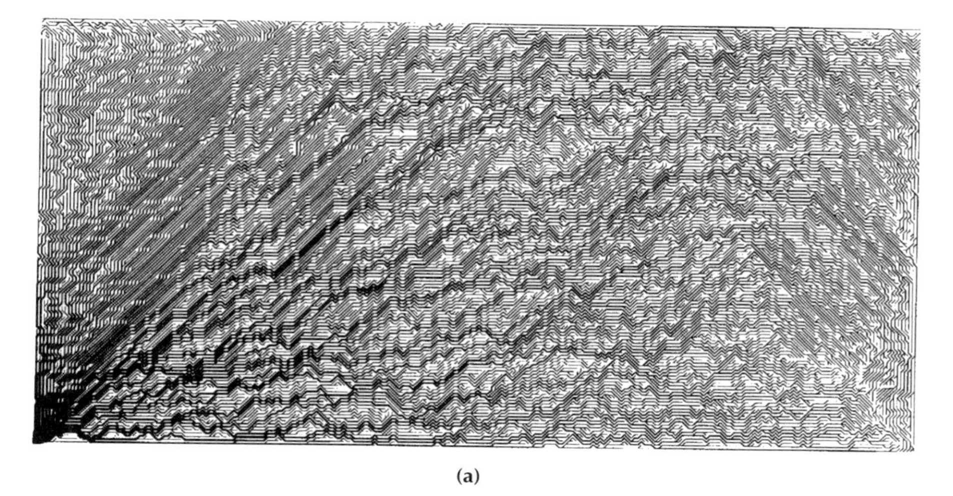

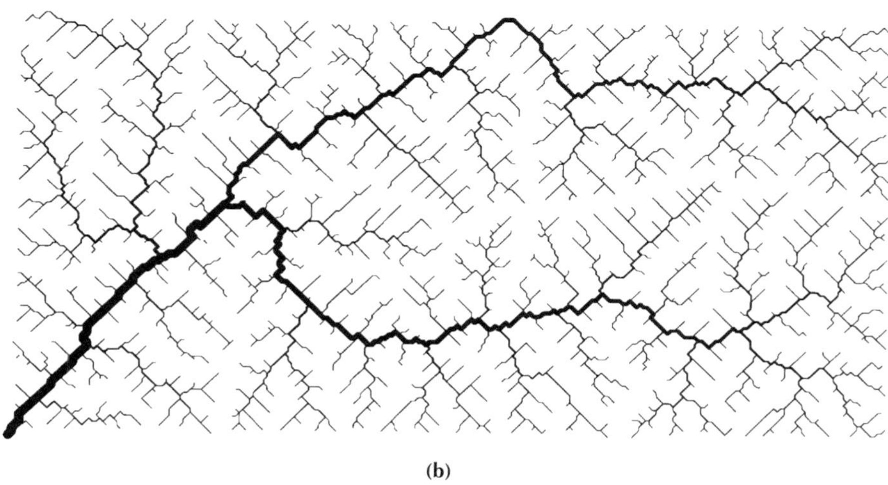

Figure 4.15. (a) Pattern obtained by minimizing $H_2(s)$ from random initial conditions with full support ($A_t = 1$); (b) OCN obtained minimizing $H_{0.5}(s)$ starting from (a). Here $A_t = 20$.

Specifically, some possible nearest-neighbor connections in the drainage plan were removed at random, and the disrupted connections remained as such throughout the whole optimization process. The salient results in these analyses, which were pursued systematically to test the threshold of appreciable changes, is that in the process of searching for OCNs, removing as many as 20% of the bonds still leads to substantially unaffected statistics, as in the case of self-organized criticality to be discussed in Chapter 5 [Bak et al., 1988]. One may thus conclude that the evolutionary process leading to OCNs is remarkably robust. A physical interpretation for quenched randomness in the context of drainage networks may reside in the effects of particular geologic features. These features, existing within the drainage basin, prevent the choice of certain drainage paths and locally deform the topological structure of the network. It is therefore reasonable that random disruptions of possible connections mimic the local topographic and geologic irregularities that do not prevent the network from attaining a structure whose statistical features are not controlled by the details of the geologic environment.

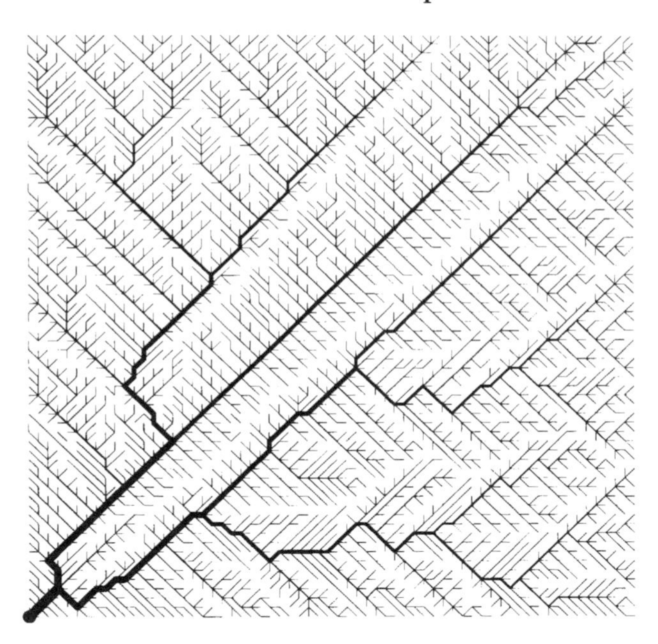

Figure 4.16. Network obtained minimizing $H_\gamma(s)$ with $\gamma = 0.9$.

4.8 Geomorphologic Properties of OCNs

A number of geomorphologic properties have been measured in OCNs, and the results have been compared against those found in natural river basins.

Two of those properties are Horton's bifurcation and length ratios, defined in Chapter 1 as

$$R_B = \frac{N(\omega)}{N(\omega + 1)} \quad \text{and} \quad R_L = \frac{\bar{L}(\omega + 1)}{\bar{L}(\omega)} \tag{4.54}$$

where $N(\omega)$ is the number of streams in Strahler's order ω and $\bar{L}(\omega)$ is the average length of streams of order ω.

Figures 4.18 and 4.19 show the evolution of Horton ratios R_B and R_L throughout several iterations in search of the pattern of minimal energy expenditure. They correspond to different initial conditions than those shown in Figure 4.7 and to a support area of one pixel in the definition of the network. In particular, Figure 4.18 shows a very robust convergence of the average (over all orders) bifurcation ratio R_B to values approximately equal to 4. It is worthwhile to observe that convergence to values close to 4 is achieved from above (the initial 'spiral'

Figure 4.17. Multiple outlet minimization of $H_\gamma(s)$ with $\gamma = 0.9$.

configuration of Figure 4.7(a) has $R_B \approx 12$) and from below (the 'exploded' initial configuration of Figure 4.7(b) has $R_B \approx 3$) with notable implications for the reliability of the result. Also, whenever the initial condition is characterized by bifurcation close to 4, OCNs remain consistently in that range.

Figure 4.19 illustrates the convergence graphs for metric statistics, that is, Horton's average (over all orders) ratio R_L. Convergence of R_L to values close to 2, which implies a definite structural relation among the parts of the configuration, is again very robust. Although the results of Figure 4.19 refer (for clarity) to the 30×60 grid because it has been widely tested, no significant difference in behavior was observed in any of the grids studied [Rinaldo et al., 1992].

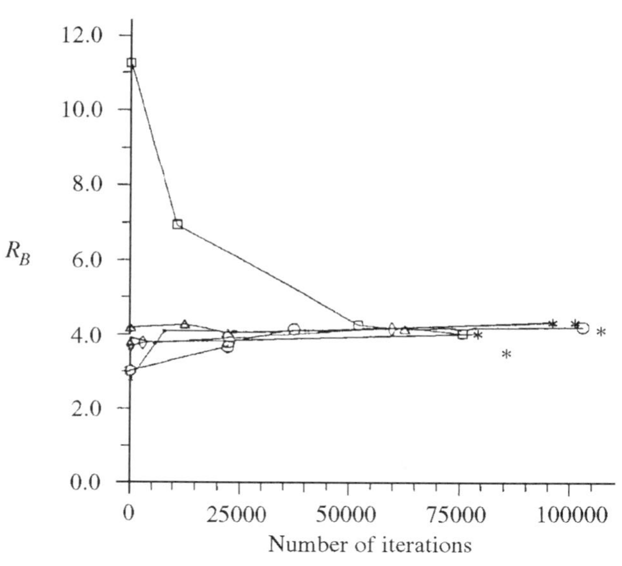

Figure 4.18. Convergence of Horton's bifurcation ratio R_B to values close to 4 regardless of initial conditions. Notice the unconnected points, representing the final state of repeated runs from the same initial conditions [from Rinaldo et al., 1992].

The values of R_B and R_L that consistently appear in OCNs are those observed in nature for mature river basin networks unaffected by strong geologic controls (Section 1.2.1). To test the goodness of Horton's equations, Figures 4.20(a) and 4.20(b) illustrate Horton's diagrams for a sample OCN (30×60 domain). These graphs are typical of the striking Hortonianity (i.e., the validity of Eq. (4.54) whatever the order ω) of OCNs.

An important geomorphologic relation is Hack's law, already discussed in Chapters 1 and 2. Hack's analysis carried out for many OCNs and some of their large subbasins shows

$$L \propto A^{0.58 \pm 0.03} \tag{4.55}$$

where L is the main stream length and A is the drainage area, in remarkably good agreement with the values reported in the literature (Section 1.2.3). The analysis was done in this case as in Hack's original study, that is, without reference to fractal issues. Later in this chapter an explanation of the elongation of river basins with size will be provided within the framework of OCNs.

Figure 4.19. Evolution of Horton's length ratio R_L along with the optimization process [after Rinaldo et al., 1992].

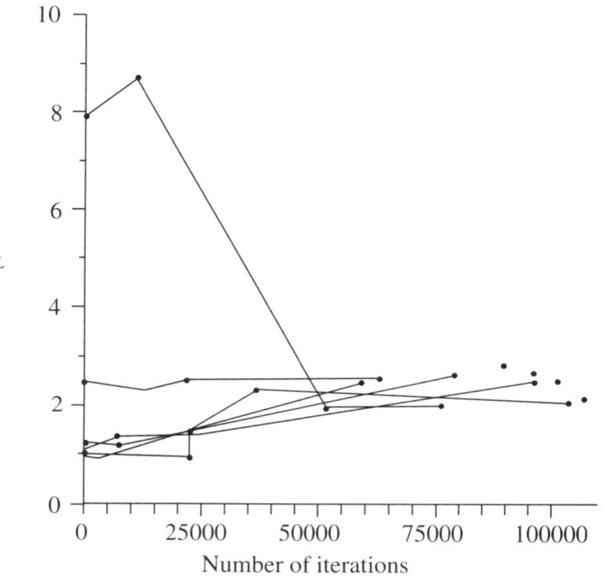

4.9 Fractal Characteristics of OCNs

An important feature of natural channel networks analyzed through DEMs lies in a large set of scaling invariance properties. These properties refer to the invariance of the probability distributions under wide changes of spatial scales, resulting from power law distributions. The power law distributions studied in Chapter 2 will now be compared with those obtained from OCNs, following Rinaldo et al. [1992].

The cumulative probability distribution $P[A \geq a]$ of a site of having a total contributing drainage area larger than a [Rodriguez-Iturbe et al., 1992a] has been found to be

$$P[A \geq a] \propto a^{-0.43 \pm 0.02} \tag{4.56}$$

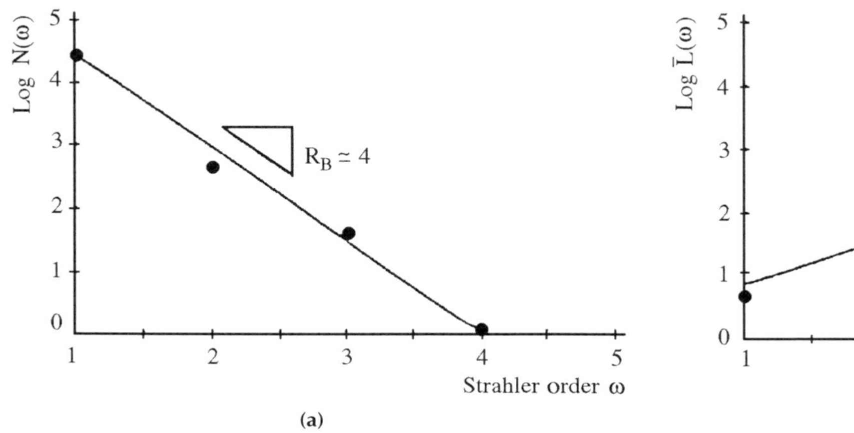

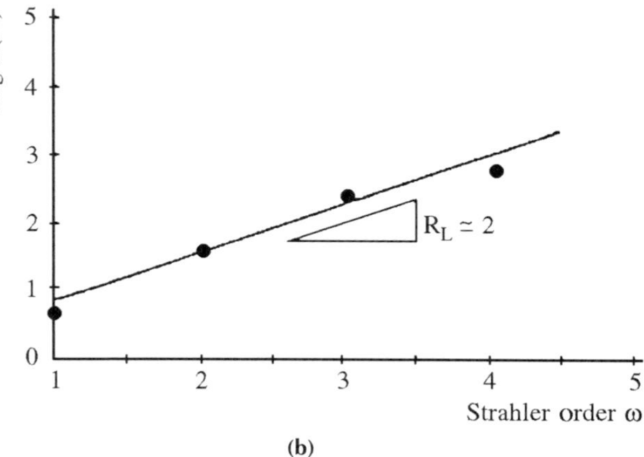

Figure 4.20. Horton's diagrams for (a) bifurcation and (b) length for a typical OCN developing in a 30 × 60 lattice [after Rinaldo et al., 1992].

(see Eq. (2.69)) from the analysis of several river basins (Section 2.5.2). The agreement of Eq. (4.56) with real data from DEMs throughout a spectrum of basins with very different geologic features is remarkable. The properties of Eq. (4.56) imply a fractal structure of the aggregated patterns, as studied in detail in Chapter 2. Finite-size scaling arguments (Section 2.10.1) have also been shown to apply in real network structures for aggregated areas.

We also saw in Section 2.2 that the geometric stream length distribution shows a hyperbolic tail with slope close to 2, while the stream length (along stream) exceedence probability follows a power law with slope in the range 1.8–1.9. Total network length was shown to be a fractal object with fractal dimension

$$D = D_l \frac{\log R_B}{\log R_L} \tag{4.57}$$

where D_l is the fractal dimension of sinuosities of individual rivers. As discussed in Chapter 2, Eq. (4.57) links scale-invariance and self-similarity to Horton's laws, which are then viewed as scaling laws.

Figures 4.21 and 4.22 illustrate the scale-invariant probability distributions of geometric stream length, say $P[L \geq l]$, and total contributing area $P[A \geq a]$ for OCNs. In all cases, the experimental or computed distributions follow good straight lines for nearly three log scales. The deviations appearing at large values of stream length or area (quite reasonably more apparent at larger values of support area) are likely to be a finite-size effect, i.e., the number of links with such large area or length are very small and the statistics lose significance. The match of the slopes of the power law distributions of OCNs with those observed in nature is indeed very good.

The values of the slopes are also relatively unaffected by the size of support threshold. Because the issue of significance of statistics is crucial for OCNs whose computation for $N \times N$ element grids carries a $O(N^2)$ growth of computational operations, in most cases studied in this book OCN statistics use full texture.

Figure 4.23(a) shows a 60 × 60 OCN developed from random initial conditions after reaching a satisfactory convergence of $H(s)$. Figures 4.23(b) and 4.23(c) show the resulting statistics for the distribution of (b) total contributing area at any point and (c) Strahler's lengths measured along the

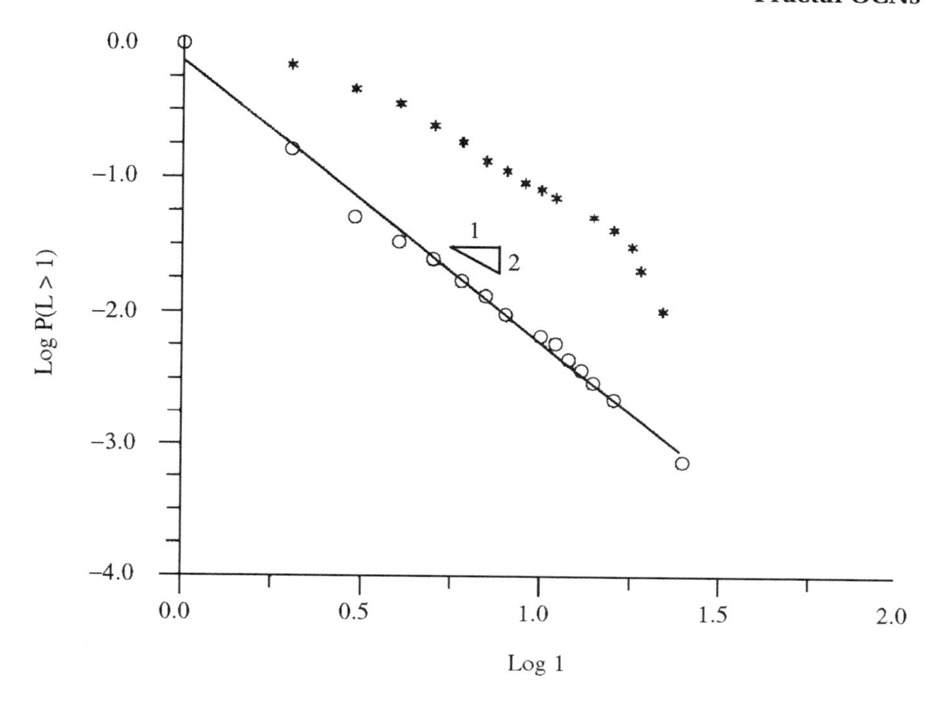

Figure 4.21. Exceedence probability of geometric stream length for a 30 × 60 OCN with different support thresholds. Upper curve, $A_t = 5$ pixels; lower curve, $A_t = 1$.

streams. We observe that although minor departures from the observational values of the slopes may occur for OCNs that get trapped into local minima (in particular, depending on the initial conditions), quite generally the OCN process tends to yield perfect statistics as in the case of Figure 4.23.

Boundary conditions may be used to prevent the minimization algorithm from getting trapped into particular configurations characterized by relatively low values of $H(s)$ and altered statistics. These are usually

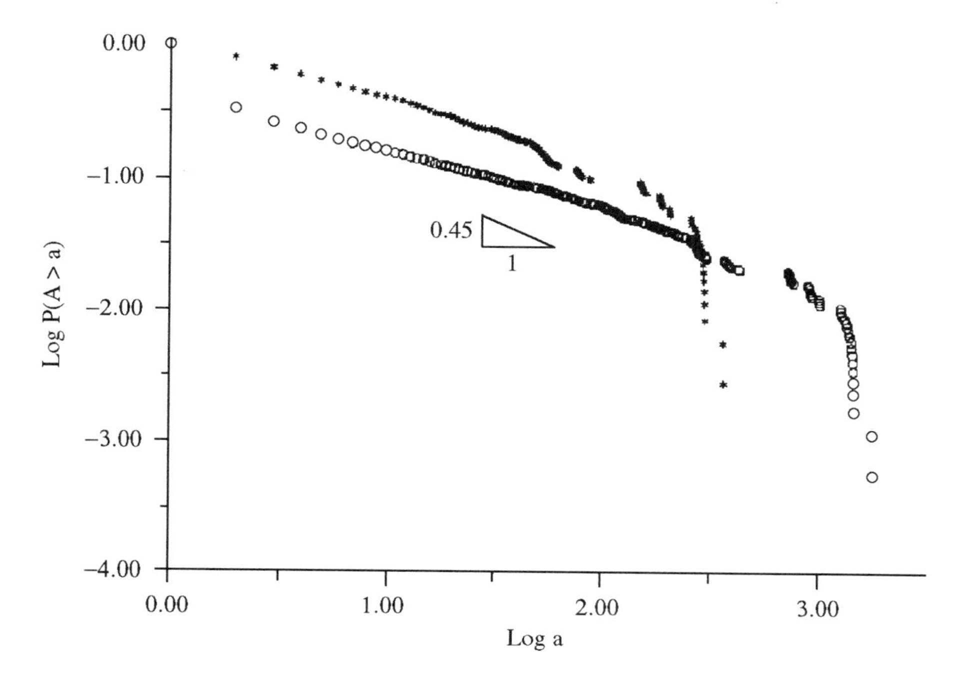

Figure 4.22. Exceedence probability of the cumulative areas with different support thresholds for a 30 × 60 OCN. The upper curve corresponds to $A_t = 5$ pixels; the lower curve to $A_t = 1$.

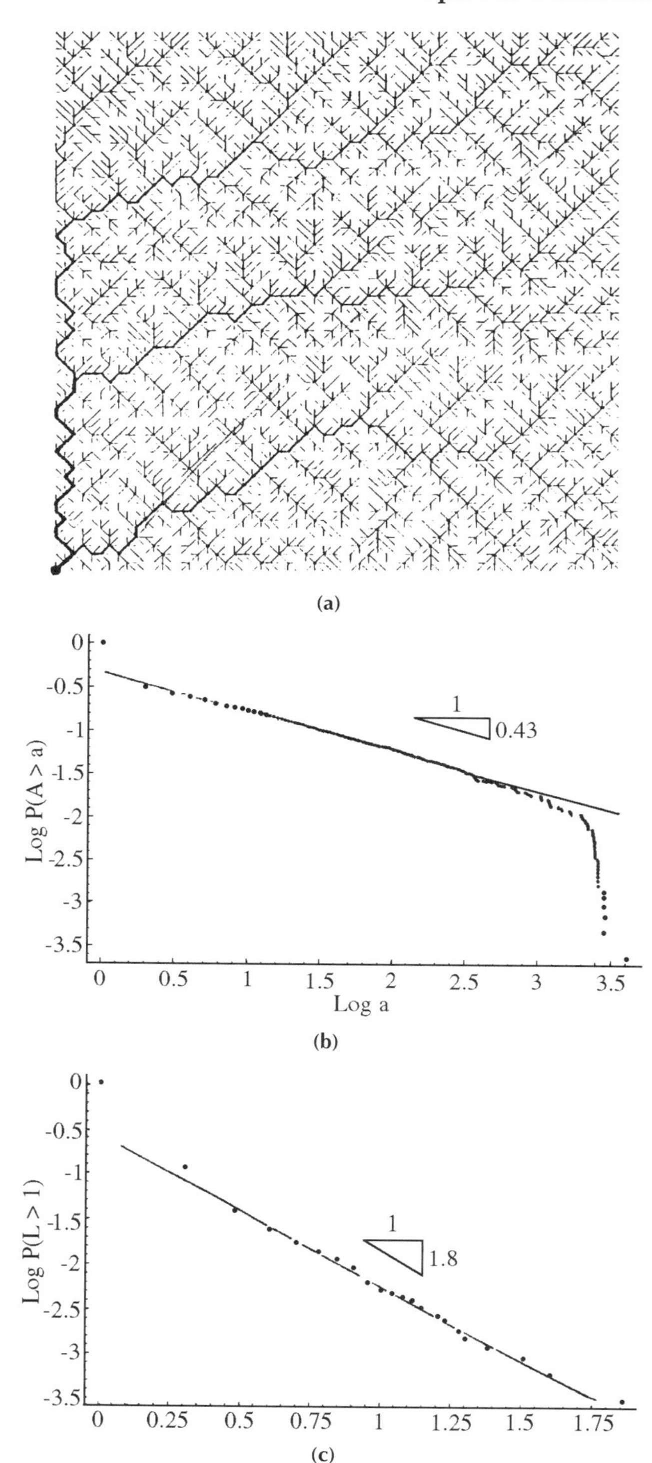

Figure 4.23. (a) 60 × 60 OCN; (b) its $P[A \geq a]$ versus a distribution; (c) its $P[L \geq l]$ versus l distribution [after Rinaldo et al., 1993].

visible as deviations from the power laws because peculiar aggregation patterns frequently exhibit characteristic lengths.

The above results prompt some speculation. It has been shown that uniform energy input in extended dissipative systems results in power law spatial distributions of energy storage and fractal energy dissipation [Mandelbrot, 1972]. Bak et al. [1988] interpreted this as a manifestation of the critical state of a self-organizing process, which acts as a stable and robust attractor of the dynamics. The reason to call it *critical* will be discussed in Chapter 5. Robustness was assessed (as in Section 4.7) with respect to changes of initial conditions, parameters (if any), and the presence of quenched randomness. Dynamic dissipative systems in which energy is injected continuously over the whole domain and that have extended spatial degrees of freedom in two or three dimensions naturally evolve toward self-organized critical states on which no preferential time or spatial scales are shown. The spatial signature of such systems is the emergence of scale-invariant (fractal) structures characterized by power law probability distributions of aggregated quantities. We will seek an explanation of these important similarities in Chapter 5.

OCNs may also be related to statistical models of pattern formation. In fact, a river network under mean annual flow conditions everywhere may be considered to be an aggregation system with a constant injection of particles. Takayasu et al. [1988] have shown that aggregation systems with injection follow a power law mass (area) distribution. The resulting cumulative probability of aggregated area satisfies

$$P[A \geq a] \propto a^{-0.465\pm0.003} \tag{4.58}$$

in two-dimensional systems (in space). The basic aggregation model used by Takayasu and Takayasu [1988] and Huber [1991], which resembles Scheidegger's [1967] model of drainage networks, is obtained by placing particles with integer units of mass on each site of a geometrical lattice. Particles ('sticky' particles) aggregate by random-walk processes with discrete time steps where injection effects operate by adding a unit mass at every site at every time step. In our notation the basic aggregation models is

$$A_i(t) = \sum_j W_{ij}(t)A_j(t-1) + 1 \tag{4.59}$$

where $A_i(t)$ is the aggregated mass (area in the analogy) at time t, and

$$W_{ij} = \begin{cases} 1, & \text{with probability } q(i-j) \\ 0, & \text{with probability } 1-q(i-j) \end{cases}$$

is a random variable denoting the realization that the particle on the jth site may jump to the ith site at time t and $q(i-j)$ is its probability of occurrence. The simplest model:

$$q(i,j) = \begin{cases} 1/2, & \text{for } i,j = 0 \text{ or } 1 \\ 0, & \text{otherwise} \end{cases}$$

is equivalent to Scheidegger's planar model of river networks, which has been shown to yield a scaling exponent $\beta = 0.33$ (Section 2.5.2). We observe the formal analogies of the aggregation model of Eq. (4.59) with the optimization process leading to OCNs with a difference related to the rule for the random change δW_{ij}. In one case, in fact, such a change is nonlocal (in OCNs a local change is accepted if the global property of energy expenditure decreases, hence reflecting a long-range interaction) and in Eq. (4.59) it is local. The question is, why do different models yield similar power laws for $P[A \geq a]$. In other words, one wonders why a local rule for q in Eq. (4.59) yields patterns with characteristics similar to those obtained from total minimum energy dissipation criteria. This points toward an interesting relation among local and nonlocal interactions, and the minimization of energy expenditure appears then to be a property arising from the aggregation process, a result of great interest and major implications.

One may thus conclude that the recurrence of similar power law exponents suggests that river basins are dissipative dynamic systems with extended degrees of freedom that naturally evolve to a critical state that is an attractor for their complex underlying dynamics.

Many other fractal properties for OCNs have been measured. Geomorphologic laws (e.g., Hack's) bearing fractal implications have already been analyzed. Fractal characters in the OCN elevation fields (i.e., obtained by superimposing on a planar OCN pattern drops in elevation consistent with the local rule $\nabla z = A^{-0.5}$, constants being immaterial) will be specifically analyzed in Chapter 5. Other properties are peculiar to the choice of boundary conditions although generated by the minimum energy dissipation requirement. For instance, Sun et al. [1994a] have measured, for multiple-outlet OCNs, the probability distribution of the areas draining at the boundaries (notice that this distribution differs from $P[A \geq a]$, the distribution of total contributing area at any point of the basin) and found that it follows a power law with slope -0.5. Box-counting analyses on the boundaries of the developing basins have also been performed as well as testing of the scaling relationships of lengths, as seen in Chapter 2.

A synthesis of the fractal and geomorphological properties obtained during the process of convergence to an OCNs is illustrated in Figure 4.24 [from Rodriguez-Iturbe et al., 1992c]. In the figure, starting from an arbitrary initial condition that does not fulfill any of the observed properties of channel networks, the optimization converges to an OCN that matches those characteristics extremely well. Figure 4.24 shows networks defined for support thresholds of one and five pixels, whereas all statistics refer to the case with full texture (one-pixel support). These include Horton's laws

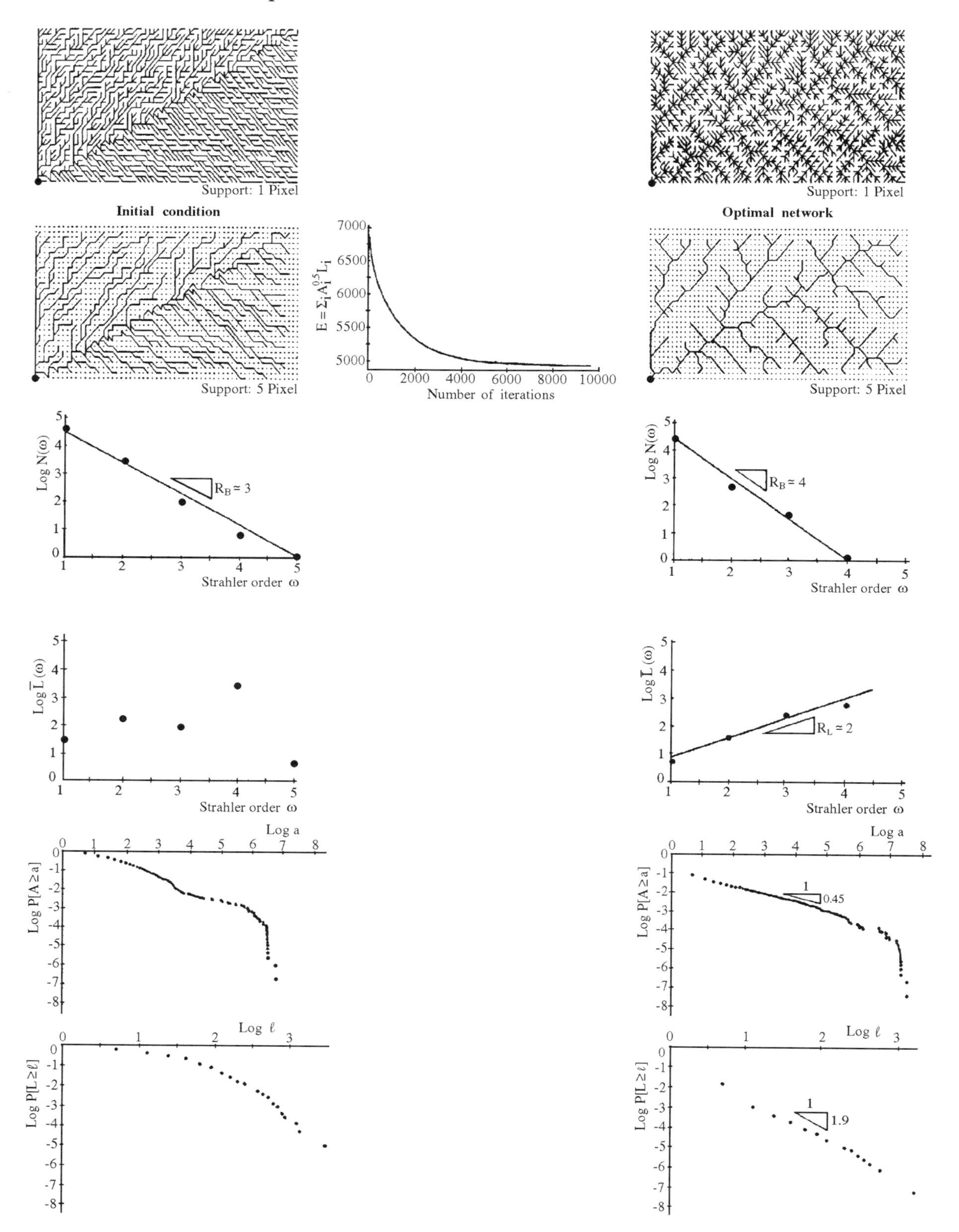

Figure 4.24. Structural characteristics of the initial network compared with those of the resulting OCN [after Rodriguez-Iturbe et al., 1992c].

and probabilities of exceedence of stream length and area. In Figure 4.24 the plots in the left column refer to the initial condition whereas those in the right column correspond to the final stage of the optimization. Also shown in the plot is the decrease in total energy expenditure of the system $\sum_i P_i$, which in the initial condition is 6,843 units and after 55,000 iterations has stabilized to 4,942 units. We note the major transformation caused by the OCN search and the fractal relationships shown by the final shape, as implied by the straightening of the probability plots.

4.10 Multifractal Characteristics of OCNs

Section 4.9 looked at the fractal characteristics of OCNs and DEMs by studying them as geometric objects. It is also interesting to study the spatial distribution of physical quantities. The multifractal formalism discussed in Chapter 3 is a useful tool for studying the spatial organization and scaling properties of variables over a set defined by the river basin structure, whatever the support chosen. In fact, multifractal analyses and measures are related to the distribution of a physical quantity on a geometric support, possibly a fractal in itself [Mandelbrot, 1977, 1990; Feder, 1988].

As discussed in Chapter 3, the important parameters for describing a multiplicative random process, needed to probe the spatial distribution of an object (like the process of dissecting the total drainage area in sparse aggregations), are the fractal dimension of the support, f, the Lipschitz–Holder exponent α of the density distribution, and their relation $f(\alpha)$, called the 'strength of the singularity α.' Although shortcomings of the procedure have surfaced [Veneziano et al., 1995], as discussed in Section 3.3, the computations of the $f(\alpha)$ singularity spectrum performed here employ the method proposed by Chhabra and Jensen [1989] (Section 3.3) for comparison of DEM data with OCNs. The purpose of this analysis, which is not a comprehensive test, is to detect possible similar behaviors of real and optimal aggregation patterns embedded in their geomorphological width functions.

We briefly recall here that the computation of the $f(\alpha)$ spectrum of a measure $\mu_i(\delta)$, where i is the ith site of an arbitrary partition of a support x, is performed by constructing a one-parameter family of normalized measures $N_i(q, \delta)$ (see Eqs. (3.43) and (3.44)) in boxes of size δ as

$$N_i(q, \delta) = \frac{\mu_i(\delta)^q}{\sum_j \mu_j(\delta)^q} \tag{4.60}$$

where q provides a microscope for exploring different regions of a singular measure. Basically, the method consists of linear fits to semilog plots of $\sum_i N_i \log N_i$ and $\sum_i N_i \log \mu_i$ versus $\log \delta$ from which one estimates $f(\alpha(q))$ and $\alpha(q)$, together with the confidence of the estimation.

With the reservations discussed in Section 3.4, we analyze the singularity spectra of width functions $W(x)$ of OCNs. Such functions define (Section 1.2.7) the proportion of drainage area placed at the same distance x from the outlet, the distance x being measured along the network and normalized by the maximum distance from source to outlet. The function $W(x)$ measures at x the number of links in the path $(x, x+dx)$ divided by the total number of network links and hence is a correct probability measure.

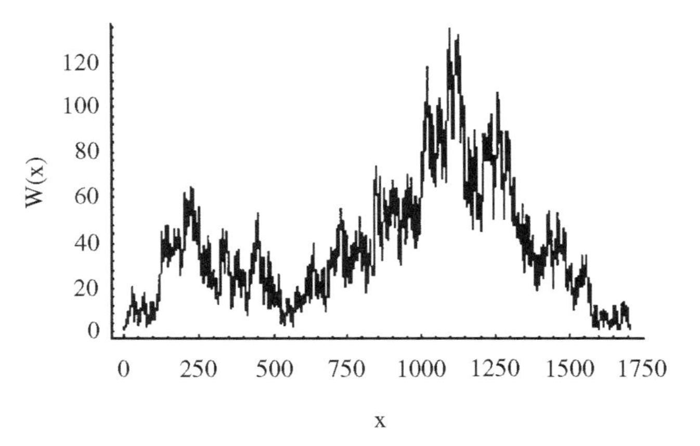

Figure 4.25. Width function of an OCN grown in a 256 × 256 lattice.

This section shows that OCNs produce width functions that consistently show multifractal spectra (though possibly affected by spurious points) indistinguishable from those observed in nature and that multiscaling induced by different cutoffs has a behavior hinting at a possible true underlying multifractal character.

Figure 4.25 shows one example of the width function produced by OCNs. As observed in Chapter 3 with reference to real river basins, the low modes (or the gross features) of the function depend on the shape of the region on which it develops. The fine structure of the function, or the high-mode irregularity, is dependent on the bifurcation structure.

Figure 4.26 shows the singularity spectra $f(\alpha)$ of the width functions $W(x)$ for a few natural river basins from DEMs (Hak, 1,260 km^2; Edel, 993 km^2; Nelk, 440 km^2; Schoharie, 2,408 km^2) compared with those of different OCNs [Rinaldo et al., 1993].

We observe that, regardless of the true nature of the underlying process, OCNs and natural networks have similar estimated multifractal spectra. This holds both for the range of values covered and for the overall form, which is heavily skewed toward the right. Notice that this was not the case in the only case known analytically, that is, Peano's width function (Section 3.2), where a symmetric form characteristic of the binomial multiplicative process was identified.

Figure 4.26. A comparison of multifractal spectra from the DEMs of real basins and several OCNs of different shape and size [from Rinaldo et al., 1993].

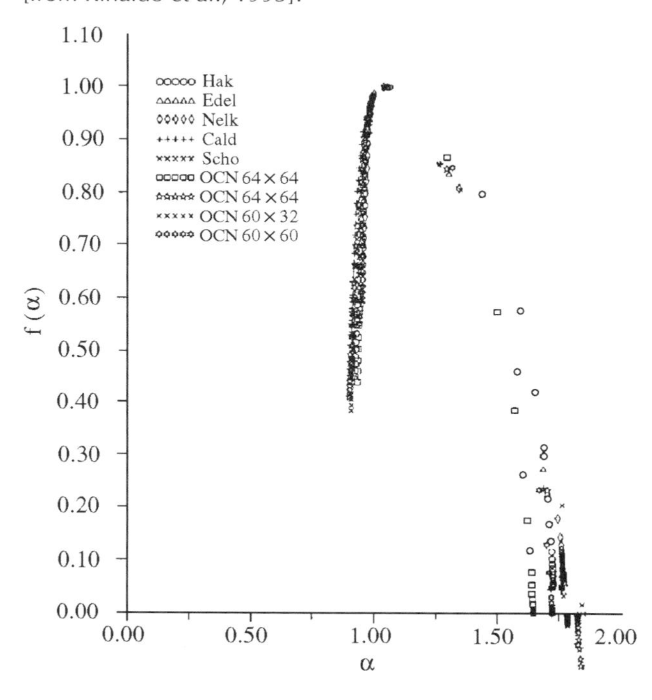

The maximum value of $f(\alpha)$ is correctly at the fractal dimension of the support ($f_{\max}(\alpha) = 1$ for $\alpha \approx 1$), and possibly spurious values of $\alpha_{\min}$ and $\alpha_{\max}$ are very close in experimental (DEM) and theoretical (OCN) results. We also note that the value $\alpha_{\min} \approx 0.8 - 0.9$ is much larger than the value $\alpha_{\min} = 0.41$ of Peano's basin [Marani et al., 1991]. This is reasonable because real-life networks and OCNs do not exhibit a regularity in their construction similar to Peano's, thus resulting in a width function spread over more length scales.

One multiscaling effect is the noteworthy impact of the support area used to identify the network on the right-hand side of the singularity spectrum, as seen in Section 3.5. This phenomenon [Jensen et al., 1991; Amitrano et al., 1991], observed in DEMs (Section 3.5) also occurs in OCNs. In fact, the evaluation of the right-hand side of the spectrum relies on the analysis of Eq. (4.60) at large negative values of q, and therefore the moments are dominated by the measures with the smallest values. This renders the computations most dependent on the lower cutoffs of the resolution. This is particularly evident for the links nearest to the outlet, where often a single channel is resolved. As one increases the value of the threshold area for extracting the network from the DEM, more channels appear, the width function grows faster than the linear function, and the right-

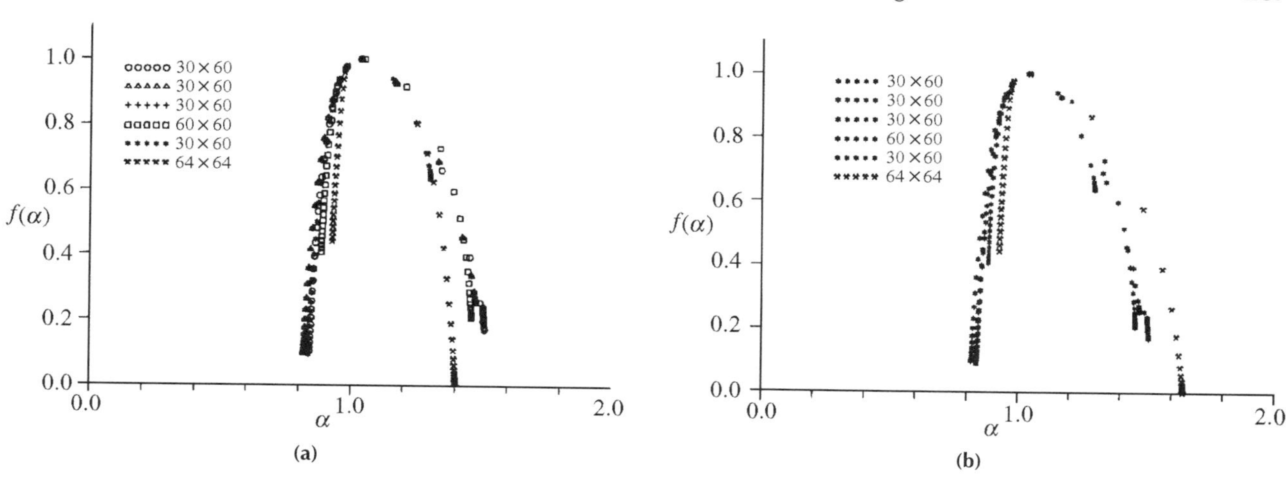

Figure 4.27. Multifractal spectra $f(\alpha)$ versus α of the width function of OCNs with (a) support equal to 5 pixels; and with (b) full support [from Rinaldo et al., 1992].

hand side of the spectrum moves to the right. Figures 4.27(a) and 4.27(b) show the effects on the computed singularity spectra of a modification in the support area used as a cutoff in the identification of the width function.

As seen in Sections 3.4 and 3.5, the genuine characters of the spectra of singularities are hard to evaluate and most likely a single scaling exponent suffices in their representation. Nevertheless, the computational procedure is used here only as a tool for comparing the embedded similarities of the resulting OCNs with nature's forms. Thus, although the multifractal spectra of the width functions of OCNs may partly be an artifact of the computational procedure, there appears to exist a common behavior exhibited by the width functions of real networks and OCNs.

4.11 Multiscaling in OCNs

As seen in Section 3.5, multifractals imply infinite sets of exponents to describe the scaling of all the moments of the distribution of some quantities defined on a (possibly fractal) geometric support.

An approach to describe the unusual and rich complexity observed in OCNs is based on scaling properties of the spatial distributions describing variables of interest. Rigon et al. [1993] extended the results concerning DLA shown in Section 3.5 by simulating OCNs of different sizes at different stages of growth L, where L is the side length of a square basin.

The networks were obtained with the usual techniques, moving from initial conditions arising partly at random and partly from previously computed OCNs. Specifically, the networks were obtained, say from the ith to the $(i+1)$th growth stage corresponding to side lengths L_i to L_{i+1}, by placing random additions to the already optimized network of size L_i in order to fill the larger available area. The procedure allows us to grow OCNs which at every characteristic size L are themselves OCNs. The networks of different sizes consistently exhibit the same statistics.

Our objective is now to look for signs of multiscaling in the width functions of OCNs. Specifically, the scaling relation investigated is of the form

$$W(x, L) = x^{-d+D(x/L)}C(x/L) \propto x^{D(x/L)} \tag{4.61}$$

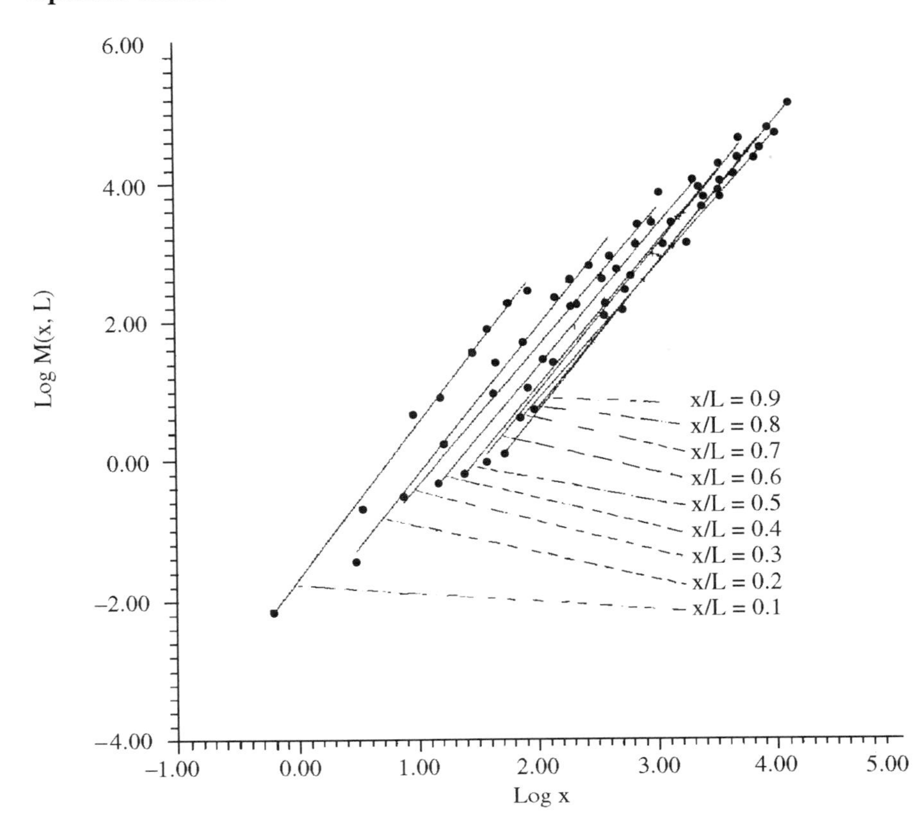

Figure 4.28. Multiscaling in the width function at different stages of growth x/L of OCNs ($A_t = 1$) [from Rigon et al., 1993].

Table 4.1. *Multiscaling induced by a cutoff in the description of the width function of OCNs*

x/L	$D(x/L)_{A_t=1}$	$D(x/L)_{A_t=5}$
0.1	2.18	1.42
0.2	2.08	1.81
0.3	1.95	1.73
0.4	1.99	1.98
0.5	2.02	2.07
0.6	2.05	1.82
0.7	2.09	1.86
0.8	1.89	1.79
0.9	2.04	2.10

Source: Rigon et al. [1993].

where Wdx^d is, in the context of DLA, equivalent to the mass $M(x, L)$ in the d-dimensional volume dx^d at distance x from the outlet, $C(x/L)$ is a scaling function, and D is the fractal dimension of the aggregate. The richness of the multiscaling structure is a function of the variability of $D(x/L)$.

The scaling properties of the width function have been calculated as shown in Figures 4.28 and 4.29. Figure 4.28 illustrates the (nonnormalized) log-mass $M(x, L) = W(x, L)dx$ contained between the distances x and $x + dx$ versus the log of the distance x at different stages of growth x/L. It shows the scaling properties of the growth of the mass at the same relative position within the network, that is, for equal values of the ratio x/L. Here the support area defining the lower cutoff is $A_t = 1$. The slopes, that is, the scaling exponents $D(x/L)$, are in the range 1.9–2.2 (Table 4.1). Figure 4.29 shows the results of the same kind of analysis when a different support area, $A_t = 5$, is used in the identification of the network. One observes that the range of fractal dimensions has been considerably altered and the overall fluctuations are much more pronounced. This points toward problems induced in the estimation of multiscaling by the introduction of lower cutoffs in the analysis of the data, a problem studied in detail by Jensen et al. [1991].

Interestingly, multifractality implies multiscaling, although multiscal-

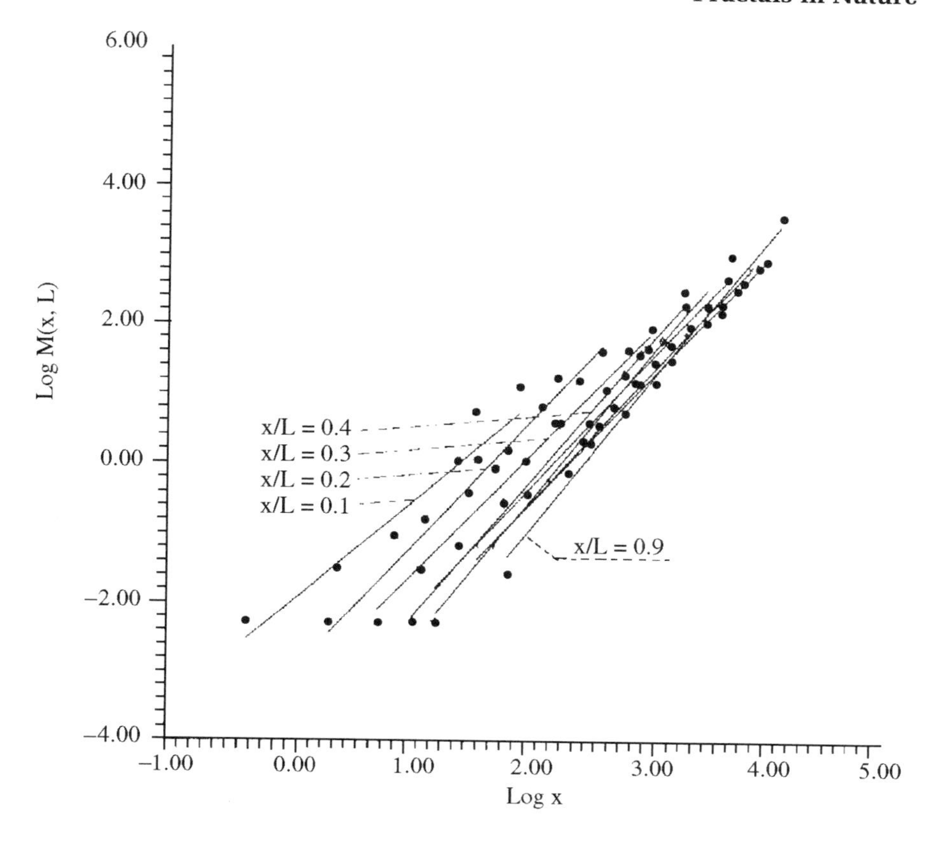

Figure 4.29. As in Figure 4.28 but with support area $A_t = 5$ [from Rigon et al., 1993].

ing does not necessarily require multifractality [Coniglio and Zannetti, 1989]. In fact, artificial multiscaling has been shown to arise when a lower cutoff impairs the correct identification of correlation functions, which in the case investigated by Jensen et al. (see Section 3.5) was known analytically.

The OCN computations of Rigon et al. [1993] concur with the theoretical conclusion of Jensen et al. [1991]. In fact, in Section 3.5 a multiscaling effect of the support area used to identify the network was detected in the evaluation of the right-hand side of the singularity spectrum (see Figure 3.23).

The results presented in Table 4.1 show that whenever larger cutoffs are used in the description of the width function of OCNs, the range of fractal dimensions is substantially altered and the overall disorder is much more pronounced, in analogy to what one observes experimentally.

4.12 Fractals in Nature: Least Energy Dissipation Structures?

The formation of complex dendritic structures like those in natural river networks (but also like colloidal aggregation, DLA, viscous fingering, or certain biological processes) presents characteristics that can be described quite well in terms of fractal geometry over a wide range of scales. Simulation and experiments generating fractals structures have been used to link the kinetics of their formation and their physical and chemical behavior to fractal geometry. A new concept was introduced in this chapter, namely, fractal structures as the result of aggregation patterns of least energy dissi-

Figure 4.30. (a) Peano's basin in a 64 × 64 lattice; (b) OCN obtained from (a) as initial condition [from Rodriguez-Iturbe et al., 1992c].

pation. The concept was validated by geomorphological empirical results and DEMs in the case of natural channel networks. A natural question is whether there can be some wider validity for this concept.

So far we have seen that the evolution from a nonfractal structure to a fractal one suggests that in the case of a natural drainage network, and likely also in other systems, the fractal structure is a least energy dissipation pattern. Moreover, this constrains natural channel patterns from the infinite variety of fractal networks to those that fulfill minimum energy dissipation requirements.

To further study this, we have imposed as an initial condition in the search for an OCN a Peano type of network as shown in Figure 4.30(a) (after Rodriguez-Iturbe et al. [1992c] and Rinaldo et al. [1992]). This idealized planar network, as seen in Section 3.2, has an important resemblance to natural networks in that, among other statistics, $R_B = 4$, $R_L = 2$, and the fractal dimension of total length is $D = 2$. By construction it is deterministically self-similar.

Figures 4.31(a) and 4.31(c) show Peano's probability distribution of (Strahler) stream lengths and of total contributing areas, respectively. We observe that neither areas nor lengths are aggregated over all length scales, certain structures being preferred by the exactly recursive scheme of its construction.

Following Rodriguez-Iturbe et al. [1992c], we now ask whether such perfect regularity meets least energy dissipation demands. Peano's basin, at 6th generation stage with $R_B^{\Omega-1} = 1{,}024$ links, has thus been adopted as the initial condition in the minimum search based on Lin's strategy, discussed in Section 4.7.

Interestingly, in the context of drainage networks Peano's basin is far from being an aggregation pattern of minimum energy. Figure 4.30(b) shows a local minimum found after a few thousand iterations. An important result (Figures 4.31(b), (d)) is that stream lengths and aggregated areas, whose dis-

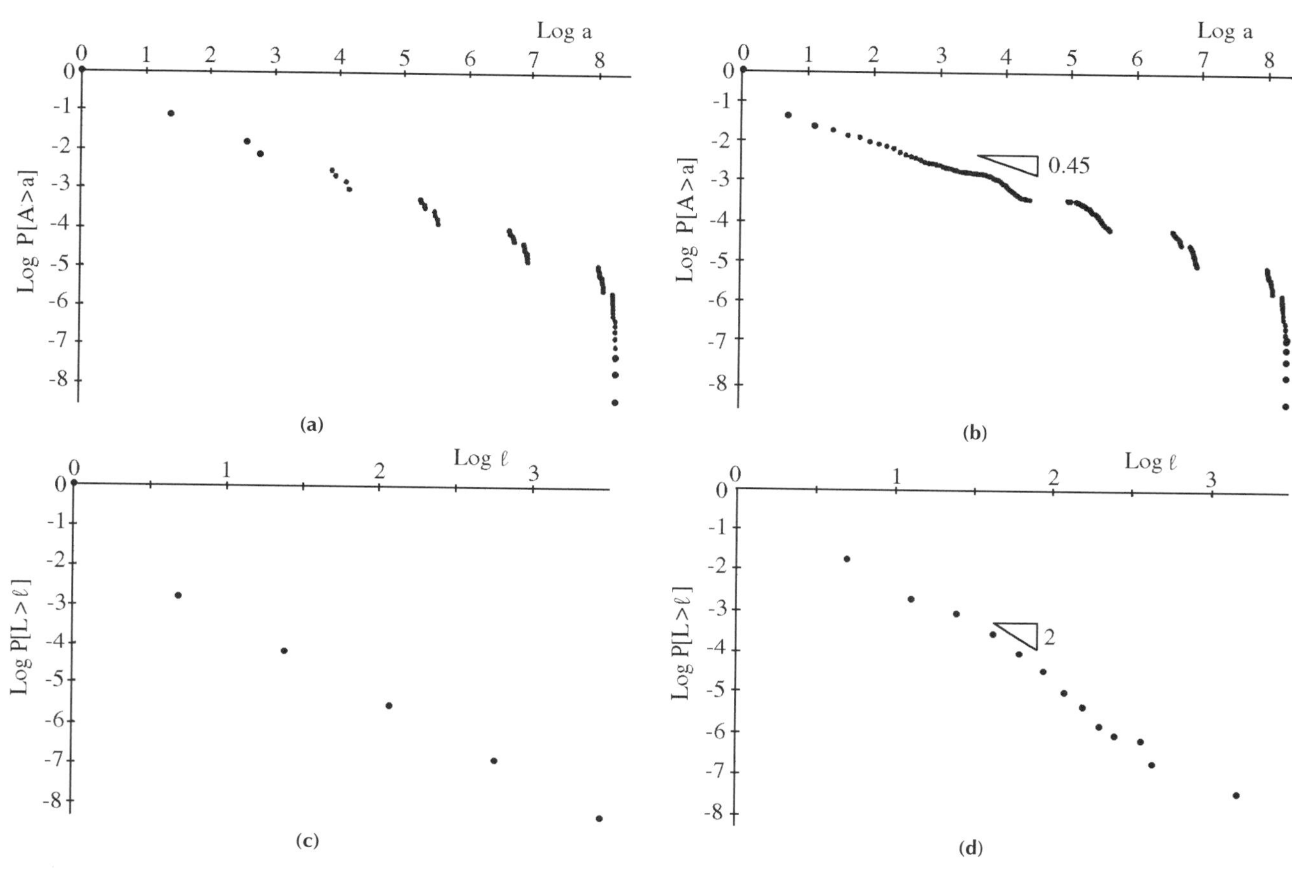

Figure 4.31. Probability distribution of (a) area and (c) length for Peano's basin, and the resulting OCN, (b) and (d) [from Rodriguez-Iturbe et al., 1992c].

tributions were initially clustered, smear clearly toward a full power law behavior in which all scales are occupied, although the statistics are not altered by the optimization process.

The basic implication of the above exercise is that minimum energy dissipation implies no preferential spatial scale. The discontinuity still present in the distribution of areas roughly at half the total area is owing to the inability of the optimization process to break the central trifurcation, where a large jump in aggregated area is observed for all downstream links. More work on the strategy for minimum search may overcome this discontinuity, and thus further smearing of the power law on a wider range of scales will most likely be obtained.

It is also important that the multifractal spectrum of the width function of the optimized network starting from Peano's initial conditions is deformed from the symmetric initial shape characteristic of an exact multiplicative process toward a form that is nearly identical to that observed in OCNs and DEMs of real river networks. The range of values of the Holder's exponent is substantially altered by the optimization and converges to a set indistinguishable from that observed in nature.

The above suggests that natural fractal structures refuse deterministic self-similarity in favor of a statistical scaling that results from the combined effects of chance and necessity. The fingerprints are power law probability distributions in the spatial structure of the main variables of interest.

The results for drainage networks presented herein suggest the possibility that other natural systems evolve toward their own fractal characterizations through least energy expenditure requirements, thus bringing out of necessity the fractal geometry commonly observed in nature.

4.13 On Feasible Optimality

In Section 4.7 we argued that different algorithms used to single out configurations s corresponding to minimum total energy dissipation $H(s)$ face a complex minimization problem, characterized by a large number of local minima. We also argued that once the system evolves from its initial configuration according to minimization of $H(s)$, it relatively quickly reaches configurations with local minima characterized by a plateau of values of $H(s)$. In this plateau the statistical properties of the configurations are virtually identical and indistinguishable from those observed in nature, and thus the question of whether the absolute minimum was reached or not was initially considered irrelevant.

In this section we analyze in detail the outcomes of different algorithms for minimization of $H(s)$ in a statistical mechanics framework. In particular, we investigate whether 'absolute' (i.e., global) minima of the type obtained by Metropolis rules are always consistent with the supposedly worse features of local minima in which less refined algorithms get trapped.

We briefly review the general procedure. An initial configuration s (loopless, spanning) is assigned. The Hamiltonian function of the system $H(s)$ is computed as $H(s) = \sum_{i \in s} A_i^{0.5}$, where $A_i = \sum_{j \in nn(i)} W_{ij} A_j + 1$ and $nn(i)$ are nearest-neighbors of i. The matrix of connections W_{ij} defines the current configuration s (see Eq. (4.52)).

To compare different procedures, we restrict our attention to two cases, that is, the Metropolis rule for simulated annealing and the basic Lin procedure (Section 4.7). Both algorithms select at random one site in the lattice, and a change in the flow directions defining W_{ij} is also randomly selected. The neighbor configuration s' obtained in this manner is first checked to see whether it contains any loops. If not, the new energy function $H(s')$ is calculated. If $H(s') < H(s)$, the change is accepted. If $H(s') \geq H(s)$ the new configuration s' is accepted with probability $\exp[-H(s') - H(s)/T]$ (Section 4.7). Otherwise, the system is returned to the old configuration s and the procedure is restarted.

The difference in the two procedures lies in the specific choice of the parameter T. Lin's basic approach, first employed by Rodriguez-Iturbe et al. [1992a], uses $T = 0$ throughout the search (i.e., it accepts only changes actually lowering $H(s)$). The Metropolis algorithm selects a schedule of temperatures T starting from a high value (i.e., accepting rather unlikely changes) and gradually lowering the values of T in a sort of cooling schedule. Cooling schedules of various types are known, and quite commonly an exponential temperature reduction scheme is adopted, as suggested by Kirkpatrick et al. [1983]. Notice that the decreasing schedule of temperatures is assigned to a sequence of stages, where within a stage with a fixed value of T a given number of change attempts are allowed.

The first observation is that generally Metropolis-like rules yield better minima than Lin's in the sense that, when convergence is assessed by a

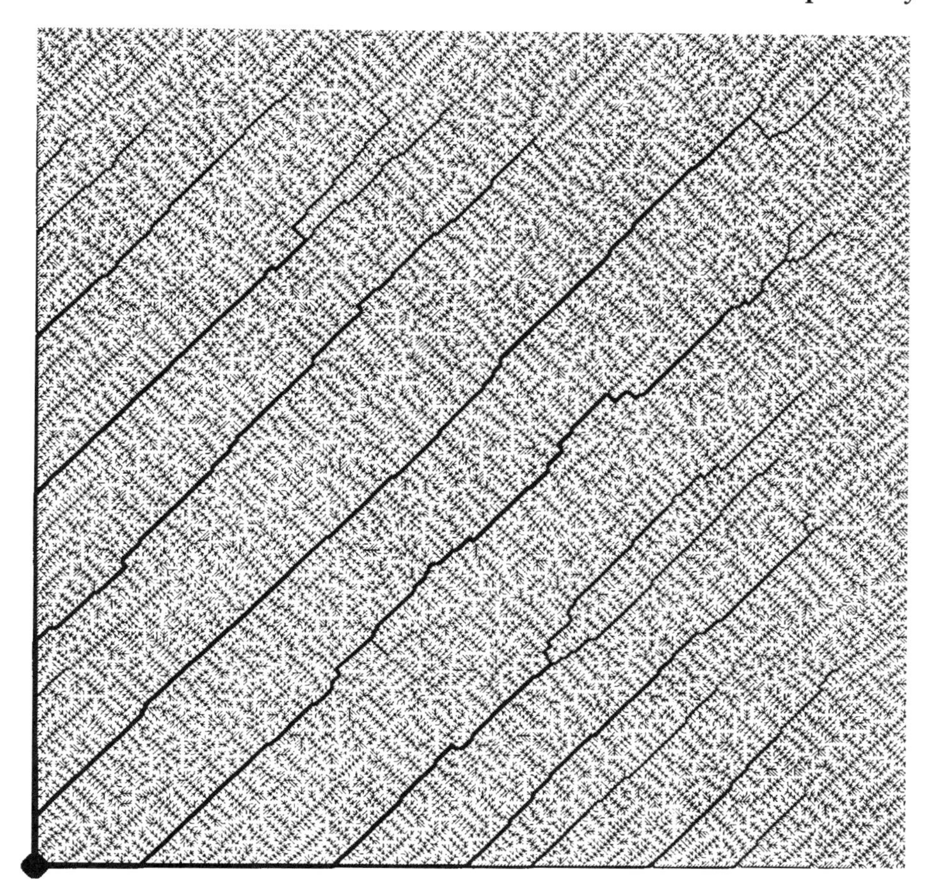

Figure 4.32. OCN obtained from comb-like, parallel drainage directions with $T = 0$.

prefixed number of void iterations, the final value of $H(s)$ for Metropolis networks is lower. The differences in the final vales of $H(s)$ from the $T = 0$ to the scheduled T approaches vary from case to case (depending on size and initial conditions). In general, the $T = 0$ procedure significantly adjusts the structure of the initial condition in the process of lowering $H(s)$ but is incapable of searching for configurations radically different than the initial condition owing to the myopic search procedure. Metropolis-like rules with variable T allow for probing the system under conditions radically different from initial ones, and thus they generally achieve lower minima if two systems (one obtained by keeping $T = 0$ and another by a cooling schedule for T) are started from the same initial conditions. In any finite-size case, the freedom allowed for the evolution of Metropolis networks is much larger than in the other case.

Figures 4.32 and 4.33 illustrate this point. Figure 4.32 shows the result of a $T = 0$ run on a 256×256 lattice starting from a regular, comb-like, nonfractal initial condition (Figures 5.15 and 5.16) composed of parallel drainage directions collected by side channels at the border. The noteworthy lowering of the energy achieved (from 1,115,000 to 250,278 in pixel units) moves the system from a nonfractal to a fractal state through about 5×10^6 changes but hardly changes the main drainage structure, composed of parallel backbones reminiscent of the initial condition. Nevertheless, all

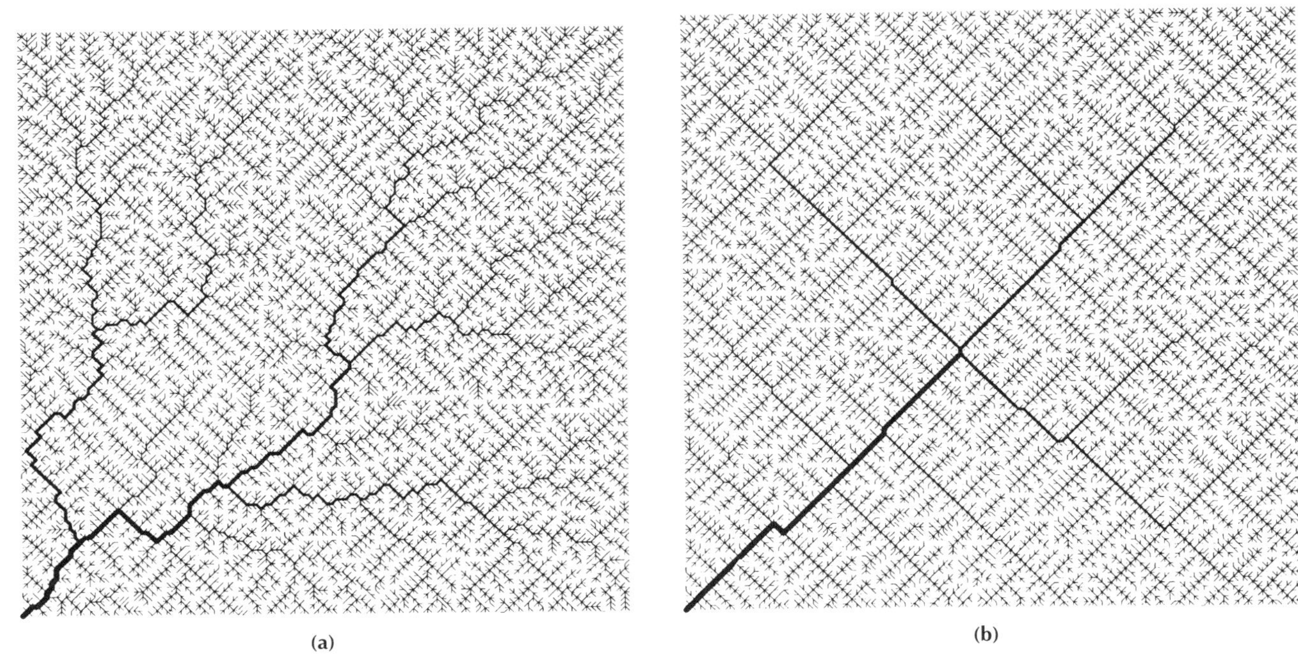

Figure 4.33. (a) $T = 0$ OCN from an initial explosion pattern; (b) variable T (Metropolis) OCN from the same initial conditions.

the statistics of the system have moved from unreasonable to quite close to those observed in nature. In particular, the probability of total contributing area $P[A \geq a]$ starts from a shape characterized by a fast decay (the upper graph in Figure 5.21), which prevents scale-free features, and then develops a power law when the network evolves toward the structure shown in Figure 4.32, corresponding to a local minimum of energy expenditure.

Figure 4.33 further illustrates our point. The OCN optimization problem is parameter-free (one only needs to choose the network whose energy dissipation function is lowest) but entails an involved selection procedure among the very large number of spanning networks in a given domain. The special feature of this problem is that the statistical attributes of the global minimum are known exactly (Section 4.23) in addition to the benchmark provided by the intensely scrutinized fluvial statistics. Here we observe that different minimum search algorithms yield different fractal structures. Figure 4.33(a) shows an example of full OCN pattern (with support area 1) grown in a square domain of size $L = 256$. To obtain (a) we employed the basic Lin approach, which uses $T = 0$ throughout the search. The initial condition for the $T = 0$ OCN (a) is a nonfractal, explosion-like pattern where channels are regular radii issuing from an outlet seeded in the left-hand corner. Figure 4.33(b) shows the outcome of a variable T algorithm (Metropolis OCN) obtained from the same initial conditions yielding (a). Note that the slowly decreasing schedule of temperatures is assigned to a sequence of stages, where within a stage with a fixed value of T a given number of change attempts are allowed. In this case the selected sequence of temperatures starts from $T = 1000$ and gradually decreases to $T = 0$. The networks in Figures 4.33(a) and (b) are single realizations obtained up to convergence over the same number of iterations and show that generally Metropolis-like rules yield better minima that Lin's, i.e., when convergence is assessed by a prefixed number of iterations the final value of $H(s)$ for Metropolis networks is lower. It is particularly relevant that the network in Figure 4.33(b) is quite symmetric and somewhat reminiscent of a Peano network. Nevertheless, relevant scaling exponents of the two networks are quite different, as discussed in the following.

An important question is therefore related to the chance that single realizations of $T = 0$ OCNs, trapped as they are into local minima of the functional $H(s)$ in the space of configurations, represent the average characters of the method. This question also relates to the assessment of such average characters, both in the $T = 0$ and variable T cases.

To answer the above questions, forty examples of 100×100 OCNs were run starting from different initial random conditions with both $T = 0$ and variable T procedures. As a synthesis of the resulting statistical characters, the probability distribution $P[A \geq a]$ was chosen as indicator. A Monte Carlo average was taken of the resulting probability distributions. Figure 4.34(a) shows the ensemble average of the forty realizations thus taken (recall that forty distinct local minima were obtained by accepting only changes lowering $H(s)$, i.e., $T = 0$), yielding an excellent power law $P[A \geq a] \propto a^{-0.43 \pm 0.02}$ for the three decades allowed by the finite size of the lattice. Figure 4.34(b) shows the plot of the ensemble average of $P[A \geq a]$ versus a and the plot of the same distribution for the single realization characterized by the highest value of $H(s)$, that is, the worst local minimum. We notice that the ensemble average characters perfectly match the observed characteristics of natural networks and that local minima attain characters close to those of the ensemble average.

Rinaldo et al. [1996b] carried out analogous computations employing a predefined cooling schedule for T. In this case, a sample of five networks was chosen. The local minima found by this method have generally lower values than the final value of $H(s)$ found by the $T = 0$ method, and the differences in the minimum values are estimated in the range 5–15%. Figure 4.35(a) shows the ensemble average for $P[A \geq a]$ obtained from the above procedure. It matches a power law with $P[A \geq a] \propto a^{-0.48}$, and the establishment of the error bar is more difficult because of the limited sample size. Figure 4.35(b) shows the comparison of the ensemble average distribution with that of the worst local minimum found. Once again, discrepancies of the single realization from the ensemble average are very small.

It is noteworthy that the better minima obtained through the Metropolis rule yield statistics significantly different from those observed in nature. Nevertheless, most structural characteristics are reproduced well.

To test the above result, we have constrained all networks to start from the same initial condition, an exploded pattern whose imprint is difficult to eradicate in the optimization process (see the example of Figure 4.33), and obtained analogous results.

Finally, to test the effects of relaxed constraints, we have run multiple-outlet OCNs with periodic boundary conditions at the sides of a square. Of all networks developed therein, we have selected the largest tree developed in each simulation. Such an arrangement yields behavior free of boundary constraints. Figure 4.36(a) shows that the $T = 0$ methods on multiple outlets yields an ensemble average behavior $P[A \geq a] \propto a^{-0.45}$ whereas an exponential schedule of T yields the 0.5 exponent characteristic of the ground state (Section 4.23) (Figure 4.36(b)).

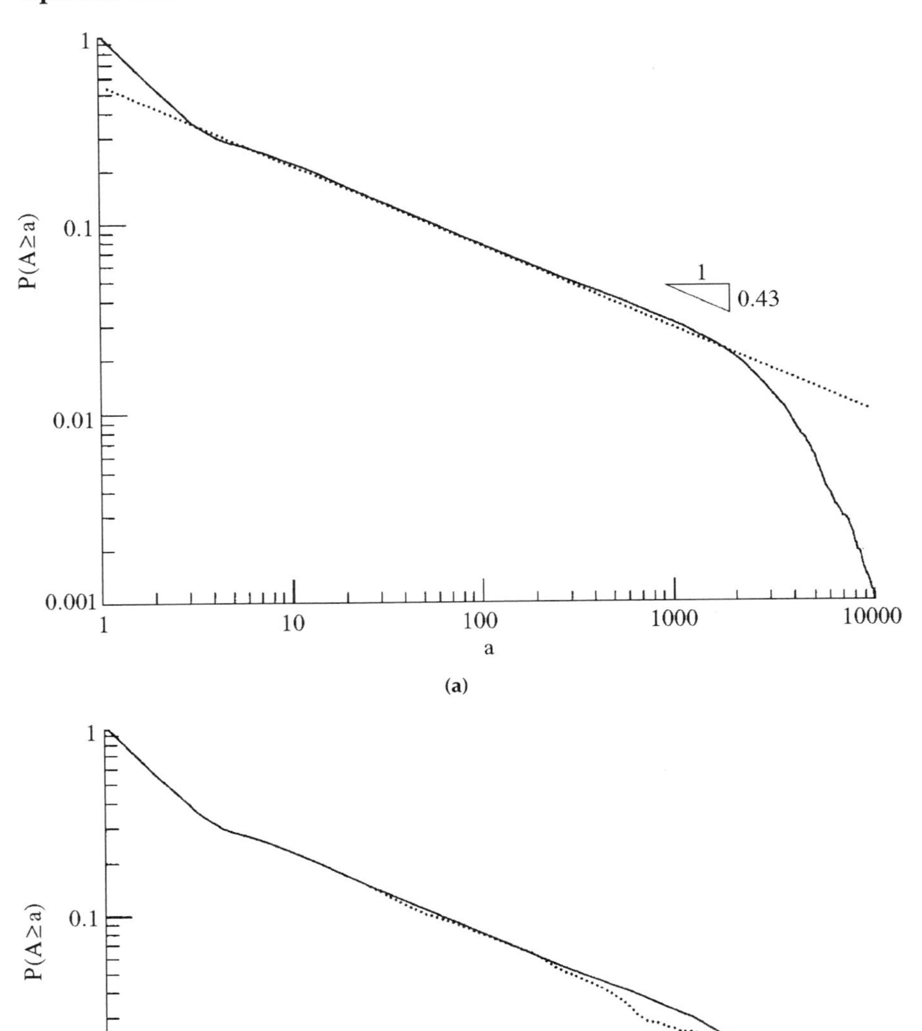

Figure 4.34. (a) Ensemble average probability distribution $P[A \geq a]$ for the $T = 0$ method; (b) ensemble average distribution and the same distribution corresponding to the single realization with the highest value of $H(s)$.

Thus boundary and initial conditions affect the feasible optimal state to different degrees depending on their constraining power. This fact matches the experimental finding (Section 2.10.1) of consistent scaling exponents (of areas, lengths, and form) that accurately describe the morphology of the fluvial basin and are directly linked, resulting in coordinated ranges of variations. Different fractal signatures embedded in linked scaling exponents

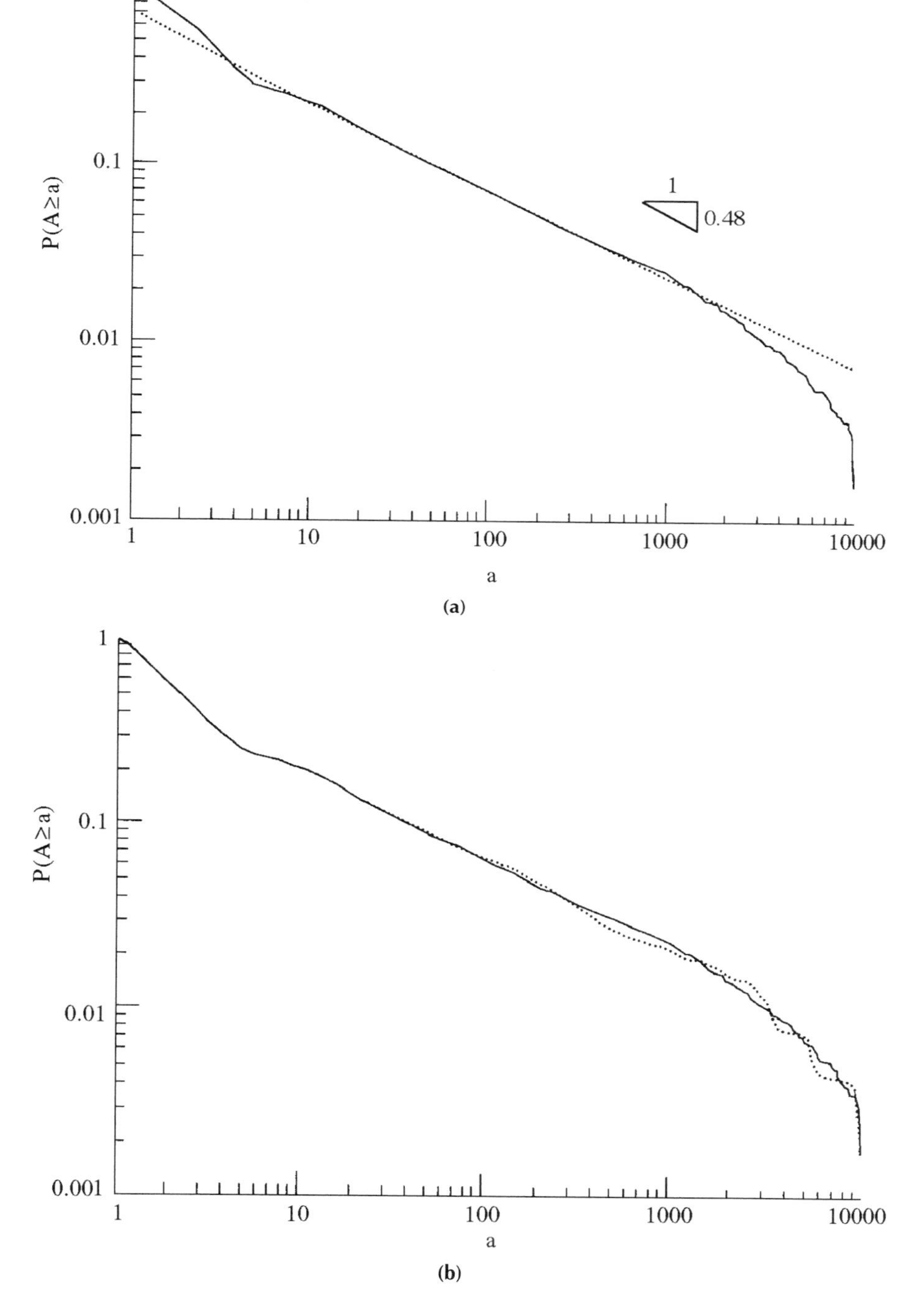

Figure 4.35. (a) Ensemble average probability distribution $P[A \geq a]$ for variable T method; (b) ensemble average, and the same distribution corresponding to the single realization with the highest value of $H(s)$.

thus suggest that the strive for fractality is adapted to the climatic and geologic environment.

Thus we conclude that OCNs show features in remarkable agreement with data *only* when the minimum is sought by procedures capable of carrying out only imperfect searches, attaining suboptimal states (local

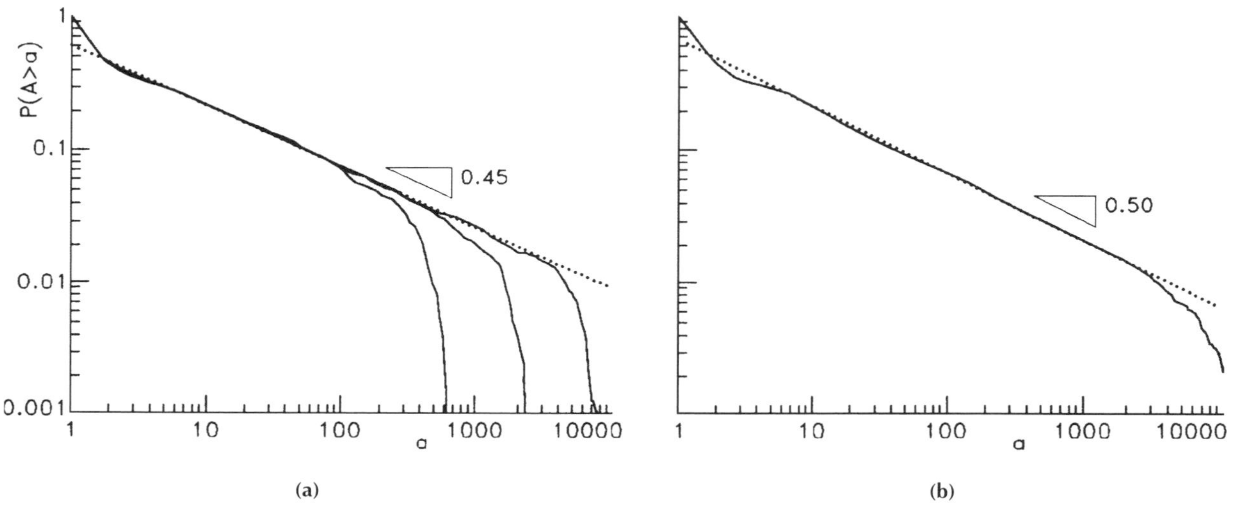

Figure 4.36. (a) A double logarithmic plot of $P[A \geq a]$ vs a obtained through ensemble averaging (10 realizations) of the largest networks developing within multiple-outlet domains (along the bottom side) with lateral periodic boundary conditions and no-flux top boundary. The three curves refer to the largest trees developed within 32×32, 64×64, and 128×128 lattices. Here a local minimum is obtained through the $T = 0$ procedure; (b) as in (a), but wilth a Metropolis procedure. Note that the exponent is 0.5, characteristic of the ground state (Section 4.23) [after Rinaldo et al. 1996b].

minima) in their quest for optimality within constraints of the initial and boundary conditions. Significantly different values of the relevant scaling exponents β ($\beta = 0.50$, which is the exact feature of the absolute minimum, see Section 4.23) are obtained through search procedures aimed at the global minimum independently of the initial condition. The worse energetic performance and the better representation of natural networks by suboptimal ($T = 0$) OCNs call for a framework of dynamic accessibility of optimal states. Feasibility implies that channel networks cannot change freely, regardless of their initial or boundary conditions, because these conditions evidently leave long-lived geomorphic signatures. Thus the optimization that nature seems to perform in the organization of the parts and the whole of the river basin cannot be farsighted, i.e., capable of evolving in a manner that completely disregards initial conditions and allowing for major migration of divides in the search for a more stable configuration, because it would necessarily involve evolution through transient unfavorable conditions. As we will discuss in Section 4.23 when referring to the exact properties of the global optimum, the experiments described before indicate that the type of optimization that nature pursues is rather myopic, that is, willing to accept changes only if their impact is favorable right after their occurrence (in the immediate) rather than in the long run. This we call *feasible optimality*.

4.14 OCNs, Hillslope, and Channel Processes

The basic implication of OCNs is that minimum energy dissipation implies no preferential spatial scale apart from boundary effects. This is reflected in the power law probability distribution of many OCN variables. The values of the scaling exponents match those found in river networks very well.

Nevertheless, a difference in the tails of the probability plots of OCNs and DEMs has been observed by Rigon et al. [1993]. The power law

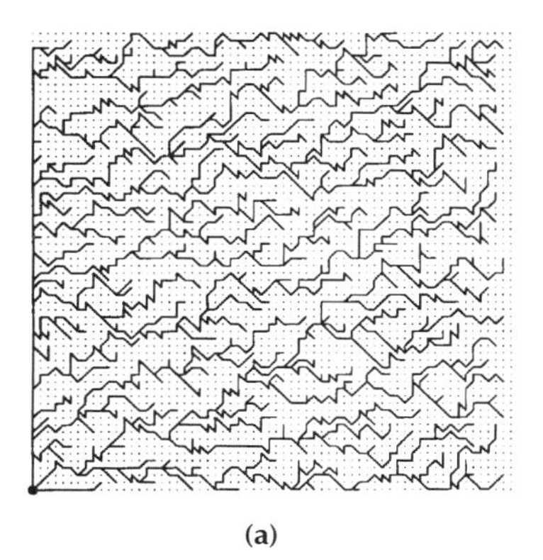
(a)

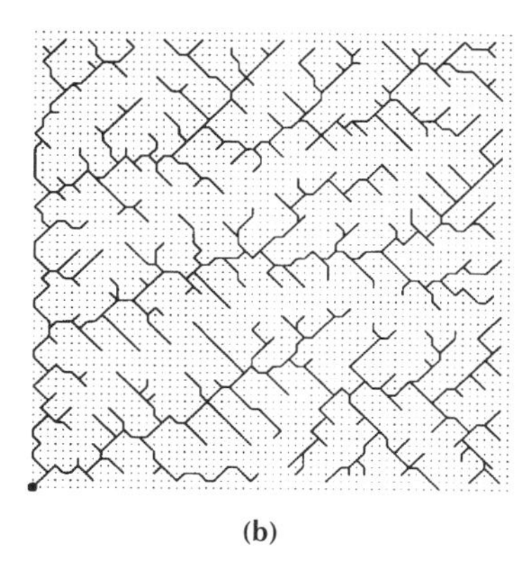
(b)

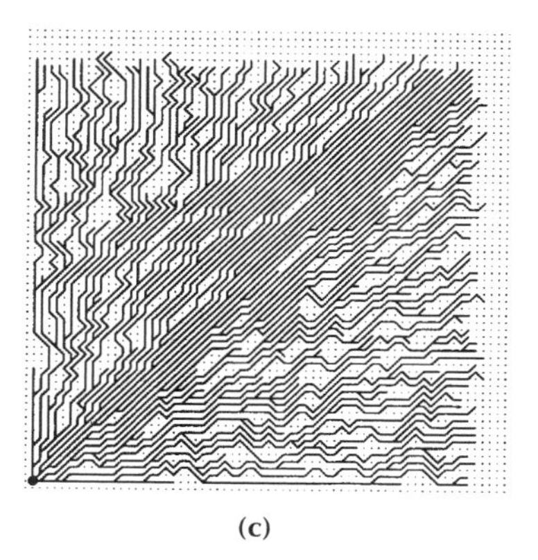
(c)

Figure 4.37. (a) A random network; (b) OCN obtained from (a); (c) pattern obtained from (a) minimizing $H_\gamma(s)$ with $\gamma = 1$.

distribution of aggregated areas for DEMs is not valid for very small and for very large areas (Section 2.5.2). At small values, hillslope behavior plays a role because its aggregation patterns differ from those of the branching networks. The failure at the largest scales is due to a finite-size scaling effect; the number of subnetworks with such a large area is small and the statistics lose significance.

OCNs do not show the small-scale deviations because they aim to describe the fluvial aggregation and do not distinguish hillslope patterns. This is an important observation, which will be discussed later in Chapter 6, where it will be shown that when diffusive processes leading to convex or parallel landforms are taken into account, the small scale hillslope behavior is reproduced. OCNs are also prone to deviations at large values of aggregated area because the constraining role of the outer ridges prevents the probability distribution from spanning over all scales.

Section 4.7 discussed the fact that optimizing networks through the minimization of the energy expenditure function

$$H_\gamma(s) = \sum_i A_i^\gamma = \text{minimum} \tag{4.62}$$

with γ very different from 0.5 could substantially change the characters of the aggregation pattern. Specifically, if $\gamma \geq 1$ no aggregation pattern is produced and the behavior is qualitatively similar to that of parallel hillslopes. Figure 4.37 illustrates this point. It shows the OCNs produced with values of $\gamma = 0.5$ and $\gamma = 1$, starting from identical (Figure 4.37(a)) random initial conditions and defined with support area equal to 4 pixels.

Figure 4.38 shows an example of the major structural modifications implied when γ approaches 1 in terms of the width functions $W(x)$.

The above observations suggest, in terms of energy expenditure at the ith pixel (i.e., $\rho g Q_i \nabla z_i \propto A_i \nabla z_i$), that if the slopes do not scale with discharge below a certain threshold, no aggregation pattern is produced at small scales. Slopes ∇z_i at such a scale may be viewed as random variables, independent of magnitude and reflecting local topographic characters. The rate of energy dissipation in the ith nonchannelized pixel (e.g., $A_i \leq A_t$,

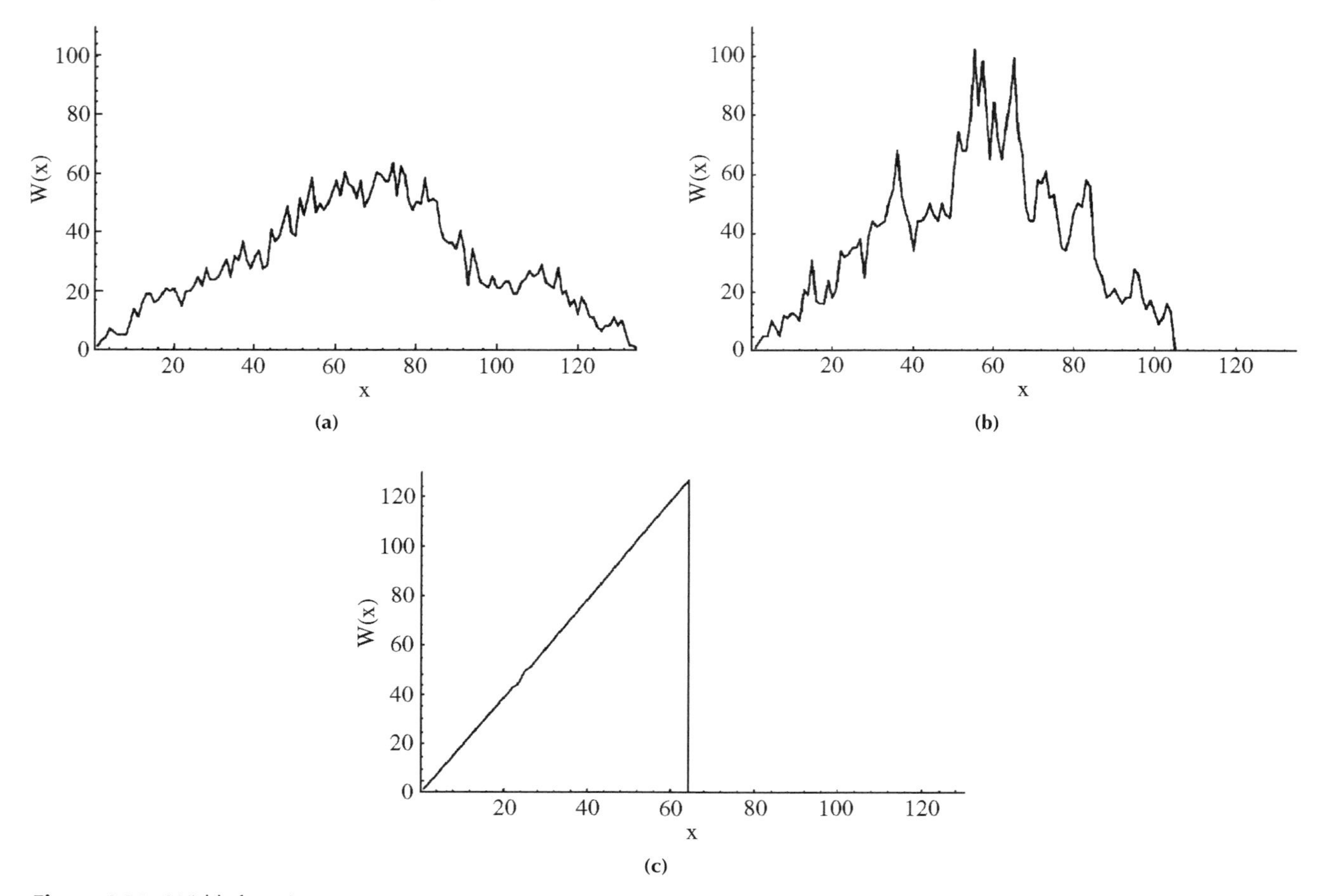

Figure 4.38. Width functions $W(x)$ of the networks in Figure 4.37(a) to 4.37(c) [from Rigon et al., 1993].

where A_t is a fixed threshold) would hence be $P_i = k_i Q_i \propto A_i$, where k_i is random but independent of A_i. One could then incorporate hillslope and channel network processes into OCN concepts through the following total energy expenditure:

$$\sum_i P_i \propto \sum_i A_i^{\eta(A_i)} L_i \tag{4.63}$$

where

$$\eta(A_i) = \begin{cases} 0.5, & \text{if } A_i > A_t \\ 1, & \text{if } A_i \leq A_t \end{cases}$$

To test this, a small DEM (Sarca di Nambrone (Italy), area 21.5 km^2, 2,151 pixels of grid size 100 × 100 m^2) shown in Figure 4.39(a) was examined [Rigon et al. 1993]. The rationale for the choice of such a small DEM is twofold. First, at this scale hillslopes dominate a large portion of small basins. Second, lacking a comprehensive renormalization approach, computational handling of OCNs larger than, say, 10,000 pixels is burdensome even for the most advanced computers. The procedure is as follows [Rigon et al., 1993]. The change in slope observed experimentally in the log-log

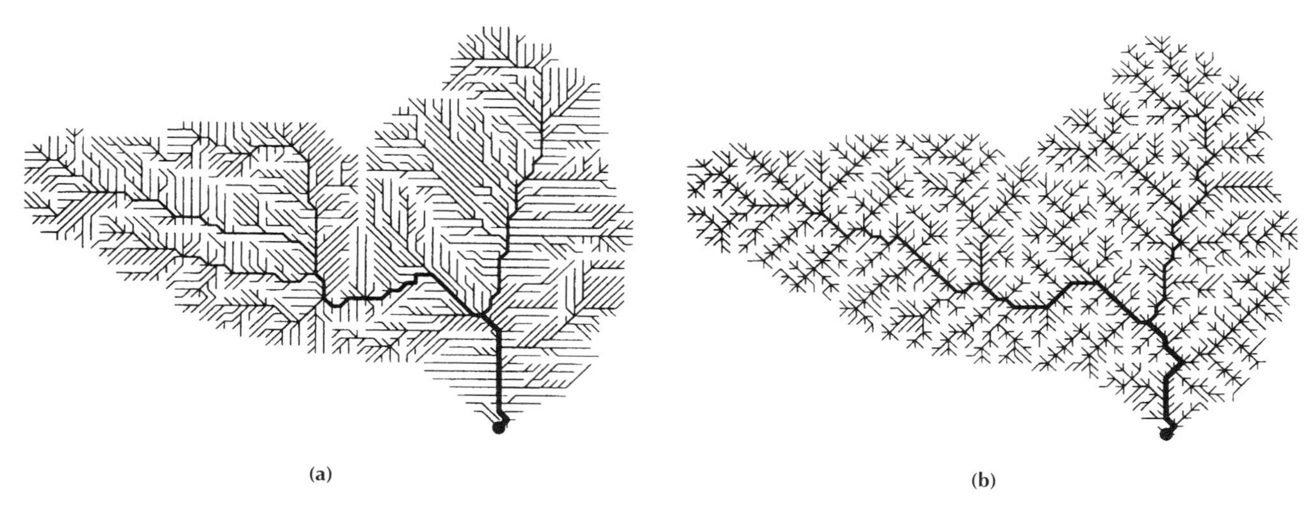

Figure 4.39. (a) DEM of the Sarca di Nambrone (Italy); (b) OCN grown within the same boundaries as the DEM in (a) [from Rigon et al., 1993].

plot of the probability of exceedence of cumulative areas in the Sarca di Nambrone basin (Figure 4.41) is interpreted as signaling the threshold area, A_t, below which hillslopes dominate. With this experimentally determined value of A_t we generated OCNs starting from random initial configurations. The results (Figures 4.39(b) and 4.40(a) and 4.40(b)) show clear differences between the organization of the network corresponding to $\gamma = 0.5$ at all scales ($A_t = 0$) and that of the network obtained through Eq. (4.63) with $A_t = 16$ (Figure 4.40(b)). Other values of A_t were studied with similar results.

This suggests that the threshold area for the hillslope regime is related to a behavior like the one described by Eq. (4.63) in terms of least energy configuration.

Implicit in Eq. (4.63) is the concept that the threshold for channel organization starts at a given area A_t. Observations and theoretical predictions, as seen in Section 1.2.12, suggest that drainage area, a surrogate for mean flow, does not uniquely determine the transition from hillslope to

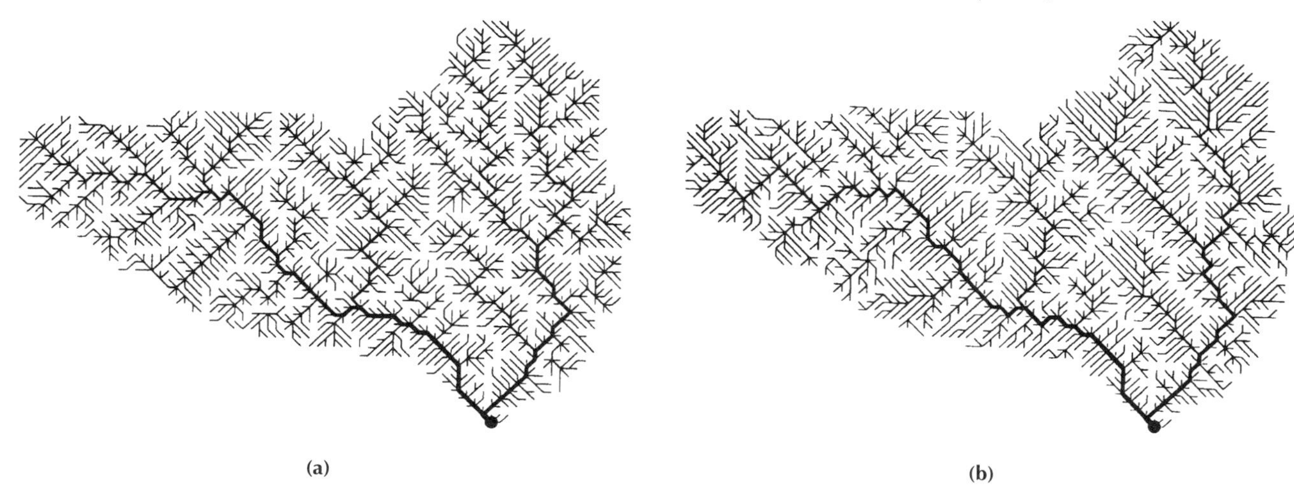

Figure 4.40. (a) OCN obtained via Eq. (4.63) with $A_t = 16$; (b) OCN with a random threshold drawn from an exponential distribution with mean $\langle A_t \rangle = 128$ pixels [from Rigon et al., 1993].

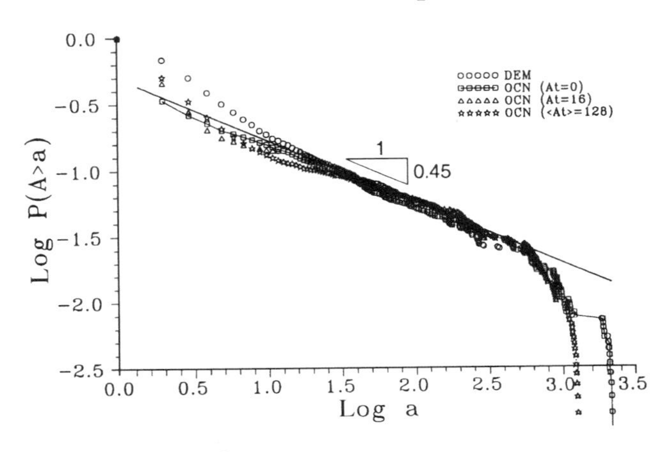

Figure 4.41. Exceedence probabilities of drainage areas for Sarca di Nambrone and the OCNs of Figures 4.39 and 4.40.

channels; many other local factors play a role. It is of interest to illustrate results from an experiment in which a random area threshold $A_t(i)$ is assigned to every pixel i. $A_t(i)$ is drawn from an exponential distribution with mean value $\langle A_t \rangle = 128$ pixels (Figure 4.40(b)). This seems consistent with the studies of Montgomery and Dietrich [1988] on channel initiation mechanisms.

In the case of a random threshold, the probability of exceedence of areas shows a smoother transition toward the steepest slope characteristic of the smallest scales and indeed is similar to that of the DEM of Sarca di Nambrone (Figure 4.41). Note that the slopes in the midrange are virtually identical and equal to 0.45.

To further study the interplay between hillslopes and channels in the exceedence probabilities of drainage areas inside a basin, an experiment in a 64 × 64 pixel network was performed, as shown in Figure 4.42. In Figure 4.42(a) an ordinary OCN is shown, that is, where the threshold $A_t = 0$ implies that at the scale of the lattice hillslope processes are not shown. Figures 4.42(b), 4.42(c), and 4.42(d) show the effect of an increased threshold for fluvial processes.

The effects on the statistics are synthesized by the exceedence probabilities of Figures 4.43(a) to 4.43(c). We have also shown in Figure 4.43(d) the results of a random threshold, here drawn from exponential distribution with mean $\langle A_t \rangle = 16$. The smoothness of the transition in the smaller scales for the random case is typical of many real-life basins, as seen in Section 2.5.2.

The hillslope threshold, whether deterministic or random, affects the distribution $P[A \geq a]$ in a manner similar to that observed in real basins. This suggests a reconciliation of the geologist's perception with OCN concepts. In fact, although at large scale OCNs represent the foremost characters of the spatial organization of the network with an appealing balance of chance and necessity, at the hillslope scales local soil cover and the nature of the exposed lithology determine (erratically) the thresholds for channel headward growth and branching. Moreover, at large scales the cumulative drainage area (and hence the topology of the connections) is a surrogate for the discharges shaping the network aggregation process, whereas at hillslope scales local heterogeneities and topographic and geologic features dominate the aggregation processes.

Sun, Meakin, and Jossang [1994c] have also addressed the explicit incorporation of hillslope processes into OCN concepts. Their approach differs from that described in Eq. (4.63) because they assume a channel initiation function that is both slope and area dependent. Also, they assume that a constant inclination characterizes hillslopes. Planar aggregation structures obtained in this manner for different channel initiation parameters are indeed suggestive. Moreover, the studies of Sun, Meakin, and Jossang [1994c] suggest that the spatial distribution of channel heads must be determined by a suitable channel initiation mechanism in combination with the development of the larger aggregation structure, and hence in combination with minimum energy dissipation requirements.

Figure 4.42. Four 64 × 64 OCNs with threshold (a) $A_t = 0$, (b) $A_t = 4$, (c) $A_t = 8$, (d) $A_t = 16$.

4.15 On the Interaction of Shape and Size

Figure 4.44 shows the results of an attempt to generate the width function of a basin by constructing an OCN of the same shape and size as the available DEM of a real-life network. The OCN (Figure 4.39(b)) was derived from a random initial network fitted to the boundaries of the DEM. In all experiments, optimality, robustness, and convergence have been tested as

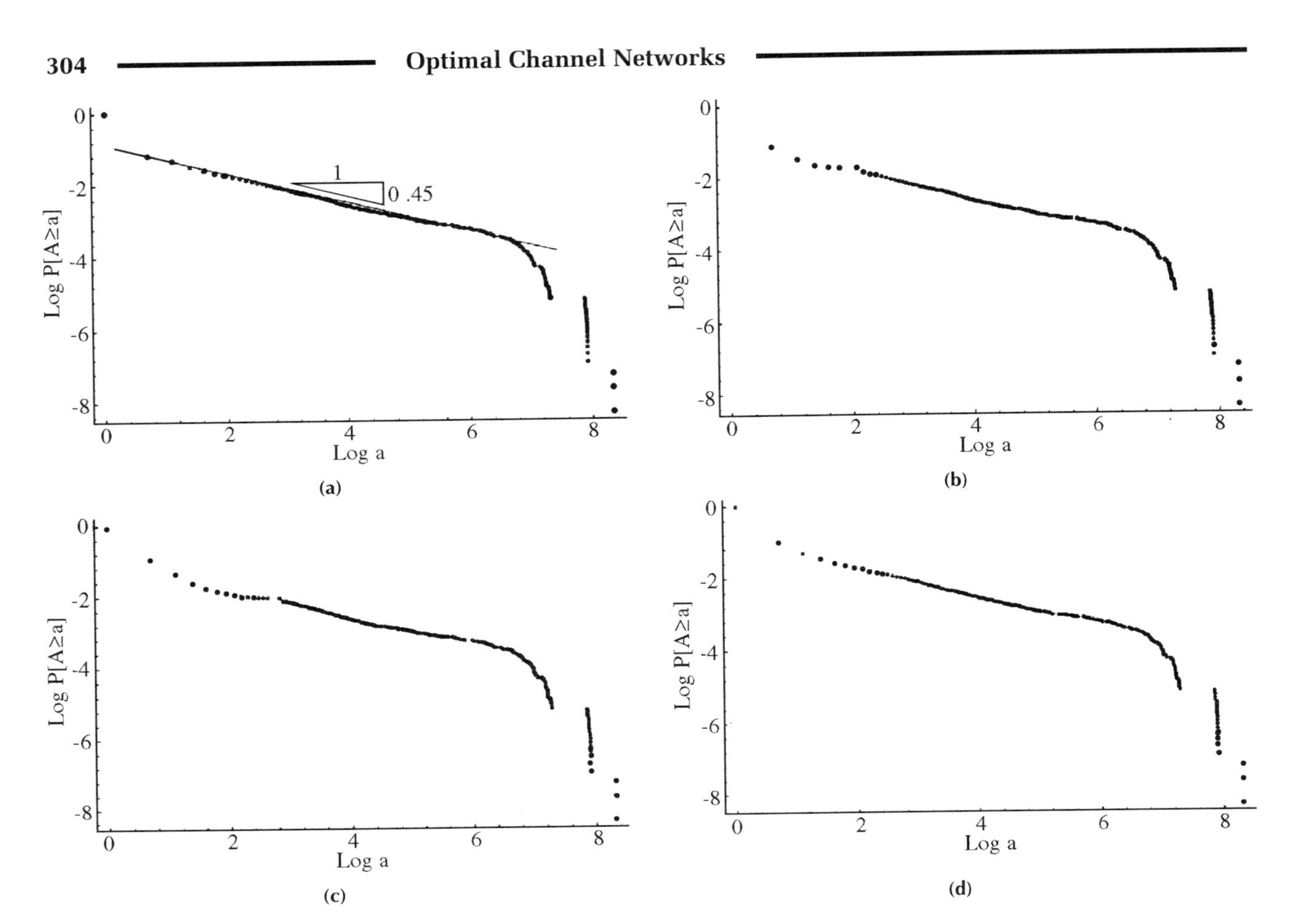

Figure 4.43. Area exceedence probability for 64 × 64 domains: (a) $A_t = 0$; (b) $A_t = 8$; (c) $A_t = 16$; (d) as in (c) but with random threshold [from Rigon et al., 1993].

described in Section 4.7. It is interesting that the real and the synthetic width functions are indeed similar, as shown in Figure 4.44.

An interesting practical consequence is that to some extent the hydrologic response of a mature basin, as reflected by its structure and spatial organization, may be predicted using OCN concepts. Because the width function contains the structure of the unit impulse–response function of the basin (see Chapter 7), the previous results may open practical possibilities of flood studies in ungauged watersheds where DEMs are not available.

Another point of geomorphologic interest concerns the range of scales over which a power law behavior is observed. We now consider OCNs grown from random initial conditions and constrained by different forms of the outer boundaries so that the domains are of equal area but different geometry.

As an example, a square domain drained in a corner is shown in Figure 4.37(b). Also shown are a square domain drained at midside and a rectangular domain drained at the midside of the shortest reach (Figures 4.45(a) and 4.45(b)).

The different constraints imposed by the boundaries are reflected in the width functions obtained and also, in a less clear manner, in the exponent of the probability distributions of total contributing areas. This is not surprising, given the results of Section 2.9.1, which linked the elongation

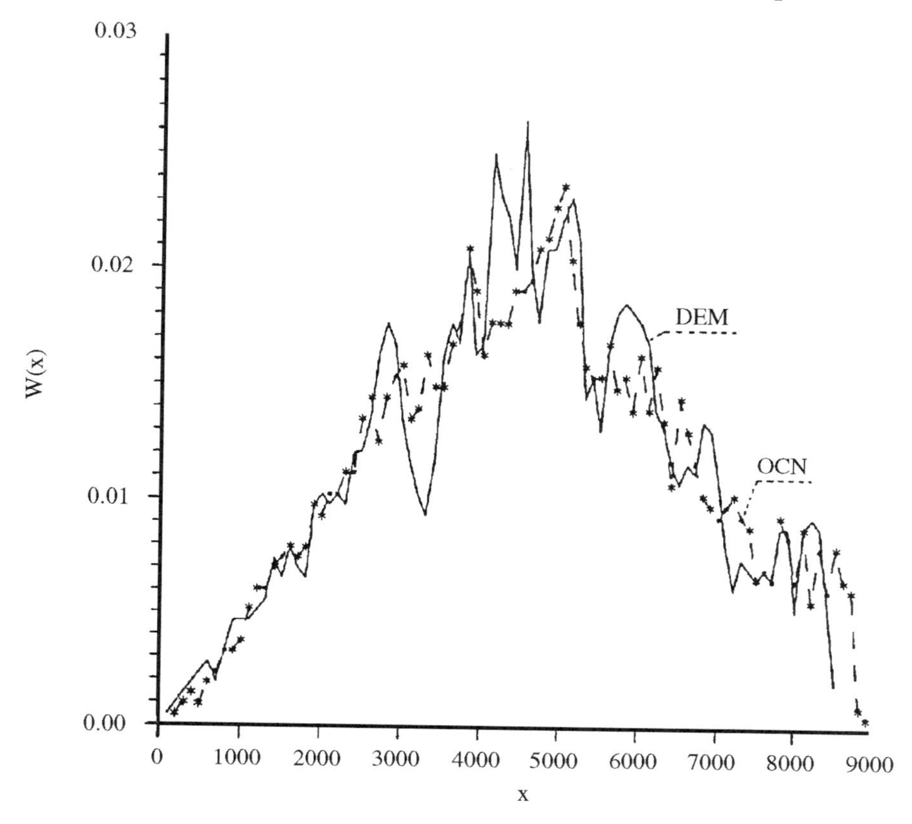

Figure 4.44. Width function for the DEM of Sarca di Nambrone and for an OCN grown within the same boundaries [from Rigon et al., 1993].

structure allowed by the imposed boundaries to the features of the aggregation. Here it suffices to notice, following Rigon et al. [1993], that the resulting OCNs have a scaling behavior close to that shown by real river basins.

Notice also that there is a preferential scale imposed by the boundaries, clearly evident in the width functions. Beyond this preferential scale the development of the network is constrained by controls which are geometric in this case and geologic in nature.

The constraints imposed on the development of the network by fixed boundaries are related to the issue of competition for drainage area within a developing network in the framework of OCNs. Two aspects of this issue are

- whether there are preferable shapes in natural basins,
- and whether subbasins within an OCN are themselves OCNs.

A large number of experiments were performed on the above topics by Rigon et al. [1993]. Figure 4.46(a) illustrates an example of a subnetwork of magnitude 335 developed within a larger 60 × 60 OCN. Here the outer boundary clearly forces part of its development, the remaining part being free to compete with other branches of the network. Its total energy expenditure (in pixel units) is 2,884. The OCN procedure was restarted using the boundaries resulting from the previous OCN and forcing a random network (Figure 4.46 (b)) within the boundaries. The random network

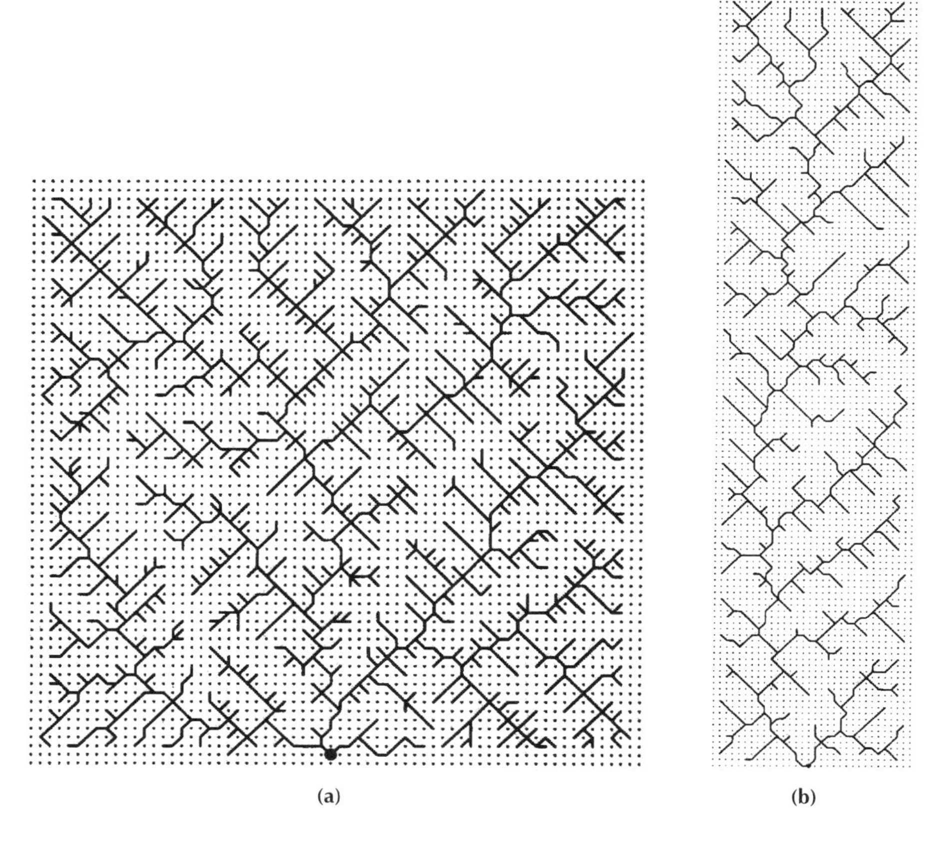

Figure 4.45. OCNs in different geometric domains [from Rigon et al., 1993].

(a) (b)

Figure 4.46. Optimality of subnetworks; (a) subnetwork of a 64 × 64 OCN ($\sum_i P_i = 2{,}884$); (b) random network ($\sum_i P_i = 3{,}568$) (c) OCN with frozen boundaries ($\sum_i P_i = 2{,}852$) [after Rigon et al., 1993].

has a total energy expenditure of 3,568. Optimization followed from this random network, yielding the OCN of Figure 4.46(c), whose energy expenditure is 2,852.

A direct comparison showed an almost perfect match of the statistics of the two OCNs, and total energy is quite similar (2,852 vs. 2,884). Indeed, the analysis of the energy evolution along the optimization process

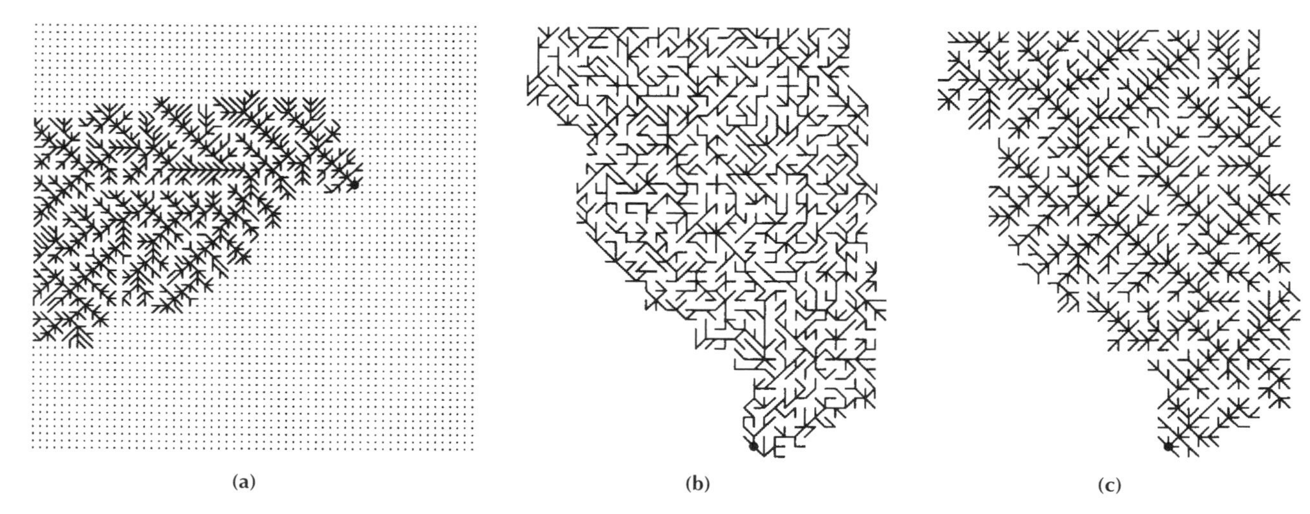

(a) (b) (c)

shows that in both cases the networks are within a plateau (of energy values) in which the features are essentially similar. The above procedure has been repeated several times on subnetworks of very different sizes with the same kind of results. This confirms that, as expected, subnetworks of an OCN are themselves optimal. Their different shapes result from the complex interrelation of competition for drainage, quenched randomness of geologic nature, outer boundary effects, and optimality in network aggregation processes, thereby preventing a unique characterization of the shape of a drainage basin.

The theoretical argument for the optimality of subnetworks of an OCN goes as follows [Rigon et al., 1993]. Let Ω be an OCN and, say, Ω_1 any of its subbasins. Ω_1 contributes to the total energy expenditure $\sum_{i(\Omega)} P_i$ both through its own area and because of the arrangement of all adjacent subnetworks. If $\sum_{i(\Omega)} P_i$ is minimum, $\sum_{i(\Omega_1)} P_i$ must also be minimum because – if this is not true – substitution of another network, Ω_2, with

$$\sum_{i(\Omega_2)} P_i \leq \sum_{i(\Omega_1)} P_i \tag{4.64}$$

would result in an overall network Ω' with

$$\sum_{i(\Omega')} P_i \leq \sum_{i(\Omega)} P_i \tag{4.65}$$

thus violating the assumption of optimality of Ω.

Note that the opposite is not true. In fact, merging optimal networks does not necessarily yield optimal larger networks. Optimization, in fact, implies migration of divides and different subbasin shapes.

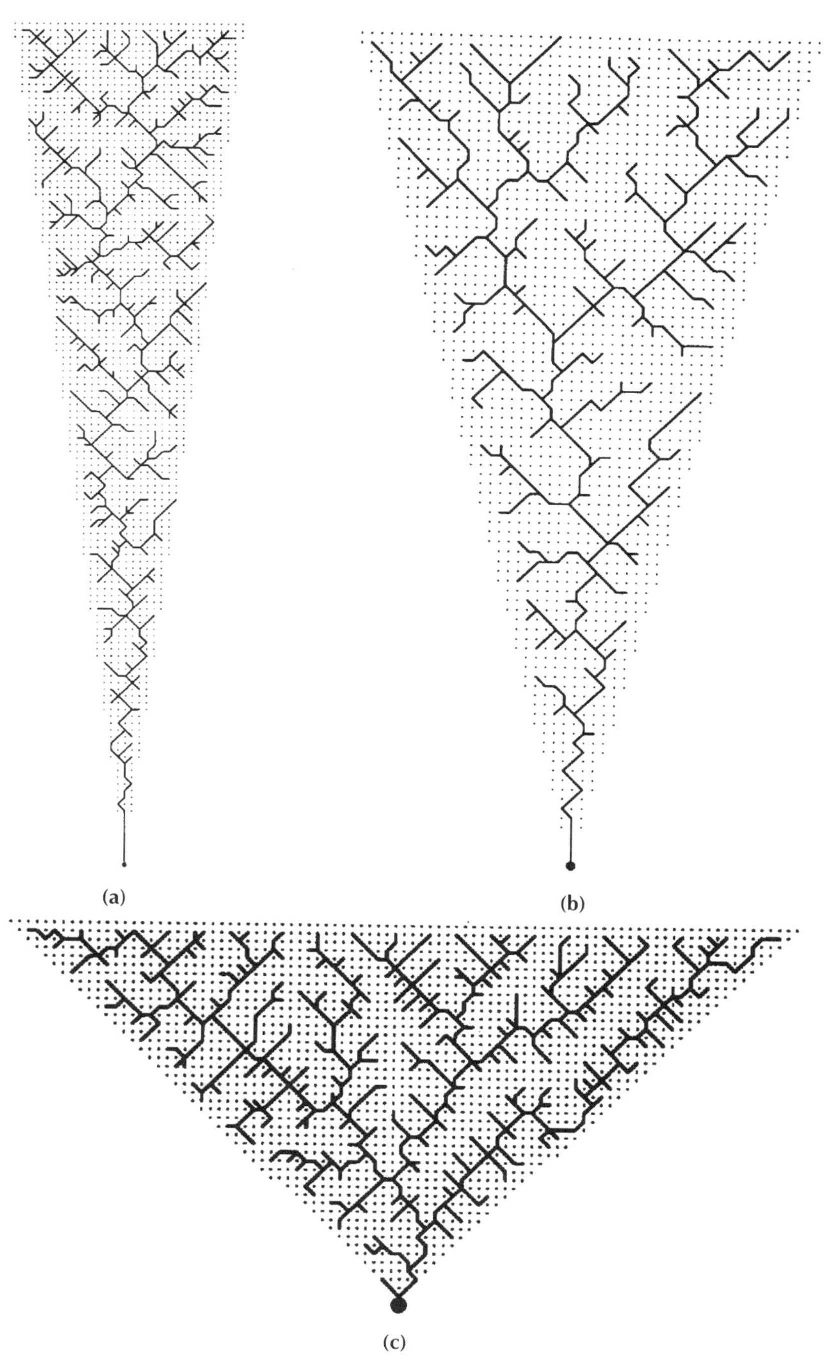

Figure 4.47. OCN in triangular domains for wedge angles: (a) $\pi/12$; (b) $\pi/6$; (c) $\pi/2$ [after Rigon et al., 1993].

To investigate the role of the shape of the basin in the optimal structure of a drainage network, Rigon et al. [1993] ran several experiments on triangular domains of different wedge angles. All OCNs resulted from repeated runs with random initial conditions. Some examples are shown in Figures 4.47(a) to 4.47(c).

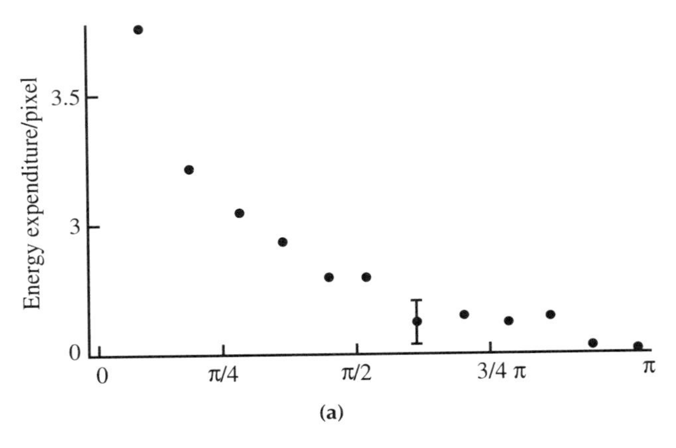

(a)

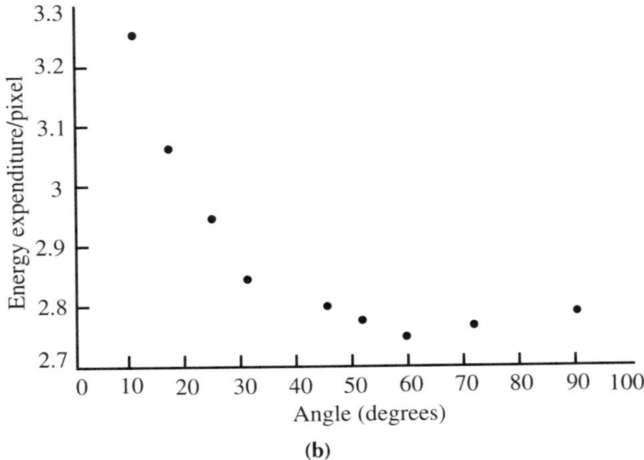

(b)

Figure 4.48. (a) Energy expenditure per unit area as a function of the opening angle of triangular domains [from Rigon et al., 1993]; (b) the same for sectors of circular domains [from Ijjasz-Vasquez et al., 1993a].

Figure 4.48(a) shows the energy expenditure per unit area computed by Rigon et al. [1993] at different values of the wedge angle. For large angles, say $\geq 3\pi/4$, more than one significant network develops in the region very close to the outlet; thus it is only for smaller openings that comparisons involving energy expenditure per unit area remain meaningful. One observes a fast decrease in the energy expenditure until the wedge angle is in the neighborhood of 60°, and then the decay is slower until it gets near 100°, where it tends to stabilize. The same kind of experiment was performed by Ijjasz-Vasquez et al. [1993a] using sectors of a circular domain rather than triangular shapes as the region of network growth. Figure 4.48(b) shows the results; one observes that indeed the shape of the domain (triangles versus sectors of circle) has an impact on the energy consumption per pixel that the OCN attains. Again, the energy expenditure decreases rapidly until the angle is near 50°, where it tends to remain in a plateau with a weak minimum at 60°.

This issue will be further pursued in this chapter with reference to the implication of Hack's law on OCNs.

4.16 Are River Basins OCNs?

The preceding sections of this chapter have shown that OCNs are able to reproduce the statistics and the most important empirical facts known to exist in the geomorphologic structure of real river networks. Nevertheless, OCNs address fluvial processes and cannot describe hillslope evolutions which dominate at small scales unless by adhoc modeling of their interplay (Section 4.14). However, the imprinting of planar fluvial processes dominates aggregation at scales from the lower cutoff (the hillslope scale) to the upper cutoff (the size that can be assumed free of geologic controls). Of course, in lattice problems the upper cutoff is the size of the domain in which the network is allowed to grow. The lower cutoff, ruled by hillslope processes, breaks the otherwise 'infinite' dissection of a landscape produced by fluvial processes (Section 1.2.12), and thus the OCN approach yielding scale invariant features is thought of as appropriate only in a limited (but large) range of scales. Thus the OCN approach cannot address the problem of where the channel begins as this phenomenon is a tradeoff of conflicting diffusive (hillslope) and erosive (fluvial) processes.

Nevertheless, a body of evidence which we have addressed in this chapter shows the similarity of real fluvial landforms and OCNs. Scaling exponents of areas, lengths, boundaries and perimeters, as well as overall patterns match those observed in nature over the wide range of scales covered by dominant fluvial processes. Notice also that, as we have discussed in Chapter 2, the matching of all scaling exponents becomes a discriminant, and a much constraining requirement.

We will first illustrate, mostly to introduce a mathematical result (the following Eq. (4.70)), a surrogate model for OCNs.

Ijjasz-Vasquez et al. [1993a] used a scheme, similar to Howard's [1990] capture model of drainage network simulation, based on the scaling relationship between slopes and areas (slope–area model), as an intermediate step in the comparison of total energy expenditure in real basins obtained from DEMs and the value predicted by the OCN formalism. The method and the results are illustrated in the following.

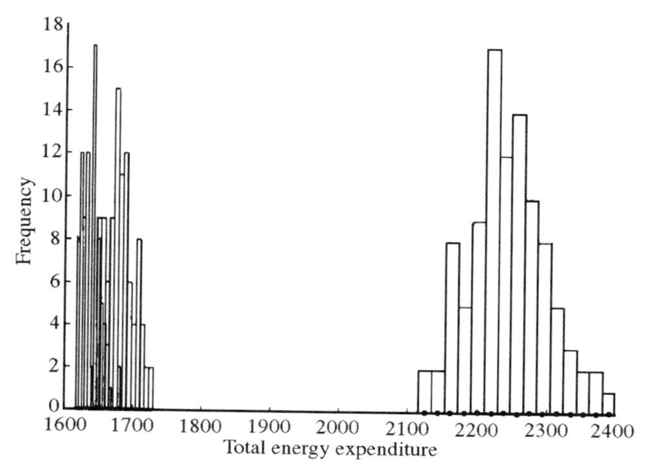

Figure 4.49. Comparison of total energy histograms for random networks (right), and for OCNs and slope–area models (left) [after Ijjasz-Vasquez et al., 1993b].

The slope–area model employed simulates river networks and their three-dimensional structure using the scaling relationship between the slopes and the drainage area surrogating the landscape-shaping flows observed in river basins:

$$\nabla z_i \propto A_i^{-\theta} \tag{4.66}$$

where ∇z_i is the slope at the arbitrary ith site, A_i is the total contributing drainage area, and θ is the scaling exponent (see Section 1.2.10).

Here we will use a value of $\theta = 0.5$, consistent with the first two principles (of local nature) of energy expenditure that constitute the physical basis of the theory of OCNs.

The slope–area model simulates the elevation field over a gridded domain starting from the outlet of the basin. At every iteration the model assigns flow directions along the steepest slope downhill. Following these flow directions, the model calculates the drainage area A_i of each pixel i. Then, the slope at pixel i at the next iteration is set to $\nabla z_i = \chi A_i^{-0.5}$, where χ is an arbitrary constant. The model keeps the elevation of the outlet (and any 'lakes', say, pixels lower than all neighbors) constant and uses them to recalculate elevations of the pixels draining through them using the actual slopes ∇z_i at each pixel. The process is iterated until an equilibrium landscape is reached.

Ijjasz-Vasquez et al. [1993b] have shown that when starting from an initial random condition in the elevation field, the final network developed by the slope–area scheme has a value of total energy expenditure comparable within a few percent of that of OCNs obtained with the random search procedure analog to the traveling salesman algorithm outlined in Section 4.7. The main difference is in the computational time, which is orders of magnitude lower in the slope–area model. Figure 4.49 shows the histogram of the final values of the total energy expenditure following many numerical experiments in the same domain (40×40 pixels) repeating searches by OCN and slope–area models. For comparison, the total energies obtained by generating random networks are also shown. In this case, a two-dimensional random Eden growth, briefly discussed in Section 5.7 [Eden, 1961], is repeatedly used in the planar terrain chosen.

We note that the total energies of slope–area models indeed almost overlap the values obtained by repeated OCN runs. Because the improvement in the total energy values was small for the regular geometries tested in order to obtain Figure 4.49, the use of slope–area models to compare the total energy expenditure of real (and large) DEMs was considered reasonable.

Here we will address two questions:

- Are real networks themselves OCNs in terms of total energy expenditure?
- What is the reason for the surprising similarities of total energy expenditure of network configurations obtained by slope–area models that enforce only the first two optimal principles of OCNs? This in turn amounts to asking why Howard's [1971b, 1990] capture model appears to minimize total energy without a specific request that this happens.

To address the first question, Ijjasz-Vasquez et al. [1993b] analyzed the total energy expenditure in DEMs and networks generated within the same domain by slope–area models. A detailed analysis carried out for four different basins of different sizes and shape across the United States revealed that indeed the similar values of energy expenditure between the real and simulated networks suggest that river networks tend toward a state of minimum energy expenditure. As an example, Ijjasz-Vasquez et al. [1993b] studied three networks with the same support area. Case (a) was the real DEM of the North Fork Cour d'Alene River Basin in Idaho, using a threshold contributing area of 50 pixels. Case (b) was the network generated by the slope–area model with the same support. Case (c) was a reference random network. Whereas the values of $\sum_i P_i$ in cases (a) and (b) are $4.2 \cdot 10^5$ and $4.0 \cdot 10^5$, respectively (in pixel units), case (c) has a total value of $6.1 \cdot 10^5$. Although there are different features in the real and synthetically generated networks, the total energy expenditure is in both cases very similar (Figure 4.49).

The different features are tied to the enforcement of the slope–area scaling relationship in this model, a condition that is known not to happen at scales comparable to hillslope and valley scales, where the scaling relationships in the elevation field are nonexistent or much more complex (Section 1.2.11). Nevertheless, at scales of thousands of square kilometers, as in the above examples, and employing support-contributing areas for the definition of channels of the order of 50 pixels (equivalent to 45,000 m^2), it is not likely that the energy computations will be critically affected by the assumption of perfect scaling $\nabla z \propto A^{-0.5}$ everywhere.

Slope–area models and OCNs tend to yield equivalent total energies whenever the initial condition for the slope–area model is chosen at random. The latter condition is indeed necessary. In fact, any initial network, even one employing parallel flow, in which the scaling is obeyed from the onset, would be unmodified by the slope–area model. A comb-like structure of the type described in Chapter 5 (Figure 5.12) would not meet any properties of real networks yet, nevertheless, would obey everywhere the scaling relationship in slope and the first two principles embedded in OCNs. Hence the slope–area model is dependent on initial conditions, a condition that does not apply to OCNs because of the random search embedded in the optimization process. Also, by construction the scheme produces an unrealistic three-dimensional structure of the network because the slope-versus-area diagrams show no scatter along the line $\nabla z \propto A^{-0.5}$ (in fact, $Var[\nabla z(A)] = 0$).

Nevertheless, the slope–area model produces networks characterized by total energy that falls within the plateau in which configurations of local minimum energy are trapped, and thus it is useful in providing reference optimal large-scale configurations.

The reason for the energy efficiency of the slope–area scheme resides in the link between the third (global) principle of optimal energy expenditure and the stability of equilibrium that results from the equivalence of total energy expenditure and the total sum of elevations [Ijjasz-Vasquez et al., 1993a]. Thus let E_p be the sum of all elevations measured, say, with respect to the outlet

$$E_p = \sum_i z_i \tag{4.67}$$

where z_i is the elevation of the ith pixel. Clearly, E_p is a measure of total potential energy of the landscape described by the field z. The elevation of the ith pixel may be thought of as the sum of all drops Δz_j from pixel to pixel along the flowpath linking the ith pixel to the outlet. If γ_i indexes the sites along the path from the outlet to the ith site, one may write

$$\sum_i z_i = \sum_i \sum_{j \in \gamma_i} \Delta z_j \tag{4.68}$$

The summations in Eq. (4.68) can be reorganized by observing that if the area ruler is unity, the total contributing area A_i to a link basically measures the number of sites connected to the ith link. Therefore A_i measures the number of times every Δz_i appears in the sum. Thus we obtain the equivalence

$$\sum_i z_i = \sum_i A_i \Delta z_i \tag{4.69}$$

We now observe that in unbiased lattices the local slope is equal to the drop in elevation; that is, $\nabla z_i = \Delta z_i$ whenever the unit length scale is the pixel size. Thus when one enforces the constraint of the scaling implied by the local principles of optimality $\nabla z_i \propto A_i^{-0.5}$, one obtains the proportionality between potential energy and total energy dissipation

$$\sum_i z_i \propto \sum_i A_i^{0.5} \tag{4.70}$$

Therefore minimizing total energy dissipation is equivalent to minimizing total potential energy under the constraint on the slopes that $\nabla z_i \propto A_i^{-0.5}$.

OCNs explicitly minimize total energy dissipation, and the slope–area model operates such that, given that the slope of each pixel comes from the preceding iteration (constrained to $\nabla z_i \propto A_i^{0.5}$), each pixel is set to drain in the steepest direction downhill. By choosing the lowest neighbor, the pixel is setting its elevation to the lowest possible value. We recall that an analogous result has been found in the analysis of stationary dendritic structures developing within a potential force field (see Section 4.5).

We thus conclude that the enforcement of the three principles results in networks characterized by stable equilibrium because the developing network tends to irreversibly lower its mean elevation once it organizes according to Eq. (4.66). The slope–area model is able to choose nearly optimal arrangements because the slope-scaling rule assumes the enforcement of the two local principles of optimality, and furthermore the iterative process, continued until one has $\nabla z_i \propto A_i^{-0.5}$ everywhere, due to the presence of randomness in the initial elevation field drives the system away from unstable landscapes. The final equilibrium condition is equivalent to the condition of minimum mean elevation under the constraint that $\nabla z_i \propto A_i^{-0.5}$ everywhere, and because of Eq. (4.68) that condition is equivalent to one of minimum total energy dissipation.

Nevertheless, the possibilities of Howard's [1990] or the slope–area [Ijjasz-Vasquez et al., 1993b] models to yield OCNs are limited to a particular choice of initial conditions, where randomness is required and specific patterns are avoided. Otherwise the final outcome could be irreversibly marked by the initial conditions. On the other hand, the beauty and generality of OCNs lie in the absence of constraints because no parameters need to be tuned, no particular initial conditions are needed, no local adjustments need to be made, and the system is robust with respect to quenched disorder in achieving a common fractal structure. This is critically linked to the enforcement of the third, global principle of minimum energy dissipation of the open system as a whole.

To further test OCNs against real river data, Rinaldo et al. [1996b] have run the following specific example of optimization. A planar map of Wolf Creek river basin (West Virginia) was taken (Figure 4.50(a)). The channel network was extracted through a slope-dependent support area, as described in Section 1.2.12, requiring that the effective channel head lies in areas of topographic concavity. Here the overall support area is of the order of 50 pixels. Figure 4.50(b) shows one example of OCN obtained within the same domain of Wolf Creek, that is, characterized by no-flux boundary conditions except for the outlet, fixed in the same position and plotted through the same support area. In this specific case, we have run an example that considers the interplay of hillslope and fluvial patterns as described by Eq. (4.63). Hence the resulting spatial distribution of channel heads is determined (as seen in Section 4.14) in combination with energy dissipation requirements. In the determination of this OCN, the optimal configuration s is chosen by minimizing the functional $H(s) = \sum_i A_i^{0.5}$ with two important constraints. First, the initial condition chosen is a random, Eden-grown, spanning network whose features are much altered by the optimization process. Second, the search procedure is 'cold' (i.e., uses $T = 0$ throughout the search, see Section 4.13). Hence the minimization is rather myopic, accepting only changes that actually lower $H(s)$. Figure 4.50(c), from the same initial condition of Figure 4.50(b), shows the OCN obtained when the developing structure is kept in a 'hot bath' (i.e., changes in the configuration s that do not lower $H(s)$ are accepted with probability proportional to a high 'temperature' T, here T =10,000, see Section 4.13) for a rather long time, thus yielding a loss of memory of the initial condition. The OCN is then stepwise frozen at $T = 0$ (i.e., accepting only changes of s lowering $H(s)$) up to convergence when no changes are further obtainable.

The statistical similarity of real and simulated networks is remarkable, all relevant scaling parameters and major structural features being indeed close. However, small scale differences arise depending on the nature of the optimization process. We notice that it is commonplace in some circles to argue that fractal statistics might not be diagnostic because the eye (unable to capture and evaluate the features of the aggregation pattern unless in a limited range of scales) captures instead small scale differences. In fact, it is interesting to notice the different tortuosity of the main streams

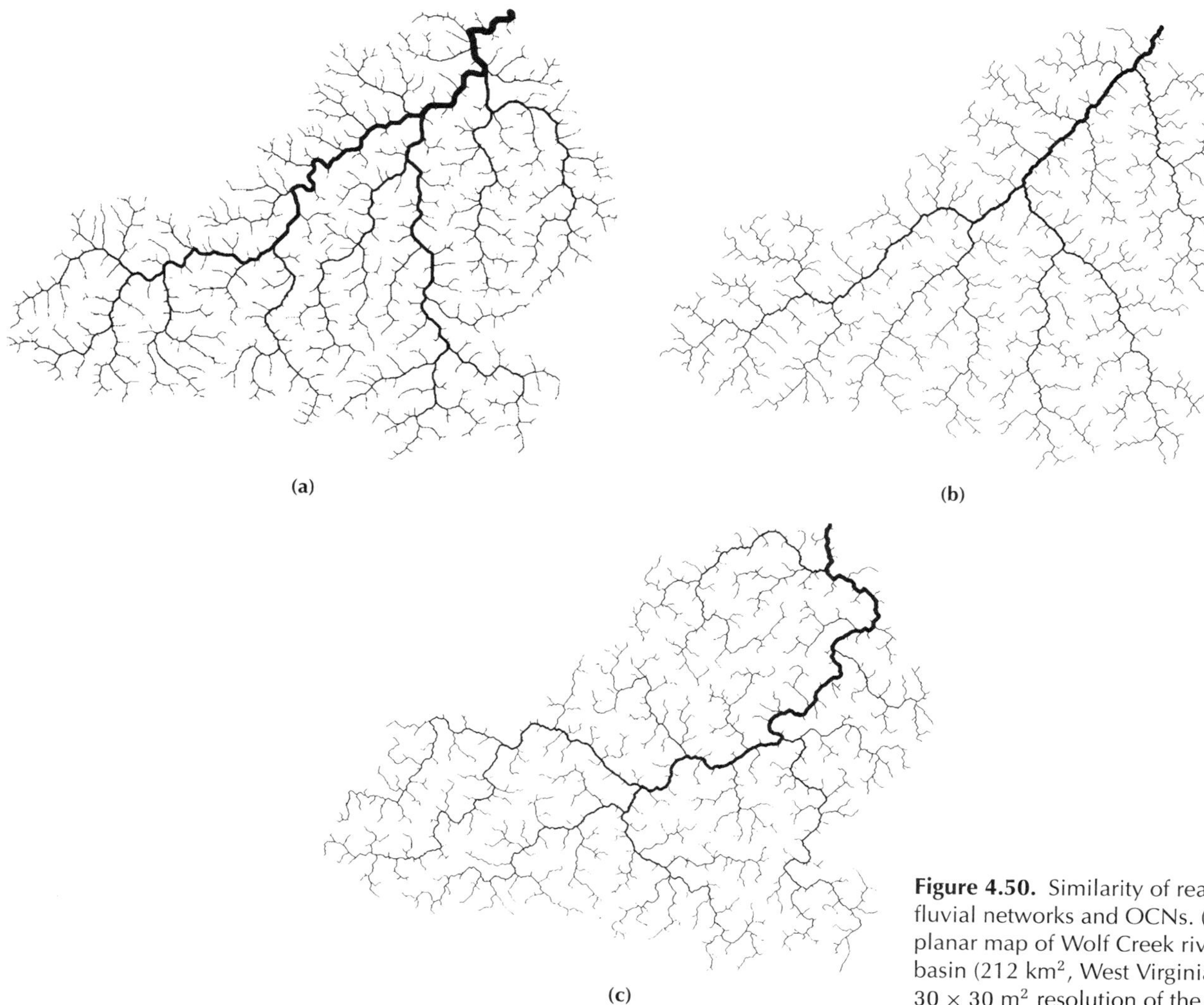

Figure 4.50. Similarity of real fluvial networks and OCNs. (a) A planar map of Wolf Creek river basin (212 km^2, West Virginia, 30×30 m^2 resolution of the digital elevation maps). Here the threshold area to support a channel head is made slope-dependent, and roughly equivalent to 50 pixels; (b) one example of cold (i.e., $T = 0$, Section 4.13) OCN obtained within the same domain of (a), that is characterized by no-flux boundary conditions except for the outlet, fixed in the same position as in (a) and plotted through the same support area; (c) from the same initial condition of (b), the OCN is kept in a 'hot bath' for a long time yielding a loss of memory of the initial condition. The OCN is then stepwise frozen at $T = 0$ up to convergence [after Rinaldo et al., 1996b].

in Figures 4.50(b) and (c), which suggests that the cooling schedule of temperatures might be related to geologic history. Nevertheless the capability of reproducing different small scale details maintaining the same aggregation structure is a remarkable feature complying with the concept of feasible optimality (Section 4.13). With reference to the debated issue of the discriminant character of the statistics adopted to compare branching patterns, once more we recall that a reliable test for comparison for large-scale structures like those in Figure 4.50 is the matching of all related scaling exponents (Section 2.10.1). We thus conclude that real rivers are feasible OCNs.

4.17 Hack's Relation and OCNs

Throughout this book we have encountered Hack's [1957] relationship, defined as

$$L \propto A^h \tag{4.71}$$

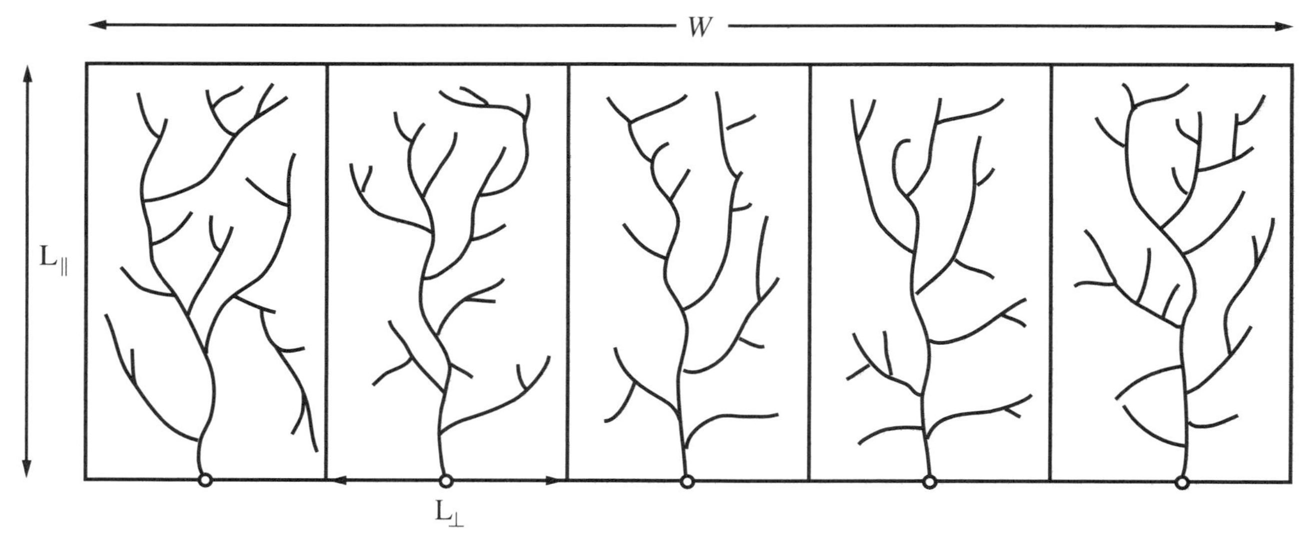

Figure 4.51. Setup for the OCN study of Hack's law [after Ijjasz-Vasquez et al., 1993b].

where L is length of the main stream channel (the longest stream measured to the drainage divide), A is the area of the basin, and h is the scaling (Hack's) exponent, commonly found in the range $h = 0.57 - 0.6$.

The above relationship, which essentially shows that basins tend to be more elongated as they increase in size, has been discussed for its geomorphological implications in Chapter 1 and for its interpretation in the context of fractal theories in Chapter 2. Following Ijjasz-Vasquez et al. [1993b], we now argue the possibility that elongation of river basins and the resulting geometrical relation, Eq. (4.71), result because fluvial systems search for a drainage configuration where energy expenditure is minimized. Thus we wish to explore the possibility that although the fractal structure of the drainage network is intimately linked to the minimization of energy expenditure, Hack's relationship may not be a direct consequence of the fractal sinuosity of river courses but rather of competition and minimization of energy in river basins.

In the previous sections we discussed OCN studies that focused on the structure of the optimal network draining a given area. The interaction of shape and size discussed in Section 4.15 showed that there are geometries that are more effective than others in relation to optimal energy expenditure. Ijjasz-Vasquez et al. [1993b] have showed that OCNs have implications for the boundary of river basins when a group of them is organized into a coherent system with the purpose of draining an area in an optimal way.

To investigate this point, one considers an area of width W and height $L_\parallel$ to be drained by OCNs constructed in subbasins of width $L_\perp$ and height $L_\parallel$ (Figure 4.51).

This configuration can be considered to be an idealization of tributaries to a main channel. The total energy expenditure for the entire area is $E = (W/L_\perp)E_w$, where E_w is the average, or typical, energy of the network in a subbasin of size $L_\perp L_\parallel$. The minimum value of E is obtained when the subbasins are such that $E_w/L_\perp$ is a minimum. In order to find this minimum, the values of total energy expenditure in OCNs constructed in areas of size $L_\perp L_\parallel$ with different values of $L_\perp$ have been calculated. The OCN

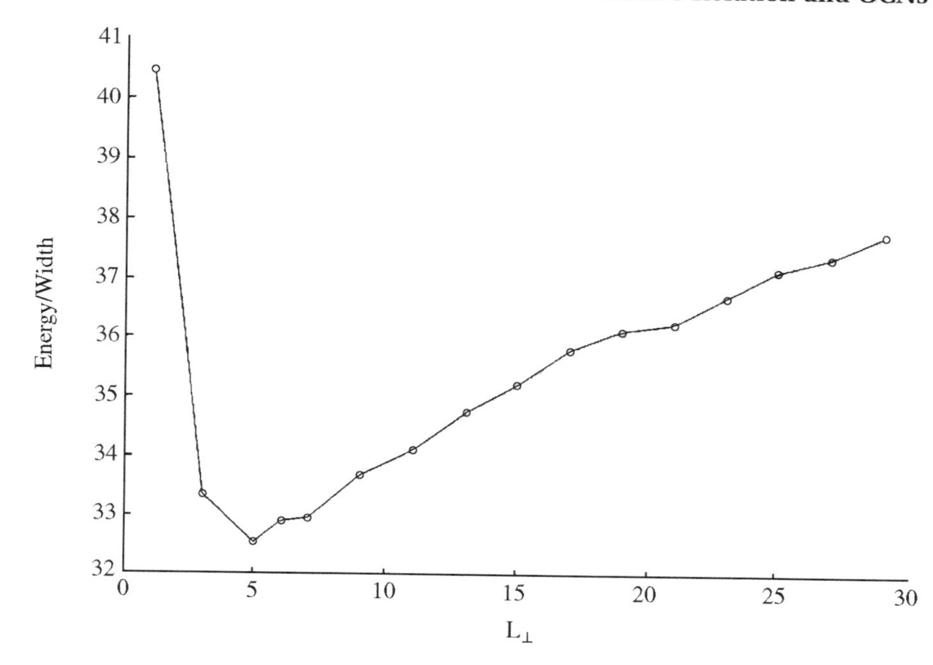

Figure 4.52. Energy per unit width ($E_w/L_\perp$) versus $L_\perp$ for $L_\parallel = 20$ pixels [after Ijjasz-Vasquez et al., 1993b].

with smallest value of $E_w/L_\perp$ yields the minimum total energy expenditure for the region and determines the optimal length/width ratio. Figure 4.52 shows a graph of the total energies per unit width as a function of the width itself for a particular value of $L_\parallel$.

A minimum is clearly identifiable at a width $L_\perp^{\text{opt}} \approx 7$ for $L_\parallel = 20$, both in pixel units. Similar minima were obtained in other similar graphs for different $L_\parallel$ studied by Ijjasz-Vasquez et al. [1993b].

Using the optimal width for a given $L_\parallel$, the relationship between length $L_\parallel$ and areas $A = L_\perp^{\text{opt}} L_\parallel$ can thus be studied in a framework that is coherent with optimality principles. The resulting scaling relationship, shown in Figure 4.53, is

$$L_\parallel \propto A^{0.57} \tag{4.72}$$

suggesting that in the search for optimal drainage configurations, basins elongate with size.

In the case of the previous derivation, the lengths used were from top to bottom of the subbasin and not along the mainstream channel, as has been the practice in the study of geometrical features of river basins. As seen in Section 2.9.1 with the aid of DEMs, the relationship between the longitudinal length $L_\parallel$ (measured by a straight line) and the mainstream length L (measured along the main channel from the outlet to the divide) is scaling, that is, $L \propto L_\parallel^{\phi_L}$ (Eq. (2.174)), with $\phi_L \sim 1.05$. At the intersection of every two links of the network the outlet of a subbasin is defined. For each subbasin the mainstream length L – measured along the main channel up to the boundary – was calculated and the longitudinal length $L_\parallel$ was also computed from that point in the boundary to the outlet. The resulting relationship in OCNs yields

$$L \propto L_\parallel^{1.04 \pm 0.01} \tag{4.73}$$

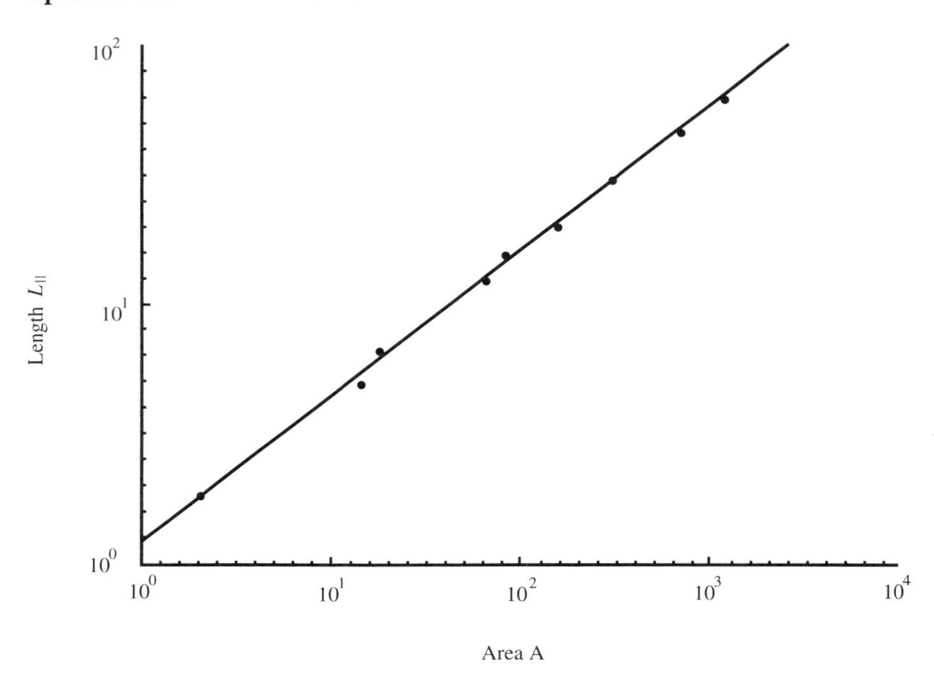

Figure 4.53. Hack's diagram for OCNs [after Ijjasz-Vasquez et al., 1993b].

Combining Eqs. (4.73) and (4.72) yields

$$L \propto A^{0.59} \tag{4.74}$$

for OCNs, a result that agrees with Hack's law.

4.18 Renormalization Groups for OCNs

The transformation due to coarse graining of the state of a given system is called renormalization in physics, and the goal of the so-called renormalization group is to study quantitatively the change that a physical quantity undergoes when the observations are taken under different degrees of coarse graining.

Fractals and renormalization procedures are intimately connected. In fact, a fractal can be defined as an object that is invariant under the transformation of a renormalization group [Takayasu, 1990].

The renormalization group method explicitly uses scale-invariance properties and has been central to many fundamental developments in physics. In the geophysical literature, it has applications for understanding the onset of fluid flow through a porous medium [e.g., Turcotte, 1992]. In fact, for fluid through porous formations the two measurable quantities are porosity (i.e., the degree to which void space may become filled with fluid) and permeability, defined as the ability of the fluid to flow through the porous formation under a pressure gradient of the fluid. Turcotte [1992] considered an idealized model that predicts the existence of a critical value of porosity beyond which an onset of permeability will appear. Problems associated with electrical conduction through a matrix of elements are essentially identical to the above "percolation" problem. A different application of the renormalization group method is to fragmentation theory [Allegre, Le Mouel, and Provost, 1982; Turcotte, 1992]. Still another geophysical application of the method deals with a fractal-tree model for the rupture of a fault [Smalley, Turcotte, and Sola, 1985; Turcotte, 1992], where the objective is to determine the probability of failure of a complex fault structure in terms of the probabilities of failure of its elements. In all the above problems the renormalization group approach is an idealization of actual geophysical processes; nevertheless, some of the results obtained are directly related to field observations. We will pursue a similar pattern here.

We will focus our interest in the behavior of energy dissipation of OCNs under coarse graining. Because optimal energy expenditure is the foundation of the OCN concept, its variation under a change in the scale of observation of the landscape is of considerable importance.

We consider the basin as a three-dimensional structure supporting an OCN. To simplify matters, let us think of a basin characterized by perfect scaling of slopes to flows (or their surrogated variables, total contributing areas), that is $\nabla z(A) \propto A^{-0.5}$ with $Var[\nabla z(A)] = 0$. This implies strict, not average, local optimality everywhere.

Notice that, in what follows, pixel lengths are a unit only at a fixed fundamental scale λ, and thus the relationship $\nabla z_i = \Delta z_i$ holds only within this same scale.

The following notation (a little involved, yet necessary for the details of the following calculations) will be employed (see Figure 4.54):

$$
\begin{aligned}
N &= A/\delta^2 \\
a &= \delta^2 \\
\bar{E} &= E/A \\
A_i &= N_i a
\end{aligned}
$$

where N is the total number of pixels making up the area of the OCN basin; A is the total area of the basin; a is the pixel area or the area ruler; δ is the length of the pixel side; E is the total energy dissipation $= k\sum_i A_i \nabla z_i L_i$; L_i is the link length; $\bar{E}$ is the energy dissipation for unit area $= E/(N\delta^2)$; A_i is the total drainage area at the (arbitrary) ith pixel, that is, $A_i = \sum_{j=1}^N W_{ij} A_j + 1$, where W_{ij} is the connection matrix defined in Section 4.7; and N_i is the number of pixels draining through the ith site.

If $\bar{z}$ is the mean elevation of the basin, then, using the same reasoning and with the same notation used in the derivation of Eq. (4.68), one has

$$
\begin{aligned}
\bar{z} &= \sum_{i=1}^{N} \frac{z_i}{N} = \frac{1}{N} \sum_{i=1}^{N} \sum_{j \in \gamma_i} \Delta z_j = \frac{1}{N} \sum_{i=1}^{N} N_i \Delta z_i \\
&= \frac{1}{NL^2} \sum_{i=1}^{N} A_i \Delta z_i = \frac{1}{NL^2} \sum_{i=1}^{N} A_i \nabla z_i L_i = \frac{1}{k} \bar{E}
\end{aligned}
\tag{4.75}
$$

where (Figure 4.54) i denotes different pixels, $j \in \gamma_i$ indexes all flowpaths from the outlet to i, ∇z_i is the local slope, L_i is the length associated with each link, and Δz_j represents the drops from pixel to pixel along the flowpath linking the ith pixel to the outlet.

Let us now group the pixels in squares of side $\lambda\delta$ so that i_λ is the new set of pixels generated. Our landscape is thus *coarse grained*. The new area ruler is $a^{(\lambda)} = \lambda^2 a = \lambda^2 \delta^2$; the new number of pixels is $N^{(\lambda)} = A/(\lambda^2 \delta^2)$.

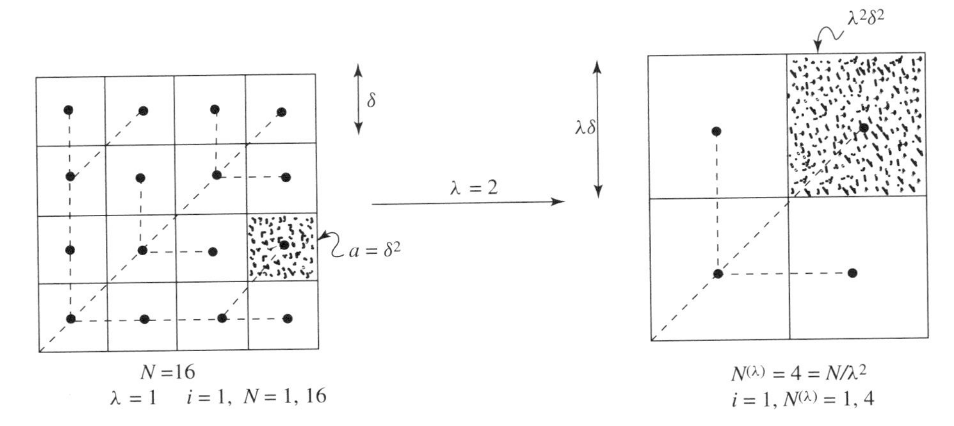

Figure 4.54. Geometric quantities for renormalization.

The new total area contributing to the ith link is thus $A_{i_\lambda}^{(\lambda)} = N_{i_\lambda}(\delta^2\lambda^2)$ (Figure 4.54). Here N_{i_λ} is the number of pixels connected by the network at the scale λ defining the contributing area at the ith link.

The renormalization now proceeds as follows. Starting from arbitrary initial conditions, an OCN is derived through the same procedure described in Section 4.5. The resulting OCN is described at a resolution of pixels of side length δ ($\lambda = 1$, that is, the fundamental structure subject to coarse graining). The three-dimensional structure of the OCN is assigned everywhere through the exact slope–area relationship $\Delta z_i(A_i) = A_i^{-0.5}$. After this the pixels are grouped in squares of side $\lambda\delta$, with $\lambda \geq 2$, so that from an initial number of pixels N one coarse grains the description of the terrain to a total number of pixels N/λ^2 (see Figure 4.54).

The elevation of each of the new larger pixels of side $\lambda\delta$, $\lambda \geq 2$, is computed as the average elevation of the λ^2 constituent pixels of side length δ that make up each particular larger pixel. From this coarse-grained three-dimensional landscape a new drainage network is drawn following as flow directions the lines of maximum steepness.

Although it is obvious that the above transformation preserves the mean elevation of the basin, we now prove it in order to become familiar with the notation being used. In fact, if

$$z_{i_\lambda} = \sum_{i \in i_\lambda} z_i/\lambda^2 \tag{4.76}$$

the renormalization rule implies

$$\bar{z}^{(\lambda)} = \sum_{i_\lambda=1}^{N^{(\lambda)}} \frac{z_{i_\lambda}}{N^{(\lambda)}} = \sum_{i_\lambda=1}^{N^{(\lambda)}} \sum_{i \in i_\lambda} \frac{z_i}{N^{(\lambda)}\lambda^2} = \sum_{i}^{N} \frac{z_i}{N} = \bar{z} \tag{4.77}$$

Hence the renormalized structure has the same mean elevation. Note that i_λ indexes the renormalized set of links, and therefore the notation $i \in i_\lambda$ indicates the arbitrary ith link that belongs to the set i_λ.

Eq. (4.75) holds also at this scale, and therefore one has

$$\bar{z}^{(\lambda)} = \frac{\bar{E}^{(\lambda)}}{k(\lambda)} \tag{4.78}$$

where in general $\bar{E} \neq \bar{E}^{(\lambda)}$ because the sum of total contributing drainage

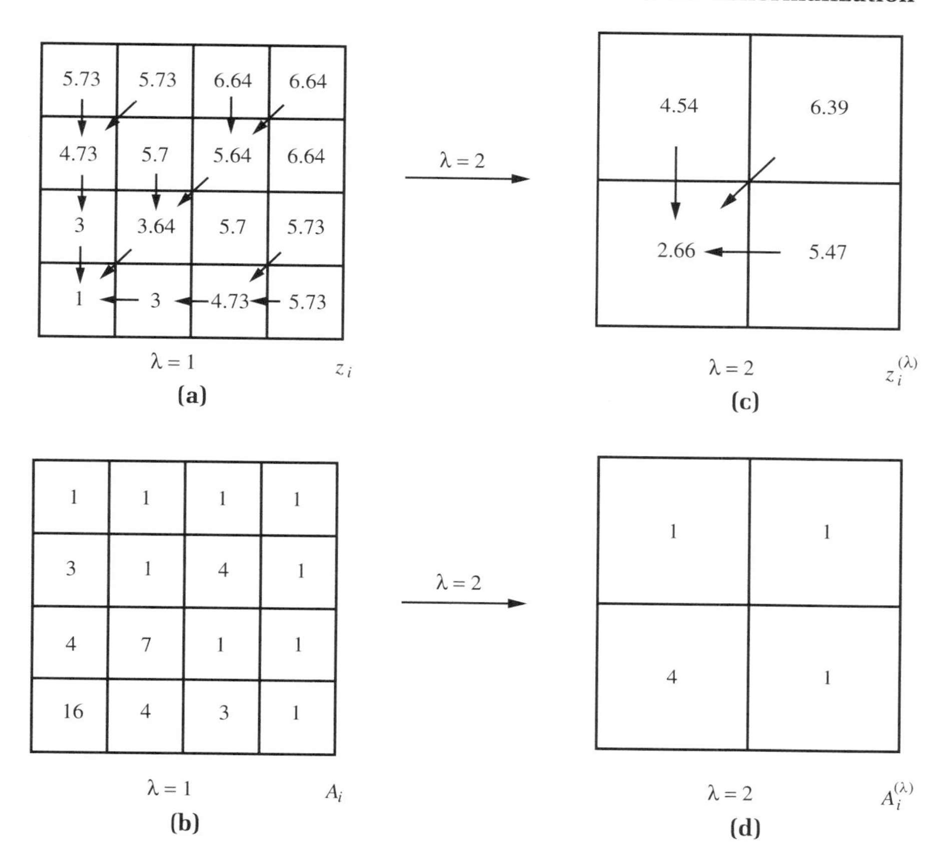

Figure 4.55. (a) Elevations of an OCN and drainage directions ($\lambda = 1$); (b) A_i ($\lambda = 1$); (c) as in (a), but for $\lambda = 2$; (d) as in (b) for but $\lambda = 2$.

areas changes under coarse graining. In fact, adding the total cumulative areas A_i in any region merged by the coarse graining procedure yields

$$A_{i_\lambda}^{(\lambda)} = N_{i_\lambda} a^{(\lambda)} = N_{i_\lambda} L^2 \lambda^2 \neq \sum_{i \in i_\lambda} A_i \tag{4.79}$$

as illustrated in Figure 4.55. Then it must be, from Eqs. (4.78) and (4.79)

$$k(\lambda) \neq k \tag{4.80}$$

Thus, on changing the scale, the coupling constant k must be renormalized by a suitable factor.

An example of an OCN progressively coarse grained according to the previous rules is shown in Figure 4.56.

A synthesis of the renormalized statistics is shown in Figures 4.57 and 4.58.

Most interesting is the fact that coarse graining induces slopes to scale imperfectly with total area, although their mean values $E[\nabla z(A)]$ scale correctly as $E[\nabla z(A)] \propto A^{-0.5}$. In other words, the perfect scaling imposed on the original condition ($\lambda = 1$), where $Var[\nabla z(A)]$ was zero for all areas, has now changed owing to coarse graining. Now $Var[\nabla z(A)] \neq 0$; furthermore, the variance induced in the slope–area relationship does not exhibit a scaling behavior. This fact has implications because we have seen in Section 1.2.10 that the slope–area relationship may have complex multiscaling characters. Aggregation, as implied by renormalization, seems to destroy

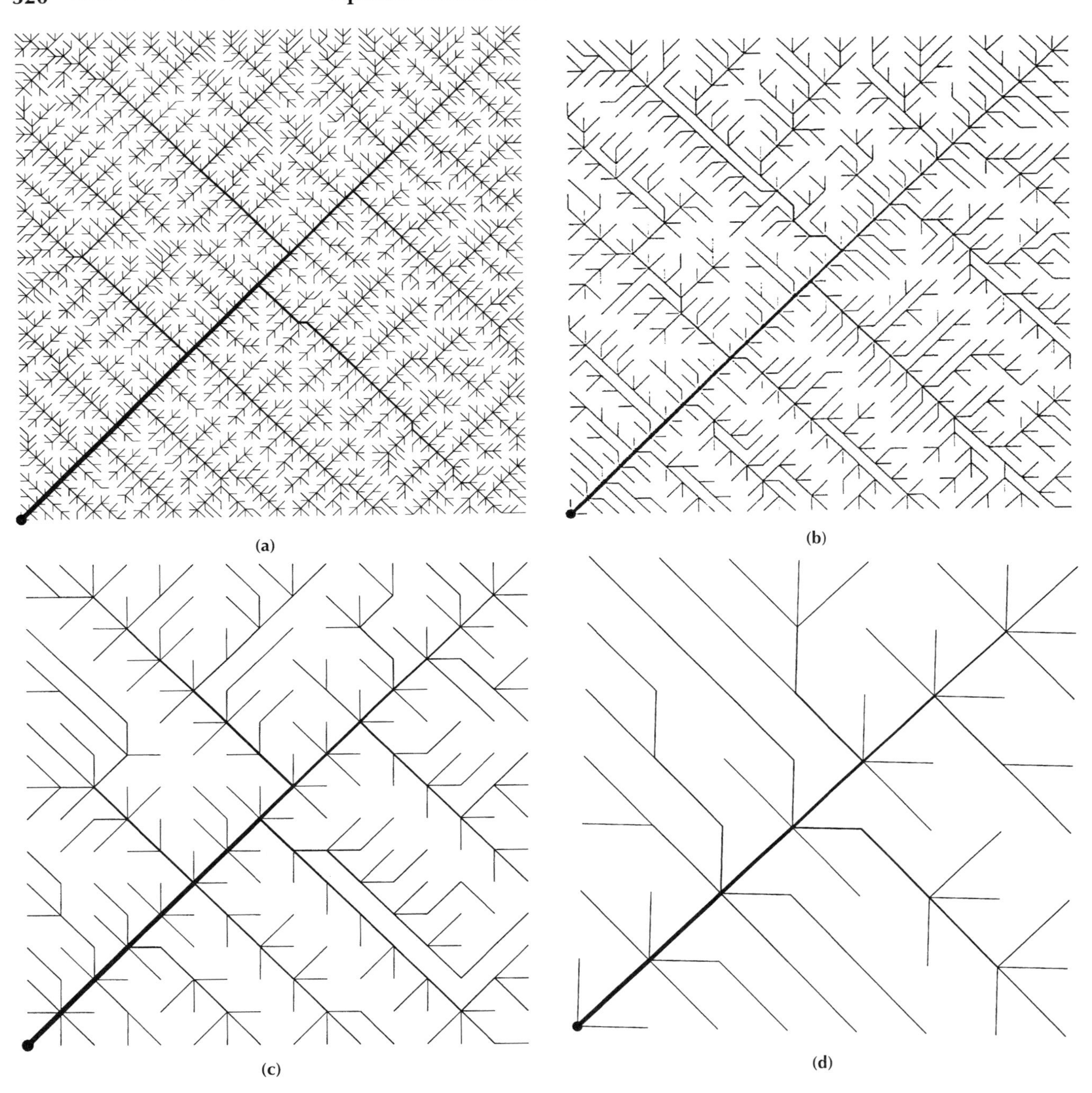

Figure 4.56. Renormalized OCNs with (a) $\lambda = 1$, (b) $\lambda = 2$, (c) $\lambda = 4$, and (d) $\lambda = 8$.

the scaling properties of moments higher than the first one. The removal of the scaling behavior in the higher moments because of aggregation suggests the use of least aggregated data (i.e., pixels of the smallest available size) for research purposes. It suffices here to observe that the above evidence suggests that the scaling properties of elevations in OCNs are respected in the mean sense of Rodriguez-Iturbe et al. [1992b] (see Section 4.6) rather than exactly. We will return to this issue when dealing with scaling and elevations in self-organized critical networks and OCNs in Chapter 5.

The experiments shown in the previous examples display a scaling of total energy expenditure as function of λ that accurately follows a power law irrespective of whether total energy is computed as $\sum_{i\in i_\lambda} A_{i_\lambda}^{0.5}$ or $\sum_{i\in i_\lambda} A_{i_\lambda}\Delta z_i$ (Figure 4.59). Notice that by construction it is only in the case where $\lambda = 1$ that one imposes the condition $\Delta z_i = A_i^{-0.5}$, and thus topological energy expenditure ($\sum_i A_i^{0.5}$) and the total energy expenditure ($\sum_i A_i \Delta z_i$) are equal. As shown by Figure 4.57 the scaling relationship $\Delta z_i = A_i^{-0.5}$ is far from exact after coarse graining, and thus we expect in general that $\sum_i A_i^{0.5} \neq \sum_i A_i \Delta z_i$.

Figure 4.59 shows that although coarse graining brings changes into the connectivity, which in turn modify the slope–area relationship imposed on the original landscape, statistically it still preserves the mean scaling rule, that is, $\langle \Delta z_i^{(\lambda)}(A)\rangle \propto A^{-0.5}$. In fact, the energy scaling relationship follows a power law:

$$\bar{E}^{(\lambda)} \propto \lambda^{-\alpha} \qquad \alpha \approx 1.1 \pm 0.01 \tag{4.81}$$

or equivalently

$$k(\lambda) = k\lambda^{\alpha} \tag{4.82}$$

The last relationship follows from the observation that at any scale λ the mean elevation is preserved and therefore we have

$$\frac{1}{k}\bar{E} = \frac{1}{k(\lambda)}\bar{E}^{(\lambda)} \tag{4.83}$$

If E_0 is the energy expenditure at $\lambda = 1$, we may write

$$E_0 = \lambda^{-\alpha}\bar{E}^{(\lambda)} \tag{4.84}$$

Thus the networks developing through OCN rules in regular isotropic lattices are suitable to renormalization of total energy.

More sophisticated questions may be posed, but we will confine our attention to an explanation of the scaling of renormalized energy.

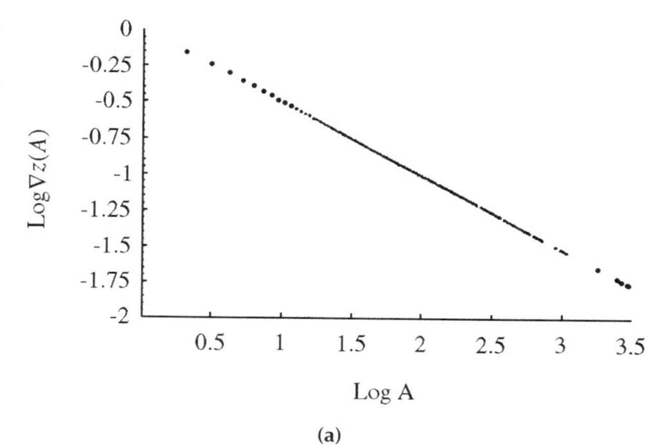

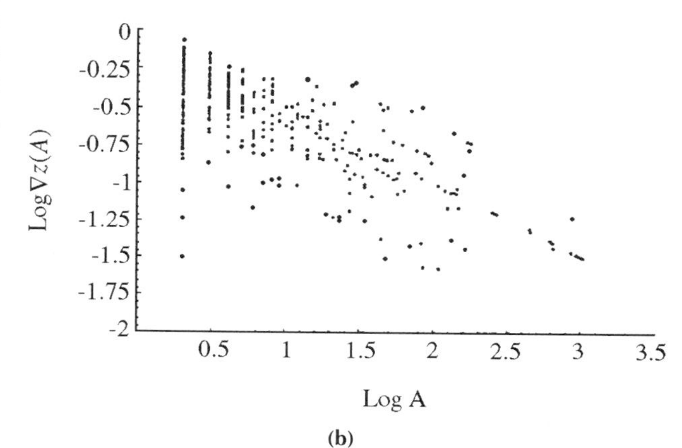

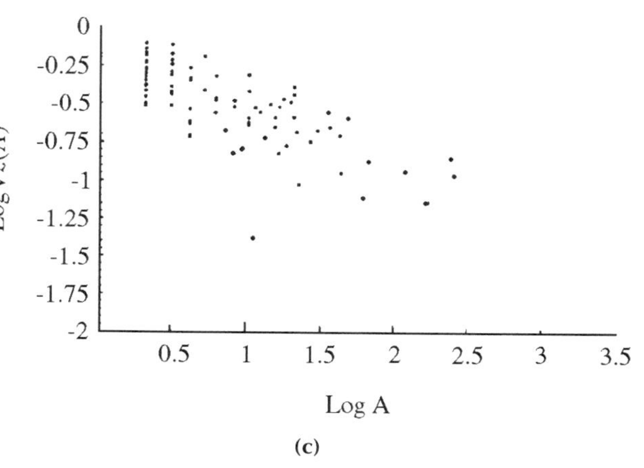

Figure 4.57. Slope–area scaling for renormalized OCNs: (a) $\lambda = 1$, slope is -0.5 and variance is zero; (b) $\lambda = 2$; (c) $\lambda = 4$.

The basic question is: Why does unit energy expenditure scale as in Eq. (4.84)? One approach to this question is to employ the probability of exceedence of total cumulative area to reproduce the observed scaling behavior. Let us rewrite the total energy expenditure $E = \sum_i P_i$ as follows (here $\lambda = 1$):

$$E = k\sum_i A_i^{0.5} L_i \propto N_{\max} \sum_{j=1}^{N_{\max}} A_j^{0.5} L_j \, p(A_j) \tag{4.85}$$

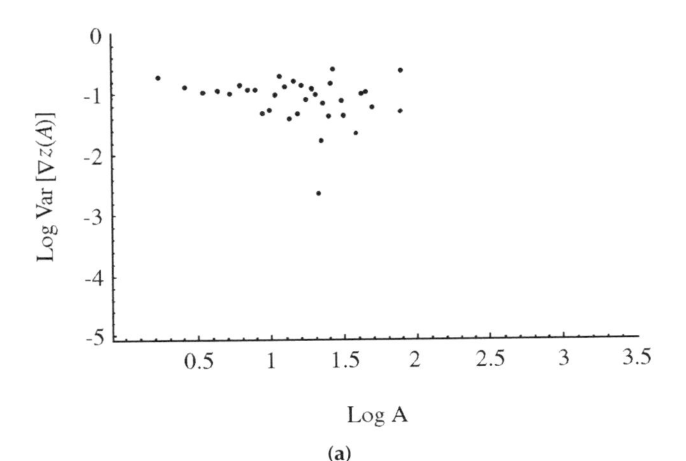

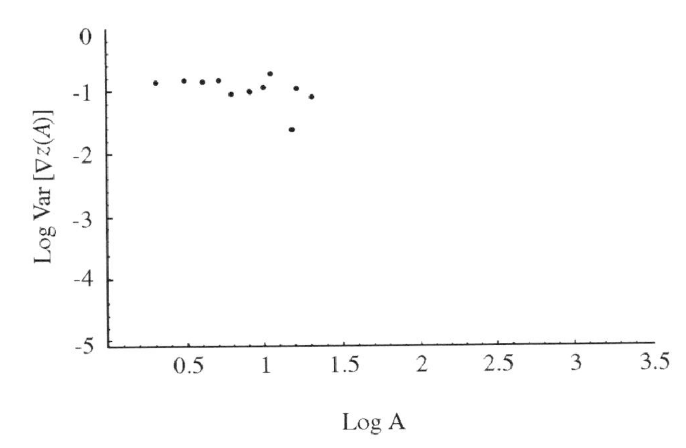

Figure 4.58. Variance of the slope–area scaling for renormalized OCNs: (a) $\lambda = 2$; (b) $\lambda = 4$.

where $N_{\max}$ is the total number of pixels and $p(A)$ is the pdf of cumulated area, that is,

$$N(A_i) = N_{\max} p(A_i) \tag{4.86}$$

where $N(A_i)$ is the number of pixels having area A_i.

We now assume that $p(A) \propto A^{-\tau}$ (Section 2.10.1), an assumption supported by experimental and theoretical evidence. Hence

$$E \propto N_{\max} \sum_{i=1}^{N_{\max}} A_i^{0.5-\tau} L_i \tag{4.87}$$

If we coarse grain via a parameter λ ($\lambda > 0$), the following quantities are generated: $N_{\max} \to N_{\max}^{\lambda} = N_{\max}/\lambda^2$; also, $L_i \to L_i^{(\lambda)} = \lambda L_i$, and $A_i \to A_i^{(\lambda)} = A_i/\lambda^2$. We therefore obtain

$$E^{(\lambda)} \propto N_{\max}^{(\lambda)} \sum_{i=1}^{N_{\max}^{(\lambda)}} (A_i^{(\lambda)})^{0.5-\tau} L_i^{(\lambda)} \tag{4.88}$$

In Eq. (4.88) we made the important assumption that the area distribution is invariant with respect to a change of scale, that is

$$p(A)^{(\lambda)} \sim p(A) \tag{4.89}$$

or $p(x)^{(\lambda)} \sim p(x)$. Figure 4.60 shows the experimental behavior of the probability of exceedence $P[A \geq a] \propto a^{-\beta}$ of A under different levels of aggregation λ. It also shows the validity of Eq. (4.89) within the limits of the cutoff imposed by the size of the system. We thus obtain

$$E^{(\lambda)} = \frac{N_{\max}}{\lambda^2} \lambda^{2(\tau-0.5)} \frac{1}{\lambda^2} \lambda \sum_{i=1}^{N_{\max}} (A_i)^{0.5-\tau} L_i \tag{4.90}$$

where all separate factors are shown for clarity. To obtain the above relationship one needs to use

$$\sum_{i=1}^{N_{\max}/\lambda^2} (A_i^{(\lambda)})^{0.5-\tau} L_i^{(\lambda)} = \frac{1}{\lambda^2} \sum_{i=1}^{N_{\max}} (A_i^{(\lambda)})^{0.5-\tau} L_i^{(\lambda)} \tag{4.91}$$

To prove Eq. (4.91), we observe that, if $I(x) = \int_0^x t^{-\gamma} dt$, coarse graining yields

$$I(x)^{(\lambda)} = I(x/\lambda^2) = \int_0^{x/\lambda^2} t^{-\gamma} dt = \lambda^{2(\gamma-1)} I(x) \tag{4.92}$$

Hence we obtain the final result

$$E^{(\lambda)} \propto \frac{\lambda^{2(\tau-0.5)}}{\lambda^3} \sum_{i=1}^{N_{\max}} (A_i)^{0.5-\tau} L_i = \lambda^{2(\tau-2)} E \tag{4.93}$$

defining analytically the scaling exponent of energy under coarse graining.

As seen in Section 2.5.2, $P[A \geq a] \propto a^{-\beta}$ with $\beta = 0.43 \pm 0.02$, and hence the related probability distribution function $p(A)$ (see Eq. (4.85)) is given by

$$p(A) \propto A^{-\tau} \qquad \tau = 1 + \beta = 1.43 \pm 0.02 \tag{4.94}$$

From Eq. (4.94) we have

$$E^{(\lambda)} \propto \lambda^{-\delta} E \tag{4.95}$$

with $\delta = 2(2 - \tau) \sim 1.1 - 1.2 > 0$. This derivation matches very well the results shown in the graphs of Figure 4.59. We will see in the next section that this result has important thermodynamic implications.

Finally if, as in the OCN computations, we compute $\hat{E} = \sum_i A_i^{0.5} = E/\delta$ (where δ is the constant link length), we obtain

$$\hat{E}^{(\lambda)} \propto \lambda^{-\delta-1} E \tag{4.96}$$

and the exponent is thus between 2.1 and 2.2.

All the above shows the statistical invariance of OCNs with the transformation group that preserves the mean elevation (and thus the total energy dissipation) of the system. This is important evidence of the fractal characters of minimum energy structures.

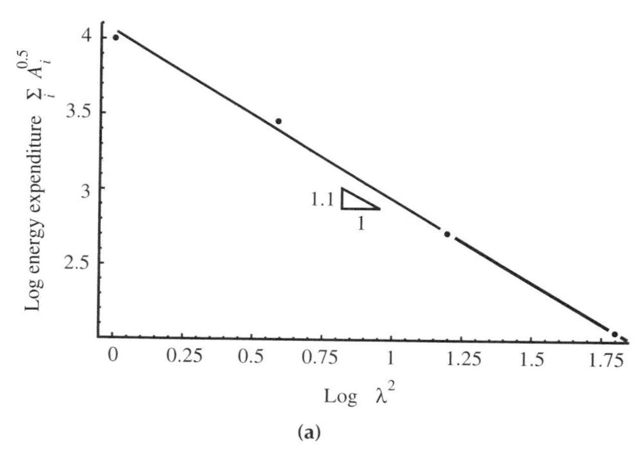

(a)

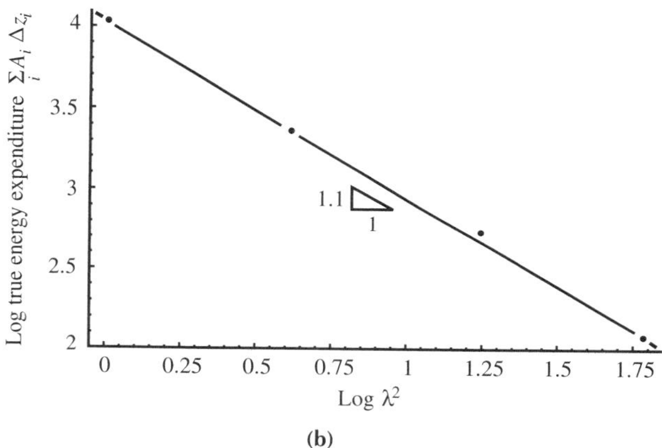

(b)

Figure 4.59. Scaling of topological total energy and true total energy with the coarse-graining parameter λ^2.

4.19 OCNs with Open Boundary Conditions

Sun, Meakin, and Jossang [1994a,b] have recently performed detailed experiments developing OCNs under the same theoretical framework devel-

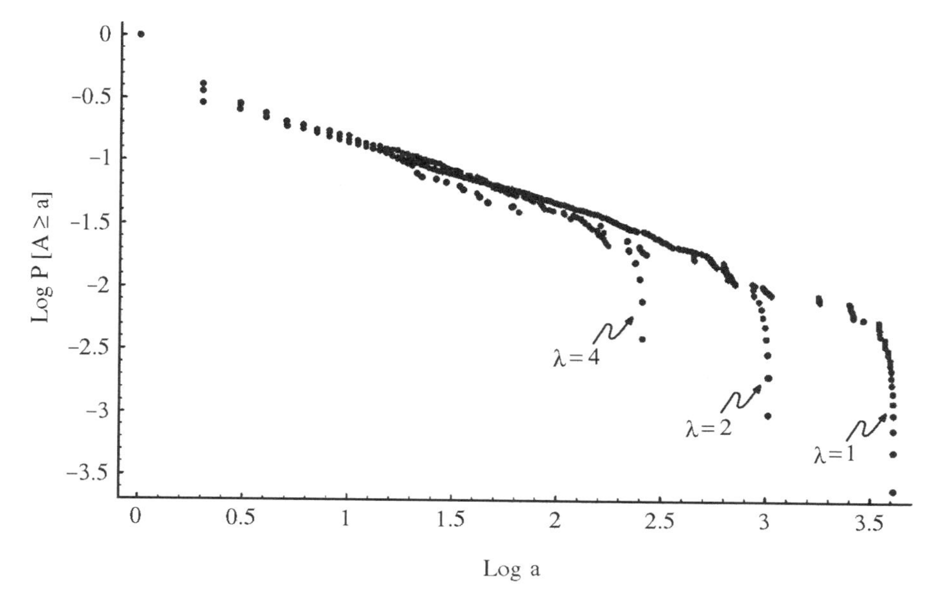

Figure 4.60. Probability of exceedence of total area A for different levels $\lambda = 1, 2, 4$ of coarse graining.

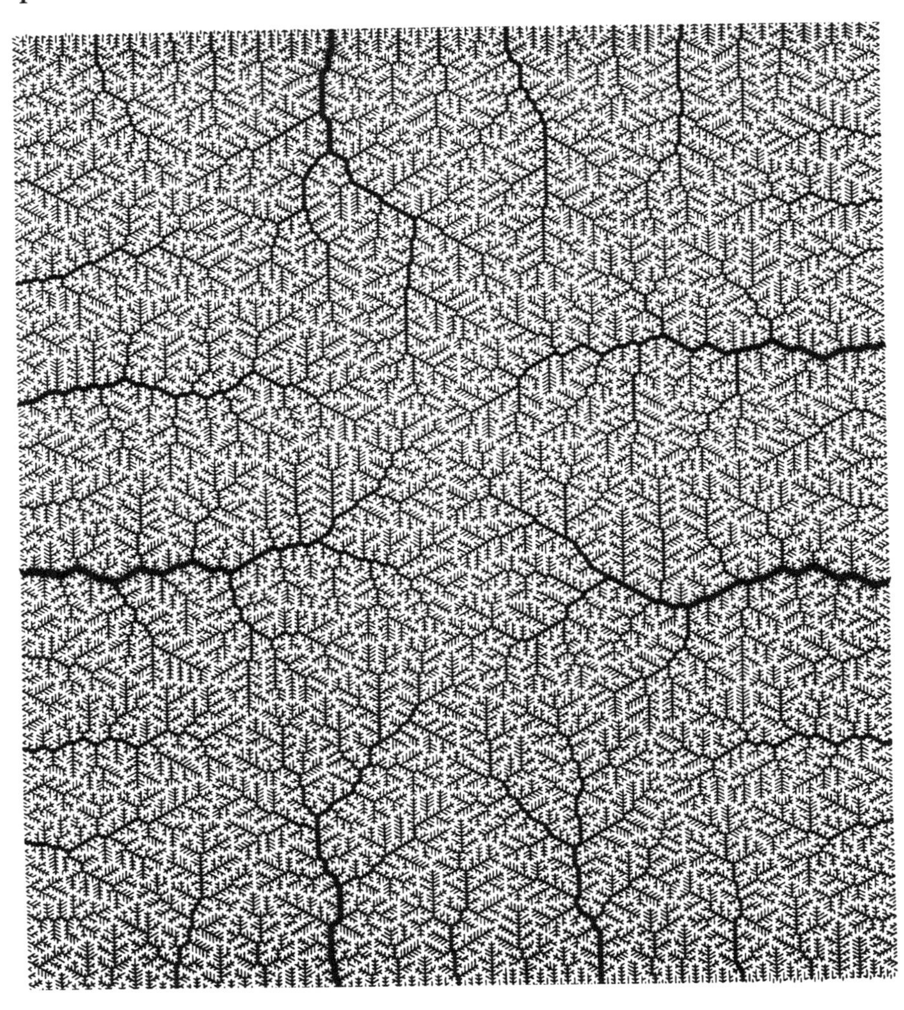

Figure 4.61. Example of OCNs with open boundary conditions in a 256 × 256 grid with a triangular lattice. The initial configuration was generated by a branched Eden growth [after Sun et al., 1994a].

oped earlier by Rodriguez-Iturbe et al. [1992b], Rinaldo et al. [1992], and Rigon et al. [1993]. Their experiments were conducted on large grids with a triangular lattice where an open boundary condition was used. Thus in the effort to simulate the processes of divide migration and competition for drainage each site on the boundary of the lattice has the possibility of being a channel network outlet.

An initial configuration is generated by performing a branched Eden growth simulation from the whole boundary of the lattice until the lattice is filled. From this configuration Sun et al. [1994a,b] effectively follow the same general problem of Rinaldo et al. [1992] using the Metropolis rule of simulated annealing as the optimization method (Section 4.7). The minimum energy expenditure type of configuration is now searched over the whole domain when many river basins have developed.

Figure 4.61 shows an example of the networks obtained, and Figure 4.62 shows the decrease of total energy dissipation with the number of annealing stages n.

The results of Sun et al. [1994a,b] match quite well those obtained by Rinaldo et al. [1992] and Rigon et al. [1993] in terms of the fractal structure

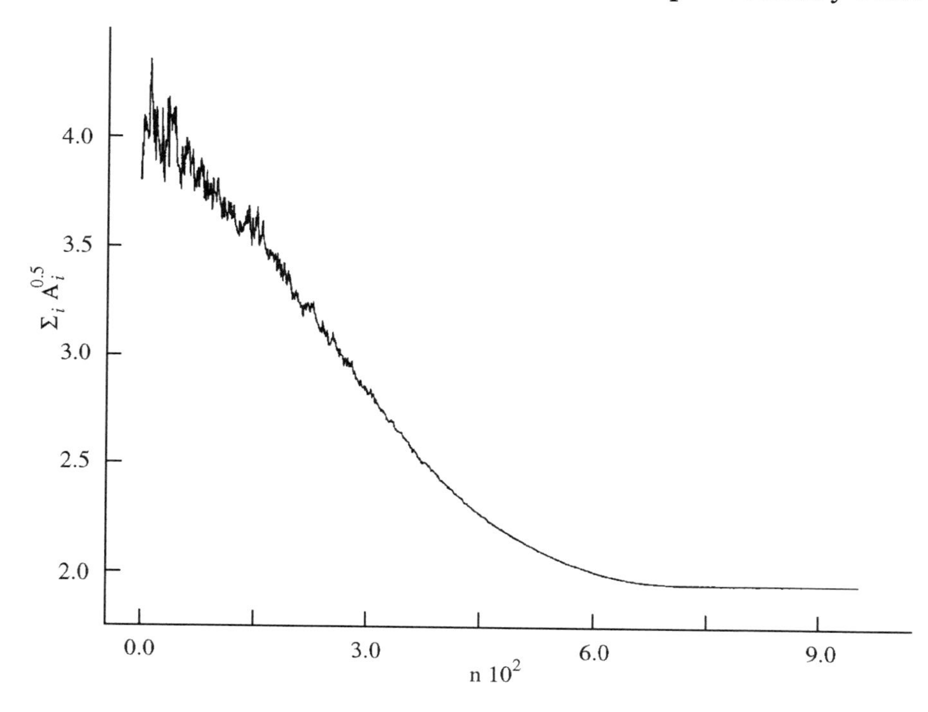

Figure 4.62. Decrease of total energy dissipation with the number of annealing stages n for OCNs with open boundary conditions [after Sun et al., 1994a].

of the obtained networks and the power laws describing the probabilistic structure of contributing drainage area at a point or the power expenditure distribution at a randomly chosen link. They were also able to study the fractal dimension of the perimeters of drainage basins enclosing the OCNs developed with the above procedure.

Sun et al. [1994a] point out that a difference between their simulations and real basins is the somewhat lower fractal dimension of the main channels of individual river basins. They obtain an effective fractal dimension of 1.03, lower than the values between 1.1 and 1.3 commonly accepted as representative of channel sinuosity of real river courses (Sections 2.2 and 2.7.7).

The simulations of OCNs with open boundary conditions confirm that the exponent γ in the energy expenditure function ($H_\gamma(s) = \sum_i A_i^\gamma$) has an important influence on the shape of the resulting basins [Sun et al., 1994b]. The basins obtained using large values of the parameter γ have much larger aspect ratios. The general structure of OCNs using different values of γ with open boundary conditions is also different.

Other experiments by Sun et al. [1994b] have dealt with the fractal structure of the landscapes obtained by imposing a perfect scaling of slopes $\nabla z(A) = A^{-0.5}$ on the planar structure of an OCN.

Recently, the open boundary OCN scheme has been employed within fractal boundaries in the search for the effects of the entire drainage area boundary on the drainage network [Sun, Meakin, and Jossang, 1995]. The results are briefly recalled here. Two types of boundaries were used for the outside boundary of the area to be drained, but here we reproduce only the case corresponding to the external perimeter of an invasion percolation cluster (Figure 4.63(a)) whose fractal dimension is $D = 4/3$. Figure 4.63(b)

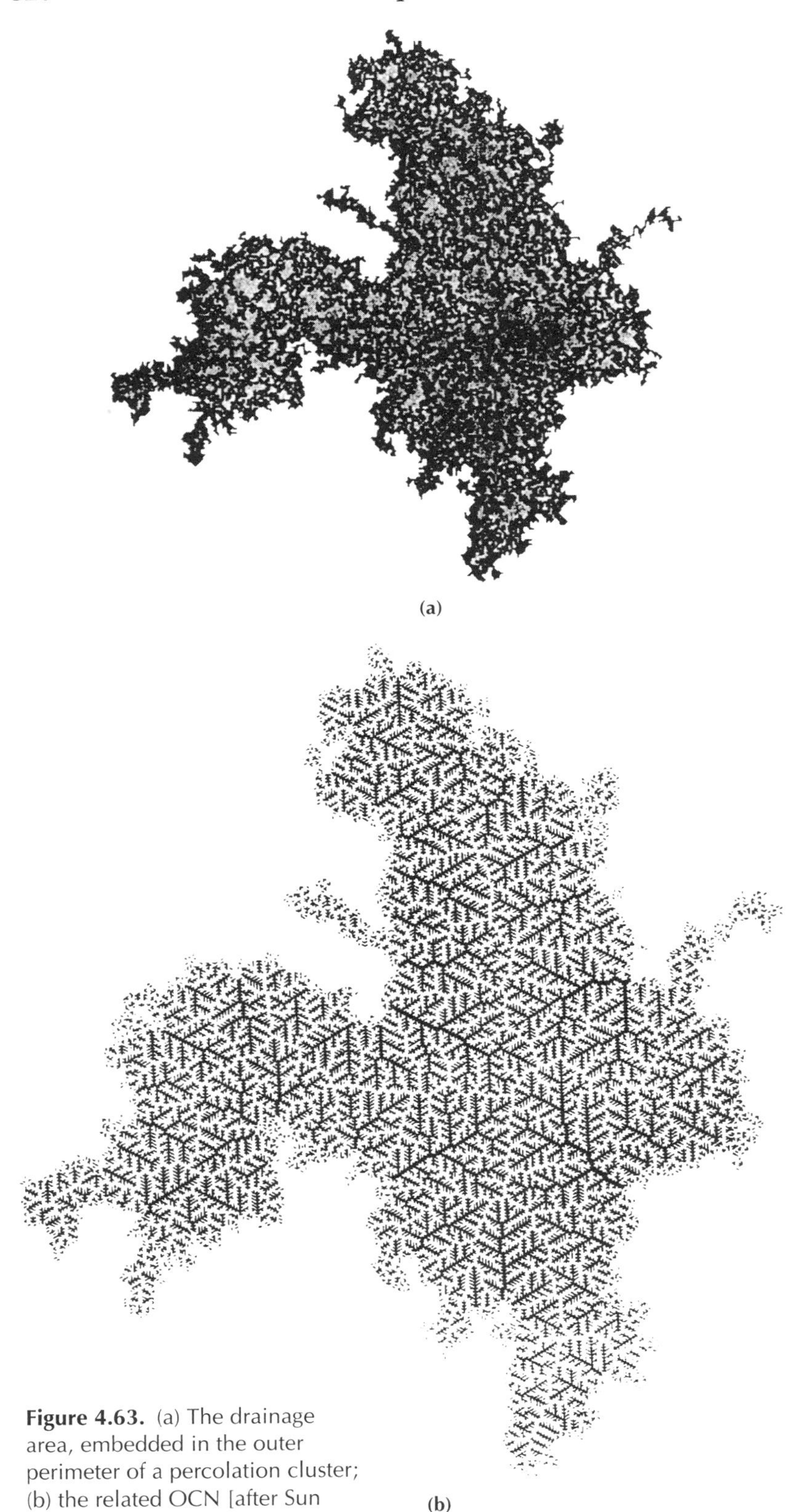

Figure 4.63. (a) The drainage area, embedded in the outer perimeter of a percolation cluster; (b) the related OCN [after Sun et al., 1995].

shows the open-boundary OCN grown within the fractal boundaries defined in this manner.

The whole area is drained by a population of river networks of different sizes. This partition of the drainage area, together with the structure within each network, minimizes total energy dissipation rate in the whole system. Individual boundaries are also fractal, as well as having many other structural characteristics. An important feature is the power law distribution of basin areas, say A^*, seeded in any outlet. Notice that the above distribution is different than the distribution of total contributing areas at any point in the basin in that the sum $\sum_j A_j^*$ over all outlet sites yields exactly the area enclosed by the fractal boundary, contrary to the sum of total contributing areas at any point that is not specified a priori. The resulting probability distribution of basin areas is

$$p(A^*) \propto (A^*)^{-1.67\pm0.02} \tag{4.97}$$

It is interesting to mention that square domains like those in Figure 4.61 yield a scaling exponent of 1.5 for $p(A^*)$.

Sun et al. [1995] argued that the scaling exponent in Eq. (4.97) is in fact equal to $1 + D/2$ where D is the fractal dimension of the boundaries. This can be explained by a scaling argument similar to that employed in Section 2.10.2. In fact, if a finite-size scaling is adopted for A^*, OCNs developing in the above systems are constrained by the plane-filling condition $\sum_i A_i^* \propto L^2$ (where i counts all basins issuing from the boundary) and obey the general scaling relationship $p(A^*, L) = (A^*)^{-\tau'} f(A^*/L^2)$. Thus (integrating as in Eq. (2.192)) $\sum_{x\in \text{boundary}} A_x^* \propto L^{D+(2-\tau')2}$ and we have, under the plane-filling condition, $D + (2 - \tau')2 = 2$ and therefore

$$\tau' = 1 + \frac{D}{2} \tag{4.98}$$

which is a relation similar to the ones established in Section 2.10.2 for real river networks. Sun et al. [1995] obtained $D = 4/3$ and $\tau' \sim 1.67$ for the patterns in Figure 4.63 whereas for those in Figure 4.61 the values are $D = 1$ and $\tau' = 1.5$.

In contrast, the results of Sun et al. [1995] also showed that other fractal dimensions (for example, of individual basin perimeters and of mainstreams, and the roughness of the resulting landscapes and contours) are independent of the fractal dimension of the entire drainage boundary. This suggests that some characters of OCNs may survive the imposition of 'geologic' constraints.

4.20 Disorder-Dominated OCNs

All previous studies have been restricted to 'homogeneous' OCNs, that is, those obtained by minimizing total energy dissipation rates where the proportionality constant in Eq. (4.13) is the same for every link, thus mimicking spatial uniformity. The scope of this section is to provide a preliminary approach for the incorporation of heterogeneity into OCN concepts and methods.

In the heterogeneous case the contributing area for the site i, A_i, can be written as

$$A_i = \sum_{j \in nn(i)} A_j W_{ij} + R_i \tag{4.99}$$

with symbols and notation as in Eq. (4.52). Recall that the meaning of the sum in Eq. (4.99) is that it adds only areas over bonds that input into i. The change with respect to Eq. (4.52) is that an injection term R_i denotes the uneven effects of rainfall at the ith site. Obviously, the above postulates the basic assumption $A_i \sim Q_i$, where Q_i is the landscape-forming discharge and Eq. (4.99) follows from continuity. R_i is commonly assumed by OCNs to be independent of i and equal to 1. The OCN is obtained by choosing the network that minimizes the energy functional

$$E = \sum_{i=1}^{N} k_i A_i^{\gamma} \tag{4.100}$$

where N is the total number of sites, k_i is a quantity (assumed to be independent of i and equal to 1 for a homogeneous OCN) related to the soil properties such as the erodibility, and γ is, as usual, an exponent that characterizes the physics of the erosional process.

Maritan et al. [1996b] have shown some properties of networks that minimize the functional (4.100) where A_i is given by Eq. (4.99) with $R_i \equiv 1$. In such cases, for $\gamma < 1$ the effects of disorder generated by heterogeneous values of k_i do not modify the features of the absolute minimum of E. Although – as discussed in Section 4.13 – OCNs do not attain the configurations of absolute minimum of E, this result suggests the robustness of the local minima to heterogeneity and thus underscores the outstanding ability of OCNs to reproduce real landforms for which heterogeneous forcings are unavoidable.

The above statement can be proved by explicitly analyzing the values of E for configurations yielding the absolute minimum, which obeys – by definition – the relationship:

$$\min E < \sum_{i \in T} k_i A_i^{\gamma} \tag{4.101}$$

where T is the tree that minimizes E in the case $\gamma = 1$. Thus one can write:

$$\min E < \left(\sum_{i \in T} A_i^{\gamma} \right) (\max k) \tag{4.102}$$

where $\max k$ is the maximum value of k_i. Notice that unless k_i has anomalously large fluctuations, $\max k$ is a real number that does not scale with the size of the system. Thus in this case, one expects the validity of the same scaling relationships that characterize homogeneous OCNs. Most important (see Section 4.21) is the scaling of minimum energy with the characteristic size, say L, of the OCN. Notice that in the rest of this chapter for simplicity we will not use the more involved notation using $L_\parallel$ and $L_\perp$ for the characteristic longitudinal and transverse size of the system. Thus for a square OCN, that is, with area L^2, in the minimum state one can write

$$\min E \propto L^2 \langle A^{\gamma} \rangle \propto L^{1+\delta(\gamma)} \tag{4.103}$$

where the average is taken through the area distribution $p(A, L) = A^{-\tau} f(A/L^{1+H})$, and $\delta(\gamma) > 0$ for $\gamma < 1$ (Section 2.10.1). Therefore Maritan et al. [1996] concluded that the exponents τ and H do not change in the presence of heterogeneity in the values of k_i.

Another exact result has been obtained by Maritan et al. [1996b] for OCNs characterized by $\gamma = 1$. It was shown that in this case even an infinitesimal disorder changes the features of the directed networks obtained by minimizing $H_1(s)$ (Eq. (4.53)), as one might expect intuitively.

We now turn to the case of random 'precipitation' R_i in Eq. (4.99) and a uniform value of $k_i \equiv 1$. Sites with area A_i have contributions from all values R_j from the j sites contributing to A_i, that is, for all $j \in \Gamma(i)$. Thus for large A_i, fluctuations in R_i that do not arise from power law distributions (yielding anomalously large fluctuations) would be expected to balance out, yielding OCNs close to he homogeneous case. Indeed [A. Maritan, personal communication, 1995] in the Scheidegger model (corresponding to a $\gamma = 1$ OCN), one can prove that this is the case. No proof is yet available for the general case $\gamma \neq 1$.

A limit case of interest is represented by disorder-dominated OCNs, that is, optimal networks grown in terrains where extreme fluctuations in k_i (Eq. 4.100) dominate the features of the aggregated structure regardless of the actual value of A_i.

Disorder-dominated river networks have been studied by Cieplak et al. [1996] as a particular case of Eq. (4.100) in the following framework. Consider a square lattice of size $L \times L$ where the links of the rivers are identified with the bonds of the lattice. Periodic boundary conditions are assumed

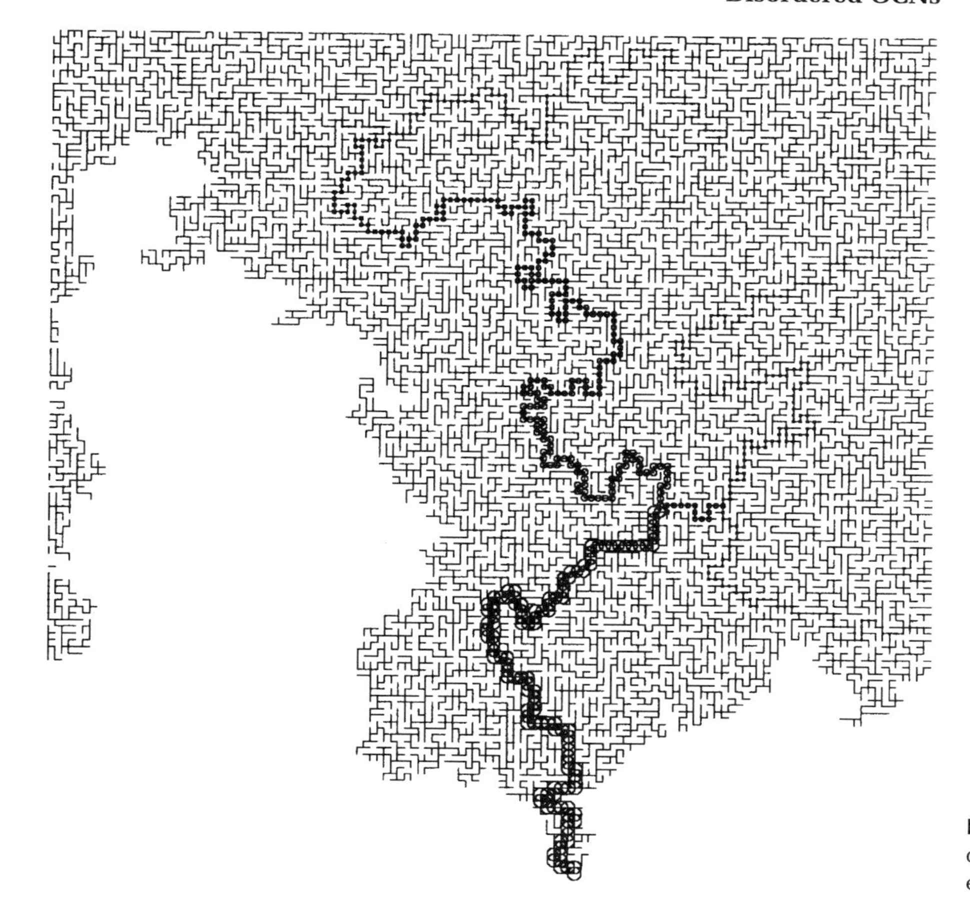

Figure 4.64. A disorder-dominated OCN [after Cieplak et al., 1996].

in the left–right direction. The bottom side of the square is defined to be the (fixed) outlet that collects the water that is flowing out. Independent random numbers in the range (0, 1) are assigned to the different bonds (i.e., links connecting the lattice of sites) representing the rank-ordering of the k_i's in Eq. (4.100).

Using invasion percolation ideas [Cieplak et al. 1996], the weakest link is selected and assumed to be a part of the network. The second-ranking weakest link is then selected and the procedure is repeated until all links are ranked in decreasing 'strength' k_i. The process is an iterative procedure progressively selecting links according to their rank and connecting them to a developing network until all sites are connected, that is, they all have a route to the outlet. Loops are excluded because once a given route is selected, alternative routes formed owing to the presence of a loop would be energetically unfavorable. Operationally, one obtains the network by incorporating the regions in order of increasing strength so that no loops are formed, yet all sites on the lattice are connected to the outlet.

A typical river obtained by this procedure is shown in Figure 4.64, in which we show the largest network grown within a 128×128 lattice where periodic boundary conditions are enforced on the sides only; that is, the bottom drains freely and the top boundary is impermeable. Notice that the size of the circles is drawn proportional to the current value of A.

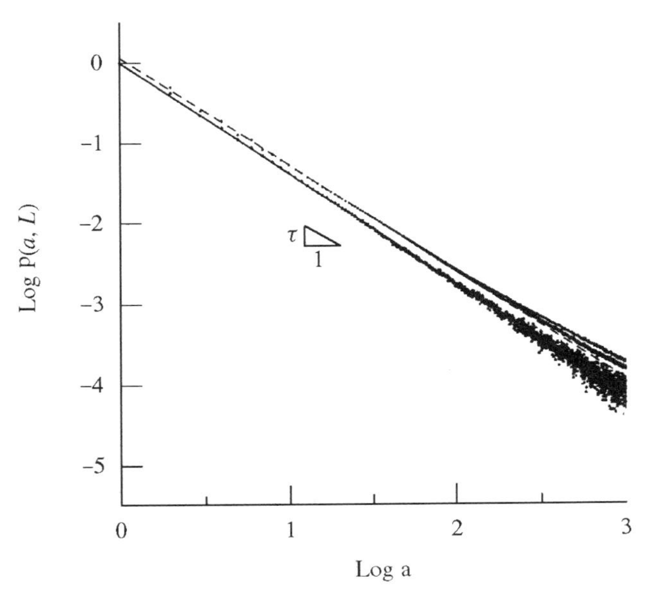

Figure 4.65. Plot of $p\,(a, L)$ versus a for disorder-dominated OCNs [after Cieplak et al., 1996].

Cieplak et al. [1996] have carried out detailed studies of the scaling properties of the resulting networks. A summary of the results of extensive numerical simulations is presented in Table 4.2 along with the results of observational data (Chapter 2). Note that the notation for the scaling exponents follows that employed in Section 2.10.1; that is, H is Hurst' exponent for the the relationship of the characteristic lengths (longitudinal and transverse) of the basin, $\tau = 1 + \beta$ is the scaling exponent of the probability distribution of total contributing area, h is Hack's exponent of the relation of mainstream length to total area, ϕ_L is the scaling exponent of mainstream length to the longitudinal size of the network, and ϱ is the scaling exponent of the distribution of upstream lengths from any point to the divide.

Figure 4.65 shows a double logarithmic plot of the total contributing area distribution in a basin of size L, $p\,(a, L)$ versus a. The lower set of data points corresponds to the ensemble average of 100 samples with $L = 128$. The solid line corresponds to a slope $\tau = 1.38$. The upper two data sets correspond to simulations of the Scheidegger model with different lattice sizes, and the broken line is the exact 4/3 slope.

Cieplak et al. [1996] demonstrate the excellent collapse of the $p\,(a, L)$ curves for different sizes for the values $\tau = 1.38$ and $H = 1$ as well as the finite-size scaling distribution of upstream lengths.

The experiments of Cieplak et al. [1996] may be viewed as a very particular case of OCNs where the fluctuations in erodibility, k_i, dominate the energy expenditure $\sum_i k_i A_i^\gamma$ to the point that the choices of different aggregated areas lose importance. Thus the minimized functional is $\sum_i k_i$. If this is the case, the sum of the k_is along any path is essentially equal to the largest k_i encountered. This equivalence derives from the properties, for instance, of an algebraic tail for the distribution of k_i, for which anomalously large values are not prohibitively unlikely. Thus the definition of disorder-dominated OCNs is appropriate.

The scaling coefficients differ from those observed in nature in a somewhat coordinated fashion. The overall sinuosity of the main patterns is larger, and the directedness of the drainage paths is considerably diminished.

Table 4.2. *Scaling exponents for disorder-dominated networks*

	Measured	River Basins
ϕ_L	1.22 ± 0.04	1.1 ± 0.02
H	1	$0.75 - 0.80$
τ	1.38 ± 0.03	1.43 ± 0.02
h	0.62 ± 0.02	$0.57 - 0.60$
ϱ	1.60 ± 0.05	$1.8 - 1.9$

Source: Cieplak et al. [1996]

The case in which correlated spatial heterogeneity dominates the distributed injection (i.e., R_j) has not yet been studied in detail. This would correspond to assuming that the flow rate at the ith link is

$$Q_i \sim \sum_{j \in \Gamma(i)} R_j \qquad (4.104)$$

where $\Gamma(i)$ indexes the upstream connected paths leading to i. Notice that if $R_i \equiv 1$, then the ordinary case where $Q_i \propto A_i$ is recovered. In the case of a stationary random field R_j with mean $\langle R \rangle$, we define a functional $H^R(s)$ as

$$H^R(s) = \sum_j \langle R \rangle_j A_j^{0.5} \tag{4.105}$$

(where $\langle R \rangle_j$ means the average taken over all sites) which suggests that the mild heterogeneity case decays to the ordinary OCN. This can be seen, for instance, for uncorrelated lognormal R_j's, that is, $R_j = \langle R \rangle e^{Y_j} \geq 0$ with $Y_j \in N(0, \sigma^2)$. In this case, one has $H^R(s) \propto e^{\sigma^2} H(s)$ and the homogeneous case is recovered as $\sigma^2 \to 0$.

Notice also that the strong disorder limit could have an interesting finite-size effect. In fact, if the distribution of the random variable R is a power law, that is, $p(R) \propto R^{-a}$, then one can determine the scaling properties of the known maximum value, say $R_{\max}$, of R drawn from a finite sample of size A_j. Hence one has

$$\left(\int_{R_{\max}}^{\infty} R^{-a} \, dR \right) A_j = 1 \tag{4.106}$$

and thus $R_{\max} \sim A_j^{1/(a-1)}$ [A. Maritan, personal communication, 1995]. Hence

$$\langle R \rangle_j = \int_1^{R_{\max}} R/R^a \, dR \propto A_j^{(2-a)/(a-1)} \tag{4.107}$$

and

$$H^R(s) \sim \sum_j A_j^{f(a)} \tag{4.108}$$

with $f(a) = 0.5 + (2-a)/(a+1)$. Therefore certain heterogeneity structures might have an effect similar to that of a perturbation of the exponent γ in the general form of the functional $H_\gamma(s)$.

We thus conclude that heterogeneous OCNs have interesting properties and that much research is expected in this area in the near future. From the preliminary results, the robustness of fractal properties to the presence of heterogeneity emerges. Thus the connection of optimality of functionals of energetic character with the growth of fractal structures is reinforced and looks promising for the establishment of a general framework for the dynamics of fractal growth.

4.21 Thermodynamics of OCNs

Central to the models of topologically random networks discussed in Section 1.3.3 is the assumption of equal likelihood of any possible tree-like configuration. However, the foundational principles for OCNs described in Section 4.3 postulate that some spanning, loopless network configurations are more likely than others, their overall likelihood being ruled by least total energy expenditure of the network structure as a whole and in its parts (recall that the set of possible configurations for the system is composed of the ensemble of all rooted loopless trees spanning a given lattice of sites defined by a complete set of oriented links among connected neighbor sites). In this section we investigate, following Rinaldo et al. [1996a], the thermodynamic rationale behind the scaling properties of the energy and entropy of OCNs.

We start by assigning a probability $P(s)$ to each particular spanning tree configuration, say s. We take $P(s)$ as a Boltzmann distribution, that is

$$P(s) \propto e^{-H(s)/T} \tag{4.109}$$

where T is a suitable parameter resembling Gibbs' temperature of ordinary thermodynamic systems and the functional $H(s)$ is the Hamiltonian of the system, that is, a global property related to energetic characters. Notice that the random topology model (Section 1.3.3), where all spanning trees are equally likely, is the limit case of the model described by Eq. (4.109) for $T \to \infty$. OCNs also belong to the class of configurations described by Eq. (4.109) where $H(s)$ is the total energy dissipation of the spanning tree configuration. In fact, OCNs are obtained by selecting the spanning network configurations, s, that maximize the probability in Eq. (4.109) by minimizing the Hamiltonian of the system, defined as

$$H_\gamma(s) = \sum_{i=1}^{L^2} A_i^\gamma \tag{4.110}$$

A slightly different notation from that in the previous sections of this chapter has been adopted for reasons that are apparent in the following, namely, i spans the L^2 sites occupied by, say, a $L \times L$ square lattice; $\gamma \sim 0.5$ is the suitable exponent describing the physics of the erosional process; and A_i is the total contributing area at the arbitrary ith site of the lattice. The outlet is taken in one of the corners of the $L \times L$ lattice and its site, say 0, therefore has $A_0 = L^2$. Thus, in principle, OCNs represent the other extreme of topologically random networks, because they constitute the maximum probability case for $T \to 0$ (recall the Metropolis algorithm for OCN determination discussed in Section 4.13). One wonders what is to be expected in ordinary thermodynamic settings characterized by a finite value of T. We will show that OCNs concepts work at any finite temperature, that is, energy minimization always maximizes the probability of a configuration, provided the network is large enough.

Before analyzing the thermodynamic implications of the Hamiltonian of OCNs, it is instructive to review a related model. Troutman and Karlinger [1992] assumed the functional $H_1(s)$ to be the difference between the sum of the flow distances of every site to the outlet computed through the drainage tree and the total flow distance to the outlet along the shortest path along the lattice. As seen in another context in Section 2.10.1, $H_1(s)$ is a measure of the sinuosity of the channels constituting the channel tree and is proportional to the mean distance to the outlet $\langle \ell \rangle$ of all sites of the lattice, such distance being measured – as for the width function $W(x)$ – along the drainage directions. The idea of finding a global measure of the likelihood of a network configuration is very suggestive, in particular because the concept of distance from a site to the outlet incorporates the structure of the three-dimensional field of elevations that controls drainage directions. Troutman and Karlinger [1992] observed that for a spanning tree s one may write $\langle \ell \rangle \propto \sum_{i \in s} \sum_{j \in x(i)} \Delta x_{ij}$ where $x(i)$ is the path along the network from i to the outlet and Δx_{ij} is the spatial step or the lattice size, taken as unit in unbiased lattices. Rearranging the sum as in Eq. (4.68) is allowed because the term Δx_{ij} is added A_i times for each site i, that is, as many times as the contributing sites upstream of i. Thus $\langle \ell \rangle \propto \sum_{j=1}^{N} A_j - 1$ for the N sites and

$$H_1(s) \propto \sum_{j \in s} A_j \tag{4.111}$$

for the configuration s where A_j is the total contributing area at the arbitrary site j, and j spans all sites within the tree s.

Troutman and Karlinger [1994] later adopted a functional form precisely like Eq. (4.110) where γ was kept as a parameter to be estimated from data. They analyzed various basins and subbasins of relatively small sizes and concluded that the observational value of γ may be quite different from 0.5. We believe that this result is much affected by the data chosen. At small scales, like the ones studied by Troutman and Karlinger [1994], convex topographies and nonaggregated patterns are commonplace and the interplay of hillslopes, valleys, and channeled parts of the landscape is dominant. Hence they certainly affect the configurations of the spanning trees and subbasins of different size. OCNs address the fluvial landscape, and thus it is likely that the particular identification procedure employed has produced spurious results.

We now turn to the scaling properties of OCNs. We recall from Section 4.7 that OCNs are obtained by selecting the optimal configuration through a traveling salesman-like algorithm starting from random initial conditions. The algorithm proceeds selecting a site i at random and perturbing the configuration ($s \to s'$) by locally assigning a change in the matrix of connections δW_{ij}. This prompts a rearrangement of A_is (all areas formerly and currently linked to i are modified). The change is accepted if $H_{1/2}(s') < H_{1/2}(s)$. The procedure stops after a prefixed number of changes are rejected. Figure 4.66 shows a 256×256 example of an OCN to be used for a specific comparison in the following.

As seen in Section 4.7, the structure of the configuration minimizing the functional H_γ strongly depends on the parameter γ but at the same time is robust to changes in γ within a certain range. We can distinguish different behaviors: For $\gamma < 0$, OCNs are attained by the so-called Hamiltonian paths, that is, the spiral-like pattern of Figure 4.1(a), in which only one stream drains all the basin area. For $\gamma > 1$ the patterns minimizing Eq. (4.110) are such that the average length of the path from each site to the outlet is the shortest (i.e., the explosion paths of Figure 4.1(c)). The range $0 < \gamma < 1$ is more interesting, and in this region the system exhibits rich scaling structures and aggregation patterns. We now analyze the above schemes in a more general manner.

For a given drainage basin B overlain with a lattice of L^2 sites, let $\mathcal{S}$ be the set of spanning loopless trees rooted in a given point, say 0. For any configurations $s \in \mathcal{S}$ we define the probability of the tree s as in Eq. (4.109) where T^{-1} is Gibbs' parameter mimicking the inverse of temperature of classic thermodynamic systems [e.g., Huang, 1963]. For a fixed γ, let $H_\gamma(\mathcal{S})$ denote the finite set of all possible values that may be taken on by $H_\gamma(s)$ for trees $s \in \mathcal{S}$. Given an energy level, say

$$E \in H_\gamma(\mathcal{S}) \tag{4.112}$$

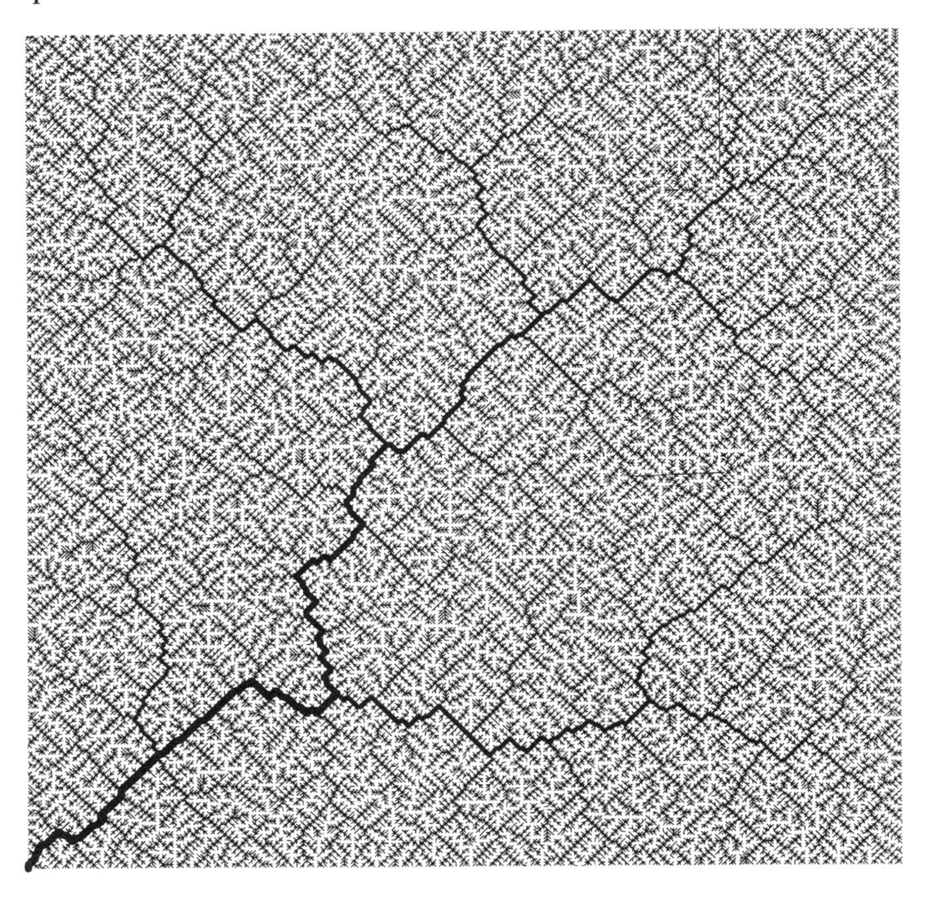

Figure 4.66. An example of a 256×256 OCN ($L = 128$) [after Rinaldo et al., 1996a].

let $N(E)$ be the degeneracy, that is, the number of different spanning trees s for which $H_\gamma(s) = E$. One therefore obtains

$$P(H_\gamma(s) = E) = \sum_{s:H_\gamma(s)=E} P(s) \propto N(E) \exp(-E/T) \tag{4.113}$$

Formally defining the thermodynamic entropy as

$$\sigma(E) = \log N(E) \tag{4.114}$$

one obtains

$$P(H_\gamma(s) = E) \propto e^{-F(E)/T} \tag{4.115}$$

where a formal free energy

$$F(E) = E - T\sigma(E) \tag{4.116}$$

has been introduced. Indeed, the most probable states correspond to an energy E that minimizes $F(E)$.

Notice that the development of the entropy concept has not taken into account the distinction between equilibrium and nonequilibrium thermodynamic processes. Here it suffices to remark, following Glansdorff and Prigogine [1971], that we address the tendency of dynamic processes with many degrees of freedom to develop stable states poised at criticality, that is, scale-free. Hence our definition of the Hamiltonian is consistent because Eq. (4.110) has the meaning of total energy dissipation.

Our central result is that entropy, σ, scales subdominantly with system size L compared to the energy, so that even for a nonzero value of Gibbs' parameter the most probable spanning tree configurations determined by minimizing the free energy can be equally well obtained by minimizing the energy, provided that L is large enough. As a result, in the thermodynamic limit the system described by the probability (4.115) always tends to operate at zero temperature; that is, the total energy in Eq. (4.110) is minimized. The proof has been given by Rinaldo et al. [1996a] and is reported here in some detail.

We first show the exact result that for the set $s \in \mathcal{S}_E$ of OCNs one has

$$E = \min H_\gamma(s) \propto L^{2+\delta} \tag{4.117}$$

with $\delta > 0$ for $\gamma \geq 1/2$.

The easiest case to prove corresponds to $\gamma = 1$ in Eq. (4.110). For a given spanning tree s we have seen (see, e.g., Eq. (4.111)) that

$$H_1(s) = L^2 \langle \ell \rangle \tag{4.118}$$

where $\langle \ell \rangle$ is the average distance of the L^2 sites from the outlet measured along the links belonging to s. Thus the minimum of $H_1(s)$ is attained by the set $\mathcal{S}_D$ of all directed spanning trees (DST), that is, the trees whose links have positive projection on the diagonal oriented in the outlet direction. DST have $\langle \ell \rangle \propto L$, that is, $H_1(\mathcal{S}_D) = \min_s H_1(s) \propto L^3$. Using this result and Schwarz' inequality, it was shown [Rinaldo et al., 1996a] that for $\gamma \in (1/2, 1]$ and any $s \in \mathcal{S}$

$$H_\gamma(s) \geq \text{ const } L^{1+2\gamma} \tag{4.119}$$

where the constant depends on lattice properties. Thus the above inequality implies that

$$\min H_\gamma(s) > \text{const } L^{2+\delta(\gamma)} \tag{4.120}$$

with $\delta > 0$ for $\gamma > 1/2$. For $\gamma = 1/2$, the global minimum has a different lower bound, that is, $E \propto L^2 \log L$, characterized by a logarithmic correction. This can be proven by explicitly constructing classes of structures that scale with this value of δ in Eq. (4.117). Nevertheless, as seen below, numerical and theoretical results for the $\gamma = 1/2$ case suggest that

$$\min H_{1/2}(\mathcal{S}_E) \propto L^{2+\delta(1/2)} \tag{4.121}$$

(where $\mathcal{S}_E$ denotes the set obtained by averaging over accessible local minima) with $\delta(1/2) = 0.1 - 0.2$. This result is consistent with a few results seen in this chapter. In fact, from the statistical properties of total contributing areas A, we recall that the probability of exceedence $P[A \geq a]$ has a density $p(a, l) = a^{1+\beta} f(a/L^\phi)$ where $\beta = 0.43 \pm 0.02$ and the finite size effect is defined by a coefficient $\phi \sim 1.8$. Then one has

$$\min_s H_{1/2} = L^2 \langle A^{1/2} \rangle = \int_1^\infty dA \quad A^{-1/2-\beta} f(A/L^\phi) \propto L^{2+\delta(1/2)} \tag{4.122}$$

with $\delta(1/2) = \phi(1/2 - \beta) \approx 0.13 > 0$ with $\phi = 1.8$ and $\beta = 0.43$ (Section 2.11). Notice that a scaling of energy with size in OCNs as $E \propto L^{2.2}$ has also been experimentally observed in multiple outlet OCNs [Sun et al.,

1994a] and is confirmed by the renormalization group argument described in Section 4.18. Hence the result in Eq. (4.117) is validated.

We will now show that for spanning loopless trees the number $N(E)$ of configurations s with given energy E scales as

$$N(E) \propto \mu^{L^2} \tag{4.123}$$

so that

$$\sigma(E) \propto L^2 \tag{4.124}$$

Here μ is a real number depending on lattice properties.

We will show the proof provided by Rinaldo et al. [1996a] only in one case, the others being analogous. For example, in a four-neighbor lattice the number N_D of DST is $N_D = 2^{(L-1)^2}$. In fact, there exist two choices of direct networks for each of $(L-1)^2$ independent sites. The total number N of spanning trees is greater than N_D but less than the number of possible ways of choosing $L^2 - 1$ links (number of links of a spanning tree) among all the $2L(L-1)$ possible links. Thus

$$2^{(L-1)^2} = N_D < N < \binom{2L(L-1)}{L^2-1} \sim 2^{2L^2}$$

Because the number $N(E)$ of configurations with a given energy is of course less than N, then $N(E) < 2^{2L^2}$ and, in general, Eq. (4.123) is satisfied with $2 < \mu < 4$ in the four-neighbor case. The general case follows directly from the same type of reasoning.

We thus conclude that for OCNs with $\gamma \geq 1/2$, entropy scales subdominantly to the energy with system size. Consequently, in the thermodynamic limit $L \to \infty$, $\min F(E) \propto L^{2+\delta}$ because $\delta > 0$. In fact, one has

$$F(E) \propto L^{2+\delta} - \text{const}\ TL^2 \tag{4.125}$$

and, for $L \to \infty$, $F(E) \propto L^{2+\delta}$ when $\delta > 0$. Hence the configuration s that minimizes H_γ also minimizes $F(E)$, whatever Gibbs' parameter T^{-1}, provided the system is large enough. Hence OCNs, which correspond to the zero-temperature assumption (i.e., the configuration yielding $\min F(E)$ is that with $\min E$ only for $T \to 0$), reproduce natural conditions for any temperature. Because fluvial networks usually develop migration of divides and competition for drainage in the absence of geologic controls over domains large with respect to the lower cutoff scale (i.e., the scale of channel initiation), it is likely that natural networks mostly operate in conditions that well approximate the thermodynamic limit $L \to \infty$. We suggest that this is the reason for the outstanding ability of OCNs to reproduce observational evidence regardless of diversities in surface lithology, geology, vegetation, or climate.

This also strongly suggests that the analysis of small size real networks may lead to spurious results. We believe that this has most likely affected observational evidence supposedly supporting the inadequacy of the low-temperature assumption [Troutman and Karlinger, 1994].

Rinaldo et al. [1996] have run an example of an entropy-dominated OCN with $\gamma = 0.5$ using the Metropolis algorithm [Metropolis et al., 1953] at a nonzero (and fixed) value of T. The example is meant to show the implications of entropy maximization; that is, if $T >$ const L^δ, entropy

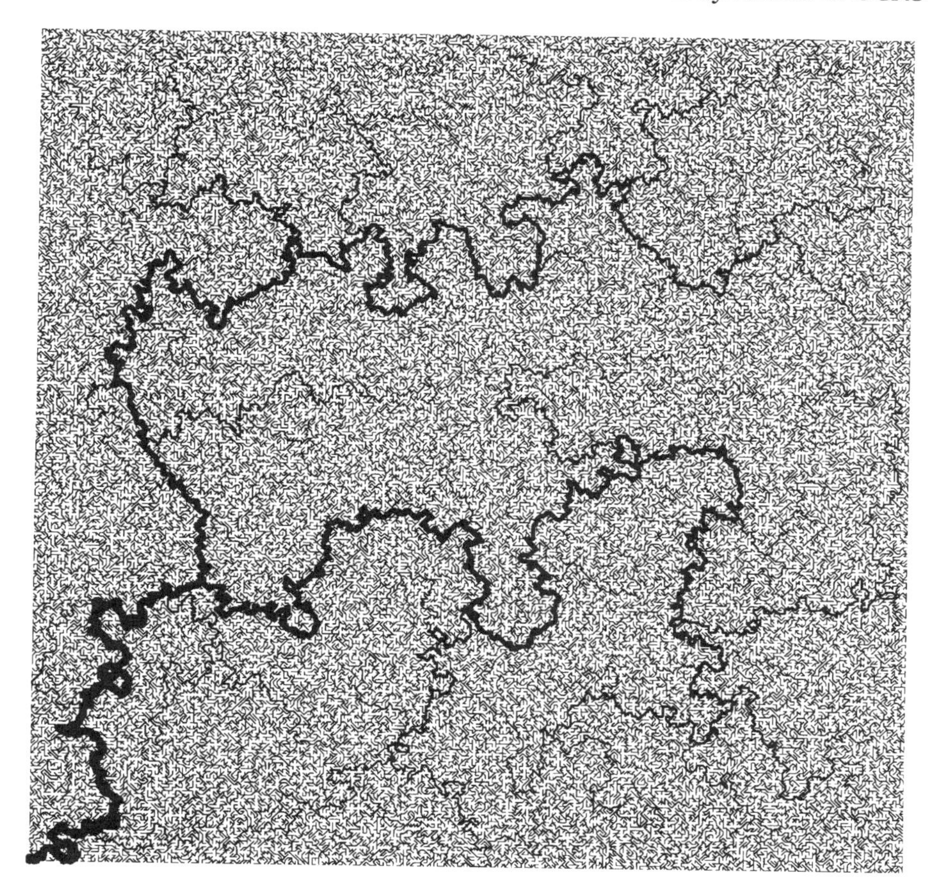

Figure 4.67. A configuration of a "hot" OCN, obtained by the Metropolis algorithm with $T = 10{,}000$ after 10^7 iterations [after Rinaldo et al., 1996a].

dominates over energy. The procedure is as follows: starting from any initial configuration, configuration changes ($s \to s'$) are accepted when they (i) lower the energy (i.e., $H_\gamma(s') < H_\gamma(s)$) or (ii) with probability $P(s') \propto \exp(-(H_\gamma(s') - H_\gamma(s))/T)$ otherwise.

Figure 4.67 shows one of the configurations, after 10^6 iterations, for a $T = 10{,}000$ OCN which, for the system size $L = 256$, is large enough to ensure the dominance of entropy. The network in Figure 4.67 has been obtained starting from the same initial conditions as the network used to produce the OCN in Figure 4.66.

The network in Figure 4.67 is clearly a fractal with well-defined self-similar and self-affine properties, none of which match those found in nature in the production zone of the fluvial system (see Figure 1.1). For example, the probability of total contributing area $P[A \geq a] \propto a^{-\beta}$ (Figure 4.68) with $\beta = 0.358$. Also, the length distribution is $P[L \geq l] \propto l^{-0.55\pm0.01}$ (Figure 4.69). Notice the highly meandering patterns, here characterised by altered scaling exponents ($\phi_L = 1.5 \pm 0.01$, Eq. (2.174); $\phi = 1.2 \pm 0.01$, Eq. (2.203)).

Also of interest is a qualitative comparison of the structures in Figure 4.68 with the fluvial patterns shown in Figures 4.70 and 4.71 or with the highly meandering patterns in the lowlands, outside the production zone of the fluvial system. We note that the network structures developed in delivery zones (Figure 1.1) depart systematically from the fluvial

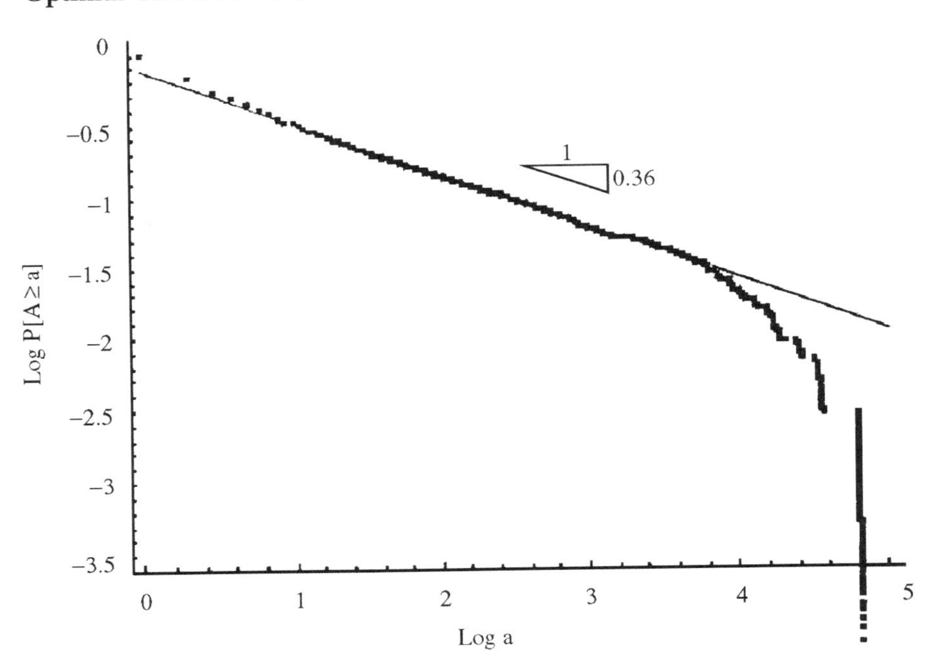

Figure 4.68. Probability of total contributing area $P(A \geq a)$ for the $T = 10{,}000$ OCN.

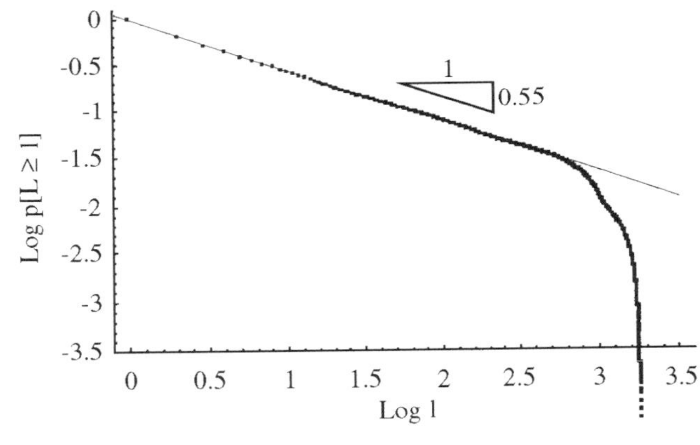

Figure 4.69. Length distribution $P[L \geq l] \propto l^{-0.55\pm0.01}$ for the $T = 10{,}000$ OCN.

morphologies observed in the production zone. Nevertheless, they do not resemble completely 'hot' patterns either, although a certain similarity exists. Understanding morphologic patterns like those shown in Figures 4.70 and 4.71 is indeed a future challenge. From the theoretical arguments proposed here, one could speculate that geologically constrained networks (that is, whose characteristic size L is relatively small) might operate under hot conditions, but at present definite experimental evidence of the existence of hot fluvial patterns is still lacking. Also, we still lack a clear geomorphologic assessment of the physical meaning of the mutational parameter T affecting the susceptibility of a network to planar changes, although one perceives hot patterns as somewhat similar to those occurring in low-relief areas.

Nevertheless, hot OCNs, which indeed obey a variational principle as they maximize the probability of a configuration given T, have important properties. It is of much interest that, after a transient, hot OCNs develop stable statistics of the spatially scale-free structures and are characterized by a preponderance of timescales once the system reaches the set of recursive states, in a process reminiscent of ordinary self-organized critical (SOC) phenomena [Bak et al., 1988] to be described in Chapter 5. We shall return to this important analogy therein. A plot of the time evolution of the total energy dissipation $H(s)$ is shown in Figure 4.72. Two features are worth discussing: (i) the rise of the mean energy level from the initial value of the random network, untenable at high temperatures, and (ii) the fluctuations of energy consequent to accepted changes that increase energy levels. It can be shown that the time signal $H_{1/2}(s(t)) = H_{1/2}(t)$, where t is time measured along the optimization process, shows signs of

$1/f$ noise. Thus the optimization process produces an entirely different, though fractal, structure both morphologically and structurally – it also evolves in time.

At lower temperatures (e.g., $T = 10$) for the chosen size $L = 256$ there is no predominance of entropy, although temporal activity is still present. The characters of the network after a given number of iterations, still much less direct than those observed in nature, nevertheless more closely resemble those of real basins. An example of these properties is given in Figure 4.73 where $P[A \geq a] \propto a^{-0.41\pm0.01}$ and $P[L \geq l] \propto l^{-0.70\pm0.01}$.

What are the implications of the above result? OCNs have by construction a much-constrained structure (loopless and spanning) and entail the minimization of total energy dissipation. These structures develop under generic conditions and do not exhibit a set of dynamically recursive states but freeze into static scale-free structures; that is, they behave like a $T = 0$ frozen system. This follows from the subdominant scaling of entropy with system size compared to the energy. One thus wonders whether this allows general conclusions of the growth of fractal structures to be drawn: In the next section we will see that the spatial characteristics of the frozen, $T = 0$, structures are still present in OCNs developed at low but nonnegligible temperatures. The important aspect is that in this case the structures are not frozen but rather display temporal structure.

It is also tempting to speculate on the differences between minimum-energy (Figure 4.66) and maximum-entropy (Figure 4.67) network structures. Is the relative scaling dominance of energy and entropy with the system size the discriminant between static (spatial) and dynamic (spatial and temporal) fractals? And conversely, are fractals the natural outcome of a variational principle embedded in the maximum probability statement of Eq. (4.109)?

No conclusion is yet at hand. The proper comparisons of the different types of fractality require the establishment of some tools, to be provided in Chapter 5, that aim at theoretical as well as speculative arguments for a new interpretation of self-organized criticality. Nevertheless, it is worth reminding that the connections of optimality with fractal growth were established for the first time following the development of the OCN concept in Rodriguez-Iturbe et al. [1992b,c] and Rinaldo et al. [1992]. All previous approaches, based on the achievement of local optimal rules [e.g., Howard, 1971b], did not address the issue of fractality, and moreover did not impose a global minimum of the structure as a whole. Followers [e.g., Sun et al., 1994a,c, 1995] have discovered important properties but essentially have only modified the boundary conditions of the main OCN scheme. One wonders if in general scaling invariance is a by-product of chance and necessity embedded in a principle of maximum fitness of open, dissipative structures that develope in nature with many degrees of freedom. If this turns out to be the case, then OCNs and their hydrologic framework will have an impact on a variety of natural structures.

4.22 Space–Time Dynamics of Optimal Networks

The space–time dynamics of a channel network depends on its susceptibility to change under the influence of perturbations, possibly of random nature, as we have seen in Section 4.21. As river networks evolve, drainage

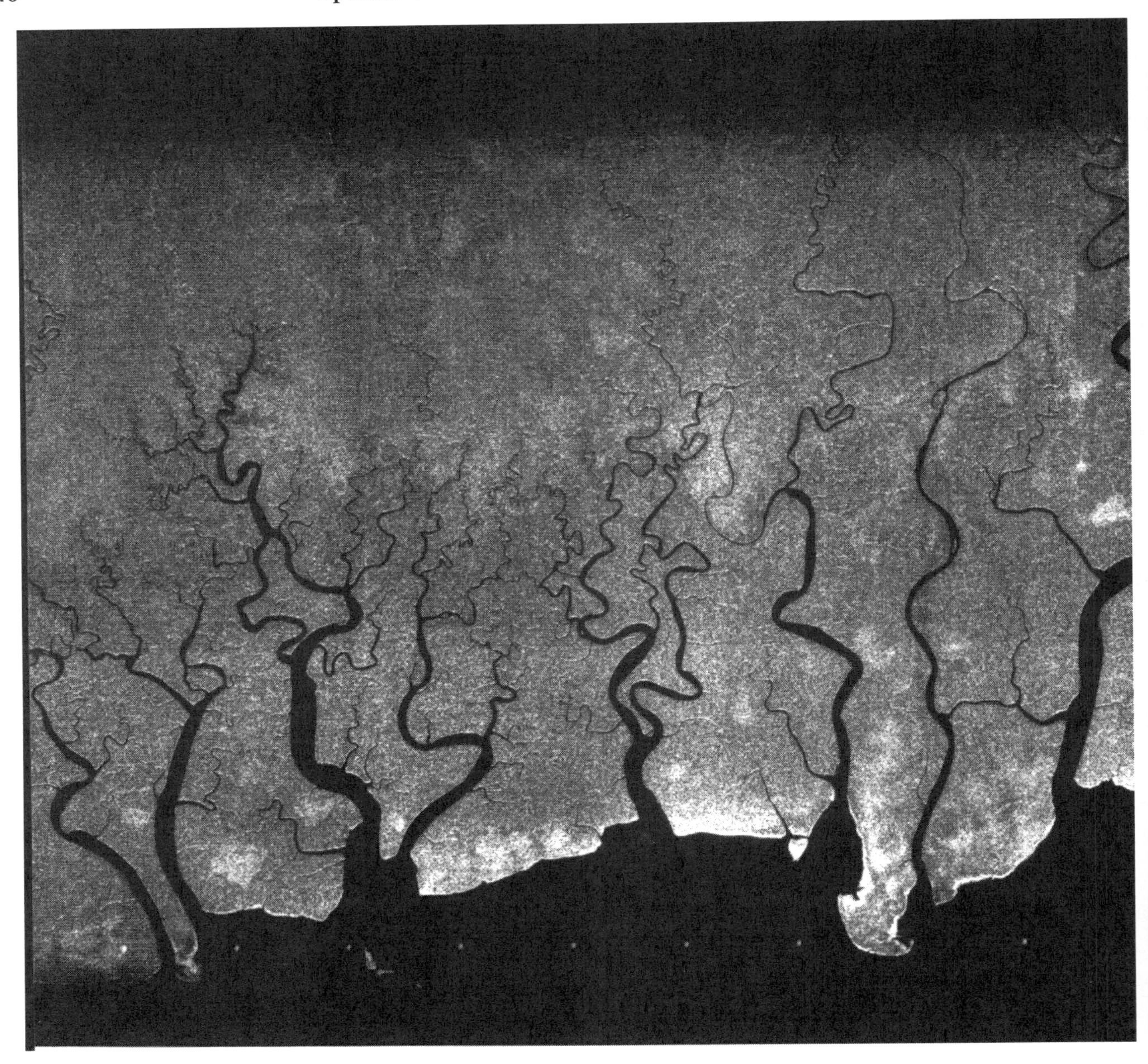

Figure 4.70. The coastal area of southwest New Guinea consists of a lowland tropical forest and an associated mangrove swamp that yield a generally uniform medium gray tone on this image. The area is impenetrable overland and is cut by the rivers that drain the region. These rivers lead 70–100 km northward in this photograph. A sluggish but powerful tidal

capture may occur, whose frequency and dynamics over the evolutionary span may result in distinctly different network attributes. Although one perceives these structures as static because the timescale of their evolution is much larger than the span of observation, a dynamics obviously exists and the legacy of drainage capture, well after it has occurred, is difficult to identify by simple inspection. Thus the importance of models in this kind of studies. Here we model the likelihood of changes in the system configuration under the thermodynamic framework implied by the Metropolis rule and described in Section 4.21. In there changes are allowed as a function of the rates of energy expenditure that they imply, depending on a mutational parameter similar to temperature in conventional thermody-

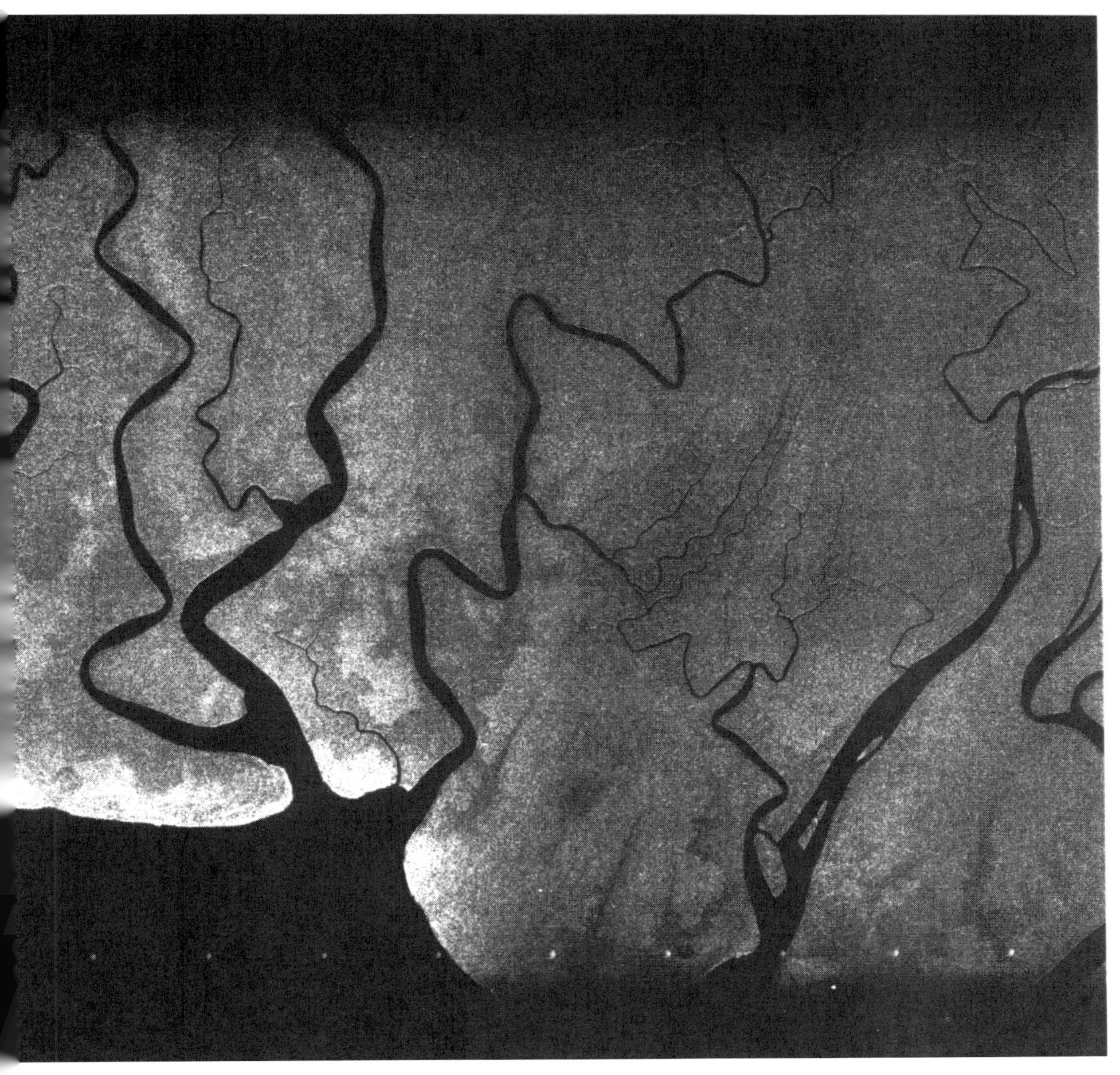

Figure 4.70. *(continued)* current in the Arafura Sea periodically inundates the coast and creates a bore that reaches 15 km upstream and causes floods of 3–6 m [Courtesy of NASA].

namic systems. This section shows that OCNs are likely to evolve with the intermittent behavior of punctuated equilibrium [Raup, 1991; Gould and Eldridge, 1993] where long periods of stasis are interrupted by bursts of activity. This temporal dynamics is coupled with spatial activity of all sizes as in self-organized critical phenomena (Chapter 5). Moreover, it is of significant interest that optimal networks may indeed typically operate under conditions characterized by low energy and at the same time be characterized by high entropy conditions that preserve the features observed in nature while embodying a rich temporal dynamics.

As the driving mechanism for network dynamics we use the reorientation at each time of a link chosen at random. Although previous models

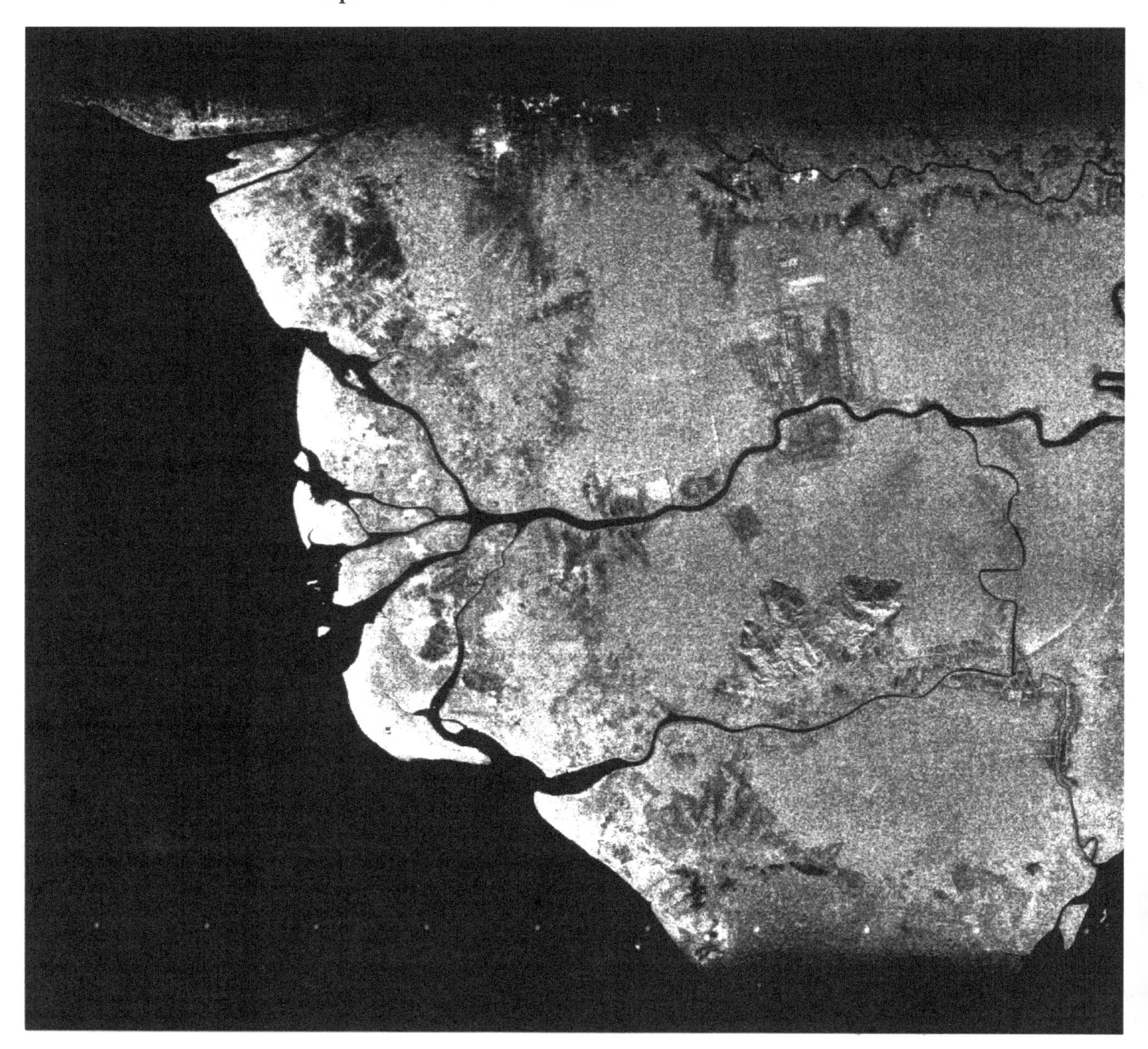

Figure 4.71. The delta and some related fluvial patterns of the Kapnas River in Indonesian Borneo [Courtesy of NASA].

[e.g., Howard, 1971] used a similar mechanism for capture, here optimality is prescribed for the global configuration of the system. The space–time dynamics of such global optimal configurations has been investigated by Rodriguez-Iturbe et al. [1996]. Notice that absolute time has no meaning in the analysis because different geologies and climates (Section 6.6) just imply different frequencies in the occurrence of perturbations. Independent of the frequency of perturbations, we embody the susceptibility of the network to change under their action in the mutational parameter, T, reminiscent of temperature in thermodynamics.

In Section 4.13 we discussed the different results one gets when applying different optimization strategies involving a temperature, T, to the OCN

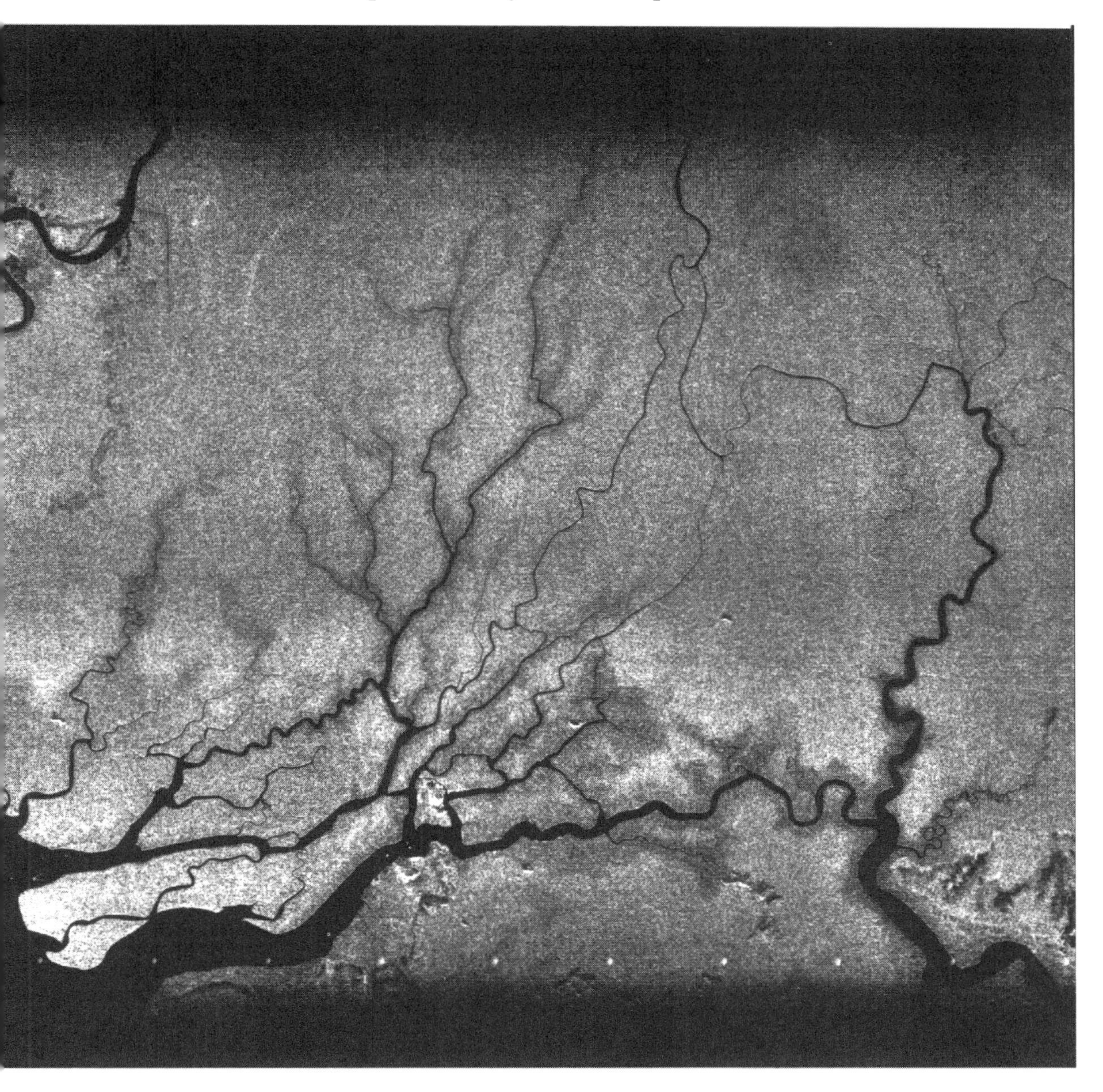

problem. The temperature parameter essentially distinguishes whether or not river networks are free to explore extended regions of their fitness landscape rather than myopically improving their present condition. Crucial to the simulated annealing procedure is the choice of a cooling schedule for T, which allows an effective search for a global optimal configuration.

Rather than using a cooling schedule, Rodriguez-Iturbe et al. [1996] have studied in depth the properties of the networks obtained with the same procedure when T is kept constant, as discussed in Section 4.21. A statistical equilibrium is obtained for the characteristics at different temperatures, which then allows the association of a value of T to any network

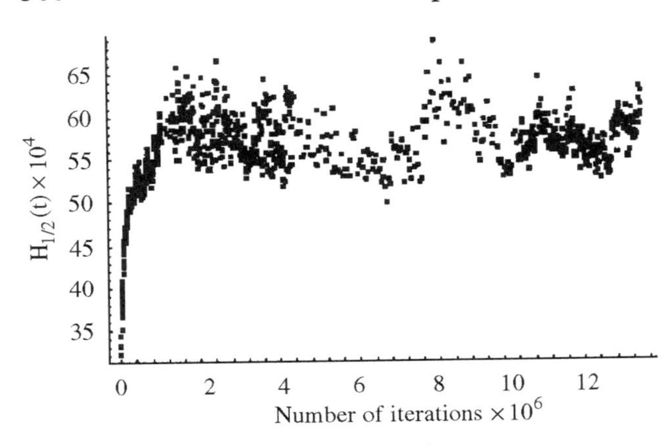

Figure 4.72. Time evolution of $H_{1/2}(t)$ for a $T = 100$ OCN in a 256×256 grid.

obtained in this manner in a domain of a given size. Keeping $T > 0$ we recognize that, during the evolution of river basins, sometimes changes drive the system in the direction of an increase in energy expenditure. It is an evolution reminiscent of controlling geological features. The higher the temperature, the larger the relative frequency with which random events change the connectivity of the network.

For a fixed value of T, the Metropolis algorithm yields the most probable configuration of the system, that is, one whose structure minimizes the free energy, as discussed in Section 4.21. Because fixing the temperature is equivalent to fixing the mean energy of the system, the Metropolis algorithm yields the most probable configuration for a given energy.

Figures 4.74 and 4.75 show typical results of the changes the network undergoes as a function of T. The networks were obtained from a random initial condition in a basin of 22,200 pixels. The basin domain was the largest one resulting from simulations with multiple outlets in a 512×512 lattice. All the networks are snapshots of the evolution after reaching statistical stationarity (except for $T = 0.001$ where the network is practically frozen).

Figure 4.76 shows the mean energy, $\langle E \rangle$, and the entropy S of the evolution. Notice that $\langle E \rangle$ is estimated from a stochastic time series where total energy fluctuates in time. The definition of entropy S is as follows. First, a time series is obtained that contains the number of active sites at each step of the process. In this context, activity refers to the number of lattice sites whose total contributing area changes upon accepting a change in the configuration. A frequency, say p_i, is associated with the number of times a value i of active sites is obtained. Informational entropy is thus defined as $S = \sum_{i=1}^{N} p_i \log p_i$ where N is the lattice size. The graphs in Figure 4.76

Figure 4.73. (a) $P[A \geq a]$ versus a and (b) $P[L \geq l]$ versus l for a $T = 10$ OCN.

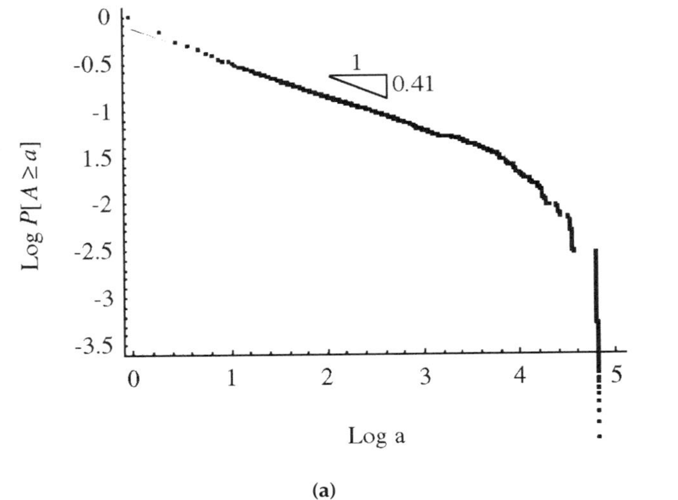

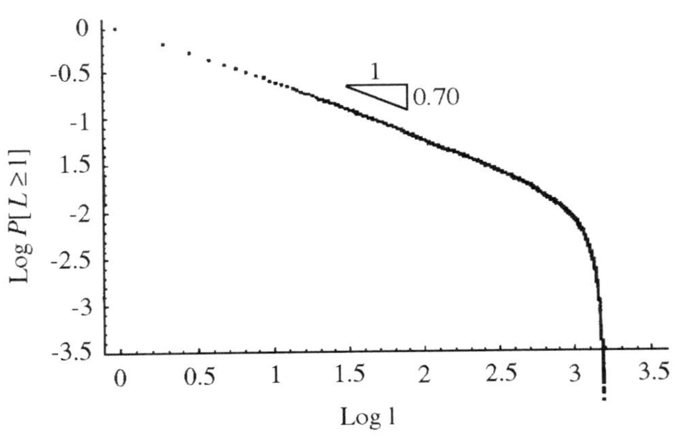

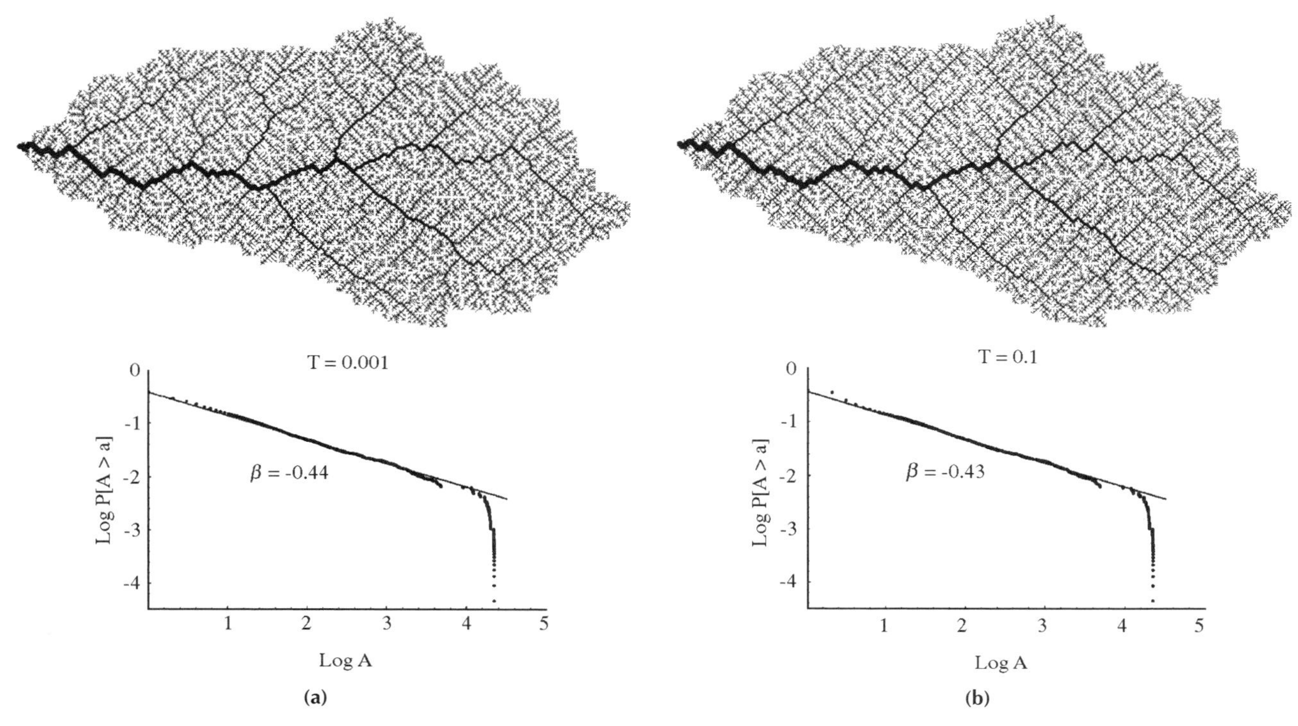

Figure 4.74. Dependence of the network structure on the temperature, or mutation parameter, T: (a) $T = 0.001$; (b) $T = 0.1$. For temperatures below a certain threshold, say $T < T^*$, the slope β of the exceedence probability for contributing area at any point, $P[A \geq a]$, is statistically indistinguishable from those most often observed in nature, that is, $\beta = 0.43 \pm 0.02$.

are obtained from the ensemble average of fifty time series of 10^5 iterations each.

A measure of the spatial extension of the changes taking place in the network is the number N, of active sites. Figures 4.77 and 4.78 study some of their relevant properties.

Two regions are clearly detected in mean energy expenditure. One observes (Figure 4.76(a)) a low plateau where $\langle E \rangle$ is stable up to a value T^* that depends on the size of the network. In this region the critical scaling exponent β fits well the behavior observed in river basin, that is, $\beta = 0.43 \pm 0.02$. For $T > T^*$ the energy expenditure increases with T and the characteristic behavior of $P[A \geq a]$ departs from the observational evidence. In this example, T^* is approximately 0.2. For high Ts we observe meandering not unlike that observed in regions characterized by small relief. The entropy of the distribution of active sites is also heavily dependent on temperature (Figure 4.76(b)). It shows a steep increase until it reaches a plateau where it is stable.

The two regions existing in the energy expenditure and entropy characteristics display different temporal dynamics controlled by the temperature or mutation parameter T. For very low values of T the system is frozen and does not react to the random changes that affect its connectivity. For somewhat larger temperatures there is a nonnegligible number of active sites whose temporal dynamics reflects the manner in which the system evolves. At temperatures near $T \sim T^*$ the system exhibits punctuated equilibrium with static periods of all lengths interrupted by bursts of activity. This intermittent evolution, typical of 1/f signals, is very different from a gradual, smooth one. Self-organized critical phenomena show this kind of dynamics where the accumulated mutational activity is described

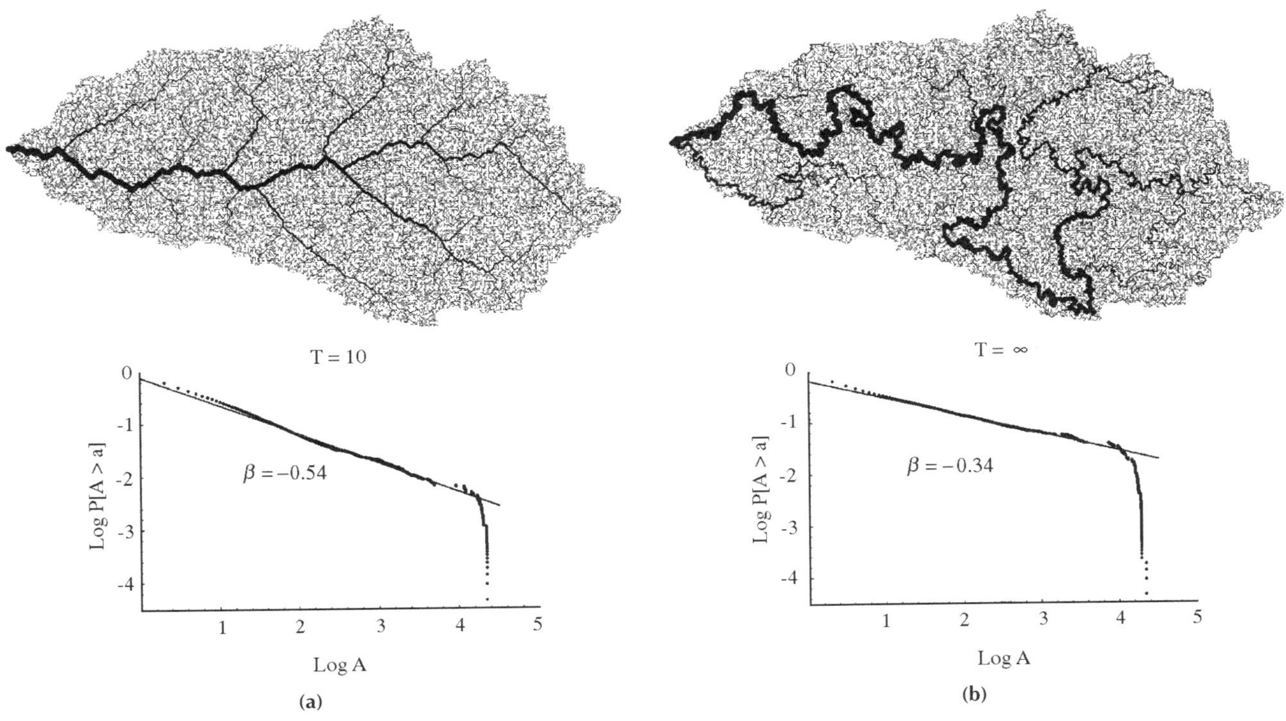

Figure 4.75. Dependence of the network structure on the temperature, or mutation parameter, T: (a) $T = 10$; (b) $T = \infty$. Above a certain threshold, say $T > T^*$, the slope β of the exceedence probability for contributing area at any point, $P[A \geq a]$, is significantly different from those observed in real basins. For very high Ts the slope approaches the theoretical value $\beta = 0.38$ valid for infinite temperature in infinitely large domains.

by a devil's staircase [Sneppen et al., 1995]. At higher temperatures the activity in the network seems more and more uniform in time, destroying the devil's staircase features. (for an introduction to the devil's staircase see Section 3.2). Figure 4.77 displays the above dynamics, which has also been observed in phenomena such as biological evolution where periods of stasis interrupted by evolutionary activity have been well documented [Raup, 1986; Gould and Eldredge, 1993]. The measure of time in geologic scales should not obscure the fact that the system is not frozen. Figure 4.78 shows the probability distribution of the number of active sites as a function of temperature. In all cases it follows a well-defined power law.

Networks with near-optimal relative energy expenditure obtained for temperatures $T \sim T^*$ are observed to have a large dynamic diversity. Thus although they may be considered cold from the energy expenditure point of view, they are hot in the characteristics of the information entropy. This latter feature is typical of self-organized critical system (see Section 4.21) and has also been observed in fluid neural networks [Solé and Miramontes, 1995]. Some animal societies like ants also appear to choose their spatial densities so as to maximize the information entropy in the spatial activity pattern. This is the point where there is the greatest variety of dynamic states.

Similar to biological evolution [Kauffman and Johnsen, 1991], many fluvial systems appear to evolve toward a regime that is between order and chaos. They exist near a transitional regime that makes them orderly enough to efficiently coordinate their activities but flexible enough to evolve in the face of change.

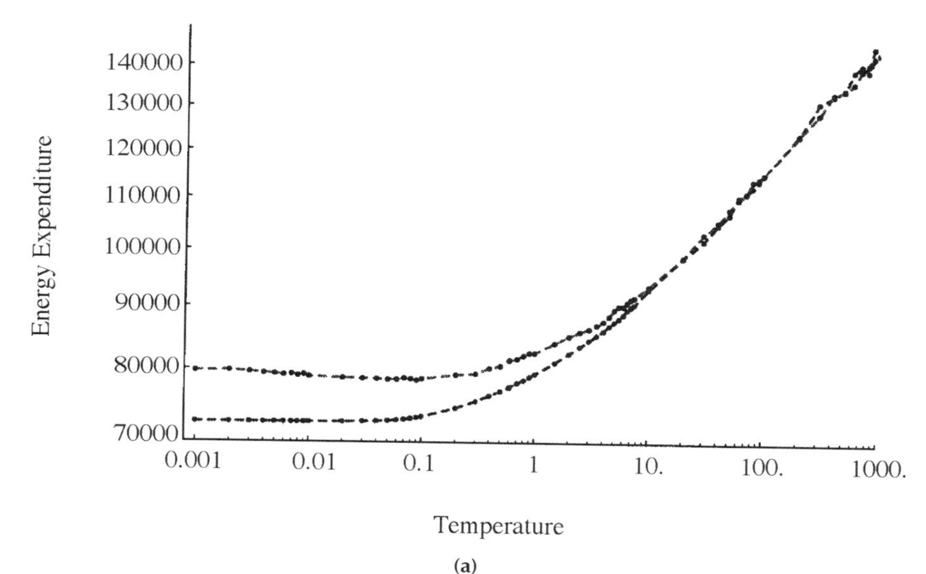

(a)

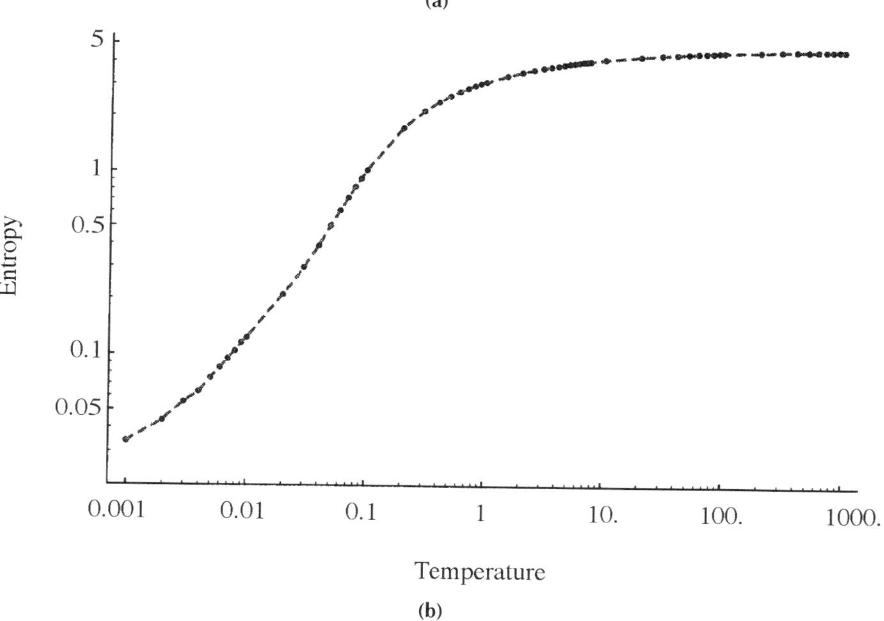

(b)

Figure 4.76. (a) Mean energy expenditure $\langle E \rangle$ and (b) information entropy S of the distribution of active sites for evolving networks – snapshots of which are shown in Figures 4.74 and 4.75. The curves were obtained from the ensemble average of fifty time series of 10^5 iterations each. The lower curve in (a), shown for comparison, refers to the energy expenditure obtained at different stages of the optimization, carried out via simulated annealing employing a careful slow-cooling schedule. As such the system evolves toward the global minimum of free energy for the given temperature (Section 4.23) [Rodriguez-Iturbe et al., 1996].

4.23 Exact Solutions for Global Minima and Feasible Optimality

Recent analytical developments complete our survey of dynamically accessible optimal states. This section covers the derivation of exact properties for the global minimum of the functional defining OCNs [Maritan et al., 1996b; Colaiori et al., 1996], and compares these properties with the numerical results of Section 4.13. We will first study the exact results. These pertain to the characteristics of the global minimum of the functional

$$E = \sum_{i=1}^{N} k_i A_i^{\gamma} \tag{4.126}$$

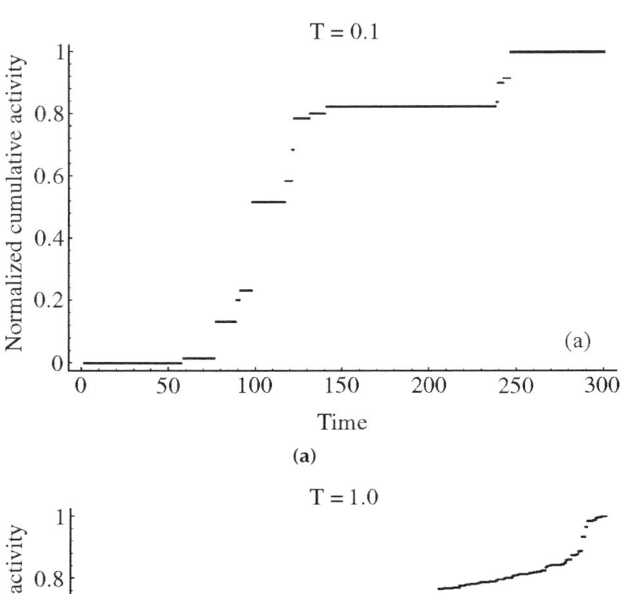

(a)

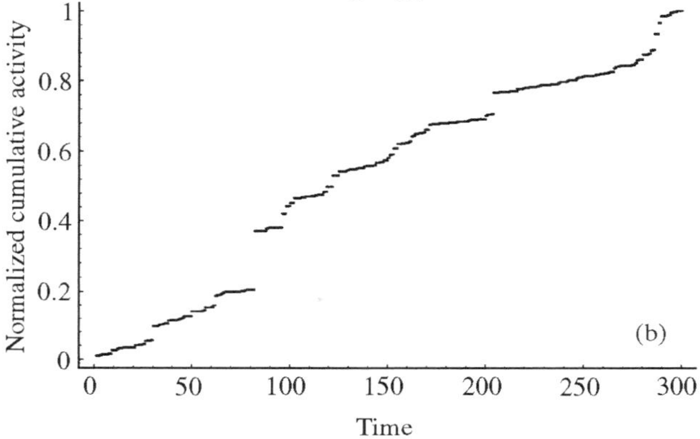

(b)

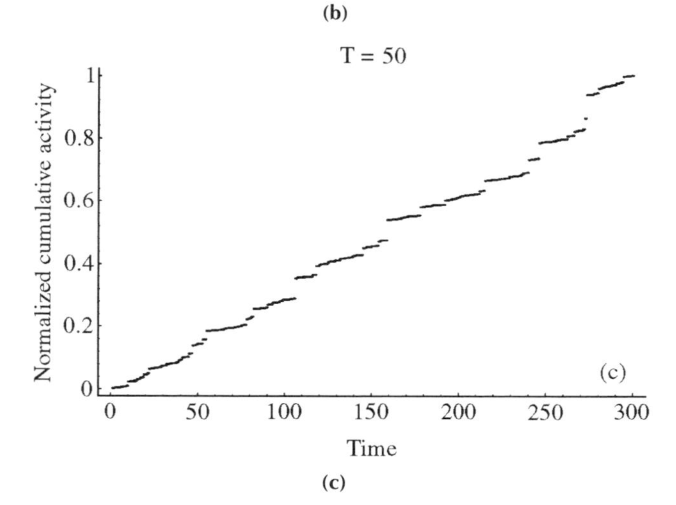

(c)

Figure 4.77. Snapshots of the temporal variability of the spatial activity for some of the networks in Figures 4.74 and 4.75: (a) $T = 0.1$; (b) $T = 1$; (c) $T = 50$. The cumulative number of sites that have shown activity at any time in the interval [0, t] is normalized by the total activity in the interval [0, 300]. The analysis is done after the system reaches a statistical steady state.

(for γ in the range [0, 1]), where N spans all sites in the lattice, k_i condenses information about the length of the ith link and the spatial value of the erosional properties, and A_i is the total contributing area at site i, defined in general as

$$A_i = \sum_{j \in nn(i)} W_{ij} A_j + R_i \tag{4.127}$$

where $nn(i)$ are the nearest neighbors to i, W_{ij} is the connection matrix defining whether j drains into i (Section 1.2.9), and R_i denotes mass injection, possibly random, at the ith site (Section 4.20). Notice that the functional E refers to the configuration yielding the minimum value of $H_\gamma(s)$ following the notation of Section 4.21. Also, the properties of E are studied either in the homogeneous case ($k_i = k$, hence is dropped from the summation) or in the heterogeneous case (where k_i varies in space) [Maritan et al., 1996c].

Let us consider first the two-limit cases ($\gamma = 0$ and $\gamma = 1$). If we denote by x_i the weighted length of the stream connecting the ith site to the outlet (calculated by assigning to each bond a length k_i), it is straightforward (see Section 4.21) to show that

$$\sum_i k_i A_i = \sum_i x_i \tag{4.128}$$

Thus the minimization of energy dissipation for $\gamma = 1$ corresponds to the minimization of the weighted path connecting every site to the outlet, that is, the mean distance from the outlet. It is a rather logical consequence that the case $\gamma = 1$ admits as a global minimum the most direct network.

The $\gamma = 0$ case, instead, implies the minimization of the total weighted length of the spanning tree:

$$E = \sum_i k_i \tag{4.129}$$

For $\gamma = 0$, every configuration has the same energy because every spanning tree has the same number of links ($L^2 - 1$ for a $L \times L$ square lattice).

For the values of $\gamma \in (0, 1)$ the search for global minima is a less trivial problem.

For $\gamma = 1$ the minimum of the energy is realized on a large subclass of spanning trees. These are all the directed trees, where every link has a positive projection along a diagonal, oriented toward the outlet and defining the preferential direction of network development. The $\gamma = 1$ case gives a minimum energy scaling $E \sim L^3$ (where L is, as usual, the characteristic linear size of the lattice) for each directed network. This follows from the observation that any directed network corresponds to the Scheidegger [1966] model of river networks where all directed trees are equally probable by construction (Section 2.5.2). Such a model can be mapped onto a model of mass aggregation with injection exactly solved by Takayasu et al. [1991] (see Section 2.5.2). From the exact mapping one recovers the important result $E \propto L^3$. Recall (Section 2.10.1) that the relevant probability distributions assume the scaling form $P[A \geq a] = a^{-\beta}\mathcal{F}(a/L^{\phi})$ for total contributing areas A, where β and ϕ are suitable scaling exponents (notice that L substitutes the more involved notation for the characteristic linear size); and (Section 2.10.2) $P[l \geq x] = x^{-\psi}\mathcal{G}(x/a^h)$, where l is the upstream length at any site and ψ and h are the relevant scaling exponents, the latter termed Hack's exponent. The corresponding exponents for the $\gamma = 1$ minimum are

$$\beta = 1/3, \psi = 1/2, \phi = 3/2, h = 2/3 \qquad (4.130)$$

Because all directed trees are equally probable, each stream behaves like a single random walk in the direction perpendicular to the diagonal through the outlet. This, among other implications, yields that its characteristic perpendicular length $L_\perp \sim L^{1/2}$ (Section 2.9), and therefore the Hurst coefficient is $H = 1/2$.

The $\gamma = 0$ case gives the same energy $E \sim L^2$ for each network, in analogy to the problem of random two-dimensional spanning trees, whose geometrical properties have been predicted in the case of a square lattice by Maritan et al. [1996b], and give, in our notation,

$$\beta = 3/8, \psi = 3/5, \phi = 2, H = 1, h = 5/8 \qquad (4.131)$$

Let us now extend the analysis to the whole range $\gamma \in [0, 1]$. Maritan et al. [1996c] have rigorously shown that in the thermodynamic limit ($L \to \infty$) the global minimum in the space, say $\mathcal{S}$, of all the spanning trees of the functional $E(\gamma, \mathcal{T}) = \sum_i A_i(\mathcal{T})^\gamma$ scales as

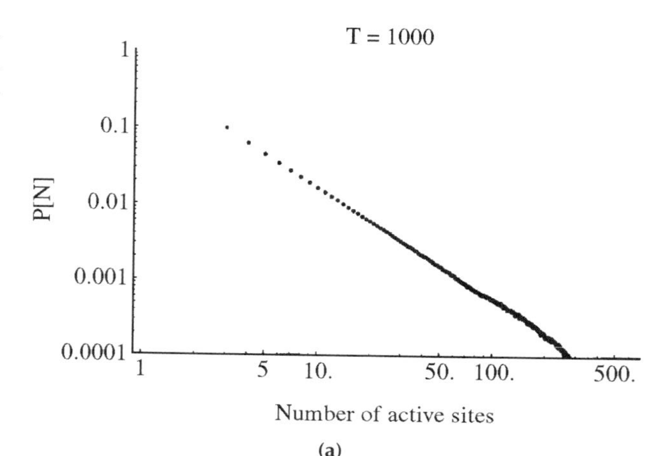

(a)

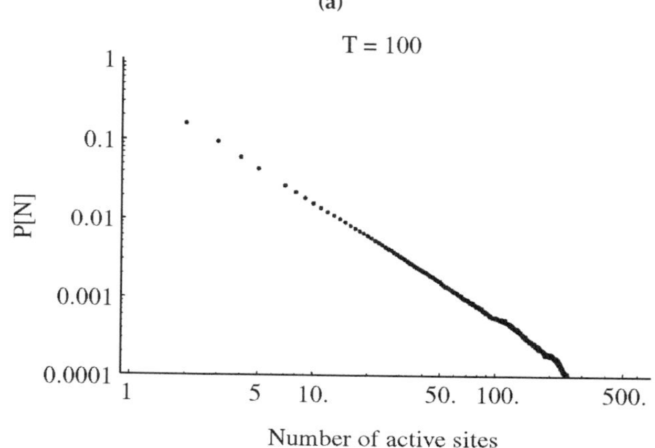

(b)

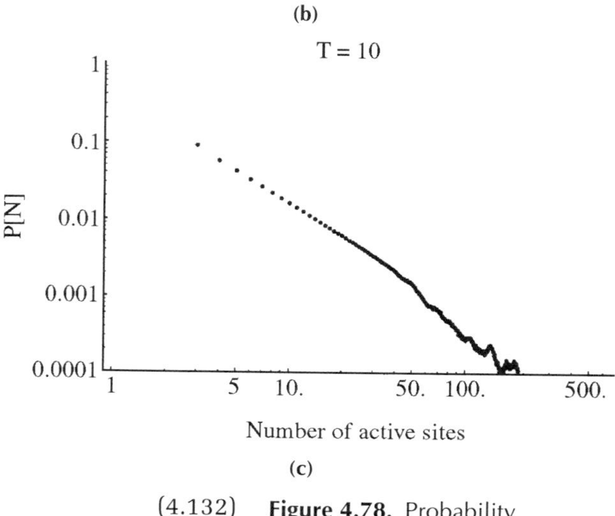

(c)

Figure 4.78. Probability distribution, $P[N]$, of having a certain number of active sites, N, at different temperatures for the networks of Figure 4.1. For temperatures lower than $T = 0.05$ the system is practically frozen and $P[N]$ loses any meaning.

$$\min_{\mathcal{T} \in \mathcal{S}} E(\gamma, \mathcal{T}) \sim \max(L^2, L^{1+2\gamma}) \qquad (4.132)$$

(where $\mathcal{T}$ is the subset of the spanning trees constituted by optimal trees) for all $\gamma \in [0, 1]$. From this result, already hinted at in Section 4.21, one can derive the scaling exponents of the global minimum. The derivation is now briefly outlined. The reader is referred to Colaiori et al. [1996] for the detailed specification of the proofs.

Because $E(\gamma, \mathcal{T})$ is an increasing function of γ and is equal to L^2 for

$\gamma = 0$, then for $\gamma \geq 0$ it is obvious that

$$E(\gamma, \mathcal{T}) \geq L^2 \tag{4.133}$$

We now observe that the sum over all the sites can be performed in two steps:

$$E(\gamma, \mathcal{T}) = \sum_{n=1}^{2L-1} \sum_{i \in \mathcal{D}_n} A_i(\mathcal{T})^\gamma \tag{4.134}$$

where $\mathcal{D}_n$ are the diagonals orthogonal to the one passing through the outlet, which we will enumerate in the following, starting from the corner farthest from the outlet. For the directed spanning trees one can observe that the sum of the areas along a given diagonal $\mathcal{D}_n$ is independent of the particular tree and is obtained [Maritan et al., 1996b] as

$$S_d(k) = \sum_{i \in \mathcal{D}_k} A_i = \begin{cases} k(k+1)/2 & k \leq L \\ L^2 - S_d(2L-1-k) & L+1 \leq k \leq 2L-1 \end{cases} \tag{4.135}$$

where we define $S_d(0) = 0$. Such quantities can be only increased considering generical "undirected" trees, thus we can write for every spanning tree

$$S(k, \mathcal{T}) = \sum_{i \in \mathcal{D}_k} A_i(\mathcal{T}) \geq S_d(k) \tag{4.136}$$

Let us observe that for $k = 0, \ldots, (L-1)$

$$S(k, \mathcal{T}) + S(2L-1-k, \mathcal{T}) \geq S_d(k) + S_d(2L-1-k) = L^2 \tag{4.137}$$

making convenient to perform the summation in Eq. (4.134) over couples of diagonals. To get a lower bound for E, we need a further inequality: for every set Γ

$$\sum_{i \in \Gamma} A_i^\gamma \geq (\sum_{i \in \Gamma} A_i)^\gamma \tag{4.138}$$

which follows easily from the Schwartz inequality, being $A_i \geq 1$ and $0 \leq \gamma \leq 1$. Now, using Eqs. (4.134), (4.137), and (4.138), we can write

$$E(\gamma, \mathcal{T}) = \sum_{n=1}^{2L-1} \sum_{i \in \mathcal{D}_n} A_i(\mathcal{T})^\gamma = \sum_{n=0}^{L-1} \sum_{i \in (\mathcal{D}_n \cup \tilde{\mathcal{D}}_n)} A_i^\gamma \geq \sum_{n=0}^{L-1} (\sum_{i \in (\mathcal{D}_n \cup \tilde{\mathcal{D}}_n)} A_i(\mathcal{T}))^\gamma$$

$$= \sum_{n=0}^{L-1} [S(n, \mathcal{T}) + S(2L-1-n, \mathcal{T})]^\gamma \geq \sum_{n=0}^{L-1} L^{2\gamma} = L^{1+2\gamma} \tag{4.139}$$

where $\tilde{\mathcal{D}}_n = \mathcal{D}_{(2L-1-n)}$. Equality in the last inequality holds for directed networks. We can thus write

$$E(\gamma, \mathcal{T}) \geq L^{1+2\gamma} \tag{4.140}$$

Eq. (4.140) together with Eq. (4.133) yields the lower bound

$$E(\gamma, \mathcal{T}) \geq \max(L^2, L^{1+2\gamma}) \tag{4.141}$$

which holds for every tree $\mathcal{T} \in \mathcal{S}$, and thus also for the minimum over $\mathcal{T}$.

Now the relevant scaling exponents can be computed. For "undirected" networks, we obtain from Eq. (2.203)

$$\langle a \rangle \sim L^{\varphi} \tag{4.142}$$

where φ is slightly bigger than one if one assumes a "quasi-direct" behavior. For a directed path one obtains instead

$$\langle a \rangle \sim L \tag{4.143}$$

Using Eq. (2.206) with $n = \gamma$ (recall that $\tau = \beta + 1$ is the exponent of the probability density of total contributing area) and the above result on the scaling of energy ($E = L^2 \langle a^\gamma \rangle \sim L^{1+2\gamma}$), one gets

$$2\gamma - 1 = (1 + H)(\gamma - \tau + 1) = \phi(\gamma - \beta) \tag{4.144}$$

which holds for $\gamma > \tau - 1$.

Equations (4.144) together with the scaling relations derived in Section 2.9 can be solved with respect to τ and H, and give

$$\gamma > 1/2 \begin{cases} \tau = \frac{3(1-\gamma)+(\varphi-1)(1+\gamma)}{2(1-\gamma)+\varphi-1} \\ H = \frac{\varphi-\gamma}{1-\gamma} \end{cases} \tag{4.145}$$

(the constraint $\gamma > \tau - 1$ becomes $\gamma > 1/2$, independently of φ). Thus, if $H \leq 1$, for all $\gamma < 1$ must be $\varphi = 1$, yielding

$$\beta = 1/2, \quad H = 1, \quad \psi = 1, \quad \varphi = 1, \quad h = 1/2 \tag{4.146}$$

for $\gamma \in (1/2, 1)$. The exponents are the same as in mean field theory [Maritan et al., 1996c] and the same as in the case of the Peano basin (Section 2.4), and indeed are different from the scaling exponents found in nature (Sections 2.9 and 2.10).

The most striking result is that the exponent of the distribution of aggregated areas is $\beta = 0.5$, quite different from the 0.43 ± 0.02 observational value consistently found in ($T = 0$) OCNs (Section 4.13). Before we draw any conclusions on the implications of the exact results, it is interesting to analyze the differences arising in the features of the global minima when heterogeneities are found, which thus affect the functional to be minimized.

Two cases have been analyzed: (i) random bonds, modeling nonhomogeneity in the local properties of the soil; (ii) random injection, modeling nonuniformity in the rainfall.

In the first case, one can show that the energy can be bounded from above with the corresponding value obtained in the absence of disorder. This gives, in the large L limit, an upper bound for the β exponent for $\gamma \in (1/2, 1)$. In the case of random rainfall, Maritan et al. [1996b] have shown that this type of randomness does not affect the energy behavior in the large L limit. All analytical results found previously for the homogeneous case, being based on the energy estimate in the thermodynamic limit, can be thus extended to this case, giving the same values for the exponents.

In the case of random bonds, we associate with each bond of the $L \times L$ square lattice (i.e., $2L(L-1)$ bonds are generated) a quenched random variable k_i, arbitrarily distributed and such that $\langle k_i \rangle = 1$. The $2L(L-1)$ variables are statistically independent and identically distributed. In the following, $b(i, \mathcal{T})$ will denote the label associated to the bond outcoming from site i within the tree $\mathcal{T}$. Let $\mathcal{T}^*(\gamma)$ denote one of the trees on which the minimum of the energy $E(\gamma, \mathcal{T})$ is realized in the homogeneous case for a given value of γ, and let $\mathcal{S}$ denote the set of all the spanning trees. Then one obtains

$$E(\gamma) = \min_{\mathcal{T}\in\mathcal{S}} E(\gamma, \mathcal{T}) = \sum_i A_i(\mathcal{T})^\gamma = \sum_i A_i(\mathcal{T}^*(\gamma))^\gamma \tag{4.147}$$

Denoting with $E_D(\gamma)$ the minimum of the energy in the heterogeneous case (and labeling all related exponents with the subscript D) averaged over the quenched disorder, the following inequality holds:

$$\begin{aligned} E_D(\gamma) &= \left\langle \min_{\mathcal{T}\in\mathcal{S}} \sum_i k_{b(i,\mathcal{T})} A_i(\mathcal{T})^\gamma \right\rangle \leq \left\langle \sum_i k_{b(i,\mathcal{T}^*(\gamma))} A_i(\mathcal{T}^*(\gamma))^\gamma \right\rangle \\ &= \left\langle \sum_i k_{b(i,\mathcal{T}^*(\gamma))} \right\rangle A_i(\mathcal{T}^*(\gamma))^\gamma \\ &= \sum_i A_i(\mathcal{T}^*(\gamma))^\gamma = E(\gamma) \end{aligned} \tag{4.148}$$

In the presence of this kind of disorder, E is thus bounded from above by the energy in the absence of disorder for any value of the γ parameter. In the large size limit this result gives bounds on scaling exponents. Also in this case Eq. (2.206) evaluated in $n = \gamma$ gives

$$\langle a^\gamma \rangle \sim L^{(1+H_D)(\gamma-\tau+1)} \tag{4.149}$$

which holds for any $\gamma > \tau_D - 1$. Eq. (4.149), compared with Eq. (4.133), leads to

$$(1 + H_D)(\gamma - \tau_D + 1) + 2 \leq 1 + 2\gamma, \qquad 1/2 \leq \gamma \leq 1 \tag{4.150}$$

In the case of self-affine behavior, Eq. (4.150) gives

$$H_D \geq 1 \qquad 1/2 \leq \gamma < 1 \tag{4.151}$$

As before, for the $\gamma = 1$ case the above inequalities are useless. In the case of self-similar behavior, inequalities for the fractal dimension φ_D can be analogously deduced:

$$\varphi_D \leq 1 \qquad 1/2 \leq \gamma \leq 1 \tag{4.152}$$

Being in general $0 \leq H \leq 1$ and $1 \leq \varphi \leq 1 + H$, Eqs. (4.151) and (4.152) give for $\gamma \in (1/2, 1)$

$$H_D = 1 \quad \text{and} \quad \varphi_D = 1 \tag{4.153}$$

The $\gamma = 1$ case has been exactly solved, and it can be mapped in an optimal path problem for which the solution has been derived in the framework of directed polymers in random media [see Maritan et al., 1996b; Colaiori et al., 1996]. The corresponding values for the exponents are

$$\beta = 2/5, \quad \psi = 2/3, \quad H = 2/3, \quad \varphi = 1, \quad h = 3/5 \tag{4.154}$$

Disorder can also be introduced in the system by replacing the constant injection in each site of the lattice with the random local injection mimicking a spatial inhomogeneity in the rainfall R_i in Eq. (4.127). It follows that [Colaiori et al., 1996]

$$A_i = \sum_j \lambda_{ij} R_j, \qquad \text{with} \quad \lambda_{ij} = \begin{cases} 1 & \text{if } i \text{ is connected to } j \\ & \text{through drainage directions} \\ & \text{or if } j = i \\ 0 & \text{otherwise} \end{cases} \tag{4.155}$$

The minimum of the energy averaged over the "random-rainfall" will be denoted by $E_R(\gamma)$ and for a given value of γ is given by

$$E_R(\gamma) = \left\langle \min_{\mathcal{T}\in\mathcal{S}} \sum_i A_i(\{R\}, \mathcal{T})^\gamma \right\rangle \tag{4.156}$$

where $\mathcal{S}$ denotes the set of all spanning trees, $\mathcal{T}$, and $\{R\}$ denotes the whole set of random variables. As in Eq. (4.147), $\mathcal{T}^*$ denotes the reference tree on which minimum energy is realized in the absence of random rainfall and for a given value of γ. One thus obtains

$$\begin{aligned} E_R(\gamma) &= \left\langle \min_{\mathcal{T}\in\mathcal{S}} \sum_i A_i(\{R\}, \mathcal{T})^\gamma \right\rangle \leq \left\langle \sum_i A_i(\{R\}, \mathcal{T}^*(\gamma))^\gamma \right\rangle \\ &= \left\langle \sum_i \left(\sum_j \lambda_{ij}(\mathcal{T}^*(\gamma)) R_j \right)^\gamma \right\rangle \leq \sum_i \sum_j \lambda_{ij}(\mathcal{T}^*(\gamma)) \langle R_j \rangle^\gamma \\ &= \sum_i A_i(\mathcal{T}^*)^\gamma = E(\gamma) \end{aligned} \tag{4.157}$$

Thus one again obtains [Maritan et al., 1996b; Colaiori et al., 1996]

$$E_R(\gamma) \leq E(\gamma) \sim \min(L^2, L^{1+2\gamma}) \tag{4.158}$$

In this case, it is also possible to bound the energy from below with an argument analogous to that used for the homogeneous case. The detailed computations are shown in Colaiori et al. [1996]. Thus one can conclude that

$$E_R(\gamma) \sim \min(L^2, L^{1+2\gamma}) \tag{4.159}$$

and all the exact results of the homogeneous case hold.

We are now in a position to further strengthen the discussion of Section 4.13: What are the implications of the suboptimal energetic performance, yet the better representation of natural networks allowed by local minima? In Section 4.13 we observed the progressive departure of the exponent β from the typical observational value of 0.43 as the minimum search procedure was refined. The values of the Metropolis minima were consistently in the range $\beta = 0.48$–0.50, where the 0.5 value, that we have shown to correspond to the global optimum, was obtained for the least-constrained arrangements (multiple outlets, periodic boundary conditions, wide schedule of temperatures in the Metropolis scheme so as to get rid of any effect of the initial condition – see Section 4.13). From the above observation, Rinaldo et al. [1996b] suggest that the adaptation of the fluvial landscape to the geologic and climatic environment corresponds to the settling of optimal structures into suboptimal niches of their fitness landscape and that feasible optimality, that is, the search for optima that are accessible to the dynamics given the initial conditions, might apply to a broad spectrum of problems of general interest.

By comparing the results of Section 4.13 with the exact results derived in this section, one observes that the fact that each constraint affects the feasible optimal state, to different degrees depending on its severity, matches the experimental finding (Sections 2.0 and 2.10) of consistent, though with minor variations, scaling exponents that describe the morphology of the fluvial basin. Different fractal signatures embedded in linked scaling exponents suggest that the strive for fractality is adapted to the climatic and geologic environment. The worse energetic performance and yet the better representation of natural networks by suboptimal ($T = 0$) OCNs imply a role of geologic constraints in the evolution of a channel network. Channel networks cannot change freely, regardless of initial conditions, because these conditions leave long-lived geomorphic signatures. An interesting question is whether the optimization that nature seems to perform in the organization of the parts and the whole of a river basin can be farsighted, that is, capable of evolving in a manner that completely disregards initial conditions and allows for major migration of divides in the search for a more stable configuration, even though it necessarily involves evolution through transient unfavorable conditions. The experiments described before seem to indicate that the type of optimization that nature performs is myopic, that is, willing to accept changes only if their impact is favorable right after their occurrence (in the immediate) rather than in the long run.

The fact that suboptimal structures differ from the ground state is neither new nor confined to fractal structures. As an example, introduced by Rinaldo et al. [1996b], the folding of an amino-acid sequence in proteins is an NP-complete problem. The native state of a protein is thermodynamically stable below a characteristic temperature, called the folding transition temperature. However, it is arguable whether a protein has its native state in a local or a global minimum of its free energy. The observation of several functionally active and significantly different conformations in a mutant T4 lysozyme and the assumption of a 'molten state', according to which the biologically active form of insulin is an intermediate state, arguably represent a case of feasible optimality.

Analogous features can be observed in another context. In a two-dimensional random ferromagnet at sufficiently low temperatures, the domain wall (i.e., the interface) between regions of different magnetization is self-affine. In a box, with + and − boundary conditions at the top and bottom faces, respectively, and antiperiodic boundary conditions at the side faces, the domain wall configuration (at zero temperature) is known to be obtainable as the global minimum of the internal energy [e.g., Huse and Henley, 1985]. In two-dimensional ferromagnets, it is also known [Huse and Henley, 1985] that the relevant scaling exponent of the self-affine domain wall of the global optimum is 2/3. With local zero-temperature spin-flip dynamics (i.e., spins are updated only if the global energy decreases), it was found that the global minimum is never reached and the domain walls are self-similar rather than self-affine [Swift et al., 1996] with a fractal dimension around 1.63. This result is robust and independent of whether one adopts a single or two spin flip dynamics and whether one uses a discrete spin model or a continuum Langevin equation [Swift et al., 1996].

The claim that river networks and domain walls in random ferromagnets are not free to explore extended regions of their fitness landscapes, suggests that nature might not search for global minima when striving for optimality. This might be true for many optimal configurations of a complex physical system with many degrees of freedom that minimize a cost function arising in a variety of contexts – for example, the classic travelling salesman problem, equilibrium thermodynamics, spin glasses and the problem of protein folding.

Clearly not all problems that can be mathematically recast in the form of a global minimum yield to the formation of fractal structures. Glasses are such an example. Scale-free features rather emerge in those optimality problems where local interactions are felt globally, through boundary conditions or through the nature of the driving forcings. Thus they constrain the system and frustrate the optimization process. Then, and only then, the strive for optimality adapts to the constraints of the specific problem, and scale invariance emerges out of feasible optimality.

The problem studied in this section (as well as in the ferromagnet) is relevant to suggest the above interpretation of the dynamic reason for fractal growth for two reasons. First, the exact features of the global minimum are known, thus allowing our conclusion that local, dynamically accessible, solutions differ from the ground state. Second, observational data match the features of suboptimal structures.

Many more examples are conceivable, and many cases where the suboptimal states differ from the ground state are known. We deem reasonable that complex proteins fold into dynamically accessible states just as river networks or domain walls of ferromagnets adjust to feasible shapes, but nevertheless the emergence of fractals would be tied, in our interpretation, to the establishment of global interactions constraining the system in a dynamic context striving for optimality.

CHAPTER 5

Self-Organized Fractal River Networks

Self-organized criticality (SOC) refers to the tendency of large dissipative systems with many degrees of freedom to build up a state poised at criticality that is characterized by a wide range of length and timescales. SOC is now a common name for a general theory of the dynamics of fractal growth, whose main features are discussed here, especially with reference to applications within the context of earth sciences. In this chapter we show that principles of critical self-organization are at work in the development of the fluvial landscape. We also show that optimal channel networks (Chapter 4) are spatial models of self-organized criticality. This reinforces the suggestion that natural fractal structures like river networks may indeed arise as a joint consequence of optimality and randomness. Specifically, we suggest that natural fractal structures in the fluvial landscape are dynamically accessible optimal states, corresponding to locally optimal niches of a complex fitness landscape where evolution can settle in a stable manner. Such relative stability is achieved with respect to perturbations and is nonetheless reminiscent if its dynamic history, including an imprinting of its initial conditions and long-lived signatures of boundary conditions, here surrogating geologic constraints.

5.1 Introduction

Two related problems have interested scientists for a long time. One is the fundamental dynamic reason behind Mandelbrot's observation that many structures in nature – such as river networks or coastlines – are fractal, that is, they look 'alike' on many length scales. The other is the origin of the widespread phenomenon called $1/f$ noise, originally referring to the particular property of a time signal, be it the light curve of a quasar or the record of river flows, which has components of all durations, that is, without a characteristic timescale. The name $1/f$ refers to the power law decay with exponent -1 of the power spectrum $S(f)$ of certain self-affine records (see Chapter 2) and is conventionally extended to all signals whose spectrum decays algebraically, that is, $S(f) \propto f^{-\alpha}$. Power law decay of spectral features is also viewed as a fingerprint of spatially scale-free behavior, commonly defined as critical. In this framework, criticality of a system postulates the capability of communicating information throughout its entire structure, connections being distributed on all scales.

The causes and possible relations between the abundance of fractal forms and $1/f$ signals found in nature have puzzled scientists for years.

Per Bak and collaborators have addressed the links between the above problems, suggesting that the abundance in nature of spatial and temporal scale-free behaviors may reflect a universal tendency of large, driven dynamic systems with many degrees of freedom to evolve into a stable critical state, far from equilibrium, characterized by the absence of characteristic spatial or temporal timescales. The key idea and its successive applications, which we will review in the following sections, address such a universal tendency and bear important implications on our understanding of complex natural processes. The common dynamic denominator underlying fractal growth is now central to our interests in landform evolution.

The resistance to Bak's idea of universality was (and still is in some circles) noteworthy. Science, and geomorphology in particular, is largely committed to the reductionist approach. The reductionist tenet is that if one is capable of dissecting and understanding the processes to their smallest pieces, then the ability to explain the general picture, including complexity, is granted. However, the reductionist approach, affected as it is by the need to specify so many detailed processes that operate in nature and for the tuning of many parameters, though suited to describe individual forms, is an unlikely candidate for explaining the ubiquity of scaling forms and the recursive characters of processes operating in very different conditions. Are the scale-free, recursive characters of the evolution of complex systems tied to the detailed specification of the dynamics? Or, on the contrary, do they appear out of some intrinsic property of the evolution itself? We believe, following Bak, that the invisible hand guiding the evolution of large interactive systems should be found in some general properties of the dynamics rather than in some unlikely fine-tuning of its elementary ingredients.

One crucial feature of the organization of fractal structures in large dynamic systems is the power law structure of the probability distributions characterizing their geometrical properties. This behavior, characterized by events and forms of all sizes, is consistent with the fact that many complex systems in nature evolve in an intermittent, burst-like way rather than in a smooth, gradual manner. The distribution of earthquake magnitudes obeys Gutemberg and Richters law [1954], which is a power law of energy release. Fluctuations in economics also follow power law distributions with long tails describing intermittent large events, as first elucidated by Mandelbrot's [1963] famous example of the variation of cotton prices. Punctuations dominate biological evolution [Gould and Eldridge, 1993] where periods when many species become extinct and new species appear interrupt long periods of stasis.

Lévy distributions (characterized by algebraic decay of tails, i.e., of the probability of large events) describe mathematically the probabilistic structure of such events. They differ fundamentally from Gaussian distributions – which have exponential decay of tails and therefore vanishing probabilities of large fluctuations – although both are limiting distributions when many independent random variables are added together. In essence, if the distribution of individual events decays sufficiently rapidly, say, with a nondiverging second moment, the limiting distribution is Gaussian. Thus the largest fluctuations appear because many individual events happen that push their action in the same direction. If, instead, the individual events have a diverging second moment – or even diverging average size – the limiting distribution could be Lévy because its large fluctuations are formed by individual events rather than by the sum of many events.

When studying large, catastrophic events in a large system with interacting agents, one can try to identify an individual event as the particular source. Rather than recognizing the achievement of a critical state, a 'Gaussian' observer may discard the event as atypical – as noted by Mandelbrot [1982] – when studying the statistics of fluctuations because the remaining events trivially follow Gaussian statistics. A rather common reaction to catastrophic concerted actions is to look for specific reasons for large events. Economists tend to look for specific mechanisms for large stock fluctuations, geophysicists look for specific configurations of fault zones leading to catastrophic earthquakes, biologists look for external sources, such as meteors hitting the earth, in order to explain large extinction events, physicists view the large-scale structure of the universe as the consequence of some particular dynamics. In essence, as Bak put it, one reluctantly views large events as statistical phenomena.

Bak noted that there is another explanation, unrelated to specific events and embedded in the mechanisms of self-organization into critical states. In such states each large event has a specific source, a particular addition of a grain of sand landing on a specific spot of a sandpile and triggering a large avalanche, the burning of a given tree igniting a large forest fire, the rupture of a certain fault segment yielding the big earthquake, or the slowing of a particular car starting a giant traffic jam. Nevertheless, even if each of the particular initiating events above were prevented, large events would eventually start for some other reason at some other place in the evolving system. In critical systems no local attempt to control large fluctuations can be successful except for directing events to some other part of the system.

What are the signatures and the origins of the process of self-organization? Bak (see, for example, Bak and Paczuski [1993]) suggested that SOC systems have one key feature in common: The dynamics is governed by sites with extremal values of the 'signal,' be it the slope of a sandpile or the age of the oldest tree in a burning forest, rather than by some average property of the field. In these systems nothing happens before some threshold is reached. When the least stable part of the system reaches its threshold, a burst of activity is triggered in the system, yielding minor or major consequences depending on its state. Complexity arises through the unpredictable consequences of the bursts of activity, suggesting that the dynamics of nature may often be driven by atypical, extremal features. This is suggestive, among other things, of Kauffman's example of biological evolution as driven by exceptional mutations leading to species with a superior ability to proliferate or to Bak's example of the introduction of program trading causing the crash of stock prices in October 1987. In both cases a new fact leads to breakthroughs propagating throughout an entire concerted system because it generates chain reactions of global size. Another feature of self-organizing processes is that, in order to have a chance to appear ubiquitously, they must be robust with respect to initial conditions or to the presence of quenched disorder and should not depend on parameter tuning.

This chapter investigates whether the dynamics of the fluvial landscape, by conforming to the above general features, may be viewed as a particular case of self-organized criticality.

5.2 Self-Organized Criticality

As seen in Section 5.1, the concept of self-organized criticality originated with reference to the search for a dynamic explanation for the behavior of many spatially extended dynamic systems with both spatial and temporal degrees of freedom. Bak et al. [1987, 1988] recognized, through an exceedingly simple dynamic model, which we describe below, that temporal effects known as $1/f$ noise and the evolution of spatial structures with fractal properties are related by a general organizing principle governing a

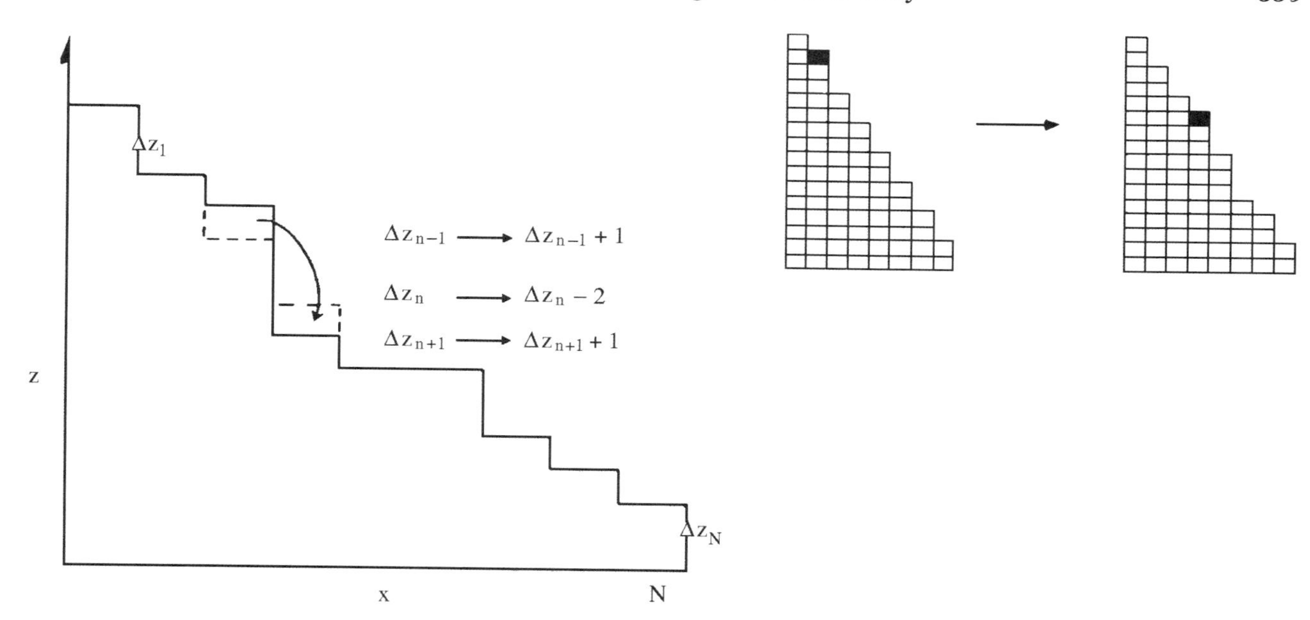

Figure 5.1. One-dimensional sandpile automaton [from Bak, et al., 1988].

wide class of dissipative coupled systems. Such a principle of organization has been named self-organized criticality because the systems subject to it evolve naturally toward a critical state with no intrinsic time or length scale.

The original model of SOC demonstrated numerically that simple dissipative dynamic systems with local, interacting degrees of freedom (the now rightly famous *sandpile* model) in two dimensions or three dimensions evolve into the final states without detailed specification of the initial conditions. These states are then viewed as attractors for the underlying dynamics.

The basic model displaying the above properties can be briefly described as follows. Figure 5.1 shows a model of a one-dimensional sandpile of length N. Boundary conditions are such that sand can leave the system at the right-hand side only. The basic field is elevation, here $z(i)$, $i = 1, N$, at site i. Height differences $\Delta z_n = z(n) - z(n-1)$ between successive positions along the sandpile are identified in this manner. The dynamics is defined from the following rules:

- sand is added at the nth position by letting $\Delta z_n \to \Delta z_n + 1$ and $\Delta z_{n-1} \to \Delta z_{n-1} - 1$;
- when the height difference becomes higher than a fixed critical value Δz_c, one unit of sand tumbles to the lower level, that is, $\Delta z_n \to \Delta z_n - 2$ and $\Delta z_{n\pm1} \to \Delta z_{n\pm1} + 1$ for $\Delta z_n > \Delta z_c$;
- closed and open boundary conditions are used for the left and right boundaries, $\Delta z_0 = 0$, $\Delta z_N \to \Delta z_N - 1$ and $\Delta z_{N-1} \to \Delta z_{N-1} + 1$ for $\Delta z_N > \Delta z_c$, respectively.

The above rules define a nonlinear discretized diffusion equation, which controls the evolution until all the Δz_n are below threshold, at which point another grain of sand is added at a random site. The discrete variable Δz_n at time $t+1$ depends on the state of the variable and its neighbors at time t

(Figure 5.1). The state of the system is specified by an array of integers representing the height difference between neighboring plateaus.

Although the one-dimensional critical states differ dramatically from the more interesting two-dimensional or three-dimensional cases, it is instructive to understand their behavior. For instance, if sand is added randomly to an empty system, the pile will build up, eventually reaching the point where all the height differences Δz_n assume the critical state $\Delta z_n = \Delta z_c$. This is the least stable of all stationary states. Any additional sand particle simply falls from site to site (in the above example from left to right) and exits the system at the end $n = N$ leaving the system in the minimally stable state. In practice, the effect of a small local perturbation may be communicated throughout the system, but the system is robust with respect to noise as it eventually returns to the globally minimum stable state. If units of sand are added randomly, the resulting sandflow is also random with power spectrum $1/f$, that is, white noise.

However appealing the behavior of the one-dimensional system may be for clarifying the basic mechanism, Bak et al. [1988] noted that the robustness of the minimally stable state is lost in two and higher dimensions. Also, they noted that the dynamic selection principle leading to the least stable stationary state is quite independent of how the sandpile is built up, as analogous results were obtained by starting with a flat sand surface and raising 'slowly' the left end of the system or, alternatively, randomly adding slope and letting the system obey the above rules. The latter case is particularly relevant to the case of river network formation because, as noted by Bak et al. [1987, 1990], it represents the dynamics of a system with a random distribution of critical height differences and a uniformly increasing slope. Another interesting case is obtained by starting with a very unstable state (i.e., $\Delta z_n > \Delta z_c$ almost everywhere) and letting the system relax. In one dimension, the minimally stable state will be reached even if the boundary conditions are such that the sand cannot leave the system.

By generalizing the above rules to higher dimensions, Bak et al. [1988, 1989] concluded that sandpile systems naturally evolve toward a self-organized critical state with spatial and temporal power law scaling behavior. The spatial scaling leads to a fractal structure and the temporal evolution, as described by the flow of grains out of the system, yields a frequency spectrum adjusting on a power law form $S(f) \propto f^{-\zeta}$, thus reflecting the dynamics of the extended system rather than noise. The resulting spatial organization of the system, which proves resilient to significant insertions of quenched randomness obtained by disrupting a priori possible sandflows, blends all the above characters and unifies seemingly unrelated features through the choice of dynamics.

Figure 5.2 shows the probability distribution of cluster sizes in two-dimensional and three-dimensional sandpile domains [after Bak et al., 1988] at criticality where averages of over 200 realizations have been employed. The dashed lines are straight lines with slopes -1.0 (two-dimensional case, Figure 5.2(a)) and -1.37 (three-dimensional case, Figure 5.2(b)). Clusters, in this case as in Section 2.1.3, are defined as regions of connected pixels where elevation is changed because of the propagation of the same random disturbance.

To investigate the temporal evolution of the developing clusters defined by the avalanche size, Bak et al. [1988] described the effect of a

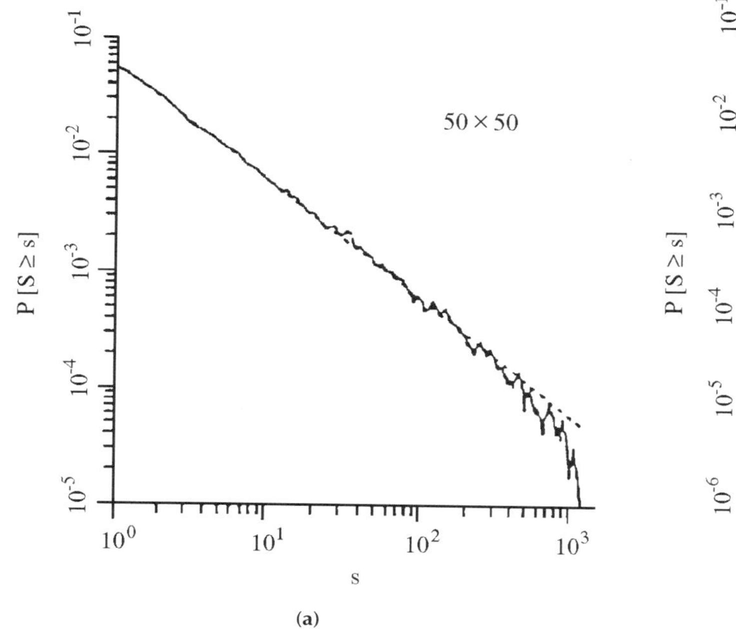

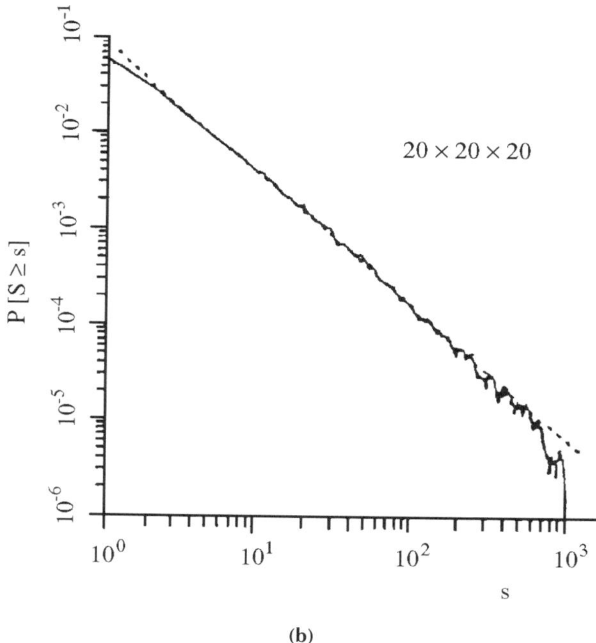

Figure 5.2. Distribution of cluster sizes at criticality in (a) two and (b) three dimensions [after Bak et al., 1988]

local perturbation at a single site on a static critical state. The perturbation spreads to some nearest-neighbor sites, then to next-nearest neighbors and so on in a *domino* effect, eventually dying out after a total time T, having triggered a total of s slidings. Figures 5.3(a) and 5.3(b) show the probability distribution of lifetimes, $P[T \geq t] \propto t^{-\alpha}$ (with $\alpha \sim 0.43$ for two-dimensional grids, $\alpha \sim 0.92$ for three-dimensional grids) [Bak et al., 1988]. For the 50×50 array, the exponent $\alpha \approx 0.43$ yields a noise spectrum $f^{-1.57}$, while for the $20 \times 20 \times 20$ array one gets $\alpha \approx 0.92$ yielding an $f^{-1.08}$ spectrum. Notice that the curves for the lifetime distribution fit a power law only over a decade or so, limiting the range of reliable data in Figure 5.3.

With reference to temporal patterns of activity, things turn out to be more complex [e.g., Hwa and Kardar, 1992]. In fact, what we have described above refers to a particular case of time activity, that is, the limit obtained by adding sand grains one at a time and allowing the system to completely relax between additions. Although many interesting scaling behaviors were found, there are a number of drawbacks associated with this particular way of probing the system. For instance, as the interval between sand additions varies, time is not well defined anymore and temporal fluctuations of transport quantities produce no sign of $1/f$ noise. Also, if the existence of criticality in a dynamic system depends sensitively on a small input rate (or driving force), it may become inapplicable to many systems in nature that exhibit $1/f$-type noise in the likely presence of arbitrary driving forces.

One important conclusion [Hwa and Kardar, 1992] is that critical scaling is not destroyed by a finite input rate. Rather, interesting temporal fluctuations such as $1/f$ noise only appear when avalanches are allowed to overlap in the presence of external driving forces. In this 'interactive' avalanche

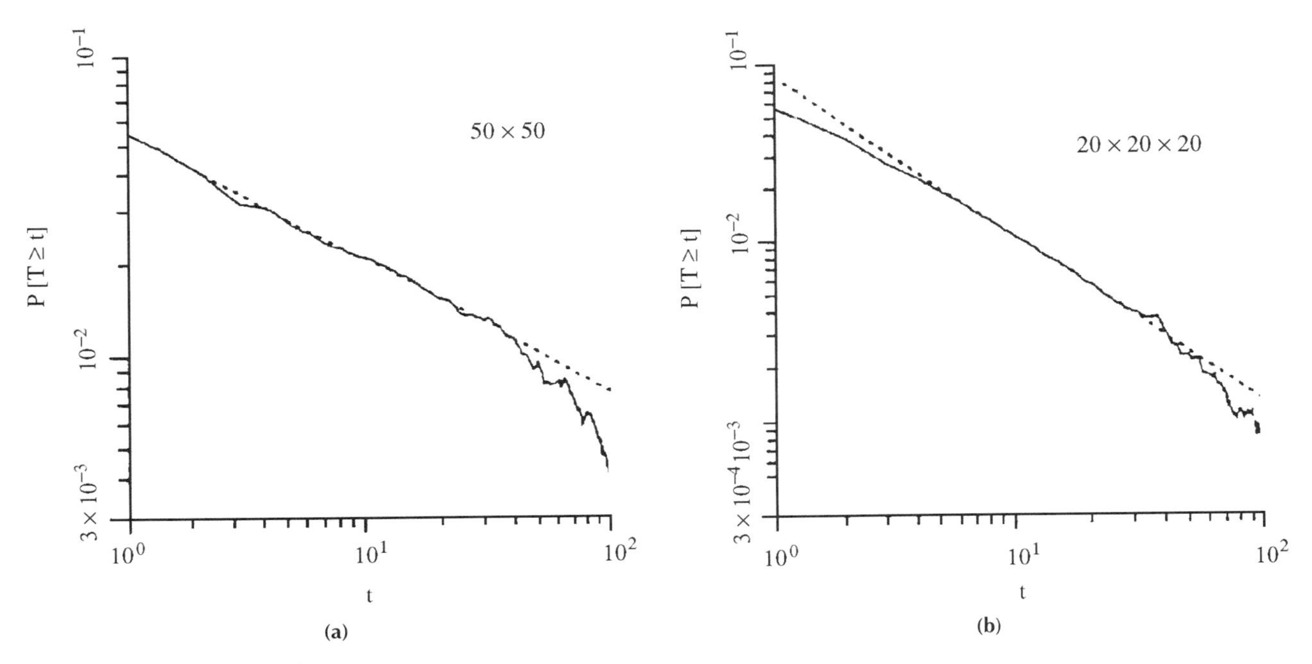

Figure 5.3. Distribution of lifetimes corresponding to the experiments of Figure 5.2 [after Bak et al., 1988].

region, the transport quantities exhibit $1/f$ signals even at timescales larger than the maximum duration of individual avalanches.

An important result [Grinstein, Lee, and Sachdev, 1990; Hwa and Kardar, 1992] concerns the compatibility of field theories of dissipative transport with self-organized criticality. As we have shown (Section 1.4.3), continuum theories have often been used to simulate landscape processes. The particular issue of whether, in the context of driven-diffusion theories, noisy nonequilibrium models with conserving deterministic dynamics may exhibit self-organized criticality, is postponed to a specific discussion in Chapter 6 in the context of coupled hillslope and fluvial processes.

The successive extensions of the original theory have been successful in explaining a number of real-life phenomena involving composite systems (whose basic structure is made up by a large number of elements that interact over a short range) in which a minor event starts a chain reaction that can affect any number of elements in the system. Experiments and models have demonstrated that many compound and complex systems at the heart of geology, economy, biology, meteorology, and fluid mechanics show signs of self-organized criticality. We will briefly review a few applications relevant to fractal river networks in the following sections of this chapter.

5.3 SOC Systems in Geophysics

Of particular relevance to the issues dealt in this book are a few applications of SOC to geophysical systems. We will describe them in some detail to take advantage of some of their properties in the following.

In the SOC framework, natural geological or geophysical systems are viewed as in marginally stable states. When perturbed from these states, they will evolve naturally back to the state of marginal stability, that is, the

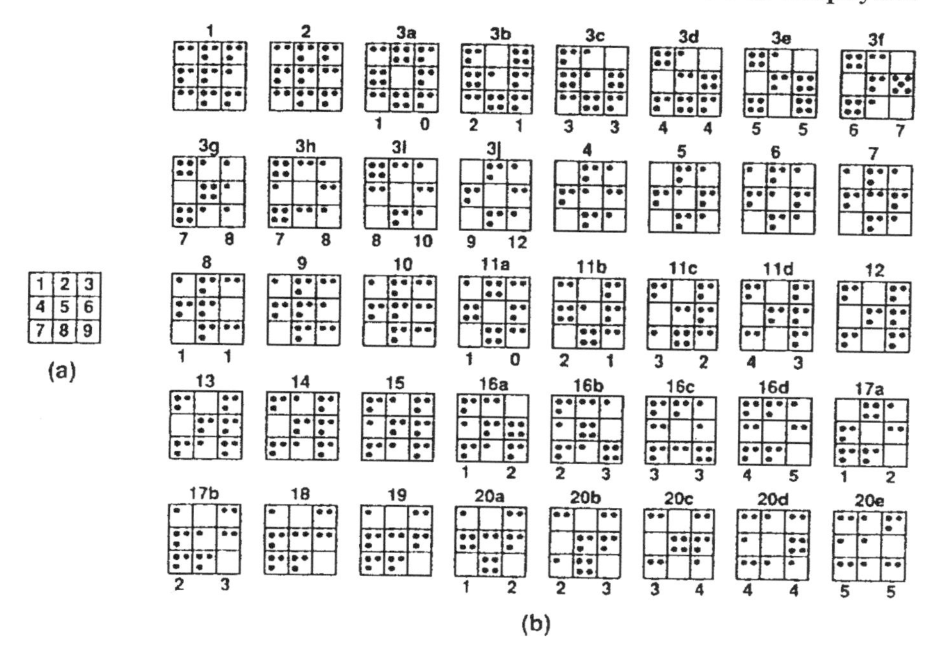

Figure 5.4. Illustration of the earthquake automaton [Turcotte, 1992] for a 3 × 3 grid.

critical state, where there is no natural length scale so that fractal statistics are applicable.

A geophysical analog of the sandpile model of Figure 5.1 is a discrete model simulating stress redistribution based on nearest-neighbor interactions, which bears striking resemblances to the seismicity associated with an active tectonic zone [Kadanoff et al., 1989; Bak and Tang, 1989; Turcotte, 1992]. This analog is shown in Figure 5.4. The model, whose implications are described later, is readily explained as follows (following Turcotte [1992]). Consider a square grid of n boxes. Particles are added and lost from the grid using the following procedure:

- A particle is randomly added to one of the boxes.
- When a box has (say) four particles it becomes unstable and the four particles are redistributed among the nearest neighbors, here chosen to be the four adjacent grid points along vertical and horizontal directions. A boundary effect is induced for edge boxes where redistribution results in the loss of one particle. Also notice that corner boxes lose two particles after undergoing redistribution.
- If, after a redistribution of particles from a box, any of the adjacent boxes has four or more particles, it becomes unstable and one (or more) further redistribution must be performed. Notice that multiple events are common occurrences for large grids. Note also that arbitrary choices may be needed once redistribution forces more than a single box into unstable conditions. Hence the model needs to decide which box should be considered first for further redistribution. This choice can be randomly made (or, instead, by choosing some specific rule) without effects on the statistical evolution of the system.

In this simple statistical nearest-neighbor model, multiple events may interact and spread over large fractions of the grid. Figure 5.4 [after

Turcotte, 1992] illustrates the model behavior in a 3×3 grid for a sequence of random additions in box 6 (step 2), box 5 (step 3a), box 5 (step 4), box 6 (step 5), box 1 (step 6), and so on. Letters indicate redistributions in the example.

It is interesting to note that the frequency–size distributions obtained by such models of self-organized criticality closely resemble the Gutemberg–Richter relationship for the distribution of earthquakes in a zone of active tectonics.

Another important implication of the possible establishment of critical states in the crust concerns long-range geophysical interactions. Specifically, one wonders whether an earthquake in one part of the planet can trigger a similar event at large distances, a possibility that classical approaches tend to rule out. In fact, stresses associated with any seismic wave are too small to trigger earthquakes. Also, no observational evidence for triggered events on a large spatial scale is available. Even if stress changes associated with the fault displacements are localized and space-damped, large-scale interactions of faults with each other (characteristic of critical phenomena) are viewed as possible in the framework of SOC, and possibly explain the (partial) success of some earthquake predictions based on SOC patterns of distributed regional seismicity in reproducing quiescence, clustering of events, and changes in aftershock statistics [e.g., Turcotte, 1992].

Interestingly, a study [Grieger, 1992] of quaternary fluctuations of the oxygen isotope abundance ratio in the ocean derived by marine sedimentary cores (a measure of the global ice volume) suggests a link with the above SOC model. In fact, a simple cellular automaton of ice-surge dynamics is constructed as follows: When the thickness of an ice sheet is greater than a certain value at a particular location, geothermal fluxes cause ice melting at the ground and surging to occur. This dynamics is simulated by a rectangular grid of ice columns, and the model is driven by the repeated addition of one unit of ice at a randomly selected position. The foremost result of Grieger [1992] is that the power spectra of the volume fluctuations in the SOC model and of the quaternary ice volume fluctuations from the oxygen isotope record are indistinguishable, suggesting that continental ice sheets may be in a self-organized critical state.

The interplay of short- and long-range interactions embedded in SOC systems is also central to the development of fluvial structures. Specifically, we wonder whether erosion activities self-organize themselves into a process that shapes the geomorphology of river networks. Along these lines, Takayasu and Inaoka [1992] proposed a model of river-like erosion embedding a different type of self-organized criticality in a model of erosion. Specifically, the effect of water erosion is analyzed with respect to the creation of river patterns and landscapes as follows. Let $z(x, y)$ be the landscape elevation in a two-dimensional triangular lattice defining sites (x, y); let also $s(x, y)$ be an analog to *flow intensity*. The following rules are then enforced:

- Rainfall with unit intensity is assumed to constantly fall on every site. Thus the overall effects of a given climate are simulated by a continuous injection of mass.
- For every site (x, y) the lowest value of the elevation among the six nearest neighbors is pointed out, $z_{\min}(x', y')$ where (x', y') denotes a nearest-neighbor

site. It is assumed that water flows in the direction of steepest descent, that is, to the lowest neighbor, if its elevation is lower than $z(x, y)$. This defines drainage directions as postulated in Section 1.2.9. The flow intensity $s(x, y)$ is defined by the sum of flows from the neighbors and the unit contribution at the site from rainfall, and therefore $s(x, y)$ equals the size of the drainage area A for the site $s(x, y)$.

- By erosion, the height at (x, y) is lowered by Δz, a (positive and monotonically increasing) function of the drop $z(x, y) - z_{\min}(x', y')$ and flow $s(x, y)$.

Iteration of the above model, jointly with an elaborate specification of the formation of pits and loops in the evolving structure, yields the development of networks.

We observe that the model never reaches a steady state, erosion being produced in a continuous fashion even for small values of the drops in elevation and the flowrates. As seen in Section 1.4.3, an evolution of the above type is characteristic of fluvial erosion models of noncohesive sediments making up the exposed surfaces in a tectonically inactive region where uplift is needed to reach a nonflat steady state. Also, the model requires some additional ad hoc modeling efforts that somewhat detract from its generality. For instance, it involves an auxiliary scalar field to treat "lakes," that is, sites lowest among their neighbors, for which water accumulation has to be accounted for until overtopping recreates a network pattern.

It is interesting to analyze the cumulative distribution of drainage area $P[A \geq a] \sim P[S \geq s]$, S being the random flow intensity at any site defined above and A the total contributing area, obtained by the above model. A power law was observed as $P[A \geq a] \propto a^{-0.41}$ in a 512×512 system size with 5,000 time steps, for all scales allowed by the lattice dimensions. We note that the computational effort to obtain the above figure is large and that the exponent is similar to the exponents observed in nature for the probability of exceedence of total contributing drainage areas (Section 2.5.2). It is interesting to note that certain modifications of the mathematical definition of the law responsible for erosion do not change (within error bars) the exponents of the power law. This robustness will be reconsidered later on in this chapter. However, we notice that, although the planar statistics obtained by Takayasu and Inaoka [1992] are very reasonable, the altitude structure of the landscape is unrealistic.

Although some geomorphologic and dynamic aspects of the patterns produced by the above model need to be revisited, it is interesting that the erosion activity here is allowed to depend on the self-organizing structure through the dependence of erosion on flow intensities $s(x, y)$ (the sum of all sites connected to x, y through drainage directions) and that the results are geomorphologically recognizable structures.

A few other studies are deemed relevant to SOC in landform evolution. Christensen, Olami, and Bak [1992] studied generic *nonconservative* models displaying self-organization. In the context of sandpile models, for example, conservation refers to the fate of the mass removed from a site because of dynamic rules and redistributed among nearest neighbors. Unit level of conservation means that all the mass is redistributed. Zero level of conservation means that it is removed from the system. Importantly, Christensen et al. [1992] showed that under generic conditions, nonconservative SOC schemes exhibit power laws in the probability distributions describing the developing structure and that the scaling exponents depend on the level of conservation. Thus the fractal dimensions associated with the structure are nonuniversal and process-dependent.

Basically, the models rely on an incomplete redistribution among nearest neighbors of the quantity associated with some critical threshold. Relaxations of the system are forced to depend on a parameter underlying the level of conservation (redistribution) yielding short-range correlations in the early stages of development and long-range interactions at criticality. The subtle and important modification of the model with respect to the original formulation is that in the latter the decrease of the value at the relaxing site was a fixed amount and the transfer to the neighbors was independent of the previous state of the system. The Christensen et al. [1992] model postulates that the relaxing amount and the transfer to the neighbors depends on the state of the system and thus on its development. The overall configuration of the model describes a much more realistic situation because the exponents of the power laws describing the power spectra have nonuniversal behavior through the choice of conservation level, a likely occurrence in nature.

5.4 On Forest Fires, Turbulence, and Life at the Edge

Other recent applications of the theory of self-organized criticality in nonconservative models of the type discussed herein [Bak, 1992; Bunde, 1992] suggest interesting connections with the material covered later in this book. Specifically, Drossel and Schwabl [1992] studied a forest fire model as an evolution of the original model proposed by Bak et al. [1988].

The basic model is a probabilistic cellular automaton defined on a d-dimensional lattice of size L^d. In the beginning each site is occupied either by a tree or by a burning tree, or it is empty. The state of the system is updated through the following rules:

- A burning tree becomes an empty site and all trees next to it become burning trees.
- At an empty site a tree grows with probability P.

It was shown that at low values of P the model is not critical but it becomes more and more deterministic with decreasing P, as it develops regular spiral-type wave fronts. The model becomes critical when a lightning parameter f is included. A third rule is therefore added:

- A tree becomes a burning tree during one time step with probability f if no neighbor is burning.

The above rules for self-organization produce power law distributions of the number N_T of tree clusters T of size s, that is, $N_T(s) \propto s^{-\tau-1}$ with $\tau \approx 1$ for $d = 2$ and $\tau = 1.5$ for large d. Also, lifetime distributions of the fire follow power laws. An example of cluster size distribution obtained with the above rules is illustrated in Figure 5.5.

Interestingly, Drossel and Schwabl [1992] observe that in the self-organized critical state the mean forest density in steady state is determined by an extremum principle, that is, it assumes its minimum possible value. In fact, if $\bar{\rho}$ is the mean overall forest density in the system at steady

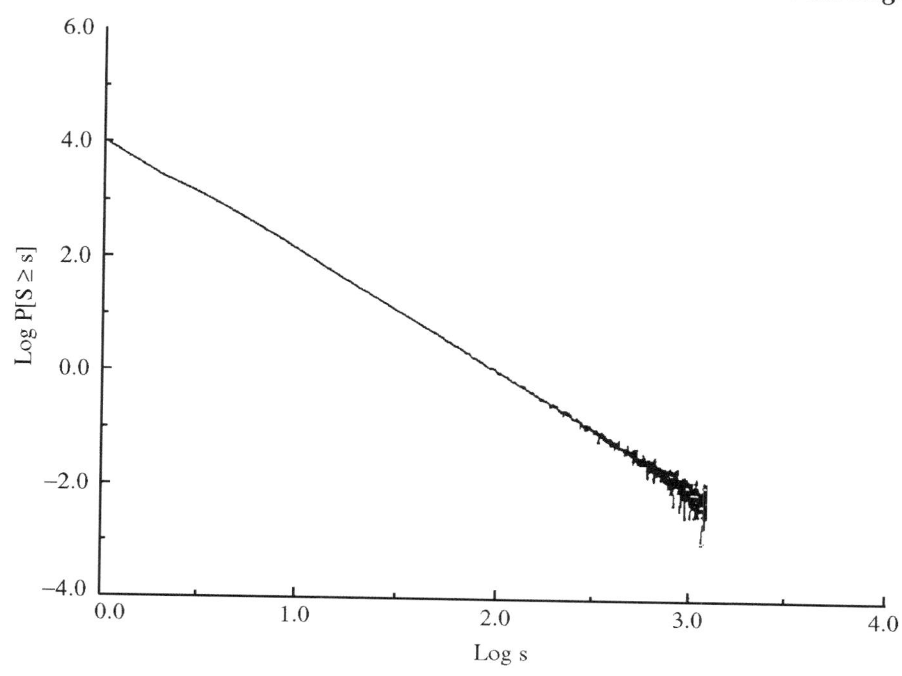

Figure 5.5. Cluster size (s) distribution for a self-organized critical forest-fire model (for $f/P = 1/200$ and $d = 2$) [after Drossel and Schwabl, 1992].

state, lightning will strike the system every $(f L^d \bar{\rho})^{-1}$ time steps on the average, and the minimum possible value of the mean forest density in a region regularly visited by a fire can be shown to be $\bar{\rho} \approx 0.39$ for $d = 2$. The latter is precisely the value obtained by SOC simulation. Thus it is of interest to investigate whether self-organized critical structures are optimal in some sense also in the context of river networks.

Recently the one-dimensional model of forest fires provided interesting analytical results [Drossel and Schwabl, 1992; Drossel, Clar, and Schwabl, 1993; Paczuski and Bak, 1993] directly related to cascade models. The basic idea is as follows. The system is defined on a line, and thus it has chains of empty sites (the hole H) and chains of trees (the forest T), initially distributed in any given manner along the line. Let $N_H(s, t)$ be the number of holes of size s at time t and $N_T(s, t)$ the number of forests of same size. Trees grow and form bigger and bigger forests, which eventually burn when hit by lightning (with assigned probability f). The analytical manipulations start with the calculation of the steady-state variations $\delta N_H(s)$, which are given by the following reasoning. A hole of size s may (a) disappear in s positions because of a growth at any of its s sites, or (b) appear in two positions from a larger size s' because of a growth, or (c) appear because of burning of a forest of size $s'' < s$ (surrounded by holes of size $s' < s$ and the remaining part of size exactly equal to $s - s' - s''$). Thus, one has exactly

$$\delta N_H(s) = -sN_H(s) + 2 \sum_{s'=s+1}^{\infty} N_H(s') + (f/P) \sum_{s''=1}^{s-2} \sum_{s'=1}^{s-s''-1} s'' N_H(s') N_T(s'') N_H(s - s' - s'') \tag{5.1}$$

Fortunately, one can show that the last term is negligible as $f/P \to 0$. Thus at steady state we have

$$sN_H(s) = 2 \sum_{s'=s+1} N_H(s') \tag{5.2}$$

which can be solved exactly to yield [Paczuski and Bak, 1994]

$$N_H(s) = \frac{1}{s(s+1)(s+2)} \tag{5.3}$$

Similarly, Drossel and Schwabl [1992] found that the distribution of forests is given by $N_T(s) = 1/(s+1)(s+2)$, matching perfectly the results of numerical analyses.

The key connection of cascade models, as in forest fires, with turbulence lies in the large-scale effects of interactions appearing on all scales. Even though an explicit link between forest fire models and fluid turbulence cannot be made [Bak and Paczuski, 1994], some similarities are remarkable, particularly because scale-free turbulent-like phenomena occur irrespective of the microscopic details of the model, and the emergent large-scale behavior has almost nothing to do with such details.

Interactions appearing at all scales are central to the statistical theory of turbulence where nonlinear interactions (the effects of convection terms) spread energy on all scales, from large to small. In turbulence this *inertial* range of scales is limited at small-scale range by the so-called dissipation scale where viscosity effects destroy high-frequency structures at exponential rates. As we will discuss in Chapter 6, fractal landscapes operate under similar conditions.

The concerted action of large numbers of cooperative factors, like the effects of energy and momentum that spread turbulent vortices throughout all scales, is crucial to critical self-organization. Bak took this concept to the edge [Bak and Sneppen, 1993] in suggesting that Darwinian biological evolution could be itself a self-organized critical process, on the *edge of chaos*. This suggestion comes from the observation that critical evolution must be the outcome of cooperative factors because of its fast pace. In fact, in noncooperative scenarios for individual mutations, large and coordinated mutations would be necessary to ensure the same effects, a prohibitively unlikely event.

The idea that life may be a SOC process is related to models of ecologies of interacting species [Kauffmann, 1993; Bak and Sneppen, 1993; Flyvbjerg, Sneppen, and Bak 1993]. These models assume that when one species mutates in order to improve its fitness, this affects other species, prompting them to mutate too. The chain reaction possibly triggered by these interactions is called a *coevolutionary* avalanche. This means that although selection happens on the scale of single individuals, evolution is considered in a coarse-grained sense where an entire species is represented by a single fitness value.

To stress the similarities with other SOC models, we briefly recall here the structure of the model of interacting species. A fitness landscape is defined following Kauffman [1993]. Each species is characterized by a genetic code of N bits, e.g., $011101100011011\cdots$. Each gene is assigned a *fitness* value, and the overall state of the system (the fitness landscape) is calculated as the sum of the fitnesses from the individual genes. Each fitness is then made dependent on K other genes in the same organism and C genes from other organisms. Kauffman [1993] shows that if there are few connections between organisms, C being small, the ecology evolves to a frozen state where the effect of a single random mutation causes at most a local disturbance in the ecology. The fitness that species acquire in this process is far from optimal. If C is large, each species depends on the state of many other species and the ecology never stops evolving. It is a chaotic state where avalanches, or bursts of activity, never stop. Also in this case the fitness that species acquire is low. Following these ideas, Bak and Sneppen [1993] suggest that the dynamics of evolution tending to optimization of fitness should drive the ecology to a critical point. This is precisely the scenario of the edge of chaos concept.

A major step forward in interpreting models of ecology of species in this framework came when Bak and Sneppen [1993] looked at conglomerates of causally connected bursts of activity. The basic model proceeds as follows. Each species is thought of as sitting in a rough fitness landscape where the height (the fitness) is a function of the genetic code. On the short time scale the favored mutations are those that make the species evolve to a local peak in the landscape defining a local fitness maximum. Typically, one finds higher peaks all over the landscape. Peaks are separated from the nearest local peak by valleys and saddles of lower fitness, and many correlated unfavorable mutations are needed in order for evolution to allow a species to travel through valleys between two peaks. Bak et al. [1994] note that this resembles evolution of wings in a grounded species since gradual evolution is very problematic because partially developed wings are less useful than no wings at all. In general, it is difficult to improve the fitness of species already fit because such species do not benefit from spontaneous mutations. Species with low fitness are suited to improve their condition because they are much more susceptible to mutation.

The key point is that species interact. When a species improves its fitness a little by the spreading of a successful mutation, the fitness of the C 'neighbor' species changes because it is functionally related to the species that experiences a mutation. Therefore a species at maximum fitness, while unable to evolve on its own, might experience a change in its fitness landscape as a result of the evolution of related species. It may also start evolving again.

The synthesis of the above considerations led to the following model [Bak and Sneppen, 1993]:

- N species are arranged on a one-dimensional line with periodic boundary conditions.
- Random fitnesses F_i, say, independent and identically distributed between 0 and 1, are assigned to each species.

At each time step the ecology is updated by the following rules:

- The site with minimum fitness is located. It is then mutated by assigning a new random number.
- The landscape of, say, the two neighbors is changed by assigning new random numbers to those sites.

In the beginning, subsequent events are quite uncorrelated in space. As fitness increases, it becomes increasingly likely that neighbors of spontaneously mutating species are next to mutate. This suggests a similarity with the sandpile system described in Section 5.2. The above model results in self-organized criticality and *punctuated equilibrium* of coevolutionary avalanches, that is, long periods of passivity interrupted by sudden bursts of activity of any size.

Interestingly, exact solutions to the statistical properties of the above model have been found by De Boer et al. [1994], Maslov, Paczuski, and Bak [1994], Paczuski, Maslov, and Bak [1994, 1995a,b] and Bak, Paczuski, and Maslov [1995].

Although the model is abstract and focuses on only a few important aspects of evolution, it provides a possible explanation for the characteristic intermittency of actual evolution, called punctuated equilibrium by Gould and Eldredge [1977, 1993], and the apparent scale invariance of extinction events described by Raup [1991]. The reference to the hypothesis of punctuated equilibria in paleontology is particularly suggestive. Through it, Gould and Eldredge contested the traditional view of evolution, that is, that species accumulate modifications at a small, constant rate. They showed convincingly that the paleontological record supports the view that a given species evolves very little, if at all, during most of its existence. The vast majority of evolutionary innovations take place in intermittent bursts that are short-lived compared with geological timescales. In essence, if life on Earth is a self-organized critical dynamic system, then intermittency and scale invariance must be universal and robust consequences, independent of the details of the dynamics and thus likely to appear even in more complex systems.

We will not pursue further the description of SOC models in this context. However, the key connotation of such systems is *complexity* in their evolution – as opposed to contingency dictated by specific events – which we also observe in the natural landscapes that form the subject of this book. Complexity is the manifestation of information carried over a wide range of scales and can occur only in large interactive dynamic systems. The principle of self-organized criticality has thus been suggested to constitute the dynamic mechanism responsible for complexity and fractal growth and seemingly it could provide a comprehensive fundamental framework to describe the ubiquity of complexity and criticality in nature.

5.5 Sandpile Models and Abelian Groups

We now return to sandpile models to explore in depth some of their remarkable features. In particular, we return to the following example formulated on a two-dimensional regular lattice of N sites with open boundaries (Figure 5.4). Integers z_i on each site i represent the local sandpile height. Addition of a sand particle to a site i is represented by increasing the value of z_i at that site by a unit. When the height somewhere exceeds a critical value z_{cr}, here taken to be 4, there occurs a toppling event wherein one grain of sand is transferred from the unstable site to each of the four neighbor sites. The value of z_i is reduced by 4 and the values of z at the four neighbor sites are increased by 1 because diagonal connections are neglected. All sites are updated simultaneously.

The state of the system can be represented by a matrix, C, whose size is that of the lattice. The elements of C are the elevations z_i of the sites and its feasible range is the set of states with all z_i values such that $z_i \geq 0$ and $z_i \leq 4$. We will use for a detailed example a simple 2×2 lattice. One example of a possible state for the system is the following:

$$C = \begin{pmatrix} 3 & 2 \\ 3 & 4 \end{pmatrix}$$

A peculiar (and unique) state is C^*, that is, the critical state defined by the above rules where all sites are at the threshold

$$C^* = \begin{pmatrix} 4 & 4 \\ 4 & 4 \end{pmatrix}$$

To explore self-organized criticality in this model, one randomly adds sand and allows the system to relax. The result of the addition quickly becomes unpredictable as the system size grows large, and one is only able to find the outcome by actually simulating the resulting avalanche. An exact result particular to this model [Bak and Creutz, 1993] is that, once in the critical ensemble, it is impossible to trigger a set of avalanches that will leave an isolated island of unaltered sites surrounded by disturbed ground. A log-log plot [Bak and Creutz, 1993] of the distribution of avalanche sizes s (described here by the number of tumblings within an avalanche) indicates

$$p(s) \propto s^{1-\gamma} \quad \gamma \approx 2.1 \tag{5.4}$$

where p is the probability density function for avalanches of given sizes s. For a random distribution of z_is one would expect the chain reaction generating the avalanche to be either subcritical (in which case the avalanche would die after a few steps and large avalanches would be exponentially unlikely) or supercritical (in which case the avalanche would explode with a collapse of the entire system). The power law, instead, indicates that the reaction is critical, that is, the probability that activity at some site branches into more than one active site is balanced by the probability that the activity dies. Thus by evolving through avalanche after avalanche, the system has *learned* to respond critically to the next perturbation.

Dhar and coworkers [Dhar, 1990; Dhar and Ramaswamy, 1989; Dhar and Majumdhar, 1990; Majumdhar and Dhar, 1992] have shown some remarkable mathematical properties that will be shown here in a condensed version of that given by Bak [1993; see also Bak and Creutz, 1993].

Operators are introduced on C by the above rules. Dhar proposed the concept of a toppling matrix Δ_{ij} with integer elements representing the change in height z at site i resulting from a toppling at site j. Under a toppling at site j, the height at site i becomes $z_i - \Delta_{ij}$. Clearly, in the simple two-dimensional example above, the toppling matrix, which has N^2 elements, is given as $\Delta_{ij} = 4$ if $i = j$, $\Delta_{ij} = -1$ if i, j are nearest neighbors, and $\Delta_{ij} = 0$ otherwise. In the above case the toppling matrix is

$$\Delta = \begin{pmatrix} 4 & -1 & -1 & 0 \\ -1 & 4 & 0 & -1 \\ -1 & 0 & 4 & -1 \\ 0 & -1 & -1 & 4 \end{pmatrix}$$

where the numbering of the sites, defining neighboring sites in the toppling matrix, is as follows:

$$\text{site number} = \begin{pmatrix} 1 & 2 \\ 3 & 4 \end{pmatrix}$$

We recall that little is unique to the specific lattice geometry chosen in the example, and the results generalize to other lattices and dimensions. The analysis requires only that under a toppling of a single site i, the site has its height decreased ($\Delta_{ii} > 0$), the height at any other site is either increased or unchanged ($\Delta_{ij} \leq 0$ for any $j \neq i$), and the total amount of sand in the system does not increase ($\sum_j \Delta_{ij} > 0$). Each site can be connected through topplings to some location at the boundaries where sand can be lost. In fact, by the above rules a toppling at an edge loses one grain of sand and at a corner a toppling loses two grains.

We now formally define various operators acting on the states C. First, the *adding sand* operator α_i acting on any C yields the state $\alpha_i C$ where $z_i \to z_i + 1$ and all other zs are unchanged. As an example, we see that the addition of sand $z_4 \to z_4 + 1$ at site 4, that is, $\alpha_4 C$, in the above case yields

$$\alpha_4 C = \alpha_4 \begin{pmatrix} 3 & 2 \\ 3 & 4 \end{pmatrix} \to \begin{pmatrix} 3 & 2 \\ 3 & 5 \end{pmatrix}$$

Second, we define the *toppling* operator t_i, which transforms C into the state with heights z'_j where $z'_j = z_j - \Delta_{ij}$. As an example, we see that $t_4 C$ in the above case yields

$$t_4 C = t_4 \begin{pmatrix} 3 & 2 \\ 3 & 5 \end{pmatrix} \to \begin{pmatrix} 3 & 3 \\ 4 & 1 \end{pmatrix}$$

Notice that a toppling in the interior of the lattice does not change the total amount of sand. A toppling on the boundary decreases this sum owing to sand falling off the edges.

Third, we define the *updating* operator U, which updates the lattice one time step. This is obtained as the product of t_i over all unstable sites:

$$UC = \prod_i t_i^{p_i} C \tag{5.5}$$

where i spans all sites and $p_i = 1$ if $z_i \geq 5$ and 0 otherwise. As an example, if the state C is

$$C = \begin{pmatrix} 3 & 5 \\ 3 & 5 \end{pmatrix}$$

the operator UC yields

$$UC = t_2 t_4 \begin{pmatrix} 3 & 5 \\ 3 & 5 \end{pmatrix} \to \begin{pmatrix} 4 & 2 \\ 4 & 2 \end{pmatrix}$$

Using U repeatedly, one can define the *relaxation* operator R. Applied to any state C this corresponds to repeating U until no more z_is change; that is, they are all subcritical. Neither U nor R have any effect on stable states. Thus $RC = UU \cdots UUC$ until all states are stable.

As an example of application of the relaxation operator R applied to the state C, that is

$$C = \begin{pmatrix} 4 & 5 \\ 3 & 5 \end{pmatrix}$$

we obtain

$$RC = UC = \begin{pmatrix} 5 & 2 \\ 4 & 2 \end{pmatrix}$$

which still has one unstable site. Thus we have three updating operators, or a product of toppling operators. In particular

$$RC = UUU \begin{pmatrix} 4 & 5 \\ 3 & 5 \end{pmatrix} = \begin{pmatrix} 2 & 3 \\ 1 & 3 \end{pmatrix}$$

We are now in a condition to define the *avalanche* operator a_i describing the action of adding a grain of sand followed by relaxation:

$$a_i C = R \alpha_i C \tag{5.6}$$

This operator a_i forms a group in algebraic language, because it modifies the system (defined by the state C) into another state through some transformation. Note that it is not obvious at all that the a_is allow the system to enter a nontrivial cycle.

It is useful to introduce the concept of recursive states. The set of recursive states, denoted by $\mathcal{R}$, includes those stable states that can be reached from any stable state by some addition of sand followed by relaxation. As the minimally stable state C^* can be obtained from any other state by adding just enough sand to each site to make z_i equal to 4, it belongs to $\mathcal{R}$. Thus one might conveniently define $\mathcal{R}$ as the set of states that can be obtained from C^* through some product of operators a_i.

It can be shown [Bak and Creutz, 1993] that there exist nonrecursive transient states. For instance, in the SOC model described here no recursive states can possibly have two adjacent heights that are both zero. Transients cancel the memory of the initial condition that may be arbitrarily characterized. In state-space the set $\mathcal{R}$ excludes transient states and describes all recursive states of the system. It therefore represents an attractor of the dynamics of the system.

The most important mathematical result is that the operators a_i, when acting on the stable states, commute and generate an Abelian group when restricted to the recursive states of the set $\mathcal{R}$. In other words, we shall now prove that the sequence of avalanches at j sites produces the same stable structure whatever the order of avalanches perturbing the system. The notion of an Abelian group postulates the existence of an inverse operator and an *identity* state whose properties also have physical relevance.

We begin by showing that the operators commute, that is

$$a_i a_j C = a_j a_i C \quad \text{for all } C \tag{5.7}$$

First we use the previous definitions to express the avalanche operators as in terms of toppling and addition operators:

$$a_i\, a_j\, C = \left(\prod_{k=1}^{n_i} t_k\right)\alpha_i\left(\prod_{k=n_i+1}^{n_j} t_k\right)\alpha_j\, C \tag{5.8}$$

where the specific numbers of topplings n_i and n_j depend on i, j, and C. The key point is that, acting on general states, ts and αs commute because they merely add or subtract constants, and thus one can shift α_i to the right in the above equation:

$$a_i\, a_j\, C = \left(\prod_{k=1}^{n_j} t_k\right)\alpha_i\alpha_j\, C \tag{5.9}$$

If α_i or α_j destabilizes the sites i, j, the product must contain t_i, t_j. We can therefore rearrange the product of topplings in the nontrivial case that either i or j (or both) becomes unstable. The product thus contains the toppling operators corresponding to those unstable sites, which, by the above reasoning, can be shifted to the right. Such operators constitute by definition the update operator U:

$$a_i\, a_j\, C = \left(\prod_{k}{}' t_k\right)U\alpha_i\alpha_j\, C \tag{5.10}$$

where $\prod_k'$ must include the remaining topplings t forming U. However, the update operator itself may leave some sites unstable. The product must therefore include additional toppling operators as

$$a_i\, a_j\, C = \left(\prod_{k}{}'' t_k\right)UU\alpha_i\alpha_j\, C = UU\cdots U\alpha_i\alpha_j\, C \tag{5.11}$$

obtained by pulling out other factors of the update operator. This procedure can be iterated until all toppling factors have been used and the resulting state is stable. Thus the product in Eq. (5.11) defines an operator that is equal to the relaxation operator R:

$$a_i\, a_j\, C = R\,\alpha_i\alpha_j\, C \tag{5.12}$$

The proof follows from the observation that $\alpha_i\alpha_j\, C$ is the same state as $\alpha_j\,\alpha_i\, C$ because they are obtained simply by addition. Thus, from the previous equation, $a_i\, a_j\, C = a_j\, a_i\, C$.

Among the consequences of this proof, the total number of tumblings occurring in the operations $a_i\, a_j\, C$ and $a_j\, a_i\, C$ are the same, although the order of the tumblings may or may not be altered by the sequence of avalanches.

It is also instructive to sketch the proof of existence of the inverse operator when the algebraic group is restricted to the set $\mathcal{R}$ of recursive states. The proof aims at demonstrating that there exists a unique operator a_i^{-1} such that

$$a_i(a_i^{-1}C) = C \quad \text{for all } C \in \mathcal{R} \tag{5.13}$$

This property defines the so-called Abelian group in algebra, and thus it is legitimate to term this class of models Abelian sandpiles.

The existence of a unique inverse in the set of recurrent configurations implies that the state in which all recurrent configurations occur with equal probability is the invariant state of its evolution [Dhar, 1990]. In this characterization of SOC systems, only recurrent configurations have nonzero probability of occurrence and this nonzero value is the same for all recurrent configurations. As an example, consider any configuration $C \in \mathcal{R}$ to which we add Δ_{ii} particles at some site one after another. Because $z_i > 0$ in C, after these additions the site will become unstable and topple, in the process adding $(-\Delta_{ij})$ particles at all other sites j where $j \neq i$. We note, with Dhar [1990], that the operators a_i $(i = 1, N)$ satisfy the equations $a_i^{\Delta_{ii}} = \prod' a_j^{-\Delta_{ij}}$, where the primed product is carried out over all $j \neq i$. Multiplying both sides by $a_i^{-\Delta_{ii}}$, we also obtain $1 = \prod_{j=1}^{N} a_j^{\Delta_{ij}}$, a result that will be useful later.

We will now show that for any recursive state C we can find another recursive state such that a_i acting on it gives C. The proof begins [Bak and Creutz, 1993] by adding a grain of sand to the state C and allowing it to relax to another recursive state, a_iC. By the definition of recursive state, there exists some way to add sand regenerating C from any given state, in particular, by some product P of avalanche operators a_i, that is

$$C = Pa_iC \tag{5.14}$$

But the a_is commute, and therefore we obtain

$$C = a_iPC \tag{5.15}$$

and thus PC is a recursive state on which a_i gives C. Hence the inverse operator exists. The proof of uniqueness will be omitted here.

The importance of the above results is related to the fact that one can count the number of recursive states. In fact, because all recursive states can be obtained by adding sand to C^*, we can define the general state $C \in \mathcal{R}$ in the following form:

$$C = \left(\prod_i a_i^{n_i} \right) C^* \tag{5.16}$$

defining in a unique manner the states C through a vector $\mathbf{n} = (n_1, \ldots, n_N)$, where the integers n_i represent the total amount of sand to be added at the respective sites moving from the critical state C^*. As an example in our 2×2 lattice, we define a vector $\mathbf{n} = (n_1, n_2, n_3, n_4) = (1, 0, 0, 0)$. Then

$$C = \left(\prod_i a_i^{n_i} \right) C^* = a_1 C^* \tag{5.17}$$

and

$$C = a_1 C^* = \begin{pmatrix} 5 & 4 \\ 4 & 4 \end{pmatrix} \to \begin{pmatrix} 1 & 5 \\ 5 & 4 \end{pmatrix} \to \begin{pmatrix} 3 & 1 \\ 1 & 6 \end{pmatrix} \to \begin{pmatrix} 3 & 2 \\ 2 & 2 \end{pmatrix}$$

which shows the equivalence of the vector $(1, 0, 0, 0)$ with the last state.

In general, there are several different ways to reach any given state. As an example, adding four grains of sand to any site forces a toppling that is precisely equivalent to adding a single grain to each of the neighbors. In operator formalism, this means

$$a_i^4 = \prod_{j \in nn} a_j \tag{5.18}$$

where nn are the nearest neighbors to site i. We can rewrite this equation by multiplying both sides by the product of inverse avalanche operators on the nearest neighbors on both sites, that is

$$\prod_{j \in nn} a_j^{-1} a_i^4 = \prod_{j \in nn} a_j^{-1} \prod_{j \in nn} a_j = I \tag{5.19}$$

where I is the identity matrix. The above, recalling the definition of the toppling matrix, can be written in the form

$$\prod_{j=1}^{N} a_j^{\Delta_{ij}} = I \tag{5.20}$$

where i refers to the arbitrary ith site and j spans the entire lattice because of the properties of Δ_{ij}.

Now we are in a condition to specify the exponents in Eq. (5.16). If we label states by the vector $\mathbf{n}$, Bak and Creutz [1993], following Dhar [1990], have observed that two states, say C_1 and C_2, are equivalent if the differences in the vectors $\mathbf{n}$ representing them is a linear combination of the elements of the toppling matrix, that is, of the form $n_i = \sum_j \beta_j \Delta_{ij}$, where the coefficients β_j are integers. In other words, if two states cannot be related by topplings, they are independent. Thus any vector $\mathbf{n}$ (or, equivalently, any recursive state C) can be translated until it lies in a N-dimensional vector space whose base is constituted by the vectors Δ_{ij}, $j = 1, \ldots, N$. The vertices of such a construction have integer coordinates, and its volume measures the number of integer coordinate points inside. We also observe that the determinant of the matrix constructed by n such vectors is the measure of the n-dimensional volume embedded by them in any vector space. This follows from geometry theorems [e.g., Spivak, 1965], which we will briefly recall here. An Euclidean (real) n-dimensional vector space is given where a $(n+1)$-dimensional simplex S is defined, that is, $S = S(P_0, P_1, \ldots, P_{n+1})$ where P_i is the ith point of the space (e.g., a triangle in two dimensions, a tetrahedron in three dimensions, etc). Let $\mathbf{v} = v_1, \ldots, v_n$ be any orthonormal basis for the vector space. If we define the application $g(v_i) = P_i - P_0$, $i = 1, 2, \ldots, n$, then the volume of S is defined by the absolute value of the determinant of g, that is, vol(S) = $|det\ g|$, where the definition of the volume is independent of the basis v adopted and of the order chosen for the points $P_0, P_1, \ldots$. Clearly, the physical meaning of the application g is to define the sides of the simplex, and thus the determinant of the matrix constructed with the vertices of the simplex measures its volume. We thus conclude that

- the volume of the object defined by the i vectors $(\Delta_{i1}, \ldots, \Delta_{iN})$ is the absolute value of the determinant of the toppling matrix Δ_{ij}; and
- such a determinant counts the number of coordinate points inside the object, that is, of distinct recursive states.

Dhar [1990] has shown that for large lattices, when we have $4N$ stable states, there are only $(3.2102\cdots)^N$ recursive states. This means that starting from an arbitrary state and adding sand, the system self-organizes into a subset of $(3.21..)^N$ states that form the attractor of the dynamics. This also postulates that, in principle, in the elementary 2×2 example if one adds $192 = (3.21)^4$ grains at the same site, one obtains the same state. In reality, the original state is obtained earlier than that because of the size of the problem. In less simple, say, 2×7 examples [Bak, 1993], 59×10^6 states are possible, and because of the rectangular lattice one needs to count all recursive states before obtaining the original state.

In Section 5.12 we will use the above properties within the context of a possible thermodynamic interpretation of self-organization. We will also examine a different form of SOC arising in the context of fractal channel network formation, whose departures from the original approach will be clarified by algebraic tools.

5.6 Fractals and Self-Organized Criticality

As we have discussed in Chapters 2 and 3, many objects in nature are best described geometrically as fractals, with self-similar features on all length scales. As Bak and Creutz [1993] put it, echoing Mandelbrot [1983], mountains have peaks of all sizes from kilometers down to millimeters, river networks consist of streams of all sizes, turbulent fluids have vortices over a variety of scales, earthquakes occur on structures of faults ranging from thousands of kilometers to centimeters. All of these phenomena are in essence scale-free in the sense that, by viewing a picture of a part of the whole, one cannot deduce its actual size unless a yardstick is shown in the same picture.

The crucial contribution of Bak and coworkers is that, moving from Mandelbrot's fundamental tenet that fractals are indeed the geometry of nature, they were successful in providing a general dynamic framework of how nature may produce them. Indeed, fractal geometry has called for a great effort in the geometric characterization of natural objects, without a parallel progress in the understanding of their dynamic origin.

Bak and Creutz [1993], in a summary of work in this area, added to this concept. They observed an inborn tendency to think of fractals like, say, the universe or the Earth's crust, as static structures because the dynamics that formed these structures is characterized by a timescale much longer than the observational period, typically a human lifetime.

The earthquakes that we observe last a few seconds, whereas the fault formations appear static because they were built up over millions of years. Nevertheless, the origin of fractals is a dynamic, not a geometric, problem. One related aspect, which we have already faced in Chapter 4, deals with the interplay of local and global interactions implied by fractal patterns that are organized over the farthest distances available for the developing structure. In OCNs a *global* rule is enforced. In SOC it is not so, and the importance of the birth of global correlations is enhanced by the fact that large systems operating near their equilibrium state tend to be only locally correlated. Only at a critical point where a phase transition takes place, are those systems fractal and correlated over all scales.

SOC, described in the earlier sections of this chapter, addresses the tendency of large dissipative systems to drive themselves to a critical state with a wide range of length and time scales. SOC thus provides a general unifying concept for large-scale behavior in systems with many degrees of freedom and, as described above, has been looked for in many different fields.

As it turned out, the sandpile model marked a fundamental step in the search for the dynamic principles leading to fractal structures. Sand, in fact, represents a state of matter, and it can exist in many stable states, almost all of which are different from the flat equilibrium state. Thus sand contains memory. If a heap of sand is perturbed, for instance by adding more sand, by tilting the whole pile, or by shaking it, the system goes from one metastable state to another in a sort of diffusion process that is very different from the relaxation of a truly diffusing system. The diffusion of sand, in fact, stops at any of many stable states and the process is threshold-dependent because nothing happens before the perturbation reaches a certain magnitude. It is therefore reasonable to expect real sandpiles to exhibit self-organized critical behavior.

The concept of migration from one metastable state to another with similar features is also central to the concept of optimal networks, where the minimization process of a global quantity, that is, total energy dissipation of the system, guides the evolution of the system into one of the many states with similar total energy dissipation. We have seen that quite different configurations of the system in fact have similar total energy dissipation, provided that their statistical and fractal features converge to certain features. This linkage of self-organized states with states of minimum energy dissipation is investigated in Section 5.7.

There are now several experiments reporting power law distributions of sand avalanches. Small heaps have been built on a precision scale and the resulting distributions of avalanches falling off the edge have been monitored. The experiments were performed using iron spheres and glass spheres of the same size. In all cases a power law distribution function was found.

Frette et al. [1996] carried out experiments on a pile of rice. Grains of rice were slowly fed into the gap between two vertical parallel plates. Two types of grains were used, one with a large aspect ratio and another with a less elongated shape. Frette et al.'s experiments show a power law distribution for the avalanches of grains with large aspect ratios but not so for the more spherical grains. The implication is that the occurrence of SOC depends on the mechanism of energy dissipation, which is quite different for the two types of grains.

The threshold dynamics of sand is a likely paradigm for many processes in nature [e.g., Bak and Creutz, 1993]. As we have discussed before, earthquakes occur only when the stress somewhere within the Earth's crust exceeds a critical value, and the earthquake takes the crust of the Earth from one stable state to another. Economic systems are driven by threshold processes, because individual agents may change their behavior only when conditions reach a certain level. Biological species are born or die when specific conditions in their environment are fulfilled. Neurons in a neural network fire when the input from other neurons reaches a threshold level. Indeed, erosion processes responsible for landscape-forming events are threshold processes. Whatever the channel initiation mechanism, an abrupt discontinuity in the basic transport mechanism is responsible for the transition from hillslopes to channels. The robustness of the statistical features of river basins and networks, regardless of climate, location, tectonics, or exposed lithology (described in Chapter 7), suggests the existence of some general self-organizing principle, which we address at the core of this chapter. Starting with Section 5.7, we describe how fractal river networks self-organize under fluvial processes. We claim that the dynamics of fluvial landscapes is driven by the exceedence of erosion thresholds and that the system is able to self-organize in a robust manner. Most important, we will also show that in the particular case of a simple threshold-dependent erosion model, self-organized critical systems obey minimum energy dissipation.

5.7 Self-Organized Fractal Channel Networks

In Section 1.4.3 we saw that fluvial landscape evolution may be described by the following threshold-dependent field equation:

$$\frac{\partial z}{\partial t} = \alpha f\,(\tau - \tau_c) \tag{5.21}$$

where $z(\mathbf{x}, t)$ is the elevation field, $f\,(x) = 0$ for $x \leq 0$; τ is the local shear stress; τ_c is a critical threshold for erosion; and α is a (dimensional) coupling constant. The shear stress produced by the flow rate Q in dynamic equilibrium is

$$\tau = \rho\, g\, y\, \nabla z \tag{5.22}$$

where ρ is the density of water, g the acceleration of gravity, ∇z the local slope, and y the flow depth (approximating the ratio of flow area and wetted perimeter) which is assumed to scale as $y \propto Q^{0.5}$. This perspective of the erosion process is justified by experimental evidence (Section 1.2.6) and by optimality principles (Section 4.3). Thus at the ith site $\tau_i \propto Q_i^{0.5} \nabla z_i$. Whenever shear stresses exceed τ_c anywhere in the network we expect activity.

The variable governing the evolution of the elevation field is the discharge Q_i in a given site in a lattice, say i. Discharge is surrogated herein (as discussed in Chapter 1) by the total contributing area, that is, $Q_i \propto A_i$. If we allow the drainage directions to coincide with the gradient of the elevation field, ∇z, areas are identified as $A_i = \sum_j W_{ij} A_j + 1$, where the sum is extended over the whole basin and W_{ij} is a transfer matrix (Section 1.2.9) assigned the unit value if j is the drainage direction in i and zero otherwise.

The crux of the matter is whether the above threshold-dependent dynamic specification yields erosion activities that self-organize themselves into a process that reproduces the geomorphology of river networks.

Rinaldo et al. [1993] suggested the equivalence of structural stability and the enforcement of the global principle of minimum energy dissipation by defining the following model of self-organization based on the threshold model of Eq. (5.21).

We choose a two-dimensional lattice. Each site i has two variables, the elevation z_i, and total flow Q_i, surrogated by total contributing area A_i. For a given initial configuration, time evolution is performed according to the following rules:

- Drainage directions are fixed by steepest descent (step (1)). A given threshold value τ_c is assigned. To simplify matters, τ_c will be initially chosen as constant in space and time and numerically small enough to affect most of the evolving landscape. Lattice effects are avoided by assigning unit length to all directions.
- The initial set of elevations is chosen indifferently at random or in a systematic way (step (2)). Random initial conditions may be generated by Eden [1961] growth or by sampling elevations and removing singularities [e.g., Takayasu and Inaoka, 1992]. For each site i the shear stress is computed as $\tau_i \propto A_i^{0.5} \Delta z_i$, Δz being the drop along the drainage direction (and hence the slope because lengths are assumed to be unit).
- The threshold exceedences possibly scattered throughout the network are computed (step (3)). The maximum exceedence is isolated, say at site j, and the elevation of the jth site is reduced to the value that yields $\tau = \tau_c$. This is the so-called greedy strategy. We will observe later that other options (e.g., a random choice of the site to be reduced, or a parallel strategy with many sites reduced simultaneously) do not affect the final statistics of the resulting structures.
- The mass released by the lowering of z_j (the avalanche in Bak et al. [1988]) is removed from the system because at equilibrium the threshold once exceeded in a site will always be exceeded downstream because of the increasing discharge (other transfer options have also been tested).
- Drainage directions are recomputed because they are altered owing to the modified elevation of site j (step (4)). Changes may be minor or extensive, depending on the actual configuration.
- Steps (3) to (4) are repeated until no exceedences are isolated. Thus at any stage the system evolves to a state where no sites are above the critical threshold (step (5)).
- The state at step (5) is perturbed at random by adding elevation to a node. The local flow convergence induced by the perturbation may yield threshold violation as in step (3) and a readjustment of the structure. The system is perturbed at random until further perturbations do not induce variations in the statistical properties of the system (step (6)). The paths leading to the final state are then seen as an analog to the transients leading to the set of recursive states in SOC models (Section 5.5).

The main difference between the model and the equations is in the specification of the rule that starts the erosion action at sites where the exceedence of the parameter is largest. One remarkable consequence of this choice is that the evolution of the network does not progress strictly by headward growth but rather is driven by the local slope instabilities appearing without preferential location in the network.

Many experiments were performed evolving from random or systematic initial conditions and the system always evolved into a final state

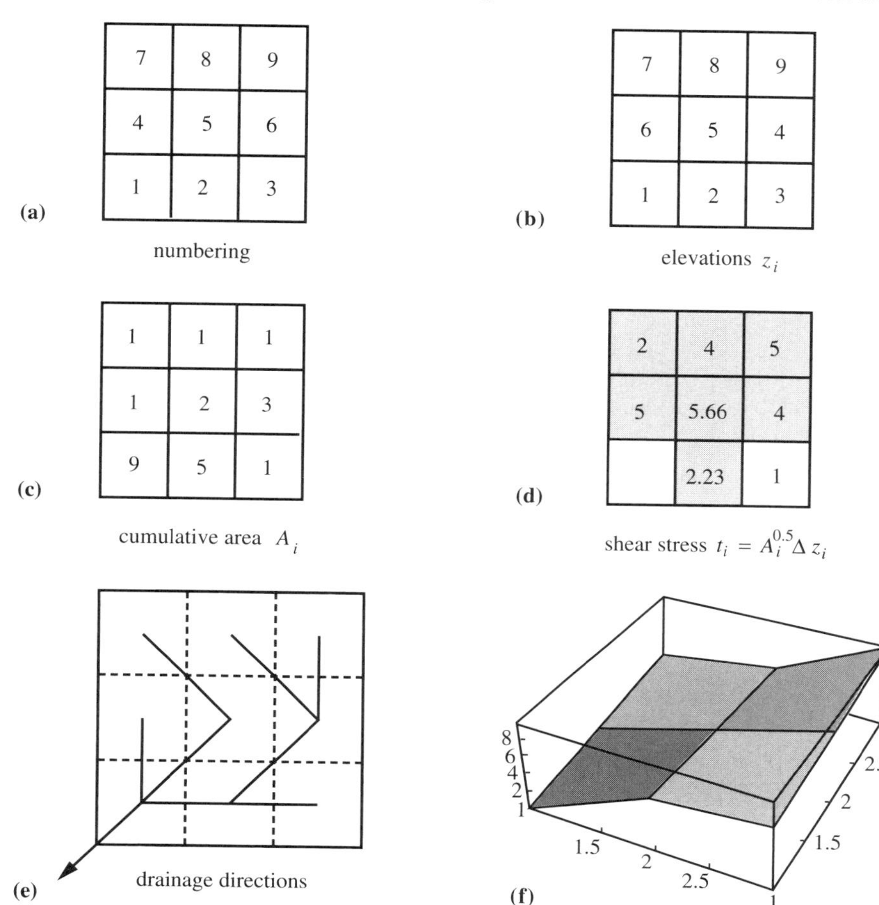

Figure 5.6. Initial conditions in the sample 3 × 3 SOC network; (a) numbering; (b) the initial field of elevations z_i; (c) cumulative areas A_i computed through, (e), drainage directions; (d) shear stresses τ_i. Note that the elevation of node 1 is kept fixed and hence τ_1 is not shown; (f) three-dimensional plot.

characterized by common features. To describe such features, we will carry out an elementary example in a 3×3 grid. Let $\tau_c = 1$ be the (constant) threshold for erosion activity. Hence at criticality, $A_i^{1/2} \Delta z_i = 1$ (recall that diagonal paths are assumed unit and hence $|\nabla z| = \Delta z$). Notice also that the elementary pixel has unit area. The initial conditions are shown in Figure 5.6. The related (topological) energy expenditure is $\sum_i A_i^{0.5} = 13.382$.

The initial elevations are chosen according to an arbitrary and regular rule, yielding a network, that is, a consistent set of drainage directions. Application of the rules pointed out above yields (a) violations of the critical shear stress (the shaded areas in Figure 5.6 (d)), where 7 pixels out of the total number of 9 yield stresses exceeding the critical unit value. Readjustment starts from the elevation of the central pixel whose shear stress ($\tau_5 = \sqrt{2}(5 - 1) = 5.66$) is largest. The new critical elevation is $z_5 = 1.707 = 1/\sqrt{2} + 1$. Drainage directions are substantially altered by the new elevation of the central node. The sequence of automatic modifications is $z_9 = 2.707$, $z_8 = 2.707$, $z_7 = 2.707$, $z_4 = 2$, $z_6 = 2.707$, $z_5 = 1.408$, $z_3 = 2.408$, $z_7 = 2.408$, $z_8 = 2.408$, $z_6 = 2.408$, and $z_9 = 2.408$. The final (symmetric) field of elevations is $z_1 = 1$, $z_2 = 2$, $z_3 = 2.408$, and $z_4 = 2$, $z_5 = 1.408$, and $z_6 = z_7 = z_8 = z_9 = 2.408$.

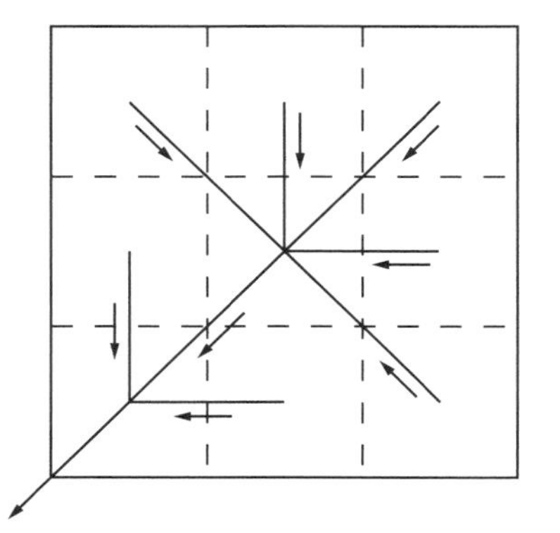

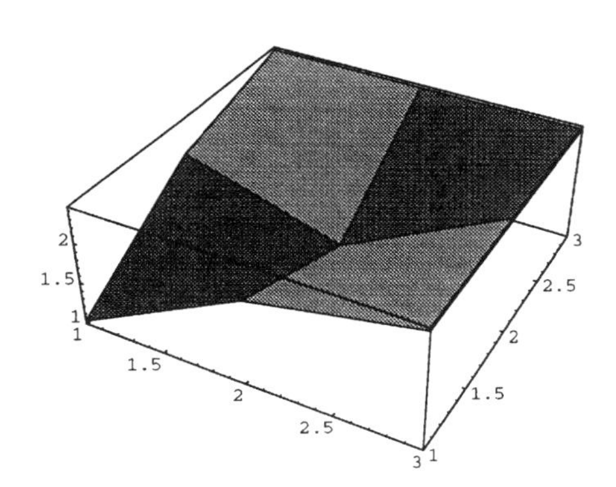

Figure 5.7. Final stage of evolution in a 3 × 3 SOC model of erosion.

Figure 5.7 shows the final configuration of the system, whose energy expenditure ($\sum_i A_i^{0.5} = 12.4495$) is the lowest topologically possible. This can be found by complete enumeration only in very small examples. Although the size of the example does not allow conclusions to be drawn about the resulting statistics, the tendency to readjust that tends to lower the total energy dissipation of the system is clear in this 3 × 3 case.

We now show the results of the basic model of self-organization in lattices of significant size.

As claimed before, in all experiments performed that evolved from random or systematic initial conditions the system always evolved into a final state characterized by some common features. To show the robustness of the features of the final state of evolution, we choose to reproduce here two significant experiments [Rinaldo et al. 1993; Rigon et al. 1994]. In one case, (i), the system is allowed to evolve from an initial surface whose realistic statistics and fractal characters already embody the required character of the final state. In another case, (ii), the landforms evolve from an initial shape that is not a fractal and that exhibits statistics radically different from those observed in nature.

Case (i) is first illustrated. Figure 5.8 illustrates a planar view and Figure 5.9 a three-dimensional plot of the SOC landscape developed from random initial conditions. The initial condition is obtained by the following steps: (a) a two-dimensional Eden growth process [Eden, 1961] is generated in the planar terrain chosen; (b) the elevation field is then enforced moving upward from the outlet via the rule $\nabla z = E[\nabla z(A)] = A^{-0.5}$; hence $Var[\nabla z(A)] = 0$; (c) an adjustment is required after step (b) because the three-dimensional structure generated in this manner is not generally consistent with the actual drainage directions. The local adjustments induce some variance in the slope versus area diagram, which otherwise shows no scatter around a perfect straight line of slope −0.5. The process is similar to the capture model of Howard [1971, 1990]. Notice that in this case, our initial condition is essentially the final stage of evolution for the capture models. Networks generated in this manner exhibit planar statistics close to those of real catchments except for the stream length versus (Strahler) order diagram, which shows some deviation from Hortonian behavior (Section 1.2.1). However, slope statistics are quite unrealistic because of the absence of dispersion (i.e., zero-variance) (see Section 1.2.10).

To analyze the landforms produced by the simple SOC rules defined above, we will now recall a few facts. In Chapters 2 and 3 it was shown

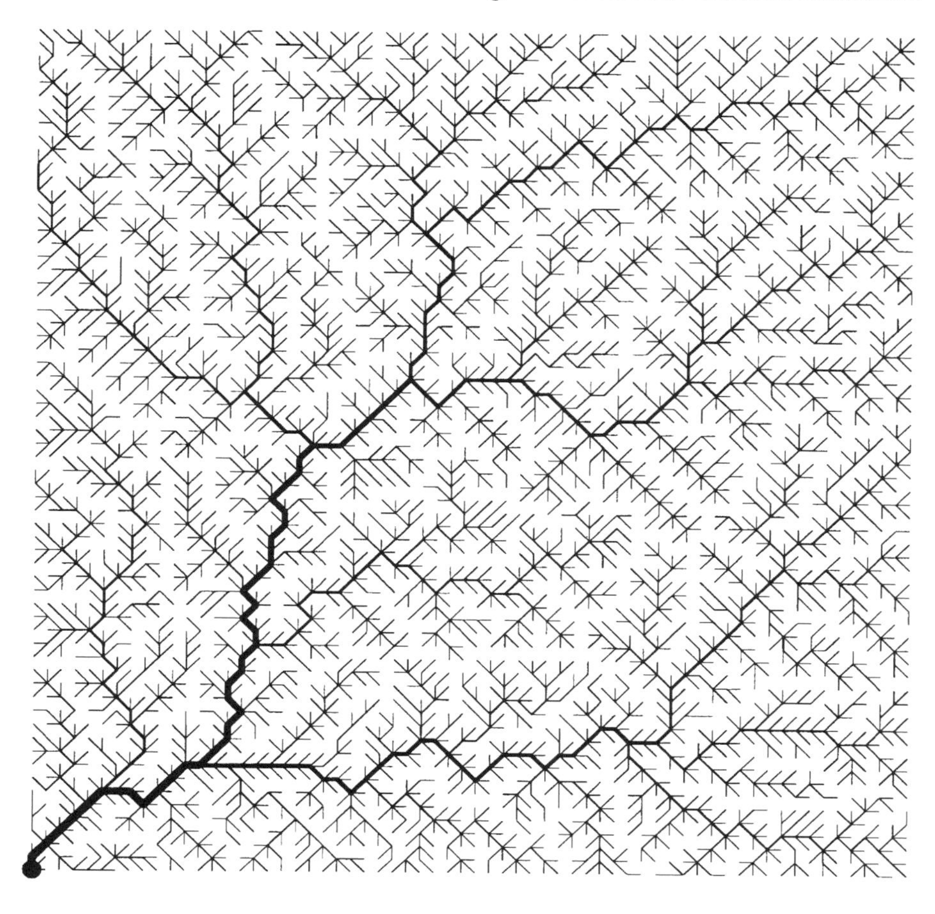

Figure 5.8. SOC network developed from random initial conditions: planar view.

through the experimental analysis of digital elevation maps (DEMs) that river patterns show consistently fractal characteristics. Power laws, which are the signature of fractal behavior, have been observed experimentally over a large range of scales in probability distributions describing river basin morphology. Most notably, cumulative total drainage area A contributing to any link follows a power law, $P[A > a] \propto a^{-0.43\pm0.03}$. The Strahler stream lengths L also follow a power law with exponent close to 2, $P[L > l] \propto l^{-1.9\pm0.1}$. Also, important statistics characterize the three-dimensional structure of the basin. The mean of the local slope ∇z of the links of a drainage network scales, once clearly in the fluvial regime (i.e., when $A > a_t$ where a_t is a suitable support area – see Section 1.2.12), as a function of cumulative area A according to $E[\nabla z(A)] \propto A^{-\theta}$ where $\theta \approx 0.5$. Importantly, the variance of the link slopes ∇z has also been studied (Section 1.2.10), and simple scaling assumptions (i.e., $E[\nabla z(A)] \propto A^{-\theta}$, $Var[\nabla z(A)] \propto A^{-2\theta}$, and, in general, the kth moment M_k scales as $M_k[\nabla z(A)] \propto A^{-k\theta}$) have been experimentally rejected. In fact, data indicate otherwise, suggesting a multiscaling behavior. The scaling exponents θ of the variances of the slopes of real basins are approximately half the corresponding exponent of the self-similar assumption, that is, $Var[\nabla z(A)] \propto A^{-\theta}$.

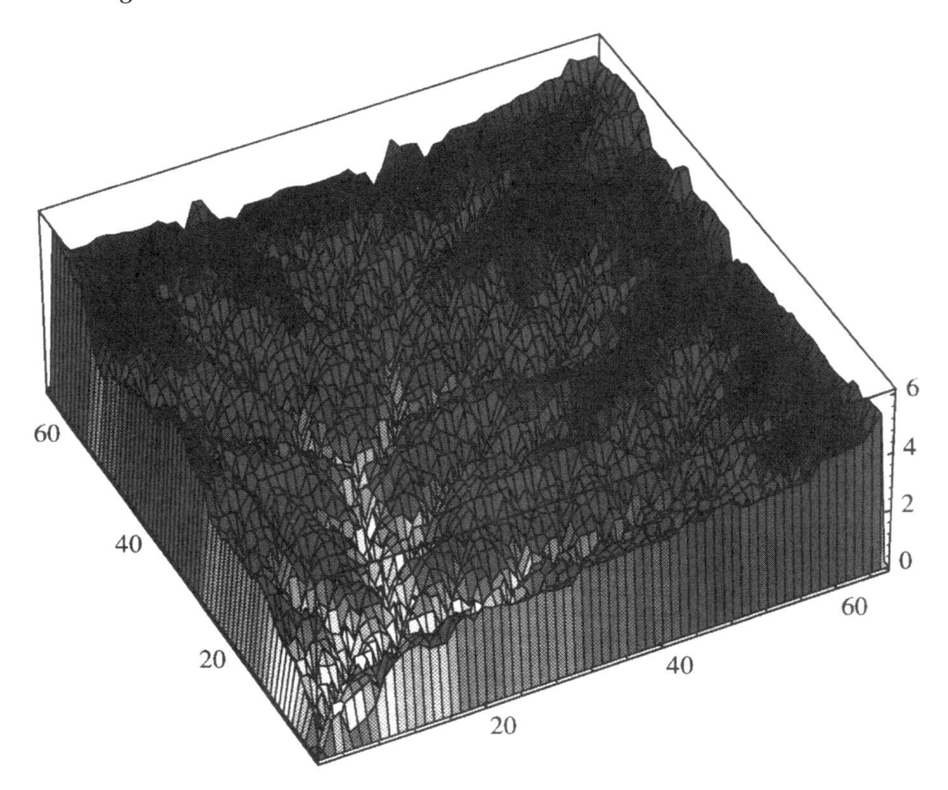

Figure 5.9. SOC network from random initial conditions: three-dimensional view.

Horton's laws of bifurcation and length – though not very distinctive – have been experimentally verified on large databases. As such, $N(\omega) \approx R_B^{\Omega-\omega}$ (Section 1.2.1), where $N(\omega)$ is the number of streams of order ω and R_B is the bifurcation ratio ($R_B \approx 4$). Horton's length ratio R_L likewise defines the average length $\bar{L}(\omega)$ of the streams of order ω, $\bar{L}(\omega) \approx \bar{L}(\Omega) R_L^{\omega-\Omega}$, with $R_L \approx 2$.

Figures 5.10 to 5.14 show some of the relevant statistics of the landscape in Figure 5.9.

Horton's plots for the SOC (i) are shown in Figures 5.10 and 5.11. Horton's numbers R_B and R_L, derived from the slopes of the plots in Figures 5.10 and 5.11, are $R_B \approx 4$ and $R_L \approx 2$, respectively, as consistently observed in nature. As discussed before (Section 1.3.2), the match of Horton's numbers with the values observed in nature is viewed as a necessary condition for outcomes of models of geomorphological structures rather than as a sufficient condition. In fact, many random networks approximately share those values – though unsuitably representing many features – and furthermore even unrealistic deterministic constructs (e.g., Peano's basin) may have perfect Hortonian statistics.

Figure 5.12 shows the probabilities of exceedence of total contributing area $P[A \geq a]$ and and Figure 5.13 shows the probability $P[L \geq l]$ for Strahler's stream lengths L. Although finite-size effects are visible, the clear power laws describing the distribution of aggregated areas and lengths are indicators of fractal structure, and the match of the resulting slopes with real data is very good.

Slope statistics $\nabla z(A)$ are shown in Figure 5.14. One realistic feature, independent of initial conditions, shown in the plot is the scatter arising in the slope–area relation. The scaling properties of the resulting slope field are indeed a relevant property of the self-organized structure. In fact, as discussed in Section 1.2.10, both the systematic and random variations in link slopes are reflected in the scaling invariance properties of their probability distributions, with drainage area serving as a scale parameter. This invariance property is referred to as statistical self-similarity of river networks parametrized by elevation. The plot in Figure 5.14 shows that the upper limit of the curve is set by the threshold as proportional to $A^{-0.5}$, and, because many sites are at the critical value, the mean scales as $\langle \nabla z(A) \rangle \propto A^{-0.50 \pm 0.01}$ as well. A rich scaling structure of the slope variability is embedded in the threshold concept because at large aggregated areas the likelihood of being below threshold is much reduced. We will return to this property in detail later in this section, since we argue that a detailed investigation of this variability yields deeper insight into the nature of landscape-forming processes.

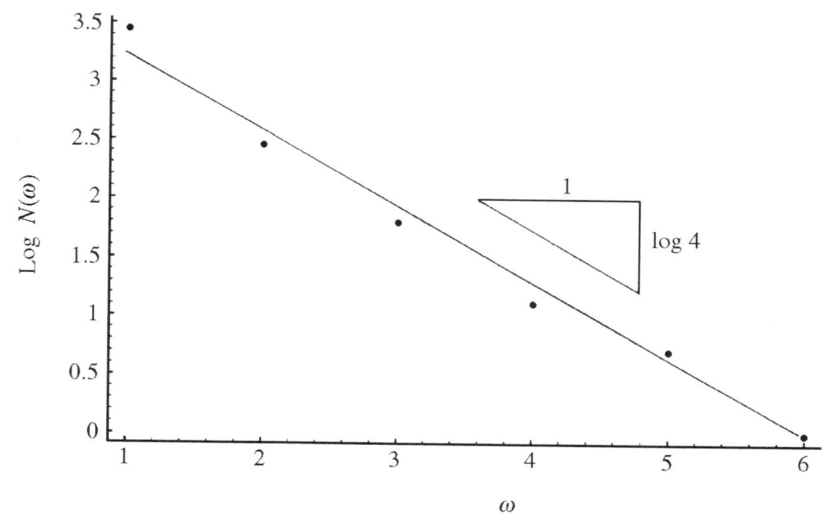

Figure 5.10. Statistics of the SOC landscape of Figure 5.9: Horton's bifurcation diagram ($R_B \sim 4$).

We finally note, for reasons to be clarified in Section 5.8, that total energy dissipation for the network in Figure 5.8 is $\sum_i A_i^{0.5} = 12,256$ in pixel units.

The above example shows the evolution of an initially random and fractal (and in the overall, realistic) network toward a fractal SOC network that did not modify the salient characters of the initial state. We will now show a second case, that is, the dramatic changes (both in plan and in the three-dimensional structure) occurring in the evolution of a deterministic nonfractal surface form, shown in Figures 5.15 and 5.16, toward a self-organized fractal channel network [Rinaldo et al., 1993]. Such evolution represents a demanding test because of the constraint imposed by the systematic arrangement of the elevation field z_i. The elevations yield parallel drainage directions almost everywhere and are exactly scaling by construction as $\Delta z_i = A_i^{0.5}$. Notice that, as a consequence, $Var[\nabla z(A)] = 0$. Hence the initial network is at the critical threshold everywhere. Thus it is a stationary critical state and the system, except for perturbations, would not evolve although the structure of the initial condition is very different from that of typical natural networks.

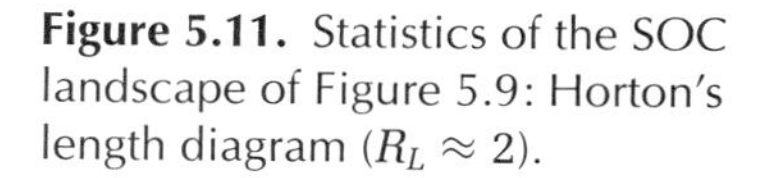
Figure 5.11. Statistics of the SOC landscape of Figure 5.9: Horton's length diagram ($R_L \approx 2$).

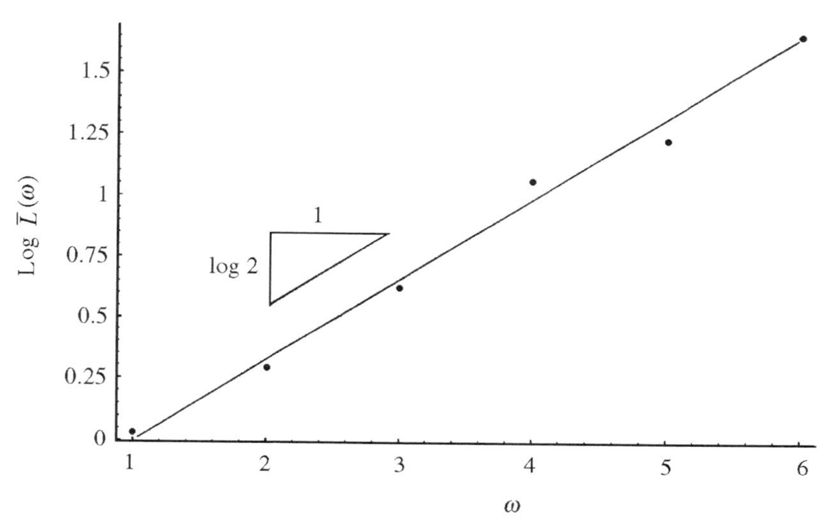

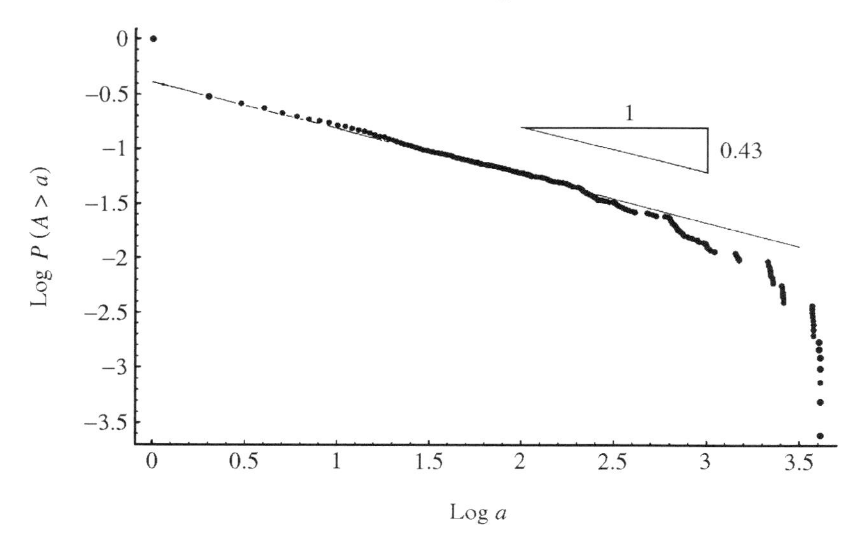

Figure 5.12. Statistics of the SOC landscape in Figure 5.9: Probability of exceedence of drainage area A, $P[A \geq a]$ versus a.

Figures 5.15 to 5.20 show the following:

- The initial condition in a 64 × 64 lattice (Figures 5.15 and 5.16) – the chosen comb-like planform yields surface slopes that generate parallel drainage directions, and at the left and bottom boundaries there are two collecting channels leading to the common outlet at the bottom left corner. The slopes are critical anywhere ($\Delta z_i = \tau_c / A_i^{0.5}$ with $\tau_c = 1$ in pixel units in this particular example), and total energy dissipation is $\sum_i A_i^{0.5} = 20{,}845$;
- an intermediate configuration after some perturbations and the related readjustments ($\sum_i A_i^{0.5} = 13{,}930$); and
- the stable stationary structure after 10^3 perturbations failed to modify the network ($\sum_i A_i^{0.5} = 12{,}234$).

From the plots one observes that the effects of a perturbation indeed depend heavily on both the site where it acts and the past history of the evolution. This important feature can be explained heuristically as follows. The first perturbation, wherever it randomly lands, acts on the maximally unstable state. It forces the merging of two formerly parallel streamlines and produces an approximate doubling of the flow rate. This creates local conditions for a pronounced erosion which, after rearranging the drainage directions, causes many connected sites to be above the erosion threshold because of the increased slopes. The burst of erosion activity may affect almost the entire network if the first perturbation touches a well-connected site near the outlet. Once complete relaxation has been achieved, a new perturbation is produced similarly to the sand addition in sandpiles. It is highly likely that a subsequent perturbation, although forcing convergence of drainage directions and local aggregation, would not achieve supercritical conditions because of the much reduced flow rate resulting from the previous capture of contributing area and the migration of divides. An example is shown in the disconnected pattern in the lower part of Figure 5.17. Thus the actual state of the system, although not its statistical characteristics, depends on the specific sequence of perturbations.

Figure 5.13. Statistics of the SOC landscape in Figure 5.9: Probability of exceedence of Strahler's stream lengths L, $P[L \geq l]$.

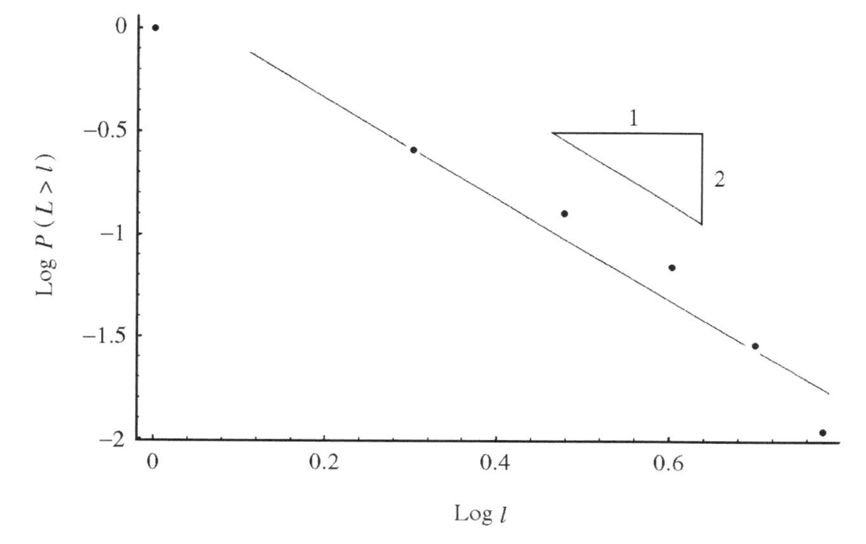

The susceptibility of the system to the order of perturbations is an

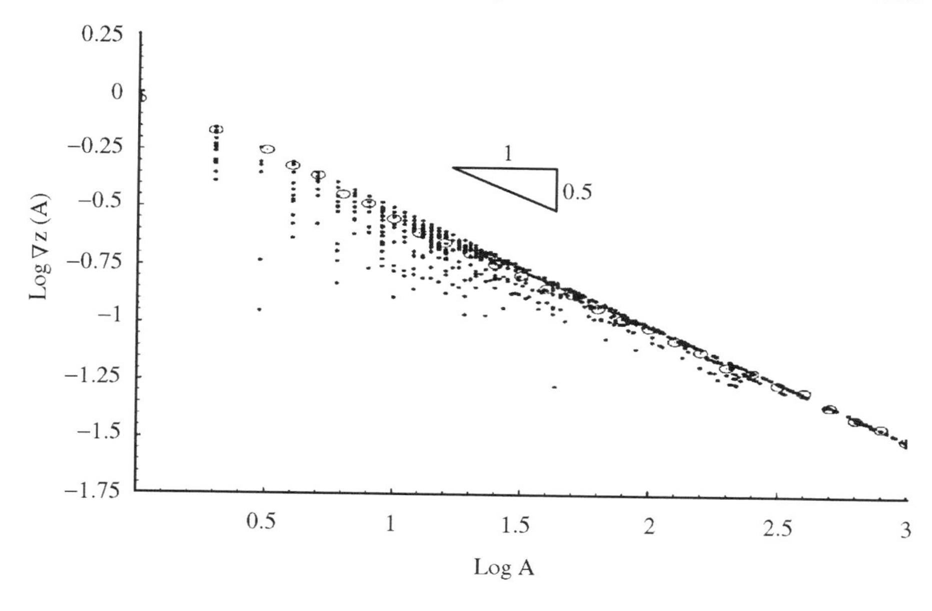

Figure 5.14. Statistics of the SOC landscape of Figure 5.9: $\nabla z(A)$ versus A relationship. Here and in the following graphs of the same type a circle indicates the mean value $\langle \nabla z(A) \rangle$.

important modification of the Abelian properties examined in Section 5.5 where a commutative behavior under a sequence of mass additions was found for the system, meaning that altering the order of a given sequence of perturbations would yield strictly the same state for the system rather than a different state with equal statistical properties. We will return to the mathematical reasons for such a property of this self-organized system in Section 5.9.

Figures 5.21 and 5.22 illustrate the plots of the probabilities of exceedence of total contributing area, $P[A \geq a]$, and stream length, $P[L \geq l]$, clearly suggesting that SOC networks evolve from a nonfractal landform (the comb-like structure) to one showing power law relationships over the whole range allowed by the lattice size. Finite-size effects are not addressed here, but indeed are described well by the methods discussed in Section 2.10.1. The slopes of power laws in the final state (when the structure proves stable to a prefixed – and large – number of perturbations) match within reasonable bounds those observed in nature and in OCNs (see Chapters 2 and 4).

Figure 5.23 illustrates the slope–area diagram for the basin shown in Figure 5.20. The same type of scatter observed in Figure 5.14 is produced.

We have already mentioned the multiscaling character observed in the slope versus area relationship of real basins. We have studied this possible feature in the SOC networks shown in this section. Figure 5.24 shows the graph of log $Var[\nabla z(A)]$ versus log A for the SOC network in Figure 5.20. From Figure 5.24 we infer that the overall scaling exponent is approximately -1.0 ± 0.1 for the case analyzed here [Rigon et al., 1994].

Figure 5.15. Evolution of SOC networks in different stages of growth: The comb-like initial condition (at the critical state everywhere) [after Rinaldo et al., 1993].

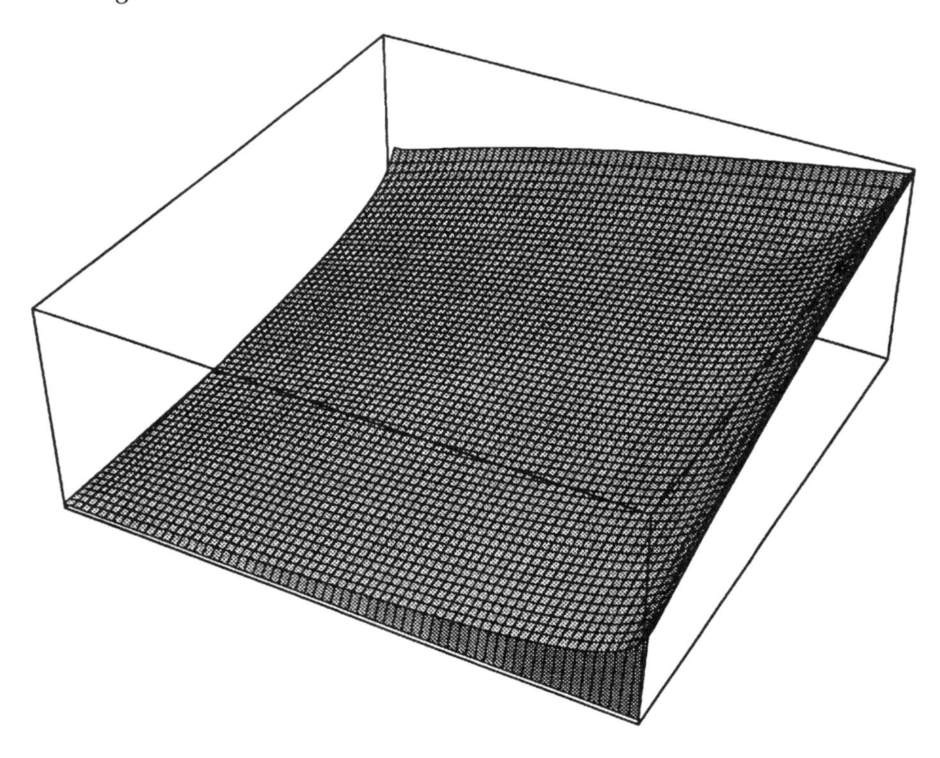

Figure 5.16. Three-dimensional plot of the comb-like initial condition.

Nevertheless, no conclusive evidence can be drawn from one example because of the difficult assessment of the impact of sample size on the variance of estimation of $Var[\nabla z(A)]$.

For the SOC model of fluvial self-organization described in this section, Rigon et al. [1994] have evaluated on a systematic basis the effect of sample size on $Var[\nabla z(A)]$ via Monte Carlo (MC) analysis of repeated 32×32 SOC networks run from identical initial conditions. The resulting diagrams of the variance of the slope versus area for a single realization, averaged over 50 and 100 MC realizations, are shown in Figure 5.25. The dots represent the variances in a single realization; the dashed line represents the average over 50 MC realizations, and the solid line represents the average over 100 realizations. The regression slopes are -1.05, -0.92, and -1.00 (and the related correlation coefficients are 0.46, 0.60, and 0.95) for 1, 50, and 100 realizations, respectively.

Figure 5.17. SOC network after a few perturbations [after Rinaldo et al., 1993].

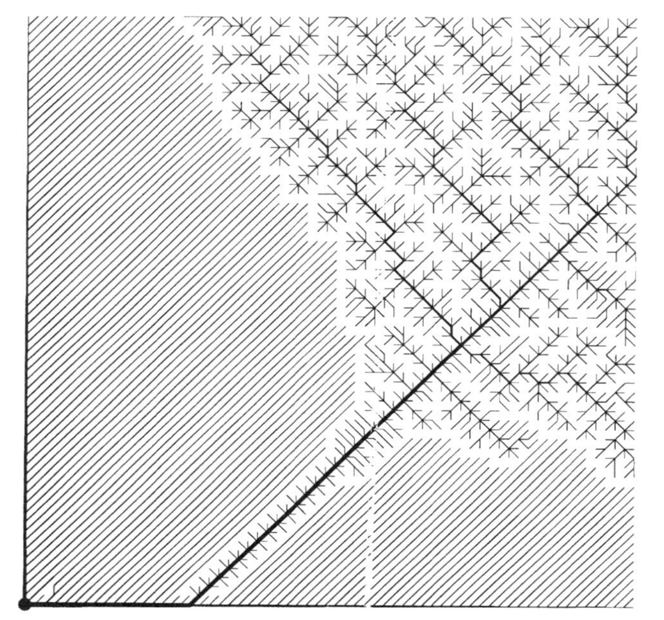

The above suggests that although finite size is an important factor in the reliability of the estimation of $Var[\nabla z(A)]$, the scaling parameter can nevertheless be studied with moderate sample sizes such as those arising from single realizations.

These results indicate that SOC models of erosion processes of the type described here are likely to yield a simple scaling structure of the type described by the self-similar assumption of Gupta and Waymire [1989] (Section 1.2.10) rather than the

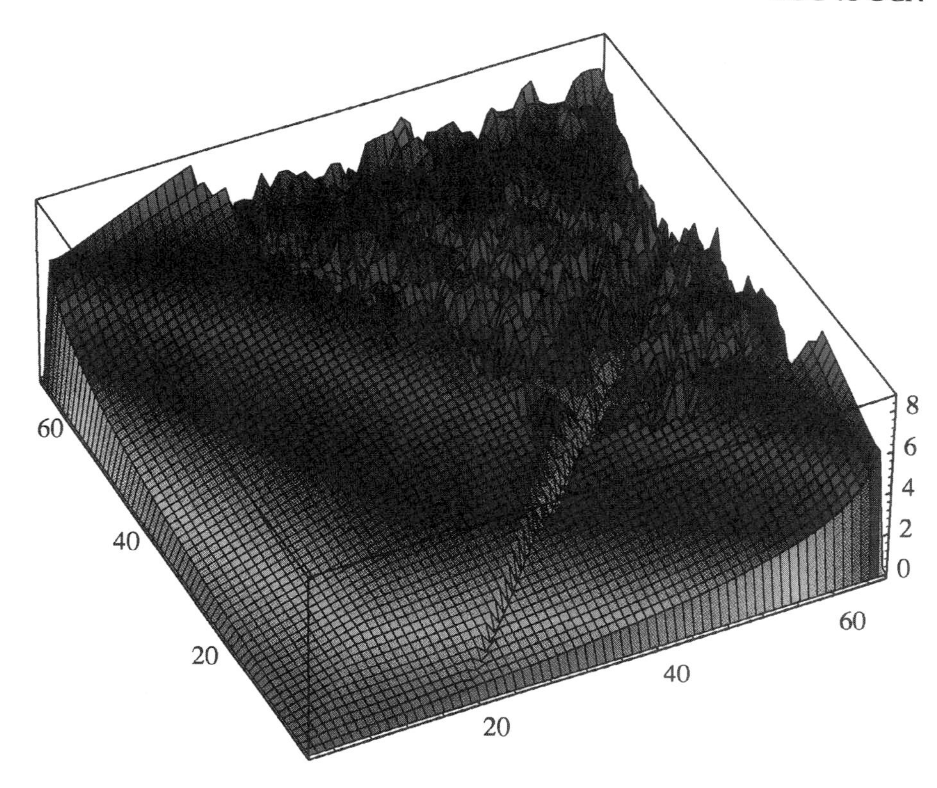

Figure 5.18. Intermediate three-dimensional configuration of the system.

multiscaling observed in nature. We will review this interpretation when examining the effects of concurrent hillslope and channel processes and the role of heterogeneity in soil mantle erosional properties (Chapter 6).

A few concluding remarks on the mechanisms of fluvial self-organization explored in this section are in order at this point:

- Principles of self-organization are likely to be at work in the development of the fluvial basin. Although strictly fluvial processes cannot be responsible for all landforms observed in nature, planar forms in particular are quite well reproduced by fluvial SOC.
- The model of self-organized criticality for river basins, defined for the first time in Rinaldo et al. [1993], produces network structures with fractal characters independent of the initial condition and with clear resemblance to real basins. This includes the demanding features required (to first-order only) by the statistics of the elevation fields.
- The self-organizing process yields a progressive lowering of the total energy dissipation of the system.

These issues will be studied in more detail in the next sections of this chapter and in Chapter 6.

5.8 Optimality of Self-Organized River Networks

The evolution of the total energy expenditure of self-organized fluvial systems is worth describing in detail.

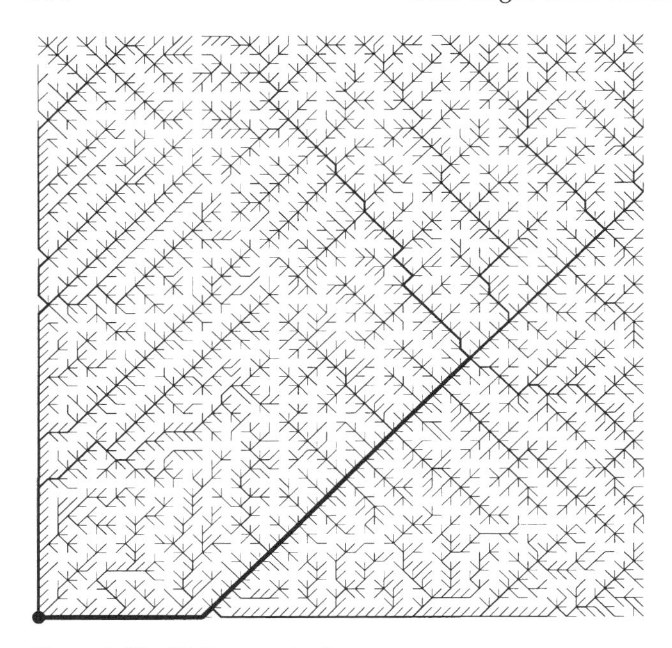

Figure 5.19. SOC network after 10,000 perturbations.

Total energy dissipation E of a given configuration s may be directly computed in this case as $E(s) = \sum_i \rho g Q_i \Delta z_i \propto \sum_i A_i \Delta z_i$ rather than by the 'topological' value $E_T(s) = \sum_i A_i^{0.5}$ because in the initial stages of the evolutionary process the height drops Δz_i usually do not scale as $\Delta z_i \propto A_i^{-0.5}$ (the latter is a property achieved in the mean, i.e., $\langle \Delta z_i \rangle \propto A_i^{-0.5}$ only once the fluvial regime is established).

In the case shown in Figure 5.15 the value of $E(s)$ decreases from the initial value of 20,845 pixel units following every readjustment, tending to the final value of $E(s) \sim E_T(s) = \sum_i A_i^{0.5} = 12{,}234$, which compares very favorably with the value of the network shown in Figure 5.8 as well as with any 64×64 OCNs (see Chapter 4).

The above behavior, which implies an embedded lowering of total energy dissipation toward an optimal value, not explicitly prescribed by the procedure, has been consistently observed in all SOC networks. From the observation that an optimal structure is one of the many configurations of local minimum energy expenditure within a plateau of relatively low total expenditure (whose shapes differ notably as a function of initial conditions and the driving perturbations but whose statistics match without exceptions), Rinaldo et al.

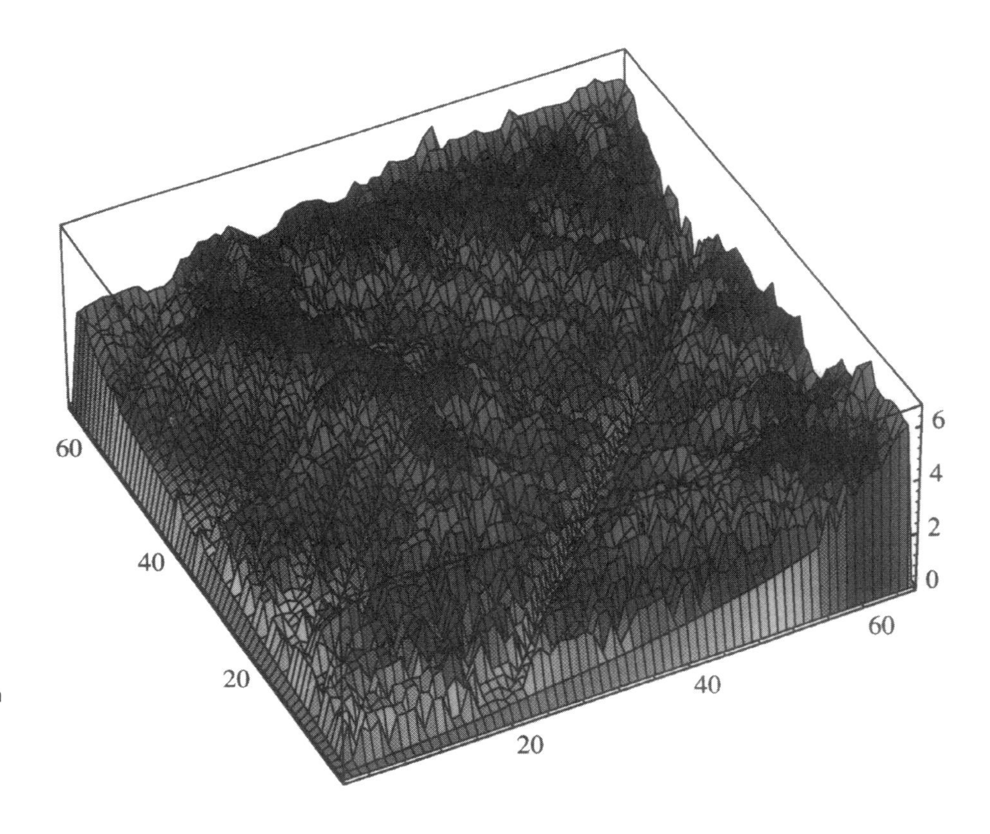

Figure 5.20. Final configuration of the self-organized fluvial landscape (three-dimensional plot).

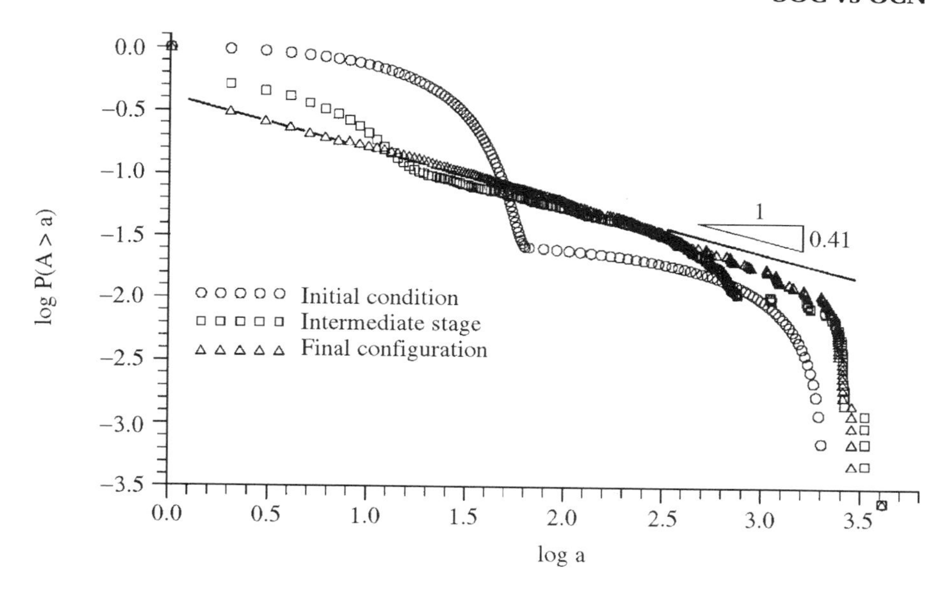

Figure 5.21. Evolution toward a SOC network from a comb-like structure: Probabilities of exceedence of total contributing area A [after Rinaldo et al., 1993].

[1993] suggested the equivalence of models of self-organized criticality with models of OCNs.

The explanation for the equivalence of the global criterion of least energy dissipation in the network and the structural stability in a self-organized model of a river basin was given by Ijjasz-Vasquez et al. [1993] (discussed in a different context in Chapter 4). The proof shows the equivalence of minimizing the total rate of energy expenditure and the total potential energy of the system constrained to obey scaling in elevations postulated by the local optimal principles discussed in Chapter 4. Such principles are the ones controlling the functional form of the critical parameter. We now return to the proof because of its relevance in the context of SOCs.

If $\sum_i z_i$ (i spans all sites) is a measure of total potential energy of the

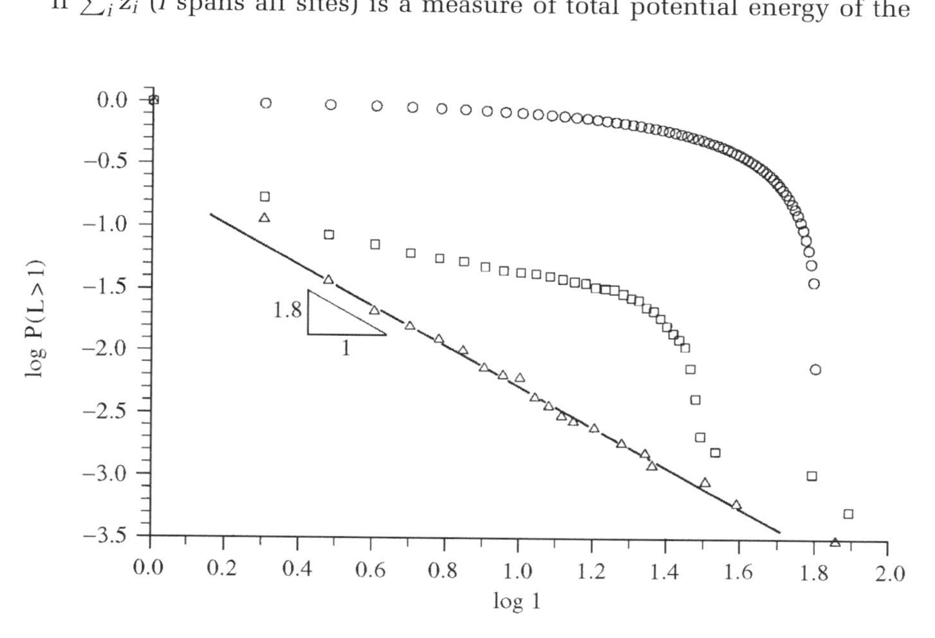

Figure 5.22. Evolution toward a SOC network from a comb-like structure: Probabilities of exceedence of Strahler's stream length. The legend is as in Figure 5.21 [after Rinaldo et al., 1993].

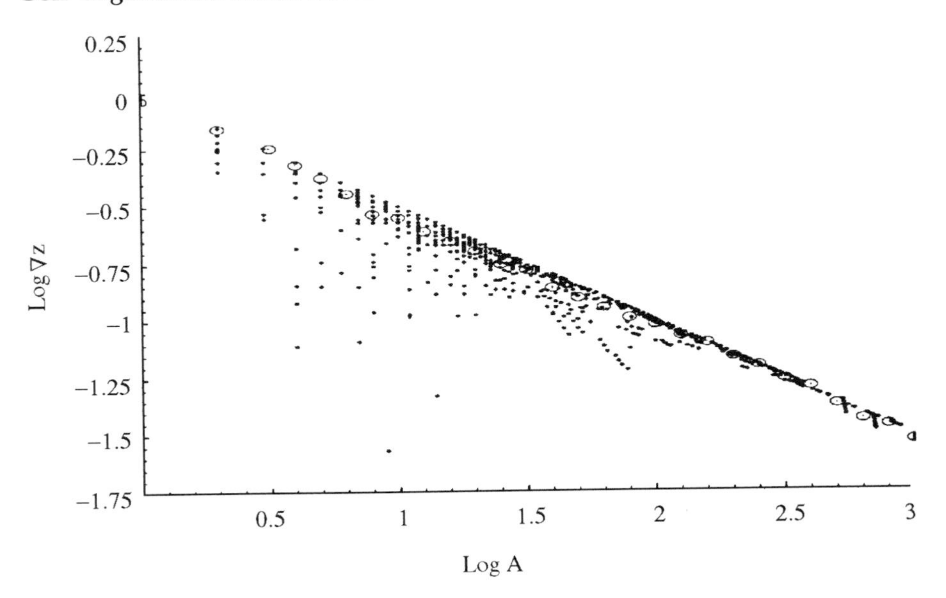

Figure 5.23. SOC network of Figure 5.20: Slope versus cumulative area. Circles represent the mean value $E[\nabla z(A)]$.

network, then one has

$$\sum_i z_i = \sum_i \sum_{j \in \gamma_i} \Delta z_j = \sum_i A_i \Delta z_i \tag{5.23}$$

where γ_i indexes the sites on the path from the outlet to the ith site. The last equality arises, as seen before, because, if the area ruler is unit, the cumulative area A_i measures the number of sites connected to the ith link and therefore the number of times every Δz_i appears in the sum. The last term on the right-hand side is proportional to the total energy dissipation, as the power P_i dissipated by the flow is $P_i = \rho g Q_i \nabla z_i L_i \propto A_i \Delta z_i$ because $Q_i \propto A_i$ and lengths are taken as unit everywhere.

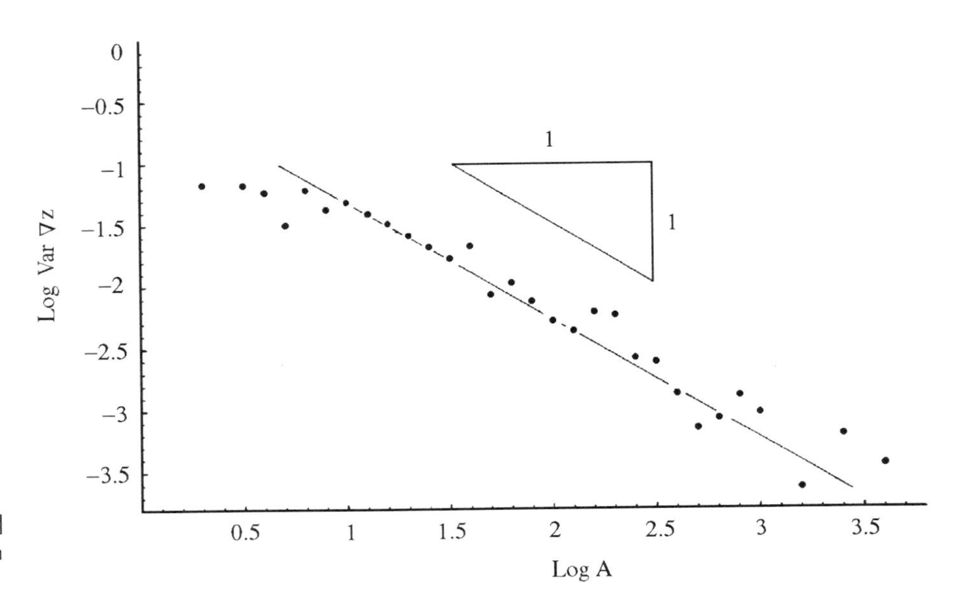

Figure 5.24. Log-log plot of the variance of the slopes $Var[\nabla z(a)]$ versus A for the SOC fluvial basin in Figure 5.20.

If we further constrain potential energy to scale as $\Delta z_i \propto A_i^{-0.5}$, we obtain

$$\sum_i z_i \propto \sum_i A_i^{0.5} \tag{5.24}$$

which is the result already shown in Chapter 4. Notice that the above implies $\Delta z_i = \langle \Delta z_i \rangle$. This result proves that minimizing the total energy dissipation in the system is equivalent to minimizing its total potential energy once the scaling of slopes is strictly $\Delta z_i \propto A_i^{-0.5}$.

Thus we observe that the system minimizing its potential energy ($\propto \sum_i z_i$) also minimizes total energy dissipation. Interestingly, this was also observed (see Section 4.5) for stationary dendritic structures developing within an open potential force field. In SOC networks, this is accomplished through random perturbations in the threshold model because the system finds a stable state when its mean elevation is low enough. Thus the resulting structures are OCNs with proper statistical scaling properties of their three-dimensional structure.

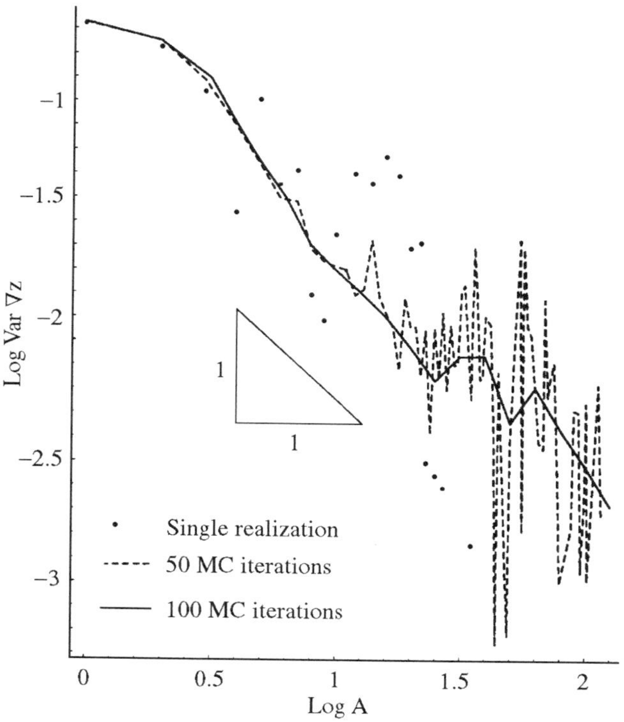

This result suggests that random perturbations (of tectonic or hydrologic origin) induce erosion activity through which the system readjusts automatically, tending to irreversibly lower its mean elevation and thus evolving toward stable configurations. Randomness, either in the initial condition or induced by perturbations, is crucial to the development of fractal scalings.

In the particular case of river networks, a model of self-organization has thus been shown to be equivalent to a model implying minimum energy dissipation for the stationary structure of an evolving open system. It will be a challenge to see whether in general self-organized criticality might be interpreted in light of optimality principles (see Section 5.12).

Figure 5.25. Sample size effect in the $Var[\nabla z(A)]$ versus A relation by averaging over Monte Carlo realizations (i.e., repeated SOC runs) from the same initial conditions. Dashed line and solid line correspond to 50 and 100 realizations, respectively. The dots correspond to a single realization [after Rigon et al., 1994].

5.9 River Models and Temporal Fluctuations

The model of self-organizing fluvial structures described in Section 5.7 may be seen as a modification of the sandpile model described in Section 5.2. It may be thought of as belonging to the set of models in which the threshold for activity, rather than being a function of a purely local variable, depends on nonlocal properties of the self-organizing structure. This represents a different type of self-organization, which is of interest in itself and worth discussing here.

In the fluvial case, the nonlocal character of the possible threshold exceedence follows from the fact that the controlling variable at the arbitrary ith site equals a shear stress, that is, $\tau_i \propto \nabla z_i A_i^{\theta}$. As such, the exceedence of τ_c depends not only on local conditions (i.e., a critical value of ∇z_i) but also on nonlocal conditions defined by the contributing area A_i computed through drainage directions. Total contributing area A_i thus depends on the entire state of the system that is self-organizing.

Notice that the above is true whatever the functional dependence of τ_i on area A_i. Notice also that the physical rationale for the nonlocal dependence lies in the fact that the system is open, that is, injected from outside, allowing flow rates to be proportional to total contributing drainage area. Thus the SOC river basin (or "river sandpile") model is a somewhat different model of self-organization because the threshold for activity is also dependent on the entire state of the system.

The long-range nature of the threshold dynamics tends to hide from the observer the temporal fluctuations that take place in the evolutionary timescale. In this sense the model described in Section 5.7 is classified as one of *spatial* self-organized criticality.

The model described before is clearly an oversimplified representation of the dynamics of the fluvial basin, especially when compared with detailed deterministic models (Section 1.4) that address a specific description of all possibly relevant processes acting in landscape evolution. The reductionist approach, where a precise description of the details is sought, is successful in the description and classification of landforms. Nevertheless, as standard in critical phenomena, the mechanisms producing scale-free structures are expected to depend only on a few key features common to all networks rather than on the details of the particular system under study. In this chapter, centered on the dynamic origins of fractal river networks, we clearly follow a nonreductionist approach based on the simplest possible, parameter-free, models capable of allowing the emergence of complexity.

Whether or not river self-organization qualifies as a more general framework of self-organized criticality remains to be seen. If SOC must necessarily refer to the occurrence of a critical state in the sense of critical phenomena, where a small local perturbation can cause a significant change in the configuration of the whole system and thus the system shows both spatial and temporal scaling, then the temporal dynamics should be specifically considered. One way to do this is through the oscillation of the threshold in time, that is, $\tau_c(t)$, simulating climatic fluctuations (see Rinaldo et al. [1995a] and also Section 6.6), through which indeed temporal activities appear, or also through perturbations of random location and strength forcing the evolution of the landscape. Temporal dynamics also occurs if the landscape-forming rainfall events are described as nonuniform in space, leading to patches of activity randomly scattered spatially (in such a case the outflow response of the system becomes a $1/f$ signal). In such a case one distinguishes flow rates Q_i and contributing areas A_i at the i-th site, as $A_i = \sum_j A_j + 1$ and $Q_i = \sum_j Q_j + R_i$ where R_i is the arbitrary injection term. Obviously, if R_i keeps fluctuating in time, a temporal activity is generated. However, regardless of any additional features, we believe that the central scope of SOC is the dynamic explanation of the growth of fractal structures of the type appearing in nature, that is, the physics of fractals.

Moreover, questioning the self-organized critical nature of the model by Rinaldo et al. [1993] on the basis of an apparent lack of temporal dynamics would also be irrelevant from a thermodynamic viewpoint because we have shown (Sections 4.21 and 4.22) that scaling properties of energy and entropy yield limit states which, depending on the constraints, may be either temporally active or frozen. Furthermore, in Section 4.22 we suggested that optimal states like the ones dynamically accessed by the model under study may exist in space–time active states precisely at the edge of chaos.

An open question, indeed more interesting than the semantics of SOC, is how to relax the constraints in the river sandpile model to produce realistic 'hot' fluvial landscapes, that is, having a geomorphologically recognizable structure and yet active like an ordinary sandpile.

A few additional features of the river sandpile are worth discussing. Dhar and Ramaswamy [1989] and Dhar [1990] have noted that the Abelian treatment described in Section 5.4 is valid when the toppling rule differs from site to site and for any dimension of the lattice but not when the toppling criteria depend on gradients of height. Thus the original Bak et al. [1988] scheme yields non-Abelian sandpile models. The model of Section 5.7 also falls under this class of models, because the procedure is greatly complicated by the nonlocal functional dependence of the threshold activity, that is, shear stresses at a site depend on total contributing area upstream of the site and thus on the self-organizing structure.

We now return to the framework of the sandpile model algebra to explore the mathematical consequences of the features described above. We recall that integers z_i on each site i represent the local landscape height. As seen in Section 5.7, addition of a sand particle to a site i is represented by increasing the value of z_i at that site by a unit. When the shear somewhere exceeds a critical value τ_c there occurs a "toppling" event in the manner described before.

The state of the system can be again represented by the C matrix described in Section 5.5 whose size is that of the lattice and whose elements represent the states, or heights, z_i of the system.

Operators are introduced on C by the rules defined earlier. To explore self-organized criticality in this model, one can randomly add sand and have the system relax as in the original SOC model. Under a toppling at site j, the height at site i becomes $z_i - \Delta_{ij}$. If we further assume, as in the case described in the previous section, that the mass released at i when activity occurs is removed from the system (i.e., does not affect the nearest neighbors), then the toppling matrix becomes diagonal, that is, $\Delta_{ij} = 0$ if $i \neq j$. Thus $\Delta_{ij} \propto \delta_{ij}$, where δ_{ij} is Kronecker's delta. Thus the toppling matrix can be represented by a *toppling vector* Δ_{ii} such that the change in height z_i at site i is produced only by its own toppling, its height being $z_i - \Delta_{ii}$. For the self-organized channel networks described in this chapter, the toppling vector is defined as

$$\Delta_{ii} = \Delta z_i - \frac{\tau_c}{\sqrt{1 + \sum_j W_{ij} A_j}} \tag{5.25}$$

Eq. (5.25) results from the following facts:

- the toppling event leaves the slopes at the critical threshold (i.e., $\tau = \tau_c$);
- the shear stress τ_i at i is proportional to total drainage area and slope via $\tau_i \propto A_i^{0.5} \nabla z_i$; and
- total drainage area A_i at site i is computed by the expression $A_i = 1 + \sum_j W_{ij} A_j$ where W_{ij} is a connection matrix made up by 1s and 0s as defined in Eq. (4.49). The definition of the update of the connection matrix W so far defies an analytical solution.

The toppling matrix, which has N^2 elements of which only N are nonzero in the simplified case studied above, needs to be updated at all times, and, remarkably, its elements become asymptotically zero, that is, $\Delta_{ij}(\infty) = 0$ for each i, j. This, in fact, is our definition of convergence. The system evolves to a stable state because no further perturbation can actually modify the state matrix C. It is clearly a completely different picture from that of the Abelian sandpile model.

In particular, the avalanche operators do not commute. In fact, the various operators acting on states C are, as before:

- The *adding sand* operator α_i acting on any C yields the state $\alpha_i C$, where $z_i \to z_i + 1$ and all other zs are unchanged.
- The *toppling* operator t_i transforms C into the state with heights z'_j where $z'_j = z_j - \Delta_{ij}$, or simply $z'_j = z_j - \Delta_j$.
- The *updating* operator U updates the lattice one time step. This is obtained as the product of t_i over all sites where the slope is unstable, that is, $UC = \prod_i t_i^{p_i} C$, where i spans all sites and $p_i = 1$ if $\tau > \tau_c$ and 0 otherwise.
- The *relaxation* operator R is defined using U repeatedly. Applied to any state C this corresponds to repeating U until no more z_i change. Neither U nor R have any effect on stable states. Thus $RC = UU \cdots UUC$ until all states are stable.
- The *avalanche* operator $a_i C = R\alpha_i C$ takes the system to another state through transformations.

The only mathematical result that we can derive is that the operators a_is, when acting on the stable states, do not commute. In other words, we shall now prove that the sequence of avalanches at j sites produces a stable structure dependent on the order of avalanches perturbing the system, that is

$$a_i\, a_j\, C \neq a_j a_i\, C \quad \text{for all } C \tag{5.26}$$

To accomplish this, we use the previous definitions to express the avalanche operators a_i's in terms of toppling and addition operators:

$$a_i\, a_j\, C = \left(\prod_{k=1}^{n_i} t_k \right) \alpha_i \left(\prod_{k=n_i+1}^{n_j} t_k \right) \alpha_j\, C \tag{5.27}$$

where the specific numbers of topplings n_i and n_j depend on i, j, and C. The key point is that, acting on general states, ts and αs do not commute because they do not act on constants, say, adding or subtracting heights, and thus one can not shift α_i to the right in the above equation. Thus the whole proof on which the Abelian construct is based does not hold.

The above is easily confirmed by the experimental results. In fact, if from the minimally stable state of the comb-like structure of Figure 5.15 we change the sequence of perturbations, we get different structures. This follows from the fact that the local flow convergence induced by the random perturbation localizes the erosion where it acts. The deep valley around which the whole network self-organizes is thus a function of the sequence of perturbations, the statistics of the system being unaltered by the following process of competition for drainage, divide migration, and erosion.

Moreover, the group does not seem to have an inverse, although we cannot provide a formal proof. It seems therefore appropriate to define our model of SOC as a semigroup.

5.10 Fractal SOC Landscapes

The self-organizing model of river basins presented in Section 5.7 also yields many interesting features related to the topography of the resulting landscapes that incorporate the drainage network.

Recently, it has been suggested that many topographies of natural geologic landscapes have a clear fractal behavior. Although recent contributions [Lavallée et al., 1993; Schertzer and Lovejoy, 1993; Rodriguez-Iturbe et al., 1994] bring new and convincing evidence for a somewhat different interpretation of the fractal characters observed in real topographies, we will first analyze SOC river landscapes as simple fractal objects.

In this section we will not consider hillslope processes and forms because they operate in a qualitatively different fashion and at different scales from that of fluvial erosion, as discussed in Section 1.4.3. Chapter 6 will explicitly consider hillslope evolution on landscape self-organization.

Relevant examples of *simple* fractal analyses dealing with topography and bathymetry measured along a linear track are well documented in the literature, as we have addressed in Section 2.7.4. Specifically, topographic tracks of height versus distance along straight lines drawn in different directions throughout the basin have been analyzed for several sets of data from real topographies. Tracks (or transects) are self-affine processes that exhibit a 'fractal' dimension, D, variable within 1.2 and 1.7 depending on many factors, chiefly the landscape-forming processes, the geologic environment, and also the orientation of the transects.

The techniques described in Section 2.7.4 have been applied to transects obtained from the topography of SOC basins. These tracks have been analyzed by a variety of techniques. One of them is the method of variations (MV) (Section 2.7.3), which proves more precise when avoiding effects due to the finite size of the data set. This method is based on the fact that the range of variation of the height z within a box of size L scales as L^H ($D = 2 - H$) if the object is simple scaling.

For comparison, we have also used the methods of Matsushita and Ouchi [1989] (MO) (Section 2.7.4) and the scale-invariance of power spectra (Section 2.7.2). MO analysis studies the deviation $\sigma^2(L)$ as a function of the horizontal length scale L as

$$\sigma^2(L) = \left\langle \frac{1}{L^2} \sum_{i,j=1}^{L} \left(z_{ij} - \langle z \rangle_L\right)^2 \right\rangle \tag{5.28}$$

where $z_{ij} = z(\mathbf{x})$ is the elevation of the lattice site $\mathbf{x} = (i, j)$ and $\langle z \rangle_L$ is the average height for the surface within a subgrid of size $L \times L$. For a self-affine fractal, σ is related to L by a power law $\sigma \propto L^H$.

All different methods have yielded a range of consistent results. Figures 5.26 and 5.27 show the results of MV and MO analyses for transect profiles of the elevation field of the SOC landscape in Figure 5.9. Fractal dimensions have also been checked by the common artifice of deriving a long sequence merging together several tracks. The corresponding fractal

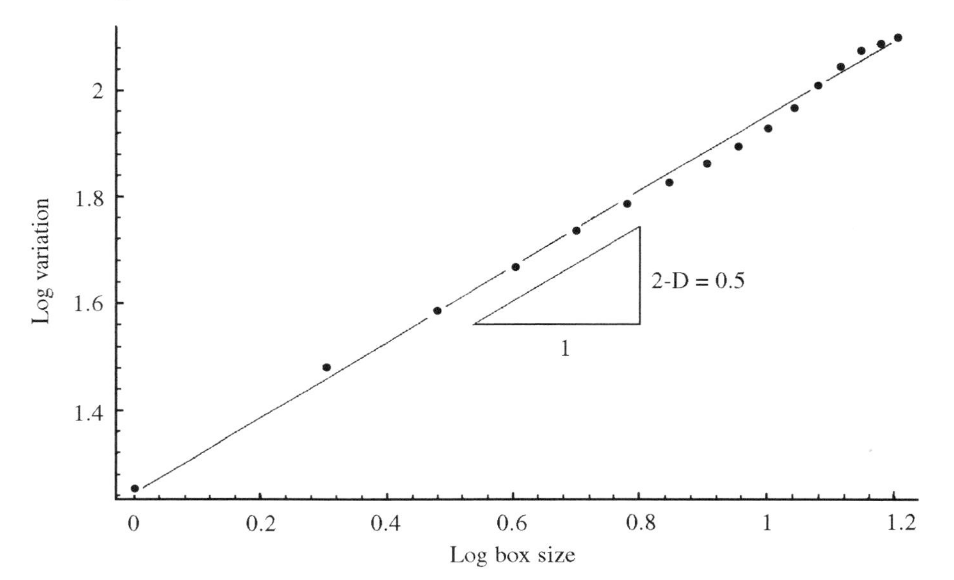

Figure 5.26. Analysis of the fractal dimension D of a single transect from the landscape shown in Figure 5.9: From the slope $H = 0.5$ of the plot for a single transect one infers $D = 2 - H = 1.5$ [after Rigon et al., 1994].

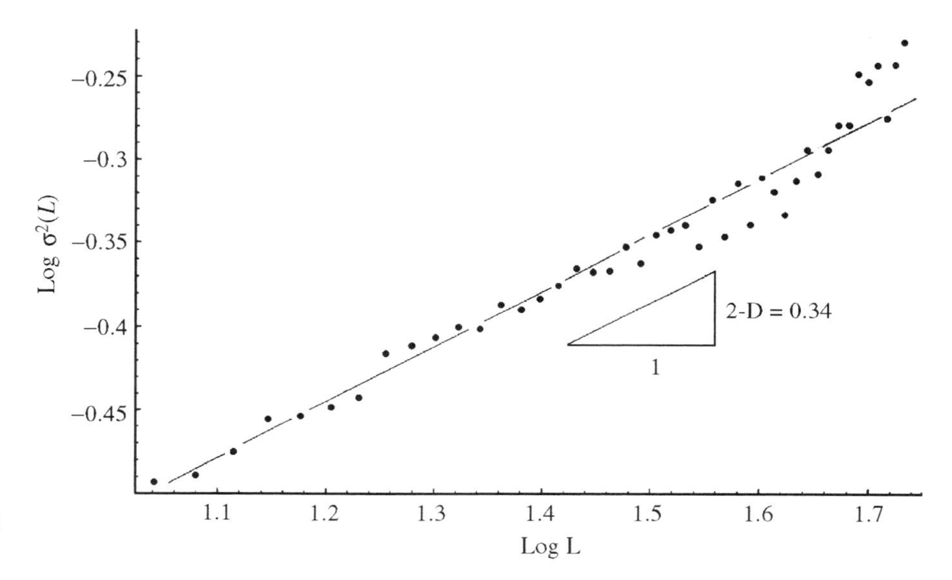

Figure 5.27. Analysis of the fractal dimension D of transects from the landscape shown in Figure 5.9. The output for the Matsushita–Ouchi method averaged over the sixty-four transects ($H = 0.34$, $D = 2 - H = 1.66$) [after Rigon et al., 1994].

analysis (carried out via the method of variations) shows two regimes, one consistent with the above results and another for large sizes of the spatial lag, induced artificially by the merging procedure.

Dimensions of the individual transects range from 1.2 to 1.7. The mean value of MV is $D = 1.64 \pm 0.05$, while the mean of MO (Figure 5.27) is $D = 1.66 \pm 0.05$, and the analyses always yield straight log-log plots for all the decades allowed by the network size. The occurrence of some very low fractal dimensions is explained by the form of some peculiar transects (typically along the boundaries).

All methods and techniques point at values of D near 1.6 for the fractal dimension of self-affine topographical transects resulting from evolution processes dominated by fluvial erosion. This holds in the absence of coupling with hillslope processes. It is thus apparent that purely fluvial landscapes are somewhat too rough with respect to many natural ones. In fact, a fractal dimension of 1.6 of the one-dimensional transect would imply a dimension of 2.6 for the landscape itself. As shown by Voss [1986], the most realistic reproductions of natural landscapes are obtained with dimensions close to 2.4. It should be emphasized that reproduction of a scaling surface with lower dimension is not simply a matter of letting the surface relax by some arbitrary mechanism. Chapter 6 explains how the coupling of fluvial and hillslope processes naturally reduces the fractal dimension observed when only fluvial processes are operating.

We now turn to more sophisticated analyses of SOC landscapes. Lavallée, et al. [1993] argued that geographical and geophysical fields are generally multifractal (characterized by an infinite hierarchy of fractal dimensions) and that inconsistencies are inevitable when they are forcedly viewed through geometric frameworks involving a single fractal dimension. We also recall, with reference to the theoretical discussion presented in Chapter 3, that multifractals arise when studying the distribution of a measure on a geometric support. The support may or may not be a fractal itself. When the mathematical measure is integrated, that is, averaged over a resolution scale, it generates a field with intensity values defined spatially. As seen in Section 3.5.1, to obtain sets one typically establishes thresholds and defines the set of interest (e.g., exceedence sets).

Consider a SOC landscape defined by the elevation $z(\mathbf{x}, t \to \infty) = z(\mathbf{x})$ at the arbitrary point $\mathbf{x}$. Given the function z, sets can be defined in many different ways. First, consider the exceedence set $S_{\geq T}$ as the set of all points x satisfying $z(\mathbf{x}) \geq T$, that is, the region whose altitude exceeds the threshold T. In general, if $T_1 > T_2$ then $S_{\geq T_1}$ is a subset of $S_{\geq T_2}$, and hence the fractal dimensions of the sets obey $D(S_{\geq T_1}) \leq D(S_{\geq T_2})$. In the case of a simple scaling (monofractal) structure, the sets $S_{\geq T}$ are characterized by a single fractal dimension $D(S_{\geq T}) = D$ irrespective of T [Lavallée et al., 1993]. This provides a means of checking the multiple scaling assumptions. Similarly, the *graph* of z, G, is defined as the set of points $(x, z(x))$ in a $(D+1)$-dimensional space, that is, $D = 2$ in the case of a topographic surface. As seen in Chapter 3, in the special case of simple scaling, the well-known result $D(G) = 3 - H$ is obtained, where the Hurst exponent H (Chapter 2) defines the scaling properties of the range exhibited by the field. Likewise, one can define the perimeter set P_T associated with $S_{\geq T}$ as the T-crossing set of S with the plane $z(\mathbf{x}) = T$, whose dimension $D(P_T)$ is the dimension of the isolines of topographic maps. We note that by definition P_T is a subset of $S_{\geq T}$ and hence $D(P_T) \leq D(S_{\geq T})$. In general, such values depend on the connectedness properties of S and may increase or decrease with T only being bounded above by the decreasing function $D(S_{\geq T})$. Isolines in simple scaling models of topography have dimension $D(P_T) = 2 - H$ independent of T [Lavallée et al., 1993].

As discussed in Chapters 2 and 3, a popular way of estimating the fractal dimensions $D(P_T)$ consists of taking qth powers of variations of the field and examining their spatial scaling behavior. In Chapter 2 we also discussed that the widely known method of estimating the fractal dimension of a surface via its power spectrum, that is, $S(f) \propto f^{-\zeta}$, yielding

$$D(P_T) = \frac{5 - \zeta}{2} \tag{5.29}$$

Figure 5.28. Planar view of a 128 × 128 SOC network.

for lines embedded in planes, is strictly valid only for simple scaling structures. In other words, Lavallée et al., [1993] argue that because in most topographic applications the sets involved are likely to be multifractals, the above relation would not hold. This also implies that there is not necessarily any contradiction between differing estimates of fractal dimensions of lines of constant altitude and spectral and variogram exponents, provided they are defined in a consistent mathematical framework.

The previous considerations should always be weighed. Nevertheless, in practice, as in the case of most of the real-world topographic data, it is difficult to assess whether or not the most significant source of deviation from expected power laws is due to sample size effects. Estimates of the necessary sample size have been obtained [Lavallée et al., 1993], which point at prohibitively large data sets for SOC analyses at the present state of the art.

A realistic representation of landscape evolution should include both threshold-dependent (TD) and slope-limited (SL) processes, and this topic will be covered in Chapter 6. Nevertheless, it is important and interesting to study whether multifractal scaling structures of the landscape are embedded in the SOC model presented in this chapter. To do so, and in view of the crucial role of size effects in this endeavor, we have constructed a large set of 128 × 128 SOC landscapes started from random initial conditions (one realization is shown in Figure 5.28). Because convergence of the SOC to the stable state in a 128 × 128 grid takes considerable computational effort and because the number of significant changes decreases steadily but ceases after a number of random perturbations proportional to N^2 (where N is the grid size), the SOC process has been stopped after the total energy dissipation $\sum_i A_i \nabla z_i$ reached a relatively stable value. We have checked the statistics of the obtained SOC network and they match the expected figures with no exception.

Figures 5.29 to 5.33 show some specific border sets P_T, where the lower cutoff scale is defined by the pixel size in the 128 × 128 grid. Every white spot corresponds to a pixel whose elevation is equal (within the accuracy of the procedure [Rodriguez-Iturbe et al., 1994]) to a fixed value T. The dimension of each set has been evaluated via box counting (see Section 2.1.2).

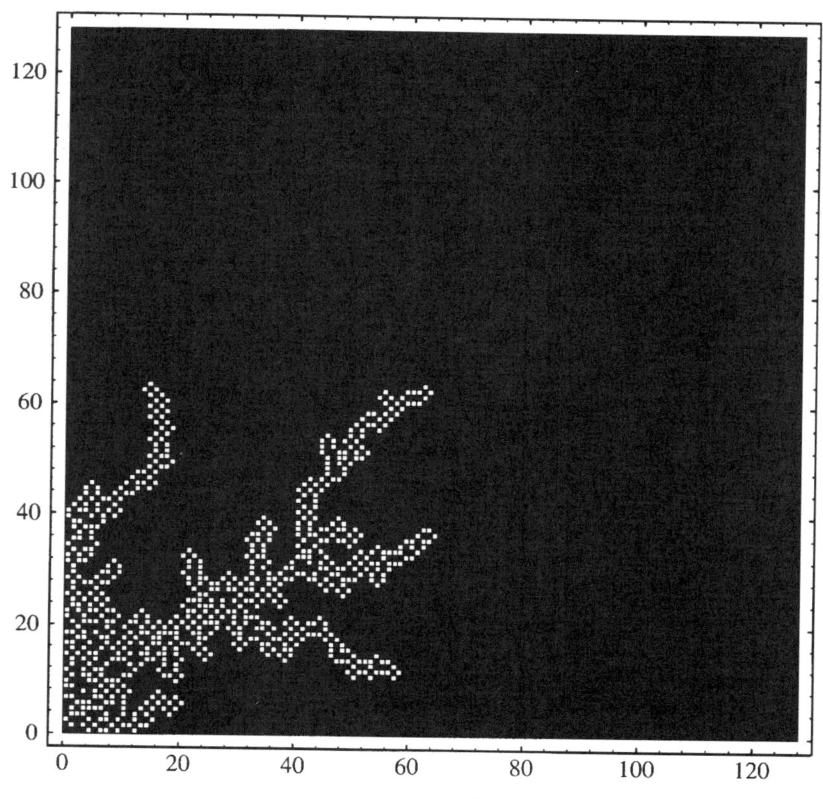

Figure 5.29. Border sets of equal elevations, $z(\mathbf{x} = 1)$ in pixel units.

For the low-elevation case ($z(P_T) = 1$), Figure 5.30 yields a fractal dimension $D = 1.38 \pm 0.01$ and a high correlation coefficient for the estimate of the slope in the double-logarithmic plot of the box-counting procedure. We thus assume that the estimation procedure is reasonably accurate notwithstanding the limited size of the sample.

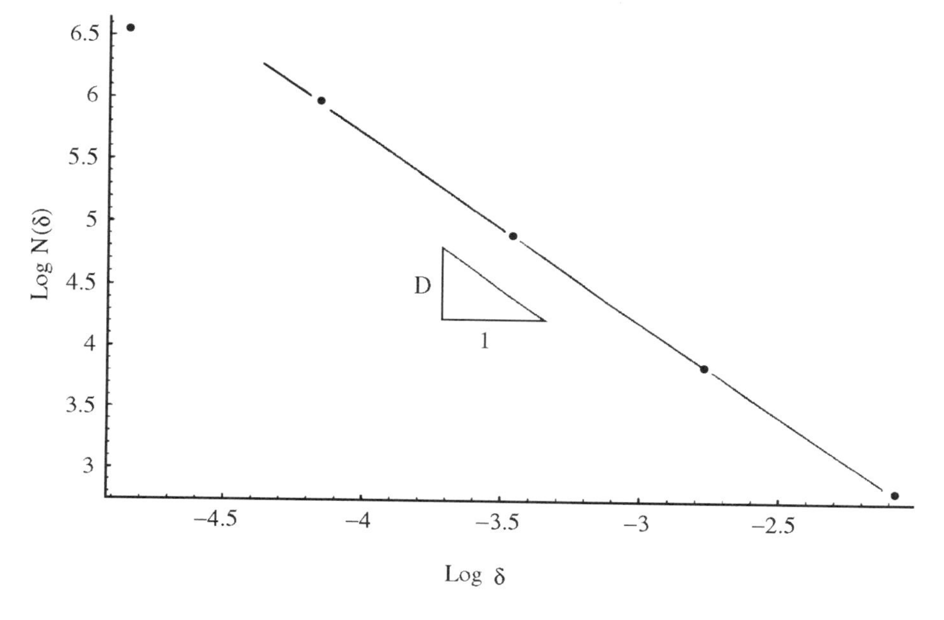

Figure 5.30. Box counting applied to determine the dimension of the contour set of elevation $z = 1$ ($D = 1.38 \pm 0.01$, correlation coefficient is $r = 0.99$).

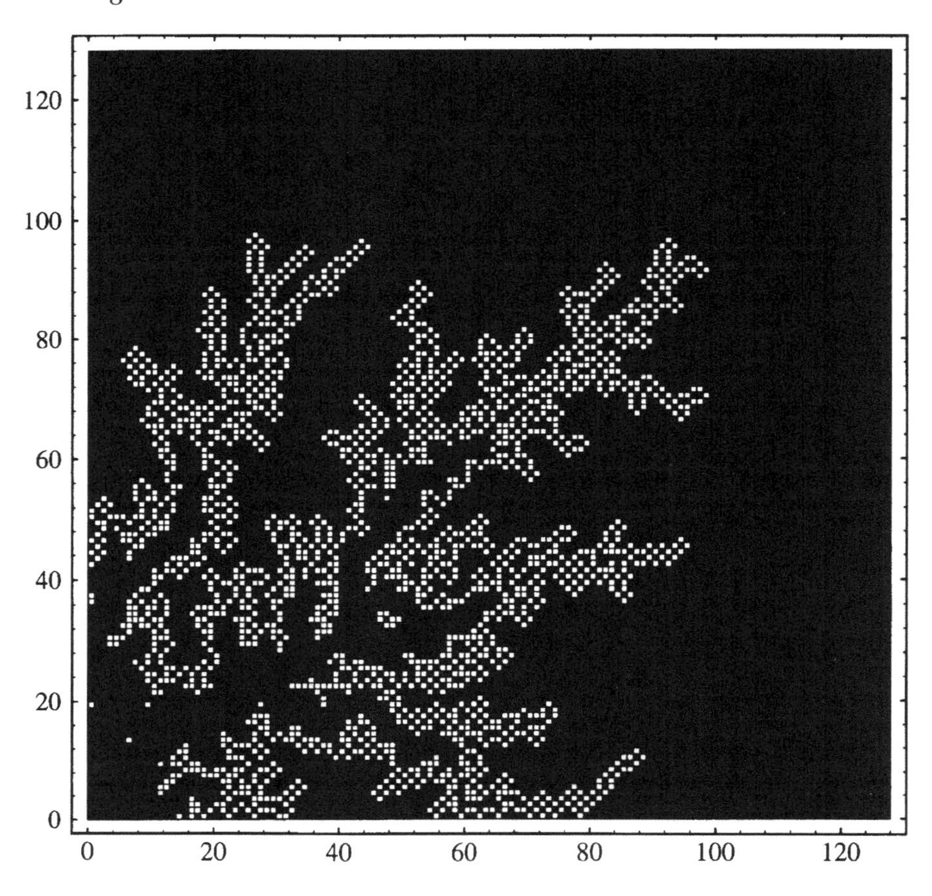

Figure 5.31. Border sets of equal elevations, $z(\mathbf{x} = 2)$ in pixel units.

A further factor partially impairing the accuracy of the computations is the presence of boundaries, which may lower the overall dimension if the boundary portion of the set becomes significant. The resulting fractal dimensions at different elevations are given in Table 5.1. The range of fractal dimensions is 1.17–1.42, and hence supposedly multifractal. However, the low values are much affected by the boundaries and, discarding the very low elevations, the real range seems to be rather 1.3–1.4. It thus seems that the SOC model produces landscapes characterized by a narrow range of fractal dimensions of lines of equal elevation and that, as a consequence, a simple scaling model may reasonably well define the features of the landscape's topographic surface. Other factors, besides size effects, render multifractal analyses difficult for the available SOC landscapes. In part these factors are related to an intrinsic nonstationarity of the field. In fact, we know that by construction the average gradients of the elevation field change in space, because when approaching the outlet not only are the elevations generally lower but the slopes are also lower. Hence simple procedures like those related to the scaling of increments may possibly prove inadequate. Also, the peculiar landscape structures produced by the threshold-dependent process of erosion induces nontrivial and anisotropic correlation structures that are not easily described within assumptions of stationarity.

Table 5.1. *Fractal dimensions of the border sets for different elevations of the landscape whose planform is shown in Figure 5.28 with the corresponding correlation coefficients*

$z(\mathbf{x}) = T$, Contour Elevation	$D(P_T)$, Fractal Dimension	r
1	1.38	0.99
1.5	1.42	0.96
2	1.43	0.98
2.5	1.43	0.94
3	1.39	0.96
3.5	1.37	0.99
4	1.35	0.92
4.5	1.30	0.90
5	1.33	0.90
5.5	1.27	0.80
6	1.17	0.80

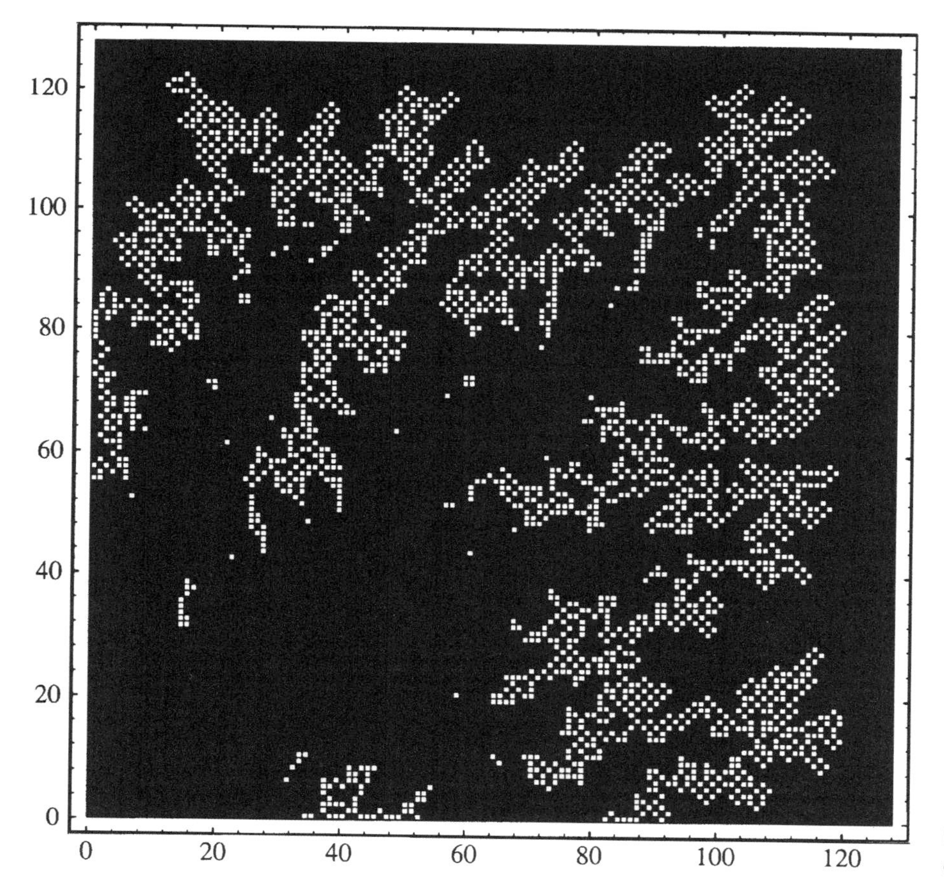

Figure 5.32. Border sets of equal elevations, $z(\mathbf{x} = 3)$ in pixel units.

In conclusion, we believe that an approximate but reliable indicator of the roughness of the SOC landscapes is represented by the single fractal dimension of the linear topographic tracks (termed, for simplicity, the fractal dimension of the landscape). Although inadequate for representing

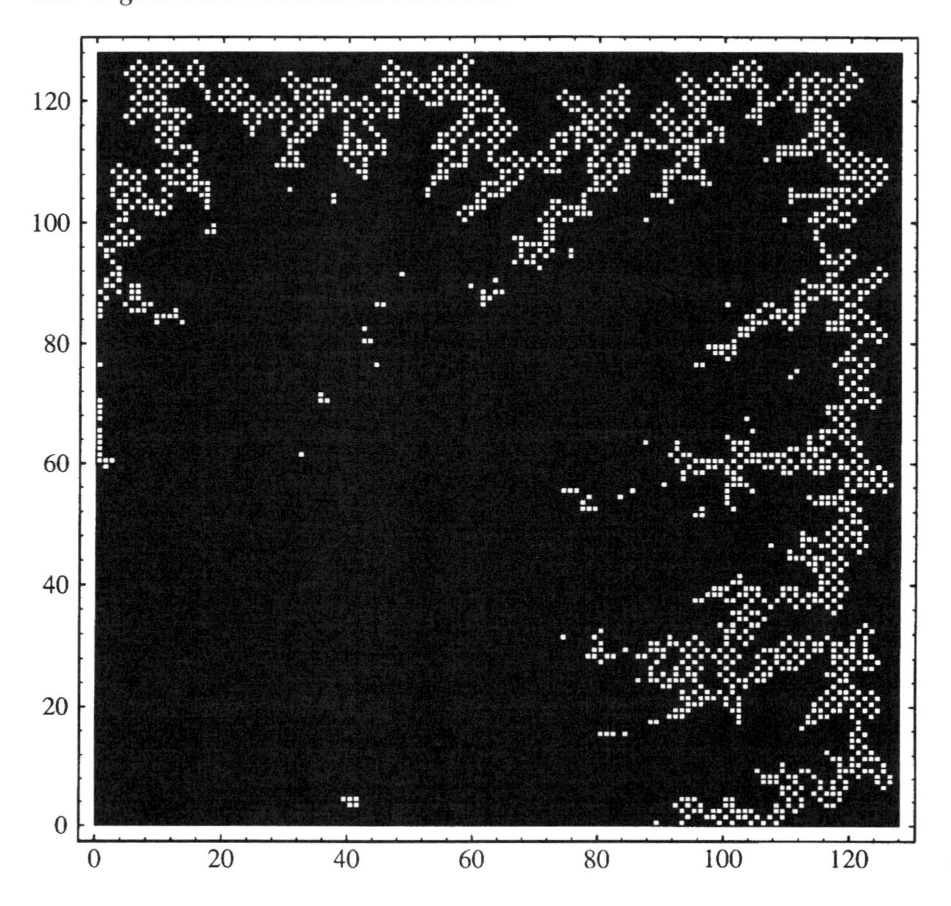

Figure 5.33. Border sets of equal elevations, $z(\mathbf{x} = 4)$ in pixel units.

probabilistically better understood fields like those generated by nonlinear superposition of random signals in the framework of generalized scaling invariance [Schertzer and Lovejoy, 1989], it gives in this case a measure of the resulting irregularity of the land surface produced by a well-defined landscape-forming process and allows a comparison (Chapter 6) with the results of the interplay of fluvial and hillslope processes.

It is worth recalling that the landscapes considered in this chapter are neither measured topographies (where it is hard to distinguish the boundaries among errors of measure, natural heterogeneities, and regions where the prevailing landscape-forming processes may differ) nor mathematically generated surfaces. They are produced by a well-defined self-organizing process, and thus it is of interest to relate to nature the resulting surfaces with some objective descriptors that have geomorphologic relevance.

The impact of other concurring processes (notably hillslope evolution processes) on the fractal appearance of the landscape will be discussed in Chapter 6.

5.11 Renormalization Groups for SOC Landscapes

In Chapter 4 we examined the transformations that coarse graining brings to an OCN. There we focused our interest in the behavior of energy dissipation under coarse graining of OCNs.

We now consider the three-dimensional SOC landscapes described in this chapter. They only include threshold-dependent processes, and the more realistic case where other types of processes are interacting is left for Chapter 6. As seen in detail before, the landscape in this case is characterized by scaling of slopes, namely, $\langle \nabla z(A) \rangle \propto A^{-0.5}$ and $Var[\nabla z(A)] \propto A^{-1}$. In the landscape of an OCN we assumed $\nabla z(A) = \langle \nabla z(A) \rangle \propto A^{-0.5}$ and $Var[\nabla z(A)] = 0$ because the slope was taken to scale proportionally to $A^{-0.5}$ everywhere in the network.

The renormalization proceeds as described in Section 4.18. Starting from arbitrary initial conditions a SOC landscape is obtained. The resulting landscape is described at a resolution of pixels of side length L. The pixels are grouped in squares of side λL, with $\lambda \geq 2$, so that from an initial number of pixels N one coarse grains the description of the terrain to a total number of pixels N/λ^2 (see Figure 4.54 for details of the notation). The elevation of each of the new larger pixels of side λL ($\lambda \geq 2$) is computed as the average elevation of the λ^2 constituent pixels of side length L that make up each particular larger pixel. From this coarse-grained three-dimensional landscape, a new drainage network is drawn following the lines of steepest descent. The transformation preserves the mean elevation of the basin as we have shown in Eq. (4.77).

An example of SOC renormalization is given in Figure 5.34. From the experiments we observe that the scaling of total energy expenditure, that is, $\sum_{i=1,N} A_i \Delta z_i$, accurately follows a power law (Figure 5.35) with the same slope (-1.1) as in renormalized energies for OCNs. Notice that here total energy is computed as $\sum_{i \in i_\lambda} A_{i_\lambda} \Delta z_i$ because the scaling properties will not grant the exact scaling $\Delta z_i = \langle \Delta z_i \rangle \propto A^{-0.5}$. As in the case of OCNs the energy-scaling relationship follows a power law (E_0 is the energy expenditure at $\lambda = 1$):

$$E_0 = \lambda^{-\alpha} \bar{E}^{(\lambda)} \tag{5.30}$$

with $\alpha = 1.1 \pm 0.01$. Here $\bar{E}^{(\lambda)}$ is the total energy expenditure at the λ level of coarse graining. Thus $E(L) \propto L^{2.2}$, where L is the linear size of the system.

The scaling of the slopes versus drainage area for the SOC landscape corresponding to Figure 5.34(a) ($\lambda = 1$) is shown in Figure 5.36. The mean values $\langle \nabla z(A) \rangle$ versus A clearly follow a -0.5 power law. The variance $Var[\nabla z(A)]$ versus A, Figure 5.36, shows a large dispersion also due to sample size effects (see Section 5.7), but it displays a tendency toward a -1.0 power relation (i.e., simple scaling) which becomes well defined when averaging over multiple realizations. Coarse graining destroys the scaling structure of the field, be it simple scaling or multiscaling. In fact, coarse graining produces a variance in the slope versus area relationship uncorrelated with the value of total area.

It would be interesting to study whether the destruction of scaling in the variance of slope versus area relationship through coarse-graining operations is also present in DEMs of natural basins.

5.12 Thermodynamics of Fractal Networks

In Section 4.21 we studied the scaling properties of energy and entropy of OCNs and showed analytically that large network development effectively

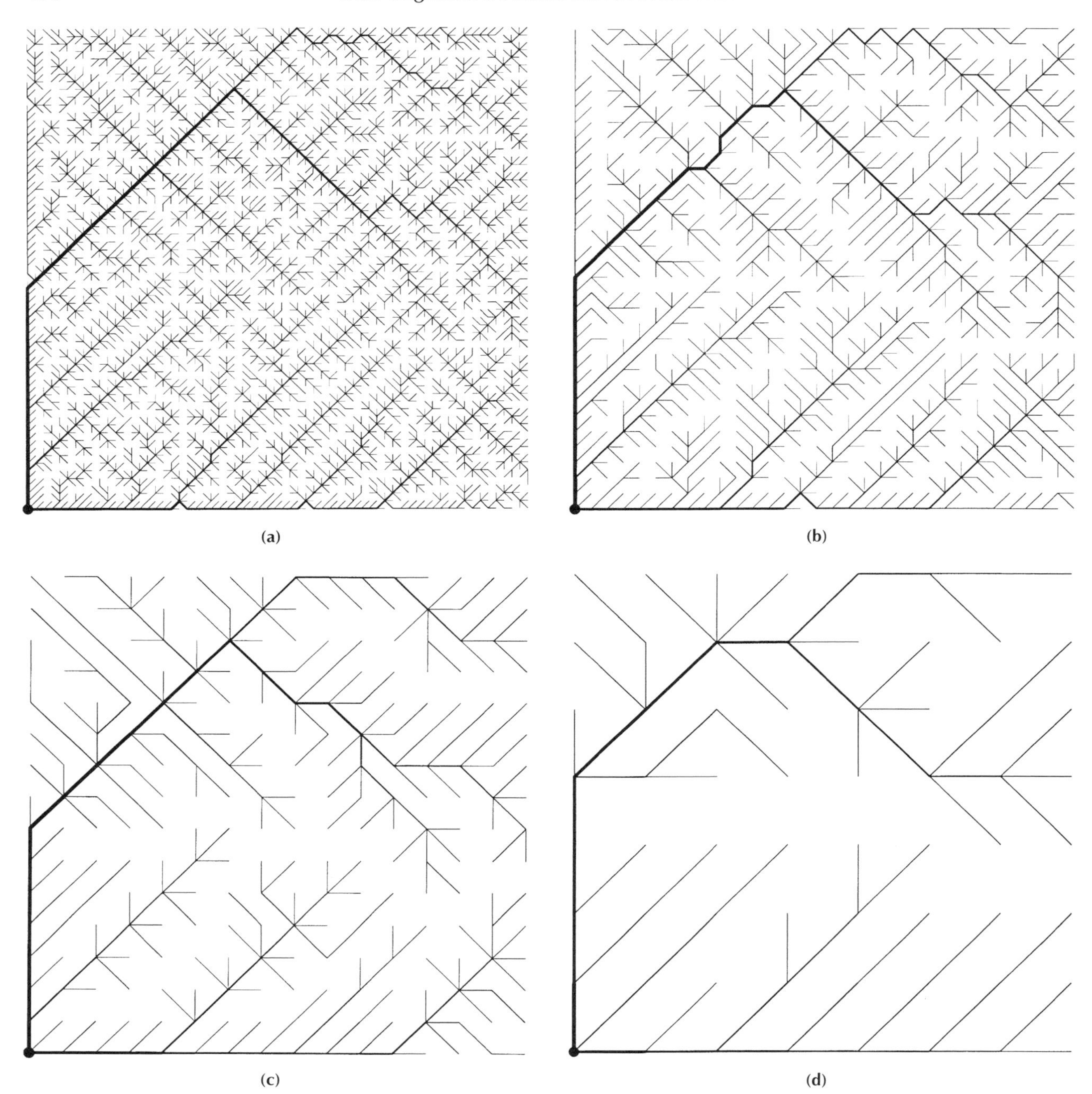

Figure 5.34. Renormalized SOC with (a) $\lambda = 1$, (b) $= 2$, (c) $= 4$, and (d) $= 8$.

occurs at zero temperature because the thermodynamic entropy scales subdominantly with system size compared to energy. As seen in Section 4.21, this does not mean that OCNs developed under generic conditions freeze into a static fractal structure. In fact we showed there that the OCN concept is compatible with a state of minimum energy expenditure and temporal dynamics described by a devil's staircase type of behavior. In fact they can share the best of two worlds: the low total energy corresponding to very low temperatures, and the high entropy – with its corresponding flexible

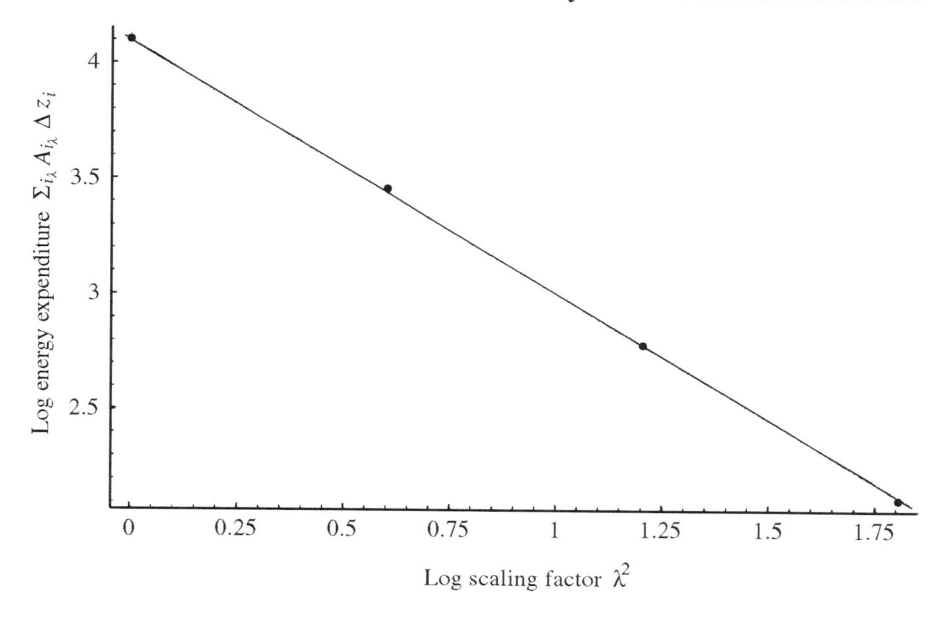

Figure 5.35. Scaling of the total energy $\sum_{i \in i_\lambda} A_{i_\lambda} \Delta z_i$ with the scaling factor λ^2.

adaptation – that characterizes high temperatures. Our closure of Chapter 4 left open the question of whether a similar rationale might also apply to ordinary self-organized critical structures.

Following the establishment of the analytical framework for the thermodynamics of OCNs, Rinaldo et al. [1996] suggested a speculative argument for a new interpretation of self-organized criticality.

As noted in Section 4.21, conventional ($T = 0$) OCNs have a highly constrained structure (loopless and spanning). They do not exhibit a set of dynamically recursive states like ordinary SOC phenomena but have a frozen structure and behave like a cold system. This follows, as shown in detail in Section 4.21, from the subdominant scaling of entropy with size compared to energy. The general physical reason for this fact is the aggregation structure of the river network and the spatial critical behavior of river networks thus originates from the correlated nature of their constrained structure.

We also recall that the construction of entropy-dominated networks (Section 4.21), by suitably defining a temperature high enough to overcome the finite-size effects, completely changes the picture, leading to hot structures with a well-defined temporal activity and a set of recursive states – though still being a fractal. Thus it seems legitimate to ask whether the generic growth of fractals both in space and time could be related to a robust achievement of scaling properties proper to the thermodynamic limit (i.e., $L \to \infty$). It is also interesting to search for SOC schemes of fluvial erosion in which the temperature of OCNs has a recognizable counterpart, possibly capable of achieving states of low energy expenditure and high entropy as attractors of the dynamics.

We suggest, following Rinaldo et al. [1996], that classic self-organized critical systems [Bak et al., 1987, 1988], for which the concept of recursive state is meaningful [Dhar, 1990], may maximize entropy in the thermodynamic limit at any temperature through a subdominant scaling of energy with system size compared to the entropy. It is also tempting to speculate

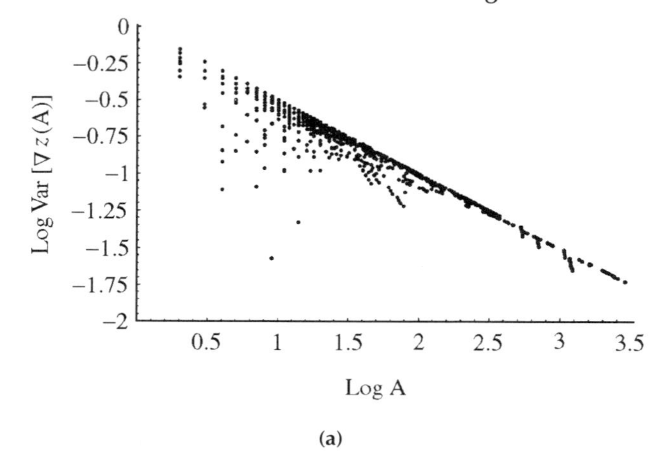

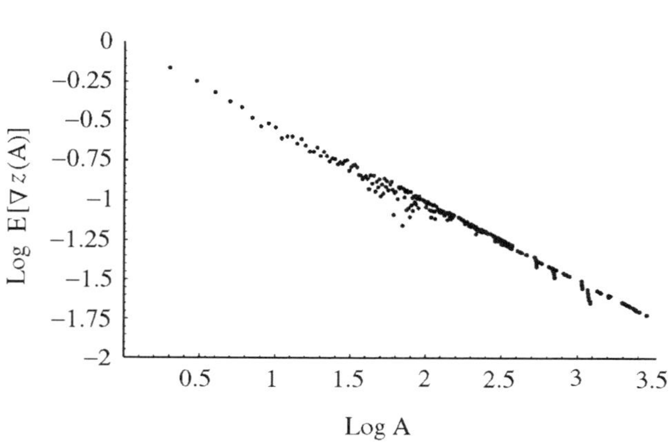

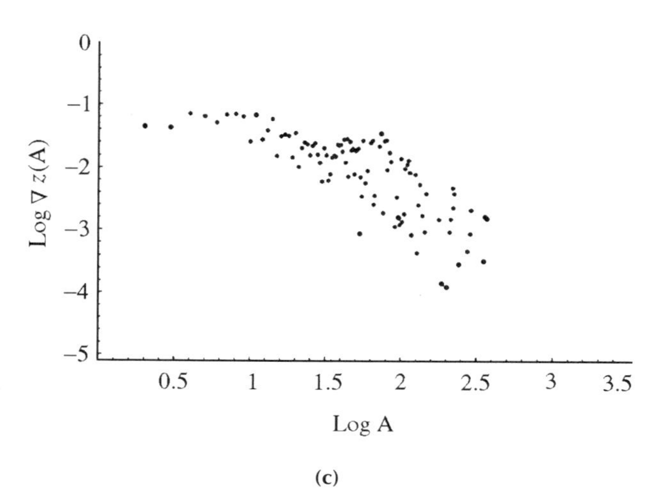

Figure 5.36. Scaling of slopes in SOC renormalization with $\lambda = 1$ (Figure 5.34 (a)): log-log plots of (a) $\nabla z(A)$ versus A; (b) $E[\nabla z(A)]$ versus A; and (c) $Var[\nabla z(A)]$ versus A.

that conventional critical phenomena, obtained by fine-tuning of a parameter (e.g., temperature) might be the outcome in which the scaling properties of energy and entropy with size L are equal. A few results, besides those outlined in Section 4.21, support this interpretation.

In classic self-organized systems, like Abelian sandpiles (Section 5.5) in two dimensions, the number of possible configurations (i.e., recursive states) N scales as

$$N \propto \mu^{L^2} \tag{5.31}$$

where μ is a suitable real number equal to 3.21.., see Section 5.5. Therefore thermodynamic entropy σ, defined as the logarithm of the number of possible states N (Section 4.21), scales like

$$\sigma \propto L^2 \tag{5.32}$$

Energy dissipation $H(s(t))$ at any time is measured by the number of active sites in the configuration $s(t)$ [Hwa and Kardar, 1992] and occurs through the boundaries. Thus its scaling is likely to be

$$H(s(t)) \propto L \tag{5.33}$$

and is certainly subdominant to the entropy in the thermodynamic limit as $L \to \infty$. Thus it is suggested, following the reasoning in Section 4.21, that in maximizing the probability of a state (i.e., minimizing free energy), entropy will be cloud total energy dissipation as L grows large. Thus the controlling principle would be the maximization of the entropy of the system subject to the constraints of its particular dynamics.

Entropy-controlled critical structures have been found [Miramontes, Solé, and Goodwin, 1994; Solé, Miramontes, and Goodwin, 1993; Solé and Miramontes, 1995] in neural networks of mobile elements with random activation – reminiscent of self-organized ant societies. Such networks operate at the edge of chaos with critical fluctuations that maximize entropy. In this case, entropy maximization and the critical behavior are obtained by tuning the density of active states. It turns out that in the case of ant societies, the density of ants thus obtained agrees with the densities observed in nature.

Finally, in this framework conventional critical phenomena would be obtained by the fine-tuning of a parameter (e.g. temperature) precisely because the scaling properties of the energy and entropy are similar, that is, they scale with the same power of L.

The SOC counterpart of hot OCNs has not been firmly established yet. However, it is being currently proposed [Rodriguez-Iturbe et al., 1997] that in SOC systems the features of the random forcing driving the system have the capability of surrogating temperature in thermodynamic systems. Specifically, one may distinguish the flow at the ith site, Q_i, from total contributing area, A_i as implied by the assumption of steady, spatially constant injection. The flow rate at the ith site thus becomes $Q_i = \sum_j Q_j + R_i$, (j spans neighbors draining into i), where R_i is an arbitrary random forcing. The simplest scheme of noise (time-varying and uncorrelated in space) is $R_i = 1 + U(0, \sigma)$, where U is a random value drawn from a uniform distribution of range σ. In the SOC model described in Section 5.7 the controlling shear stresses are then computed as $\tau_i = \sqrt{Q_i}\Delta z_i$ and the procedure described therein applies without further modification. At increasing strengths of the noise (large σ), the system should attain the features of hot systems, while for $\sigma \to 0$ the behavior decays into the ordinary cold one. Thus a relationship between temperature and the strength of the driving noise would allow one to observe under what conditions one expects a punctuated equilibrium-type of dynamics for a fluvial landscape.

Rinaldo et al. [1996] concluded that it could be possible that SOC systems are not obtainable in general through free-energy minimization principles. Nevertheless, an intriguing alternative is that the natural evolution of fractal structures in open, dissipative systems with many degrees of freedom is the by-product of chance and necessity, the latter being embedded in the strive for optimality that we see everywhere in natural forms.

This, of course, is speculative at this moment, but could lie at the core of what Kauffman [1995] calls ".. the search for a theory of emergence." Kauffman [1993] developed the concept of the fitness landscape as the terrain on which biological evolution driven by random forcings searches for feasible optimality. One could speculate that the fitness landscape of hot systems, like those so frequently found, for example, in economics, is basically the entropy that characterizes the different configurations of the system. Evolution, in this case, will proceed in the search for the peaks of such landscapes. On the other hand, in the case of cold systems, the fitness landscape may be one of energy expenditures under the different configurations. The system then evolves, driven by random forcings toward the valleys of such landscapes. Geophysical systems may develop at the edge of the two states, retaining the properties of both hot and cold systems. In fact, a realistic interpretation of how nature works looks at time-scales as measured in relation to the frequency of the random events that drive the system. In this framework a temporal activity of fluvial SOC systems is compatible and significant.

Profound order is being discovered in large and coordinated dissipative systems, as we have repeatedly tried to show throughout this book for the case of river basins. In the search for an explanation, we have seen that one does not need to know all the details of the processes that take place to explain the generic properties of this order. Simple models that capture the most relevant features of this complex cooperative scenario are probably our best hope for a unifying synthesis of the recursive patterns that we observe. Once again it comes to mind that models are fun, and sometimes even instructive [Kadanoff, 1995].

5.13 Self-Organized Networks and Feasible Optimality

In this concluding section we will address the issue of feasible optimality in the context of self-organization, that is, the structure of dynamically accessible optimal states.

One important issue concerns the imprinting of external conditions on the characters of self-organized structures. A detailed account of the inference of finite-size effects and of boundary and initial conditions on fluvial self-organization has been provided by Caldarelli et al. [1996]. A bidimensional square lattice for sizes up to $L = 200$ was cast with the usual no-flux boundary conditions in the direction transverse to the main flow and possibly open boundary conditions (i.e., multiple outlets) in the parallel one. Notice that a mean flow direction is defined by tilting the initial surface uniformly, that is, the initial surface $z_0(\mathbf{x})$ is a sloping plane where the gradient ∇z is constant, defining a preferential flowpath. To such a tilted structure, random disorder in the form of irregularities in the surface elevation $(z_0(\mathbf{x}) + \eta(\mathbf{x}))$ may or may not be applied depending on the specific experiment, as discussed below. The specific initial and boundary conditions studied are

Model A: An initial comb-like structure with a single outlet is chosen. The single outlet is maintained in one corner throughout the simulation as in the cases in Section 5.7;

Model B: The initial condition is a smooth inclined plane where all sites at the bottom of the plane are maintained as possible outlets. This condition was selected with the aim of investigating the inference of an arrangement with multiple outlets as opposed to the constrained arrangement characterized by a single outlet. The assumption relaxes the constraints on the developing structure and allows the competition for drainage among rivers in the same basins, which is an effect actually active in nature;

Models C and D: Models A and B were rerun with the addition of a random uncorrelated noise $\eta(\mathbf{x})$ to the field $z_0(\mathbf{x})$ (characterized by a significant amplitude, in general less than 10 percent of the average height) to test the effects of disorder on the developing structure.

A Monte Carlo ensemble average over a few (up to ten) configurations was taken for the resulting statistics to avoid noise corresponding to singularities of single realizations. This number proved sufficient to portray ensemble averages, due to the self-averaging nature of the random perturbation.

Many landscapes sculpted by the above dynamic process have been produced for the four models (A, B, C, and D). An example of four typical networks that developed under the different assumptions is shown in Figure 5.37. Two features are immediately captured from the above figure. First, models A and B exhibit a strong memory of the initial configuration, despite the fact that the dynamics of the erosion process was expected to be sufficiently elaborate so as to lose memory of its initial state. Second, the single outlet restriction imposed in models A and C appears to be a severe constraint because it seriously affects the wandering of the river and the overall aggregation, especially in the lower parts of the basin where aggregation is forced. However, this holds only in the case of smooth initial conditions (models A and B). Instead, the outlet arrangement appears

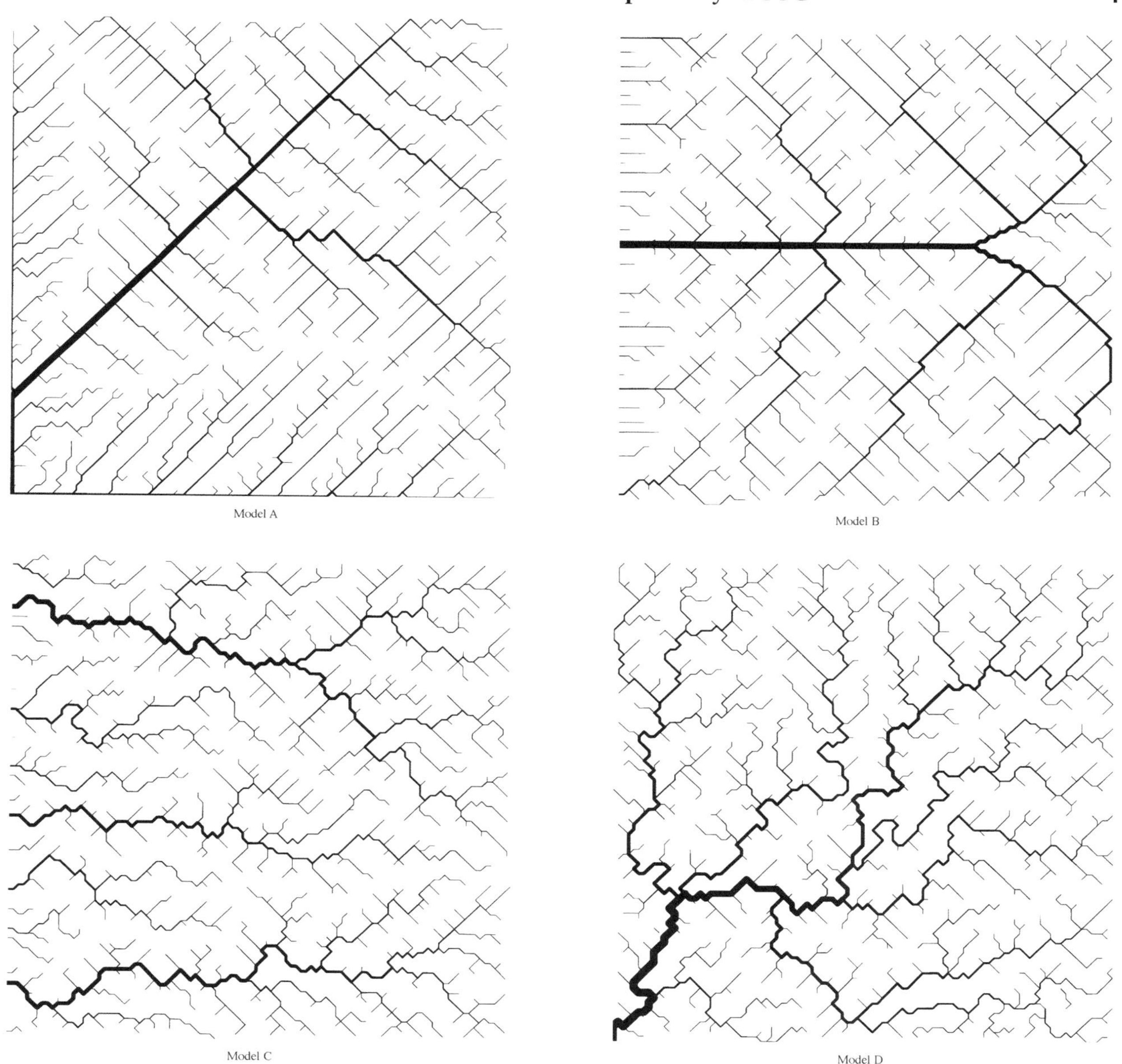

Figure 5.37. In clockwise order from the upper left corner, the network structure corresponding to the landscapes for model A, model C, model D, and model B. Models A and D have a single outlet, models B and C have multiple outlets. The initial condition is deterministic for A and B, whereas C (D) has a small random noise superimposed on A (B). Only sites with area ≥ 5 are shown [after Caldarelli et al., 1996].

to be quite irrelevant as soon as a significant initial perturbation is added to the corresponding initial condition. This is clearly seen through scaling exponents, as described below.

We choose, as a significant morphological indicator, the scaling exponent of the distribution of total contributing areas. Let us make use, once more, of $P(a, L) = P[A \geq a, L]$, the cumulative probability distribution of the drainage area, A, in a given basin of finite size L. We have seen in Section 2.10 that quite generally the scaling form holding in the river basin is $P(a, L) = a^{-\beta}F(a/L^{\phi})$ [Maritan et al., 1996a], where β and ϕ are related scaling coefficients. We recall that careful examination of data shows (Section 2.10.2) minor and related variations in the observed values of the scaling exponents, suggesting an adaptation of natural fractal structures

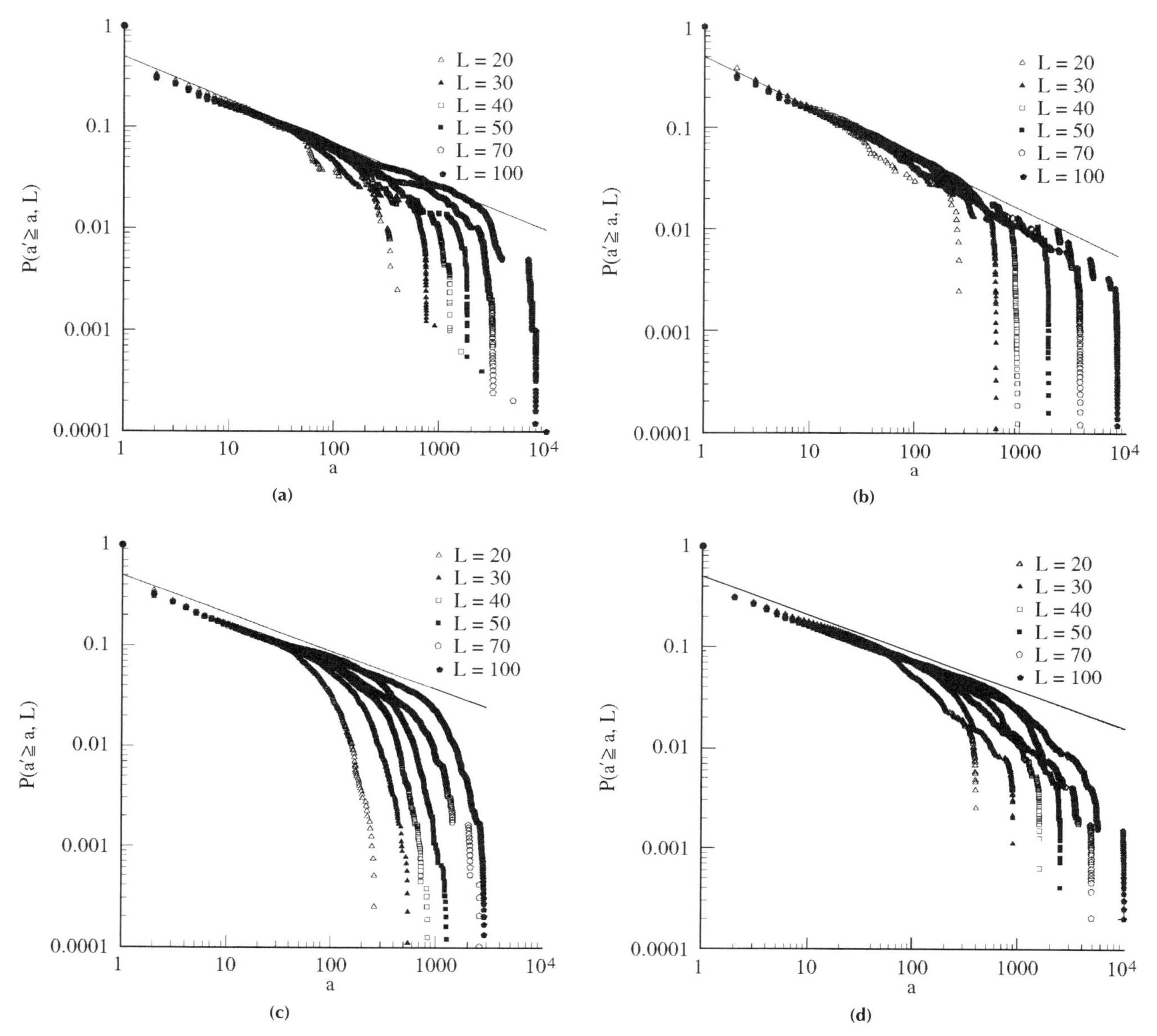

Figure 5.38. Log-log plot of the area cumulated distribution $P(a, L)$ versus a for models A, B, C and D, in the same clockwise order as in Figure 5.37. The full line has a slope corresponding to $\beta = 0.43$, $\beta = 0.50$, $\beta = 0.38$, and $\beta = 0.38$, respectively.

to the geologic and climatic environment. This leads to the conclusion that no 'universal' exponents are expected in the fluvial system. Rather, the roles of geology and tectonics act in concert, effecting a coordinated scaling structure that reflects the strive for fractality adapted to its geologic environment. The results of the model described in this section yield to this view.

The results of the four models for the area distributions are shown in Figure 5.38. It is apparent that, owing to the somewhat pathological initial conditions, the scaling behavior for models A and B is noisier than for models C and D. Figure 5.39 contains the collapse plot (i.e., a plot of $a^{\beta} P(a, L)$ vs a/L^{ϕ}, see Section 2.10.1) for case C. Case D (not shown) is almost undistinguishable from case C. A summary of the relevant results is

Table 5.2. *Values of the scaling exponents of total contributing area, β, through Monte Carlo average of several runs of Models A, B, C, and D*

	Model A	Model B	Model C	Model D
β	0.43 ± 0.03	0.50 ± 0.03	0.38 ± 0.02	0.38 ± 0.02

Source: Caldarelli et al. [1996].

included in Table 5.2. It should be added that all scaling exponents have been studied and a consistent picture of related variations has been observed [Caldarelli et al., 1996] in analogy to the results obtained through careful examination of data (Section 2.10).

Similar results can be obtained for many other significant geomorphic indicators, chiefly upstream length distributions and Hack's relationship. In Figure 5.40 we show, as an example, the results for the cumulative length distribution.

It is somewhat surprising that the numerical value of the critical exponents for models C and D appears to be close to that of disordered spanning trees (Section 4.20). Rather large values of β have indeed been obtained, that is, $\beta \sim 0.5$, implying that the dynamics may be trapped in states with much reduced aggregation. In any case, we note that the SOC model is not robust with respect to boundary and initial conditions, contrary, for example, to sandpile models. This can possibly be ascribed to the impact of long-range interactions embedded in the definition of a critical shear stress. It is also noteworthy that in any case scaling structures are produced that are adapted to the external conditions, here epitomized by initial and boundary conditions.

This result is reminiscent of the concept of feasible optimality (Section 4.10), which portrays fractal structures as dynamically accessible optimal configurations. In this case, locally optimal configurations are obtained consistently, as we deduce from the monitored decrease of total energy dissipation in the SOC systems (in all models A, B, C, and D) to different plateaus in each different run. Once the dynamics is attracted toward a relatively stable configuration, only minor adjustments can occur because of the constraints of the dynamics, as discussed in Section 5.8.

A further important result is that the robustness of the SOC procedure, and the recurrence of the common value $\beta = 0.43$, are recovered when quenched randomness is forcing aggregation. We will examine this issue following Caldarelli et al. [1996], who have shown that, regardless of boundary and initial conditions, geomorphologic constraints in the form of quenched randomness (i.e., sites where erosion is not allowed by pinning them at their original elevation values) play a key role for the robust emergence of aggregation patterns that bear a striking resemblance to real river networks. The added ingredient, random pinning,

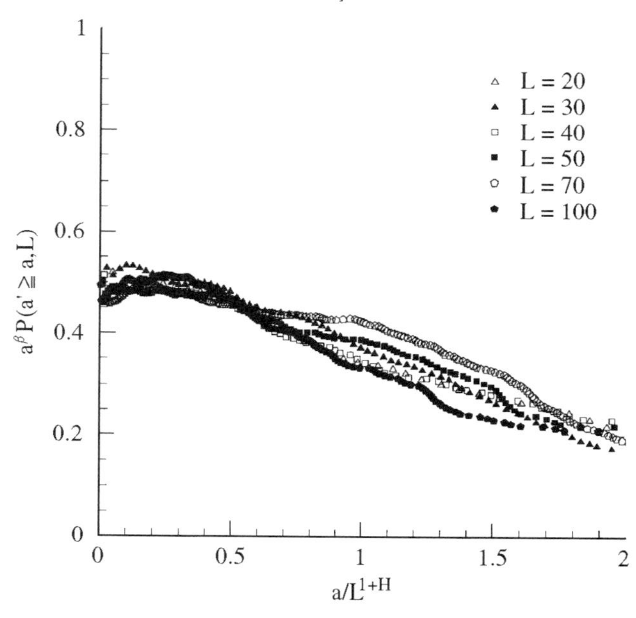

Figure 5.39. Scaling function $a^{\beta}P(a, L)$ versus a/L^{ϕ} for model C. The values used to obtain the collapse were $\beta = 0.38$ and $\phi = 1.6$, corresponding to a self-affine (Section 2.10.1) basin with Hurst coefficient $H = 0.6$. Notice that the sensitivity of the collapse plot much enhances our ability to estimate scaling coefficients [after Caldarelli et al., 1996].

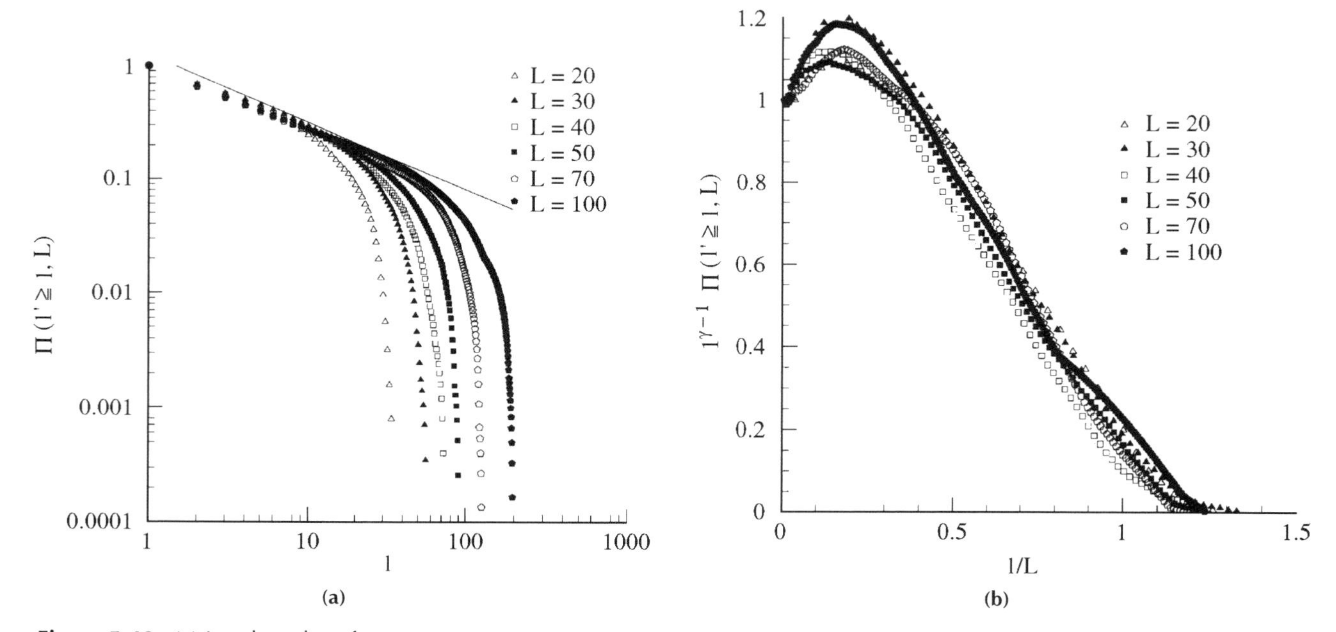

Figure 5.40. (a) Log-log plot of the distribution of upstream lengths $\Pi(l, L)$ ($\propto l^{-\gamma}$) versus l for model C. The solid line has a slope corresponding to $\gamma = 1.6$ [after Caldarelli et al., 1996]; (b) scaling function $l^{1-\gamma}\Pi(l, L)$ versus l/L for model C. The value of γ used to obtain the collapse is 1.6.

proves to be effective and of physical significance because heterogeneities of the terrain and geologic constraints are ubiquitous and expected to play a definite role in the development of the fluvial basin.

The procedure is readily explained. A small percent (typically 5 percent) of the heights is chosen (randomly and in uncorrelated manner) to be pinned at their initial value. Thus these sites do not take part in the landscape evolution process. Logically, one expects this effect to increase the meandering of the main river structures and thus to provide an additional mechanism for aggregation.

After a suitable number of perturbations, the system reaches a steady state that is insensitive to further perturbations and where all statistics of the networks are stable. Notice that also in this case this resulting steady state proves scale-free, that is, characterized by power law distributions of the physical quantities of interest. It is reasonable that the above holds for relatively minor dilution of quenched randomness. In Figure 5.41 the typical network obtained using the above dynamics is shown without (left) and with (right) random pinning, starting with the same initial configuration.

The results for the cumulative area distributions are shown in Figure 5.42. A value of the scaling exponent equal to $\beta = 0.43 \pm 0.01$ was found irrespective of the choice of model. Figure 5.42 also contains the corresponding collapse plot from finite-size scaling analysis. It is worth mentioning that all numerical results are fully consistent both with the experimental values observed in nature and with the intertwined relations provided by scaling analysis (Section 2.10).

It is noteworthy that in the absence of pinning the scaling exponents prove much more sensitive to the boundary and initial conditions. It is tempting to speculate about the role of pinning in annihilating the effects of disorder in the initial conditions that would otherwise direct the evolution toward a less pronounced aggregation (and thus to lower values of $\beta \sim 0.38$) irrespective of the type of outlet arrangement allowed. Once

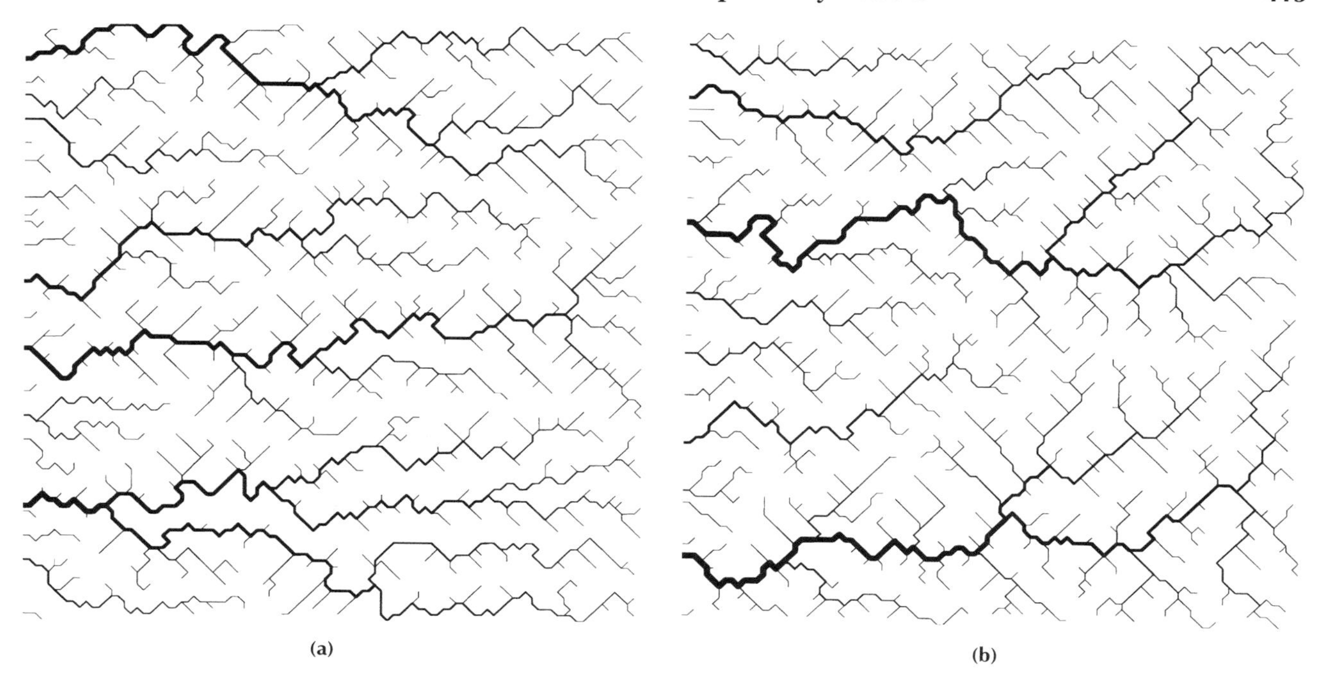

Figure 5.41. Comparison of the evolution of two identical initial configurations (model D) with size $L = 100$ (left) without and (right) with pinning. The pinning dilution was 5 percent, meaning that a 5 percent proportion of the lattice sites, chosen at random, is pinned to the initial value of the elevation field. Thus erosion circumvents the pinned sites and different aggregation patterns are obtained [after Caldarelli et al., 1996].

again, this sustains our view of the accessibility of optimal states by the dynamics, a feature critical to self-organized fluvial landscapes.

Finally, the robustness of the results with respect to the pinning dilution has been demonstrated. Even if a significant part of the landscape is frozen (i.e., pinned) to the original elevation value, SOC fluvial evolution produces a scaling behavior much resembling that of natural landforms. These results prove rather robust to boundary and initial conditions of the system, and thus we suggest that quenched randomness is an important ingredient of the self-organization, at least in its simplest mechanical formulation described herein.

A few concluding remarks on the mechanisms of fluvial self-organization explored in this section are in order at this point:

- Principles of self-organization are likely to be at work in the development of the fluvial basin. Although strictly fluvial processes cannot be responsible for all landforms observed in the river basin, the imprinting of the fluvial process, addressed by this crucially simple SOC scheme, seems plausibly close to natural planar landforms.
- The model of self-organized criticality for river basins produces network structures always endowed with fractal characters, in specific cases closely resembling real basins. This similarity includes, in particular, the demanding features required by the statistics of the elevation fields.
- The sensitivity of key scaling exponents to boundary and initial conditions in ordinary (i.e., not quenched) conditions is reminiscent of the features of feasible optimality (Section 4.7) for a fluvial network adapting to a geologic and climatic environment. The possibility for SOC dynamics to get trapped into unrealistic aggregation patterns highlights the importance of local minima of the optimal configuration of the system.

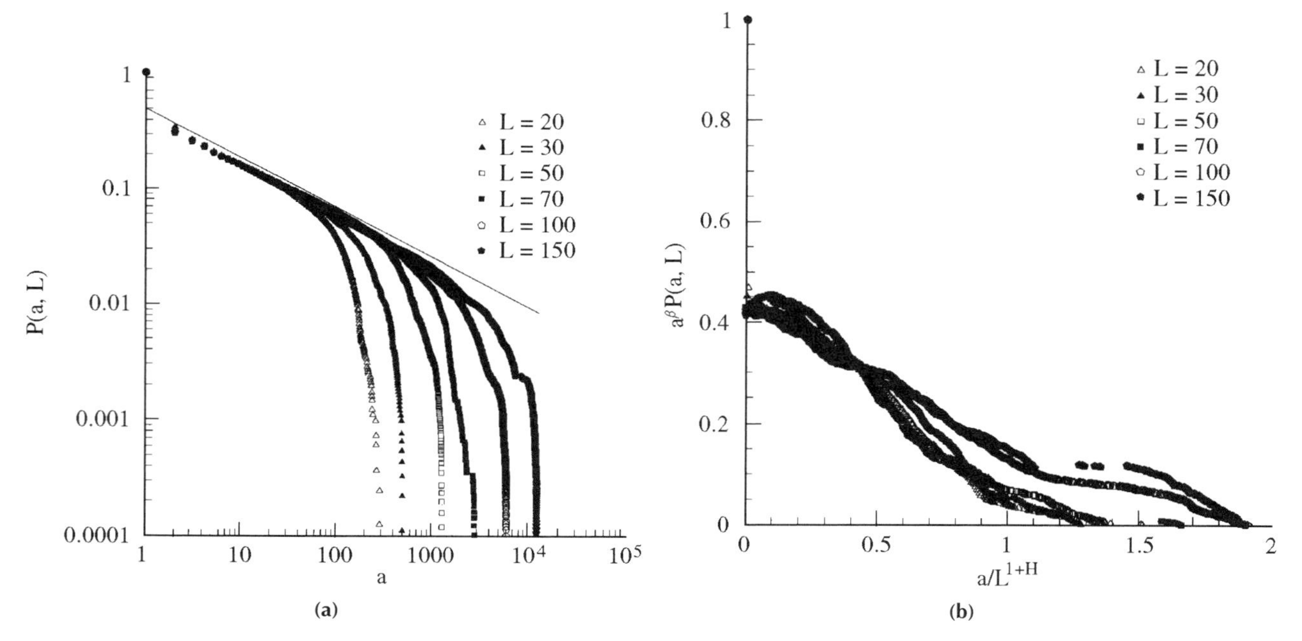

Figure 5.42. Log-log plot of the area distribution $P(a, L)$ versus a (left) and (right) corresponding collapse plot for the networks in Figures 5.41(b).

- The robustness of aggregated patterns granted by quenched randomness (that is, random pinning) and the fact that, once pinned, the fluvial landscapes obtained reproduce well the scaling exponents of real rivers suggest a mechanism of clear physical significance, which is possibly responsible for the recurrence of common patterns in nature.
- The self-organized process always yields a progressive lowering of the total energy dissipation E of the system, and different values of E (stable to perturbations) are obtained depending on the constraints imposed on the dynamic process. This reinforces the idea that natural fractals are dynamically accessible optimal states.

These issues will also be studied in the context of more realistic landscape evolution processes in Chapter 6.

CHAPTER 6

On Landscape Self-Organization

This chapter examines the morphology of landscape systems in the search for process-based explanations of fractal landforms in which the dynamic interplay of fluvial and hillslope processes is embedded. The coupled processes considered are slope-dependent hillslope mass wasting and threshold-limited fluvial erosion. The chapter also addresses the presence of geomorphologic signatures of varying climate and attempts to show that geomorphologic thresholds, coupling of different erosional processes, and concepts of self-organization play a key role in the basic general mechanisms that govern landscape evolution.

6.1 Introduction

A landscape consists of a great number of individual geomorphologic forms, such as slopes, river segments, valleys, peaks.... Scheidegger [1991] argued that it is impossible to build a classification of landscapes on the basis of the morphology of their elements, as the latter can combine to form systems characterized by extraordinary complexity. Attempts at such classifications must therefore be based on features of a larger scale and of statistical character.

Indeed, we have already attempted to describe the extraordinary diversity and yet the deep statistical symmetry that characterize river networks (Chapter 1). We now attempt a generalization of the concepts underlying the evolution of the landforms described in Chapter 5 (which we will term the *fluvial* landscape) aimed at including coupled erosional processes of different nature. The generalization presented herein includes processes that describe mass wasting and slope evolution of various origins that are substantially different from fluvial processes and mostly operate at smaller scales.

The need for a distinction between different landscape-forming processes has long been recognized. A few remarks of general nature seem appropriate here.

Gilbert [1877] first enunciated some simple qualitative principles of landscape formation: the law of uniform slope, through which rapid erosion of steeper slopes tends to reduce the landscape to low relief; the divide law, which states that streams steepen toward headwaters; and the law of geologic structure, which observes that hard rocks erode less and in different timescales and hence stand out from the surrounding landscape.

An old attempt at a description of landscape systems was based on a hypothetical descriptive natural history of the systems' genesis. Davis [1924] proposed that each landscape represents a certain stage of an evolutionary cycle. Because degradations by water, ice, and wind (the *exogenic* agents) are powerful destructive processes in a landscape, Davis argued that a buildup must have occurred prior to the onset of the destructive actions (the *endogenic* processes). Thus the famous cycle theory was proposed: a geomorphologic cycle has its beginning soon after an endogenic geodynamic process has finished creating an uplifted area such as a mountain range. Weathering, erosion, and detrition thus begin to act on the uplifted area and gradually reduce it to a baselevel, therefore completing the cycle. A new cycle is started when a new endogenic diastrophism occurs.

Davis [1924] recognizes three distinct stages in the geomorphologic cycle termed youth, maturity, and old age. In a humid climate, these are defined as follows:

- In youth, rudimentary trunk streams are present but not many large tributaries. Valleys are strongly V-shaped, their depth being basically a function of their mean elevation above sea level. Variations in surface lithology cause waterfalls and chutes that are bound to disappear as time elapses.
- In maturity, the drainage system becomes more integrated. Waterfalls and chutes eventually disappear and most rivers are in dynamic equilibrium between sediment production and transport. In maturity, the extent of the landscape relief is at maximum.
- In old age, valleys become very broad and most of the landscape relief has disappeared because of continental planation. The level of the drainage basin approaches the baselevel of erosion. The final stage of the cycle is reached when all relief has been reduced to the baselevel, leading to a gently undulating plain (the *peneplain*).

Although historically quite important, cycle theory has been extensively questioned, especially in relation to the lack of a clear definition of a geomorphic cycle, as youthful forms (in Davisian sense) may locally exist even in very late stages and vice versa.

Also, climate plays a major role in the development of a landscape. There is a wide body of literature [see, e.g., Bloom, 1991; Summerfield, 1991] on the dependence of slope, mountain development, and geomorphology on climate. Normally, we deal with humid climates where water and wind are the leading exogenic agents. In cold climates, ice and snow become the key geomorphic agents that produce the so-called glacial or periglacial forms.

One should also note that in many cases there is no simple distinction of climates. For instance, the Pleistocene experienced several periods during which large continental glaciations existed, and thus remnants of glacial forms are frequently found in areas that are subject to a normal humid climate at the present time.

Arid climates produce still different landforms. These are characterized by rare but very intense rainfalls which, concurrent with wind actions, become the key geomorphic factors. Later in this chapter we will explore the consequences of a hypothetical climate change within the framework of our theoretical landform development schemes.

It is now clear that landscape development, and thus its possible fractal characters, is not a temporally sequential affair with degradation following buildup, but rather a by-product of degradation and uplift occurring concurrently. A given landscape thus represents a dynamic equilibrium between the antagonistic action of endogenic and exogenic processes [Scheidegger, 1991]. Youth, maturity, or old age simply represents given ratios of the uplift and degradation velocities defining the age-stage of a landscape.

Spatially nonuniform rates of degradation or uplift, especially in relation to plate tectonic phenomena, may be important in the analysis of large-scale morphologies. However, these issues are not deemed central to this book, focused as it is on the river basin scale. Even though tectonic controls may also impose important constraints (i.e., characteristic scales) on the river basin, we will not cover this subject. As mentioned in the preface, this is not meant to be a comprehensive book on geomorphology. The reader is referred to the literature [e.g., Scheidegger, 1991; Bloom, 1991; Summerfield, 1991] for comprehensive reviews of general geomorphologic character.

This chapter describes the impact of various landscape-forming processes on the resulting fractal characters of the topographic surface. The role of self-organization in such a context is studied and a comparison with experimental results is described.

6.2 Slope Evolution Processes and Hillslope Models

A number of theories of landscape evolution have been summarized by Scheidegger [1991]. We will not review here the many mechanisms of rock reduction by exogenic agents (chemical, physical, or biological) nor the effects of spontaneous mass movements on slopes. We are interested in mathematical models of denudation to infer the basic characters of actions that are concurrent with fluvial erosion.

We will first review some models that have been commonly used to study hilllsope evolution. As discussed in Section 1.4.3, the general form of the governing equation for elevation changes in an open dissipative system is basically

$$\frac{\partial z(\mathbf{x}, t)}{\partial t} = -\nabla \cdot \mathbf{F} + U + \eta(\mathbf{x}, t) \tag{6.1}$$

where, as usual, z is the landscape elevation in the arbitrary site of geographic coordinates $(x_1, x_2) = \mathbf{x}$ belonging to the catchment; $\mathbf{F}$ defines the flux of sediments; U is the arbitrary uplift rate; η is some external random forcing; and t is time.

In the case of nonactive uplift ($U = 0$), the unforced system is described by

$$\frac{\partial z(\mathbf{x}, t)}{\partial t} = -\nabla \cdot \mathbf{F} \tag{6.2}$$

where $\mathbf{F}$ is the sediment flux vector, reasonably oriented in the direction of the gradient of the topographic field z (the steepest topographic descent), that is

$$\mathbf{F} = -F\frac{\nabla z}{|\nabla z|} \tag{6.3}$$

where the fraction is the unit downslope vector.

The early quantitative studies on slope evolution were one-dimensional profile evolutions ($\mathbf{x} \rightarrow x$). The simplest theory reduces Eq. (6.2) to a one-dimensional profile evolution where the net flux of mass waste is

Table 6.1. *Typical values of m, n in the transport law*

Process	m	n
Soil creep	0	1
Rainsplash	0	1–2
Soil wash	1.3–1.7	1.3–2
Rivers	2–3	3

Source: after Kirkby [1971].

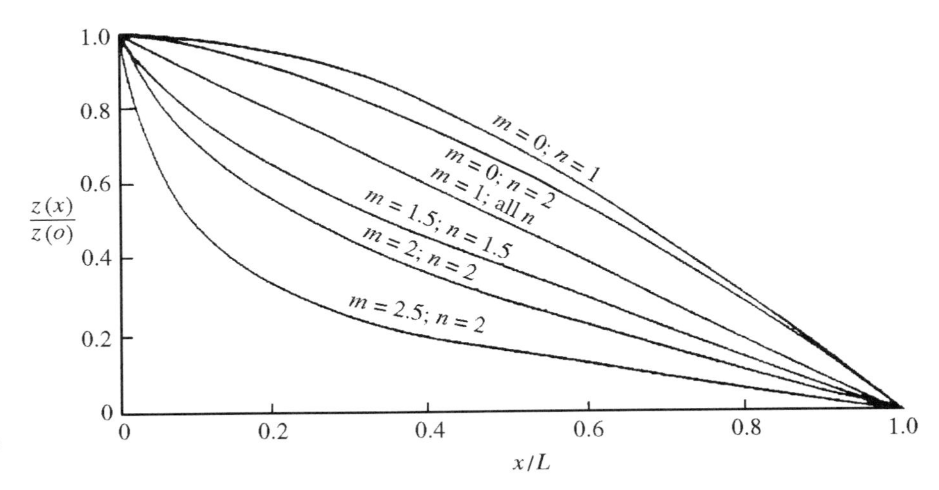

Figure 6.1. Slope profiles $z(x) = z(x, t \to \infty)$ for different values of the transport coefficients [after Kirkby, 1971].

proportional to the elevation z. One therefore obtains

$$\frac{\partial z(x,t)}{\partial t} = -Kz \tag{6.4}$$

where K is dimensionally the inverse of a characteristic timescale for erosion. With the initial condition $z(x, 0) = z_0(x)$ at $t = 0$, the solution is trivially

$$z(x,t) = z_0(x)e^{-Kt} \tag{6.5}$$

which gives an exponential decay of topography with time but no information about the spatially evolving landscape. This is a scheme of little interest for our purposes because of the embedded scale built a priori into the process.

Kirkby [1971] studied the one-dimensional formulation of the sediment transport law as

$$\nabla \cdot \mathbf{F}(Q, |\nabla z|) \propto Q^m |\nabla z|^n \tag{6.6}$$

where the exponents m, n are a function of various processes (See Table 6.1). The solutions to the above one-dimensional problem in different scenarios of transport yield the profiles shown in Figure 6.1. Nevertheless, the limitations of the one-dimensional scheme are evident with respect to a fundamentally multidimensional process where spatial interactions play a dominant role.

The most commonly adopted model for hillslope evolution is a diffusion analogy for the deterministic relaxation first proposed by Culling [1960, 1963, 1965]. The diffusion analogy assumes sediment fluxes ($\mathbf{F}$) to be proportional to local slope ∇z in the formulation of the mass balance. The basic diffusion assumption therefore yields

$$\mathbf{F} = \mu \nabla z \tag{6.7}$$

where μ plays the role of a diffusion coefficient. Thus the basic equation of landscape evolution becomes

$$\frac{\partial z(\mathbf{x}, t)}{\partial t} = \nabla \cdot \mu \nabla z \approx \mu \nabla^2 z \tag{6.8}$$

The last equality results from the assumption of a spatially constant value of the diffusion coefficient. Eq. (6.8) has been used in a number of studies of landscape evolution. In particular, the form of many alluvial fans and prograding deltas can be approximated by solutions of Eq. (6.8) that have also been employed to estimate the age of faults and shoreline scarps. Measures of the size of colluvium regions have also helped in the field validation of diffusive models of hillslope evolutions [e.g., Dietrich et al., 1993].

The observation that deterministic landscape evolution models of linear characters cannot possibly yield scale invariance is crucially important. This observation is well-known in other contexts [e.g., Whittle, 1962] and surfaced in the geomorphologic literature owing to Turcotte [1992]. However, as we will show in this section, one does not necessarily need a nonlinear theory to produce (or maintain) scaling topographies.

One example, which we will review in detail, refers to solutions of Eq. (6.8) where an initial landscape $z(\mathbf{x}, 0)$ is allowed to evolve. The main result of Newman and Turcotte [1990] is that the exponential decay of high spatial frequencies prevents the growth and the stabilization of fractal structures.

Assume that topography has (at any time) a Fourier representation, and let $\hat{z}(\mathbf{k}, t)$ be the discrete Fourier transform of $z(\mathbf{x}, t)$:

$$z(\mathbf{x}, t) = \sum_{\mathbf{k}=-N/2}^{N/2} \hat{z}(\mathbf{k}, t) e^{i2\pi \mathbf{k}\cdot\mathbf{x}} \tag{6.9}$$

where $\mathbf{k} = (k_1, k_2)$ is a wave vector and N is a suitable cutoff (notice that we will deliberately skip any detailed description of problems related to aliasing in the discrete Fourier representation in order to avoid clouding the central idea with details). Substitution of Eq. (6.9) into Eq. (6.8) yields (recall that $\partial z/\partial x_1 = \sum_{\mathbf{k}} (2i\pi k_1)\hat{z} \exp(2i\pi \mathbf{k}\cdot\mathbf{x})$ because we assume – somewhat reasonably because of the continuity of z – that differentiation by series is allowed by our transform representation)

$$\sum_{\mathbf{k}} e^{2\pi i \mathbf{k}\cdot\mathbf{x}} \left(\frac{\partial \hat{z}(\mathbf{k}, t)}{\partial t} + 4\mu\pi^2 k^2 \hat{z}(\mathbf{k}, t) \right) = 0 \tag{6.10}$$

where $k = |\mathbf{k}| = \sqrt{k_1^2 + k_2^2}$ is the modulus of the wave vector. The completeness of the orthogonal basis $e^{2\pi i \mathbf{k}\cdot\mathbf{x}}$ [e.g., Mei, 1995] ensures that the quantity in parentheses is identically zero, yielding the solution

$$\hat{z}(\mathbf{k}, t) = \hat{z}(\mathbf{k}, 0)e^{-4\pi^2\mu k^2 t} \tag{6.11}$$

The amplitudes decay exponentially and the rate of decay is proportional to the square of the wave number. Short-wavelength topography, characterized by large wave numbers, is flattened much faster than long-wavelength topography. The loss of fractal characters can be best described in this context by the departures of the spectra from power laws (see Chapter 2). Significantly, the power spectral density of topography $S(k, t)$ (let $k = |\mathbf{k}|$) is proportional to the square of the spectral amplitude, that is

$$\frac{S(k, t)}{S(k, 0)} = e^{-8\pi^2\mu k^2 t} \tag{6.12}$$

Introducing a reference wave number k_0 and a time parameter $\mathcal{T} = 4\pi^2\mu k_0^2 t$, we can write

$$\frac{S(k, \mathcal{T})}{S(k, 0)} = \exp\left(-\frac{2\mathcal{T}k^2}{k_0^2}\right) \tag{6.13}$$

Newman and Turcotte [1990] assumed, as an example, an initial topography with spectral density inversely proportional to the square of the wave number (i.e., Brownian noise, see Chapter 2) so that

$$\frac{S(k, 0)}{S(k_0, 0)} = \left(\frac{k_0}{k}\right)^2 \tag{6.14}$$

The resulting evolution equation is

$$\frac{S(k, \mathcal{T})}{S(k_0, 0)} = \left(\frac{k_0}{k}\right)^2 \exp\left(-\frac{2\mathcal{T}k^2}{k_0^2}\right) \tag{6.15}$$

The dependence of the spectrum on the wave number k/k_0 for several values of the dimensionless time $\mathcal{T}$ is shown in Figure 6.2.

Figure 6.2 shows that the suppression of landforms characterized by large wave numbers rapidly destroys the initial scaling behavior. This, of course, holds for any initial landscape characterized by spectral density in the form $S(k, 0) \propto k^{-\beta}$ as the evolution of the spectrum via Eq. (6.12) is given by

$$S(k, t) \propto k^{-\beta}e^{-8\pi^2\mu k^2 t} \tag{6.16}$$

We conclude that the deterministic diffusion equation, Eq. (6.8), destroys spatial scaling properties of the landscape topography by the intrinsic embedding of a characteristic length scale $\sqrt{\mu t}$. Whether or not we should expect that hillslope processes produce scale-invariant landforms remains to be seen. Nevertheless, it is important to analyze the broad characters that slope evolution models should – or should not – possess in view of the interpretation of the observational evidence discussed in Chapter 1.

Although many other refinements of hillslope models are available from the literature, it seems appropriate here to recall a class of conceptually different slope evolution models that still make use of a diffusion component as in Eq. (6.8). Dietrich et al. [1995] proposed the explicit incorporation within the landscape evolution equation of a term describing soil production from underlying bedrock. The basic evolution equation in this case becomes

$$\rho_s \frac{\partial h}{\partial t} + \rho_r \frac{\partial (z - h)}{\partial t} = \mu_s \nabla^2 z(\mathbf{x}, t) \qquad (6.17)$$

where ρ_s, ρ_r are the bulk densities of the soil and the underlying bedrock, respectively, and $h(\mathbf{x}, t)$ is the soil depth. Thus the first term is the change of soil thickness with time t. The second term is the rate of conversion of bedrock to soil resulting in the lowering of the soil–bedrock interface. The last term is the net soil transport, assumed as diffusive in the original work by Dietrich et al. [1995]. Soil production functions defining $\partial(z - h)/\partial t$ are estimated from field data. The importance of the class of models that includes soil production terms lies in its intrinsic ability to describe rock emergence (when soil thickness disappears) and other geomorphic factors that have an impact on channel initiation, on one hand, and on the roughness of the resulting elevation field on the other.

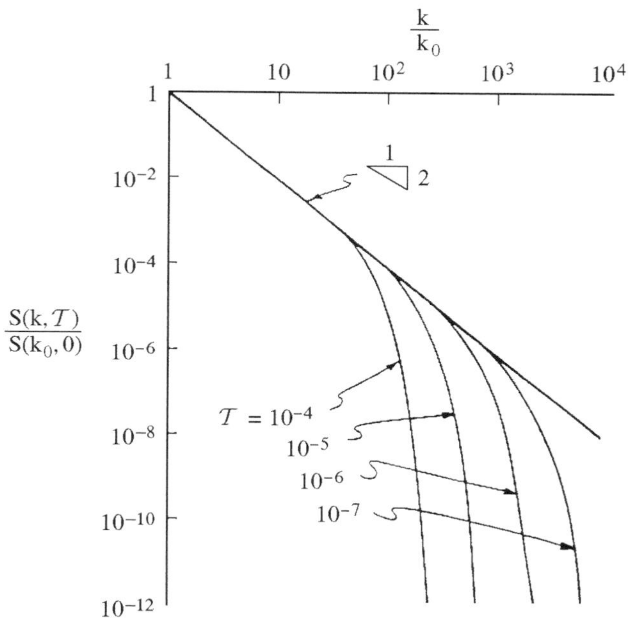

Figure 6.2. Power spectral density as a function of wave number for initial power law topography evolved through a diffusive model [after Newman and Turcotte, 1990].

A further important question is related to the intrinsic ability of dynamic models to produce scaling structures, either in their spatial or temporal properties. Although, of course, we do not claim that all landforms should bear the signatures of fractality, as stated in many places throughout this book, the central interest here lies precisely in the identification of the dynamic reasons and the observational evidence for scale-invariant landforms. Thus it seems appropriate to briefly review some mathematical mechanisms that entail the growth of fractals.

Two mechanisms are known to produce scaling structures in continuum approaches of the type in Eq. (6.1): nonlinearities in the divergence of fluxes ($\nabla \cdot \mathbf{F}$) and the effects of a driving noise η. We will briefly discuss them separately, focusing only on the power law characters of spectral densities in space and time as the signature of scaling structures.

6.2.1 The Effects of Nonlinearity

Nonlinear mechanisms are naturally suited to produce scaling structures. From spectral analysis, the basic ingredient is the property of convolution sums of spreading spectral energy across different modes, thus possibly neutralizing the dissipation effect of high-order linear terms. Many nonlinear mechanisms are known, especially from the literature on the morphology of growing interfaces [Kardar et al., 1986].

We will discuss here one example related to landform evolution [Newman and Turcotte, 1990]. Topography will be measured relative to its mean height $\langle z(\mathbf{x}, t)\rangle$ and considered a stationary field. The equal-time autocovariance function is defined by

$$\rho(\mathbf{X}, t) = \left\langle z(\mathbf{x} + \mathbf{X}, t) z(\mathbf{x}, t) \right\rangle \qquad (6.18)$$

which is assumed spatially homogeneous (i.e., independent of x). The autocovariance function and the power spectral density are related by

$$\rho(\mathbf{X}, t) = \sum_{\mathbf{k}=1}^{N/2} S(\mathbf{k}, t)e^{i2\pi\mathbf{k}\cdot\mathbf{X}} \tag{6.19}$$

Newman and Turcotte [1990] introduced a modified discrete Fourier transform associated with topography $z(x, t)$ on a linear track x:

$$\rho(x, t) = \sum_{n=0}^{N} S_n(t)e^{-i2\pi 2^n k_0 x} \tag{6.20}$$

where N is a suitable cutoff. The above representation compresses all the information contained between wave numbers $k_0 2^{n-1/2}$ and $k_0 2^{n+1/2}$ into the Fourier amplitude coefficient S_n.

It is of interest to relate the coefficients S_n in the modified Fourier series (6.20) to the spectrum $S(k, t)$ in the standard Fourier integral representation. In one dimension we have

$$S_n(t) = \int_{k_0 2^{n+1/2}}^{k_0 2^{n-1/2}} S(k, t)dk \tag{6.21}$$

in the region $k_0 2^{n-1/2} < k < k_0 2^{n+1/2}$. Newman and Turcotte [1990] studied the conditions for the maintenance of a fractal structure described by a power law spectral density of the type

$$S(k, t) = S(k_0, t)\left(\frac{k}{k_0}\right)^{-\zeta} \tag{6.22}$$

which, once substituted into Eq. (6.21), yields

$$S_n(t) = \frac{S(k_0, t)}{1-\zeta} k_0^{\zeta}(2^{-(1-\zeta)/2} - 2^{(1-\zeta)/2})k_n^{1-\zeta} \tag{6.23}$$

where $k_n = k_0 2^n$. Thus the scaled coefficients are found to satisfy the recurrence relation:

$$\frac{S_{n+1}}{S_n} = 2^{1-\zeta} = \lambda \tag{6.24}$$

because $k_{n+1}/k_n = 2$. Using the relation between the exponent ζ of the power spectral density and the (local box counting) fractal dimension D (valid for a simple fractal object) given in Section 2.7.4, that is, $\zeta = 5 - 2D$, we obtain

$$\lambda = 2^{2(D-2)} \tag{6.25}$$

The evolution of fractal surfaces may be obtained from a dynamics such that the Fourier coefficients S_n result from an equation of the type

$$\frac{dS_n}{dt} = -R_n S_n \tag{6.26}$$

where R_n is a rate coefficient dependent on n, such that Eq. (6.24) holds. Notice that if $R_n = 2\mu$ for all n, the model reduces to the diffusion model Eq. (6.8). This result follows from Eq. (6.10) and the substitution of Eqs. (6.18) and (6.19) after some technicalities.

A scaling behavior may arise because of the key dependence of the rates R_n on the wave number n. This implies that erosion rates at every length scale depend on the topographic distribution of features at related length scales. This is a basic spectral definition of nonlinearity.

In well-known turbulence studies [e.g., Kolmogorov, 1941; Landau and Lifschitz, 1959] equivalent cascade models are defined as having the rates R_n dependent on the Fourier coefficients at that scale, S_n, and the immediate adjacent scales, S_{n-1} and S_{n+1}. For the actual erosional problem, expressions for the rates R_n are based on the observation that large gullies generate gullies on the next smaller scale in a scale-invariant manner. The problem requires a closure. The scheme proposed by Newman and Turcotte [1990] is as follows:

- At the largest scale, $n = 0$ and $k = k_0$, one assumes $R_0 = \alpha_0$, a constant. Thus $dS_0/dt = -\alpha_0 S_0$, and consequently it is assumed that the largest features have an exponential decay.
- At the other scales, it is assumed that features on a given scale are influenced only by the next *larger* scale, roughly implying that gullies of a specified scale generate smaller gullies by a factor of about 2 in a scale-invariant process. Thus R_n depends only on S_n and S_{n-1}. The proposed model relation is thus

$$R_n = \gamma \left(\frac{S_n}{\lambda S_{n-1}} \right) \tag{6.27}$$

where γ is a constant. Eq. (6.27) when inserted in Eq. (6.26) yields the desired recurrence relation contained in Eq. (6.24).

The description of the families of models that can derived by the above assumptions is beyond the scope of this book. Although the model contains interesting implications concerning the birth and evolution of geomorphic fractal structures, a physical basis has yet to be provided.

6.2.2 The Effects of a Driving Noise

We now proceed to study the second mechanism known to produce scaling structures in the framework of Eq. (6.1), namely, the effects of driving noise.

It was established long ago [Whittle, 1962], with reference to the interpretation of evidence from agricultural uniformity trials, that equal-time homogeneous covariance functions in the plane, say $\Gamma(r)$, decaying ultimately as the inverse of distance r, that is, as r^{-1}, could be theoretically derived through mathematical models of driven diffusion in three dimensions. Covariance functions at large distances r, and the related spectral density functions at small frequencies, were examined through a deterministic diffusion model in two or three dimensions into which random noise is continuously being injected. As such, random variations that diffuse in a deterministic fashion through physical space were labeled as random diffusion processes.

Within our context, the relevance of Whittle's [1962] analysis is due to the basic equation discussed, which is the following:

$$\frac{\partial z}{\partial t} = \mu \nabla^2 z + \eta(\mathbf{x}, t) \tag{6.28}$$

where $\eta(\mathbf{x}, t)$ is an *additive* random component whose statistical structure we discuss below. Eq. (6.28) is a particular case of Eq. (6.1) where a diffusive approximation for the fluxes and the assumption of nonactive tectonics ($U = 0$) are adopted.

Notice that in the presence of a preferred direction, say $x_\parallel$, corresponding to anisotropic fluxes, say $\mu_\parallel \partial z/\partial x_\parallel$ and $\mu_\perp \partial z/\partial x_\perp$ (where $x_\perp$ is the transverse direction), from the conservation Eq. (6.8) one has

$$\frac{\partial z}{\partial t} = \mu_\parallel \frac{\partial^2 z}{\partial x_\parallel^2} + \mu_\perp \frac{\partial^2 z}{\partial x_\perp^2} + \eta(\mathbf{x}, t) \tag{6.29}$$

Later in this section we will suggest conditions through which noise, driving anisotropic systems, will have effects on the emergence of generic scale invariance.

The effects of other types of noise, for example, the presence of vectorial noise within the divergence term as $\nabla(\mu \nabla z + \eta)$, have also been investigated in the literature [e.g., Grinstein et al., 1990]. Nevertheless, the mathematical formalism becomes much more involved, and thus we will describe the foremost results without proof.

The presence of the random term (here reminiscent of the introduction of any source of variation, say by weather, vegetation, or rainfall heterogeneity) substantially modifies the properties of the scaling-destroying mechanisms embedded in the deterministic diffusion part. If the process is kept going, a statistical balance can ultimately be achieved between the disturbances caused (at all scales if the noise is white) by η and the selective smoothing effects of the diffusion term. The process will then become stationary with respect to both $\mathbf{x}$ and t (the correlation functions will depend only on the modulus of spatial or temporal lags).

In the simplest case, Whittle [1962] referred to spatially uncorrelated noise, for which

$$\langle \eta(\mathbf{x}, t)\eta(\mathbf{x}', t) \rangle \propto \delta(|\mathbf{x} - \mathbf{x}'|, t) \tag{6.30}$$

but we notice, following Hwa and Kardar [1992] and Grinstein et al. [1990], that the specific structure of the noise bears important consequences on the scale invariance properties of the dynamic model in Eq. (6.28). Specifically, noise can be strictly conservative or conservative in the mean. In the simple case shown in Eq. (6.28) the noise is additive and the implied addition of mass from outside destroys the local conservation rule. In steady state, erosional transport to the outside implies $\langle \eta(\mathbf{x}, t) \rangle = 0$. Thus acting random forcings in the landscape evolution equation may be mimicked by uncorrelated Gaussian noise with the leading moment given by

$$\langle \eta(\mathbf{x}, t)\eta(\mathbf{x}', t') \rangle = 2D\,\delta(\mathbf{x} - \mathbf{x}')\,\delta(t - t') \tag{6.31}$$

where D is a measure of the strength of the noise.

The mean and the covariance function of the noise define, to second order, the random fields $\eta(\mathbf{x}, t)$. Without mathematical details, we simply notice that through the dynamics of the process one can manipulate the transform of the covariance function, Eq. (6.31), to yield the so-called conservative noise, that is, one that acts in the evolution equation and does not violate the conservation of the variable $z(\mathbf{x}, t)$ even in single realizations. Additive Gaussian noise, in fact, is conservative only in the mean.

Scale-invariance properties can be investigated through the two-point correlation function:

$$\left\langle (z(\mathbf{x}, t) - z(\mathbf{x}', t'))^2 \right\rangle = \Gamma\left(|\mathbf{x} - \mathbf{x}'|, t - t'\right) \tag{6.32}$$

(or, alternatively, $\langle (z(\mathbf{r}, t) - z(0, t)^2 \rangle_{|\mathbf{r}|=r} = \Gamma(r)$).

Through Green's approach, the general solution of Eq. (6.28) is readily obtained in the form

$$z(\mathbf{x}, t) = \int_{-\infty}^{\infty} d\mathbf{y} \int_{0}^{\infty} d\tau \;\; G(\mathbf{x}, \mathbf{y}, \tau)\; \eta(\mathbf{y}, t - \tau) \tag{6.33}$$

where Green's function G for Eq. (6.28) is

$$G(\mathbf{x}, \mathbf{y}, \tau) = \frac{1}{\sqrt{2\pi\mu\tau^2}} \exp\left(-\frac{\left(|\mathbf{x} - \mathbf{y}|\right)^2}{2\mu\tau} \right) \tag{6.34}$$

From the above exact definition of $z(\mathbf{x}, t)$ one can compute equal-time covariance functions $\Gamma(r)$ in one, two, or three dimensions. The covariances can be obtained by direct integration using the properties of Green's functions and the uncorrelated nature of the noise. In fact

$$\begin{aligned} \Gamma(r) &= \left\langle \left(z(\mathbf{x}, t) - z(\mathbf{x} + \mathbf{r}, t)\right)^2 \right\rangle_{|\mathbf{r}|=r} \propto \left\langle z(\mathbf{x} + \mathbf{r}, t) z(\mathbf{x}, t) \right\rangle_{|\mathbf{r}|=r} \\ &= \int_{-\infty}^{\infty} d\mathbf{y} \int_{0}^{\infty} d\tau_1\, G(\mathbf{x}, \mathbf{y}, \tau_1) \int_{-\infty}^{\infty} d\mathbf{z} \\ &\times \int_{0}^{\infty} d\tau_2\; G(\mathbf{x} + \mathbf{r}, \mathbf{z}, \tau_2) \left\langle \eta(\mathbf{y}, t - \tau_1) \eta(\mathbf{z}, t - \tau_2) \right\rangle \end{aligned} \tag{6.35}$$

In the case of uncorrelated noise one employs Eq. (6.31) thus obtaining

$$\Gamma(r) = \int_{-\infty}^{\infty} d\mathbf{y} \int_{0}^{\infty} d\tau\; G(\mathbf{x}, \mathbf{y}, \tau) G(\mathbf{x} + \mathbf{r}, \mathbf{y}, \tau) \tag{6.36}$$

in one, two, or three dimensions. For the three-dimensional case, Whittle [1962] has shown, by substituting Eq. (6.34), that $\Gamma(r)$ decays as a power law, that is

$$\Gamma(r) \propto r^{-1} \tag{6.37}$$

Notice that the divergence of the variance $\Gamma(0)$ for zero lag is produced by the extreme irregularity of the uncorrelated noise given as input to the process.

Whittle [1962] also showed that if η has a small amount of autocovariance (i.e., spatial correlation, which is quite plausible from the physical viewpoint), then $\Gamma(r)$ is finite at $r = 0$ and still behaves like r^{-1} for large r. To show this, it is expedient to derive correlation properties from spectral density functions. The spectral density function has been defined in Eq. (6.12). Notice that here we generalize the Fourier transform (6.9) to the case where the function is also periodic in time even though periodicity is not entirely justifiable in the absence of uplift.

Let ω denote frequency and $\mathbf{k}$ denote the wavenumber vector that is complementary to t and $\mathbf{x}$, respectively, that is

$$z(\mathbf{x}, t) = \sum_{|\mathbf{k}|} \hat{z}(\mathbf{k}, \omega) \exp^{i\pi(\mathbf{k}\cdot\mathbf{x} + \omega t)} \tag{6.38}$$

The same notation is employed for the random noise η.

If z and η, regarded as processes both in space and time, have space–time spectral density functions $S(\mathbf{k}, \omega) = \langle \hat{z}(\mathbf{k}, \omega)\hat{z}^*(-\mathbf{k}, \omega)\rangle$ (where $*$ denotes complex conjugate) and $\Phi(\mathbf{k}, \omega) = \langle \hat{\eta}(\mathbf{k}, \omega)\hat{\eta}^*(-\mathbf{k}, \omega)\rangle$, respectively, then it follows from Eq. (6.28) and standard methods that

$$(i\omega + k^2\mu)\hat{z}(\mathbf{k}, \omega) = \hat{\eta}(\mathbf{k}, \omega) \tag{6.39}$$

and

$$S(\mathbf{k}, \omega) = \frac{\Phi(\mathbf{k}, \omega)}{(i\omega + k^2)(-i\omega + k^2)} \tag{6.40}$$

where $k = |\mathbf{k}|$ [Whittle, 1962]. If we consider $z(\mathbf{x}, t)$, for fixed t, as a stationary process in space alone, then this spatial process will have a spectral density function

$$S(\mathbf{k}) = \frac{1}{2\pi} \int_{-\infty}^{\infty} S(\mathbf{k}, \omega)\, d\omega \tag{6.41}$$

In particular, if η is uncorrelated in time – so that $\Phi(\mathbf{k}, \omega)$ will be independent of ω and equal, say, to $\Phi(\mathbf{k})$ – then from Eq. (6.28) one concludes that

$$S(\mathbf{k}) = \frac{\Phi(\mathbf{k})}{k^2} \tag{6.42}$$

The above equation is related to the covariance function $\Gamma(r)$ previously considered by the Fourier transformation:

$$\Gamma(r) = \frac{1}{(2\pi)^2} \int \exp(i\mathbf{k} \cdot \mathbf{r})\, S(\mathbf{k})\, d\mathbf{k} \tag{6.43}$$

Whittle's [1962] foremost result is that if one assumes η to be uncorrelated both in time and space, than $\Phi(\mathbf{k})$ is constant and

$$S(\mathbf{k}) \propto k^{-2} \tag{6.44}$$

which is consistent with the results obtained above. The process still retains infinite variance because the diffusion mechanisms do not smooth η sufficiently. It can be shown that when the process has finite variance η, it may still be uncorrelated in space. In such a case, the noise term has in time a covariance function $\phi(t)$, its space–time covariance function being $\phi(t)\delta(r)$, maintaining the asymptotic power law characters.

Finally, if η is uncorrelated in time but has a spatial spectral density function $\phi(\mathbf{k})$, one obtains $S(\mathbf{k}) = k^{-2}\phi(\mathbf{k})$. Suppose now that $\phi(\mathbf{k})$ is finite and nonzero at the origin but decays fast enough for large k. Then a mild degree of spatial correlation in the input will also ensure that $z(\mathbf{x}, t)$ has a finite variance.

We now turn to the general case for which a few general results are available [e.g., Grinstein et al, 1990; Hwa and Kardar, 1992]. In the aforementioned references the field theory of dissipative transport, especially with reference to the occurrence of generic scale invariance, has been addressed. By generic scale invariance we mean a concept central to self-organized criticality, that is, the robustness of the consistent growth of fractal structures regardless of initial conditions and parameter values. This is in contrast to the typical behavior of critical phenomena where scale invariance is typically obtained by fine tuning of some parameters.

The dynamic ingredients relevant to the growth of generic scale invariance consist of (i) a driving action and (ii) a subsequent relaxation. As such, a good candidate model is Eq. (6.28) where the left-hand side takes on the form of the conservative, deterministic relaxation that follows perturbation. The right-hand side represents the external sources and sinks in terms of a random input function η.

Grinstein et al. [1990] and Hwa and Kardar [1992], among others, have studied the possible effects of anisotropy in the constitutive equation (6.28) and in the driving noise. It has been shown that in the context of noisy, nonequilibrium models of the type in Eq. (6.29), systems characterized by strictly conservative deterministic dynamics and conservative noise require spatial anisotropy to exhibit spatial and temporal correlations that decay algebraically under generic conditions. In other words, they have infinite correlation lengths for any set of parameters as indicated by the power law decay of equal-time correlations $\Gamma(r) \propto r^{-a}$, $a > 0$, for large r. Instead, systems with conservative dynamics and additive noise like Whittle's do not require anisotropy for generating spatial and temporal correlations that decay algebraically in three dimensions under generic conditions.

It is important to realize that random noise coupled to a diffusive relaxation allows us to evolve a complex system through states that are scaling and thus can be described by synthetic descriptors like a fractal dimension. The above result is important in view of the applications of landscape evolution models pursued in the next sections.

6.3 Landscape Self-Organization

Let $z(\mathbf{x}, t)$ be the elevation field whose *fluvial* evolution is specified by the dynamic equation:

$$\frac{\partial z}{\partial t} \propto f(\tau - \tau_c) \qquad (6.45)$$

where $f(x) = 0$ for $x \leq 0$, τ is the local shear stress, and τ_c is a critical threshold for erosion. As discussed in Chapter 5, the shear stress produced by the flow rate Q in dynamic equilibrium is

$$\tau = \rho g y \nabla z \qquad (6.46)$$

where ρ is the density of water, g is the acceleration of gravity, ∇z is the local slope, and y is the flow depth (approximating the ratio of flow area and wetted perimeter), which is assumed to scale as $y \propto Q^{0.5}$. This perspective of a threshold-limited (TL) erosion process is justified by experimental evidence [e.g., Leopold et al., 1964] and by optimality principles [Rodriguez-Iturbe et al., 1992b]. Thus at the ith site, $\tau_i \propto Q_i^{0.5} \nabla z_i$. Whenever shear stresses exceed τ_c anywhere in the network, we expect erosional activity.

The variable governing the evolution of the elevation field is the discharge Q_i at any given site in a lattice, say i. Discharge is surrogated, as in Chapter 5, by the the total contributing area, that is, $Q_i \propto A_i$.

The crux of the matter in the evolution of *fluvial* landscapes was to show that threshold-dependent dynamic specifications yield erosion activities that self-organize themselves into a process reproducing significant characters of the geomorphology of river networks.

Nevertheless, we noticed in the previous section that besides fluvial erosion, other processes act in the river basin, especially at the scale where discharge is small and ephemeral, that is, the hillslope scale. At this scale the critical shear stress is normally below threshold and therefore every evolution of the landscape elevation $z(\mathbf{x}, t)$ is threshold-independent and essentially related to the gradient of the elevation field (slope-dependent (SD) processes). The contribution of SD processes to the dynamic equation (6.45) yields additional terms of the type $F(\nabla z)$ where F represents modes of transport referred to in Section 6.2. As noted in Section 1.4.3, transport can eventually be enhanced by the presence of water flow that acts as catalyst for physical or chemical reactions and may lead to a very complicated functional structure embodied in the notation $F(Q, \nabla z)$.

The fact that fluvial erosion processes start only when a threshold is exceeded means that in those regions where $\tau < \tau_c$ only SD terms act in reducing elevations. Eq. (6.45) can be thought of as a particular case of the following continuous model when $\tau > \tau_c$ (see Eq. (1.56)):

$$\frac{\partial z}{\partial t} = \alpha f(\tau - \tau_c) + \beta F(\nabla z) + U \tag{6.47}$$

where F models SD transport and α, β are two coupling constants (dimensionally the inverse of time) that, together with the uplift U, determine the relative strength and the timescales of the processes. At first we will neglect tectonic forces responsible for U to avoid clouding the first issue addressed, that is, the self-organized effects of the coupling of different processes within a landscape system.

It can be argued that $\alpha \ll \beta$ and thus when erosion forces are concurrently acting, Eq. (6.45) plays the central role because the timescale of fluvial erosion is much shorter than the timescale of hillslope evolution. Technically, we assume that at prefixed times, say t_α, landscape-forming events trigger intense fluvial erosion and thus

$$\frac{\partial z}{\partial t} = \alpha f(\tau - \tau_c) \qquad \text{for} \quad t \simeq t_\alpha \tag{6.48}$$

because we assume that all TD changes triggered by the same event may be assumed to occur in the same geological instant t_α. Notice that the particular choice of interarrival times t_α of landscape-forming events is irrelevant, as random or equal-time arrivals would yield equivalent results (see Section 6.6). Thus time is allowed to elapse only between subsequent erosion events. Landscape relaxation between subsequent erosion events is described by the following evolution equation:

$$\frac{\partial z}{\partial t} = \beta F(\nabla z) \qquad t_\alpha < t \le t_{\alpha+1} \tag{6.49}$$

where we identify as t_β the relaxation time allowed for diffusive processes, $t_\beta = t_{\alpha+1} - t_\alpha$. Within the above framework the two types of processes may be decoupled. Although the temporal coupling of diffusion and fluvial erosion will be the subject of a deeper study in Section 6.6, in the current section (following Rigon, et al. [1994]) we simply alternate the action of the two processes acting independently, that is, alternating cycles of diffusive and fluvial character. This is justified for $\alpha \gg \beta$.

In this simplified but realistic perspective we use the lattice model described in Section 5.7 for obtaining SOC landscapes and their embedded networks from any arbitrary initial condition.

After a SOC fractal network and its corresponding landscape are obtained as the result of the first cycle in fluvial erosion, they become the initial condition for the action of a *diffusion* process that operates as follows:

- Sites, say i, are ordered by the magnitude of the largest drop Δz_{ij} from the elevation z_i to the elevation z_j of one of the j neighbors; the elevation z_i of the site with the largest drop is decreased by a quantity proportional to Δz_{ij}. The mass removed from z_i is redistributed among the neighbors whose elevation is lower than z_i (unit level of conservation). Mass redistribution on the lower neighbors is done according to the following scheme (let z_j^{new} be the new elevation in j):

$$z_j^{\text{new}} - z_j = \frac{\nu}{\sum_{k \in P} z_i - z_k}(z_i - z_j) \qquad 0 < \nu < 1 \tag{6.50}$$

  where $k \in P$ indicates that the sum is extended to the points lower than i in its neighborhood. Thus the point i does not change its drainage direction as its elevation is lowered; ν plays the role of a diffusion coefficient.
- The points draining into i change their ∇z whereas the points drained by i change both z and ∇z. Pits cannot be formed in the neighborhood of i.
- As a result, we expect a whole body of adjustments following a SD activity in a point. The above process is iterated an arbitrary number of times. The slope structure thus reflects the age and evolution of the diffusive processes.

Notice that the diffusion mechanism employed in the above scheme (Eq. (6.50)), is an anisotropic one in which material is transported only toward pixels with lower altitude. This is important because, as seen in the previous section, a driven isotropic diffusion mechanism in two dimensions would tend to destroy the scaling characteristics produced by the TL processes. On the other hand, a randomly driven anisotropic diffusion is compatible with the generic evolution through scaling states.

The resulting landscape and its drainage network become the initial condition for a new cycle of fluvial (TL) erosion, and so on indefinitely.

The results of many erosive-diffusive cycles in the network shown in Figure 5.9 are reported in Figures 6.3 and 6.4. Here t_β is defined by 10,000 allowed relaxations, that is, diffusive changes of elevation, as clarified below. A value of $\nu = 0.4$ in Eq. (6.50) was used in this particular example.

The concept of *cycle* needs some clarification. A cycle, in the context of SD actions, means a prefixed number of events in which diffusion is allowed to occur in the time span, $t_\alpha < t \leq t_{\alpha+1}$, through the above rules. The prefixed number of events mimics the interplay between the strength of the diffusive processes and the time of exposure to these processes before

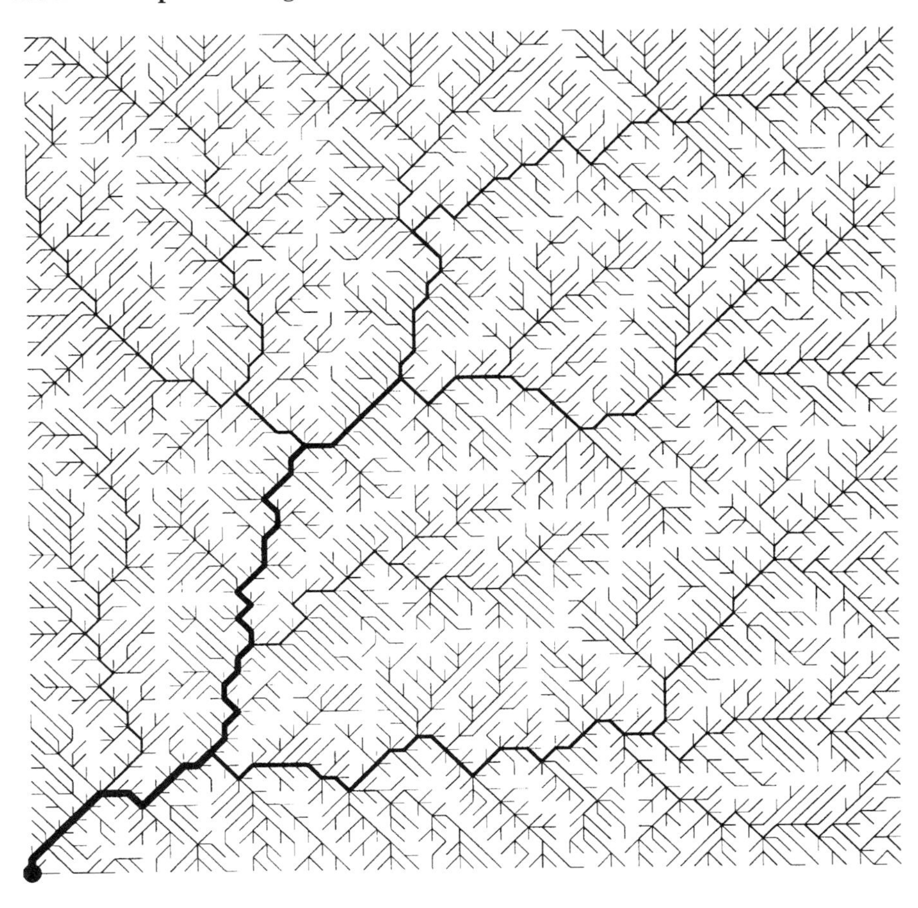

Figure 6.3. The action of SD processes on the network of Figure 5.8 [after Rigon et al., 1994].

TL erosion is set in action again. This is an obvious simplification of a complex reality because both TL and SD processes act concurrently and their decoupling is justified for modeling purposes only because $\alpha \gg \beta$.

The term cycle is also used for TL actions, defined as the number of avalanches required to reach equilibrium, as described in Chapter 5. Nevertheless, this is assumed to happen instantaneously with respect to the much longer exposure to SD actions. In fact, because diffusion modifies the structure of the landscape, in particular bringing some sites above the critical threshold for erosion, self-organized TL erosion will further modify the landforms in a never-ending process. Whether or not in the long range other factors, for example of climatic origin, play a role will be addressed in Section 6.6.

Some significant statistics of the landscape in Figure 6.4 are shown in Figures 6.5 to 6.8.

One important feature that is readily observed is that for landscapes resulting from the interaction of TL and SD erosional processes, planar statistics are almost unaffected. This is an important result, first reported in Rigon et al. [1994]. It suggests the reason why OCN and SOC models of drainage networks – which were essentially aimed at the sole fluvial evolution – reproduce planar features observed in nature. Although the

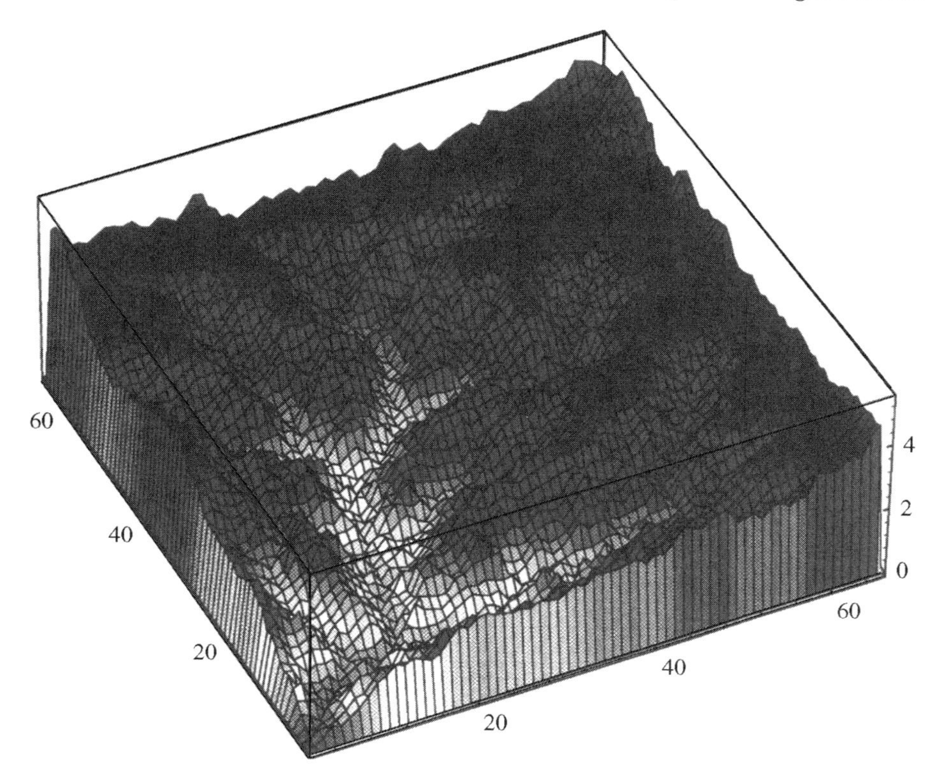

Figure 6.4. Three-dimensional view of the landscape associated with the network in Figure 6.3.

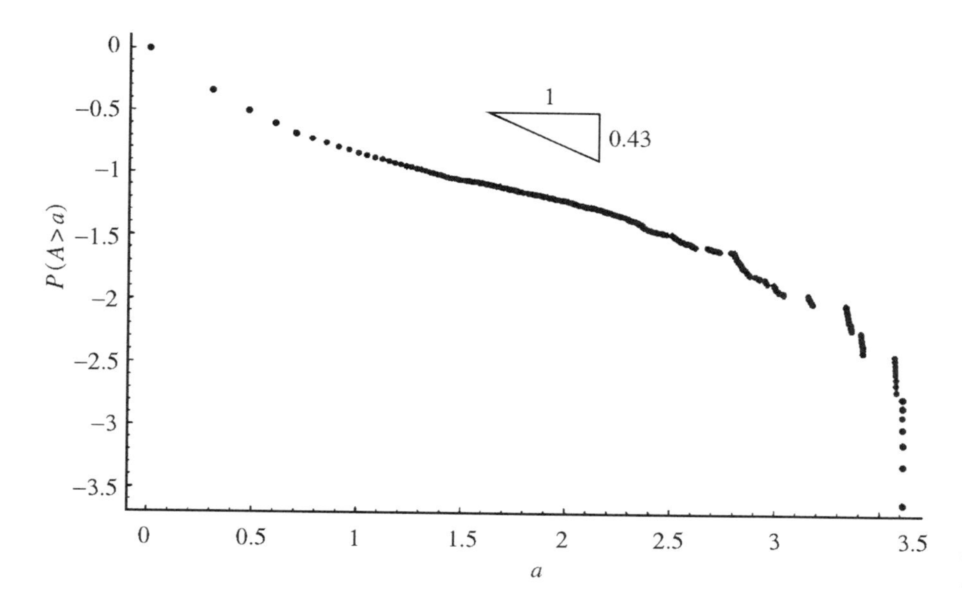

Figure 6.5. Probability of exceedence of total contributing area of the network in Figure 6.3.

statistics are robust in all respects, the sole visible planar signatures of the SD process are a certain elongation of the sources (Figure 6.3) and a modified structure of the probability of contributing area A, $P[A > a]$, for small values of A (Figure 6.5). Both of these are well-known geomorphologic features of real basins.

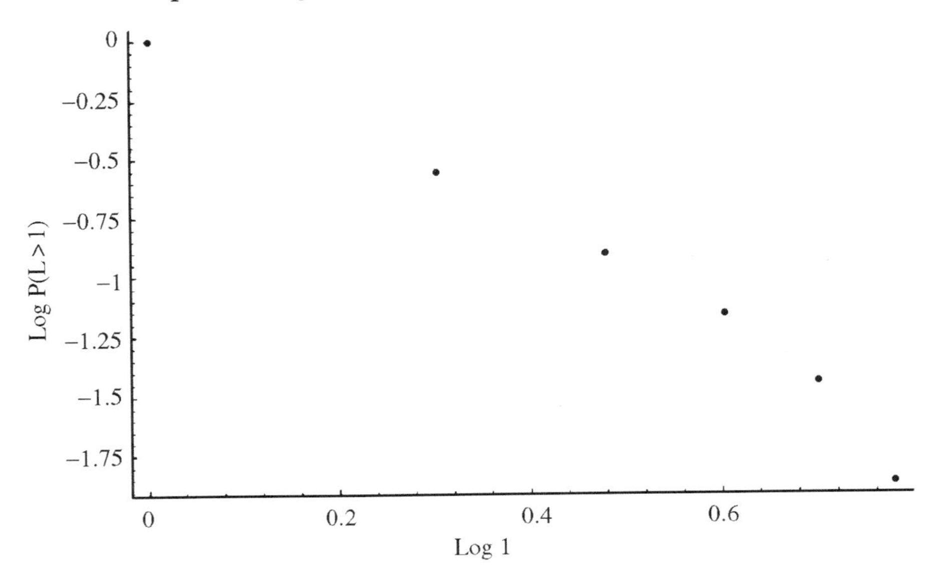

Figure 6.6. Probability of exceedence of Strahler stream length of the network in Figure 6.3.

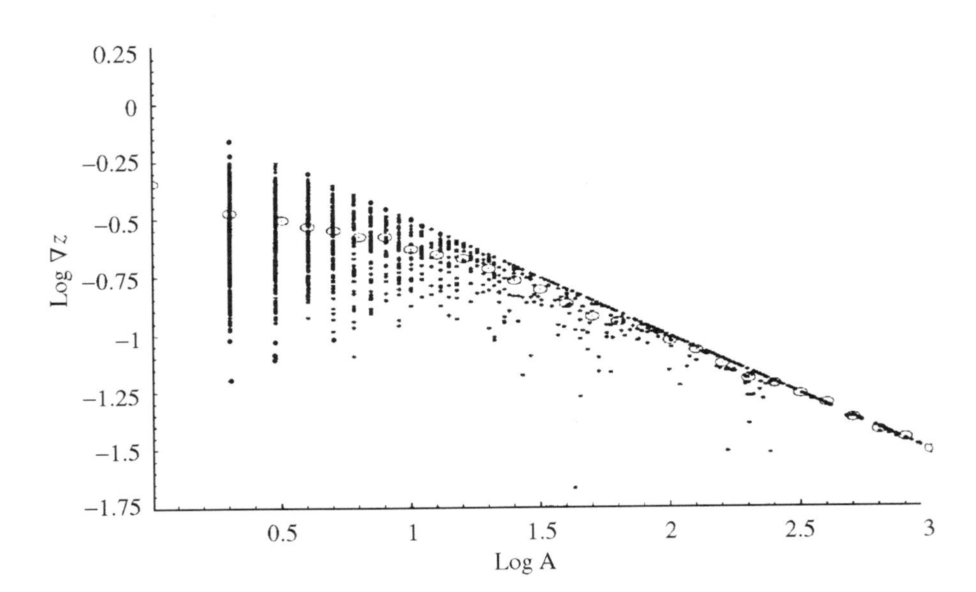

Figure 6.7. Double-logarithmic plot of the slope versus area for the landscape in Figure 6.4.

The upper curvature of the graph of $P[A \geq a]$ at small contributing areas, already observed from DEMs [Tarboton et al., 1992; Rodriguez-Iturbe et al., 1992a] is now more pronounced than, say, in Figure 5.12, where the straight line continued undeflected throughout minimal areas. The corresponding probability plot of Strahler's stream lengths is shown in Figure 6.6.

Interestingly, statistics involving elevations and areas (i.e., log ∇z vs log A) are significantly affected (Figure 6.7), even though the mean values approximately scale, as in SOCs without diffusion ($E[\nabla z(A)] \propto A^{-0.5}$). For smaller values of A the relationship is now like the one found in real basins.

A particular new feature is of key importance. In fact, it was observed in Section 1.2.11 that real basins from DEMs show multiscaling in the slope versus area relationship. This was detected from the fact that the

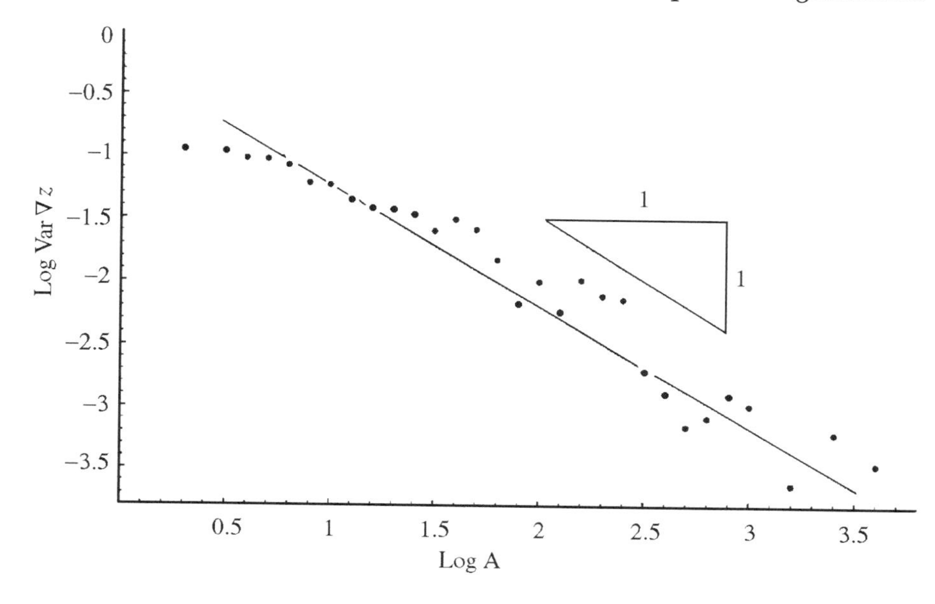

Figure 6.8. Double-logarithmic plot of the variance of the slopes $Var[\nabla z(A)]$ versus the contributing area A relative to the landscape in Figure 6.4.

mean of the slopes ($E[\nabla z(A)]$) scales with $A^{-0.5}$, but the variance scales approximately as $Var[\nabla z(A)] \propto A^{-\theta}$ with $\theta \sim 0.5$ rather than A^{-1}, as would be expected in the simple scaling case (Section 1.2.11). We recall that in Chapter 5 we studied the above features in SOC fluvial networks. We concluded that SOC *fluvial* topographies are essentially simple-scaling structures. It is of interest to investigate whether SD processes modify the simple-scaling behavior.

Figure 6.8 shows the graph of log $Var[\nabla z(A)]$ versus log A for the landscape in Figure 6.4, relative to coupled TL + SD processes. It can be seen that the overall scaling exponent is still approximately −1, a value that remains approximately the same for different values of the coefficient ν.

The previous results suggest that the coupling of TL and SD processes yields a simple-scaling structure of the type described by the self-similar assumption of Gupta and Waymire [1989], regardless of whether SD diffusion is acting or not. This result is confirmed by a detailed average through Monte Carlo realizations of the type performed in Section 5.7. It is only through the effects of spatial heterogeneities in the erosional processes that multiscaling is attained in the slope versus area relationship, at least according to the models we have employed. We will return to this topic in the next section.

One observation that follows from the previous description is that SD processes can by themselves produce (i.e., without any random disturbance in elevations or in discharge) the exceedence of the threshold and trigger events of considerable spatial extension.

Once a system has reached an equilibrium state of low total rate of energy dissipation, equivalent to a low mean elevation of the basin compatible with the existing erosion thresholds, the diffusion process will only produce minor changes in the landscape that keep the system near the low energy level already attained. Obviously, if the erosion thresholds are taken as zero, the equilibrium landscape would be a peneplain. It is only through geologic phenomena like uplift or through climatic changes that may increase the shear stresses τ (or, what is computationally equivalent, lower the threshold τ_c) that the landscape may evolve to a new equilibrium state significantly different from that previously attained. Throughout this evolutionary process the landscape may experience considerable changes in the roughness features, from rugged and steep hillslopes (typical of purely fluvial erosional processes) to the gentle and smooth characteristics of prolonged SD erosional processes. As explained before, the previous cycle is neverending, and the concepts of equilibrium and, say, fractal dimension, are intimately related to the duration of the process that drives the cycle (e.g., uplift, climate change) and to the TL and SD processes that actually built the landscape but that operate at different timescales.

6.4 On Heterogeneity

In natural river basins, fluvial erosion thresholds are likely to be characterized by spatial heterogeneity, which may reflect both the variations in the exposed lithology or vegetation and (or) the conditions favorable to different prevailing erosional mechanisms. As an example, we may recall that vegetation, by colonizing an initially bare soil surface, may increase by orders of magnitude the erosion threshold by overland flow. A source of heterogeneity lies also in the interplay of different mechanisms responsible for channel initiation, a synthesis of which has been given in Section 1.2.12. Regions of prevailing landslides, subsurface saturation without erosion, diffusion-dominated and combined transport mechanisms are thus observed in different places across the basin. Finally, when assuming that landscape-forming flow rates are directly proportional to total contributing area, one assumes that driving rainfall rates are uniform. On a finer level of detail, intense rainfall fields embed significant spatial and temporal fluctuations.

The above considerations make clear that the critical stress triggering fluvial erosion varies throughout the basin. However, if the observational scale for the developing network is large with respect to some characteristic scale of hillslope–channel transition, a constant threshold may constitute a reasonable representation. As such, we have shown that it gives plausible interpretations of observational evidence.

If the hillslope scale is significant with respect to the basin scale of observation, a reasonable assumption [Rigon et al., 1994] is to view channel initiation as a random spatial process where at the arbitrary (critical) site $\mathbf{x}$ one has

$$A(\mathbf{x})^{1/2}\nabla z(\mathbf{x}) = \tau_c(\mathbf{x}) \tag{6.51}$$

$\tau_c(\mathbf{x})$ being a random field with spatial correlation. Although no limitations are posed to the distribution of the threshold model adopted, Rigon et al. [1994] assumed a threshold field lognormally distributed with $\tau_c(\mathbf{x}) = \langle\tau_c\rangle e^{Y(\mathbf{x})}$, where $Y(\mathbf{x}) \in N(0, \sigma_\tau^2)$ (where $\langle\rangle$ denotes, as usual, ensemble averaging exchanged with spatial averages), as commonly adopted in the geophysical literature [e.g., Dagan, 1989] for nonnegative random variables. The spatial correlation is defined by the two-point covariance function

$$C_y(\mathbf{x}_i, \mathbf{x}_j) = \langle Y(\mathbf{x}_i)Y(\mathbf{x}_j)\rangle \tag{6.52}$$

and, as a reasonable first-order assumption, a statistically homogeneous and stationary field is assumed (i.e., $C_y(\mathbf{x}_i, \mathbf{x}_j) = C_y(|\mathbf{x}_i - \mathbf{x}_j|) = C_\tau(r)$, where r stands for the modulus of the two-point separation vector $\mathbf{r}$). A common choice of correlation is the exponential type:

$$C_y(r) = \sigma_\tau^2 e^{-r/\lambda_\tau} \tag{6.53}$$

where λ_τ is the correlation scale, that is, a characteristic length at which the thresholds are statistically uncorrelated. For a given variance of the field, it controls the smoothness of the spatial changes that exist in τ_c.

Efficient numerical techniques are available to generate such random spatial fields, for instance, the turning bands method [e.g., Mantoglu and Wilson, 1988] or spectral methods. As an example, an efficient tool uses fast Fourier transform (FFT) generated fields [Gutjahr, 1989] where the spectrum of the function $X(\mathbf{x})$ (say, $\hat{X}(\mathbf{k})$) is generated through the iterated use of the following expression in the wave number space:

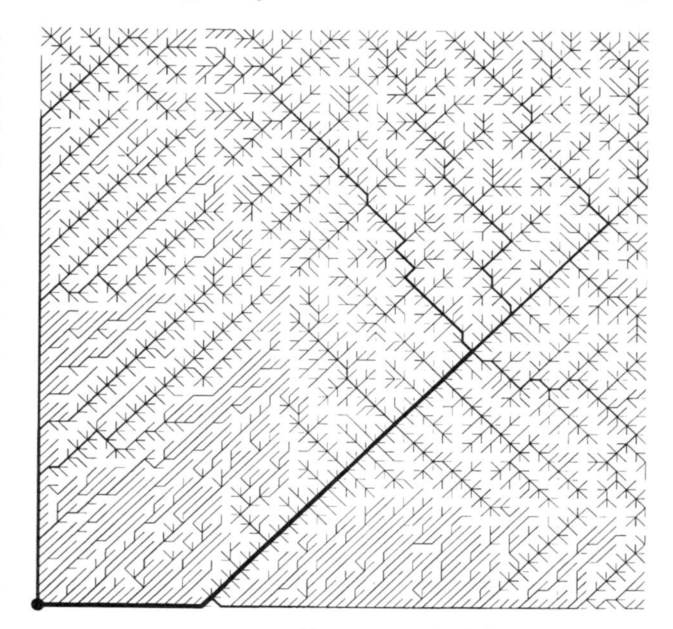

Figure 6.9. If $\tau_c(\mathbf{x})$ is a random space variable, the resulting SOC network before a diffusion cycle looks more irregular. Here the initial condition is the landscape of Figure 5.16, $\langle \tau_c(x) \rangle = 1$, $\sigma_\tau = 0.1$ and $\lambda_\tau = 32$ [after Rigon et al., 1994].

$$\hat{X}(\mathbf{k}) = \hat{C}_X(\mathbf{k})^{1/2}[R_\mathbf{k} + iT_\mathbf{k}] \tag{6.54}$$

where $\hat{C}_X(\mathbf{k})$ is the FFT of the covariance function $C_X(r)$, and R, T are pairs of independently generated Gaussian random numbers with zero mean and variance one-half. One may prove that inverse transformation of the above equation yields (apart from numerical accuracy) a stationary field with zero mean, variance σ_X^2 and correlation structure $C_X(r)$.

The SOC procedure described in Section 6.3 is applied without modifications once the threshold field is assigned.

We show here the results of two experiments in a 64×64 lattice [after Rigon et al., 1994] in which (1) the initial condition is that given in Figure 5.16, with $\langle \tau_c \rangle = 1$, $\sigma_\tau^2 = 0.1$, and $\lambda_\tau = 32$ pixels; (2) the initial condition is again that in Figure 5.16, but $\langle \tau_c \rangle = 1$, $\sigma_\tau^2 = 0.3$, and $\lambda_\tau = 2$ pixels are used in the computations.

Figures 6.9 to 6.12 show the resulting landscapes for case (1), before (Figures 6.9 and 6.10) and after a diffusion cycle (Figures 6.11 and 6.12).

Randomness in the distribution of the erosion threshold tends to alter the regularity possibly embedded in the initial conditions. Nevertheless, the effects of heterogeneity are related to its strength (i.e., the variance of the field) and to the ratio of the correlation scale to the size of the domain. Little tortuosity is still observed in the backbone of the networks in Figures 6.9 and 6.10, as a result of the initial comb-like structure and of the mild heterogeneity. Other computational results suggest that to produce a marked sinuosity of the main channel

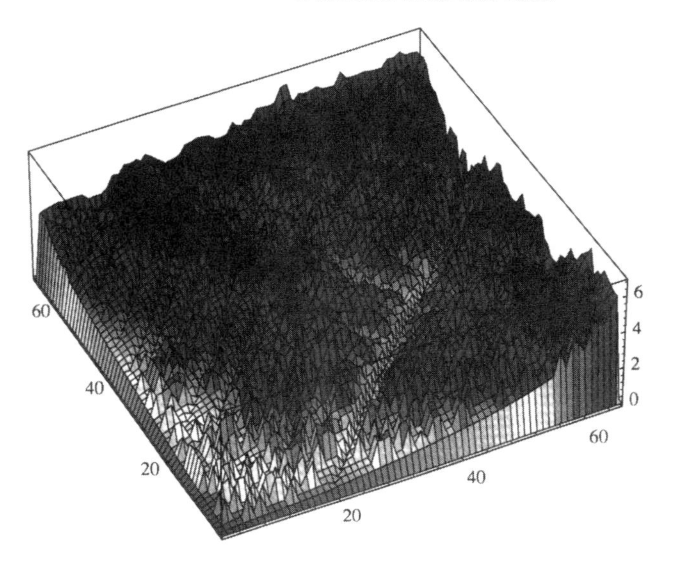

Figure 6.10. Three-dimensional network after the SOC cycle with a random threshold field.

Figure 6.11. Planar view of the SOC with random threshold field after a diffusion cycle.

starting from parallel drainage directions, one has to increase significantly the variance of the threshold ($\sigma_\tau^2 \approx 1$) while lowering the correlation scale, thus enhancing the differences in erodibility between adjacent pixels.

The statistical properties of the network reproduce quite well those observed in nature.

Interestingly, not much is changed for the planar statistics, such as the slope of the power law part corresponding to the regions where fluvial processes are clearly established. However, as somewhat expected, in the case of random thresholds for erosion the slope statistics ($\nabla z(A)$ versus A) show quite a different behavior than they do in the case of homogeneous processes and closely resemble the behavior in DEMs (Figure 6.13).

It is interesting to examine in detail the combined effects of heterogeneous erosion and diffusion on the scaling structure of slopes because it is only in the three-dimensional structure of the basin, rather than in the planar morphology, that the coupling of different processes shows a definite signature. In particular, we will observe that to second order (i.e., log $Var\nabla z(A)$ versus log A) a scaling structure very close to that observed by Tarboton et al. [1989b] appears.

In fact, the mean slope scales as $E[\nabla z(A)] \propto A^{-0.5\pm0.05}$ in all cases. Figure 6.13 shows one example of scaling of slopes for the basin in Figure 6.11. It was found that the variance scales with exponents 0.4 and 0.5, respectively, for the cases where (1) $\sigma_\tau^2 = 0.1$ and $\lambda_\tau = 32$ pixels, and (2) $\sigma_\tau^2 = 0.3$ and $\lambda_\tau = 2$ pixels (Figures 6.14 and 6.15).

The analysis thus suggests the existence of a multiscaling structure for the elevation fields. The features of the heterogeneous threshold field (σ_τ and λ_τ) have an impact on the scaling exponents of the variance but not on the mean of the slopes. This suggests a linkage between the structure of heterogeneities of the landscape and the multiscaling structure type, a result consistent with the theoretical argument of Rodriguez-Iturbe et al. [1992b] where the multiscaling behavior of the slopes of the network (as a function of the contributing area) was explained on the basis of arguments that randomized the second principle of energy expenditure (Section 4.4). In their study the energy expenditure per unit area was assumed to be a random variable whose mean is the same everywhere rather than a constant value. In fact, when the product $A^{0.5}\nabla z(A) = \tau$ is regarded as a random variable with constant mean value, $E[\tau]$, the variance of the slopes scales as $Var[\nabla z(A)] \propto A^{-1}Var[\tau]$. If $\tau = \tau_c$ everywhere, $Var[\tau] = Var[\nabla z(A)] = 0$. In the case of TL+SD processes, quite a few sites are left

Figure 6.12. Three-dimensional landscape with random threshold field after a diffusive cycle.

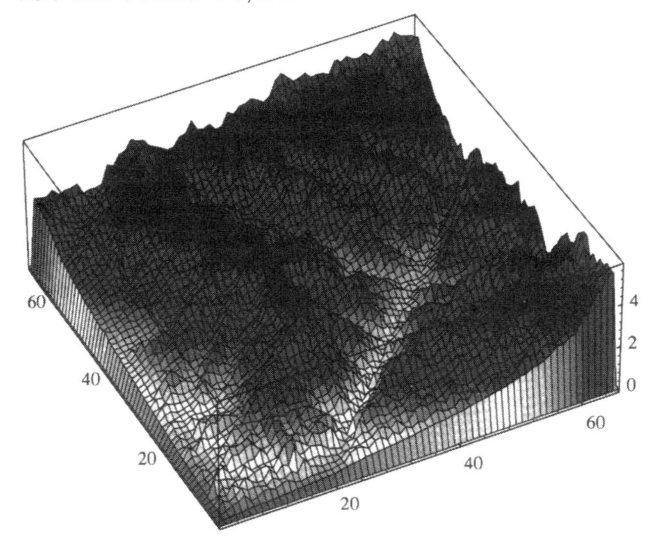

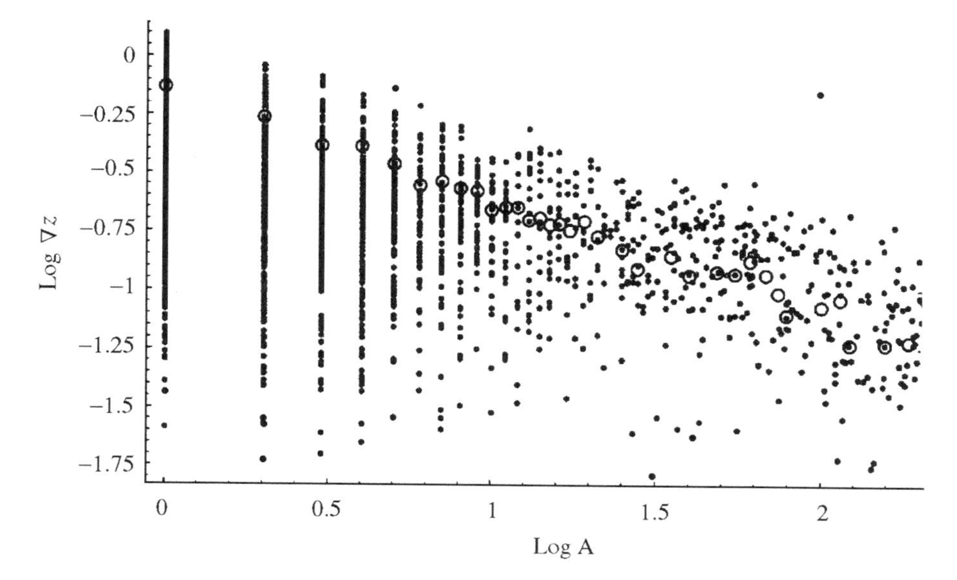

Figure 6.13. Double-logarithmic plot of the slope versus area relationship of the SOC with random threshold field after a diffusion cycle.

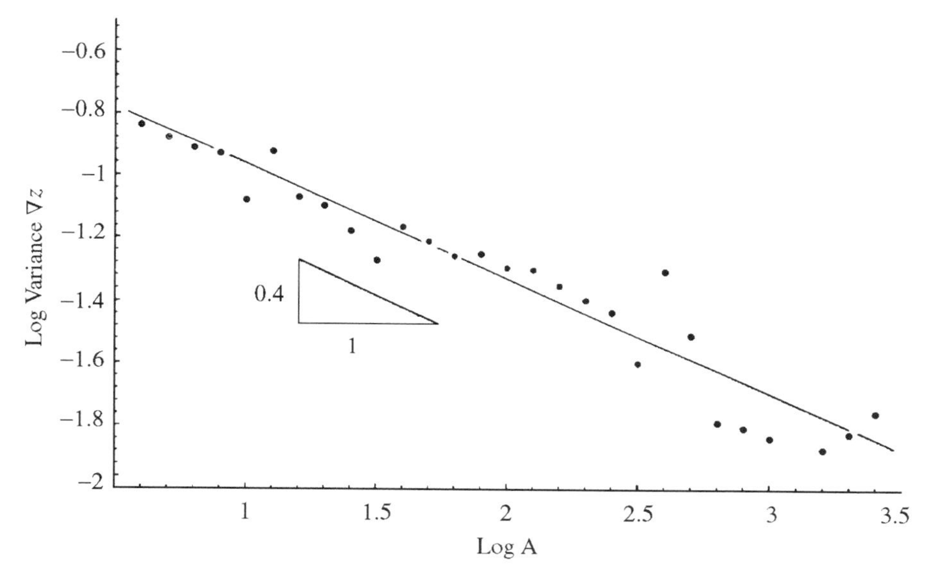

Figure 6.14. Double-logarithmic plot of $Var[\nabla z(A)]$ versus A for the case of a SOC network after a sequence of erosion–diffusion cycles with random threshold ($\sigma_\tau^2 = 1$, $\langle \tau_c \rangle = 1$ and $\lambda_\tau = 32$ pixels).

below the critical threshold for erosion and therefore $\tau \leq \tau_c$ and $Var[\tau] \neq 0$, quite possibly independent of the particular value of A.

The robustness of planar features is noteworthy. It turns out that other indicators, chiefly related to the three-dimensional structure of the basin, are much more sensitive to climatic, geologic, or vegetational changes. The main structure of the fluvial network is relatively insensitive (very much so in a statistical sense and less so in different individual realizations) to effects of spatial disorder or to the mixing of different processes. The dominance of fluvial processes over a large set of scales endows the fluvial basin with recurrent fractal characters that, at large scale, tend to obscure the deviations, rooted at small scale, produced by the concerted actions of disorder and nonfluvial processes.

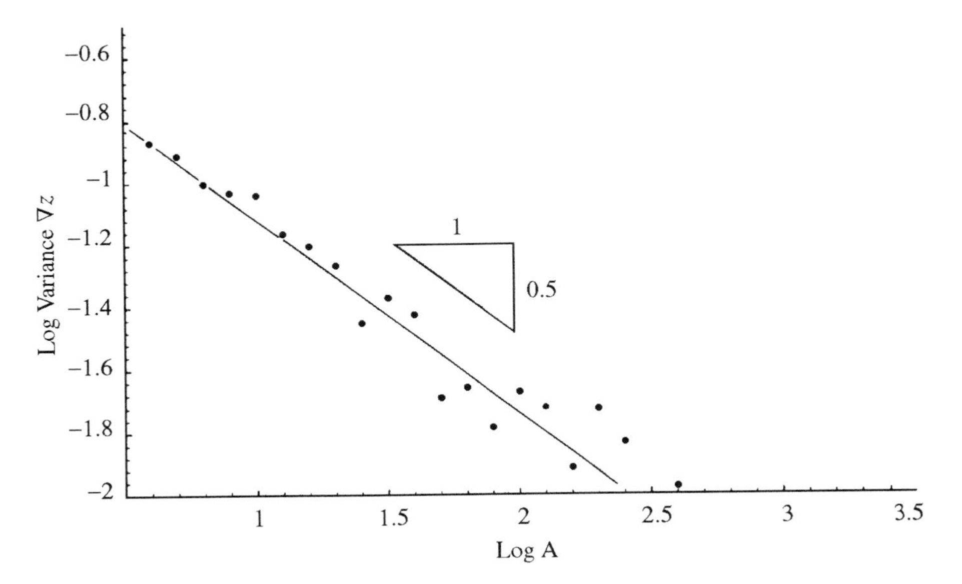

Figure 6.15. Double-logarithmic plot of $Var[\nabla z(A)]$ versus A for the case of SOC network (after a sequence of erosion–diffusion cycles) with random threshold ($\sigma_\tau^2 = 0.3$, $\langle \tau_c \rangle = 1$ and $\lambda_\tau = 2$ pixels).

We now proceed to a refinement of the analysis of the important effects of disorder.

Figures 6.16, 6.17, and, 6.18 show larger (128 × 128) SOC landscapes produced via (i) a uniform threshold $\tau_c = 1$; (ii) random threshold fields, after fluvial (TL) processes alone; and (iii) after a cycle of diffusive (SD) actions, respectively.

Figure 6.19 shows the slope–area statistics relative to the evolving landscapes shown in Figures 6.17 and 6.18. It is clear that the combined effects of TL and SD erosion have the ability to modify significantly the scaling relationship for small contributing areas. This is consistent with the field observations discussed in Chapter 1. Figure 6.20 shows the resulting variances. Values relatively close to unity for the slopes of the double logarithmic plots (i.e., 0.85 ± 0.05 and 0.95 ± 0.05) are obtained. When subject to ensemble averaging over a few realizations, the scaling of the variance tends consistently to a unit value, as observed in Chapter 5 for fluvial processes.

A more complex situation is observed when the threshold field τ_c is not constant but is described by a spatially random field, $\tau_c(\mathbf{x})$, with a prescribed spatial correlation. In the lognormal case, the degree of heterogeneity is controlled by the variance of the field and by the correlation scale. Figures 6.21 and 6.22 show the scaling properties of two sample SOC fields among thirty simulated landscapes generated from the same initial conditions with statistically different characterizations of $\tau_c(\mathbf{x})$. The only difference between the two cases in Figures 6.21 and 6.22 is the correlation scale, λ_τ; the latter is much more correlated in space than the former, thus resulting in a reduced degree of spatial irregularity in the properties determining soil erodibility.

From the above simulations, we observe that the scaling of the mean value $\langle \nabla z(A) \rangle$ preserves the power law in a robust manner, that is, $\langle \nabla z(A) \rangle \propto A^{-0.5 \pm 0.01}$, with the exception of small contributing areas where the hillslope processes reproduce the typical crossover effect shown in real basins [Montgomery and Dietrich, 1992].

Figure 6.16. A 128×128 self-organized landscape generated from random initial conditions with $\tau_c = 1$ and $\nu = 0.2$. Channeled sites are defined in areas of convergent topography ($\nabla^2 z \geq 0$).

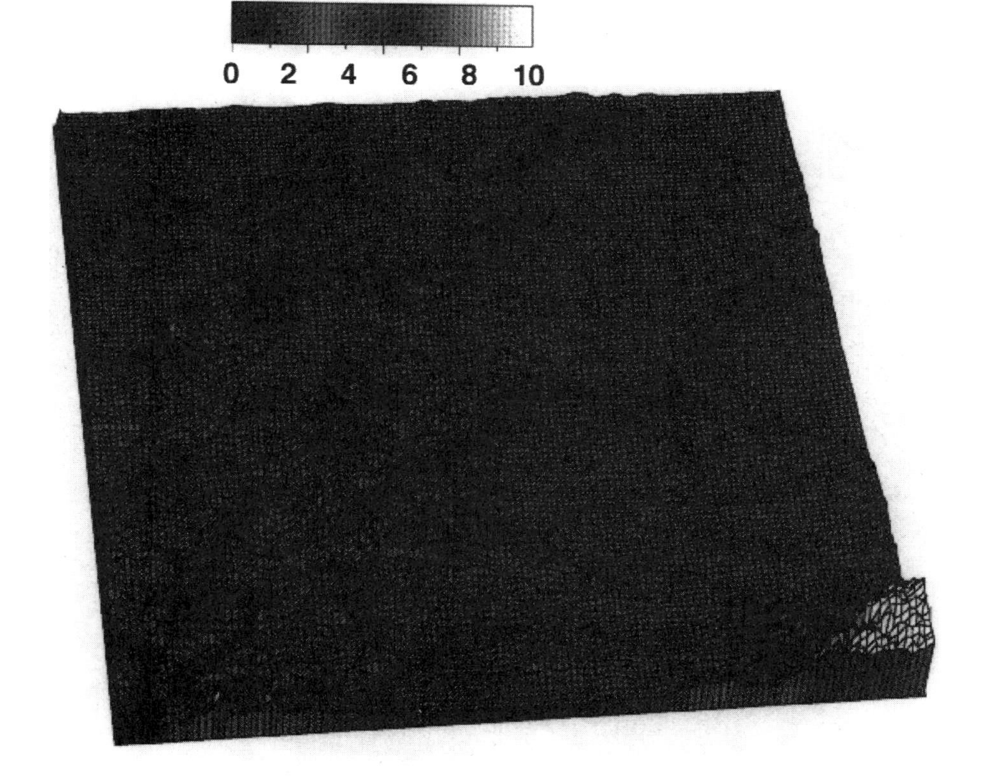

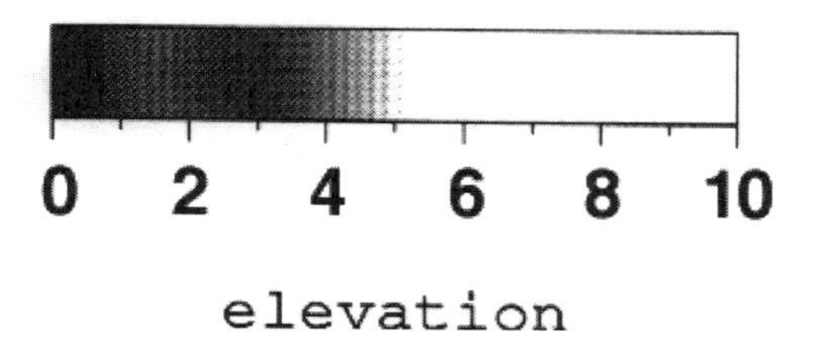

Figure 6.17. Fluvial landscape produced by a random threshold field drawn from a lognormal distribution where only TL processes are allowed to occur ($\langle\tau_c\rangle = 1, \sigma_\tau^2 = 0.2, \lambda_\tau = 10$ in pixel units).

Figure 6.18. Landscape produced by relaxing the z field in Figure 6.17 by a diffusive cycle.

The variances scale with an exponent dependent on the degree of heterogeneity. In the case of Figures 6.21 and 6.22, the exponents are in the range 0.4–0.6. A summary of the results of these experiments for lognormal $\tau_c(\mathbf{x})$ fields with different variances and correlation scales λ_τ is presented in Table 6.2. In the table we give, for a significant choice of heterogeneous cases (described by the correlation scale λ_τ, the mean $\langle \tau_c \rangle$, and the variance σ_τ), the exponents derived from the scaling of the variance (a) after fluvial erosion only (θ_{TL}, i.e., $Var[\nabla z(A)] \propto A^{-\theta_{TL}}$) and (b) after both fluvial and hillslope erosion ($Var[\nabla z(A)] \propto A^{-\theta_{TL+SD}}$). One infers that the multiscaling behavior does not seem so sensitive to the degree of heterogeneity. This can be deduced from the values of θ for all cases when $\sigma_\tau \neq 0$. The role of disorder is important with respect to multiscaling behavior. When $\sigma_\tau = 0$, θ is considerably larger than 0.5 and, in fact, not very different from the unit value even after pronounced SD actions.

Although the above results depend on particular assumptions (e.g., the mathematical description of heterogeneity or the fact that we have not considered heterogeneous hillslope processes owing to their smaller scale of influence), the above conclusions might indeed have general character.

We will now briefly discuss the effects of coarse graining on scaling properties of the elevation field, following the ideas presented in Section 5.11. In that section we suggest that poor resolution in DEM measurements (corresponding to a coarse-grained version of an original field) induces scatter in the scaling relationships of elevations. Such scatter selectively affects the different moments. Figure 6.23 shows the scaling of the mean slopes for coarse-grained versions of the landscape in Figure 6.16. In spite of the increasing scatter, the slope is substantially maintained even for major degrees of coarse graining.

Figure 6.24(a) shows the effects of coarse graining on the scaling of the variances for the same data analyzed in Figure 6.23. Clearly, the effects of coarse graining applied to a basin where τ_c is spatially constant alter the original scaling but seem to preserve some scaling structure. The scaling exponents depend on the stage of the coarse-graining procedure. However, when they are applied to a landscape generated with spatially random $\tau_c(\mathbf{x})$ (Figure 6.18), the change in the scaling exponent is much less evident (Figure 6.24(b)). This suggests that coarse graining produces effects equivalent to those due to the introduction of random heterogeneities and thus reinforces the suggestion that simple scaling slope–area statistics are unlikely occurrences for the fluvial landscape for which heterogeneities are unavoidable.

Finally, a possible effect of macroscopic heterogeneities is shown in Figure 6.25. To generate such landforms a bimodal threshold field, basically reproducing two homogeneous zones with different erodibility, was

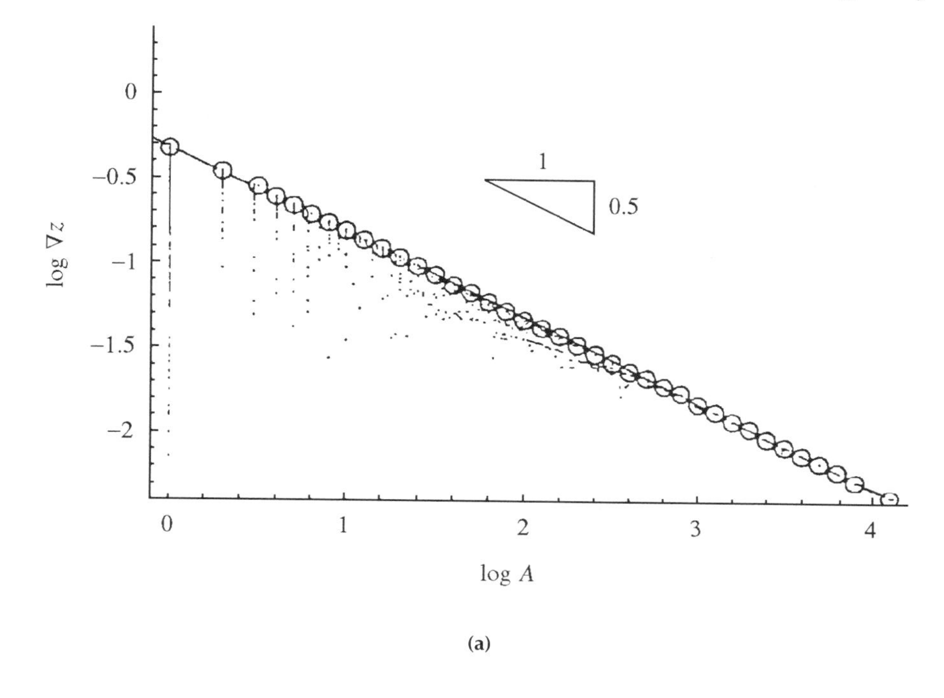

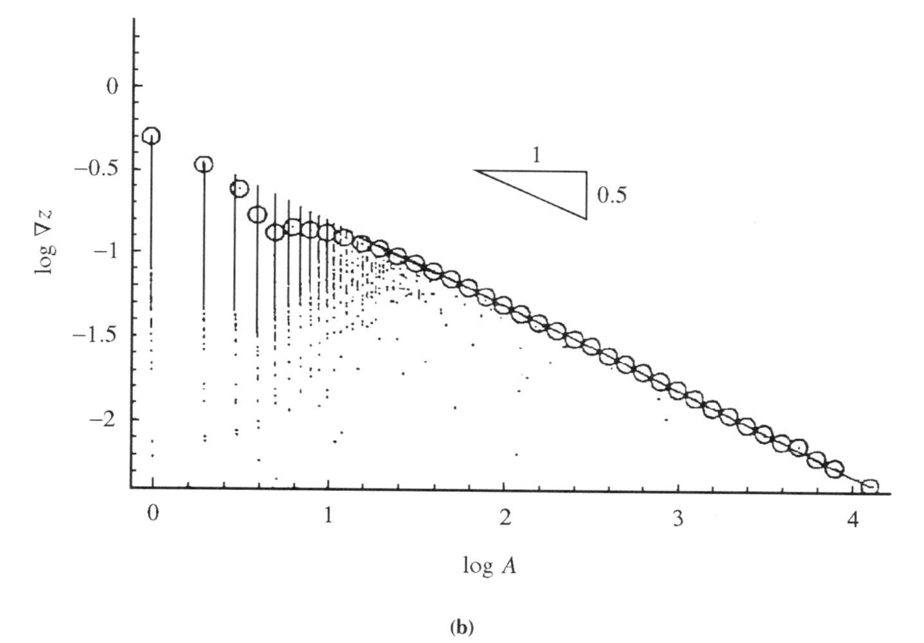

Figure 6.19. Scaling of the mean local slope $\nabla z(A)$ with A: (a) after fluvial (TL) erosion (Figure 6.17); and (b) after both TL and SD processes (Figure 6.18).

assigned by dividing the domain into two distinct parts along the diagonal issuing from the outlet. The upper part of the basin is characterized by a high threshold, $\langle \tau_c \rangle = 2$, and the lower part by a low value, $\langle \tau_c \rangle = 0.5$. Figure 6.25 shows the result of a complete cycle of TL and SD erosion. Interestingly, the arrangement of zones with different erodibility causes the formation of a cliff that stands out, as in Gilbert's law.

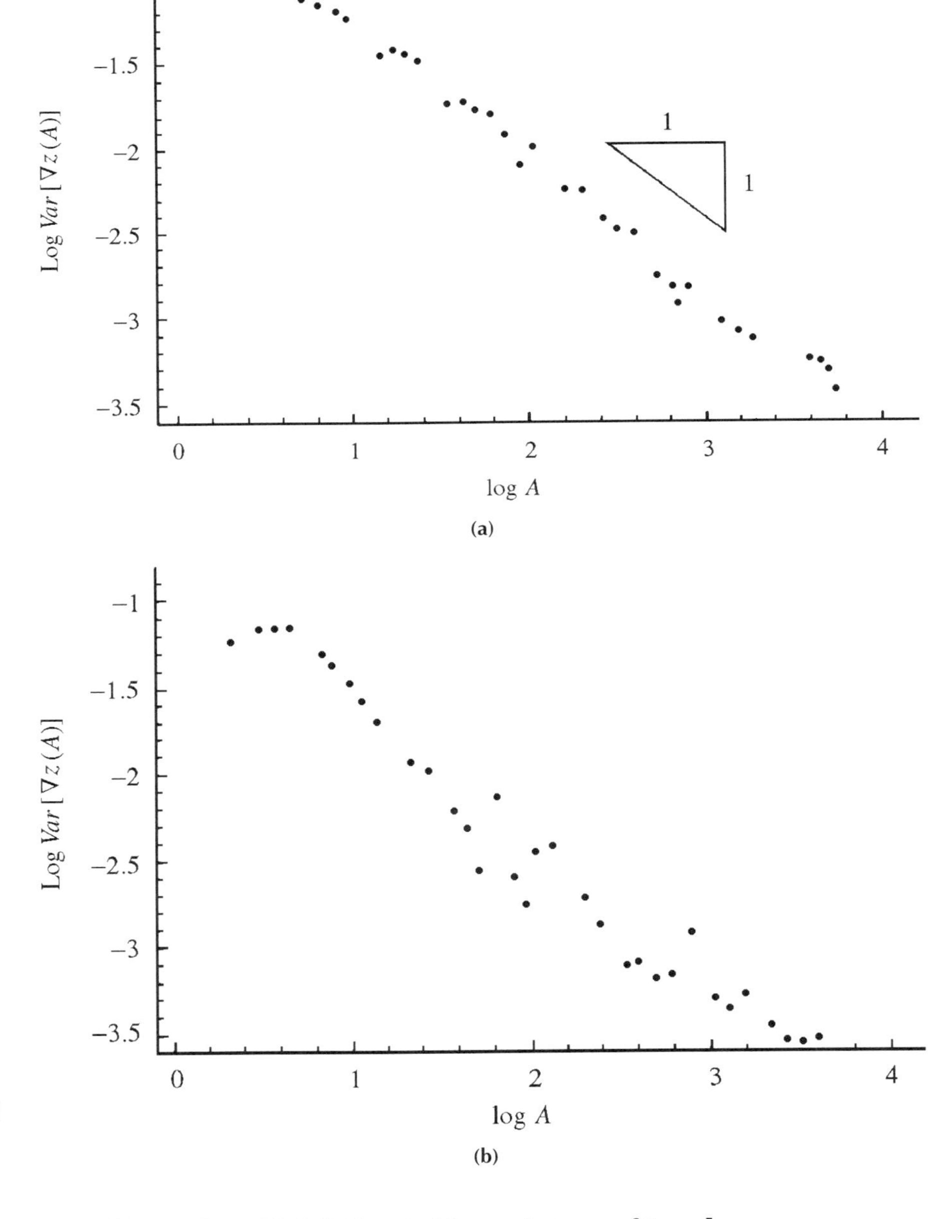

Figure 6.20. Scaling of the variances of the local slopes: (a) after homogeneous fluvial (TL) erosion; and (b) after both TL and SD processes.

6.5 Fractal and Multifractal Descriptors of Landscapes

This section describes a systematic analysis of the fractal properties of the topographies resulting from coupled TL and SD processes and a comparison with measures of the roughness of natural terrains.

In Chapter 5 we studied the fractal characters of the fluvial landscape. Among the salient conclusions, we noted that the structure of the elevation field (i.e., the *landscape*) resulting from a TL self-organized process can be reasonably well described by a single fractal dimension. Such a fractal dimension may be deduced from the self-affine transects (obtained

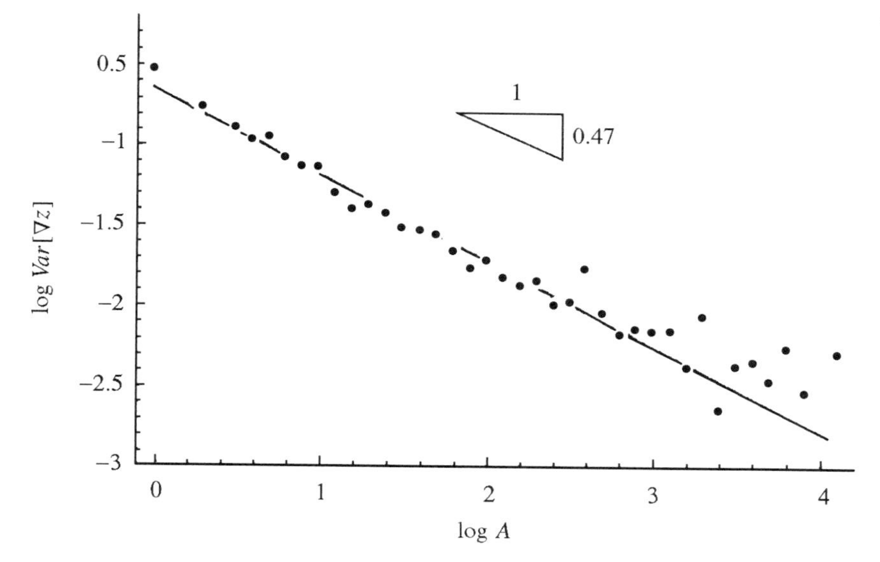

Figure 6.21. Double-logarithmic plot of $Var[\nabla z(A)]$ versus A from a landscape with heterogeneous threshold $\tau_c(\mathbf{x})$ ($\langle \tau_c \rangle = 0.5$, $\sigma_\tau = 0.6$ and λ_τ is 4 pixels).

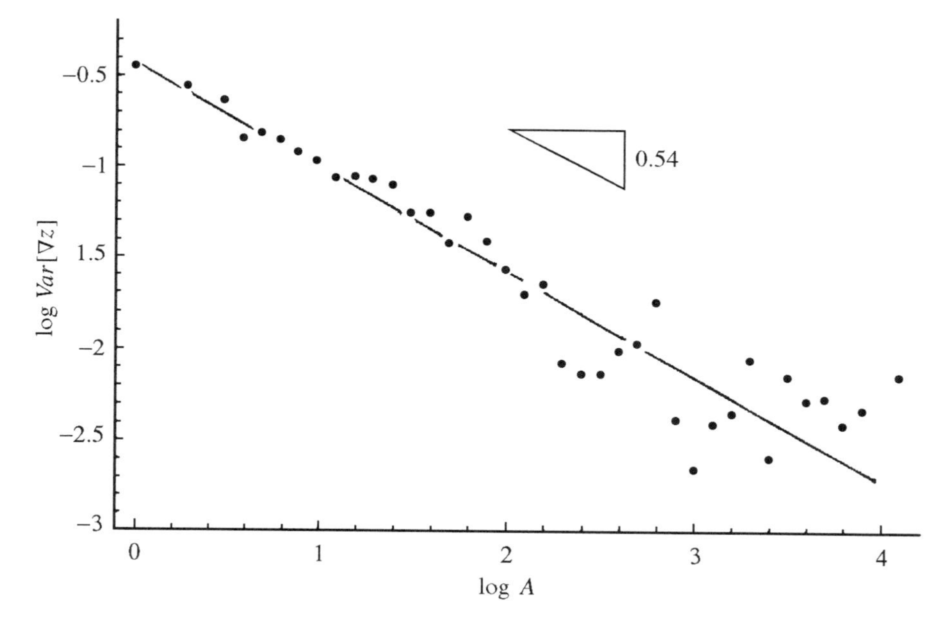

Figure 6.22. Double-logarithmic plot of $Var[\nabla z(A)]$ versus A from a landscape with heterogeneous threshold $\tau_c(\mathbf{x})$ ($\langle \tau_c \rangle = 0.5$, $\sigma_\tau = 0.6$ and λ_τ is 32 pixels).

by intersections of the surfaces with parallel vertical planes) through different algorithms. It is now of interest to examine the impact of coupled fluvial and hillslope processes on the fractal characters of the resulting landscapes.

One of the relevant effects of the hillslope processes is the lowering of the overall fractal dimension of the landscape. More precisely, we will show that the relaxation of the fluvial (TL) landscape produced by the mass-conserving diffusive scheme, Eq. (6.50), allows the system to evolve through fractal states characterized by a progressively lower dimension. This is a significant achievement of Eq. (6.50) in view of the fact that a simple linear diffusive scheme would have destroyed the scaling properties of

Table 6.2. *The exponents of the relationship* $Var[\nabla z(A)] \propto A^{-\theta}$ *for a set of numerical experiments*

λ_τ	$\langle \tau_c \rangle$	$\sigma_\tau / \langle \tau_c \rangle$	θ_{TL}	θ_{TL+SD}
∞	0.5	0	0.80	0.90
∞	1	0	0.84	0.95
∞	2	0	0.84	–
32	0.5	0.6	0.54	0.44
32	1	0.4	0.54	0.48
32	2	0.2	0.56	–
16	0.5	0.6	0.55	0.51
8	0.5	0.6	0.62	0.56
4	0.5	0.6	0.47	0.43
4	0.5	0.4	0.49	–
4	0.5	0.2	0.52	–

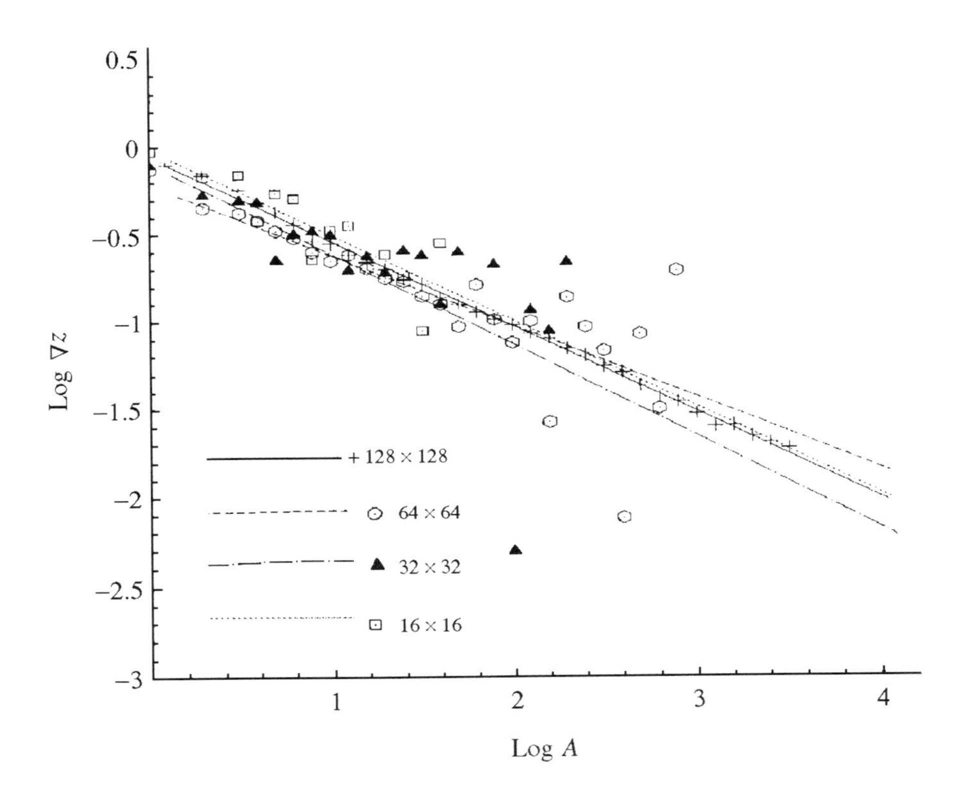

Figure 6.23. Expected values of $\langle \nabla z(A) \rangle$ versus A during coarse graining of the landscape in Figure 6.16.

the elevation field unless it was anisotropic and driven by additive random noise (Section 6.2).

We will first assume a simple-scaling framework for which the use of a local fractal dimension of a self-affine track is a meaningful concept. Later in this section we will relax the simple-scaling constraint via a generalized procedure.

We will first discuss the fractal properties of fluvial and coupled landscapes by analyzing the properties of their (possibly self-affine) transects. Table 6.3 [after Rigon et al., 1994] shows the fractal dimensions computed

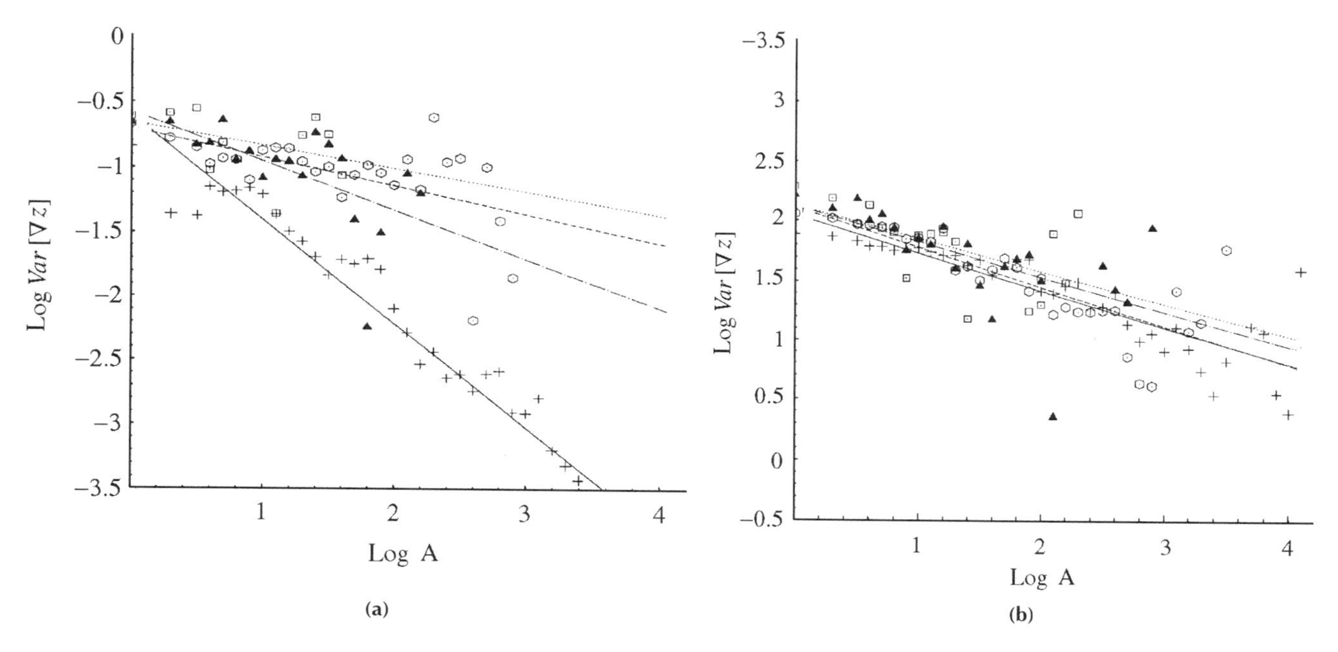

Figure 6.24. Scaling of $Var[\nabla z(A)]$ under coarse graining: (a) landscape generated by constant τ_c (Figure 6.16); (b) landscape generated by spatially random $\tau_c(\mathbf{x})$ (Figure 6.18). Legends are as in Figure 6.23.

by two different procedures for a sample of sixty-four transects taken in the west–east direction in the basins of Figures 5.9 and 6.4. These basins correspond to a purely fluvial (TL) landscape and to a coupled (TL+SD) landscape, respectively. The TL+SD landscape, as shown in Section 6.3, was obtained by relaxing the fluvial landscape of Figure 5.9 via many diffusive cycles. The numbering of the tracks indexes the transects taken from the southern boundary (track # 1) to the northern boundary (track # 64).

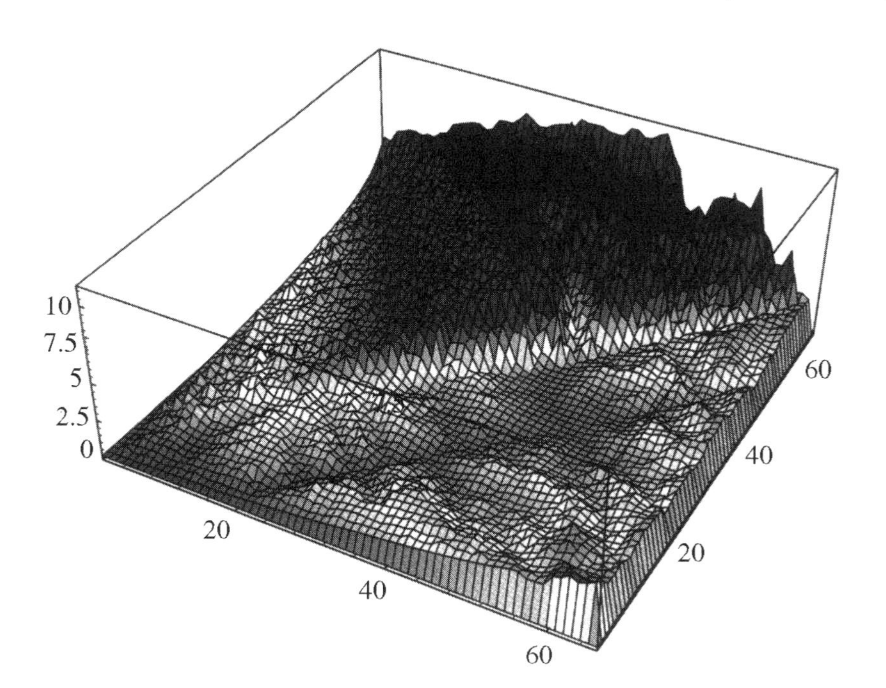

Figure 6.25. SOC with diffusion obtained from the initial condition in Figure 5.16 when a bimodal threshold field is employed. Less erodible parts stand out and a cliff is created [after Rigon et al., 1994].

Table 6.3. *Fractal dimensions for sample tracks of a TL landscape (Figure 5.9) and a TL+SD landscape (Figure 6.4)*

Track #	$D_{MV} \pm 0.05$ TL landscape	$D_{MO} \pm 0.25$ TL landscape	$D_{MV} \pm 0.05$ TL+SD landscape	$D_{MO} \pm 0.25$ TL+SD landscape
1	1.3	1.5	1.3	1.5
2	1.4	1.6	1.3	1.5
3	1.5	1.5	1.3	1.4
10	1.5	1.6	1.4	1.4
20	1.5	1.5	1.4	1.4
30	1.7	1.8	1.6	1.8
40	1.6	1.8	1.6	1.7
41	1.7	1.8	1.6	1.8
42	1.7	1.8	1.6	1.7
43	1.7	1.8	1.6	1.7
44	1.7	1.8	1.6	1.7
45	1.7	1.8	1.6	1.7
50	1.7	1.8	1.6	1.7
60	1.7	1.8	1.6	1.7
64	1.5	1.7	1.5	1.6
$\bar{D}$	1.6	1.7	1.5	1.6
$Var[D]$	0.1	0.2	0.1	0.2

Note: D_{MV} is the estimate by the method of variations, D_{MO} by that of Matsushita and Ouchi.
Source: Rigon et al., 1994.

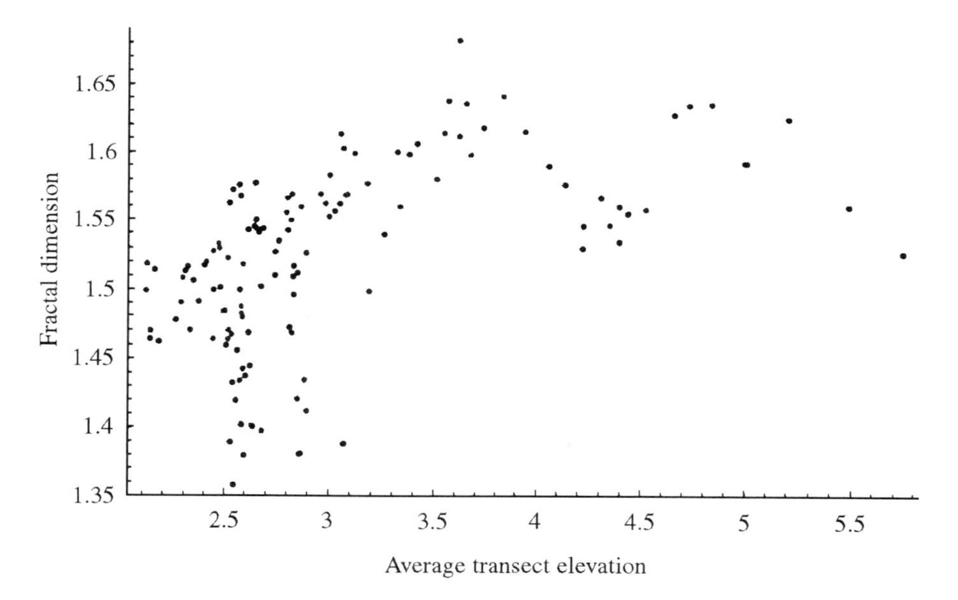

Figure 6.26. Fractal dimensions of individual tracks plotted against their mean elevation.

The estimates of fractal dimensions are obtained through the variations method (see Section 2.7.4) and Matsushita and Ouchi's method (see Section 2.7.6).

Notice that the values of fractal dimensions estimated for tracks at (or near) the boundaries are significantly lower than the dimensions of those relatively unaffected by them. We also note (Figure 6.26) that no clear

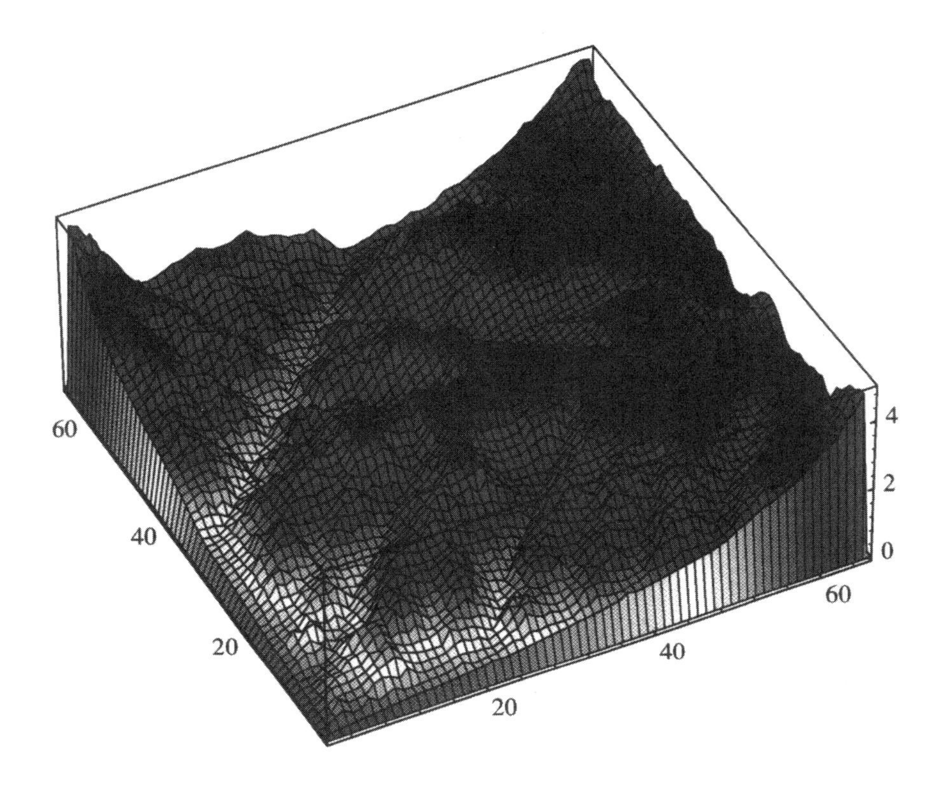

Figure 6.27. The smoothed landscape as it appears after a long cycle of diffusion–erosion processes, starting from the comb-like initial condition of Figure 5.16. The fractal dimensions of the transects are generally close to 1.4.

correlation seems to exist between the roughness of the track and its average elevation.

The results in Table 6.3 suggest that slope-dependent diffusion processes render the landscapes more realistic by lowering the fractal dimensions of the surfaces produced by purely fluvial processes. The reduction caused by SD processes points toward values similar to those observed in nature for both fluvial and diffusive landscapes which were described in Chapter 2.

The result of a longer sequence of diffusion–erosion cycles acting on the comb-like initial conditions (Figures 5.15 and 5.16) is shown in Figure 6.27. The final fractal dimension in this case is 2.4 ± 0.1.

Figure 6.28 shows, for the landscape in Figure 5.9, the sets of points that belong to different ranges in elevation for a SOC landscape obtained when only TL actions are operating, that is, $z(\mathbf{x}, t_\alpha)$, where t_α represents the time of occurrence of an erosion event, assumed to be instantaneous (see Eq. (6.48)). The initial condition $z(\mathbf{x}, 0)$ was obtained by Eden growth. Notice that for the final state we assume the notation $z(\mathbf{x}, t \to \infty) = z(\mathbf{x})$. Here $\tau_c = 1$ on a 128×128 lattice. The solid black area in Figure 6.28 refers to the set $S_{\leq 2}$, that is, the set of points whose elevation is lower than 2 ($z(\mathbf{x}) \leq 2$). Different levels of grey indicate sets S where $2 \leq z(\mathbf{x}) \leq 3$ and so on.

Figure 6.29 shows contour lines of the landscape $z(\mathbf{x}, t_\beta)$ (with t_β defined by 10,000 diffusive relaxations) obtained from a cycle of SD processes acting from the initial condition $z(\mathbf{x}, t_\alpha)$, whose contour lines are shown in Figure 6.28. The resulting field is also represented through border sets in Figure 6.30.

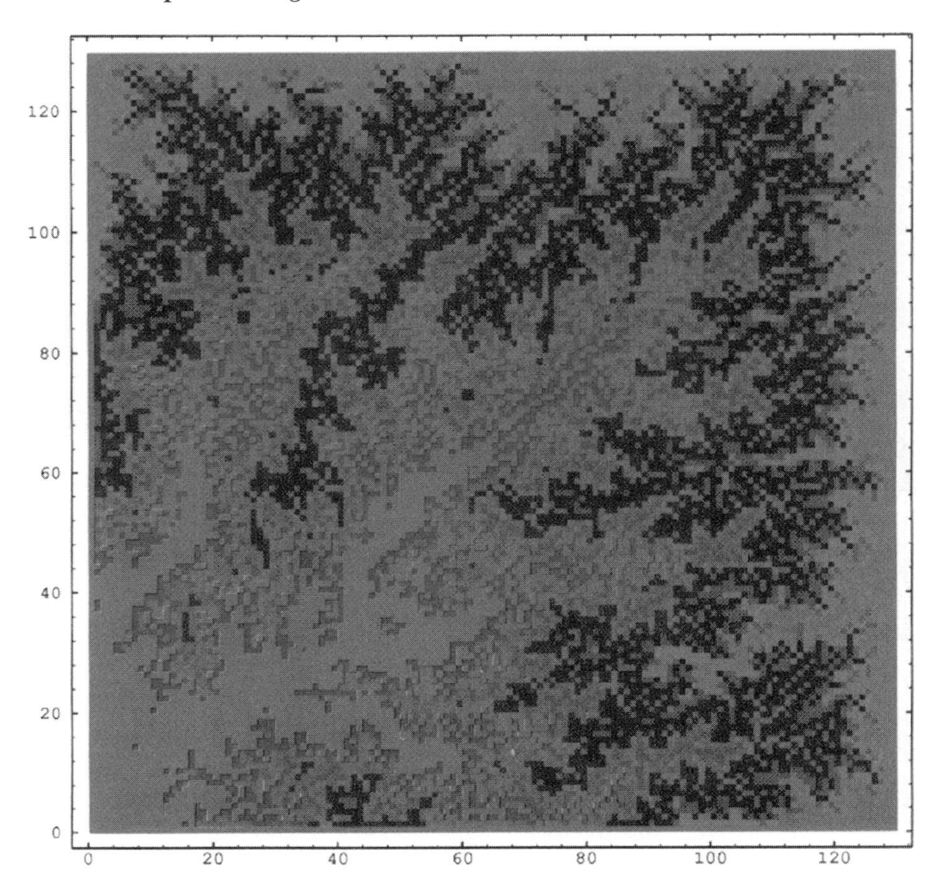

Figure 6.28. Plot of the set of points S that belong to predetermined ranges of the TL landscape in Figure 5.9 [after Rodriguez-Iturbe et al., 1994].

Box counting has been applied to the sets shown in Figures 6.28 and 6.29 to determine the fractal dimension of each contour line. An example of the related results is shown in Figure 6.31. Box-counting data are represented by a straight line; that is, the slope in a double-logarithmic plot of the number of boxes $N(\delta)$ that cover the whole contour versus the box size δ is the fractal dimension $D(P_T)$. This indicates that any contour of a self-organized landscape may be considered a self-similar fractal. The main results of the analysis are illustrated in Table 6.4.

Rodriguez-Iturbe et al. [1994] noted that the fractal dimensions of the isolevel lines obtained for different contour elevations are slightly different. Similar results were observed for the optimal topographies obtained by Sun et al. [1994b] by enforcing an artificial perfect scaling of slopes on an optimal channel network. The contours near the average height have very similar fractal dimensions, and overall the surfaces produced seem to show the signs of a relatively complex self-affine fractal. However, in the above analysis sample size effects are important both at the higher and at the lower elevations, and thus partially impair the related results. In the range of elevations where the sample size is constant we observe a consistent behavior analogous to that observed in the topographies obtained by Sun et al. [1994b], which only involve a fluvial behavior.

Other methods of analysis also agree with the previous results. Figure 6.32 shows the double-logarithmic plot of the power spectrum $\langle S(f)\rangle$

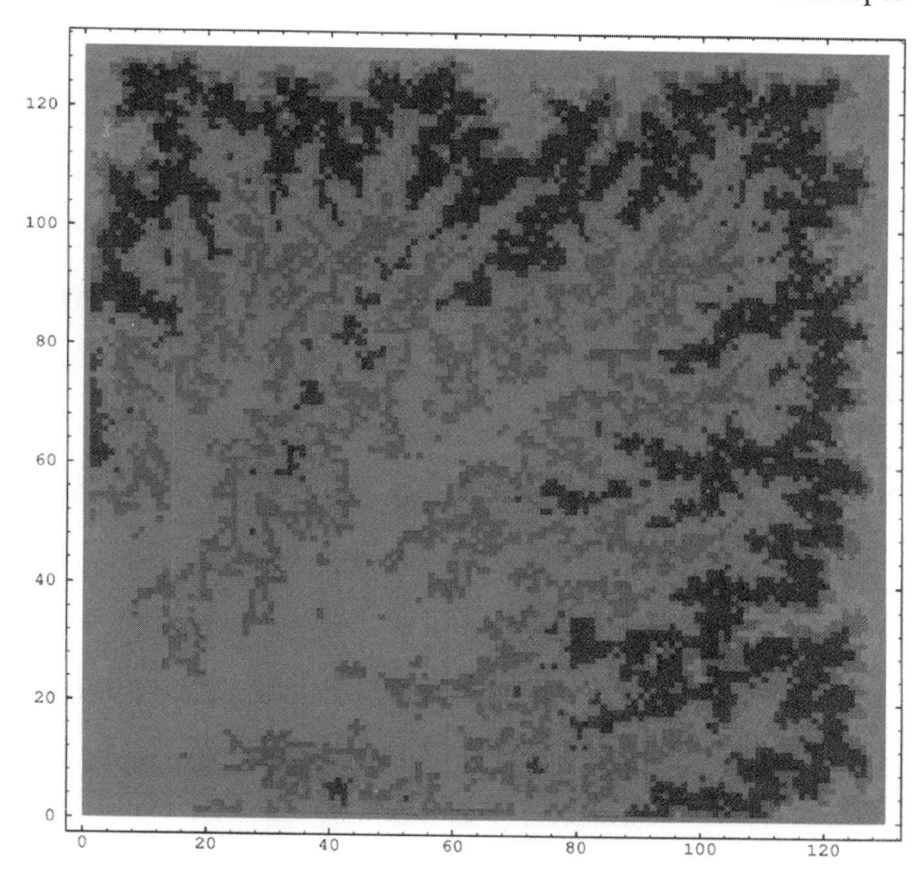

Figure 6.29. Plot of the set of points S that belong to predetermined ranges of elevation of the TL + SD landscape in Figure 6.4 [after Rodriguez-Iturbe et al., 1994].

averaged over all south–north (S–N) and east–west (E–W) topographic profiles obtained from vertical transects through a SOC landscape produced by TL processes. The data can be fitted to a straight line (i.e., $S(f) \propto f^{-\zeta}$) with slope $\zeta = 1.6$, giving $D = (5 - \zeta)/2 = 1.7 \pm 0.1$ for the mean fractal dimension of the transects.

The range of variation of the height z within a box of size L scales as $\propto L^H$ $(D = 2 - H)$ if the object is simple scaling. After averaging over many transects is performed, the resulting fractal dimension in a 128×128 landscape computed through the variation method is $D = 1.65 \pm 0.05$.

Mean deviation analysis, first introduced for self-affine fractals in Section 2.7.4, studies the deviation $\sigma^2(L)$ as a function of the horizontal length scale L as

$$\sigma^2(L) = \left\langle \frac{1}{L^2} \sum_{i,j=1}^{L} (z_{ij} - \langle z \rangle_L)^2 \right\rangle \tag{6.55}$$

where z_{ij} is the elevation at the lattice sites $\mathbf{x} = (i, j)$ and $\langle z \rangle_L$ is the average height of the surface within a grid of size $L \times L$. For a self-affine fractal, σ is related to L by a power law $\sigma \propto L^H$. The dependence of σ on L for a TL landscape yields $D = 1.66 \pm 0.05$.

We have repeated the above analyses for the landscape in Figure 6.4 which incorporates TL + SD processes and have obtained consistent results for the mean fractal dimensions. Nevertheless, we observe a

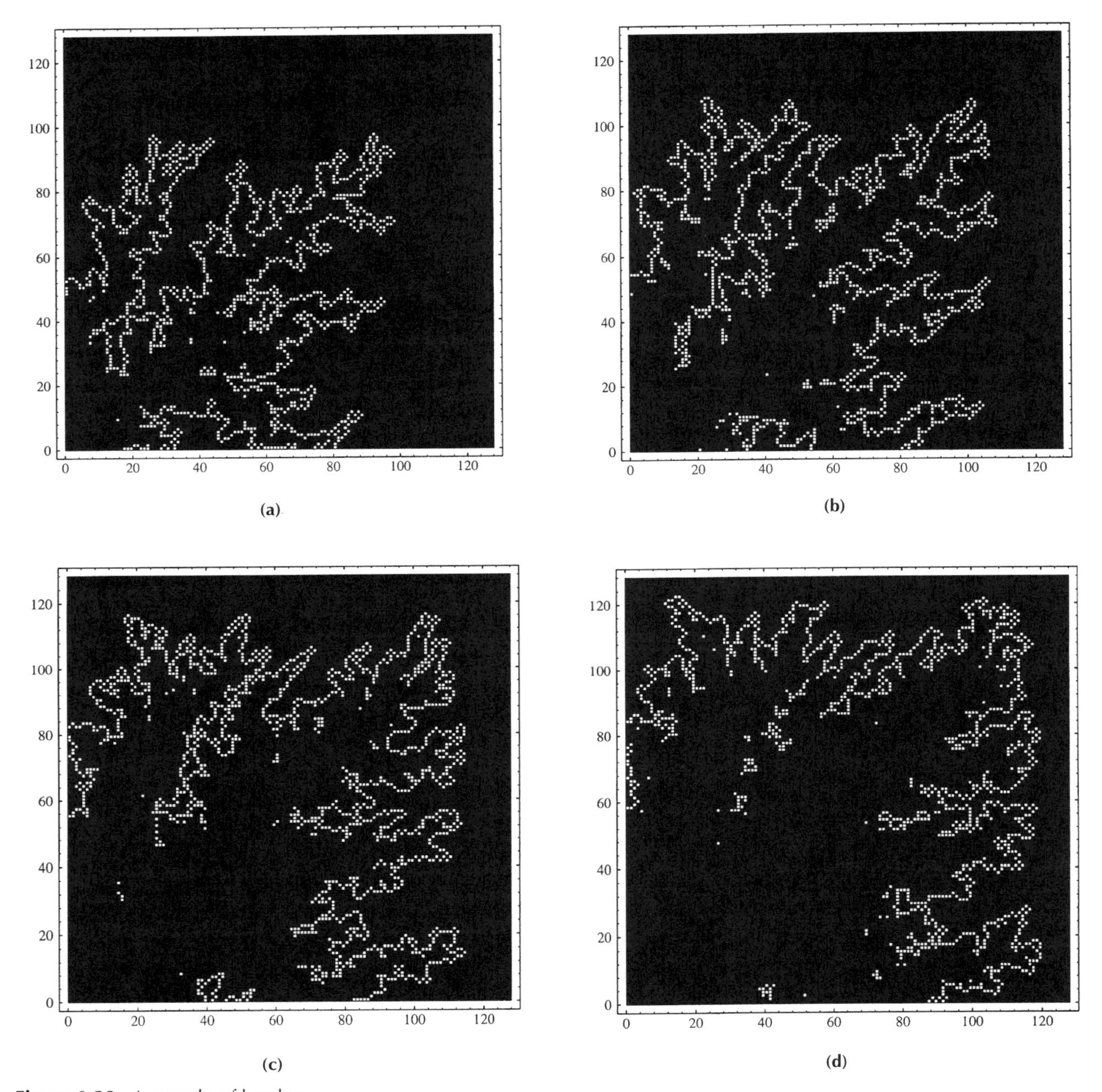

Figure 6.30. A sample of border sets in Figure 6.4 for $z = 2, 2.5, 3$, and 3.5, respectively. The resulting fractal dimensions are 1.30, 1.29, 1.26, and 1.22.

nonnegligible range in the individual fractal dimensions of the topography transects and in the contour lines at different elevations. Figure 6.33 shows a plot of the fractal dimensions for different E–W tracks of TL and TL + SD landscapes. One notices that the self-affine characteristics of the TL + SD surfaces are obviously smoother than those of the TL landscapes.

As already seen in Section 3.6.2, the qth moment of a range, or generalized variogram, $C_q(\lambda)$, is defined as

$$C_q(\lambda) = \langle |z(\mathbf{x}) - z(\mathbf{x} + \mathbf{r})|^q \rangle_{|\mathbf{r}|=\lambda} \tag{6.56}$$

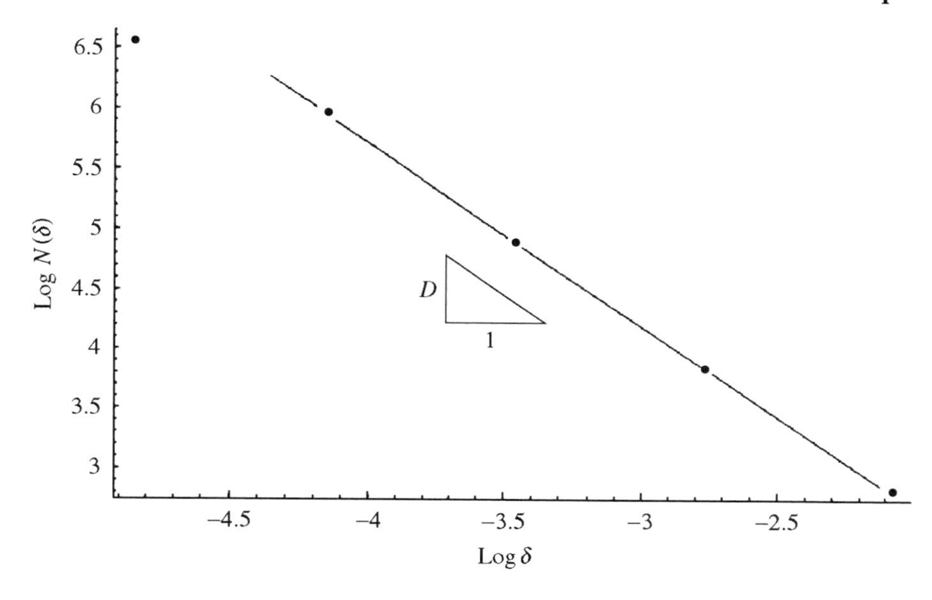

Figure 6.31. Typical dependence of $N(\delta)$ on δ found in the box-counting procedure applied to one of the isolines of Figure 6.30 [after Rodriguez-Iturbe et al., 1994].

Table 6.4. *Elevations, fractal dimensions of the border sets, and correlation coefficients (r) of box-counting analysis for TL and TL+SD SOC landscapes*

$z(\mathbf{x}) = T$, contour elevation	$D(P_T)$ (TL actions)	r	$D(P_T)$ (TL + SD actions)	r
1	1.38	0.99	1.30	0.99
1.5	1.42	0.96	1.30	0.98
2	1.43	0.98	1.30	0.98
2.5	1.43	0.94	1.29	0.97
3	1.39	0.96	1.26	0.97
3.5	1.37	0.99	1.22	0.96
4	1.35	0.92	1.21	0.94

Source: after Rodriguez-Iturbe et al. [1994]

where λ is the separation distance. The qth moment will be said to be scaling if the following relation holds:

$$C_q(\lambda) = \lambda^{K(q)} C_q(1) \quad \propto \lambda^{K(q)} \tag{6.57}$$

In the case of simple scaling, the exponent $K(q)$ is linear, that is, $K(q) = qH$.

We have applied the analysis of moments to self-organized landscapes, computing qth moments from averages along both S–N and E–W transects. The scaling properties of the moments obtained from the 128 × 128 TL landscape in Figure 6.16 are shown in Figure 6.34. The lowest correlation coefficient was found to be 0.99. The related $K(q)$ versus q graph is shown in Figure 6.35 where the best linear fit (with zero intercept) has been plotted to make the simple-scaling behavior more evident. The exponent that characterizes the simple scaling was found to be $H = 0.303 \pm 0.001$ in the case of fluvial (TL) processes and $H = 0.368 \pm 0.004$ for TL+SD processes, showing again that SD processes are responsible for consistently lowering the fractal dimension of landscapes (recall that $D = 2 - H$). The results

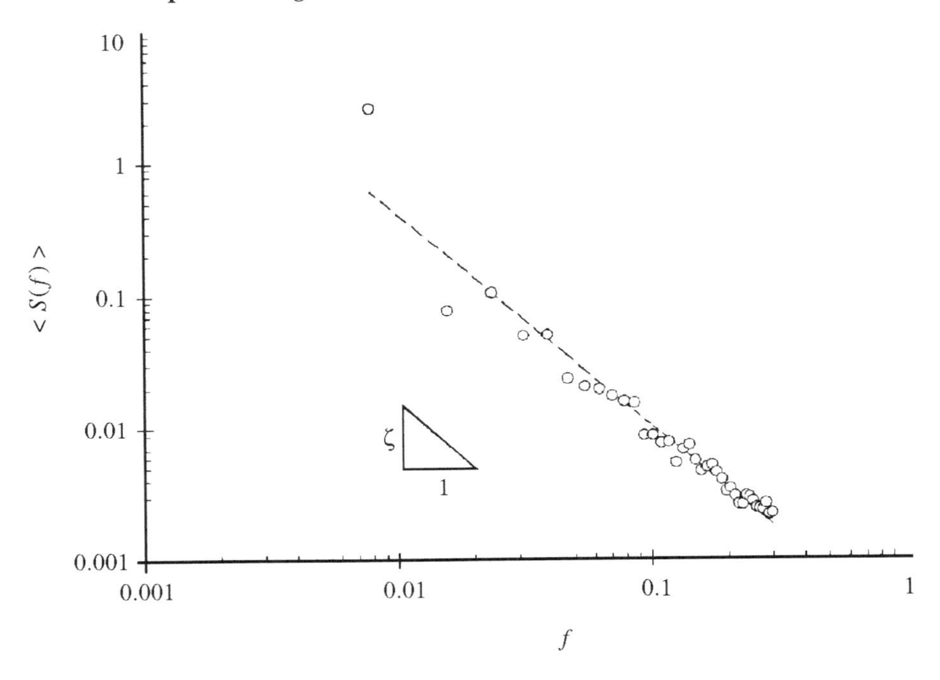

Figure 6.32. Power spectrum averaged over all S–N and E–W transects of the 128 × 128 SOC landscape of Figure 6.16 [after Rodriguez-Iturbe et al., 1994].

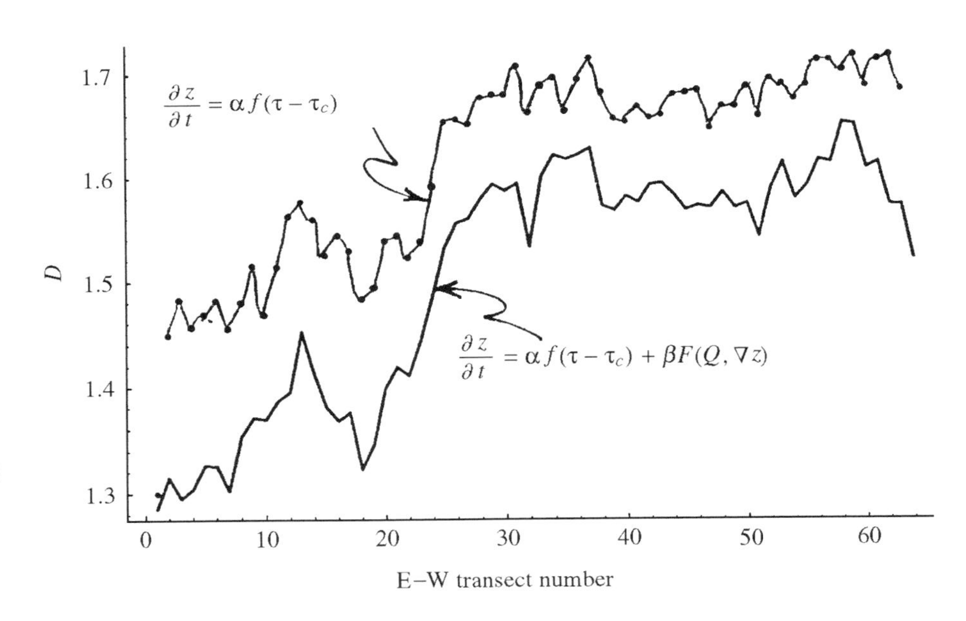

Figure 6.33. Values of the fractal dimension for different E–W transects of TL and TL + SD topographies [after Rodriguez-Iturbe et al., 1994].

obtained through the study of qth moments are consistent with those arising from the analysis of contour lines.

To compare the features described above for theoretical landscapes with those of real data, Rodriguez-Iturbe et al. [1994] have studied a 2,200-km^2 portion of the digital terrain map (DTM) of northern Italy produced by Servizio Geologico Nazionale (pixel size: 250×310 m^2). The results of the literature [Lavallèe et al., 1993] and of our own computations (Figure 6.36) show, quite convincingly, that real topographic fields indeed contain the signatures of multiscaling, that is, a nonlinear relation between $K(q)$ and

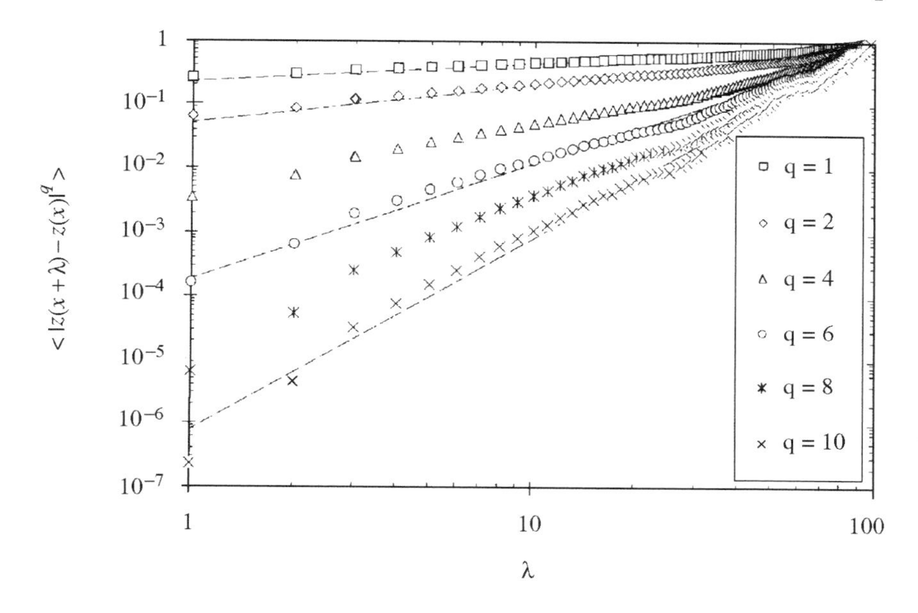

Figure 6.34. Scaling of qth moments for the fluvial elevation field of Figure 6.16 where only TL processes (with a constant τ_c) are allowed to occur [after Rodriguez-Iturbe et al., 1994].

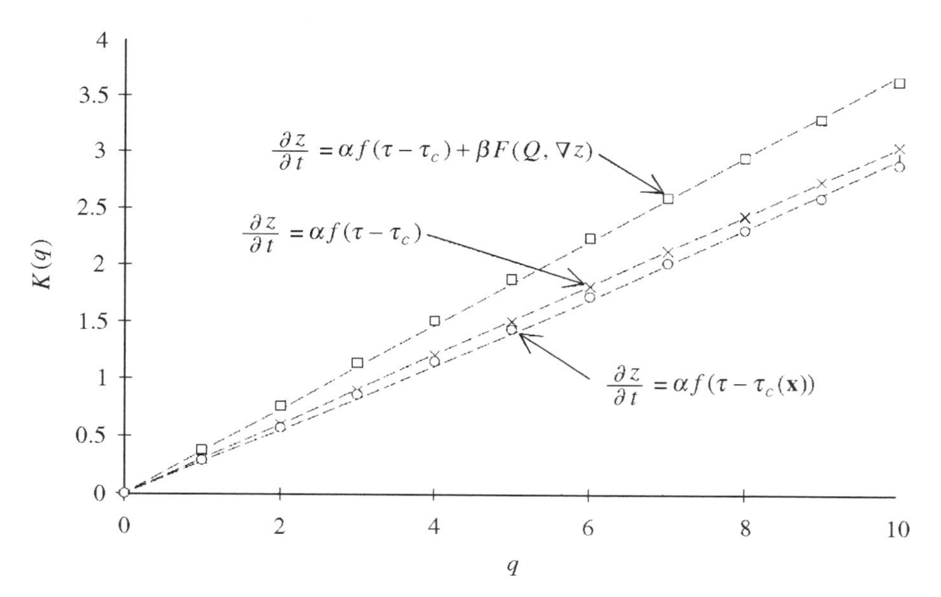

Figure 6.35. $K(q)$ versus q curves for landscapes resulting from TL, TL+SD, and heterogeneous TL processes, respectively [after Rodriguez-Iturbe et al., 1994].

q. This fact implies that the geometric properties of natural landscapes cannot be described in general by specifying a single fractal dimension.

Rodriguez-Iturbe et al. [1994] investigated further the assumption that multiscaling in the field of elevations of the landscape may be a by-product of heterogeneity in the erodibility of the soil mantle. Thus they consider, as in Section 6.4, landscapes in which heterogeneity is modeled through a random spatial distribution of the field parameters. The field equation is in this case

$$\frac{\partial z}{\partial t} = \alpha f\left(\tau - \tau_c(\mathbf{x})\right) \qquad (6.58)$$

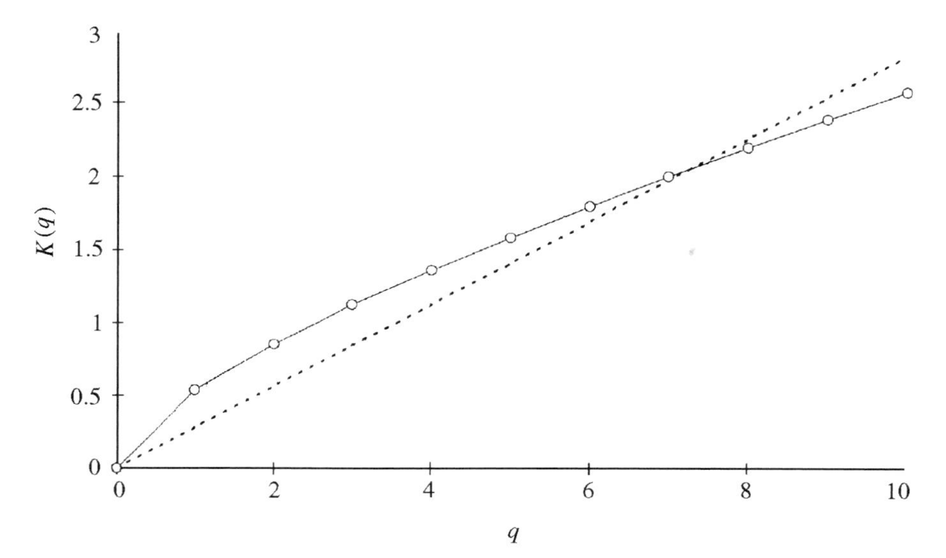

Figure 6.36. $K(q)$ versus q curve for the natural elevation field in northern Italy. A linear best fit (with zero intercept) has been plotted to visualize the multiple-scaling properties evidenced by the field [after Rodriguez-Iturbe et al., 1994].

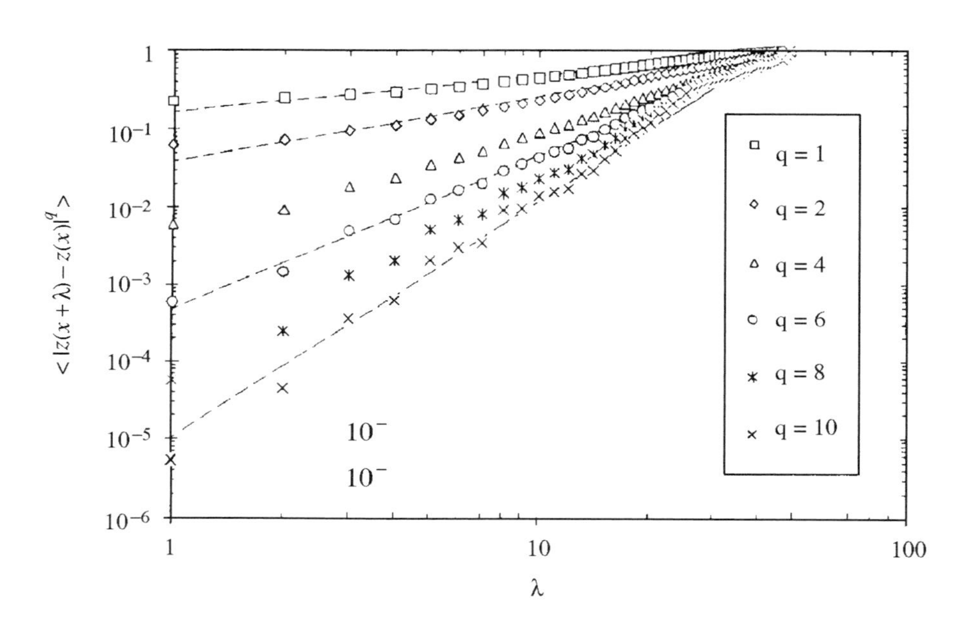

Figure 6.37. Scaling of qth moments for the SOC landscape shown in Figure 6.25 [after Rodriguez-Iturbe et al., 1994].

where the spatial variability of τ_c is described, say, by a lognormal field with an exponential correlation structure. The results of the generalized covariance analysis applied to the heterogeneous landscape in Figure 6.17 are also shown in Figure 6.35.

The effects of mild heterogeneities result in significantly rougher landscapes ($H = 0.280 \pm 0.002$ from Figure 6.35) whose structure may still be considered simply scaling.

To test the effects of greater heterogeneities, the same analysis has been performed on the landscape of Figure 6.25, which is characterized by a bimodal distribution of the critical shear stress. The scaling of the moments and the $K(q)$ versus q relation are shown in Figures 6.37 and 6.38, respectively. It can be seen that moments match a power law scaling behavior

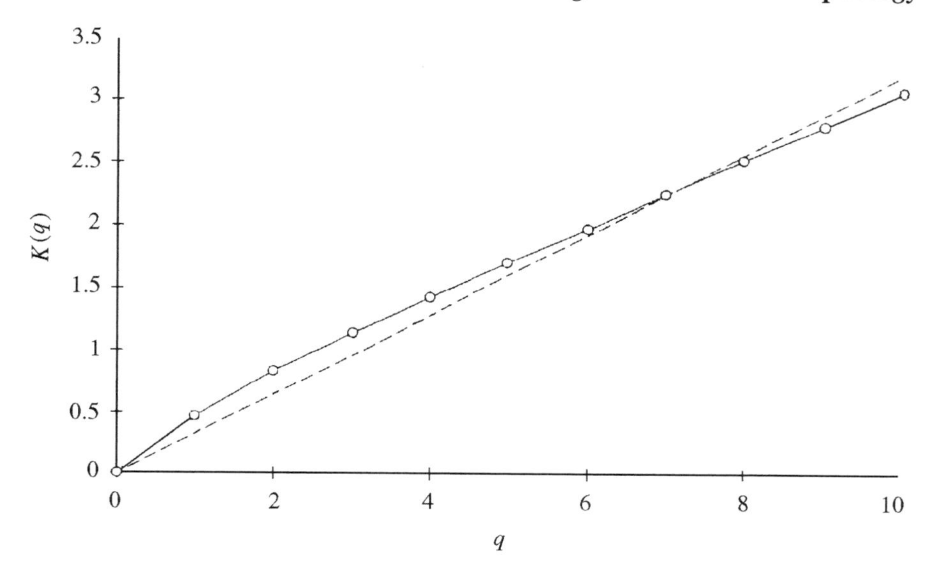

Figure 6.38. $K(q)$ versus q curve for the heterogeneous field of Figure 6.25. A linear best fit (with zero intercept) has been plotted to visualize multiscaling properties.

(the minimum correlation coefficient of the linear regression being 0.96) and that a departure from the simple-scaling relationship is revealed by the form of $K(q)$ versus q. Whether or not this signature of multiscaling is artificially induced by the bimodality remains to be seen.

In conclusion, a spatially variable threshold field creates rougher but still substantially simple-scaling structures. Multiscaling occurred only when a highly nonstationary threshold field (e.g., bimodal) was used. It is likely that pronounced disorder might favor multiscaling behavior. In any case, the type of multiple-scaling behavior experimentally found in natural landscapes cannot be explained in terms of spatially homogeneous processes, and thus heterogeneity of field properties is needed for multiple scaling to emerge.

Also, from the study of the scaling properties of self-organized landscapes elevation fields we observe that their topography appears to be more complex than that of a simple self-affine fractal, although in some cases a simple fractal framework may be adequate for their description. The fact that multiple scaling may indeed emerge as a result of strong heterogeneity in the field properties reflecting spatially variable climate and geology suggests a field of interest for future research.

6.6 Geomorphologic Signatures of Varying Climate

When we stand on a ridge and look across a landscape, do we see landforms that are a consequence of processes acting in the current climate? Sometimes this is obviously not the case, as in the glaciated landscapes of the northern latitudes and high mountains. But what of those areas where climate-driven processes have only changed in intensity? Are there relicts or a mixture of signatures from previous climatic fluctuations left in the developing landscape for periods of time? In this section (following Rinaldo et al., [1995a]) we address the important issue of the geomorphological response to varying climate and suggest, through the mathematical model of landscape self-organization presented in this chapter, that both cases –

contemporary balance and relict features – are indeed possible. We show that, for a sinusoidal climatic fluctuation, all climate states leave geomorphic signatures only when no geologic uplift affects the basin landscape. With active uplift and intense runoff and erosion (humid conditions), however, geomorphic features track the current climate. In the case of arid climate and active tectonics, relict features reflect the wettest climate experienced in the past.

Rinaldo et al. [1995a] also show that with or without uplift the temporal evolution of the landscape in response to cyclic climate forcing is complex and leads to the unexpected conclusion that valley density is largest during periods dominated by slow downslope movement of sediments rather than during times of strong fluvial incision, as we would have anticipated from steady-state models (Section 1.4.5).

The interpretation of the presence of climatic signatures on geomorphology has been a longstanding argument in the field [e.g., Summerfield, 1991; Bloom, 1991], leading to what we believe is a false distinction between those who study processes and those who study history of landscapes. The distinction is false because to correctly interpret any landscape, understanding of processeses and historical analysis must be married in order to establish causality.

Many have argued, in a manner similar to Bloom [1991], that landscapes are like open books on which nature keeps leaving signatures through climatic fluctuations. The capability of seeing and interpreting such signatures is important both in the analysis of real landscapes, nowadays much improved through digital mapping techniques [e.g., Dietrich et al., 1992], and in relating processes and forms [e.g., Kirkby, 1993]. The latter has gained broad interest after the discovery of the fractal characters of the geometry of nature.

As seen in Chapter 2, the existence of many fractal geomorphologic structures in the fluvial landscape has been documented in a wide range of scales that is limited by a lower cutoff dictated by channel initiation processes. We will again focus our attention on the fractal crossovers defined by the channel head because the interplay of fluvial and hillslope processes epitomized by channel initiation plays a central role in reading the signs of climate.

Climatic records (e.g., constructed from oxygen isotope records in deep cores or benthic foraminiphera) show cycles in all timescales examined, from thousands of years to millions of years, with peaks of spectral density for the basic cycles of 22,000, 41,000, and 100,000 years. Over such long timescales all empirical analysis must rely on inference, either by interpretation of features or projection of contemporary process through long periods of changing conditions. Furthermore, decoding the massive statistical information about geometry and topology available through large digital elevation data sets requires assumptions about processes and forms as played out over the long timescales of landscape evolution. Under these conditions one can hardly underestimate the role of theory.

This section quantitatively addresses the problem of landform inheritance by simulating the effects of climatic fluctuations through models of landscape self-organization. The basic structure of the model and its rationale, with reference to theories of general dynamics of fractal growth, have been described in Sections 5.7 and 6.3. The specific mathematical model used by Rinaldo et al. [1995a] incorporates a minor but significant modification with respect to that described in the previous sections. It displays, as usual, an interplay of fast threshold-limited (TL) erosion by running water, resembling fluvial geomorphologic processes, and slow slope-dependent (SD) movement characteristic of hillslopes, possibly affected by uplift. TL actions are triggered by landscape-forming events at scattered times, say, T_k, $k = 1, 2, \ldots$ and their response lasts a short time, say t_α, in the geologic timescales.

In this version of the model the slow component of transport, depending on the actual slopes ∇z, acts in conjunction with geologic uplift U.

The characteristic timescales of TL fluvial erosion are again assumed to be much smaller than those of SD hillslope processes, and thus, as seen before, we decouple the instantaneous TL landscape self-organization from the evolution of SD landform development. The basic outcome of the model is, as usual, a topographic field $z(\mathbf{x}, t)$ on a lattice, where z is the point elevation of the landscape, $\mathbf{x}$ is a Cartesian coordinate, and t is time. Landscape evolution is thus described by the following equations:

$$\frac{\partial z(\mathbf{x}, t)}{\partial t} = \alpha f\big(\tau - \tau_c(t)\big) \qquad T_k \leq t \leq T_k + t_\alpha \tag{6.59}$$

whenever TL processes are acting (notice the time dependence of τ_c, which will be discussed in detail). All sediment mobilized by this process is removed from the system – no redeposition can occur. In long timescales the field equation is

$$\frac{\partial z(\mathbf{x}, t)}{\partial t} = \beta F(|\nabla z|) + U \qquad T_k + t_\alpha \leq t \leq T_{k+1} \tag{6.60}$$

when SD transport (defined by the net flux $\beta F(|\nabla z|)$) and uplift (U) are the dominating processes. Mobilized sediments are redistributed in a mass-conserving scheme.

If one assumes that $t_\alpha \sim 0$, all readjustments following TL erosion are considered instantaneous. This is a reasonable assumption, as seen in Section 6.3.

Rinaldo et al. [1995a] have used regular arrival times T_k for TL landscape-forming events, but other choices, for instance, randomly selected arrivals, do not change the results with reference to the issues dealt with here. Also, it is important to note that it is assumed that changes in slope-dependent transport due to climate changes are negligible compared to those occurring in rates of threshold-limited transport owing to varying critical shear stress.

The basic assumption proposed by Rinaldo et al. [1995a] is the simulation of climatic fluctuations by imposing a cycle on the threshold value $\tau_c = \tau_c(t)$. Of course, climatic oscillations are not simple cycles and landscape response to, say, varying precipitation can have significant lags and complexities associated, for instance, with vegetation response. We suggest, however, that the very simplicity of this model enables us to identify causal relationships in landscape response and to capture the essence of the combined effects of climatic changes. Here we interpret low τ_c and high τ_c as representing wetter and drier conditions, respectively.

Though immaterial to our conclusions because an oscillation of the threshold by climatic fluctuation is generally accepted, the task of refining

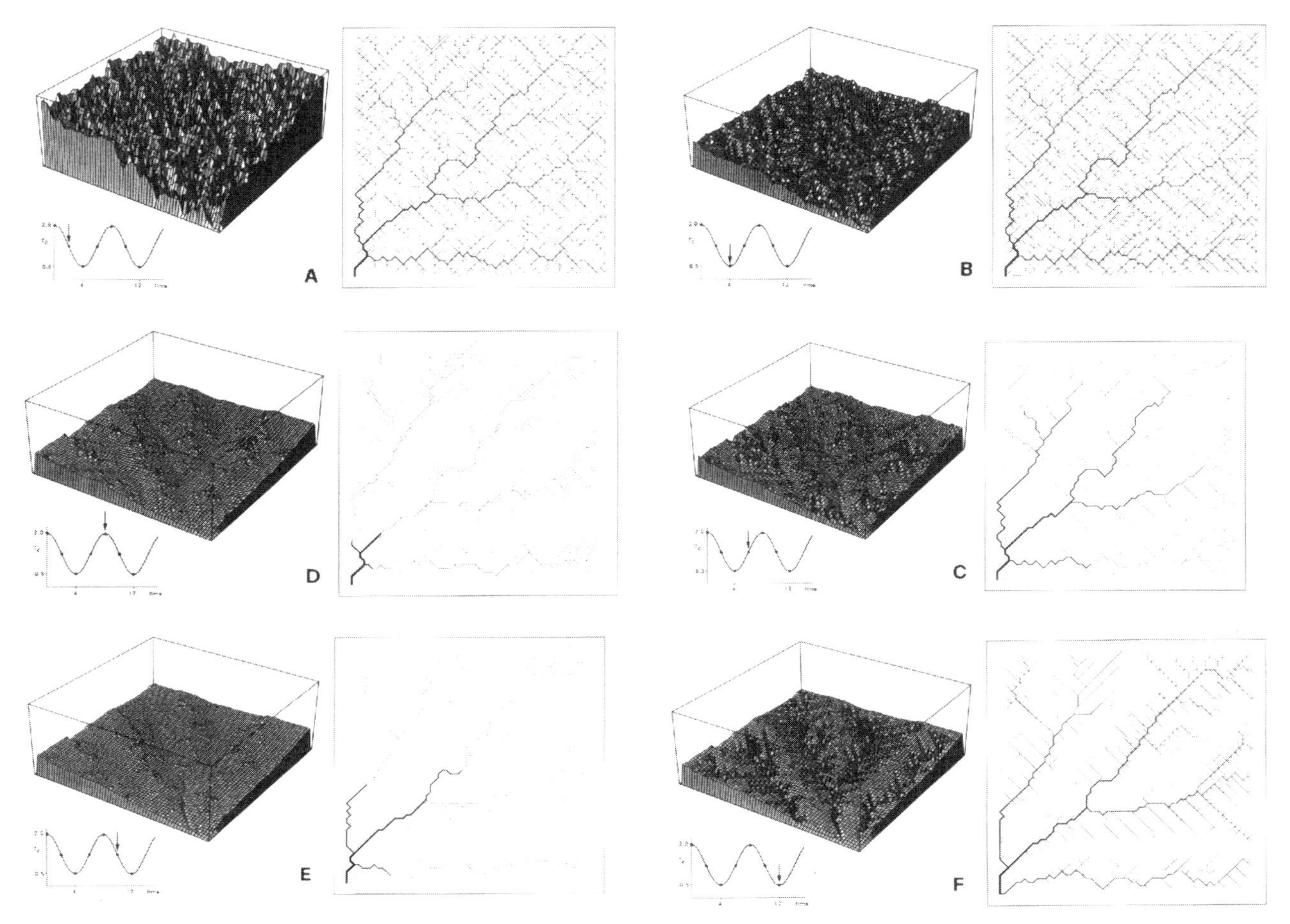

Figure 6.39. An example of the effects of climatic fluctuations on landscape self-organization (the case $U = 0$) [after Rinaldo et al., 1995a].

the above interpretation is a complex one – and worth some discussion. Higher rainfall rates corresponding to wetter climates result in an increase of the shear stress τ because of the proportionality of flow rates and cumulated areas through rainfall rates. However, a lower threshold τ_c does not represent simply, nor directly, wetter climates (higher τ). Drier climates modify surface features affecting τ_c (through, say, the density and extent of vegetation cover), and many geomorphic patterns are also modified by increased rainfall rates. The different lags in the response to climate changes of hillslope potential for erosion and of adaptation of the vegetation system also support a time-dependent scheme for $\tau_c(t)$. It is suggested that a cycle captures the essence of the combined effects of climatic changes and, as we will discuss later in this section, that complexity in the cycle structure does not substantially alter the main findings.

Figures 6.39 and 6.40 show an example of the effects on landscape evolution of a sinusoidal climatic fluctuation reflected by a large variation in the critical shear stress. In the example of Figures 6.39 and 6.40 the uplift rate U is set to zero. To describe the landscape evolution, for every time chosen in Figure 6.39 we show (for the case $U = 0$) the current value of τ_c used to update TL erosion at regular time intervals, the resulting elevation

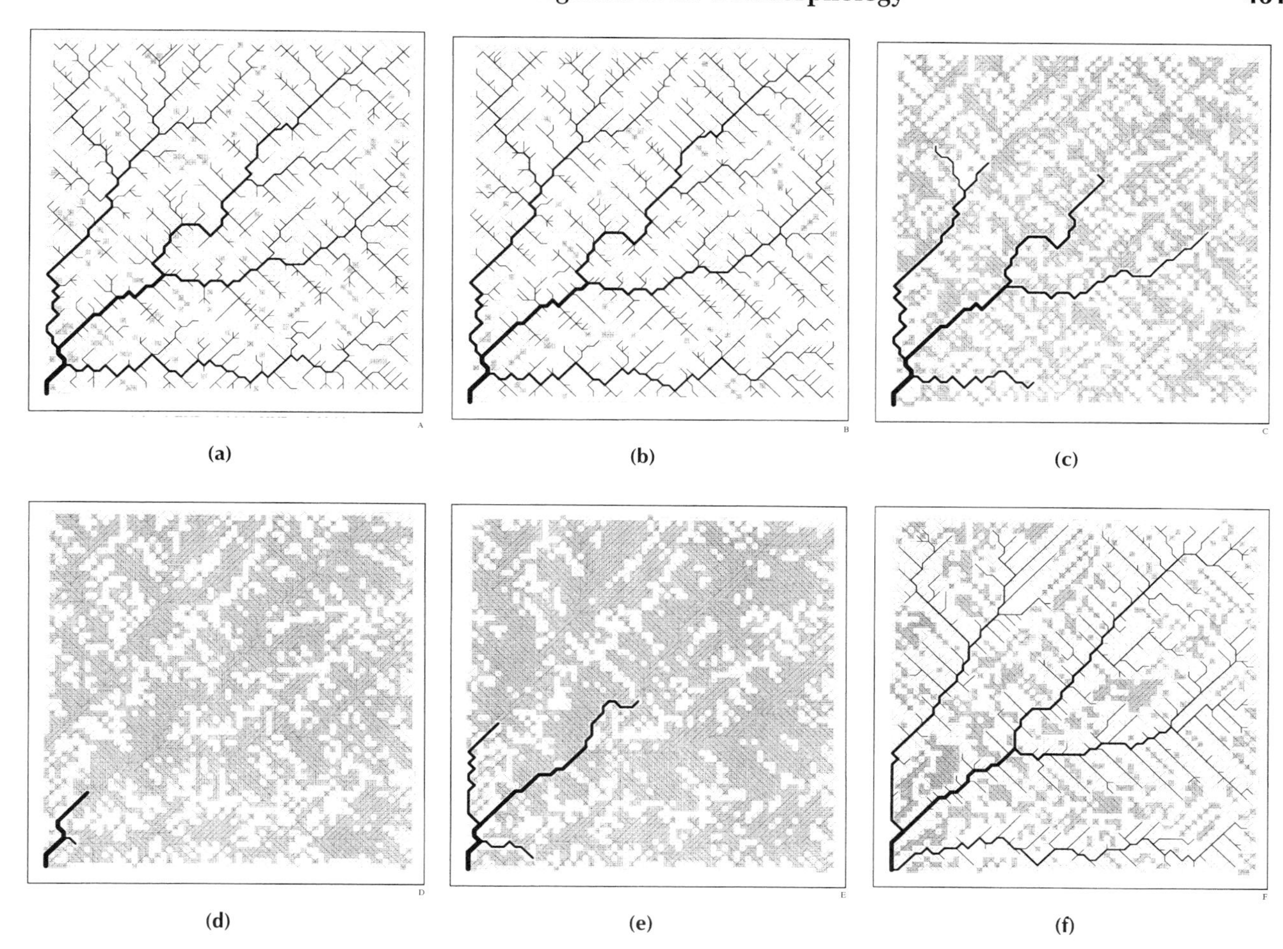

Figure 6.40. Channeled and unchanneled valley sites corresponding to the landscapes A–F in Figure 6.39.

field, and the planar view of the resulting drainage directions. Note that the dotted lines indicate drainage directions in unchannelized regions and the solid lines (whose thickness is proportional to A) indicate channels. In Figure 6.40 shaded pixels indicate valleys defined, at time t, by the condition $\nabla^2 z \geq 0$ (see Section 1.2.12). Channeled valleys (defined by $\nabla^2 z \geq 0$ and $\tau \geq \tau_c(t)$) are represented by solid lines.

We observe that landscape states corresponding to equal thresholds (say A and E, and B and F) differ significantly as the system evolves (Figure 6.41) because the progressively declining mean slope causes the critical shear stress to be less frequently exceeded. Unchanneled valleys (Figure 6.40) are created by this oscillation in a manner suggested by field observations [Dietrich and Dunne, 1993].

Figure 6.42 shows a three-dimensional color plot of a sequence of the resulting landscapes where each color corresponds to a different elevation. Notice that the entire pattern of climate changes is likely to leave geomorphic signatures when no geologic uplift is affecting the basin landscape. As the landscape undergoes a complete cycle in the case of no uplift ($U = 0$), it never recovers its original degree of channelization.

We have repeated the cycle shown in Figure 6.39 for two different uplift rates U, here defined as 0.4 and 0.8 height units per climatic cycle.

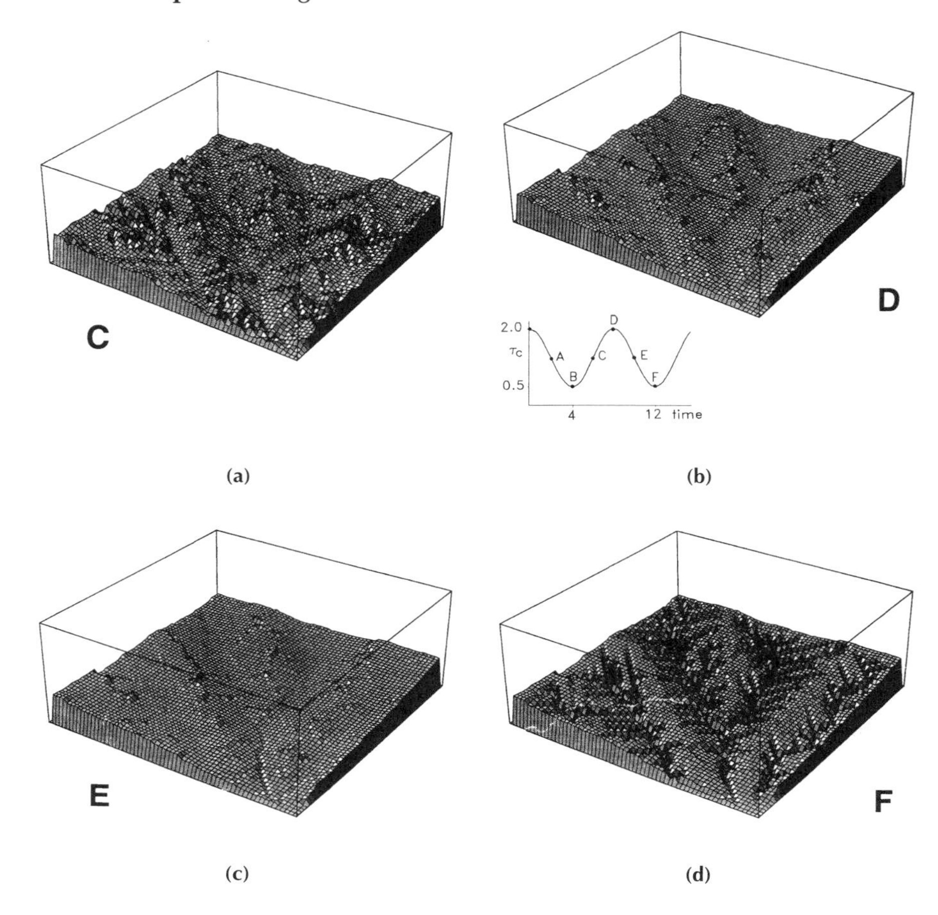

Figure 6.41. A comparison of four states of landscape evolution [after Rinaldo et al., 1995a].

Figure 6.43 shows a plot analogous of that of Figure 6.39 for the case $U = 0.8$ height units per cycle.

Other simulations based on more complex fluctuations built on different climatic frequencies, not reported here, confirm our present conclusions.

A summary of our results is illustrated in Figures 6.44 and 6.45. We use three geomorphic indicators to describe the effects of climatic oscillations on landscapes: a global fractal dimension of the elevation field; the drainage density of the resulting basins; and the related valley density.

Fractal dimensions of the elevation field measure the roughness of the landscape and are determined, as usual, by generalized covariance analysis. In Section 6.5 we saw that in the case of spatially constant τ_c the fractal dimension is a significant and synthetic geomorphologic measure for self-organized river basins and that the simple-scaling assumption is a viable one for their topography.

Drainage density is a measure of the degree to which the basin is dissected by channels. It is usually defined by the ratio of the total length of channels to the total area of the basin. Here we employ a dimensionless ratio defined as the ratio of the number of pixels where a channel occurs divided by the total number of pixels. As discussed in Chapter 1, experimental and theoretical evidence strongly suggests that a site i is occupied

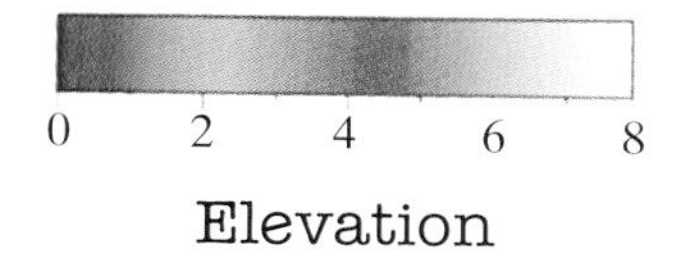

Figure 6.16. A 128×128 self-organized landscape generated from random initial conditions with $\tau_c = 1$ and $\nu = 0.2$. Channeled sites are defined in areas of convergent topography ($\nabla^2 z \geq 0$).

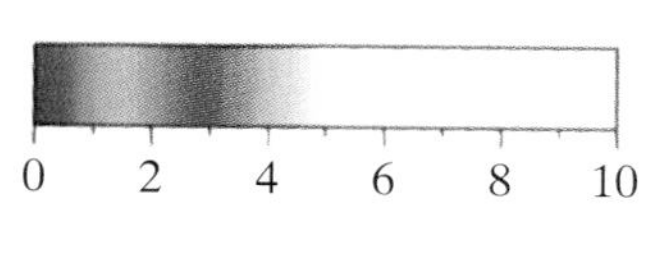

Figure 6.17. Fluvial landscape produced by a random threshold field drawn from a lognormal distribution where only TL processes are allowed to occur ($\langle \tau_c \rangle = 1, \sigma_\tau^2 = 0.2, \lambda_\tau = 10$ in pixel units).

Figure 6.18. Landscape produced by relaxing the z field in Figure 6.17 by a diffusive cycle.

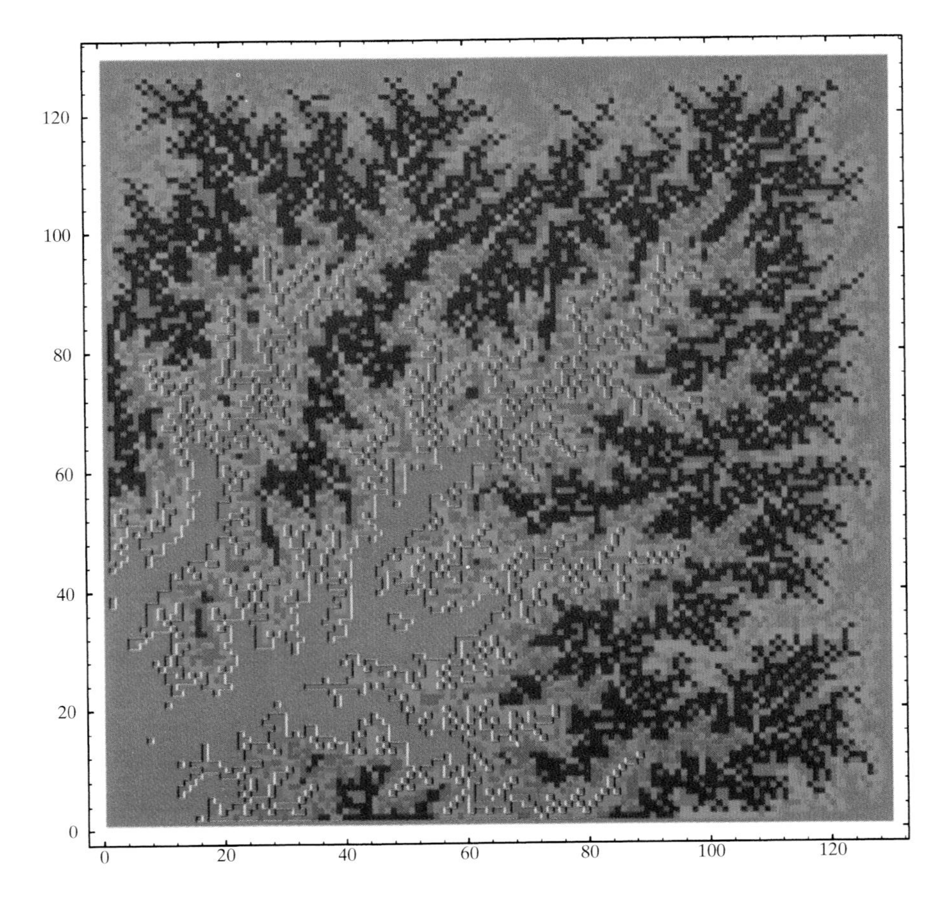

Figure 6.28. Plot of the set of points S that belong to predetermined ranges of the TL landscape in Figure 5.9 [after Rodriguez-Iturbe et al., 1994].

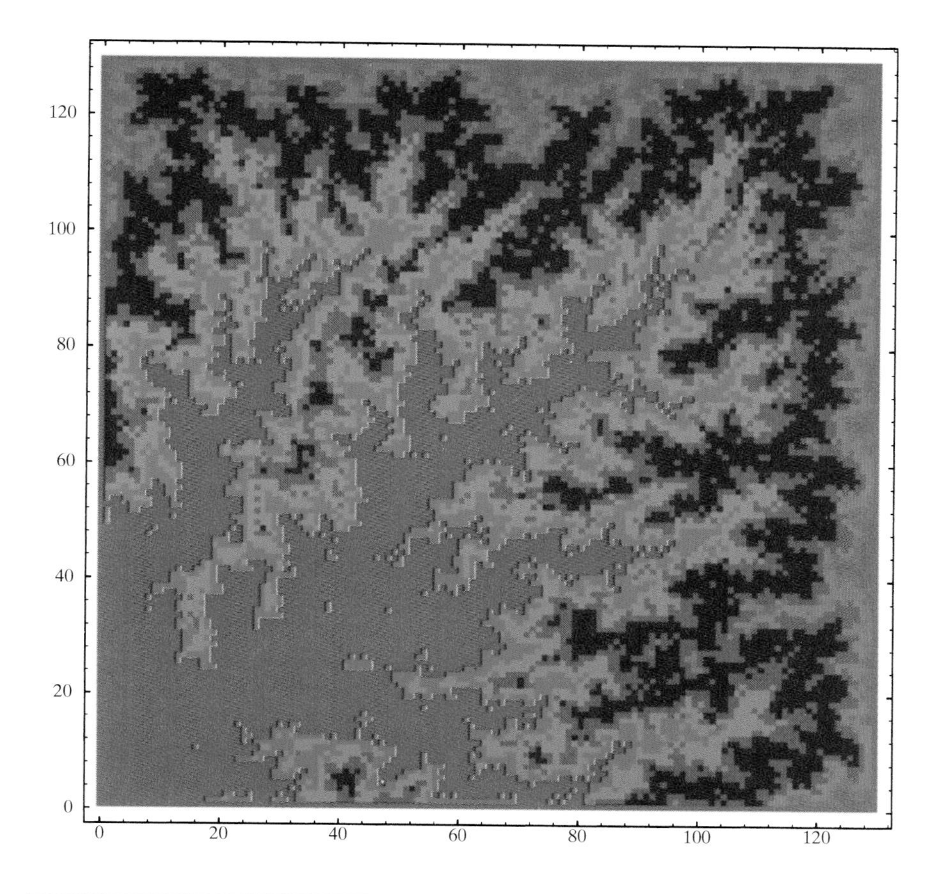

Figure 6.29. Plot of the set of points S that belong to predetermined ranges of elevation of the TL + SD landscape in Figure 6.4 [after Rodriguez-Iturbe et al., 1994].

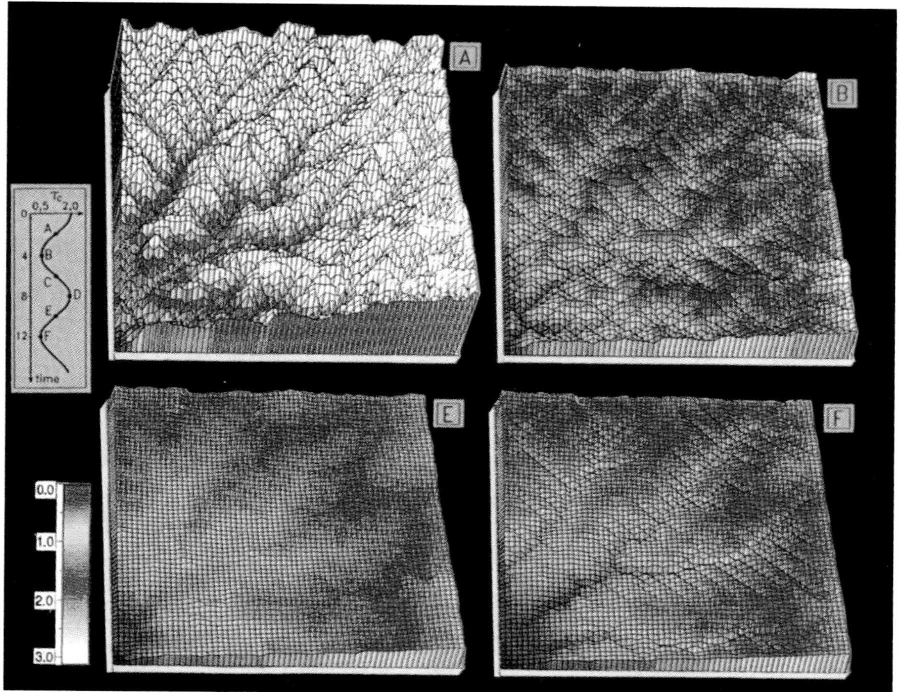

Figure 6.42. Clockwise from top, A, B, F, D [after Rinaldo et al., 1995a].

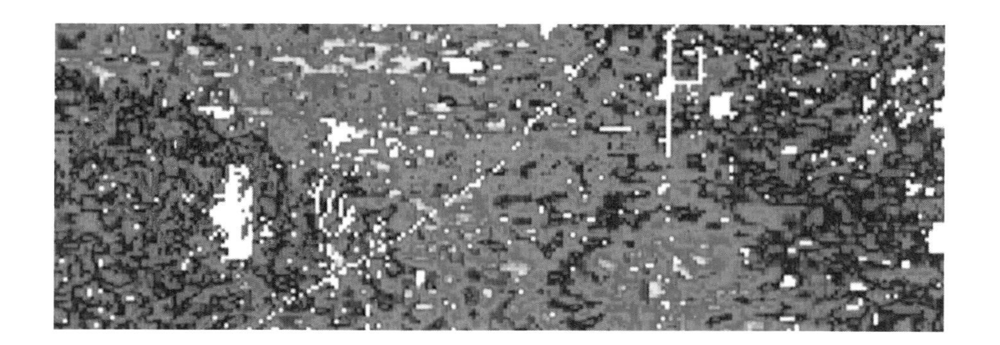

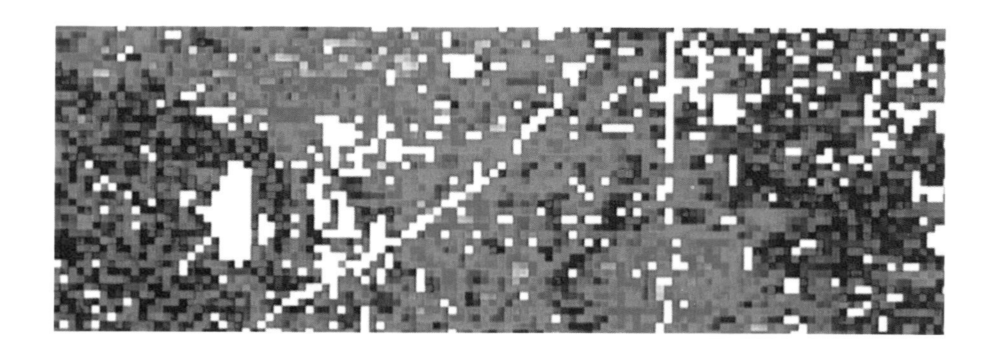

Figure 7.26. Relative soil moisture field $S(\mathbf{x})$ on day 10, at different levels of aggregation ($200 \times 200\ \mathrm{m}^2$, $400 \times 400\ \mathrm{m}^2$, and $600 \times 600\ \mathrm{m}^2$) . Red pixels are saturated ($S = 1$); green and yellow pixels have low values of S.

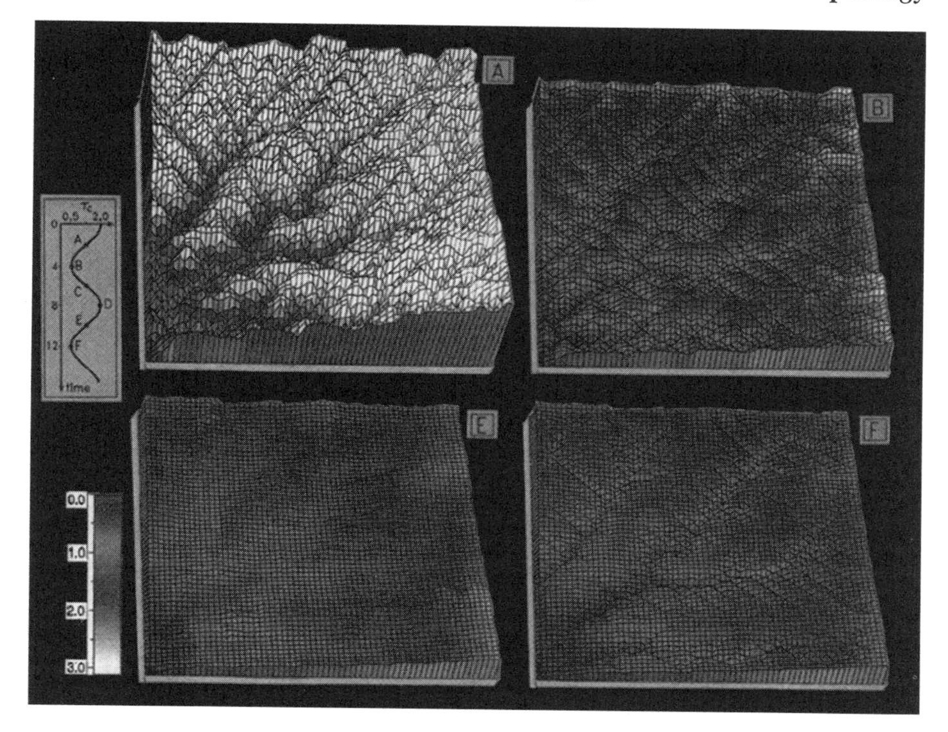

Figure 6.42. Clockwise from top, A, B, F, D [after Rinaldo et al., 1995a].

by channels if $\tau_i = \sqrt{A_i}\nabla z_i \geq \tau_c$ and the site is also located in a concave valley region where $\nabla^2 z_i \geq 0$.

Valley density is defined as the ratio of the number of concave sites (i.e., where $\nabla^2 z \geq 0$), defining valley areas of convergent topography usually identified with colluvium regions, to the total number of sites in the basin.

Important results emerge. During periods of weakened river incision but of still active diffusion processes, the stable base levels of the valley bottoms cause the concave area to expand by diffusive processes. Although it is well established that concavity results from threshold-limited processes (and convexity originates from diffusive slope-dependent processes), when the climate forcing is periodic, as will occur virtually anywhere, we actually see periods of greater concavity associated with periods of weakened river incision. Importantly, the expansion of the concave region is not due to hillslope sheet wash, but rather to diffusive processes.

In general, during rising values of critical shear stress, hillslope processes tend to smooth the landscape; and during falling values, fluvial processes tend to roughen it. In the absence of uplift ($U = 0$) the roughness of the landscape (and thus its fractal dimension) shows a damped oscillation that is not observed when uplift occurs (Figure 6.44(b)). Without uplift, progressive flattening of the landscape causes the TL processes to become less active while the diffusion-like transport continues; this prevents restoration of the higher fractal dimensions during periods of low critical stress.

During rising critical shear stress values (equivalent to aridification in this model), SD processes may infill the previous channel networks, creating an extensive network of unchanneled valleys (where shear stress is below critical), but not altering substantially the overall valley planform

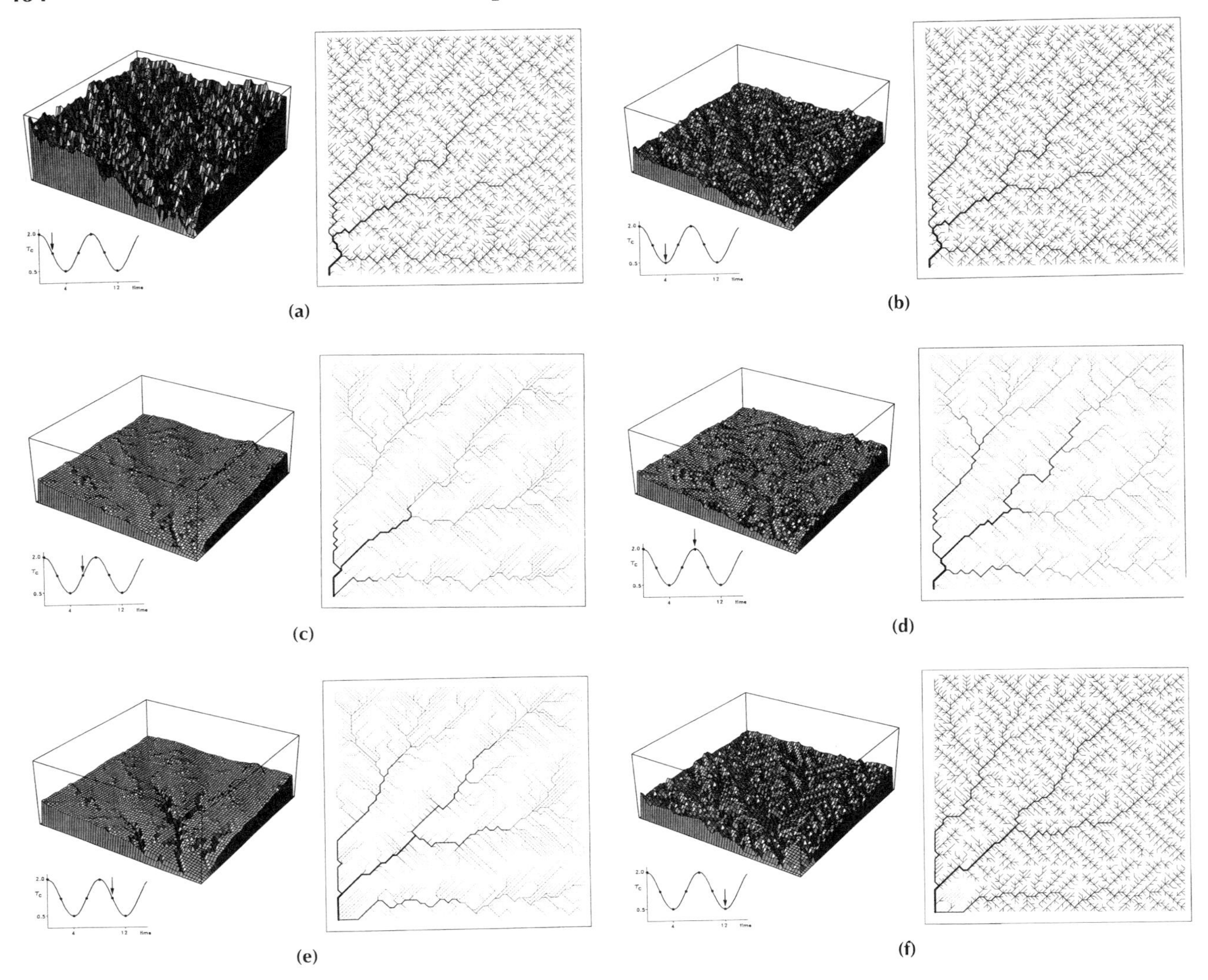

Figure 6.43. An example of the effects of climatic fluctuations on landscape self-organization for the case $U = 0.8$ height units per cycle.

created during the previous low critical shear stress (humid) cycle. For either the case of no uplift or active uplift, the valley density (Figure 6.44 (d)) is less variable than the drainage density. In this model, then, it is the timing and magnitude of drainage density change in response to climatic oscillations that lead to landform inheritance.

Figure 6.44 (c) shows that drainage density increases (more of the basin becomes channelized) as the climate becomes more humid (lower τ_c) and decreases as the climate becomes more arid (higher τ_c) when the channels nearly disappear. With active uplift, when the critical shear stress is periodically lowered to similar values in successive cycles, all past drainage paths are recut. Without tectonic uplift ($U = 0$), however, drainage density during successive wet periods progressively declines.

The response function of valley density to cyclic forcing (Figure 6.44 (d)) shows an intriguing asymmetry reflected in the short lag occurring between the maximum and the minimum values. If this is similar to what happens in nature, it suggests how significant time lags between climate

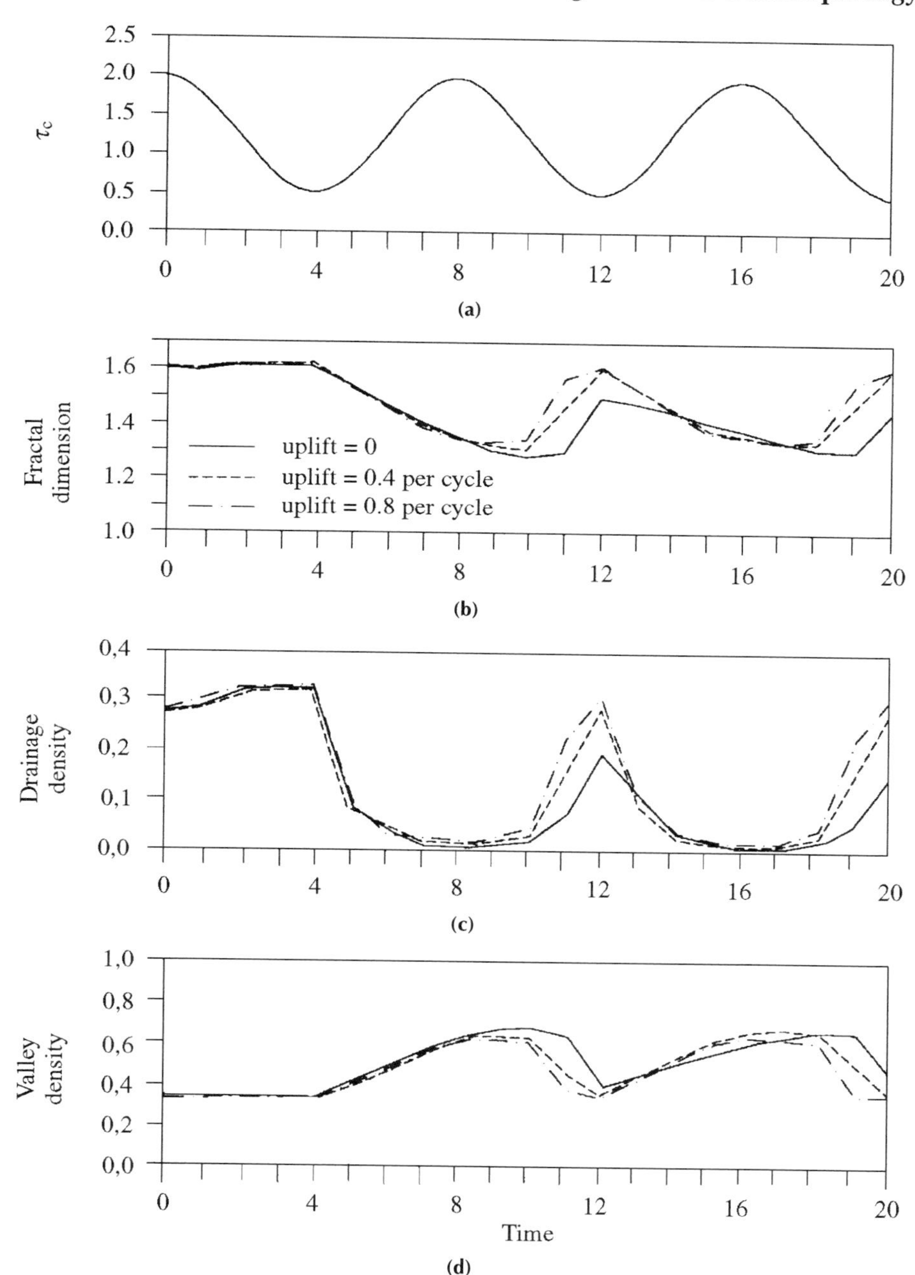

Figure 6.44. Time evolution of (a) the critical shear stress $\tau_c(t)$ defining the climatic cycle; (b) fractal dimensions of the elevation field; (c) drainage density; and (d) valley density ($U = 0$, 0.4 and 0.8 height units/cycle) [after Rinaldo et al., 1995a].

change and landform adjustment might occur. With the cyclic reestablishment of conditions favorable to threshold-limited erosion after the peak of the dry phase (the descending limb of $\tau_c(t)$, Figure 6.44(a)),valleys again become active with channels that advance headward and cut downward. During this advancement phase, diffusive processes keep wearing the slopes and expanding the area of concave topography, effectively increasing the valley area. The valley density increases only when the channels extend up to their former maximum extension and begin cutting downward, because only then do the convex hillslopes associated with basal incision expand

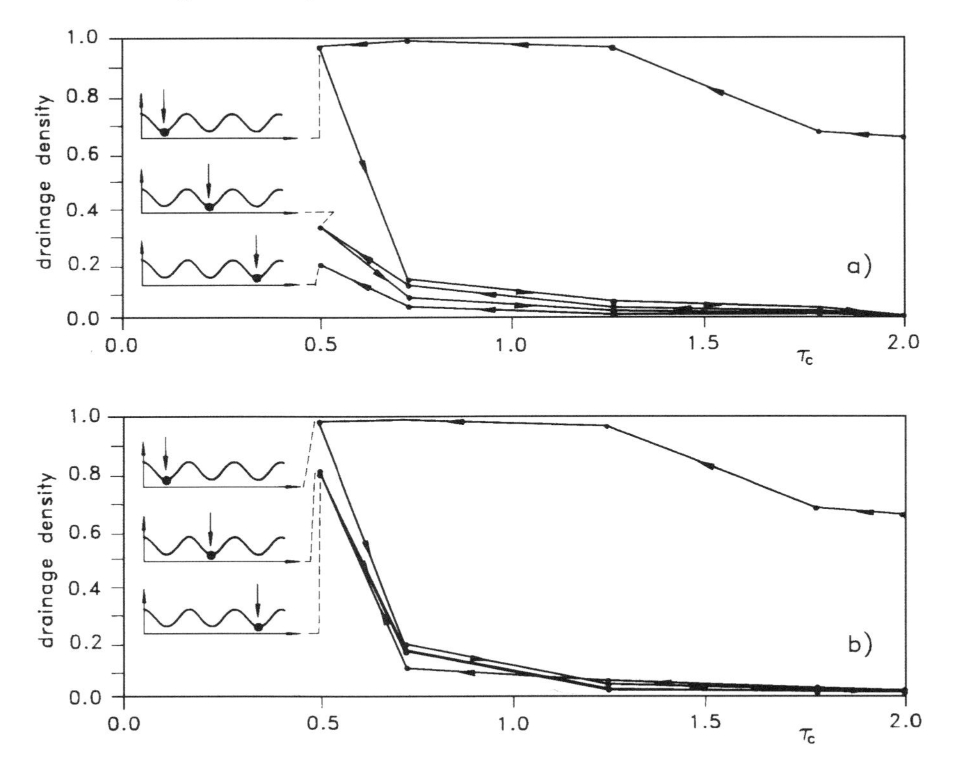

Figure 6.45. The variation in drainage density with changes in critical shear stress for (a) a landscape without uplift, and (b) for one with rapid uplift.

downslope. This explains the unforeseen lag and the short time incurred between maximum and minimum valley density. Thus the extent of convexity of the topography is positively correlated with channel incision, rather than being negatively correlated as one would otherwise expect by looking at competition of processes under steady-state conditions. This means that a landscape that is, on the average, in steady state (that is, the average uplift and erosion are equal) but that is experiencing periodic climate forcing may be radically different in morphology than one that is strictly in steady-state conditions (no variation in uplift or climate forcing). We observe that morphological differences are especially manifest in the extent of unchanneled valleys, because valley and drainage densities are sometimes both increasing and at other times are changing in opposite directions. Thus channel initiation patterns not only set a finite scale to the landscape (and a lower cutoff to the growth of fluvial fractal structures) but also play a complex role in the interpretation of the geomorphologic signatures of past climates.

The change of drainage density with the climate variable $\tau_c(t)$ clearly shows hysteresis in the case $U = 0$ (Figure 6.45(a)), suggesting that the entire climatic pattern affects the current geomorphology. Notice that the initial point ($t = 0$) is in the upper right part of the graph ($\tau_c = 2$ and drainage density is 0.40). The arrow in the insert shows the relative value of critical shear stress for the hinge points on the graph. The time variation in drainage density with critical shear stress is shown by the direction of the arrows defining the curves in each figure.

An analogous behavior was detected for fractal dimensions and valley densities.

In the $U = 0$ case, the drainage density at any particular time depends on the effectiveness of erosion during past climates in lowering the average slope. Each climatic cycle produces a different oscillation in drainage density and a systematic addition of unchanneled valleys (the difference between valley density and drainage density), which are only slowly eliminated by SD processes.

With active tectonics, however, once the initial conditions are erased the particular drainage density is purely a function of the specific climate (Figure 6.45(b)). The extent of unchanneled valleys reflects just the previous threshold minimum and does not tend to increase through successive cycles. After one cycle in which the initial conditions matter, the value of drainage density is determined solely by the current value of critical shear stress; past values do not matter.

We conclude that the current geomorphologic structure of a basin depends on the entire pattern of climatic fluctuations experienced whenever tectonic actions are negligible. In areas subject to tectonic uplift (in excess of isostatic recovery) the geomorphic *memory* of climatic change is limited to the previous threshold minimum, and we observe a repeated cycle in landscape roughness and drainage density in a sinusoidal climatic oscillation. The nature of the climatic fluctuations and the geology of a region, then, determine whether landscapes have few inherited landforms (memory-limited) or are dominated by inherited forms (memory-dominated).

It is also suggested that, given the apparent lack of preferential scale in the climatic record (a power law behavior of climatic fluctuation rather than a smooth sinusoidal pattern), extreme climatic excursions that cause accelerated threshold-limited erosion seem likely to occur and to leave very long-lived morphologic evidence where slope-limited processes are ineffective. This may explain rock pediment growth and the persistence of ancient fluvial landforms in hyper-arid climates.

CHAPTER 7

Geomorphologic Hydrologic Response

In order to capture the relation between the structure of channel networks and the characteristics of runoff processes, this final chapter presents the geomorphologic theory of the hydrologic response within the framework of the fractal geometry of the basin. A brief introduction to the observational evidence is also provided that suggests a scaling structure in the spatial organization of soil moisture.

7.1 Introduction

The geomorphologic theory of the unit hydrograph (GIUH) was first formulated by Rodriguez-Iturbe and Valdes [1979], who interpreted the runoff hydrograph as a travel time distribution to the basin outlet, embedding the dynamics of the processes of rainfall–runoff transformation and, most importantly, geomorphology. From that point on, several contributions have appeared in surface and groundwater hydrology that have considerably expanded the perspectives and the potential of the original idea. Rather than referring to a specific model, we now refer to a theory of the hydrologic response that embeds transport of chemical species and soil–climate interactions affecting rainfall–runoff transformations. To focus on the impact of fractal topologies and geometries on the hydrologic response (the purely geomorphologic effect), we will narrow our attention to such effects, arguing that the geomorphologic characters dominate in many cases of interest.

According to GIUH theories, the hydrologic response of a river basin is studied by decomposing the process of runoff formation into distinct contributions: (i) one that accounts for the mechanisms of runoff generation and travel times within the hillslopes; (ii) another that involves flow propagation as the foremost process within individual channel reaches (which we will term hydrodynamic dispersion to underline the many analogies with transport processes of flood routing in which both storage and dynamic effects are simultaneously operating); and (iii) a third contribution, of chief geomorphologic interest, that accounts for the morphology of the network structure (geomorphologic dispersion). This chapter reviews the formulation of transport processes through travel time distributions from the perspective of fractal geometries in order to stress the relation between the structure of river networks and the characteristics of their geomorphologic hydrologic response.

We will place a special emphasis on dispersive effects of geomorphologic nature. When assuming an instantaneous uniform injection of particles over a contributing area drained by a tree-like network, different lengths and connections of the drainage paths to the control section imply a distribution of travel times to the basin outlet even in the case of constant alongstream drift assigned to flow ‘particles’ (notice that we somewhat freely use the concept of flow particle, which will be precisely defined in Section 7.2). The heterogeneity of travel lengths and paths (hence times) to the outlet results in a diffusive effect on runoff response, which is imprinted in the landforms.

An entirely new chapter of geomorphologic hydrology has been originated by fractal theories as widely discussed throughout this book. Since Mandelbrot's [1983] early and seminal conjectures, empirical and digital mapping evidence has suggested that departures from Euclidean measures are the norm rather than the exception. We will examine the impact of these theories on the characterization of the hydrologic response. This will be pursued by exploring the statistical regularity of drainage networks with a view to data from real basins and also to exactly self-similar Hortonian networks. Many connections between the geometry of fractal networks and the characteristics of their underlying hydrologic response will thus be established. A revised width-function formulation of the GIUH is also discussed along with the results of related work.

The final part of this chapter focuses on the scaling nature of the spatial organization of soil moisture fields. This is a new field of research, which holds great interest and implications for a general theory of the hydrologic response.

7.2 Travel Time Formulation of Transport

We will first review the formulation of transport by travel time distributions, a tool ideally suited to deal with complex processes such as those involved in the hydrologic response at the basin scale. This complexity stems from the fact that natural environments are intrinsically heterogeneous. In particular, the set of all physical media involved in rainfall–runoff transformation processes is a collection of naturally heterogeneous formations: the surficial features of soils where precipitation occur, subject to variations in both space and time; the textures and features of formations hosting infiltration, interflow, seepage, and groundwater flow; and the characteristics of the hillslopes and their accompanying drainage networks. In addition to physical processes, there is also the uncertainty in the space/time distributions of the forcing rainfall fields and of parameters (like, for instance, pH or organic matter content) controlling chemical or biological reactivity of fixed (the heterogeneous media) and mobile (solutes in the carrier hydrological flow) phases. The possible interaction and correlation of physical and chemical heterogeneities adds further interest and complication to the description of environmentally relevant transport problems. Here our interest is confined to the case of conservative matter, the focus being on the impact of fractal geometries on hydrologic dynamics. Elsewhere [Rinaldo and Rodriguez-Iturbe, 1996], the geomorphologic theory of the hydrologic response has been shown to embed reactive solute transport and geomorphoclimatic interactions as well.

Under these highly heterogeneous conditions it is accepted that deterministic models are too gross a tool for simulation and prediction of processes at the scales of hydrologic interest. Stochastic models have therefore been engineered to embed uncertainty in the definition of transport variables. In fact, the aim of stochastic theories is to relate measurable properties of heterogeneous quantities to the uncertain characters of the predictions. The uncertainties in the processes are surrogated by the introduction of probability statements. In other words, we admit that a particle moving within the control volume and driven by a hydrologic carrier flow has a trajectory which, at time $t \neq 0$, is only partially known, that is, with a certain probability.

Let $m(\mathbf{x}_0, t_0)$ be the initial mass of a water particle injected at time t_0 in the (arbitrary) initial position $\mathbf{X}(t_0) = \mathbf{x}_0$. Each trajectory is defined by its Lagrangian coordinate $\mathbf{X}(t) = \mathbf{X}(t; \mathbf{x}_0, t_0) = \mathbf{x}_0 + \int_0^t \mathbf{u}(\mathbf{X}(\tau), \tau) d\tau$, where $\mathbf{u}(\mathbf{X}, t)$ is the point value of the velocity vector. The notation is meant to emphasize the Lagrangian character of the analysis where the properties depend on the trajectory.

The spatial distribution of mass concentration in the (arbitrary) volume $\mathcal{V}$ as a result of the injection of a *single* particle is given by [Taylor, 1921]

$$C(\mathbf{x}, t; \mathbf{x}_0, t_0) = \frac{m}{\Theta} \delta(\mathbf{x} - \mathbf{X}(t; \mathbf{x}_0, t_0)) \tag{7.1}$$

where $\delta(\cdot)$ is Dirac's delta distribution, $\int_{\mathcal{V}} C \Theta d\mathbf{x} = m$, and Θ is the possibly existing 'porosity' of the transport volume, that is, the coefficient defining the active portion of the transport volume. The δ distribution (or *delta function* as it is sometimes referred to) is a generalized function that we will define here simply by two operational properties:

$$\int_{-\infty}^{\infty} d\mathbf{x}\, \delta(\mathbf{x}) = 1 \tag{7.2}$$

and

$$\int_{-\infty}^{\infty} d\mathbf{x}\, a(\mathbf{x}) \delta(\mathbf{x} - \mathbf{x}_0) = a(\mathbf{x}_0) \tag{7.3}$$

Eq. (7.1) states that, in the one-particle world, concentration (mass per unit transport volume) is nonzero only at the site where the particle is instantaneously residing (i.e., at its trajectory). Thus uncertainty in the dynamic specification of the particle is reflected in the transport process.

Notice that the one-particle, one-realization picture needs to be generalized for applications relevant to a theory of the hydrologic response, which is typically characterized by large injection areas and pronounced time variability. The nonpoint source, one-realization picture is now defined. More involved formalism is not needed for a general description of the main results and thus, after the proper definitions, we will abandon the general notation without loss of generality. Nonpoint sources characterized by the initial distribution $M_0(\mathbf{x}_0) d\mathbf{x}_0$ yield the following relation for the spatial concentration distribution:

$$C(\mathbf{x}, t; t_0) = \int_{\mathcal{V}} d\mathbf{x}_0 \quad M_0(\mathbf{x}_0) C(\mathbf{x}, t; \mathbf{x}_0, t_0) \tag{7.4}$$

and likewise an input mass flux defined as $\dot{M}(\mathbf{x}_0, t_0) d\mathbf{x}_0 dt_0$ yields

$$C(\mathbf{x}, t) = \int_{\mathcal{V}} d\mathbf{x} dt_0 \quad \dot{M}(\mathbf{x}_0, t_0) C(\mathbf{x}, t; \mathbf{x}_0, t_0) \tag{7.5}$$

From now on, indices will be omitted, thereby implicitly assuming that any of the Eqs. (7.1), (7.4), or (7.5) may be referred to without affecting the results.

All mass exchange processes (gain/loss processes of physical, chemical, or biological nature, as seen from the mobile phase) are defined by the space/time evolution of the mass $m(\mathbf{X}, t)$ attached to the single particle studied herein. It is important to note that, in the general case, a functional dependence is shown for m on both the spatial position (postulating that the paths undertaken have a bearing on the result) and the time available to mass transfer processes.

Hydrologic processes define the evolution in time and space of the trajectory $\mathbf{X}(t; \mathbf{x}_0, t_0)$ of the particle that is therefore seen as a random function.

Let $g(\mathbf{X})d\mathbf{X}$ be the probability that the particle is in $(\mathbf{X} - d\mathbf{X}, \mathbf{X} + d\mathbf{X})$ at time t (notice that the functional dependence $g(\mathbf{X})$ implies $g(\mathbf{x}, t)$ in terms of Cartesian coordinates because of the evolution of the trajectory with time). The uncertainty cast on the particle's position in time and space is related to the difficulty of characterizing completely the complex chain of events making up hydrologic transport processes (and, in particular, rainfall–runoff generation processes). Convection processes of interest in this book are, in fact, strictly related to hydrologic transport phenomena and, as such, they are affected by the natural heterogeneity of the media at the basin scale. The ensemble average (i.e., many realizations) concentration $\langle C(\mathbf{x}, t)\rangle$ over all possible paths is then given by the classic relation [Taylor, 1921; Dagan, 1989]

$$\langle C(\mathbf{x}, t)\rangle = \int_{-\infty}^{\infty} \frac{m(\mathbf{X}, t)}{\Theta} \delta(\mathbf{x} - \mathbf{X}) g(\mathbf{X}) d\mathbf{X} \qquad (7.6)$$

A particular case of Eq. (7.6) may be thought of as significant for the transport processes involved with the hydrologic response, that is, that of the transport of conservative matter where no distinction is drawn between the carrier and the carried particles in terms of travel times. This will be the case if the convection field is unaffected by the concentration of solute and the solute does not undergo mass exchange phenomena with fixed phases.

Mathematically, the assumption of passive solute postulates that the mass m with which each particle is labeled is conserved throughout the transport process, that is, $m(\mathbf{x}, t) \approx m$. The passive behavior is further described by the lack of dependence of the velocity field characterizing $\mathbf{X}$ on the actual values of the concentration of mass in the field. Mixing and dispersion mechanisms are due to the combined effects of the geometry of the transport volume, of molecular diffusion, and of the heterogeneous velocity fields.

Integration of Eq. (7.6) yields Taylor's [1921] relation (conceived for stationary turbulence, although of general validity)

$$\langle C(\mathbf{x}, t)\rangle = \frac{m}{\Theta} g(\mathbf{x}, t) \qquad (7.7)$$

which links the displacement mechanics of the particles (epitomizing hydrology and hydraulics) and the concentration of solutes carried by the flow fields. The distribution $g(\mathbf{x}, t)$ is called the displacement probability density function.

Decoupling the flow field $\mathbf{u}(\mathbf{x}, t)$ from the properties of dispersing matter is allowed in the above framework of passive transport. The Lagrangian coordinate of a single particle is defined by the relationship

$$\mathbf{X}(t; \mathbf{x}_0) = \langle \mathbf{X} \rangle + \mathbf{X}'(t; \mathbf{x}_0) + \mathbf{X}_B(t) \tag{7.8}$$

Taking the velocity field as $\mathbf{u}(\mathbf{x}, t) = \langle \mathbf{u} \rangle + \mathbf{u}'(\mathbf{x}, t)$ where $\mathbf{u}'$ stands for velocity fluctuations, we have

$$\langle \mathbf{X} \rangle = \langle \mathbf{u} \rangle t \qquad \mathbf{X}'(t, \mathbf{x}_0) = \int_0^t \mathbf{u}'(\mathbf{X}_\tau, \tau) d\tau \tag{7.9}$$

In general, $\mathbf{X}_B(t)$ is an isotropic Brownian motion component either of molecular origin or condensing information from smaller scales (mathematically, one has $\langle \mathbf{X}_B(t) \rangle = 0$ and $\langle \mathbf{X}_B(t)^2 \rangle = 2D_B t$, where D_B is a diffusion coefficient). Notice that the technical literature uses both σ and D as commonly accepted notations for the diffusion coefficient. We do not employ a strictly consistent notation here and in Chapter 2, as confusion should not arise.

An important model [see, e.g., Dagan, 1989] to characterize displacement distributions, $g(\mathbf{x}, t)$, or $\langle C(\mathbf{x}, t) \rangle$ (from Eq. (7.7) $g \propto \langle C \rangle$) is Fokker–Planck's equation

$$\frac{\partial g(\mathbf{x}, t)}{\partial t} + \sum_i \langle u \rangle_i \frac{\partial g(\mathbf{x}, t)}{\partial x_i} = \sum_i \sum_j D_{ij}(t) \frac{\partial^2 g(\mathbf{x}, t)}{\partial x_i \partial x_j} \tag{7.10}$$

where D_{ij} is the general dispersion tensor defined below. Before proceeding with the characterization of the components of Eq. (7.10) it is instructive to discuss the physical and mathematical meaning of the above classic scheme. The distribution $g(\mathbf{x}, t)$ characterizes the probability of a particle's trajectory being at $\mathbf{x}$ at time t. It is highly likely that the maximum probability lies about the mean displacement whose coordinates are $\langle X \rangle_i = \langle u \rangle_i t$. Thus Fokker–Planck's model embeds the second term of the left-hand side, which tends to shift the maximum probability along the mean trajectory. The term on the right-hand side of Eq. (7.10) explains how likely it is to find departures from the mean trajectory. This is due to the effect of fluctuations $\mathbf{u}'$ about the mean velocity of the carrier. Notice that mathematically the *diffusion* term is generally anisotropic and time dependent but not space dependent (in contrast to standard diffusion, or Fickian, models where D is constant in space and time and direction independent). This has important mathematical implications whose analysis is beyond the scope of this book.

We now return to the nature of the dispersion tensor D_{ij}. The key feature of the Fokker–Planck's model is that the actual dispersion is related to the fluctuations of velocity that generate departures of trajectories from the mean displacement. Thus diffusion is not simply a mechanism whose features are superimposed on the convection process, but rather is embedded in the heterogeneous structure of the convection field. With reference to the ith component of the residual displacements X'_i in Eq. (7.8), a key role is played by the correlations of such residual displacements generated by velocity fluctuations. We now introduce the component X_{ij} of the covariance tensor of the residual displacements $\mathbf{X}'$ (notice that for simplicity of notation we assume statistical homogeneity, i.e., $\mathbf{X}'(t, \mathbf{x}_0) = \mathbf{X}'(t)$):

$$X_{ij}(t) = \langle X'_i(0) X'_j(t) \rangle \qquad i, j = 1, 2, 3 \tag{7.11}$$

The physical meaning of the component X_{ij} of the displacement tensor is the square of the length characteristic of the spreading of trajectories around the mean; that is, $X_{ij} = 0$ implies no diffusion at all. Notice that in the Taylor picture we may think of different realizations of a random trajectory as equivalent to different positions of particles of equal mass. In fact, techniques for solution of the mass balance equation based on the study of statistics of trajectories are called particle-tracking schemes. Thus the centroid of the distribution of particles at a given time coincides with the mean trajectory at the same time, and the ij th component of the covariance of the residual displacements is the square root of the ij th moment of inertia of the many-particle system. Before linking the correlations X_{ij} of the residual displacements to the statistics of the (random) velocity field, we notice that the coefficient of dispersion in Eq. (7.10) is related to the above covariance tensor by the relationship [e.g., Taylor, 1954]

$$D_{ij}(t) = D_B + \frac{1}{2}\frac{dX_{ij}}{dt} \tag{7.12}$$

Eq. (7.12) is quite general in nature. It states that the dispersion process is defined by the time evolution of the covariance of displacements from the mean value. The displacement covariance is a robust measure that characterizes the important class of *Gaussian* dispersion processes [Cox and Miller, 1965].

The dynamics of the process is thus defined by fluctuations of the displacements about that of the centroid ($\langle \mathbf{u} \rangle t$). The velocity field determines the dispersive character of transport via the following relation:

$$\frac{d^2 X_{ij}}{dt^2} = 2\langle u_i(0, 0) u_j(\mathbf{X}(t), t)\rangle = 2u_{ij}(\mathbf{X}(t), t) \tag{7.13}$$

where an important assumption is that of Lagrangian stationarity for the statistics, u_{ij}, of the velocity field [Cramér, 1946; Batchelor, 1953; Landahl and Mollo-Christensen, 1987]. The derivation of Eq. (7.13) is not straightforward because differentiation and integration of random functions differ from traditional calculus. A synthetic but rather complete discussion of differentiation and integration of random functions in the hydrologic context may be found in Dagan [1989, Chapter 1], which also contains the derivation of Eq. (7.13).

Eq. (7.13) has been solved in important classes of linear stochastic problems, most notably in the context of transport in heterogeneous porous formations [e.g., Dagan, 1989] where $\mathbf{u}(\mathbf{x}, t) \sim \mathbf{u}(\mathbf{x})$. The linear solution assumes $\mathbf{X}(t) \sim \langle u \rangle t$ in the argument of the covariance function, and the solution to Eq. (7.13) yields

$$X_{ij}(t) = 2\int_0^t (t - \tau) u_{ij}(\langle u \rangle \tau)\, d\tau \tag{7.14}$$

(where $u_{ij}(r) = \langle u_i(0) u_j(r)\rangle$) whenever the mean flow $\langle u \rangle$ is stationary and uniform. Asymptotic solutions of Eq. (7.14) are $X_{ij}(t) \propto t^2$ (and hence $D_{ij} \propto t$) for $t \to 0$ (the Taylor regime) and $X_{ij}(t) \propto t$ (hence D_{ij} reaches a constant values because of Eq. (7.12)) for $t \to \infty$ (the Fickian regime).

All diffusive models of transport phenomena are derived from Eq. (7.10), via different degrees of approximation or averaging. A most common model contained in Eq. (7.10) is the so-called convection–dispersion scheme that

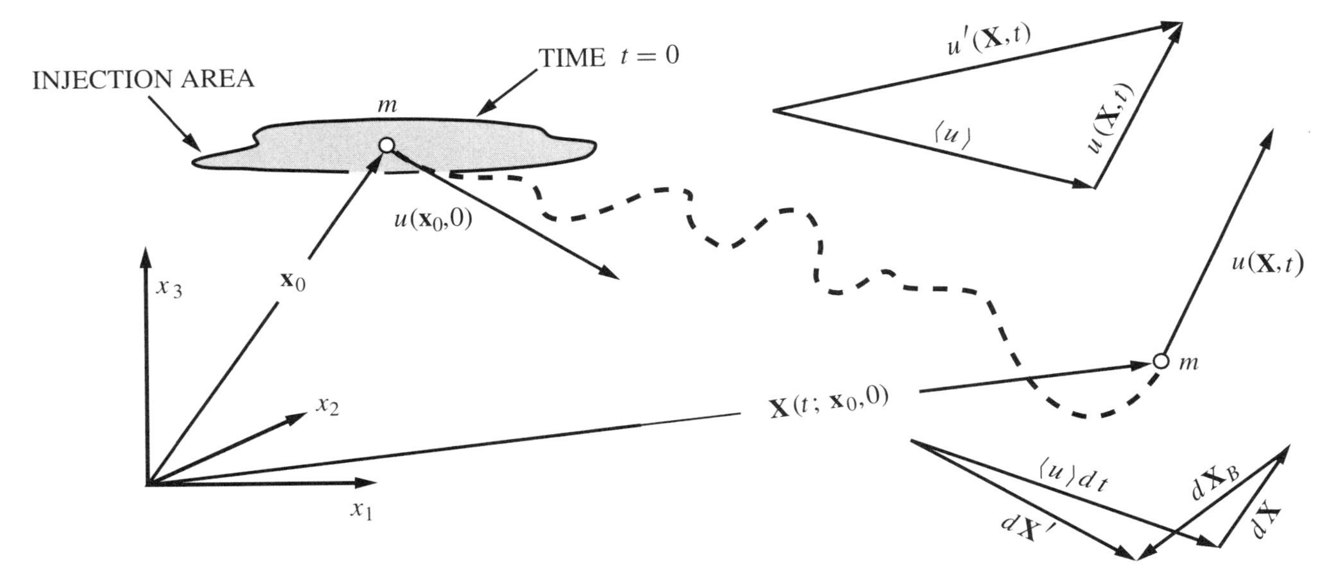

Figure 7.1. Sketch of the transport volume and relevant quantities.

is quite successful in incorporating the interplay of diffusion and heterogeneous convection. In the case of one coordinate (x) aligned along the mean flow we have

$$\frac{\partial C}{\partial t} + U\frac{\partial C}{\partial x} = D_{\parallel}\frac{\partial^2 C}{\partial x^2} + D_{\perp}\left(\frac{\partial^2 C}{\partial y^2} + \frac{\partial^2 C}{\partial z^2}\right) \tag{7.15}$$

where the average fluid velocity U in the x direction and the longitudinal and transverse dispersivities $D_{\parallel}, D_{\perp}$ depend on both U and the geometry of the system. The quantities represent suitable statistical averages (e.g., $U = \langle u \rangle$ and $C = \langle C \rangle$).

A very important connection of classic Eulerian approaches (i.e., based on the solution to Eq. (7.10)) and of travel time approaches is now described [e.g., Dagan, 1989]. One can readily establish the relation between the displacement pdf $g(\mathbf{x}, t)$ due to the kinematics of the carrier flow field and the first-passage (or travel time) distribution at a fixed control section. Let $\mathcal{V}$ be the transport volume in which a control section is defined (see Figure 7.1). We assume that the time t in which a particle crosses the control section is unique and, most importantly, that all particles injected in $\mathcal{V}$ ensuing from $\mathbf{x}_0 \in \mathcal{V}$ must transit the predefined cross section (river sections at the closure of basin-scale transport volumes are good examples of the trapping state described above).

Owing to the uncertainty characterizing $\mathbf{X}$, the arrival time T is a random variable characterized by a probability $P(T < t) = P(t; \mathbf{x}_0, t_0)$, that is, the probability that the particle originated at $\mathbf{x} \in \mathbf{x}_0$ at $t = t_0$ has already crossed the trapping state at time t. The link of the Eulerian and the Lagrangian approaches is defined by the following relationship:

$$P(T < t) = 1 - \int_{\mathcal{V}} g(\mathbf{x}, t; \mathbf{x}_0, t_0) d\mathbf{x} \tag{7.16}$$

which, upon substitution of Eq. (7.7), yields the fundamental relation

$$P(T \geq t) = \frac{\Theta}{m} \int_{\mathcal{V}} \langle C(\mathbf{x}, t) \rangle d\mathbf{x} = \langle M(t) \rangle$$

$$f(t) = \frac{dP(T < t)}{dt} = -\frac{d\langle M(t) \rangle}{dt} \tag{7.17}$$

where $\langle M(t) \rangle$ is the (ensemble mean) mass still in the transport volume at time t normalized by the initial mass-injected m. The control section acts as an absorbing barrier or trapping state. From continuity of mass, $d\langle M \rangle/dt = -Q(t)$ (where $Q(t)$ is the outflow from the control volume) and therefore one obtains

$$f(t) = Q(t) \tag{7.18}$$

The probability density of travel times (herein called without distinction travel time or arrival time distribution) is therefore the instantaneous mass flux at the control section and as such has a clear physical meaning.

In surface hydrology, when the input is a unit of effective rainfall, this pdf is called the *instantaneous unit hydrograph.*

We now derive the travel time distribution for the transport problem where the mean convection ($\langle u \rangle$) and the dispersion coefficients ($D_{\parallel}$ and $D_{\perp}$) are taken as their constant asymptotic value. Such a solution proves useful in some geomorphologic applications. In this case, the probability $g(\mathbf{x}, t)d\mathbf{x}$ for a particle being in $(x, x + dx)$ at t is computed by the general model of longitudinal dispersion described by Eq. (7.15) where the (constant) mean convection is directed along the direction x. Eq. (7.15) requires specification of two conditions for every spatial coordinate $\mathbf{x}$. A reference solution (matters of boundary conditions particular to each problem will be discussed later in this chapter) involves a Dirac delta input at $x = 0$ (say, the displacing particle is at $\mathbf{x} = 0$ at $t = 0$ with probability one) and $g(\pm\infty, t) = 0$. Also, $g(\mathbf{x}, 0) = 0$ except at the origin where a unit pulse of flux at $x = 0$ is enforced as

$$\left| \langle u \rangle g(x, t) - D_L \frac{\partial g}{\partial x} \right|_{x=0} = \delta(t) \tag{7.19}$$

The above condition specifies that the total flux (convective and diffusive) at the input control surface is impulsive. The general solution in the above conditions is the standard Gaussian model

$$g(\mathbf{x}, t) = \frac{1}{\sqrt{2\pi D_{\parallel} D_{\perp}^2 t^3}} \exp\left(-\frac{(x - \langle u \rangle t)^2}{4D_{\parallel} t} - \frac{y^2}{4D_{\perp} t} - \frac{z^2}{4D_{\perp} t} \right) \tag{7.20}$$

with $\mathbf{x} = (x, y, z)$. We notice that the Gaussian solution is separable. If we define $g(x', t)$ as the part of g containing the variable $x' = x - \langle u \rangle t$, that is, $g(x', t) = 1/(\sqrt{2\pi D_{\parallel} t}) \exp(-(x - \langle u \rangle t)^2/4D_{\parallel} t)$, and likewise with the other variables, we obtain $g(\mathbf{x}, t) = g(x', t)g(y, t)g(z, t)$. Eq. (7.17) can thus be written as

$$f(t) = -\frac{d}{dt} \int_{\mathcal{V}} g(\mathbf{x}, t) d\mathbf{x} \tag{7.21}$$

We now turn to the control volume $\mathcal{V}$ as in Figure 7.2 with $x = x_1, y = x_2$, and $z = x_3$. We consider $x \in (-\infty, L)$, L being the x coordinate of the

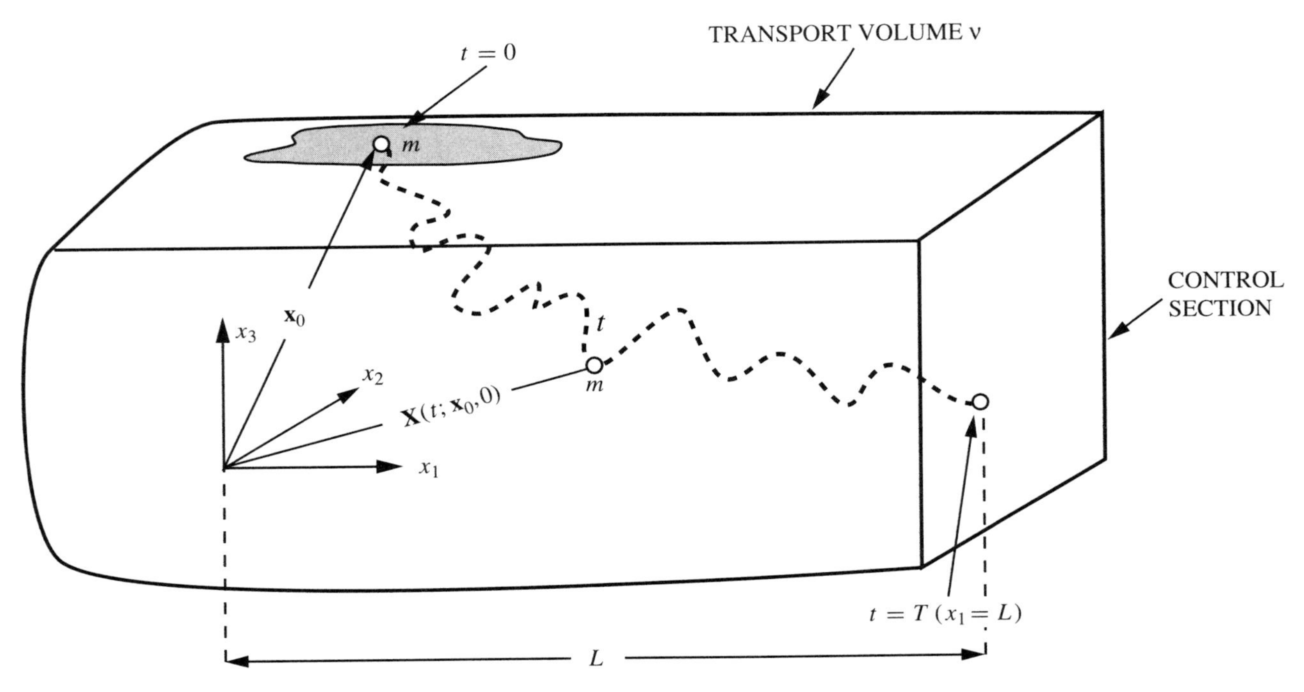

Figure 7.2. Transport volume and the trapping state.

control section that acts as an absorbing barrier and $y, z \in (-\infty, +\infty)$. The fact the displaced particle cannot return into the transport volume once its coordinate x exceeds $x = L$ is defined as the *absorbing barrier* assumption. A *reflecting* barrier is assumed at $x = -\infty$, describing the fact that the particle cannot leave the transport volume except through the control section at $x = L$. Thus one has

$$f(t) = -\frac{d}{dt}\int_{-\infty}^{L} dx \int_{-\infty}^{\infty} dy \int_{-\infty}^{\infty} dz\, g(\mathbf{x}, t)$$

$$= \int_{-\infty}^{L} dx\, g(x, t) \int_{-\infty}^{\infty} dy\, g(y, t) \int_{-\infty}^{\infty} dz\, g(z, t) \tag{7.22}$$

The second and third integral are unity, and thus, as noticed by Dagan [1989], transverse dispersion $D_\perp$ does not play a role in the determination of travel times under the above assumptions. The general solution for the case of an absorbing barrier at $x = L$ and a reflecting barrier at $x = -\infty$ is thus obtained by integrating the marginal Gaussian distribution $g(x', t)$ in space from $x = -\infty$ to $x = L$ and taking its time derivative by Leibnitz' rule [Dagan, 1989, p.166]:

$$f(t) = \frac{1}{2}\left(\frac{L}{t} + \langle u \rangle\right) g(L, t) \tag{7.23}$$

with

$$g(L, t) = \frac{1}{\sqrt{2\pi D_\parallel t}} \exp\left(-\frac{(L - \langle u \rangle t)^2}{4D_\parallel t}\right) \tag{7.24}$$

A more general solution can be obtained by the above procedure whenever $D_\parallel$ depends on time t (recall that we set $t_0 = 0$ for simplicity of notation without loss of generality). In this case, if we define

$$X_{\parallel}(t) = 2\int_0^t D_{\parallel}(\tau)d\tau \tag{7.25}$$

Rinaldo et al. [1991] showed that the travel time distribution $f(t)$ with the same set of boundary conditions becomes

$$f(t) = \left(\frac{L - \langle u \rangle t}{2X_{\parallel}(t)}\frac{dX_{\parallel}}{dt} + \langle u \rangle\right) g(L, t) \tag{7.26}$$

which reduces to Eq. (7.23) if $D_{\parallel}$ becomes constant. It can be shown, by numerical comparisons, that Eq. (7.23) and Eq. (7.26), with a constant reference value $D_{\parallel}(\infty)$, yield similar travel time distributions even for a wide range of variation in $D_{\parallel}(t)$.

The travel time distribution blends all sources of uncertainty into a single curve that depends on all the transport processes that are taking place. In using the travel time formulation of transport in surface hydrology, two courses have been pursued: one course assumes the form of the pdf, characterizing it by some parameters of clear physical meaning like, say, mean travel time. An example of this is the exponential pdf used to describe travel times of water particles in the original approach by Rodriguez-Iturbe and Valdes [1979] to derive the geomorphologic unit hydrograph. A second approach exploits the analogy of fluxes and pdfs to deduce travel times from the equations of motion. Eulerian, Lagrangian, or travel time approaches therefore may differ formally even though they are derived from the same assumptions. The prejudice of considering one approach superior to the other is therefore incorrect, although rather common in the literature. The relative balance of merits for the different approaches has been discussed by Dagan [1989]. Applications of travel time concepts to relevant transport problems are found in Rinaldo and Marani [1987], Shapiro and Cvetkovic [1988], Rinaldo et al. [1989a,b, 1991], and Dagan et al. [1992].

The travel time approach allows us to account for the effects of the morphology of the network in the response of a basin by combining analytically the travel times within the individual parts of the whole.

7.3 Geomorphologic Unit Hydrograph

The response of the system to an instantaneous unit volume of effective rainfall uniformly distributed in space is central to the understanding of hydrologic dynamics. Such a response is termed an instantaneous unit hydrograph. When related to the morphologic parameters of the network, this shock response, defining the travel time distribution of the water particles to the outlet of the basin, is called the geomorphologic instantaneous unit hydrograph (GIUH or GUH) [Rodriguez-Iturbe and Valdes, 1979].

The pdf of random travel times, controlled by the dynamics of the system, defines the hydrologic response because, as shown in Eq. (7.17), $f(t) \propto d\langle M(t)\rangle/dt$, where $\langle M(t)\rangle$ is the expected mass of matter within the transport volume at time t. In the case at hand, this flux of mass is simply the discharge $Q(t)$ at the control section.

The determination of the GIUH was initially accomplished by following an analysis of the detailed motion of water 'particles' in space and time over a channel network. Recall from Chapter 1 that Ω is the order of the basin; c_i, $1 \leq i \leq \Omega$, denotes a channel state of Strahler order i; and o_i, $1 \leq i \leq \Omega$, denotes an overland region of the basin that drains into a channel of order i. It is assumed [Rodriguez-Iturbe et al., 1979] that initially the particles are only located in the overland regions and thus the amount of rainfall that initially falls directly onto channels is neglected. Consequently, the particles, initially in any one of the regions o_i, undergo transitions according to the following rules:

- The only possible transitions out of the state o_i are those of the type $o_i \to c_i$, $1 \leq i \leq \Omega$.
- The only possible transitions out of the state c_i are those of the type $c_i \to c_j$ for some $j > i$, $i = 1, 2, .., \Omega$.
- A state $c_{\Omega+1}$ is defined as a trapping state; that is, all particles have to cross the ideal compliance surface defined by the trapping state. In other words, transitions out of the state $\Omega + 1$ are impossible. It is defined as the outlet of the basin.

The above rules define a collection Γ of paths γ that a particle may follow through to the trapping state at the basin outlet. As an example, consider the network shown in Figure 7.3.

The collection of paths, $\Gamma = \gamma_1, \gamma_2, \gamma_3, \gamma_4$, consists of the following routes to the outlet:

$$o_1 \to c_1 \to c_2 \to c_3 \to c_4(\text{outlet})$$
$$o_1 \to c_1 \to c_3 \to c_4(\text{outlet})$$
$$o_2 \to c_2 \to c_3 \to c_4(\text{outlet})$$
$$o_3 \to c_3 \to c_4(\text{outlet})$$

Note that the above proposition may be extended to an arbitrary numbering of the states. If o_i, $i = 1, N$, is the number of overland states pertaining to each link of the network (N is the total number of links), one may think of collecting all paths $o_i \to c_i \to \cdots c_{\Omega+1}$ following the actual connections defined by the channel network and its ordering and numbering procedure. Strahler's ordering induces averaging over states (e.g., over all first-order streams draining into second-order streams and so on) with equal ordering features and this may (or may not, depending on specific processes) be a desired feature.

The above rules specify the spatial evolution of a particle through a network of channels and surface regions. During the travel time of a particle along any one of the above paths, it spends a certain amount of time in each of the states that actually compose the path. The time T_x that a particle spends in state x ($x = o_i$ or $x = c_i$) is a random variable that can be described by probability density functions $f_x(t)$. Obviously, for different states x and y, T_x and T_y can have different pdf's $f_x(t) \neq f_y(t)$, and we assume that T_x and T_y are statistically independent for $x \neq y$ [Rodriguez-Iturbe and Valdes, 1979]. The latter assumption is not deemed entirely necessary to the mathematical construction, but it is indeed physically plau-

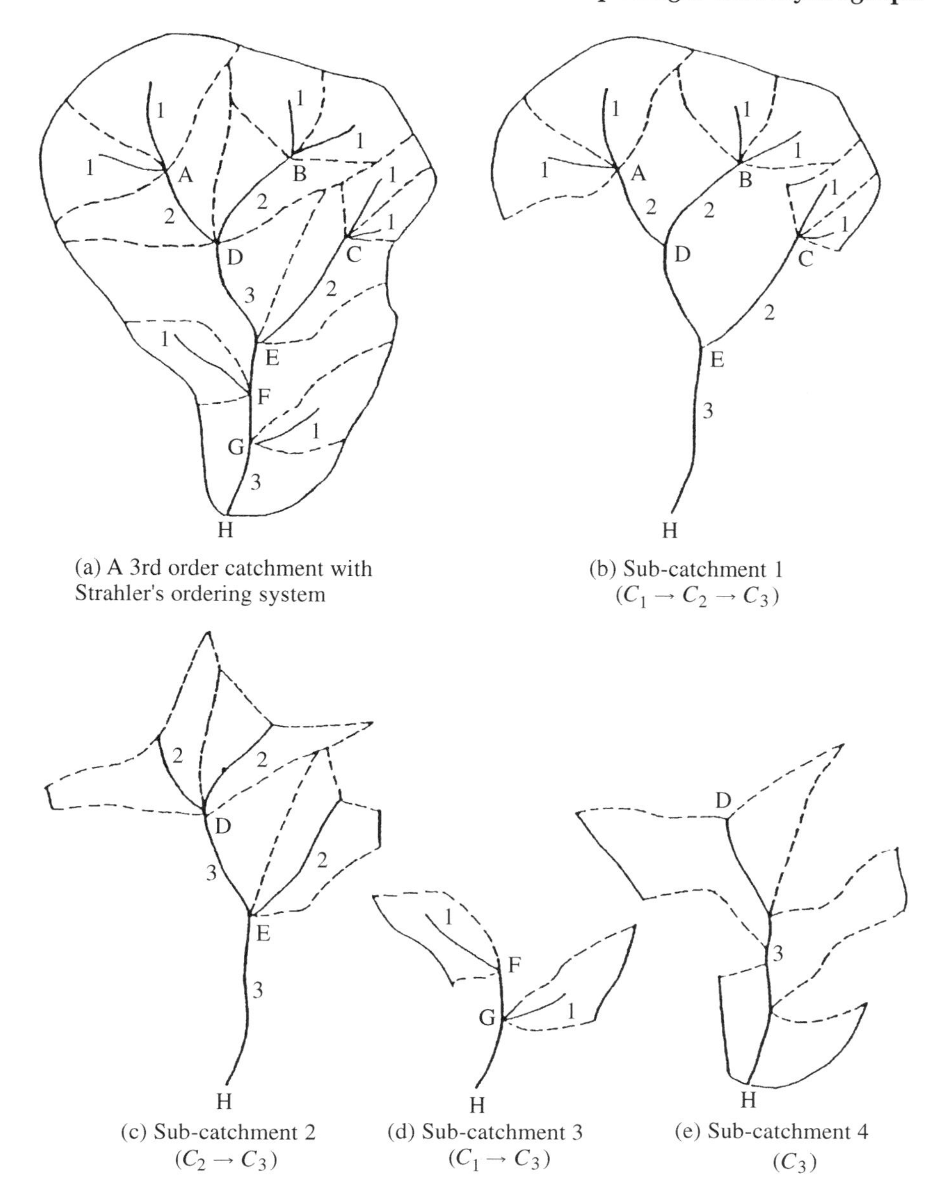

Figure 7.3. Third-order basin [after Rodriguez-Iturbe and Valdes, 1979; and Jin, 1992].

sible. For a path $\gamma \in \Gamma$ defined by the collection of states $\gamma = \langle x_1, \ldots, x_k \rangle$ (where, in turn, $x_1, \ldots, x_k \in (o_1, .., o_\Omega, c_1, .., c_\Omega)$) we define a travel time T_γ through the path γ as

$$T_\gamma = T_{x_1} + \cdots .. + T_{x_k} \tag{7.27}$$

From the statistical independence of the random variables T_{x_i} it follows that the derived distribution $f_\gamma(t)$ of the sum of the (independent) residence times T_{x_i} is the convolution of the individual pdfs:

$$f_\gamma(t) = f_{x_1} * \cdot \cdot * f_{x_k} \tag{7.28}$$

where the asterisk denotes the convolution operator.

In the case of the path γ_2 in Figure 7.3 the transitions are defined by the sequence $o_1 \to c_1 \to c_3$ and the residence time is $T_\gamma = T_{o_1} + T_{c_1} + T_{c_3}$. The probability $P(T_\gamma \leq t) = \int_0^t f_\gamma(t)dt$ is uniquely determined by its pdf:

$$f_\gamma(t) = \int_0^t \int_0^{t''} f_{o_1}(t')f_{c_1}(t'' - t')dt' f_{c_3}(t - t'')dt'' \tag{7.29}$$

Travel time distributions $f(t)$ at the outlet of a system whose input mass is distributed over the entire domain are obtained by randomization over all possible paths [Rodriguez-Iturbe and Valdes, 1979; Gupta et al., 1980]:

$$f(t) = \sum_{\gamma \in \Gamma} p(\gamma) f_{x_\omega} * .... * f_{x_\Omega}(t) \tag{7.30}$$

where γ is the arbitrary path constituted of states $\langle x_\omega, \ldots, x_\Omega \rangle$ such that a water particle experiences transitions as $x_\omega \to \cdots \to x_\Omega$; Γ is the collection of all possible paths from the source to the outlet; and $p(\gamma)$ is the path probability.

We now define the types of path probabilities. In the simplest case of uniform spatial distribution, the path probability is simply defined by $p(\gamma) = A_{x_\omega}/A$ where A_{x_ω} is the contributing area draining into the channel state x_ω of the given path γ; A is the total area drained by the channel network. Notationwise, here x_ω is viewed as any element of the network and f_{x_ω} as its residence time distribution.

If the states x_ω are defined as geomorphologic (e.g., Strahler's) states, the path probability may be written as [Rodriguez-Iturbe and Valdes, 1979]

$$p(\gamma) = \pi_{x_\omega} p_{x_\omega, x_{\omega+1}} \cdots p_{x_{\omega+i}, x_{\omega+r}} \cdots p_{x_{\Omega-1}, x_\Omega} \tag{7.31}$$

where π_{x_ω} is the probability that the particle starts its travel in a hillslope segment draining into a stream of order ω; and $p_{x_\omega, x_{\omega+1}}$ is the transition probability from the state x_ω to state $x_{\omega+1}$ (note that in geomorphologic descriptions of river networks, transitions from x_ω to $x_{\omega+2}, x_{\omega+3}$ are admissible). In general, $p_{x_j, x_{j+k}}$ is equal to the number of streams x_j draining into state x_{j+k} divided by the total number of streams x_j. Notice that if states x_ω are identified by Hortonian rules, then the area A_{x_ω} appearing above is defined by $A(\omega, \Omega)$, that is, the area directly draining into streams of order ω embedded in a network of order Ω.

To be more specific, Rodriguez-Iturbe and Valdes [1979] show, that the initial state probabilities π_i and the transition probabilities $p_{i,j}$ are functions only of the geomorphology and the geometry of the river basin. The physical interpretation of the probabilities is as follows:

$$\pi_i = \frac{\text{total area draining directly into the stream of order } i}{\text{total basin area}}$$

$$p_{i,j} = \frac{\text{number of streams of order } i \text{ draining into streams of order } j}{\text{total number of streams of order } i}$$

It is interesting to use results related to Strahler's ordering, that is, computing transition probabilities calculated as a function of the number N_i

Table 7.1. *Initial and transition probabilities for Hortonian third-order basins*

$\pi_1 = \frac{R_B^2}{R_A^2}$	$p_{1,2} = \frac{R_B^2 + 2R_B - 2}{2R_B^2 - R_B}$
$\pi_2 = \frac{R_B}{R_A} - \frac{R_B^3 + 2R_B^2 - 2R_B}{R_A^2(2R_B - 1)}$	$p_{1,3} = \frac{R_B^2 - 3R_B + 2}{2R_B^2 - R_B}$
$\pi_3 = 1 - \frac{R_B}{R_A} - \frac{R_B^3 - 3R_B^2 + 2R_B}{R_A^2(2R_B - 1)}$	$p_{2,3} = 1$

Source: after Rodriguez-Iturbe and Valdes [1979].

of Strahler streams of each order i [Gupta et al., 1980]:

$$p_{i,j} = \frac{(N_i - 2N_{i+1})E(j,\Omega)}{\sum_{k=1+1}^{\Omega} E(k,\Omega)N_k} + \frac{2N_{i+1}}{N_i}\delta_{i+1,j} \tag{7.32}$$

where δ is Kronecker's delta distribution (unit and different from zero only if $j = i + 1$); and $E(i, \Omega)$ is the mean number of interior links of order i in a finite network of order Ω, whose expression is given by [e.g., Smart, 1972]

$$E(i,\Omega) = N_i \prod_{j=2}^{i} \frac{(N_{j-1} - 1)}{2N_j - 1} \tag{7.33}$$

for $i = 2, \ldots, \Omega$. Interior links have been defined in Chapter 1. Likewise, the probability that a drop falls in an area of order ω is approximated by the following expressions [Rodriguez-Iturbe and Valdes, 1979]:

$$\pi_1 = \frac{N_1 \bar{A}_1}{A(\Omega)} \tag{7.34}$$

$$\pi_\omega = \frac{N_\omega}{A(\Omega)} \left(\bar{A}_\omega - \sum_{j=1}^{\omega-1} \bar{A}_j \frac{N_j p_{j,\omega}}{N_\omega} \right) \tag{7.35}$$

for $\omega = 2, \ldots, \Omega$, where $\bar{A}_i$ means mean area draining into state i. Bras [1990] noted that slight inaccuracies are embedded in the above equations because it is assumed that stream transitions from streams of order i to streams of order j are uniformly distributed between different drainage pathways $i \to j, j = 1, \ldots, \Omega$. Consequences are minor, and one just needs to adjust the initial probabilities to avoid the possible occurrence of very small negative probabilities.

The complete list of initial and transition probabilities is given in Table 7.1 for the third-order basin of Figure 7.3, where use of Horton's bifurcation and area laws, described in Chapter 1, is made to substitute for each N_ω and $\bar{A}_\omega$ as functions of Horton's ratios R_B and R_A.

In the representation of the GIUH given by Eq. (7.30), the path probability functions $p(\gamma)$ can be completely specified on the basis of basin morphology. However, the pdfs f_{x_i} of the random holding times appearing in Eq. (7.30) need additional consideration.

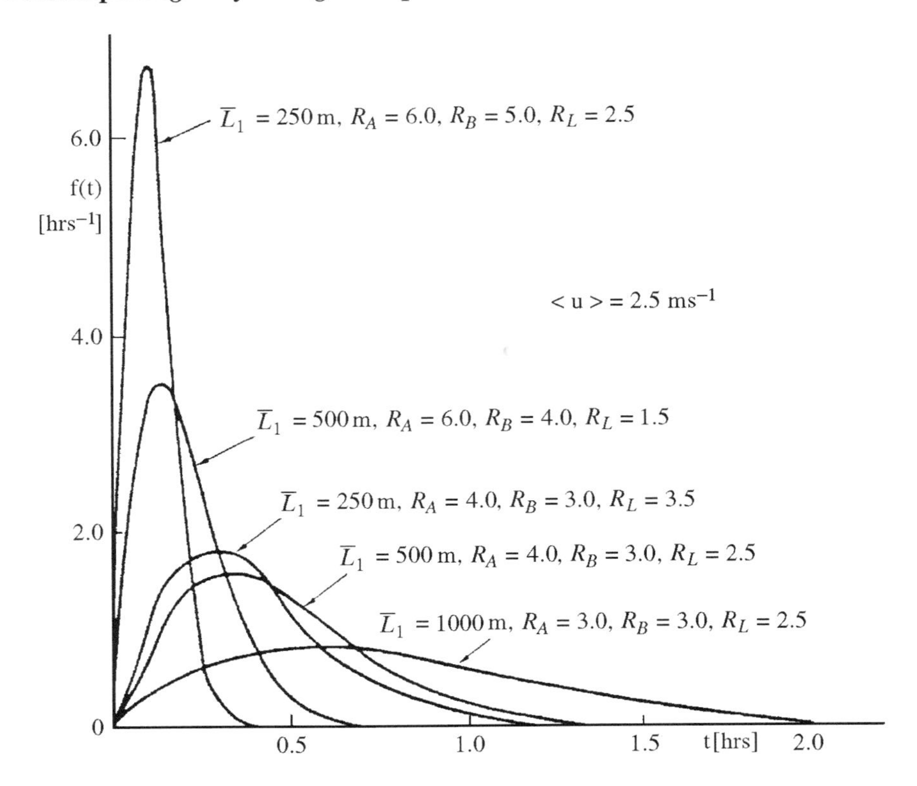

Figure 7.4. GIUHs for different geomorphologic properties [after Rodriguez-Iturbe and Valdes, 1979].

Early models of GIUH considered particular forms for the pdfs in Eq. (7.30). In the original version of Rodriguez-Iturbe and Valdes [1979] the pdfs f_{x_i} were assumed to be exponential defined by characteristic timescales $\lambda_{x_i} > 0$ ($f_{x_i}(t) = \exp(-t/\lambda_{x_i})/\lambda_{x_i}$). Rodriguez-Iturbe and Valdes also offered physically based support for the exponential assumption. In this case, the term $f_{x_1} * \cdots * f_{x_k}$ in Eq. (7.30) becomes the k-fold convolution of independent nonidentically distributed exponential random variables given by the following expression:

$$f_{x_1} * \cdots * f_{x_k}(t) = \sum_{j=1}^{k} C_{jk} e^{-\lambda_{x_j} t} \tag{7.36}$$

where the coefficients C_{jk} are given by

$$C_{jk} = \frac{\lambda_{x_1} \cdots \lambda_{k_1}}{(\lambda_{x_1} - \lambda_{x_j}) \cdots (\lambda_{x_{j-1}} - \lambda_{x_j})(\lambda_{x_{j+1}} - \lambda_{x_j}) \cdots (\lambda_{x_k} - \lambda_{x_j})} \tag{7.37}$$

In such a case, the GIUH is given by the following expression:

$$f(t) = \sum_{\gamma \in \Gamma} \sum_{j=1}^{k} p(\gamma) C_{jk} e^{-\lambda_{x_j} t} \tag{7.38}$$

which allows the derivation of important relationships.

Examples of derived GIUHs are shown in Figures 7.4 and 7.5 for cases of Hortonian networks of order $\Omega = 3$ where the states x_i are identified through Strahler's orders i. In these examples, as well as in many practical formulas derived from the theory of the GIUH, a characteristic velocity

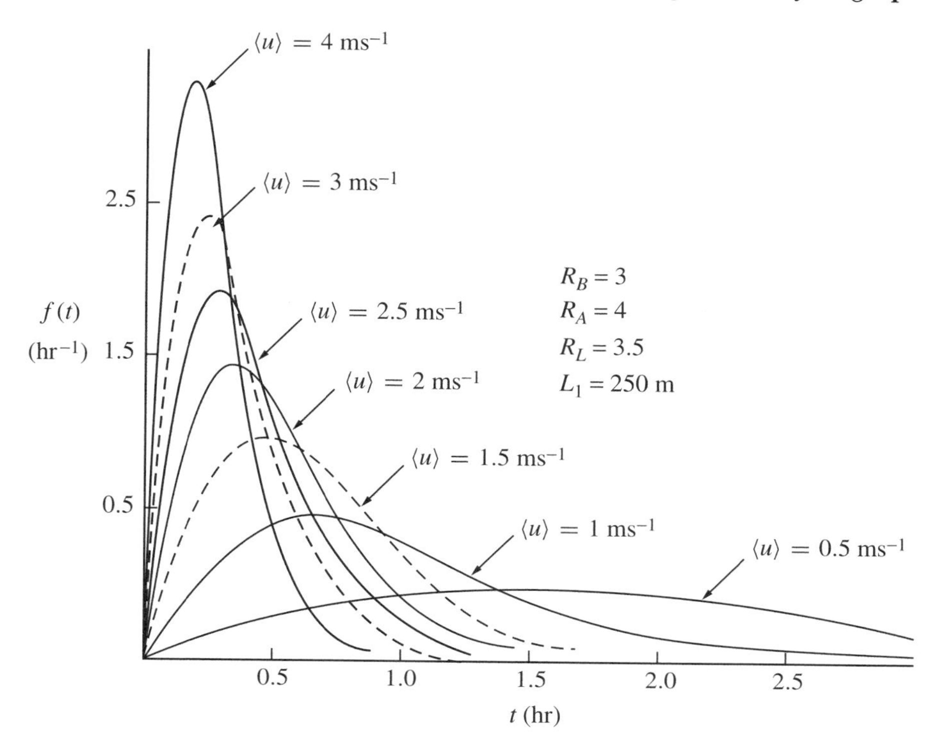

Figure 7.5. GIUHs for different drift velocities [after Rodriguez-Iturbe and Valdes, 1979].

$\langle u \rangle$ is assumed to be representative for the flow everywhere in the basin throughout the whole occurrence of the runoff hydrograph. The assumption of a spatially constant velocity was discussed under a different context in Chapter 1. Its validity as a practical approximation for the description of the dynamics of the unit impulse response function was established through the experiments of Valdes et al. [1979].

The assumption of constant drift velocity in each stream, $\langle u \rangle$, yields characteristic residence timescales $\lambda_{x_i} = \bar{L}_i/\langle u \rangle$ where $\bar{L}_i$ is the average length of streams of order i. The different λ_{x_i}s are all related through Strahler's length ratio, R_L, and the average length of any set of streams of a given order. Figure 7.4 shows the impact of different geomorphologies in the GIUH where the same drift velocity is assumed for all cases. Figure 7.5 shows the effects of different drift velocities when the same geomorphologic description is used in all cases. In these examples the path probabilities $p(\gamma)$ were computed from Eq. (7.35) with the terms π_i and p_{ij} as given in Table 7.1.

Rodriguez-Iturbe and Valdes [1979] also gave the expression for the most important characteristics of a GIUH, the peak Q_P and the time to peak T_P. Unfortunately, the sum of exponential functions appearing in Eq. (7.38) does not lend itself to exact mathematical manipulations. Thus Rodriguez-Iturbe and Valdes [1979] resorted to numerically analyzing the variations of the above quantities in a large number of combinations of geomorphologic parameters to derive the following relationships:

$$T_P = 0.44 \frac{L(\Omega)}{\langle u \rangle} \left(\frac{R_B}{R_A} \right)^{0.55} R_L^{-0.38} \tag{7.39}$$

The choice of units for $L(\Omega)$ [km] and $\langle u \rangle$ [m/s], yields T_P in hours. The peak discharge in $[h^{-1}]$ is

$$Q_P = \frac{1.31\langle u \rangle}{L(\Omega) R_L^{0.43}} \tag{7.40}$$

These results have a good deal of practical importance and will be used later for different kinds of comparisons. Interestingly, the product $Q_P T_P$ is independent of the velocity $\langle u \rangle$ and of the scale $L(\Omega)$ and is approximated by $Q_P T_P \approx 0.58(R_B/R_A)^{0.55}$. This suggests that in practice the GIUH may be determined by only one parameter, either T_P or Q_P, and moreover it allows an approach to the elusive problem of hydrologic similarity [Rodriguez-Iturbe et al., 1979].

Eqs. (7.39) and (7.40) can be rearranged so as to be measured in coherent units, resulting in [Rosso, 1984]

$$Q_P = 0.36 R_L^{0.43} \langle u \rangle L(\Omega)^{-1} \tag{7.41}$$

$$T_P = 1.58 \left(\frac{R_B}{R_A} \right)^{0.55} \frac{L(\Omega)}{\langle u \rangle} \tag{7.42}$$

Rosso [1984] also fitted Eqs. (7.41) and (7.42) to the peak and time to peak of a gamma probability density function defined as

$$f(t) = \left(\frac{t}{k} \right)^{\alpha - 1} \frac{\exp^{-t/k}}{k\Gamma(\alpha)}$$

(where $\Gamma(x)$ is the complete Gamma function of argument x). The two parameters α and k are then given by

$$\alpha = 3.29 \left(\frac{R_B}{R_A} \right)^{0.78} R_L^{0.07} \tag{7.43}$$

$$k = 0.70 \left(\frac{R_A}{R_B R_L} \right)^{0.48} L(\Omega) \langle u \rangle^{-1} \tag{7.44}$$

where k, $\langle u \rangle$, $L(\Omega)$ should be measured in coherent units.

Among other related results of interest, we recall the geomorphoclimatic extension of the original idea by Rodriguez-Iturbe et al. [1982a,b]. In their extension the dynamics of direct runoff formation embedded in the flow velocity was related to the effective rainfall intensity $\bar{i}$. Peak discharge Q_P and the time to peak T_P of the GIUH for a storm with intensity $\bar{i}$ were obtained as

$$Q_P = \frac{0.871}{\mathcal{P}^{0.4}} \qquad T_P = 0.585 \mathcal{P}^{0.4} \tag{7.45}$$

where $\mathcal{P} = L(\Omega)^{2.5} / (\bar{i} A(\Omega) R_L \Theta_0^{1.5})$; Θ_0 is a parameter, $\Theta_0 = \sqrt{S_0(\Omega)} / (n\, w(\Omega)^{2/3})$; $\bar{i}$ is the mean intensity of the effective rainfall; $L(\Omega)$, $S_0(\Omega)$, and $w(\Omega)$, are the average length, slope, and width of the highest stream Ω in a basin, respectively; and, finally, n is a suitable Manning's coefficient for uniform flow resistance [e.g., Henderson, 1966]. The units are the same as those for Eqs. (7.39) and (7.40) with $\bar{i}$ in centimeters per hour and $A(\Omega)$ in square kilometers. The randomization of the storm characteristics also allows for a probabilistic description of the unit hydrograph characteristics [Rodriguez-Iturbe et al., 1982a,b]. An extensive review of the geomorphologic unit hydrograph theory and its geomorphoclimatic extension may be found in Rodriguez-Iturbe [1993].

Within the above general framework for the GIUH, Gupta et al. [1980] assumed the f_{x_i}'s were uniform over intervals $[0, \tau_{x_i}]$ where the parameters are the timescales τ_{x_i}. The resulting convolution of Eq. (7.30) reads

$$f_{x_1} * \cdots * f_{x_k}(t) = \frac{t_+^{k-1}}{\prod_{i=1}^{k} \tau_{x_i}(k-1)!} + \sum_{\nu=1}^{k}(-1)^{\nu} \sum_{x_\nu \in (x_1,\ldots,x_k)} \left(t - \sum_{i=1}^{\nu}\right)_+^{k-1} \tag{7.46}$$

where $_+$ means the positive part. Thus the GIUH under the above assumption is built by polynomial splines of degrees no greater than the order of the basin.

Gupta et al. [1980] observed that the above expressions provide geomorphologic insight into the fictitious but widely used construct for obtaining the IUH of a basin in terms of responses of linear reservoirs and linear channels [Nash, 1957a,b; Dooge, 1973]. For instance, it has been shown [e.g., Rinaldo and Marani, 1987] that modeling the hydrologic response by a cascade of equal reservoirs implies only one path with n states whose mean residence time $\lambda_{x_i} = \lambda$ is the same. This is a very particular case of Eq. (7.30).

The theory of the GIUH, through the incorporation of the morphology of the network, allows for a realistic description of the dynamics and the effects of the geometrical characteristics. In classical conceptual models all dynamic and morphological effects are blended into parameters – at times of very unclear physical meaning – that cloud distinctions of the role of each component.

Many other papers have addressed the characterization of the GIUH. We will not review these here because their main concern was unrelated to the impact of fractal structures on the features of the hydrologic response. However, as we will discuss in the next sections of this chapter, the theoretical framework of the GIUH is indeed of very general character. It allows important analytical derivations in cases of complex fractal structures and links transport problems of different nature.

To close this section, we look at a few other interesting questions that concern the differences between travel time for nodal injections or distributed arrivals along the streams and the role of fast and slow flow components. Bras [1990, Chapter 12] describes several applications of the GIUH theory, including critical reexaminations of some of the original modeling assumptions. Particular reference is made to the incorporation of hillslope travel times through incorporating a fast and a slow component into total travel times [Gupta et al., 1980; Gupta and Waymire, 1983; Kirshen and Bras, 1983; Mesa and Gupta, 1987; Van der Tak and Bras, 1988; Wyss, 1988; Jin, 1992].

The issue of hillslope travel times is of importance. In fact, hillslope residence times are responsible not only for lags in the overall routing but are also important to the understanding of derived transport processes, chiefly, solute generation and transport to runoff waters [Rinaldo et al.,

1989a,b; Rinaldo and Rodriguez-Iturbe, 1996]. Apart from mass transfer processes that chiefly occur within hillslope states, it has been stressed [Troendle, 1985; Bras, 1990] that the flow velocity increases by at least an order of magnitude once the water reaches the channel by whatever pathway it has accessed.

As observed by Bras [1990], to characterize residence times outside the networks states, that is, in the ith *hillslope* state h_i, all that is needed is the introduction of a proper travel time pdf $f_{h_i}(t)$. Wyss [1988] suggested that the travel distance from any point in a hillslope to the closest stream may be taken to be exponentially distributed with mean distance equal to $\bar{L}_h = 1/2\mathcal{D}$ where $\mathcal{D}$ is the drainage density of the network (Chapter 1), that is

$$f_{h_i}(t) = \lambda_{h_i} e^{-\lambda_{h_i} t} \tag{7.47}$$

with $\lambda_{h_i} = 1/(2\mathcal{D}V_h)$ where d is the (local) drainage density and V_h is a characteristic scale of velocity within the hillslopes. Although at the cost of added parameters, the incorporation of the above mechanism of travel time seems indeed relevant at basin scales not exceedingly larger than the hillslope scales.

Finally, the response for a time-varying effective rainfall time series $i(t)$, $t \geq 0$, uniformly injected in space onto the watershed can be obtained in the simplest case of linear invariant hydrologic response by a further convolution integral:

$$Q(t) = \int_0^t i(\tau) f(t-\tau) d\tau \tag{7.48}$$

where $f(\cdot)$ is the GIUH. The convolution integral Eq. (7.48) can be easily computed by discretizing it into appropriate convolution sums or by using appropriate transforms. The foremost practical problem is the determination of the net or effective fraction of rainfall $i(t)$ and its temporal pattern that contributes to direct runoff. This problem is outside the main aim of this book and will not be addressed here. A variable of key importance in this transformation is the soil moisture whose spatial structure continuously evolves through time owing to hydrologic forcing such as rainfall and evapotranspiration. An introductory perspective of the fractal character of soil moisture spatial structure will be given in Section 7.10.

Within the framework of this chapter, geomorphoclimatic schemes may be derived through the coupling of residence time concepts with soil moisture dynamics to produce nonempirical evolutions of net rainfall distributions. An interesting application is derived from Eq. (7.48) by using defective kernels, that is

$$Q(t) = \int_0^t i(\tau) f(t-\tau) \mathcal{S}(t-\tau, \tau) \, d\tau \tag{7.49}$$

where $\mathcal{S}$ represents a loss function. This and other applications are examined in Rinaldo and Rodriguez-Iturbe [1996].

Examples of use of GIUHs for the computation of the direct runoff response are shown in Figure 7.6 for a 8.6-km^2 mountain watershed subject to intense storms [after Rinaldo et al., 1989a,b].

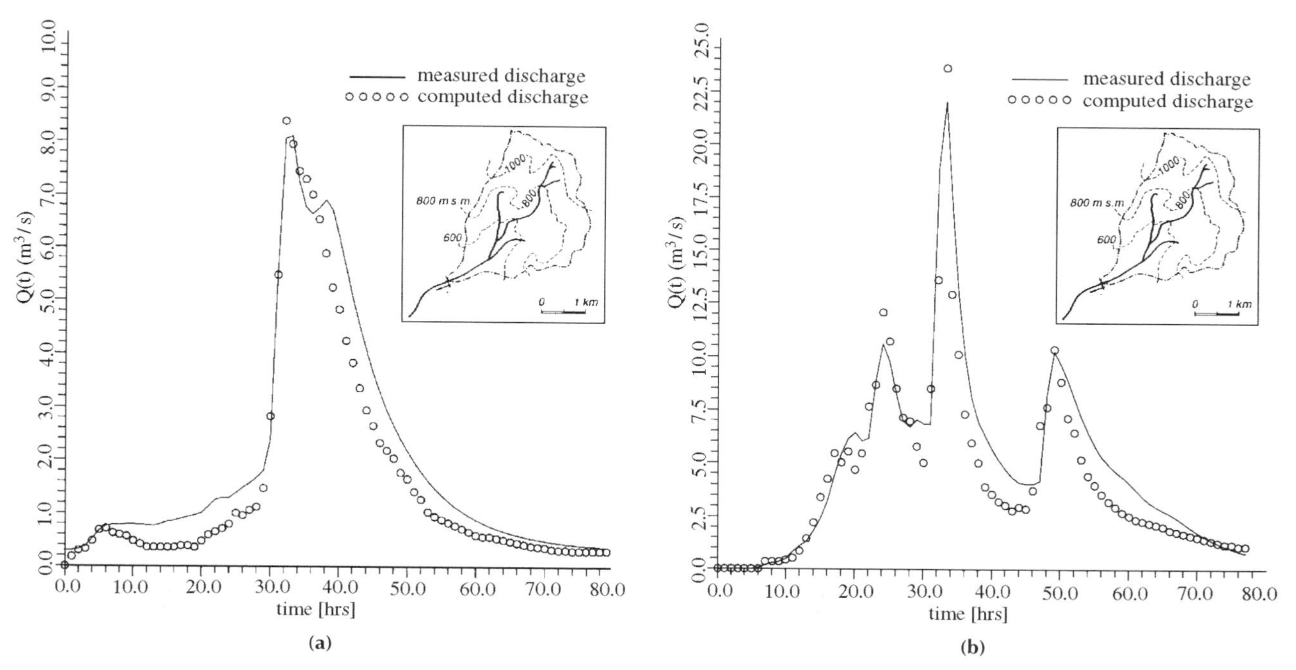

Figure 7.6. Example of use of GIUHs to compute real-life hydrologic responses [after Rinaldo et al., 1989a,b].

7.4 Travel Time Distributions in Channel Links

The previous examples of GIUH assumed particular forms of travel time distributions within channel links, namely, the exponential [Rodriguez-Iturbe and Valdes, 1979] and the uniform [Gupta and Waymire, 1983] distributions. Nevertheless, the formulation of transport by travel time distributions presented in Section 7.2 allows us to link in a general manner the choice of distribution with general momentum balance equations. In this section we will discuss how to derive residence time distributions in which both convective and diffusive effects are present. The results derived will be instrumental in deriving analytical results in the case of Hortonian (and fractal) channel networks (see Section 7.7).

Our main goal here is to derive from first principles the pdfs $f_X(t)$ of residence times within individual reaches X constituting the network. In fact, the geomorphologic framework first proposed by Rodriguez-Iturbe and Valdes [1979] allows the incorporation of the topological properties of a network as a whole into the hydrologic response regardless of the detailed specification of travel times within parts of the whole.

We notice here that in the following treatment we will not consider fractal sinuosity of the individual reaches. This is done in order to keep the formalism simple enough and yet able to capture the impact of fractal characters of the network structure such as its space-filling total length (see Chapter 2). As such, for instance, the average length of streams of Strahler order ω embedded in a network of order Ω will be denoted by $L(\omega, \Omega)$. Where the scale of observation defining the order Ω is defined unambigously, we will turn to the simpler notation $L(\omega)$. Also, we drop the use of the overbar, that is, $\bar{L}(\omega)$ to denote average length of streams of a given order ω for simplicity of notation. We seek a theoretical framework that enables us to describe the effects of fractal characters of the network on its hydrologic response.

Individual reaches are open channels whose transport capacity depends on hydraulic factors. The control volumes are viewed as one-dimensional in the coordinate x, $x \in (0, L)$ (where L is the length of the reach) along the longitudinal direction, so that the flux at the outlet is the flow rate $Q(L, t)$. Geomorphologic unit hydrograph approximations using linear flow through topologically random or perfectly Hortonian channel networks have been studied by Troutman and Karlinger [1985] and Rinaldo et al. [1991], among others. In these studies the rule for motion of matter in channel reaches is described by average convection and an analog to hydrodynamic dispersion, sometimes termed the parabolic model. We will now summarize the basic implications of such studies in order to exploit their features in the perspective of a fractal framework.

Within a channel reach, say X, of length L, discharge $Q(x, t)$ (which at its endpoint yields the flux $Q(L, t) \equiv Q(t)$), cross-sectional area $S(x, t)$, topwidth w, hydraulic radius R_H, and flow depth $y(x, t)$, mass continuity yields

$$\frac{\partial Q}{\partial x} + \frac{\partial S}{\partial t} = 0 \tag{7.50}$$

where we neglect the distributed injection of mass along the reach, thus concentrating it at the nodes. Integration on the transport volume $\mathcal{V}$ yields

$$Q(L, t) \propto \frac{d}{dt} \int_{\mathcal{V}} dx\, S(x, t) \tag{7.51}$$

Let us now turn to the probabilistic interpretation of the flux at the control section given in Eq. (7.17). Clearly, following the Taylorian theorem Eq. (7.1), the displacement probability $g(x, t)$ is proportional, in the simple one-dimensional scheme above, to the cross-sectional area $S(x, t)$. In a rectangular channel characterized by constant width w, such an area is simply $S(x, t) = w\, y(x, t)$. In general, if the gradients of w are negligible

$$g(x, t) \propto y(x, t) \tag{7.52}$$

Thus the time evolution of the free surface y gives information on the structure of the travel time distribution because the sought-after pdf $f_X(t)$ equals the (properly normalized) flux $Q(L, t)$.

From the complete De Saint Venant equations of momentum balance [e.g., Henderson, 1966] different approximations can be derived. Following Rinaldo et al. [1991], we will employ the so-called linear channel, which basically yields a simplified Fokker–Planck probability propagation. In fact, a convenient writing of the dynamic equation for the flow velocity $u(x, t)$ at the arbitrary site x [Henderson, 1966, Eq. (9.29)] is

$$u(x, t) = \chi \sqrt{R_H \left(S_0 - \frac{\partial y}{\partial x} - \frac{u}{g} \frac{\partial u}{\partial x} - \frac{1}{g} \frac{\partial u}{\partial t} \right)} \tag{7.53}$$

where χ is Chezy's friction factor and S_0 is the average slope in the reach. The nonlinear character of the previous equation is removed by dropping some of its terms. Specifically, Henderson [1966] contends that from experimental analyses, the relative weight of the four terms S_0, $\partial y/\partial x$, $(u/g)\partial u/\partial x$, and $(1/g)\partial u/\partial t$ are 26, 1/2, 1/8 – 1/4, and 1/20, respectively. The previous figures hold even for fast-rising floods. Thus influences of the third and fourth terms are viewed as negligible. In this case, dropping the related terms in Eq. (7.53) and assuming that reference uniform-flow conditions y_0 and u_0 are meaningful, the dynamic equation reduces to

$$\frac{\partial y}{\partial t} + \langle u \rangle \frac{\partial y}{\partial x} = D_L \frac{\partial^2 y}{\partial x^2} \tag{7.54}$$

where

$$\langle u \rangle = \frac{3}{2} u_0 = \frac{3}{2} \chi \sqrt{y_0 \left(S_0 - \frac{\partial y}{\partial x} \right)} \approx \frac{3}{2} \chi \sqrt{y_0 S_0} \tag{7.55}$$

and

$$D_L \approx \frac{\langle u \rangle y_0}{3 S_0} \tag{7.56}$$

D_L is defined in this context as a hydrodynamic dispersion coefficient. The physical meaning of $\langle u \rangle$ is that of the mean kinematic celerity of the traveling wave. Experimental evidence [e.g., Pilgrim, 1976; 1977] supports the validity of models of hydrologic response employing a basin-constant drift $\langle u \rangle$. In practical cases, we also assume that D_L is a constant. A reasonable range for dimensional D_L values is 10^2–10^3 m^2/s.

In the framework of travel time distributions the probability $g(x, t)dx$ for a particle being in $(x, x + dx)$ at t is described by the so-called Wiener process (or *biased* diffusion) where the probabilistic process is described by particles displaced by a mean drift $\langle u \rangle t$ to which a Brownian motion X_B is superimposed. Recall – see Section 7.2 – that the stochastic process $X_B(t)$ implies $\langle X_B \rangle = 0$ and $\langle X_B^2 \rangle = 2D_L t$, that all odd moments $\langle X_B^{2n+1} \rangle = 0$, $n = 1, 2, \ldots$, and that all even moments $\langle X_B^{2n} \rangle$ are a function of $\langle X_B^2 \rangle$. In such a case, the displacement pdf $g(x, t)$ can be computed by the general model of longitudinal dispersion described by the one-dimensional equation [Cox and Miller, 1965]:

$$\partial g/\partial t + \langle u \rangle \partial g/\partial x = D_L \partial^2 g/\partial x^2 \tag{7.57}$$

The solution of the above equation for particular boundary conditions was given in Section 7.2.

As noted in Section 7.2, Gaussian solutions, like, for example, Eq. (7.23), are not suited to network models owing to matters of boundary conditions. In fact, Eq. (7.57) requires specification of two conditions which, in the solution Eq. (7.23), are given by the character of a reflecting barrier at $x = 0$ and the Gaussian decay at $x = \infty$, respectively. The appropriate boundary conditions for network models are as follow: (i) initial conditions are (let $x_0 = 0$, $t_0 = 0$) $g(x, 0) = 0$ except at the origin where a unit pulse of flux at $x = 0$ is enforced as in Eq. (7.19):

$$\left| \langle u \rangle g(x, t) - D_L \frac{\partial g}{\partial x} \right|_{x=0} = \delta(t) \tag{7.58}$$

(ii) as shown by Cox and Miller [1965, pp. 219–221], the only appropriate condition when the endpoint of a reach is the source node for a subsequent propagator is that a sink be placed at each tube end $x = L$, i.e., $g(L, t) = 0$.

To obtain manageable solutions, it is convenient to solve Eq. (7.57) using Laplace transform techniques. As such

$$\hat{g}(s, x) = \mathcal{L}(g) = \int_0^\infty g(x, t)e^{-st}\, dt \tag{7.59}$$

and the transformed equation reads

$$s\hat{g}(s, x) + \langle u \rangle \frac{\partial \hat{g}(s, x)}{\partial x} - D_L \frac{\partial^2 \hat{g}(s, x)}{\partial x^2} = 0 \tag{7.60}$$

where obviously two boundary conditions are required. The above follows from the continuity of the function g and the operational property of the Laplace transform

$$\int_0^\infty e^{-st} \frac{\partial g(x, t)}{\partial t}\, dt = s\hat{g}(s, x) - g(x, 0) \tag{7.61}$$

after integration by parts and making use of $g(x, 0) = 0$.

The Laplace transform, say $\hat{f}_X(s)$, of the travel time distribution $f_X(t)$ follows from Eq. (7.16) as

$$\hat{f}_X(s) = -s \int_0^L \hat{g}(s, x)\, dx \tag{7.62}$$

The general solution for the Laplace transform of first-passage distributions $\hat{f}_X(s)$ in the reaches X of length L is obtained by solving Eq. (7.60):

$$\hat{f}_X(s) = \bar{A}e^{L\theta_1(s)} + \bar{B}\, e^{L\theta_2(s)} \tag{7.63}$$

where the arguments of the exponentials follow from the roots of the characteristic equation of Eq. (7.60):

$$\theta_{1,2}(s) = (\langle u \rangle \pm \sqrt{\langle u \rangle^2 + 4sD_L})/2D_L \tag{7.64}$$

The general solution for $\bar{A}, \bar{B}$ yields a rather opaque expression [De Arcangelis et al., 1986, Eq. (6)]. Cox and Miller [1965, Chapter 5] give a discussion of the different solutions of Eq. (7.63) as a function of the type of boundary conditions.

Regarding a choice of constants in Eq. (7.63), two points are relevant when referring to channel networks. On one hand, in fact, one wonders whether the imposition of the reflecting barrier at $-\infty$ rather than close to the injection point has any effect. This is important because the mathematics is considerably simpler. Intuitively, one would tend to think that such an assumption may be valid especially for long channels and large velocities. This was studied by Rinaldo et al. [1991] through the values of what they call basin Peclet number, $Pe = \langle u \rangle L(\Omega)/D_L$, which measures the relative importance of convective over diffusive terms. In the dynamic conditions arising in real channels the average basin Peclet number is indeed large.

Furthermore, with reference to the role of runoff-contributing areas, one also wonders to what extent uncertainty on the injection point $x = 0$ affects travel time distributions.

To address the first point, we observe that general solutions are available in the Laplace transform domain [Cox and Miller, 1965, Chapter 5]. Straightforward numerical Laplace inversion shows that at $Pe = 10$ (here

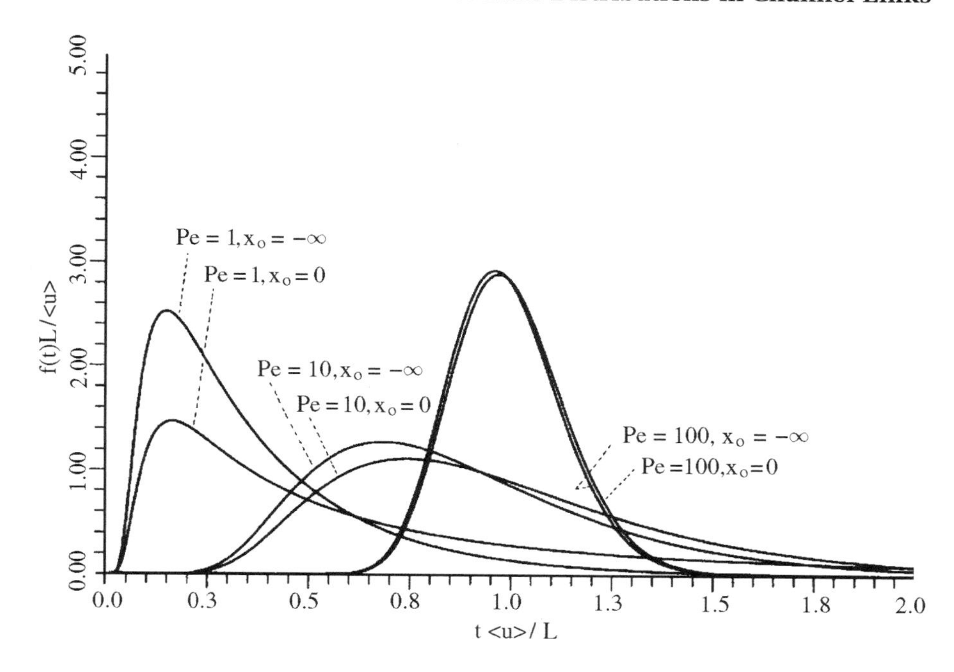

Figure 7.7. The effect of a reflecting barrier on travel time distributions within individual reaches at differing *Pe* numbers [after Rinaldo et al., 1991].

the spatial scale is defined by the distance L to the absorbing barrier located at the end point of the reach), setting a reflecting barrier at $x = 0$ or at $-\infty$ yields very similar pdfs (Figure 7.7). Therefore the solution with a reflecting barrier at infinity, which is considerably simpler, is well approximated by the solution with a reflection at the injection point already at $Pe = 10$.

Thus for practical purposes in channel network hydrodynamics one can neglect the effect of upstream boundaries and use a reflecting barrier at $x = -\infty$. This considerably simplifies the Laplace transform of travel time distributions within individual reaches. In fact, recalling that $\hat{f}_X(0) = \int_0^\infty f(t)dt = 1$, we have, for a travel distance L

$$\hat{f}_X(s) = e^{L\theta(s)} \tag{7.65}$$

where $\theta(s) = (\langle u \rangle - \sqrt{\langle u \rangle^2 + 4sD_L})/2D_L$ (the positive root is discarded for the requirement of boundedness). Eq. (7.65) is suited to analytic inverse transformation [Rinaldo et al. 1991] to yield

$$f_X(t) = \left(\frac{L}{4\sqrt{\pi D_L} t^{3/2}} \right) \exp(-(L - \langle u \rangle t)^2 / 4D_L t) \tag{7.66}$$

which is an expression in which the relative effects of convective and storage-diffusion effects are accounted for. Moreover, the parameters define not only the mean travel time but also the overall effects of storage and friction in deforming traveling waves within the reach.

Figure 7.8 shows the dimensionless travel time distributions for individual reaches obtained from Eq. (7.66) at various *Pe* values for fixed travel length L:

$$f_X(t) = \frac{\langle u \rangle}{L} \frac{\sqrt{Pe}}{4\sqrt{\pi t^3}} e^{-Pe(1-t)^2/(4t^2)} \tag{7.67}$$

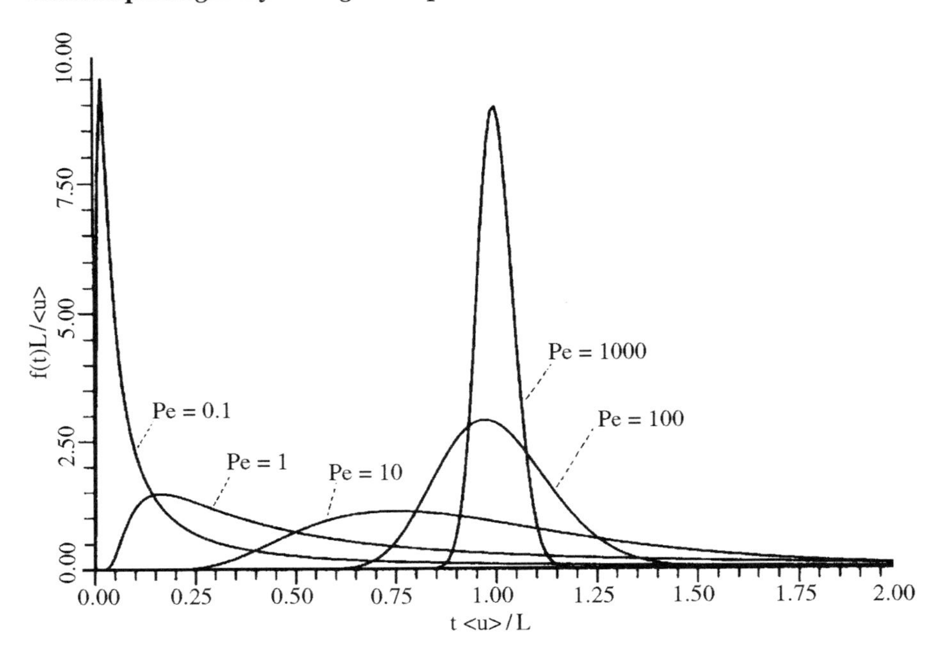

Figure 7.8. Travel time distributions at various *Pe* numbers [after Rinaldo et al., 1991].

with time being dimensionless as $t\langle u\rangle/L$. Note that when increasing the predominance of convection the solution tends to the Dirac distribution characteristic of pure convection. The parameter of the curves is the basin Peclet number *Pe* qualifying the ratio of propagation and storage effects. It is clear that the travel time pdf tends to linear storage routing for $Pe \to 0$ and to pure convection for $Pe \to \infty$.

The travel time distribution in the more restrictive case of absorbing barrier at $x = L$ and of a reflecting barrier at $x = -\infty$ (let $x_0 = 0$, $t_0 = 0$ for simplicity of notation), with constant D_L, is given by Eq. (7.23):

$$f_X(t) = \frac{1}{2}\left(\frac{L}{t} + \langle u\rangle\right) g\,(L, t) \tag{7.68}$$

with

$$g\,(L, t) = \frac{1}{\sqrt{4\pi D_L t}} e^{-\frac{(L-\langle u\rangle t)^2}{4D_L t}} \tag{7.69}$$

as discussed in Section 7.2.

The role of the transitions from overland areas to channel states deserves a brief mention. If we treat the injection point $x = x_0$ as a random variable because uncertainty affects its location, we obtain

$$\hat{f}_X(s) = \int_0^L e^{(L-x_0)\theta(s)} \tilde{p}\,(x_0) dx_0 \tag{7.70}$$

where $\tilde{p}\,(x)dx$ is the marginal probability for the particle to be in $(x+dx, x-dx)$ at the initial time. In the important case of p uniformly distributed [Kirshen and Bras, 1983] $(p\,(x_0) = 1/L)$, one obtains

$$\hat{f}_X(s) = (e^{L\theta(s)} - 1)/\theta(s)L \tag{7.71}$$

Hence in the general formulation we let $\hat{f}_X(s)$ represent all mechanisms of travel time in the network link. This will be the subject of further analysis in Section 7.8.

7.5 Geomorphologic Dispersion

Let $\hat{f}(s)$ be the Laplace transform of $f(t)$. Then from Eq. (7.30) one obtains [Rinaldo et al., 1991]

$$\hat{f}(s) = \sum_{\gamma \in \Gamma} p(\gamma) \prod_{x_\omega \in \gamma} \hat{f}_{x_\omega}(s) \tag{7.72}$$

which generalizes the first passage probability density obtained for random networks [e.g., De Arcangelis et al., 1986] in that the sum over all the paths Γ from the inlet to the outlet is weighed by the path probability.

The question of whether Eq. (7.72), which gives an exact rule for computation of first passage distributions, is amenable to analytic solution rests on the definition of $f_{x_\omega}(t)$, which in turn is the travel time probability density that matter injected at the inlet of the state x_ω reaches the endpoint of the reach in a time t. Eq. (7.72) is derived through the property that the convolution operator reduces to the product of Laplace transforms:

$$\mathcal{L}\left(f_1 * f_2 * \cdots f_n(t)\right) = \hat{f}_1(s)\hat{f}_2(s)\cdots\hat{f}_n(s) \tag{7.73}$$

This operational rule is needed in the following.

The transform of individual travel time distributions in Eq. (7.73) is obtained through Eq. (7.65) with the present notation:

$$\hat{f}_{x_\omega}(s) = e^{L(\omega)\theta(s)} \tag{7.74}$$

where $\theta(s) = (\langle u \rangle - \sqrt{\langle u \rangle^2 + 4sD_L})/2D_L$. Obviously, the reach x_ω has length $L(\omega)$.

The above represents a dynamic model of routing in individual reaches that embeds models of the linear storage and the kinematic wave as particular cases. Substitution into Eq. (7.72) yields

$$\hat{f}(s) = \sum_{\gamma \in \Gamma} p(\gamma) e^{-\sum_{x_\omega \in \gamma} \theta(s) L(\omega)} \tag{7.75}$$

which is valid whatever the variations in hydrodynamics (e.g., $\langle u(\omega) \rangle$, $D_L(\omega)$) from headwaters to the outlet of the basin.

Eq. (7.75) can be exactly inverted, yielding a close-form GIUH, when $\langle u \rangle$ and D_L are independent of the geomorphologic order ω. The corresponding analytic expression for the hydrologic response (Eq. (7.30)) is the following [Rinaldo et al., 1991]:

$$f(t) = \frac{1}{4\sqrt{\pi D_L t^3}} \sum_{\gamma \in \Gamma} p(\gamma) L(\gamma) e^{-((L(\gamma) - \langle u \rangle t)^2 / 4D_L t)} \tag{7.76}$$

where $L(\gamma) = \sum_{x_\omega \in \gamma} L(\omega)$ is the length to the outlet through the γ path, and D_L is given by Eq. (7.56). Eq. (7.76) allows the computation of travel time distributions in structured networks for which basin-constant mechanisms are meaningful.

We observe that the analytical formulation of the GIUH, Eq. (7.76), allows the direct calculation of many statistics, as shown later in this section.

Furthermore, the assumption of a known convection $\langle u \rangle$ may be relaxed by randomization of the adopted velocity. Let $\tilde{p}(\tilde{u})$ be the marginal distribution describing the uncertainty in the estimation of $\langle u \rangle$ ($\tilde{u}$ is the estimate of $\langle u \rangle$). The unconditional distribution is given by

$$f(t) = \int_0^\infty f(t|\tilde{u})\tilde{p}(\tilde{u})d\tilde{u} \tag{7.77}$$

where $f(t|\tilde{u})$ is represented by Eq. (7.76). In particular cases, say the uniform distribution, close-form solutions for Eq. (7.77) are available. In general cases in which the marginal distribution is given (as in the noteworthy case of experimental measurements), numerical solutions to Eq. (7.77) may be easily computed. Eq. (7.77) considers the impact of the error of estimation of the average velocity upon $f(t)$. If we assume a Gaussian distribution of $\tilde{u}$ characterized by the two moments, $\langle \tilde{u} \rangle$ and $\sigma_{\tilde{u}}^2$, one can estimate the interval of confidence (estimated through the variance $\sigma_f(t)$) for $f(t)$ via

$$\sigma_f^2(t) = \left(\frac{\partial f}{\partial \tilde{u}}\right)^2\Bigg|_{\langle \tilde{u} \rangle} \sigma_{\tilde{u}}^2 \tag{7.78}$$

which depends on time t. In the case of the analytical GIUH, Eq. (7.76) (to simplify matters let $\partial D_L/\partial \tilde{u} \approx 0$) the variance $\sigma_f^2(t)$ is

$$\sigma_f^2(t) = \frac{\sigma_{\tilde{u}}^2}{2D_L t^3}\left(\sum_{\gamma\in\Gamma} p(\gamma)L(\gamma)\frac{L(\gamma)-\langle \tilde{u} \rangle t}{2D_L} e^{-\frac{(L(\gamma)-\langle \tilde{u} \rangle t)^2}{4D_L t}}\right)^2 \tag{7.79}$$

The dotted line of Figure 7.9 [from Rinaldo et al., 1991] shows the $\pm\sigma$ interval of confidence for the travel time distribution of a single reach (Eq. (7.66)). The single reach is assumed of unit length, and $\langle u \rangle$ and D_L are also assumed unit.

Application of the preceding equation yields the confidence interval in the rather broad range $0.3 \le \sigma_{\tilde{u}}/\langle \tilde{u} \rangle \le 1$ for the coefficient of variation of the estimated velocities. Notice that the shaded area in Figure 7.9 is the range of uncertainty for the GIUH $f(t)$ in the case of $\sigma_{\tilde{u}}/\langle \tilde{u} \rangle = 0.3$. The upper and lower dotted curves denote the interval of confidence in the case $\sigma_u/\langle \tilde{u} \rangle = 1$. The curves $f(t) - \sigma_f(t)$ and $f(t) + \sigma_f(t)$ are assumed to be the limits of confidence.

As suggested by Figure 7.9, the range of uncertainty for the peak value of the GIUH $f(t)$ is 27 percent in the case of a coefficient of variation of the velocities of 0.3. It is likely that real-life cases exhibit smaller variations and that one may thus expect that $f(t) \approx \langle f(t) \rangle$ (where the brackets denote ensemble averaging).

It is interesting to recall that Eq. (7.75) is a moment-generating function as

$$\langle t^n \rangle = (-1)^n \frac{d^n \hat{f}(s)}{ds^n}\Big|_{s=0} \tag{7.80}$$

where $\langle t^n \rangle = \int_0^\infty t^n f(t)dt)$ is the nth moment of the arrival time distribution.

The above basic property of the moment-generating (or characteristic) function is derived by expanding in power series the exponential into the

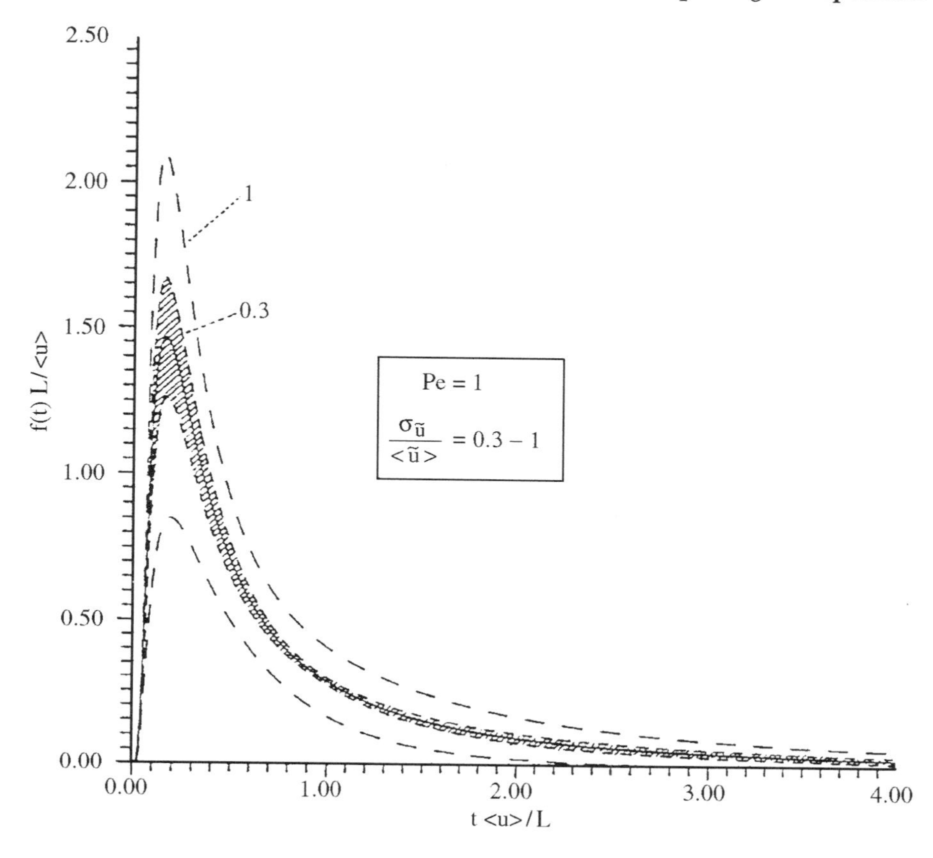

Figure 7.9. Intervals of confidence on travel time distribution for a single reach. The solid line is the GIUH. The pairs of dotted lines denote the intervals of confidence [after Rinaldo et al., 1991].

Laplace transform:

$$\hat{f}(s) = \int_0^\infty e^{-st} f(t)\, dt = \int_0^\infty \left(1 + st + \frac{(st)^2}{2!} + \frac{(st)^3}{3!} + \cdots\right) f(t)\, dt \tag{7.81}$$

Assuming that the series within the integral is uniformly convergent, one can differentiate under the integral sign. Thus

$$\frac{d^n \hat{f}(s)}{ds^n} = \int_0^\infty \left(t^n + st^{n+1} + \frac{s^2 t^{n+2}}{2!} + \cdots\right) f(t) dt \tag{7.82}$$

from which, setting $s = 0$, Eq. (7.80) is recovered.

Straightforward computations yield the mean travel time $E(T) = \langle t \rangle$ and its variance $Var(T) = \langle t^2 \rangle - \langle t \rangle^2$:

$$E(T) = \sum_{\gamma \in \Gamma} p(\gamma) \sum_{x_\omega \in \gamma} \frac{L(\omega)}{\langle u \rangle} \tag{7.83}$$

$$Var(T) = 2 \sum_{\gamma \in \Gamma} p(\gamma) \sum_{x_\omega \in \gamma} \frac{L(\omega) D_L}{\langle u \rangle^3} + \sum_{\gamma \in \Gamma} p(\gamma) \left(\sum_{x_\omega \in \gamma} \frac{L(\omega)}{\langle u \rangle} \right)^2 - \left(\sum_{\gamma \in \Gamma} p(\gamma) \sum_{x_\omega \in \gamma} \frac{L(\omega)}{\langle u \rangle} \right)^2 \tag{7.84}$$

In the important case where $\langle u \rangle$ and D_L are constant, we obtain

$$E(T) = \frac{1}{\langle u \rangle} \sum_{\gamma \in \Gamma} p(\gamma) L(\gamma) \tag{7.85}$$

and

$$Var(T) = 2\frac{D_L}{\langle u \rangle^3} \sum_{\gamma \in \Gamma} p(\gamma) L(\gamma) + \frac{1}{\langle u \rangle^2} \left(\sum_{\gamma \in \Gamma} p(\gamma) L^2(\gamma) - \left(\sum_{\gamma \in \Gamma} p(\gamma) L(\gamma) \right)^2 \right) \tag{7.86}$$

where $L(\gamma) = \sum_{x_\omega \in \gamma} L(\omega)$.

Notice also that when using DEMs every pixel may identify a path. Thus the path probability may be substituted by the relative proportion of pixels at isochrone distances from the outlet [Marani et al., 1991], that is, the path length $L(\gamma)$ is replaced by the arbitrary distance x to the outlet measured along the network and $p(\gamma)$ is replaced by the width function $W(x)$ (Section 1.2.7). One may thus write

$$E(T) = \sum_{x=1}^{N} W(x) \frac{x}{\langle u \rangle} \tag{7.87}$$

and

$$Var(T) = 2 \sum_{x=1}^{N} W(x) \frac{x D_L}{\langle u \rangle^3} + \sum_{x=1}^{N} W(x) \left(\frac{x}{\langle u \rangle} \right)^2 - \left(\sum_{x=1}^{N} W(x) \frac{x}{\langle u \rangle} \right)^2 \tag{7.88}$$

where N indexes the maximum length from source to outlet.

Higher-order moments can be calculated by straightforward but lengthy derivations.

The variance of travel times at the control section is made up by two contributions: one, owing to the first term of Eq. (7.85), involving hydrodynamic dispersion, and the other, dispersion-free, computed by the last two terms in Eq. (7.85). We will discuss them separately here to underline their physical meaning. Let $\bar{L}(\Omega) = \sum_{\gamma \in \Gamma} p(\gamma) \sum_{x_\omega \in \gamma} L(\omega)$ be the mean length of the collection of all paths γ leading to the outlet.

The first contribution to the total variance involves hydrodynamic dispersion, and it involves the weighted sum of the variances of the arrivals generated within every individual link for every path γ. Assuming D_L and $\langle u \rangle$ are constant, this term reduces to the variance $2\bar{L}(\Omega) D_L / \langle u \rangle^3$, which is simply the variance of the arrivals induced by a Gaussian hydrodynamic dispersion along a mean path of length $\bar{L}(\Omega)$.

The two remaining terms do not involve hydrodynamic dispersion and are related to the morphology of the network. They physically determine what part of the overall variance of the arrival times is due to the heterogeneity of the paths leading to the outlet. We observe that this contribution (i) is zero only when $p(\gamma)$ reduces to a single value, that is, only one path, say γ_0, is available to the control section – in any other case Schwartz' inequality proves that this contribution is positive; (ii) defines the contribution to the variance of the travel time distribution owing only to transport along paths of different length.

To quantify the individual roles of the different variance-producing processes, we define a geomorphologic diffusion coefficient D_G as follows (where D_L and $\langle u \rangle$ are basin constants):

$$\begin{aligned}\frac{\langle u \rangle^3 Var(T)}{2\bar{L}(\Omega)} &= D_L + D_G \\ &= D_L + \frac{\langle u \rangle}{2\bar{L}(\Omega)}\left(\sum_{\gamma\in\Gamma} p(\gamma)(L(\gamma))^2 - \left(\sum_{\gamma\in\Gamma} p(\gamma)L(\gamma)\right)^2\right)\end{aligned} \tag{7.89}$$

obtained by rearranging Eq. (7.84). Hence the geomorphologic contribution to the total variance is

$$D_G = \frac{\langle u \rangle}{2\bar{L}(\Omega)}\left(\sum_{\gamma\in\Gamma} p(\gamma)(L(\gamma))^2 - \left(\sum_{\gamma\in\Gamma} p(\gamma)L(\gamma)\right)^2\right) \tag{7.90}$$

which can be generalized as [Marani et al., 1991; Snell and Sivapalan, 1994]

$$D_G = \frac{\langle u \rangle Var[L]}{2E[L]} \tag{7.91}$$

expressed in terms of the moments of the flow distances.

We note that, as in the Taylor [1954] theory of diffusion of conservative matter by continuous movements, different dispersion mechanisms are additive.

It is interesting that, as in the theory of diffusion by continuous movements hydrodynamic dispersion soon obscures molecular diffusion [Taylor, 1921], at the macroscale geomorphologic diffusion tends to overwhelm the dispersion mechanisms operating at smaller scales whenever $D_G \gg D_L$.

The mechanisms contributing to the variance of travel times may be described as follows: (i) Part of the variance of the particles' travel times is explained by the weighted sum of the variances along the individual routes to the network outlet. In particular cases (like, e.g., radial equal streams converging at the end) this would constitute the total variance as the rate of arrivals through different paths would coincide. (ii) A significant contribution to spreading of the arrival rates affecting the variance is derived from the heterogeneity of the paths. The difference between the two types of mechanisms is indeed important because one may assess when the second mechanism is bound to prevail, thereby addressing the modeling effort toward the description of the network characteristics rather than the particular detailed dynamics that takes place at the individual elements of the network.

Snell and Sivapalan [1994] provided an interesting experimental assessment of geomorphologic dispersion through three approaches by which geomorphology is described, namely, (i) using Horton's order ratios to derive pathway parameters, (ii) extracting $p(\gamma)$ from a Strahler order superimposed to the network, and (iii) using a width function (as seen in

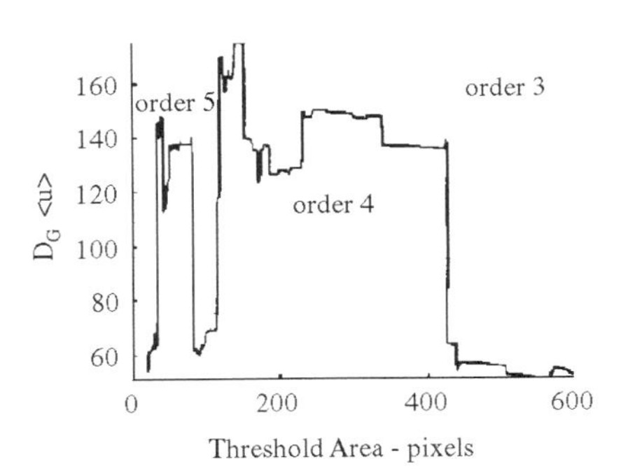

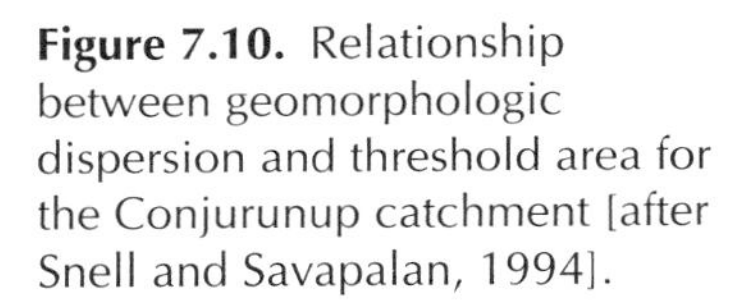

Figure 7.10. Relationship between geomorphologic dispersion and threshold area for the Conjurunup catchment [after Snell and Savapalan, 1994].

Section 1.2.7 identified by a contributing area–flow distance function) extracted directly from accurate DEMs. Networks were extracted automatically using the criterion of constant support areas (Section 1.2.9) and the related width and area functions were thus obtained.

Figure 7.10 shows the wide variability exhibited by the relationship between the geomorphologic dispersion coefficient Eq. (7.90) and the threshold area for channel extraction in the Conjurunup catchment whenever the path probabilities $p(\gamma)$ are forced into Hortonian schemes.

A more stable evaluation of D_G is accomplished through the width function, whose main characters are not heavily affected by the threshold area used. In fact, the geomorphologic diffusivity $D_G/\langle u\rangle$ estimated through the width function, that is, by Eq. (7.91), are $D_G/\langle u\rangle = 859$, 842, and 861 pixels for 425, 83, and 23 pixel area thresholds, respectively.

Snell and Sivapalan [1994] thus conclude that the most reliable estimation of the variance of the GIUH is based on the mean and variance of the area function because it does not reduce the probability distribution into a small number of generic pathways.

7.6 Hortonian Networks

In Chapters 2 and 3 we saw how Horton's laws can be interpreted as signs of the fractal structure of the underlying network. We will now examine the impacts of such fractal characters on the geomorphologic response of a basin embedding a Hortonian network.

We will assume that Horton's laws of bifurcation and length hold strictly, not in an approximate sense, and that the link lengths are strictly constant. With notation like that in Chapter 2, let

$$N(\omega, \Omega) = R_B^{\Omega-\omega}, \quad L(\omega, \Omega) = L(\Omega, \Omega)R_L^{\omega-\Omega} \tag{7.92}$$

where, in a network of order Ω, $N(\omega, \Omega)$ is the number of streams of order ω whose length is $L(\omega, \Omega)$. Use of the relation

$$\sum_{j=\omega}^{\Omega} L(j, \Omega) = L(\Omega, \Omega) \sum_{j=\omega}^{\Omega} R_L^{j-\omega} = L(\Omega, \Omega)\frac{R_L^{\omega-\Omega} - R_L}{1 - R_L} \tag{7.93}$$

is made in the following.

Our starting point is the analytic result for the characteristic function of the travel time distribution, Eq. (7.75), which we reproduce here:

$$\hat{f}(s) = \sum_{\gamma\in\Gamma} p(\gamma) e^{-\sum_{x_\omega \in \gamma} \theta(s) L(\omega,\Omega)} \tag{7.94}$$

where $\theta(s) = (u - \sqrt{u^2 + 4sD_L})/2D_L$ and u is a constant drift. Notice that, from this point, no use will be made of a variable drift velocity. Hence we will drop the bracketed notation $\langle u\rangle$.

A first step in the analysis is taken from Eq. (7.94) when $\theta(s)$ does not depend on ω. Thus

$$\hat{f}(s) = \sum_{\gamma \in \Gamma} p(\gamma) e^{-L(\gamma)\theta(s)} \tag{7.95}$$

where different paths $\gamma = \langle x_\omega, \ldots, x_\Omega \rangle$ (implying transitions $x_\omega \to \cdots \to x_\Omega$) are identified by different distances $L(\gamma)$ to the outlet of the basin.

The distances $L(\gamma)$ can be calculated exactly because the link lengths are assumed constant. Then, from Eq. (7.92), the constant link length, $L(1, \Omega)$, is equal to $L(\Omega, \Omega)/R_L^{\Omega-1}$. As seen in Chapter 2, the link length becomes the natural ruler for metric computations because it tends to zero as the order $\Omega \to \infty$ and allows the order Ω to increase exactly by one unit with a scale change. Under such conditions, the sum over all possible paths γ becomes a sum over all multiples of the link length $L(\Omega, \Omega)/R_L^{\Omega-1}$ ranging from 1 to the maximum distance from source to outlet $L_{\max}(\gamma)$. The maximum distance is obtained by adding the stream lengths of all consecutive orders [Marani et al., 1991]:

$$L_{\max}(\gamma) = \sum_{\omega=1}^{\Omega} L(\omega, \Omega) = L(\Omega, \Omega) \frac{R_L^{1-\Omega} - R_L}{1 - R_L} \tag{7.96}$$

As discussed in Chapter 2, it is interesting to study the limit results for $\Omega \to \infty$. We thus obtain $\lim_{\Omega\to\infty} L_{\max}(\gamma) = R_L L(\Omega, \Omega)/(R_L - 1)$. Notice that $L(\Omega, \Omega)$ is kept fixed while $L(1, \Omega) = L(\Omega, \Omega)/R_L^{\Omega-1} \to 0$ as $\Omega \to \infty$.

The resulting moment-generating function may thus be written in either of the following forms [Rinaldo et al., 1991]:

$$\begin{aligned} \hat{f}(s) &= \sum_{k=1}^{R_L^{\Omega}/R_L - 1} p(k) e^{k\theta(s) L(\Omega,\Omega)/R_L^{\Omega-1}} \\ &= \sum_{k=1}^{R_L^{\Omega}/R_L - 1} p(k) e^{k\theta(s) L(1,\Omega)} \end{aligned} \tag{7.97}$$

where $p(k)$ is the probability of the paths characterized by a common distance $kL(1, \Omega)$ (i.e., k-multiples of the ruler) from the outlet. Notice that the upper limit of the sum is an integer because of the assumption of constant link length. The integer multiples of the link length $kL(\Omega, \Omega)/R_L^{\Omega-1}$ with $k = 1, 2, \cdots, R_L^{\Omega}/R_L - 1$ cover all distances from any node to the outlet. It follows from the explanation leading to Eq. (7.87) that

$$p(k) = W\left(k \frac{L(\Omega, \Omega)}{R_L^{\Omega-1}}\right) = W(k) \tag{7.98}$$

where $W(x)$ is the normalized width function and the notation in the second equality underlines the integer k, which indexes all distances x, that is, $x = kL(1, \Omega)$.

The following relationship may be derived by inverse Laplace transformation:

$$f(t) = \frac{L(\Omega, \Omega)}{R_L^{\Omega-1}\sqrt{4\pi D_L t^3}} \sum_{k=1}^{R_L^{\Omega}/R_L - 1} kW(k) \exp\left[-\left(\frac{kL(\Omega, \Omega)}{R_L^{\Omega-1}} - \langle u \rangle t\right)^2 \Big/ 4D_L t\right] \tag{7.99}$$

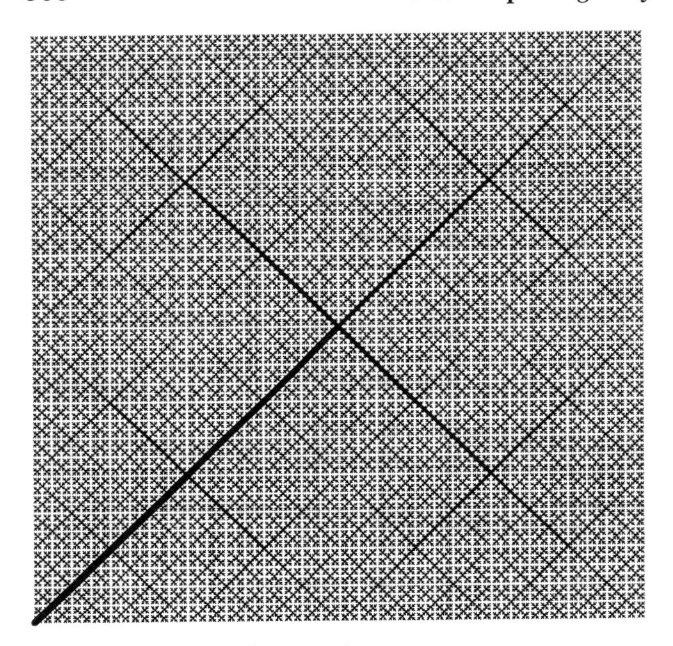

Figure 7.11. Peano's basin at the $\Omega = 6$ stage of generation.

The above relationship is valid for perfectly Hortonian basins with links of equal lengths and constant drift throughout the basin.

An analytical solution for the moments of the hydrologic response has been provided by Rinaldo et al. [1991] for the particular case of Peano's basin (Figure 7.11). It results from Eq. (7.98) through complete mathematical induction over all Strahler's orders. The following results hold exactly for the reference case where $L(\Omega, \Omega) = 1 : R_L = 2$, $R_B = 4$, $L(1, \Omega) = L(2, \Omega) = 1/2^{\Omega-1}$, $L(\omega, \Omega) = 2^{\omega-\Omega} L(\Omega, \Omega)$, with $\omega > 2$.

The formulation for $\hat{f}(s)$ follows from Eq. (7.97) and from the use for $p(k)$ of the values of the width function in Table 3.1. In fact, at the $\Omega = 2$ stage of the multiplicative process we have $\hat{f}(s) = 1/4e^{\theta(s)} + 3/4e^{2\theta(s)}$. At the $\Omega = 3$ stage of generation we have $\hat{f}(s) = 1/16e^{\theta(s)/2} + 3/16e^{\theta(s)} + 3/16e^{3\theta(s)/2} + 9/16e^{2\theta(s)} = 1/16e^{\theta(s)/2}(1 + 3e^{\theta(s)/2})(1 + 3e^{\theta(s)})$. In the general case one has

$$\hat{f}(s) = \frac{1}{2^{\Omega-1}} \prod_{\omega=2}^{\Omega} (1 + 3e^{\theta(s)2^{\omega-\Omega}}) e^{\theta(s)/2^{\Omega-1}} \tag{7.100}$$

The moments of the GIUH can be computed straightforwardly through the derivatives of the characteristic function Eq. (7.100) as described in Section 7.5. Rinaldo et al. [1991] provided the resulting analytic results

$$\frac{E(T)u}{L(\Omega, \Omega)} = \frac{3}{2} \tag{7.101}$$

$$\frac{Var(T)u^2}{L(\Omega, \Omega)^2} = \frac{1}{4} + 3Pe \tag{7.102}$$

and the ratio of geomorphologic dispersion to hydrodynamic dispersion results (see Eqs. (7.90) and (7.91)):

$$\frac{D_G}{D_L} = 1 + \frac{Pe}{12} \tag{7.103}$$

where Pe is the basin Peclet number, $Pe = uL(\Omega, \Omega)/D_L$ as defined in the discussion of Figure 7.7.

It is interesting to compare the above asymptotic moments with the values of the moments of the GIUHs computed by Troutman and Karlinger [1985] and discussed further by Gupta and Mesa [1988]. The assumptions built in their results are the following:

- The travel time pdf in individual links is purely convective with a constant drift everywhere. Thus first-passage time distributions at the endpoint of the link are Dirac-delta distributed. This is a particular case of Eq. (7.99) when $D_L = 0$.
- Link lengths are gamma-distributed. Thus a dispersive effect in the rate of arrivals at the outlet of the network is induced by the heterogeneity of the lengths.

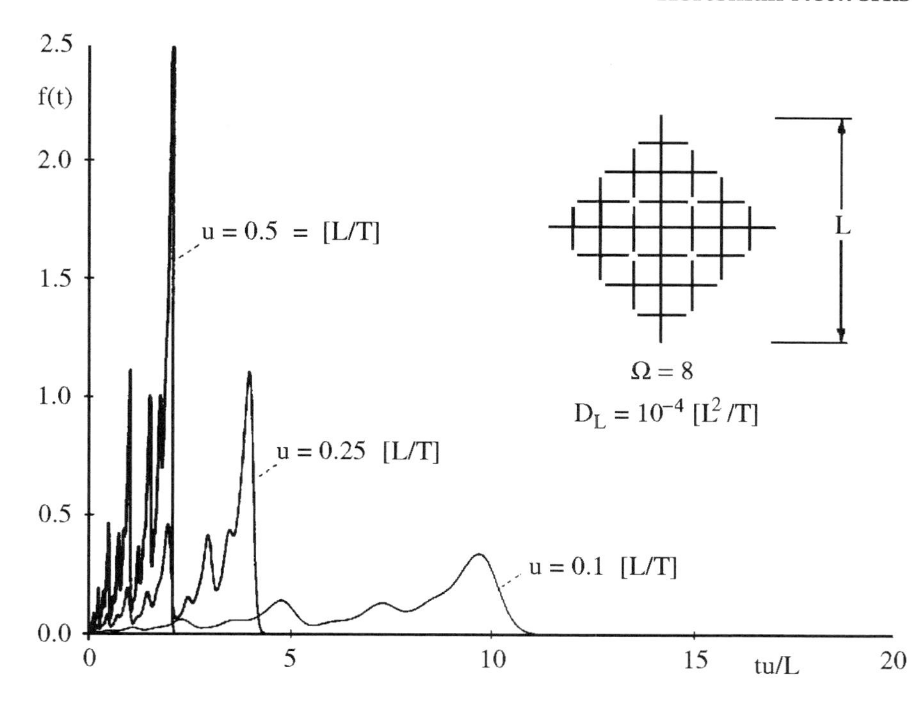

Figure 7.12. The GIUH for a Peano basin in the case $D_L = 10^{-4}[L^2/T]$ and different average velocities [after Rinaldo et al. 1991].

- The network is a topologically random network (see Chapter 1) with magnitude large enough to allow limit results, in the same sense as the Hortonian calculations above for $\Omega \to \infty$.

In our notation, Troutman and Karlinger's asymptotic moments are given by

$$\frac{E(T)u}{L(\Omega, \Omega)} = \sqrt{\pi} \tag{7.104}$$

$$\frac{Var(T)u^2}{L(\Omega, \Omega)^2} = 4 - \pi \tag{7.105}$$

where use has been made of the constant link length $L(1, \Omega) = L(\Omega, \Omega)/R_L^{\Omega-1} = 1/2^{\Omega-1}$ and of the magnitude of the network at the Ω stage of generation, $4^{\Omega-1}$. Notice that Eq. (7.105) compares with $Var(T)u^2/L^2 = 1/4$ in Eq. (7.102) because $D_L = 0$. Also, as explained in Chapter 1, in infinite topologically random networks the expected values of Horton's ratios are $R_B = 4$ and $R_L = 2$, and therefore results can be consistently compared.

The difference in the mean travel time is of the order of 10 percent. The larger variance of the geomorphologic contribution in the formulation of Troutman and Karlinger, Eq. (7.105), with respect to that of Eq. (7.102) is due to the effects of the distribution of link lengths. Nevertheless, the results are of the same order although derived by completely different approaches.

Figures 7.12 to 7.14 illustrate the GIUH, Eq. (7.99), in the case of the Peano basin. The order of the basin for the computations is $\Omega = 8$. The length scale is normalized by the longest path L to the control section. Time is normalized by the time to the control section from the farthest source at unit velocity (arrival from a distance L is at $t = 1$ for $u = 1$ [L/T]).

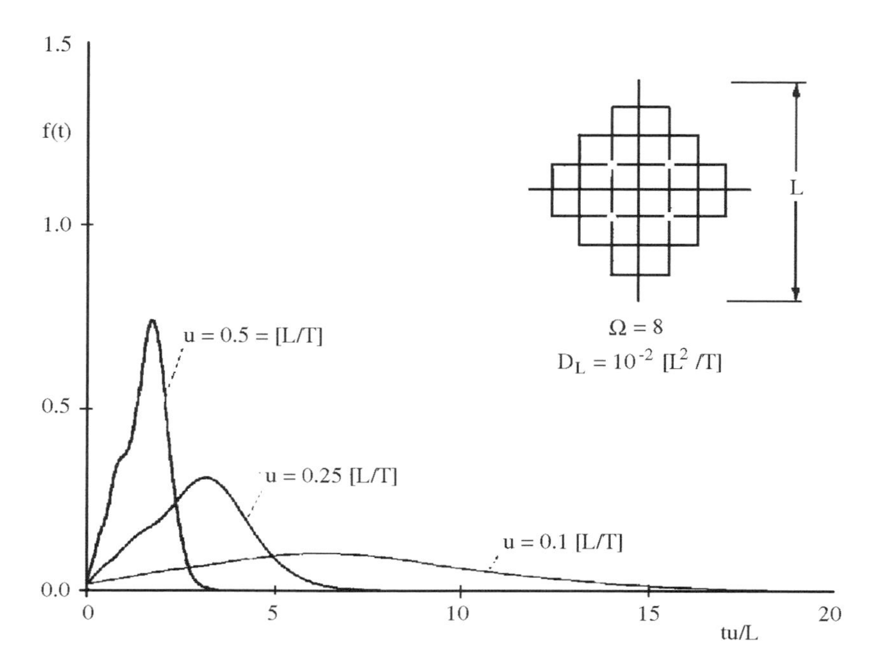

Figure 7.13. The GIUH for a Peano basin in the case $D_L = 10^{-2}[L^2/T]$ and different average velocities [after Rinaldo et al., 1991].

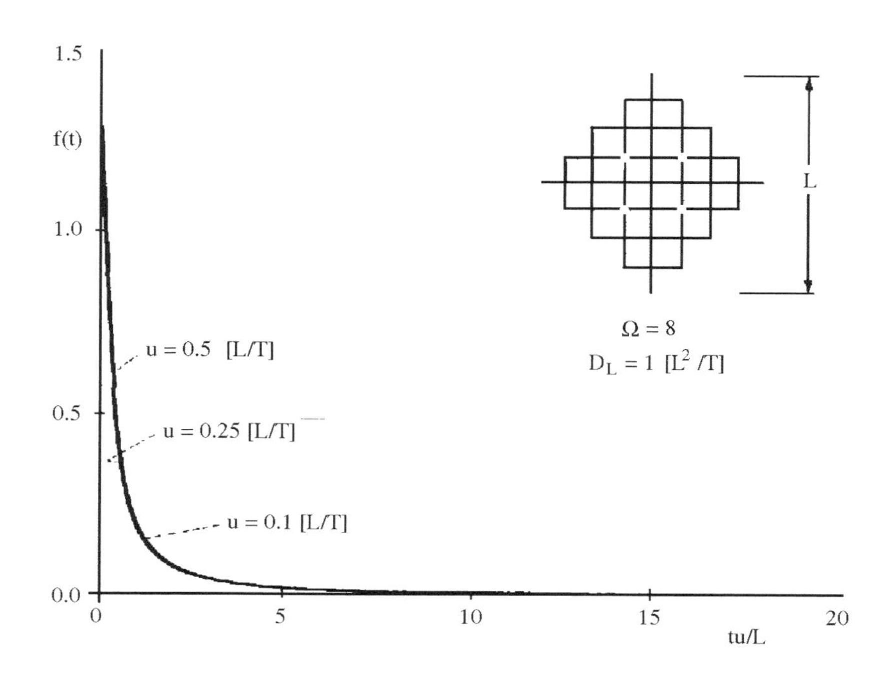

Figure 7.14. The GIUH for a Peano basin in the case $D_L = 1[L^2/T]$ and different average velocities [after Rinaldo et al., 1991].

We note that the cases characterized by small values of D_L (Figure 7.12) tend to reproduce the characters of the width function to which the GIUH reduces as hydrodynamic dispersion becomes negligible. Such character is smeared out progressively by the effects of diffusion. Nevertheless, the foremost characters of the travel time distribution are not affected by D_L.

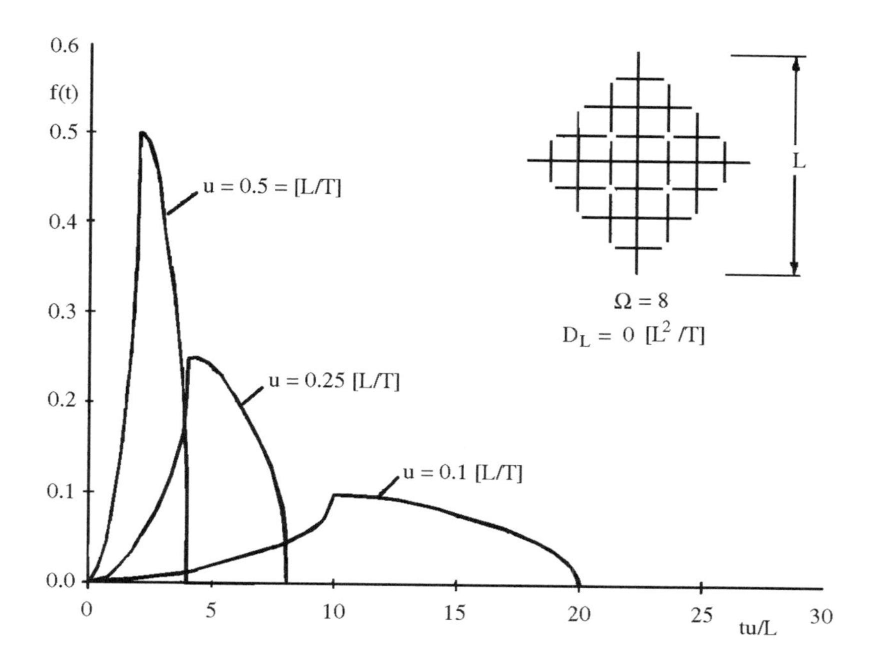

Figure 7.15. Purely convective ($D_L = 0$) hydrologic response of Peano's basin after an input pulse lasting a time L/u [after Rinaldo et al., 1991].

A wide range of $Pe = u/D_L$ values (recall that $L(\Omega, \Omega) = 1$) was tested by changing the values of u and D_L. We observe that for decreasing Pe values the behavior tends to be that characterized by predominant linear storage effects.

Figures 7.15 to 7.18 illustrate the response of Peano's basin under different dynamic conditions for pulses of unit volume, uniformly applied in time. Their duration is fixed by the time to reach the control from the farthest source. Thus pulses of effective rainfall last a duration equal to the time of concentration (i.e., $L_{\max}/u = 2L(\Omega, \Omega)/u$) in the different cases. The response is obtained by numerical convolution of the unit rainfall pulse with the GIUH given by Eq. (7.99). In this case, the order of the basin for the computations is $\Omega = 8$. The length and velocity scales are as in Figure 7.12.

The relative weight of hydrodynamic dispersion and geomorphology can be seen in the different cases by comparison with the case of pure convection, that is, $D_L = 0$ (Figure 7.15). The smoothing of the curves is due to the effects of hydrodynamic dispersion. One can, even visually, detect the small change in the overall variance of travel times with respect to the case of pure convection. The variance of the GIUH is still predominantly due to geomorphologic contributions, that is, to the geometry of the system.

We finally notice that the estimation of the moments of the GIUH for Peano's basin can be linked to a more general problem of diffusion in disordered media, at times termed anomalous diffusion, where the cause of dispersion is the geometric disorder imposed on the paths leading to the control section [Havlin and Ben-Avraham, 1987; Bouchaud and Georges, 1990].

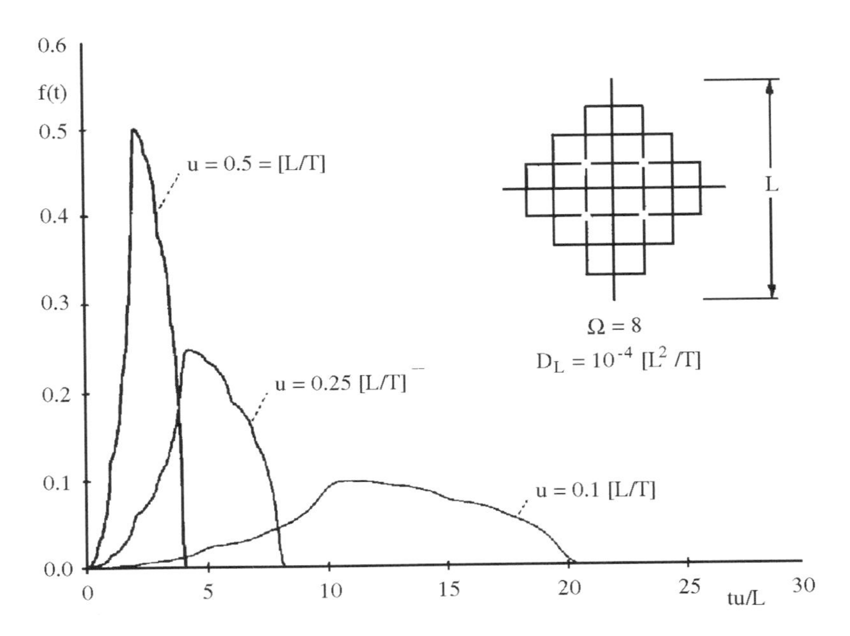

Figure 7.16. Hydrologic response for the Peano basin in the case $D_L = 10^{-4}$ [after Rinaldo et al., 1991].

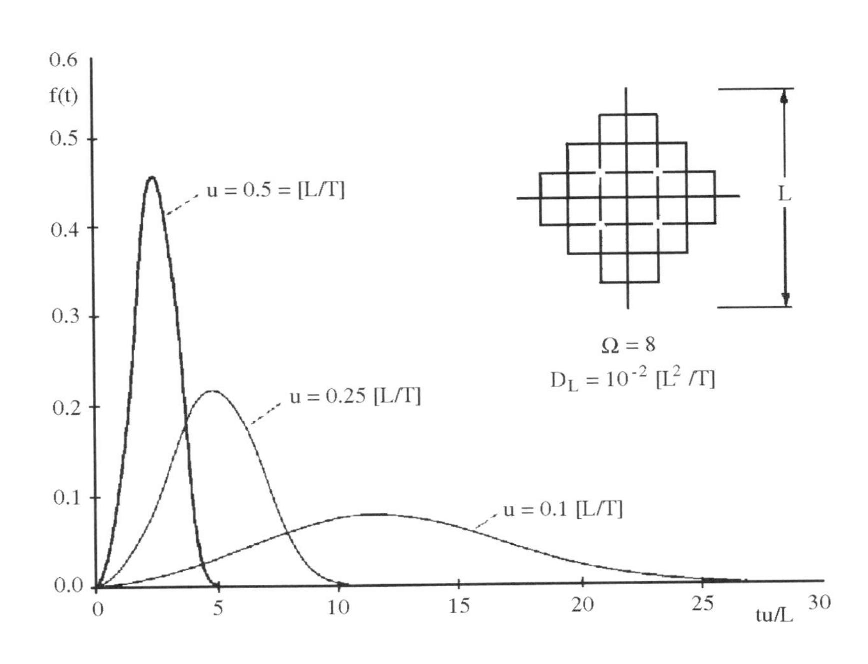

Figure 7.17. Hydrologic response for the Peano basin in the case $D_L = 10^{-2}$ [after Rinaldo et al., 1991].

7.7 Width Function Formulation of the GIUH

As seen in Chapters 1 and 3, the width function $W(x)$ is defined as the probability measure obtained when dividing the number of links at given distance x from the outlet by the total number of links in the network (the distance is measured along the network and normalized by the maximum distance along the streams from source to outlet). We will not consider here

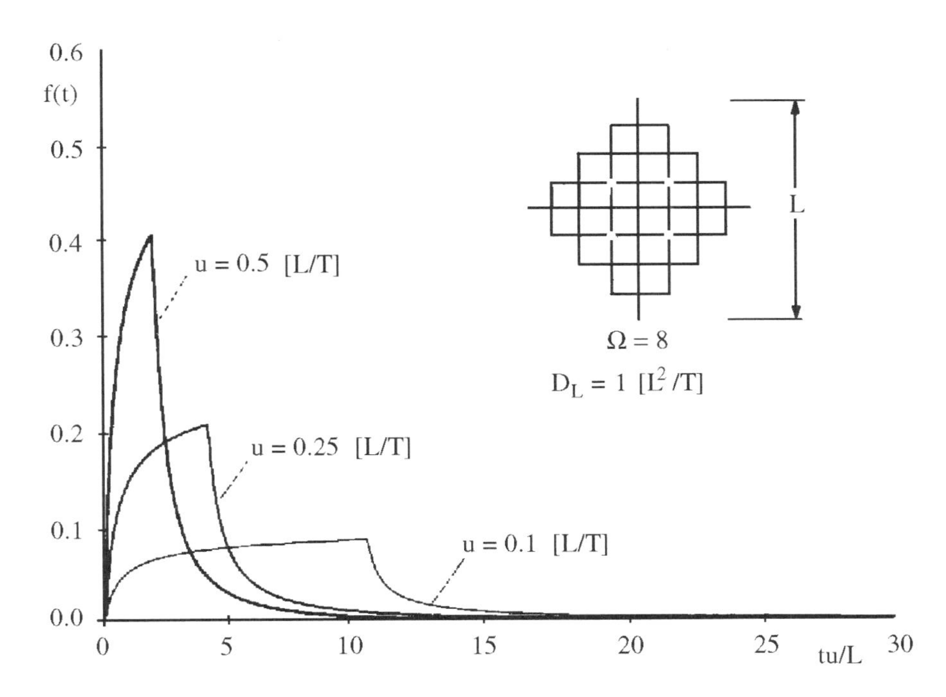

Figure 7.18. Hydrologic response for the Peano basin in the case $D_L = 1$ [after Rinaldo et al., 1991].

fractal sinuosity of the individual links, and therefore we assume that they have a defined finite scale for travel times.

As seen in Sections 2.7.5, 3.4, and 3.7.2, the width function of natural river networks shows common characters that are possibly derived from the multiplicative nature of the embedded dissection process of the total drainage area along the network. $W(x)$ is also directly linked with the runoff response of a basin because it embodies the average of isochrones of travel times in a basin. The first linkage of the hydrologic response with the width function came from Kirkby [1976]. Later the concept was refined by Gupta et al. [1986], Troutman and Karlinger [1984, 1985], Mesa and Mifflin [1988], Gupta and Mesa [1988], and Marani et al. [1991].

Mesa and Mifflin's [1988] formulation of the GIUH is

$$f(t) = \int_0^\infty dx\, f_x(t)\, W(x) \tag{7.106}$$

where $f_x(t)$ is the travel time distribution from the arbitrary distance x, which can be described, for example, by Eq. (7.68), and $W(x)$ is the width function. This follows from the observation that the probabilities defining the number of distinct ways available to reach the outlet at a given time (given that all paths initiated at time $t = 0$) are proportional to the rates of contributing area at time t. The above holds in the important case of uniform injection, which is instrumental to the GIUH.

The width function precisely defines the rate of total drainage area available at a given distance l from the outlet, so that, in general terms:

$$f(t) \propto \int_{l(t_i)}^{l(t_i+\Delta t)} W(x)dx \tag{7.107}$$

where $l(t)$ is the distance from the outlet to the border of the contributing area; $0 \leq l(t) \leq \ell$ indicates that the contributing distances evolve with time as dictated by the dynamics of the process, and ℓ is the alongstream maximum distance from source to outlet.

Consider a particle reaching the outlet at time t. The underlying equation of motion defines its trajectory, and the initial position is at distance $x = l$. Under these assumptions, our basic problem is to determine the probability $P(l; t)$ that the particle, found at the outlet at time t, was in the position $x = l$ at time $t = 0$. Thus in general the width function formulation of the hydrologic response is

$$f(t) \propto \int_0^\infty W(x) dP(x; t) \tag{7.108}$$

Notice that we will use u as the average convection defining the mean travel length at time t as $x = ut$, and that $dP(x; t) = f(x|t)dx$, where f is the probability of a given travel distance x, conditional on the occurrence of the particle at the outlet at time t.

In the simplified but important case of constant dynamics (pure convection with constant velocity u) where $dP(x; t) = \delta(x - ut)dx$ (implying that the particle at the outlet at time t was at distance ut at time $t = 0$ with probability one) the GIUH is simply the width function rescaled as [Gupta et al., 1986]

$$f(t) \propto W(ut) \tag{7.109}$$

One can try to characterize $P(x; t)$, that is, the probability that the arbitrary particle, found at the outlet at time t, was at x at time $t = 0$. One way is to assume a basic rule of motion

$$X(t; l_i, 0) = l_i - \int_0^t u(X(\tau; l_i, 0), \tau) d\tau + X_B(t) \sim l_i - ut + X_B(t) \tag{7.110}$$

where u is the drift, X the Lagrangian position of the particle initially at l_i, and $X_B(t)$ a Brownian motion component, which we define for convenience as characterized by $\langle X_B \rangle = 0$ and $\langle X_B^2 \rangle = 2D_L t$. In such a case, the probability $P(x; t)$ has a density satisfying the Kolmogorov backward equation [Gupta et al., 1986; Rinaldo et al., 1991] whose solution is

$$dP(x; t) = \frac{x\,dx}{\sqrt{4\pi D_L t^3}} e^{-((x-ut)^2/4D_L t)} \tag{7.111}$$

where (see Section 7.3 and Eq. (7.106)) D_L defines the basic hydrodynamic dispersion mechanism. The corresponding expression for the GIUH in Eq. (7.108) is given by [Rinaldo et al., 1991; Marani et al., 1991]

$$f(t) = \frac{1}{\sqrt{4\pi D_L t^3}} \sum_{i=1}^{N_{\max}} W(l_i)\, l_i\, e^{-((l_i-ut)^2/4D_L t)} \tag{7.112}$$

where $l_i = i\Delta l$, Δl is the spatial step size, and $N_{\max}$ is the total number of links in the longest path from source to outlet.

The above formulation of the GIUH, Eq. (7.112), has been applied to real catchments. Figure 7.19 (a) shows the width functions of the Schoharie River Basin (2,408 km^2) whose DEM features are shown in Table 1.1. In the figure, as many as 100,000 pixels define the relative distribution of drainage area along the maximum path from source to outlet. Figures 7.19 (b) to 7.19 (d) show the effects of introducing a hydrodynamic dispersion

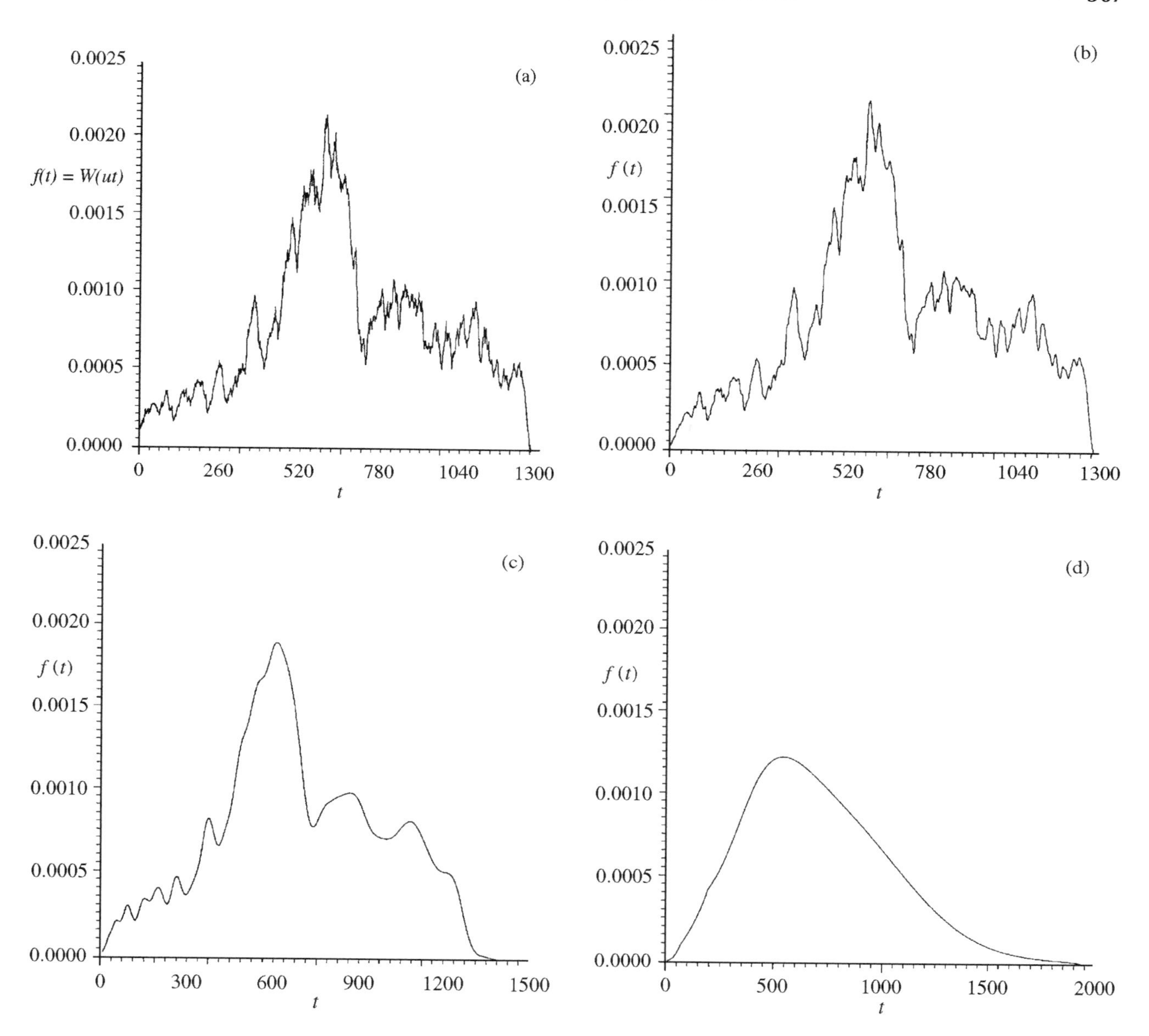

Figure 7.19. The width function of the Schoharie River Basin (2,408 km^2) and the width function formulation (Eq. (7.112)) of the hydrologic response (a) $D = 10^{-6}$ [L^2/T]; (b) $D = 10^{-4}$; (c) $D = 10^{-2}$; (d) $D = 1$. Notice that t is in pixel units, where a unit length and a unit value of u are assumed [after Rinaldo et al., 1995b].

of the type contained in Eq. (7.112) superimposed on a unit drift u, applied to the width functions of the Schoharie River Basin. Here in all cases $u = 1$ [L/T]. The cases shown in Figure 7.19 are (a) $D = 10^{-6}$ [L^2/T]; (b) $D = 10^{-4}$ [L^2/T]; (c) $D = 10^{-2}$ [L^2/T]; and (d) $D = 1$ [L^2/T]. The distribution of the arrivals tends toward a skewed form typical of the hydrologic response.

Throughout this book we have been showing that landscape and network geometries display many scale-independent characters. It is interesting to ask why their hydrologic response does not retain this important character. In other words, although one cannot a priori distinguish even the approximate scale of a topographic map, the hydrologic response of a large network is distinguishable from that of a small watershed in that it is smoother.

The result in Eq. (7.112) suggests that uncertainty in the dynamic specification of the motion within a river network destroys the scaling characters of the hydrologic response. The physical interpretation of this result can be inferred from Eqs. (7.107) and (7.108). In fact, the smoothing of the hydrologic responses reflects the wider bands, $l(t+\Delta t)-l(t)$, on which the width functions $W(x)$ are averaged as time elapses. Only Dirac-delta distributed travel time pdfs reproduce the original self-similar scale-independent structures in a time series of arrival at the outlet of a basin. Also, the effects of uncertainty are selective owing to the nature of the Kolmogorov equation. Thus high-frequency peaks are smoothed much more rapidly than low-frequency modes. As a result, the hydrologic response selectively modifies the main characters of the width function rapidly smoothing its high-frequency features. This is particularly evident for the skewnesses induced in the width functions of Figure 7.19. The selective character of the filtering operated by the dynamics in a fractal network is thus the cause of the lack of fractal characters in its response.

It is important to point out that the dynamic effects described in this section are far from being solely responsible for the typical skewness observed in real life hydrographs or for their characteristic smoothness. Even more important than these effects are those resulting from the dynamics of the hillslopes in the response hydrographs. This topic will be the subject of Section 7.8.2.

7.8 Can One Gauge the Shape of a Basin?

This section investigates the effects of the geometric factors that characterize the shape of a river basin on the features of its hydrologic response. In particular, following Rinaldo et al. [1995b], we wonder whether by measuring the hydrologic response, the salient geomorphic features of the basin can be recovered. We argue that the basic structure of the channel network tends to some universal characters of the width function $W(x)$ that defines the relative proportion of contributing area at distance x from the outlet. $W(x)$ exhibits low-frequency features that are geometry-dominated and high-frequency features that are determined by local aggregation patterns. It is suggested that given the shape of the basin one can indeed forecast in a rational manner the main characters of the hydrologic response because they are imprinted in the width functions, which in turn have dominant features dependent on the basin shape. However, the inverse problem (i.e., the determination of the shape from the measure of the hydrologic response) is less solidly defined because of the possible loss of irretrievable information induced by the dynamics of runoff processes.

In a classic paper, entitled "Can one hear the shape of a drum?," Kac [1966] wondered whether someone with perfect pitch can recover the precise shape of a drum just by listening to its fundamental tone and all the overtones. This question identified an inverse spectral problem and motivated numerous works on the subject. The question, far from trivial, is linked to the asymptotics of the infinite sequence of eigenvalues arising in the drum vibration context and is also linked to whether one can indeed derive complete inference of the geometric boundary conditions from such a sequence. Lapidus [1989] revisited the problem using the context of fractal domains. Borrowing from Kac's title, we investigate how geometric factors characterizing the shape of a river basin affect the features of the basin's hydrologic response.

Besides their intrinsic interest, these questions bear practical consequences for hydrology. In fact, flood prediction in ungauged basins is a crucial engineering problem for disaster prevention. Because the shape of the basin can be obtained objectively, say from space, the link of shape and response is relevant to the above questions as the knowledge of some geomorphic shape factors and spatial scales coupled with climatic observations could surrogate costly and, at times, complex measures.

7.8.1 Estimation of Basin Shape from the Width Function

In this section we show that the information contained in the width function allows for the decoding of the main features of boundary profiles of river basins.

We have already noticed that the geomorphologic width function $W(x)$ has recurrent characters, reflecting common mechanisms of the drainage structure organization regardless of climate, geology, vegetation, etc. These recurrent characters pertain to the high frequencies of their spectra. The low frequencies are quite variable from basin to basin as they reflect the bulk of the contributing area that is available to drainage and therefore grossly reflect the width of the basin from ridge to ridge measured along a direction orthogonal to a diameter. At a larger scale the availability of drainage area and the overall fractal shape of the boundaries are regulated by the competition for drainage and the migration of divides. Within this context we assume that the boundaries are time-invariant.

One may proceed with a simple exercise in which one constructs a network by successive random additions from the outlet. The random additions have to be exactly equal to the width function, which is given a priori. Although there is no unique tree with respect to this rule, the main characters are reproduced as shown below. The reader is referred to Groff [1992] and Rinaldo et al. [1995b] for the details of the method. Here we give only a brief heuristic account of the procedure.

Starting from the outlet (the seed), one constructs a layer of neighbors. Among those neighbors, one wishes to select $W(\Delta x)$ sites equally distant Δx from the seed. The neighboring layer is constituted of $M > W(\Delta x)$ pixels among which one chooses, at random, exactly $W(\Delta x)$ sites, assigning them a drainage direction (i.e., a unit value of the connection matrix W_{ij}, Eq. (1.14), where i and j are the connected pixels). An adjacent layer is then constructed by choosing all the neighbors of the cluster generated by the previous additions. Again one searches, as possible new growth sites, all the pixels whose distance to the seed is $2\Delta x$. The new adjacent layer offers yet another choice of pixels, from which one draws at random exactly $W(2\Delta x)$ pixels. Drainage directions are assigned to these pixels and a larger cluster is created. A new layer is added and the procedure is repeated for $W(3\Delta x), \ldots, W(N\Delta x)$. The procedure is iterated until the length from the seed, $N\Delta x$, is the maximum length from source to outlet specified by the width function.

However, technicalities arise because the above procedure needs to be constrained. In fact, nonphysical realizations (e.g., basins looping onto

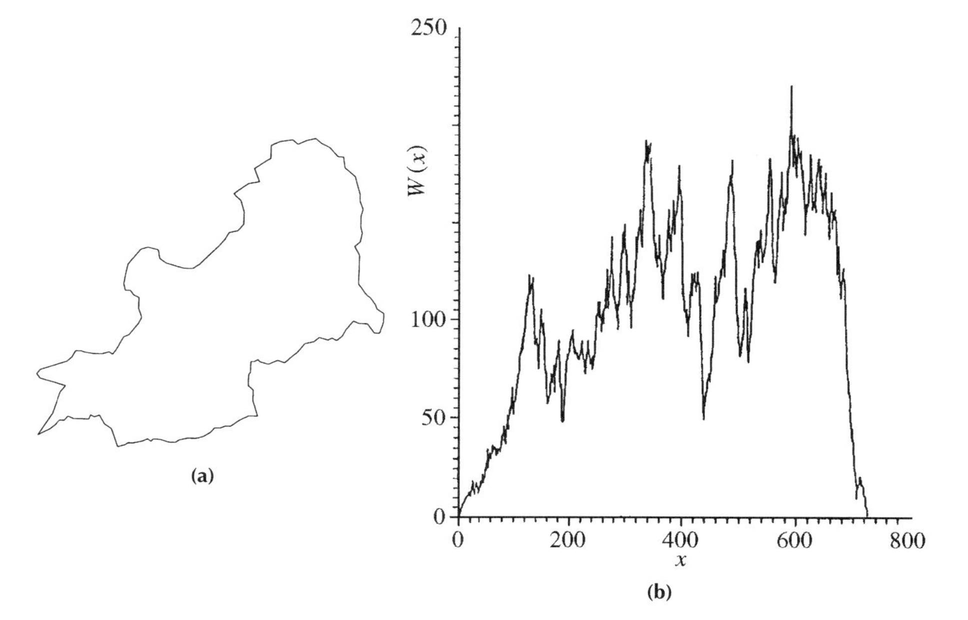

Figure 7.20. (a) The boundary and (b) the width function, $W(x)$, (in pixel units) of the Nelk River Basin [after Rinaldo et al., 1995b].

themselves) are discarded a posteriori, while nonphysical choices of drainage directions (looping, crossing) are not accepted a priori. One also faces the complex problem of developing a plane-filling network. This, at times, forces the choice of certain additions in order to avoid 'holes' in the developing basins.

The development of the structure of the network therefore follows from the availability of area ruled by $W(x)$. Thus the resulting shape is the outcome of a sequence of compatibility controls on the developing structure. Through the above procedure, any realization of the random process is characterized by the prescribed width function $W(x)$.

Figure 7.20 shows the boundaries and the width function of the Nelk River Basin, extracted from a DEM with pixels of size 62.6×92.2 m^2 [after Tarboton et al., 1988; Ijjasz-Vasquez et al., 1993a]. Figures 7.21(a) to 7.21(b) show two realizations of the boundaries of the structure resulting from the above procedure once the width function of the Nelk River Basin is imposed. Figure 7.21(c) shows the reconstructed shape as the ensemble average of ten realizations of the shapes obtained according to the previous approach.

The individual configurations of the basin boundaries obtained are indeed remarkably robust. In fact, in almost all individual realizations the overall shape of the Nelk River Basin is well reproduced. An objective boundary representation (i.e., obtained by unrolling the boundary profiles via a polar plot and expressing the resulting function as a Fourier series [Russ, 1994]) of the different results has been used to substantiate the visual impression.

We have also tried to simulate the same shape by using a filtered version of $W(x)$ where only low frequencies had been retained. The same robustness is shown by the shape of the basin, and one may conclude that the information contained in the low frequencies of the width function suffices in reconstructing the shape of the basin.

Figure 7.21. (a and b) Two realizations of the boundaries of the Nelk River Basin; (c) reconstructed shape as the ensemble average of ten realizations [after Rinaldo et al., 1995b].

The inverse procedure, that is, the possibility of an adequate reproduction of the width function from knowledge of the outer shape of the basin, has been studied by Rigon et al. [1993] using optimal channel networks (OCNs) concepts (see Chapter 4). It was shown in Section 4.14 that OCNs developed within given basin boundaries have similar width functions among themselves and also to those observed in real basins with the same shape. Because the link of width functions to the hydrologic response has also been established (see Section 7.8), the above result suggests that by gauging the flow at the control section of a watershed, one could pursue the inverse pattern, from the flow to the width function and from the widths to the boundaries to effectively gauge the shape of a basin. This is the subject of Section 7.8.2.

7.8.2 Geomorphologic Hydrologic Response

In Section 7.7 we observed that the scaling character of basin morphology cannot be transferred to the dynamic response $f(t)$. In fact, when the mean travel distance is large with respect to mean dispersion displacements, the self-affine characters observed for real width functions are progressively destroyed. Thus the larger the basin, the smoother the gauge trace. Therefore the mechanisms and the timescale imposed by dispersion break the one-to-one relationship of width functions and unit hydrograph responses. We also notice that the observed width function of real basins is not always skewed to the left with positive third central moments of the distribution. Nevertheless, the hydrologic response is invariably so, showing a relatively fast rising limb and a long tail.

It was argued in Section 7.7 that an analytical explanation could be embedded in the moment-generating function Eq. (7.75), because geomorphologic dispersion has a selective and pronounced skewing effect. However, the above exact analyses were missing some important parts of the picture, chiefly related to nature and extent of hillslope patterns and their effects on the general dynamics. Rather than modeling the delay in the access from overland flow regions to the network, as in Eq. (7.71), we will use the technology of DEMs to extract more detailed geographic information and a suitable dynamic framework.

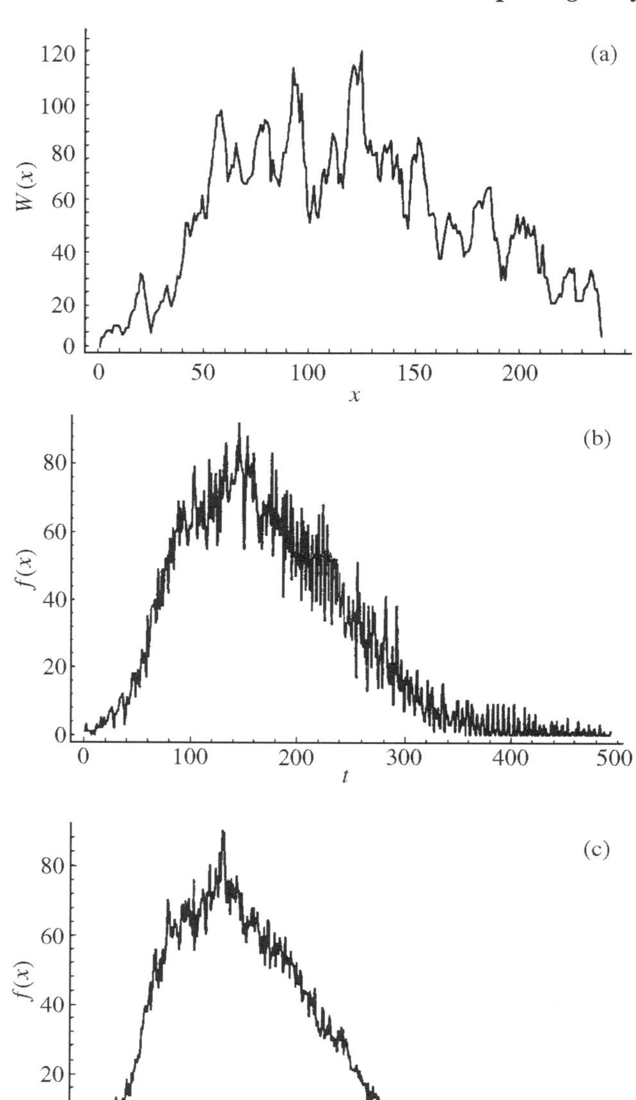

Figure 7.22. GIUH of the Fella River: (a) $f_T(t) = W(ut)$ $(u = u_h = u_c)$; (b) $f_T(t)$ for the deterministic case where $u_h/u_c = 0.1$ and $u_c = 1$; (c) randomized lognormal u_h. The curve is obtained as the ensemble average of five realizations [after Rinaldo et al., 1995b].

In this section we will focus on direct runoff and therefore bypass the estimation of effective rainfall. Also, we neglect the spatial variability of the rainfall input although both factors could in principle be accounted for in the geomorphologic framework.

The appearance of a skewness not contained in the original width functions and the smoothing effect of the dispersion suggested that the pronounced skewness of real-life hydrographs could result solely from dynamic considerations. The above suggestions were incomplete because the processes involved neglect the role of hillslope patterns. It turns out that these are of foremost importance.

To describe the former, one needs to objectively distinguish hillslopes, valleys, and channels. Hillslopes are seen as areas of topographic divergence, and valleys are areas of topographic convergence. Channels appear within areas of topographic convergence but are not defined by curvature alone. In Chapter 1 (Section 1.2.12) we discussed the extraction of channel networks, as well as the identification of valley and hillslope regions from DEMs. With this type of characterization one can objectively compute, for any path originated at site i, a measure of the hillslope paths $L_h(i)$ required to reach the first channel site, say j_i, and the related length of channel path $L_c(j_i)$ from j_i to the outlet, in both cases by following local drainage directions defined by ∇z.

Following the idea of Van der Tak and Bras [1988], Rinaldo et al. [1995b] rescaled the width function by differentiating the drift velocity depending on whether the moving particle is in a hillslope or a channel. Let u_h denote the mean drift in hillslopes and u_c be analogous for channels. Mean travel time T_i for the particle originating at site i is therefore

$$T_i = \frac{L_h(i)}{u_h} + \frac{L_c(j_i)}{u_c} \tag{7.113}$$

and the distribution $f_T(t)$ follows simply from sampling all sites i within a DEM. Figures 7.22(a) to 7.22(c) shows the result of the above computation when applied to the Fella River Basin shown in Figure 1.33. Figure 7.22(a) shows the width function $W(x)$ (the particular case of response $f_T(t) = W(ut)$ where $u_h = u_c = u = 1$; here the maximum length from source to outlet is taken as $\ell = 1$ for convenience). We observe that $W(x)$ is not particularly skewed. Figure 7.22(b) shows (for the case where the threshold is as in Figure 1.33) the response function computed by an arbitrary ratio $u_h/u_c = 0.1$ and $u_c = 1$. Figure 7.22(c) shows the results when a Monte Carlo procedure has been applied to the process to mimic the effects of heterogeneities. In Figure 7.22(c) the drift u_h has been spatially random-

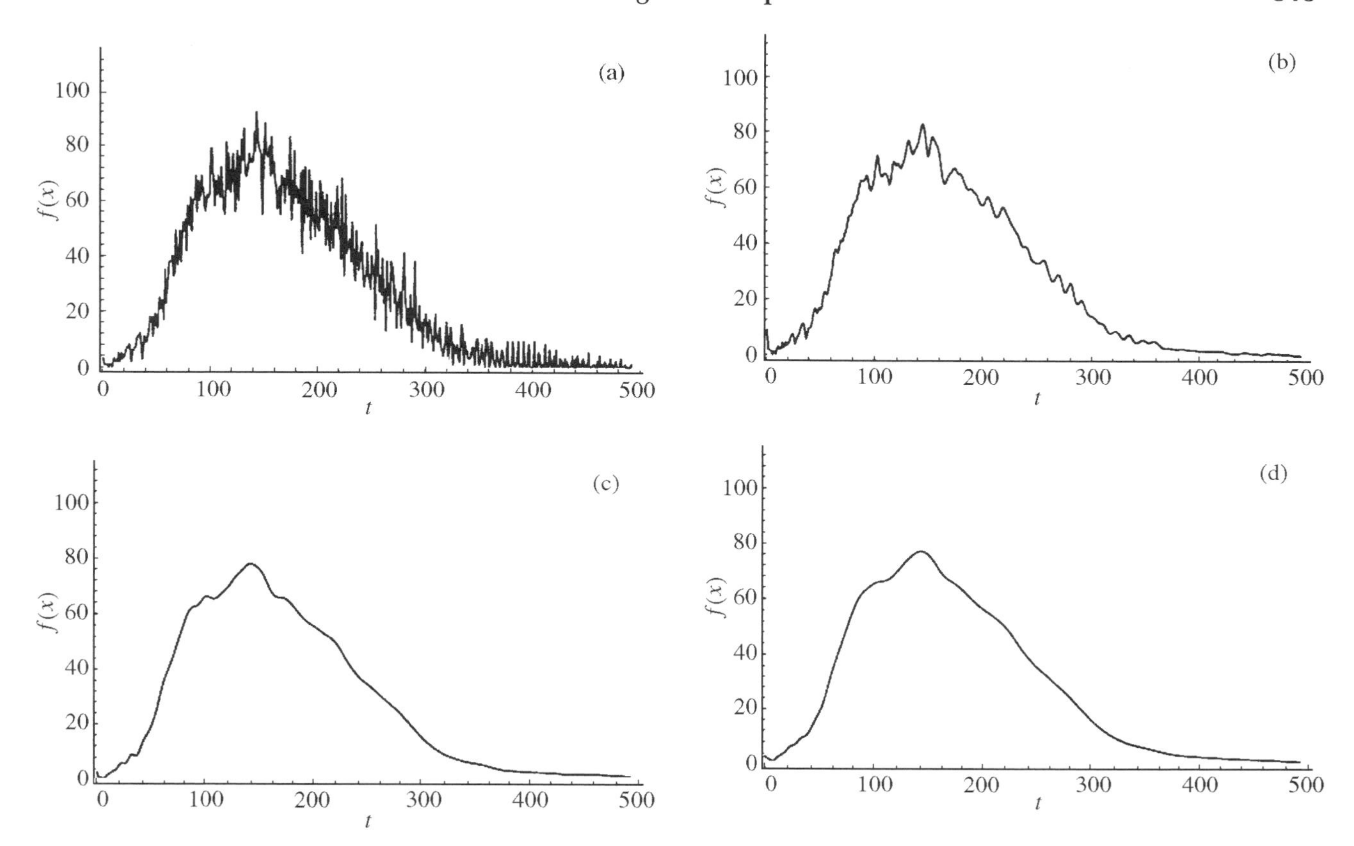

Figure 7.23. The effects of hydrodynamic dispersion on the responses of Figure 7.22 [after Rinaldo et al., 1995b].

ized by assuming a lognormal distribution with mean 0.1 and variance 0.1. Five realizations were generated with this procedure and an ensemble average was then taken through the realizations.

The results in Figure 7.22 clearly suggest that the different dynamic characterizations for hillslopes and channels, even in the gross deterministic framework of Figure 7.22(b), drastically enhance the positive skewness of the response. The nonphysical degree of roughness of the response of Figure 7.22(b) is significantly reduced by simulating heterogeneity through Monte Carlo techniques. It is not uncommon to find width functions with negative skewness appearance that the above mixed dynamics translates into a smooth response with a positive skewness.

Figures 7.23(a) to 7.23(d) show the addition of hydrodynamic dispersion of the type described by Eq.(7.76) to the response in Figure 7.22(b), where the related dispersion coefficients have been arbitrarily set to $D = 0$, 0.05, 0.1, and 0.2 (Figures 7.23(a)–7.23(d), respectively), all of them in pixel units (L^2/T). Notice that no Monte Carlo average has been implemented.

The effects of progressive Monte Carlo averaging are tested in Figure 7.24, where $\bar{u}_h/u_c = 1/8$, $u_c = 1$, and the variance of the (mean 1/8) lognormal distribution for u_h is 0.5. A smoothing of the response is progressively achieved while the main features remain unchanged, as they are most sensitive to the ratio u_c/u_h. Figure 7.25 shows the dramatic changes

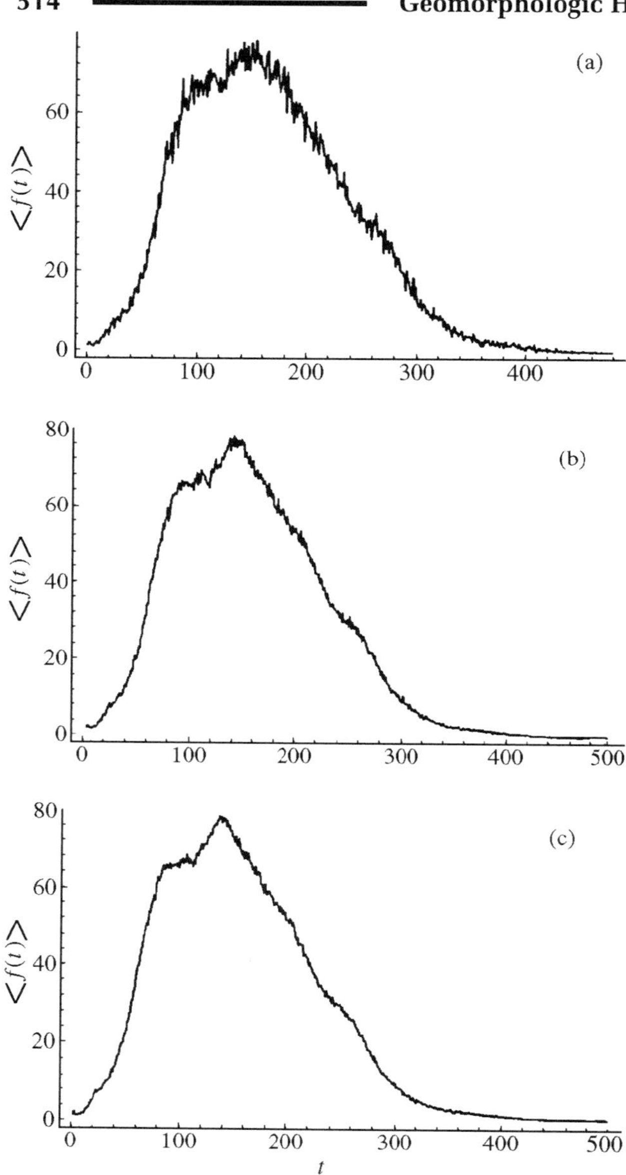

Figure 7.24. The effects of progressive Monte Carlo (MC) averaging on $\langle f(t)\rangle$: (a) 10 MC realizations; (b) 50 MC realizations; (c) 100 MC realizations [after Rinaldo et al., 1995b].

that mean, variance, and dimensionless third moment of the response undergo as a function of the ratio $\bar{u}_h/u_c$ in the deterministic case.

Many other cases have been tested for river basins in northern Italy and comparisons with real runoff data have been performed. What matters always is that different dynamics can greatly mix the geometric and topologic information contained in the width functions. Because the dynamic characters can vary considerably in time and space (e.g., depending on initial moisture conditions, vegetation state, and rainfall intensity), the fact that they have considerable impact on the response seems to considerably limit the capabilities of the inverse procedure, that is, from gauging to morphology.

7.9 On the Spatial Organization of Soil Moisture Fields

7.9.1 Introduction

All our analyses of the geomorphologic structure of the hydrologic response have assumed the input as effective rainfall once losses have been subtracted from the spatial and temporal patterns of total precipitation. Indeed, a crucial hydrologic problem is the determination of those 'losses' (not necessarily a physical exclusion from the control volume) whose temporal evolution is marked by an extreme spatial heterogeneity. A key variable in this analysis is the soil moisture state of the basin.

Soil moisture in space and time constitutes a most important process in the study of hydrologic phenomena and soil–atmosphere interaction [Delworth and Manabe, 1988, 1989]. Its multiple links with soil and vegetation processes, with topography, and with the atmosphere are made up of a large number of interacting elements and information over a wide range of spatial and temporal scales.

In this section we are interested in exploring if, in the apparent disorder and diversity that seem to characterize the spatial and temporal structure of soil moisture, there exists a well-defined organization of statistical character. This section closely follows Rodriguez-Iturbe et al. [1995]. These studies are still in their infancy but the importance of the topic and its relevance to the hydrologic response of river basins make at least a preliminary discussion advisable. In fact, if a statistical organization exists, then its characterization becomes a critical piece of information for the modeling of the soil moisture field and for the incorporation of this variable in the study of the hydrologic response under a general scientific framework.

Probably the most challenging and fascinating aspect in the study of soil moisture and the soil–atmosphere interaction is the wide range of scales through which the phenomena manifest themselves and affect other processes. To address this issue, Rodriguez-Iturbe et al. [1995] report on the experimental collection of soil samples in cubes of 5 cm per side, as through the difference between wet and dry weight it is possible to estimate the gravimetric soil moisture. In a different scale range one can study the spatial patterns of soil moisture in fields that cover up to one or two hundred kilometers in typical length with resolutions of the order of several tens or hundreds of meters. At much larger scale one can also focus on fields of typical lengths of the order of one to several thousand kilometers. From a hydrologic point of view one is interested in the latter two scales.

Processes and patterns occurring over regions of several tens or hundreds of kilometers in typical length are very important from the point of view of regional water balance, hydrologic response to varying precipitation input, dynamics of soil–water–vegetation systems, etc. In this scale the feedback mechanisms between soil and atmosphere are already present and could be an important factor in determining the spatial and temporal structure of the rainfall input into the area and the soil moisture inside the region.

When going to larger scales (for example, scales whose characteristic length is of the order of 1,000 km), the soil–atmosphere interaction is crucial in the structure of the spatial and temporal patterns. Thus there is a continuous spectrum of temporal and spatial scales, from tens of meters to thousands of kilometers and from several hours to several months, which are embedded one into another. The phenomena in these scales are not independent, but the structure of the spatial and temporal patterns is affected by very different variables and mechanisms.

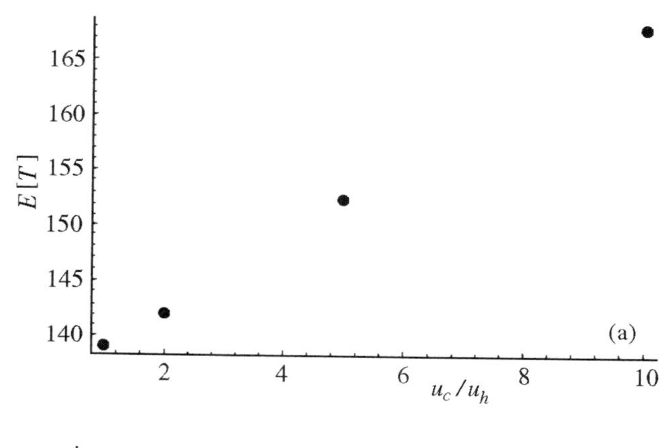

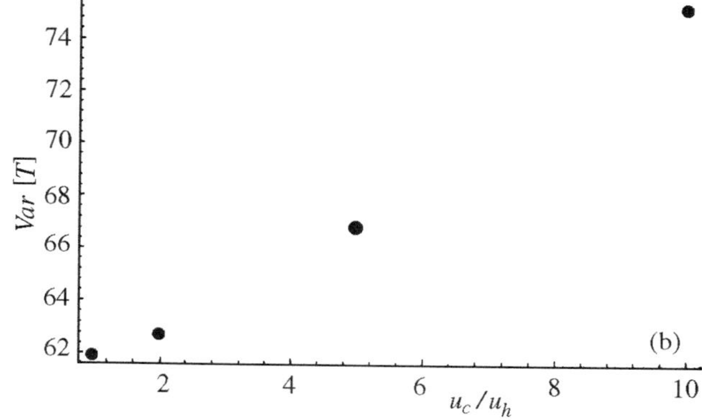

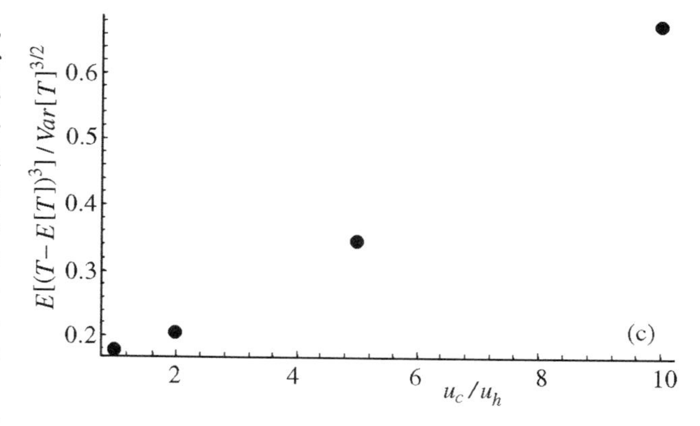

Figure 7.25. Mean, $E[T]$, variance, $Var[T]$, and dimensionless third moment of travel times versus u_c/u_h for the Fella River Basin [after Rinaldo et al., 1995b].

The discussion in this section will focus on spatial scales of tens of meters to hundreds of kilometers with temporal scales of the order of one day. We are interested in the fluctuations of soil moisture occurring in those scales of analysis. More specifically, we want to study the links between the properties of the soil moisture process when the process is studied as a field that is observed and analyzed at different degrees of resolution. The emphasis is on the spatial, field-like, characters of the fluctuations of soil moisture. The temporal aspect will not be a structural part of the analysis, and it will only play a role in the time variability that the field undergoes when its evolution is followed throughout several days.

To our knowledge, the best set of data on soil moisture and soil properties at the spatial scale of interest is the one collected by the U.S. Department of Agriculture (USDA) during the so-called Washita experiment. A brief description of some of the characteristics of this experiment, judged to be most relevant for our purposes, will now be provided.

Washita '92 was a cooperative experiment between NASA, USDA, and several other government agencies and universities conducted with the primary goal of collecting a time series of spatially distributed data focusing on soil moisture and evaporative fluxes [Jackson et al., 1993]. Data collection was conducted from June 10 to June 18, 1992. The region received heavy rains over a period before the experiment started with the rain ending on June 9, 1992, and zero precipitation occurring during the experiment period. This is quite fortunate because it allows us to study the spatial organization during a period when the system is not gaining moisture and is only losing it through drying conditions.

A complete summary of the characteristics of the region and the research data is contained in Allen and Naney [1991]. The area is located in southwest Oklahoma and the main body of soil moisture information was retrieved from ESTAR – Electronically Steered Thinned Array Radiometer – image data. ESTAR and many other operational sensors were carried by a C-130 NASA aircraft during the Washita '92 Experiment. The details of the C-130 flightlines, altitudes, etc., can be found in Jackson et al. [1993]. The data sets studied by Rodriguez-Iturbe el al. [1995] and used here were directly derived from those provided in the high-density PC diskette "Washita '92 Data Sets" (T. J. Jackson, version 12/20/93). The data sets provide, besides the ESTAR passive microwave data and the soil moisture data, information on soil texture and land use that were relevant for the purposes of the research. All the image files are 228 pixels by 93 lines, covering an area of 45.6 km by 18.6 km (848 km^2) with a pixel grid of 0.04 km^2. Practically all the topography is of very gentle character with no major orographic features. The area belongs to the Little Washita watershed and has a channel system that provides adequate drainage.

The soil moisture files record volumetric soil moisture (percentage of soil moisture in the total volume of soil) estimated mainly on the top 5 cm of soil. The values represent areal averages over pixels of 200 $\times$ 200 m, and there is an image for each day of the experiment. Rodriguez-Iturbe et al. [1995] converted these files of volumetric soil moisture to relative soil moisture by dividing them by the effective soil porosity of the characteristic soil of the pixel. Thus Jackson's files provided soil texture data for each pixel, classifying its soil in one of the nine categories shown in Table 7.2. Each class was assigned an effective porosity as described in Rodriguez-Iturbe et al. [1995]. Figure 7.26 gives the field of relative soil moisture $S(\mathbf{x})$ integrated on pixels of 200 $\times$ 200 m^2, 400 $\times$ 400 m^2, and 600 $\times$ 600 m^2. There are some missing data and urban pixels that are treated as missing data. Fortunately, there are not many but they bring an important constraint when aggregating data at larger spatial scales, as will be seen in the discussion.

The Washita '92 Experiment also provided a different type of soil moisture data in addition to that given above. In fact, with the primary purpose of providing calibration and verification for the microwave remote sensing instruments, ground observations of surface soil moisture were collected in the region. Spatially distributed sites were selected so as to represent

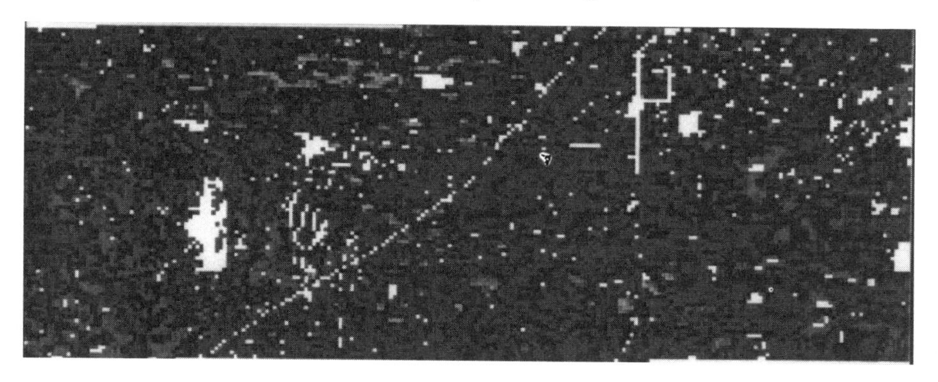

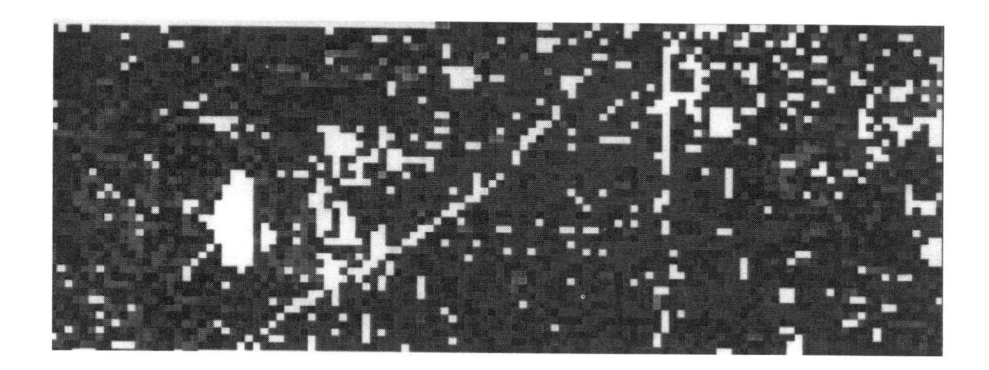

Figure 7.26. Relative soil moisture field $S(\mathbf{x})$ on day 10, at different levels of aggregation (200×200 m^2, 400×400 m^2, and 600×600 m^2) . Red pixels are saturated ($S = 1$); green and yellow pixels have low values of S.

the various soil textures and land cover conditions. Of special interest are the so-called small-site ground samples. In the small sites the area sampled was 30 m by 30 m with nine samples collected on a square grid basis (10 m apart). Each sample was made up of a 5-cm cube of soil. All samples were weighed, dried for 24 hours, and weighed again to obtain a dry weight. The gravimetric soil moisture was then computed by dividing the difference between the wet and dry weight by the dry weight. Data are available comprising the individual nine samples for each day, of each of sixteen 30 m by 30 m sites. The soil characteristics of each site were characterized by the bulk density and effective porosity corresponding to the soil type of the site. The volumetric soil moistures of each 5-cm cube soil sample were

Table 7.2. *Classification of soils in the Washita '92 experiment area*

Indicator	Classification	Effective porosity
1	Sand	0.417
2	Loamy fine sand	0.401
3	Fine sandy loam	0.412
4	Loam	0.434
5	Silt loam	0.486
6	Silty clay loam/clay loam	0.432
7	Pits, quarries, urban	–
8	Gypsum	–
9	Water	–

Note: Porosity values are taken from Rawls, Brakensiek, and Miller [1983].

obtained by multiplying the gravimetric soil moisture by the soil bulk density and dividing by the density of water. The relative soil moisture was then obtained by dividing the volumetric soil moisture by the effective porosity of the soil.

The systematic sampling of 30 m by 30 m sites across the whole region allows the study of the spatial patterns of soil moisture over an area much smaller than the area of the pixel resolution from the remote measurements (e.g., 900 m^2 versus 40,000 m^2). Nevertheless, it is important to remark that the remote measurements are a direct measurement of the integrated process over the pixel area. In the case of the small 30 m by 30 m sites, the inferences are based on nine discrete samples taken 10 m apart on each site. From the statistical point of view this implies an additional variance of estimation when going from the discrete samples to the field properties, which is not present in the remote measurements. This needs to be taken into account when analyzing the characteristics of soil moisture fluctuations integrated over different spatial scales.

Figure 7.27 shows the mean of the 200 m by 200 m pixel resolution field of relative soil moisture for each of the nine days of the experiment.

Figure 7.28 shows histograms of relative soil moisture for two days where the effect of drying is clearly seen.

7.9.2 The Effect of Aggregation on the Statistics of the Soil Moisture Field

The field of relative soil moisture in every day of the experiment was aggregated at larger and larger spatial scales. Thus from the scale corresponding to 200 m × 200 m pixels, one can aggregate to scales of pixels with side length of 400 m, 600 m, 800 m, and 1000 m. Pixels with missing data or with urban inferences were discarded in the analysis, thus severely limiting the possibility of aggregating at larger pixel sizes. At each scale of aggregation larger than the original one, the relative soil moisture assigned to the pixel is estimated as the arithmetic mean of the values corresponding to the 200×200 m^2 grid that make up the larger pixel. The variance and correlation of the fields were then estimated at the different aggregation levels. Figure 7.29 shows how the variance changes during the different days as a function of the pixel size. In all cases there is a remarkable power law with very little dispersion in the fit. The point in the extreme left of the power law corresponds to the variance of the 30 m × 30 m plots. As explained earlier, the estimates of soil moisture for these plots are based on nine discrete samples taken 10 m apart. Thus one wishes to estimate $Var[(1/A)\int_A S(\mathbf{x})d\mathbf{x}]$, where A is the 30 m × 30 m region and the term in brackets is a random variable. The estimation is carried out by first computing the mean relative soil moisture of each of the sixteen 30 m × 30 m sites, $\bar{S}_i = (1/N)\sum_{j=1}^{N} S_i(j)$, where N is equal to 9 and $i = 1, \ldots, M$, with $M = 16$. The variance of the $\bar{S}_i$s ($i = 1, \ldots, M$) is then a first approximation to $Var[(1/A)\int_A S(\mathbf{x})d\mathbf{x}]$. This estimate can be further corrected by

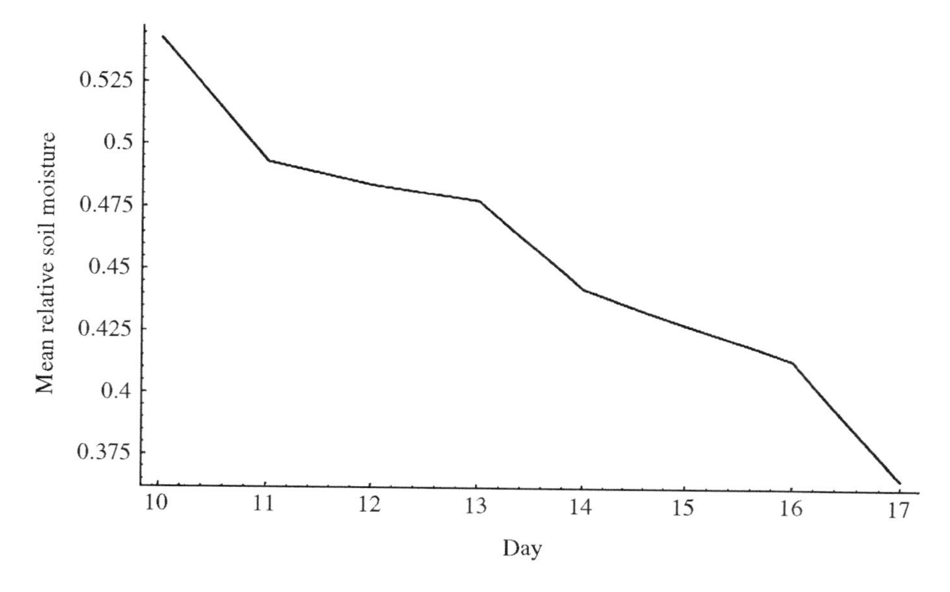

Figure 7.27. Mean relative soil moisture in the field on days 10 to 18 [after Rodriguez-Iturbe et al., 1995].

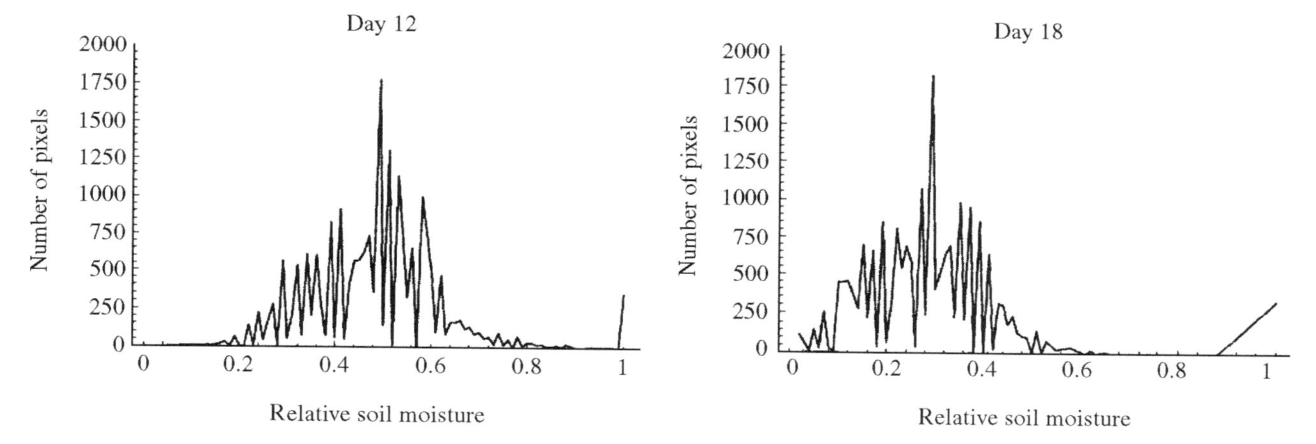

Figure 7.28. Distribution of relative soil moisture on days 12 and 18 [after Rodriguez-Iturbe et al., 1995].

subtracting from it the variance of estimation of the $\bar{S}_i$s. In other words, $\bar{S}_i$ from point samples is just itself an estimate of $(1/A)\int_A S(\mathbf{x})d\mathbf{x}$, and the variance of the $\bar{S}_i$s is an overestimation of the desired variance. To evaluate the variance of estimation of $\bar{S}_i$, Rodriguez-Iturbe et al. [1995] first calculated the correlation between the point samples (5 cm × 5 cm) using all sixteen sites. The correlation was effectively zero between point samples, implying that the variance of estimation of $\bar{S}_i$ is simply σ_P^2/N where σ_P^2 represents the point variance of the soil moisture field $S(\mathbf{x})$ and $N = 9$. The point variance was estimated by repeated sampling of one point value from each of the sixteen 30 × 30 m^2 fields, estimating $\sigma_P^2(i)$, repeating the procedure several times, and averaging over the $\sigma_P^2(i)$.

It is interesting to notice that the slope of the graphs in Figure 7.29 increases monotonically with time. This is clearly seen in Figure 7.30. The increase in slope is expected because the process of drying will produce a larger degree of small-scale heterogeneity than that existing at higher degrees of saturation.

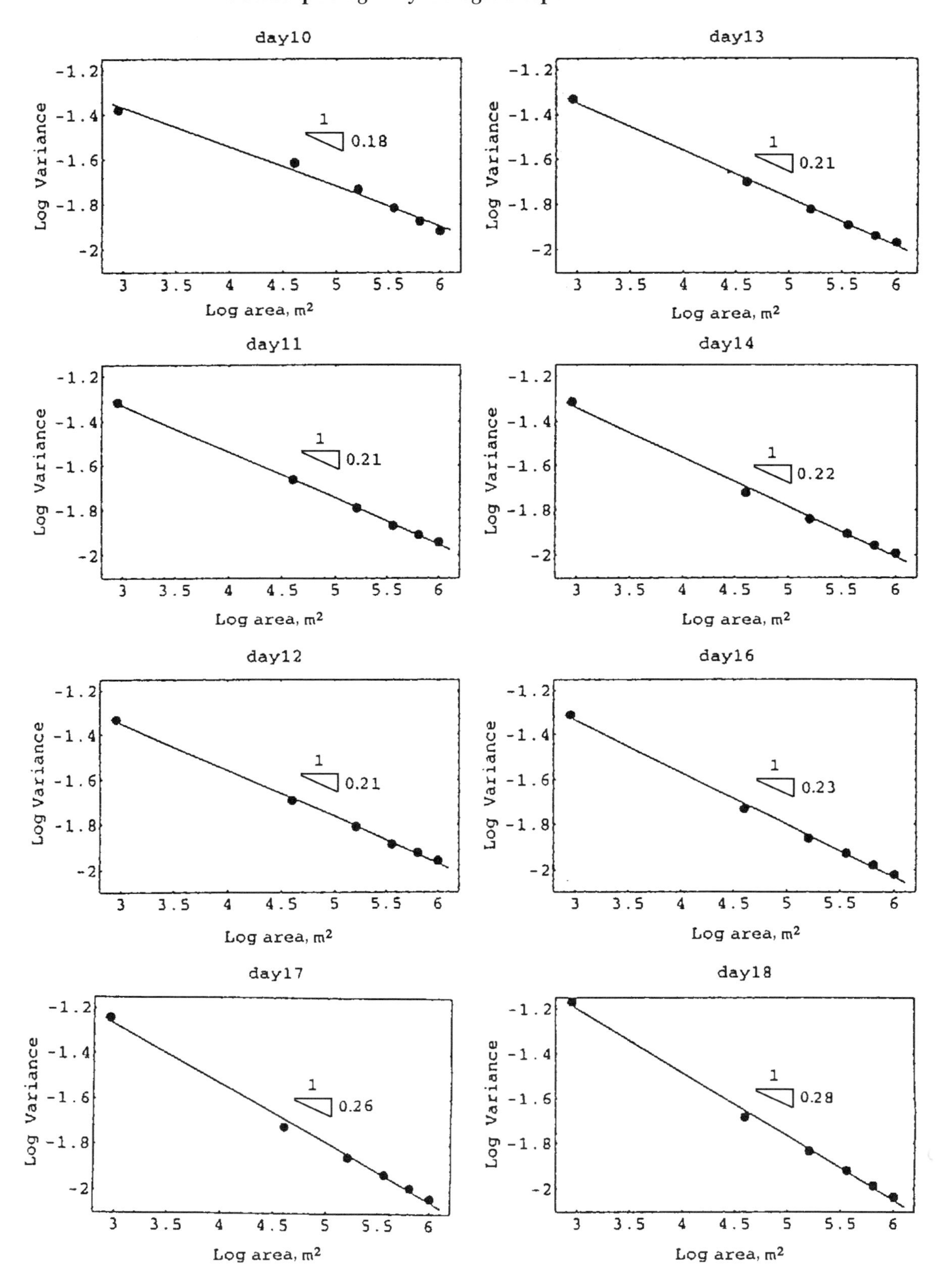

Figure 7.29. Variance of the soil moisture field $S(\mathbf{x})$ versus pixel area on days 10 and 18 [after Rodriguez-Iturbe et al., 1995].

The power law decay of the soil moisture variance as a function of the size of the averaging area implies that the spatial correlation function, $\rho(x)$, also falls off as a power law for large distances [Whittle, 1962]. Thus with $S(A) = A^{-1} \int_A S(\mathbf{x})d\mathbf{x}$, the scaling of the variance $Var[S(A)] \propto A^{\mu-2}$ implies that $\rho(x) \propto x^{2(\mu-2)}$ for large distances. The correlation structure of the soil moisture field characterized by pixels of 200 m × 200 m in the Washita'92 Experiment was estimated assuming stationarity in the field and averaging over all pixels separated by a fixed lag in the longitudinal direction. The results in all cases gave power laws with decay exponents that monotonically decrease from −0.31 for day 10 to −0.48 for day 18.

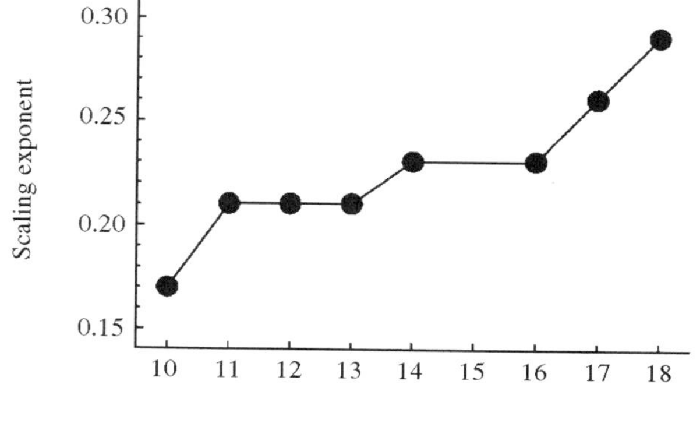

Figure 7.30. The slope of the variance versus pixel area shown in Figure 7.29 as a function of time after the major rainstorm. No data were available for day 15.

Figure 7.31 shows typical examples of $\rho(x)$ for some days of the experiment. The process is described in 200 m × 200 m pixels, and the correlation is estimated at distance multiples of 200 m. The slopes of the fitting lines in Figure 7.31 are day 11, −0.33; day 14, −0.35; and day 18, −0.48.

The variance and correlation structure of the aggregated soil moisture field are typical of scaling processes that do not change the statistics of their fluctuations when studied at different levels of resolution. The scaling properties previously detected suggest the existence of similar characteristics in the clustering spatial patterns. This has theoretical and practical importance. From the theoretical point of view it opens the door to a unifying – across scales – type of analysis of the spatial shapes present in soil moisture. From the practical point of view it allows the quantitative probabilistic assessment of the patches of different soil moisture levels.

Figure 7.31. Correlation function of the relative soil moisture field [after Rodriguez-Iturbe et al., 1995].

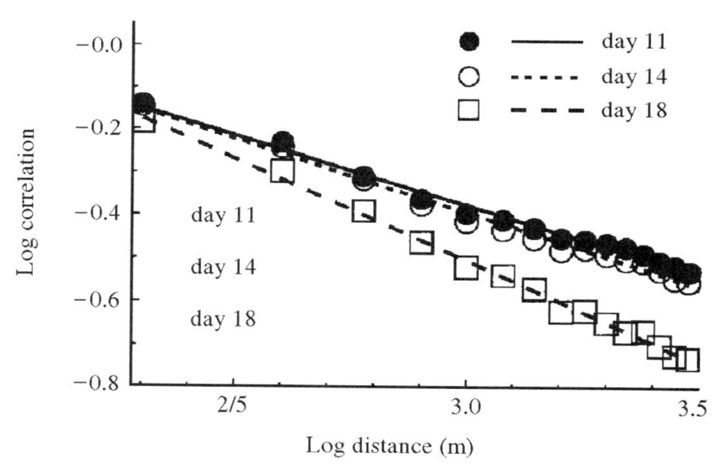

A soil moisture island may be defined as a group of pixels all with soil moisture at or above a certain level that are connected so that each pixel in the group shares one full side with another pixel in the cluster. The field of soil moisture islands at or above a fixed level was studied for different levels in each day of the Washita '92 Experiment by Rodriguez-Iturbe et al. [1995].

The analyses were carried out using the original field of 200 × 200 m^2 pixels. Figure 7.32(a) shows examples of the power laws obtained for the size distributions of the soil moisture islands. The slopes of the fitting lines are day 11 ($S > 0.60$), −0.92; day 14 ($S > 0.50$), −0.79; and day 18 ($S > 0.35$), −0.79. The moisture level is decreased when advancing in time in order to keep an adequate sample size because of the drying effect. In all cases the fit is excellent.

In geography it is well known that the size of islands, say A, follows a power law distribution, that is, the so-called Korcak [1940] law: $P[A > a] \propto a^{-b}$. Korcak's exponent b is related to the fractal dimension of the coasts of the islands through $b = 0.5D_{\text{coast}}$ [Mandelbrot, 1975] and, for islands, it was found to range from 0.5 in Africa to 0.75 in Indonesia. The

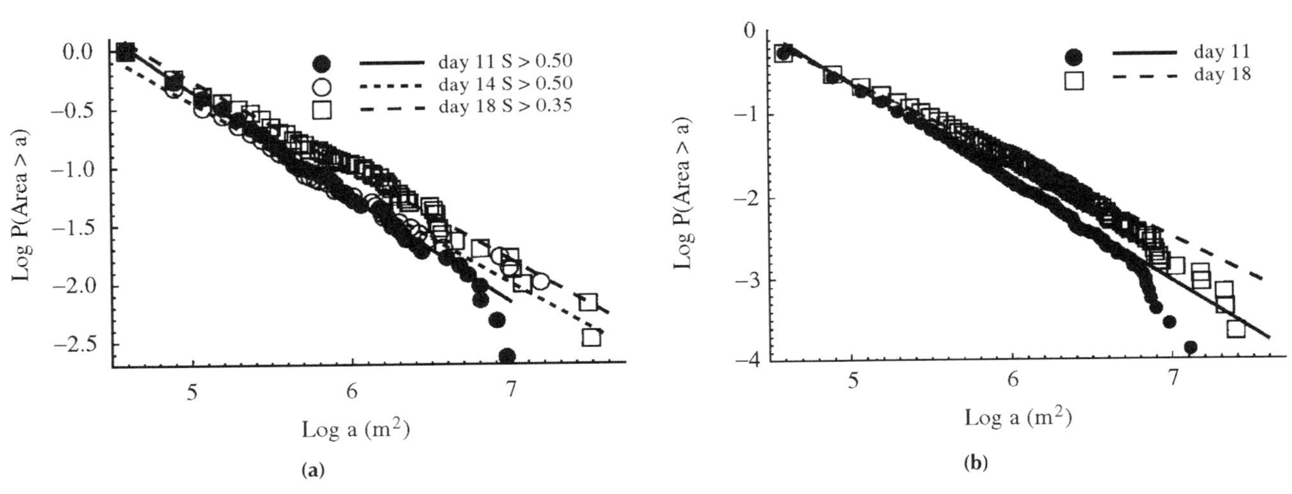

Figure 7.32. (a) Probability distribution of the size of soil moisture islands above different thresholds; (b) probability distribution of the size of soil moisture clusters [after Rodriguez-Iturbe et al., 1995].

values of the Washita data imply a very rough fractal perimeter for the soil moisture islands. The changes in the fractal dimension of the perimeters of soil moisture islands when the level of reference is changed point toward a multifractal character of the field. This characteristic is similar to that observed in the isolines of the landscape of the river basin, as seen in Section 6.5.

Other types of clustering patterns were studied for the Washita data. For instance, clusters may be defined as connected pixels of equal soil moisture (± 0.05). Other kinds of clusters where the soil moisture is required to remain at a fixed level everywhere (also ± 0.05) were also studied. An example of cluster analysis is shown in Figure 7.32(b). The slopes of the fitting lines are day 11, -1.22; and day 18, -0.97. In all cases the power law fit is excellent, confirming the scaling nature of the spatial patterns.

The size of the area involved in the Washita '92 Experiment makes it unrealistic that the scaling detected in the soil moisture fields has any relation to the dynamics of soil–atmosphere interaction phenomena or with any appreciable space–time organization of rainfall [Entekhabi and Rodriguez-Iturbe, 1994]. Also, from the surface runoff viewpoint, there is not much redistribution except through the channel network, which will take less than one day to move the water out of the region once it reaches the network. Additionally, from the viewpoint of significant moisture redistribution through underground dynamics, the timescales involved make that mechanism ineffective for the type of data at hand [Entekhabi and Rodriguez-Iturbe, 1994].

The above reasoning suggests that the spatial scaling of soil moisture at the scales of this study is a consequence of the existence of spatial organization in the soil properties that command the infiltration of moisture. Other authors [Burrough, 1983, 1989; Wood, 1994] have studied some aspects of this organization but – to our knowledge – not the features analyzed by Rodriguez-Iturbe et al. [1995].

To investigate the connection with soil properties, porosity may be used as a soil characteristic likely to be related to relative soil moisture in terms of the possible scaling properties of both fields. Rodriguez-Iturbe et al. [1995] have studied the statistical properties of the porosity field under different scales of resolution in a manner analogous to the methodology used for soil moisture. Figure 7.33 shows the power law obtained for the decrease of variance with area of aggregation. The fit is excellent with a slope close to −0.21. This value is in the middle range of the exponents found for the different days in the case of relative soil moisture. Notice that because the porosity values were assigned on the basis of soil type, there are no missing values in this field.

The estimation of the correlation coefficients for different spatial lags at different scales of observation was also carried out. Again, it was found that the correlation structure remains practically unchanged regardless of the scale at which the porosity is described. These results show that porosity is also a scaling spatial field at the scales of the study.

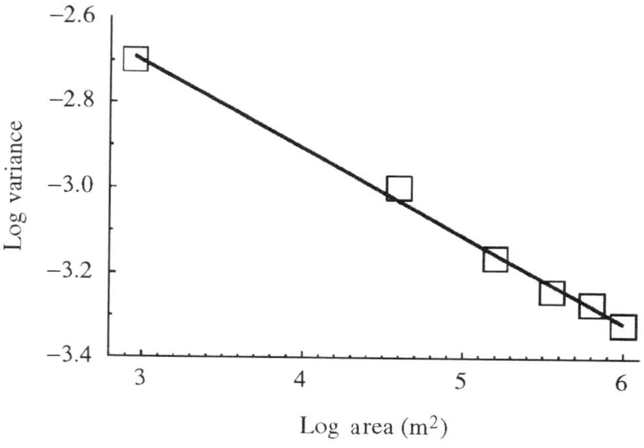

Figure 7.33. The variance of the porosity field as a function of the area of the pixel over which the process is averaged [after Rodriguez-Iturbe et al., 1995].

The above analyses show a spatial scaling structure of soil moisture fields at the scales considered in the study. They also suggest that such a structure may arise from the existing scaling in soil properties (e.g., porosity). The scaling properties detected have direct implications for sampling of the soil moisture field, for modeling of its spatial fluctuations, and for its linkage with large-scale precipitation phenomena.

There is indeed a need for much research in this area. Moreover, the mechanism of runoff production makes it likely that the spatial structure of soil moisture is related to the organization of the drainage network and its accompanying landscape. The linkage of the fractal and multifractal properties of the river basin with those of the soil moisture field is an exciting frontier that will bring much new insight into the problem of the geomorphologic basis of the hydrologic response.

A first step in this direction has recently been taken by Woods and Sivapalan (unpublished manuscript, 1996). Runoff generation can be modeled using a topographic wetness index; the fraction of the catchment where the wetness index exceeds some threshold is assumed to be saturated and thus controls runoff-generation processes. In humid climates in particular, the terrain surrounding the drainage network (usually colluvium regions characterized by topographic concavity) is the one that becomes more easily saturated. Thus one would expect that some of the scaling properties of the network may also be present in the probabilistic characterization of the runoff-producing terrain.

Woods and Sivapalan (unpublished manuscript, 1996) used the wetness index $\omega_i = a_i/\nabla z_i$ of Beven and Kirkby [1979 and Beven and Wood, 1983], computed at the ith site of a DEM, to characterize the runoff-producing areas of a basin. In the definition of the index, a_i is the total contributing area per unit contour length at site i and ∇z_i is the local slope. The index ω constitutes a random field whose characteristics were connected to the properties of the underlying channel network. They show that the distribution of ω follows a power law whose exponent is different from that of the distribution of contributing area. They also find that this is true only for catchments with areas above some critical size, very reminiscent of the theoretical result of Rinaldo et al. [1996] (Section 4.21), where it is shown that a critical size exists for scaling properties to hold.

The possibility of determining a unifying statistical structure with scaling properties describing the runoff–producing regions of basins of different size and topography is indeed an exciting challenge for hydrologic research. At the core of such a structure will surely lie the circuits of reciprocal control between the hillslopes and the drainage network. It is in the scaling properties of the river basin and through the dependence of these properties on chance and self-organization that hydrologists hope to find unifying theories for the hydrologic response.

References

Abrahams, A. D. 1977. The factor of relief in the evolution of drainage basin networks with a digital computer, *Am. J. Sci.*, **277**, 626–630.

Abrahams, A. D. 1984. Channel networks: A geomorphological perspective, *Water Resour. Res.*, **20**(2), 161–168.

Abrahams, A. D., and A. J. Miller. 1982. The mixed gamma model for channel link lengths, *Water Resour. Res.*, **18**, 1126–1136.

Aharony, A. 1986. Percolation. In *Directions in Condensed Matter Physics*. G. Grinstein and G. Mazenko editors, World Scientific, Singapore, pp. 1–50.

Aharony, A. 1989. Measuring multifractals, *Physica D*, **38**, 1–4.

Ahnert, F. 1976. Brief description of a comprehensive three-dimensional process–response model for landform development, *Z. Geomorphol. Suppl.*, **25**, 29–49.

Ahnert, F. 1984. Local relief and the height limits of mountain ranges, *Am. J. Sci.*, **284**, 1035–1055.

Ahnert, F. 1987. Process response models of denudation at different spatial scales. In *Geomorphological Models Theoretical and Empirical Aspects*. F. Ahnert editor. *Catena*, Suppl. **10**, 31–50.

Allegre, C. J., J. L. Le Mouel, and A. Provost. 1982. Scaling rules in rock fracture and possible implications for earthquake prediction, *Nature*, **297**, 47–49.

Allen, P. B., and J. W. Naney. 1991. Hydrology of the Little Washita river watershed: Data and analyses, *USDA Tech. Rep. ARS-90.*

Alstrom, P. 1990. Self-organized criticality and fractal growth, *Phys. Rev. A*, **41**(12), 7049–7052.

Amitrano, C., A. Coniglio, P. Meakin, and M. Zannetti. 1991. Multiscaling in diffusion-limited aggregation, *Phys. Rev. B*, **44**(10), 4974–4977.

Andrews, D. J., and T. C. Hanks. 1985. Scarp degraded by linear diffusion: Inverse solution for age, *J. Geophys. Res.*, **90**, 10193–10208.

Andrews, D. J., and R. C. Bucknam. 1987. Fitting degradation of shoreline scarps by a nonlinear diffusion model, *J. Geophys. Res.*, **92**(12), 857–867.

Andrle, R. 1992. Estimating fractal dimensions with the divider method in geomorphology, *Geomorphology*, **5**, 131–142.

Andrle, R., and A. D. Abrahams. 1989. Fractal techniques and the surface roughness of talus slopes, *Earth Surf. Proc. Landforms*, **14**(3), 197–209.

Armanini, A., and G. Di Silvio, editors. 1992. *Hydraulics of Mountain Regions*, Springer–Verlag, Berlin.

Athelogou, M., B. Merte', P. Deisz, A. Hubler, and E. Luscher. 1989. Extremal properties of dendritic patterns: Biological applications, *Helvetica Physica Acta*, **62**, 250–253.

Athelogou, M., B. Merte', P. Poschl, A. Hubler, and E. Luscher. 1990a. Extremal properties of dendritic patterns: Biological applications, II, *Condensed Matter H. P. A.*, 908–909.

Athelogou, M., B. Merte', P. Deisz, A. Hubler, and E. Luscher. 1990b. Extremal properties of dendritic patterns: Investigations on systems of blood-vessels, *Herbsttagung der SPG/SSP*, **63**, 539–540.

Avissar, R., and R. Pielke. 1989. A parametrization of heterogeneous land surfaces for atmospheric numerical models and its impact on regional meteorology, *Mon. Weather Rev.*, **117**, 2113–2120.

Bak, P. 1992. Self-organized criticality in non-conservative models, *Physica A*, **191**, 41–46.

Bak, P. 1996. *How Nature Works*, Copernicus/Springer–Verlag, New York.

Bak, P. 1994. Lectures on Self-Organized Criticality, Lecture Notes, Summer School on Environmental Dynamics, Istituto Veneto di Scienze, Lettere ed Arti, Venice, Italy, preprint.

Bak, P., and K. Chen. 1989. The physics of fractals, *Physica D*, **38**, 5–12.

Bak, P., and K. Chen. 1991. Self-organized criticality, *Sci. Am.*, **46**, 52–61.

Bak, P., and C. Tang. 1989. Earthquakes as a self-organized critical phenomenon, *J. Geophys. Res.*, **94**, 15635–15637.

Bak, P., and M. Creutz. 1993. Fractals and self-organized criticality. In *Fractals and Disordered Systems.* A. Bundle and S. Havlin editors, Springer–Verlag, Berlin, pp. 93–101.

Bak, P., and K. Sneppen. 1993. Punctuated equilibrium and criticality in a simple model of evolution, *Phys. Rev. Lett.*, **71**(24), 4083–4986.

Bak, P., and M. Paczuski. 1993. Why nature is complex, *Physics World*, **12**, 39–43.

Bak, P., C. Tang, and K. Wiesenfeld. 1987. Self-organized criticality: An explanation of $1/f$ noise, *Phys. Rev. Lett.*, **59**, 381–385.

Bak, P., C. Tang, and K. Wiesenfeld. 1988. Self-organized criticality, *Phys. Rev. A*, **38**(1), 364–374.

Bak, P., K. Chen, and M. Creutz. 1989. Self-organized criticality in the 'Game of Life', *Nature*, **342**, 780–784.

Bak, P., K. Chen, and C. Tang. 1990. A forest-fire model and some thoughts on turbulence, *Phys. Lett. A*, **147**, 297–313.

Bak, P., H. Flyvbjerg, and K. Sneppen. 1994. Can we model Darwin? *New Scientist*, **12**, 36–42.

Bak, P., M. Paczuski, and S. Maslov. 1995. Complexity and extremal dynamics, Brookhaven Natl. Lab. preprint no. 62678.

Band, L. E. 1986. Topographic partition of watersheds with digital elevation models, *Water Resour. Res.*, **22**(1), 15–24.

Barenblatt, G. I., A. V. Zhivago, Y. P. Neprochnov, and A. A. Ostrovskij. 1984. The fractal dimension: A quantitative characteristic of ocean bottom relief, *Oceanology*, **24**, 695–697.

Batchelor, G. K. 1953. *The Theory of Homogeneous Turbulence*, Cambridge University Press, Cambridge.

Beer, T., and M. Borgas. 1993. Horton's laws and the fractal structure of streams, *Water Resour. Res.*, **29**(5), 1475–1487.

Bell, T. H. 1975. Statistical feature of sea-floor topography, *Deep Sea Res.*, **22**, 883–892.

Bellin, A., P. Salandin, and A. Rinaldo. 1992. Dispersion in heterogeneous porous formations: Statistics, first-order theories, convergence of computations, *Water Resour. Res.*, **28**(9), 2211–2227.

Bennett, J. G. 1936. Broken coal, *J. Inst. Fuel*, **10**, 22–39.

Benzi, R., G. Paladin, G. Parisi, and A. Vulpiani. 1984. On the multifractal nature of fully developed turbulence and chaotic systems, *J. Phys. A*, **17**, 3521–3531.

Berkson, J. M., and J. E. Mathews. 1983. Statistical properties of sea floor roughness. In *Acoustics and the Sea-Bed.* N. G. Pace editor, Bath University Press, Bath, pp. 215–223.

Beven K. J., and M. J. Kirkby. 1979. A physically based variable contributing area model of basin hydrology, *Hydrol. Sci. Bull.*, **24**(1), 43–49.

Beven, K., and E. Wood. 1983. Catchment and geomorphology and the dynamics of the runoff contributing areas, *J. Hydrol.*, **65**, 139–158.

Beven, K., and I. D. Moore, editors. 1993. *Terrain Analyses and Distributed Modelling in Hydrology*, J. Wiley, New York.

Beven, K., and M. J. Kirkby, editors. 1993. *Channel Network Hydrology*, J. Wiley, New York.

Bloom, A. L. 1991. *Geomorphology: A Systematic Analysis of Late Cenozoic Landforms*, Prentice Hall, New Jersey.

Blumenfeld, R., Y. Meir, A. Aharony, and A. B. Harris. 1987. Resistance fluctuations in randomly diluted networks, *Phys. Rev. B*, **35**, 3524–3535.

Bouchaud, J. P., and A. Georges. 1990. Anomalous diffusion in disordered media: Statistical mechanisms, models and physical applications, *Phys. Rep.*, **195**(4-5), 127–293.

Bounds, D. 1987. New optimization methods from physics and biology, *Nature*, **329**, 215.

Boyd, M. J. 1978. A storage-routing model relating drainage basin hydrology and geomorphology, *Water Resour. Res.*, **14**(5), 921–928.

Bras, R. L. 1990. *Hydrology*, Addison–Wesley, Reading, MA.

Bras, R. L., and I. Rodriguez-Iturbe. 1985. *Random Functions in Hydrology*, Addison–Wesley, Reading, MA.

Breyer, S. P., and R. S. Snow. 1992. Drainage basin perimeters: A fractal significance, *Geomorphology*, **5**, 143–158.

Brush, L. M. 1961. Drainage basins, channels and flow characteristics of selected streams in central Pennsylvania, *U.S. Geol. Surv. Prof. Paper* no. 282.

Bunde, A., editor. 1992. *Fractals and Disorder*, North-Holland, Amsterdam.

Burrough, P. A. 1983. Multiscale sources of spatial variation in soil. I, The application of fractal concepts to nested levels of soil variation, *J. Soil Sci.*, **34**, 577–591.

Burrough, P. A. 1989. Fractals and geochemistry. In *The Fractal Approach to Heterogeneous Chemistry*, D. Avnir editor, J. Wiley, New York, pp. 383–406.

Cabral, M., and S. J. Burges. 1994. Digital elevation model networks (DEMON): A model of flow over hillslopes for computation of contributing and dispersal areas, *Water Resour. Res.*, **30**(6), 1681–1692.

Caldarelli, G., A. Maritan, J. R. Banavar, A. Giacometti, I. Rodriguez-Iturbe, and A. Rinaldo. 1996. Cellular models for river networks, preprint.

Calver, C. W. 1978. Modelling headwater development, *Earth Surf. Proc.*, **3**, 223–241.

Carlston, C. W. 1969. Downstream variations in the hydraulic geometry of streams: Special emphasis in mean velocity, *Am. J. Sci.*, **64**(2), 241–256.

Carrara, A. 1988. Drainage and divide networks derived from high fidelity digital terrain models. In

Quantitative Analysis of Mineral and Energy Resources, NATO ASI Series C, Mathematical and Physical Sciences, vol. 223, C. F. Chuna et al. editors, D. Reidel, Himgham, MA, pp. 215–223.

Cayley, A. 1859. On the analytical forms called trees, *Phil. Mag.*, **18**, 374–378.

Chase, C. G. 1992. Fluvial landsculpting and the fractal dimension of topography, *Geomorphology*, **5**, 39–57.

Chen, K., P. Bak, and S. P. Obukhov. 1991. Self-organized criticality in a crack-propagation model of earthquakes, *Phys. Rev. A*, **43**(2), 625–630.

Chhabra, A. B., and R. V. Jensen. 1989. Direct determination of the $f(\alpha)$ singularity spectrum, *Phys. Rev. Lett.*, **62**, 1327–1331.

Chorley, R. J., editor. 1969. *Water, Earth and Man*, Methuen, London.

Chorley, R. J., R. P. Beckinsale, and A. J. Dunn. 1973. *The History of the Study of Landforms*, Methuen, London.

Chorley, R. J., S. A. Schumm, and D. E. Sugden. 1985. *Geomorphology*, Methuen, London.

Chow, V. T. 1959. *Open-Channel Hydraulics*, McGraw–Hill, New York.

Christensen, K., Z. Olami, and P. Bak. 1992. Deterministic $1/f$ noise in nonconservative models of self-organized criticality, *Phys. Rev. Lett.*, **68**(16), 2417–2420.

Churchill, R. V., and T. Brown. 1979. *Fourier Series and Boundary Value Problems*, 2nd edition, McGraw–Hill, New York.

Cieplak, M., A. Giacometti, A. Maritan, A. Rinaldo, I. Rodriguez-Iturbe, and J. R. Banavar. 1996. Disorder-dominated river basins, preprint.

Colaiori, F., A. Flammini, A. Maritan, and J. R. Banavar. 1996. Analytical solutions for optimal channel networks, *Phys. Rev. E*, in press.

Coniglio, A. 1986. Multifractal structure of clusters and growing aggregates, *Physica A*, **140**, 51–62.

Coniglio, A., and M. Zannetti. 1989. Multiscaling and multifractality, *Physica D*, **38**, 37–40.

Coniglio, A., C. Amitrano, and F. Di Liberto. 1986. Growth probability distribution in kinetic aggregation processes, *Phys. Rev. Lett.*, **57**(8), 1016–1019.

Cordova J. R., I. Rodriguez-Iturbe, and P. Vaca. 1982. On the development of drainage networks, Proc. Exter Symposium on Recent Developments in the Explanation and Prediction of Erosion and Sediment Transport, IAHS Publ. 137, 239–249.

Cox, D. R., and H. D. Miller. 1965. *The Theory of Stochastic Processes*, Methuen, London.

Cox, D. R., and V. Isham. 1980. *Point Processes*, Chapman and Hall, London, 182 pp.

Cramér, H. 1946. *Mathematical Methods of Statistics*, Princeton University Press, Princeton, NJ.

Culling, W. E. H. 1960. Analytical theory of erosion, *J. Geol.*, **68**, 336–344.

Culling, W. E. H. 1963. Soil creep and the development of hillside slopes, *J. Geol.*, **71**, 127–161.

Culling, W. E. H. 1965. Theory of erosion on soil-covered slopes, *J. Geol.*, **73**, 230–254.

Culling, W. E. H. 1986. On Hurst phenomena in the landscape, *Trans. Jpn. Geomorphol. Union*, **7**(4), 221–243.

Dagan, G. 1984. Solute transport in heterogeneous porous formations, *J. Fluid Mech.*, **145**, 151–177.

Dagan, G. 1989. *Flow and Transport in Porous Formations*, Springer–Verlag, Berlin-Heidelberg.

Dagan, G., V. Cvetkovic, and A. Shapiro. 1992. A solute flux approach to transport in heterogeneous formations. 1, The general framework, *Water Resour. Res.*, **28**(5), 1369–1376.

Davis, W. N. 1899. The geographical cycle, *Geogr. J.*, **14**, 481–504.

Davis, W. N. 1924. *Die erklarende Beschreibung der Landformen*, 2nd edition, Teubner, Leipzig.

De Arcangelis, L., S. Redner, and A. Coniglio. 1985. Anomalous voltage distribution of random resistor networks and a new model for the backbone at the percolation threshold, *Phys. Rev. B*, **31**, 4725–4727.

De Arcangelis, L., J. Koeplik, S. Redner, and D. Wilkinson. 1986. Hydrodynamic dispersion in network models of porous media, *Phys. Rev. Lett.*, **57**(8), 996–999.

Delworth, T., and S. Manabe. 1988. The influence of potential evaporation on the variabilities of simulated soil wetness and climate, *J. Clim.*, **1**, 523–547.

Delworth, T., and S. Manabe. 1989. The influence of soil wetness on near-surface atmospheric variability, *J. Clim.*, **2**, 1447–1462.

De Boer, J., B. Derrida, H. Flyvbjerg, A. D. Jackson, and T. Wettig. 1994. Simple model of self-organized biological evolution, *Phys. Rev. Lett.*, **73**(6), 906–909.

De Vries, H., T. Becker, and B. Eckhardt. 1994. Power law distribution of discharge in ideal networks, *Water Resour. Res.*, **30**(12), 3541–3543.

Dhar, D., and S. N. Majumdar. 1990. Abelian sandpile model of the Bethe lattice, *J. Phys. A: Math. Gen.*, **23**, 4333–4350.

Dhar, D. 1990. Self-organized critical state of sandpile automaton models, *Phys. Rev. Lett.*, **64**(14), 1613–1616.

Dhar, D., and R. Ramaswamy. 1989. Exactly solved model of self-organized critical phenomena, *Phys. Rev. Lett.*, **63**(16), 1659–1662.

Dietler, G., and Y. Zhang. 1992. Fractal aspects of the Swiss landscape, *Physica A*, **191**, 213–219.

Dietrich, W. E., and T. Dunne. 1993. The channel head, In *Channel Network Hydrology*, K. Beven and M. J. Kirby editors, J. Wiley, New York, pp. 176–219.

Dietrich, W. E., C. J. Wilson, and S. L. Reneau. 1986.

Hollows, colluvium, and land slides in soil-mantled landscapes. In *Hillslope Processes*, E. D. Abrahams editor, Allen and Unwin, New York, pp. 361–388.

Dietrich, W. E., S. L. Reneau, and C. J. Wilson. 1987. Overview: "Zero order basins" and problem of drainage density, sediment transport and hillslope morphology, *IAHS Publ.* no. 165, 49–59.

Dietrich, W. E., D. R. Montgomery, S. L. Reneau, and P. Jordan. 1988. The use of hillslope convexity to calculate diffusion coefficients for a slope dependent transport law, *EOS Trans. AGU*, **69**(16), 316–317.

Dietrich, W. E., C. Wilson, D. R. Montgomery, J. McKean, and R. Bauer. 1992. Erosion thresholds and land surface morphology, *J. Geol.*, **20**, 675–679.

Dietrich, W. E., C. J. Wilson, D. R. Montgomery, and J. McKean. 1993. Analysis of erosion thresholds, channel networks and landscape morphology using a digital terrain model, *J. Geol.*, **3**, 161–180.

Dietrich, W. E., R. Reiss, M. L. Hsu, and D. R. Montgomery. 1995. A process-based model for colluvial soil depth and shallow landsliding using digital elevation data, *Hydrol. Processes*, **9**, 383–400.

Dolan, R., L. Vincent, and B. Hayden. 1974. Crescentic coastal landforms, *Z. Geomorph. N. F.*, **18**(1), 1–12.

Dooge, J. C. I. 1973. Linear theory of hydrologic systems, *USDA Tech. Bull. 1468.*

Drossel, B., and F. Schwabl. 1992. Self-organized criticality in a forest-fire model, *Physica A*, **191**, 47–50.

Drossel, B., S. Clar, and F. Schwabl. 1993. Exact results for the one-dimensional self-organized critical forest fire model, *Phys. Rev. Lett.*, **71**(23), 3739–3742.

Dubuc, B., J. F. Quiniou, C. Roques-Carmes, C. Tricot, and S. W. Zucker. 1989a. Evaluating the fractal dimension of profiles. *Phys. Rev. A*, **39**(3), 1500–1512.

Dubuc, B., S. W. Zucker, W. Tricot, J. F. Quiniou, and D. Wehbi. 1989b. Evaluating the fractal dimension of surfaces, *Proc. Roy. Soc. London*, **a425**, 113–127.

Dunne, T. 1978. Field studies in hillslope flow processes. In *Hillslope Hydrology*, M. J. Kirkby editor, J. Wiley, New York, pp. 227–294.

Dunne, T. 1980. Formation and controls of channel networks, *Progr. Phys. Geogr.*, **4**, 211–239.

Dunne, T., and W. E. Dietrich. 1980a. Experimental study of Horton overland flow on tropical hillslopes. 1, Soil conditions, infiltration and frequency runoff, *Zeits. fur Geomorph.*, Suppl. Bd., **35**, 40–58.

Dunne, T., and W. E. Dietrich. 1980b. Experimental study of Horton overland flow on tropical hillslopes. 2, Hydraulic characteristic and hillslope hydrographs, *Zeits. fur Geomorph.*, Suppl. Bd., **35**, 59–80.

Dunne, T., and B. F. Aubrey. 1986. Evaluation of Horton's theory of sheetwash and rill erosion on the basis of field experiments. In *Hillslope Processes*, A. D. Abrahams editor, Allen and Unwin, Winchester, pp. 31–53.

Eagleson, P. S. 1970. *Dynamic Hydrology*, McGraw–Hill, New York.

Eagleson, P. S. 1978. Climate, soil and vegetation. 6, Dynamics of the annual water balance, *Water Resour. Res.*, **5**, 749–755.

Eagleson, P. S. 1994. The evolution of modern hydrology (from watershed to continent in 30 years), *Adv. Water Resour.*, **17**, 3–18.

Eden, M. 1961. A two-dimensional growth process. In Fourth Berkeley Symposium on Mathematical Statistics and Probability, Volume IV: *Biology and Problems of Health*, J. Neyman editor, University of California Press, Berkeley, pp. 223–239.

Eigen, M., and R. Winkler. 1993. *Laws of the Game*, Princeton University Press, Princeton, NJ.

Entekhabi, D., and P. S. Eagleson. 1989. Land surface hydrology parametrization for atmospheric general circulation models including subgrid scale spatial variability, *J. Clim.*, **2**, 816–831.

Entekhabi, D., and I. Rodriguez-Iturbe. 1994. Analytical framework for the characterization of the space–time variability of soil moisture, *Adv. Water Resour.*, **17**, 35–46.

Fairfield, J., and P. Leymarie. 1991. Drainage networks from grid digital elevation models, *Water Resour. Res.*, **27**, 709–717. (Correction, *Water Resour. Res.*, **27**, 2809, 1991.)

Falconer, J. 1985. *The Geometry of Fractal Sets*, Cambridge University Press, Cambridge, U.K.

Falconer, J. 1990. *Fractal Geometry: Mathematical Foundations and Applications*, J. Wiley, New York.

Family, F. and T. Vicsek, editors. 1991. *Dynamics of Fractal Surfaces*, World Scientific, Singapore.

Feder, J. 1988. *Fractals*, Plenum, New York.

Feller, W. 1971. *An Introduction to Probability Theory and Its Applications*, vol. 2, 2nd edition, J. Wiley, New York.

Fisher, M. E. 1971. In *Critical Phenomena*, M. S. Green editor, Academic, New York.

Flint, J. J. 1973. Experimental development of headward growth into channel networks, *Geol. Soc. Am. Bull.*, **84**, 1087–1095.

Flint, J. J. 1974. Stream gradient as a function of order, magnitude, and discharge, *Water Resour. Res.*, **10**(5), 969–973.

Flyvbjerg, H., K. Sneppen, and P. Bak. 1993. Mean field theory for a simple model of evolution, *Phys. Rev. Lett.*, **71**(24), 4987–4090.

Fox, C. G. 1989. Empirically derived relationship between fractal dimension and power law from frequency spectra, *Pure Appl. Geophys*, **131**, 307–313.

Fox, C. G., and D. E. Hayes. 1989. Quantitative methods for analyzing the roughness of the seafloor, *Rev. Geophys.*, **23**, 1–48.

Freeman, T. G. 1991. Calculating catchment area with

divergent flow based on a regular grid, *Comp. Geosci.*, **17**, 413–422.

Frette, V., K. Christensen, A. Malthe-Sørenssen, J. Feder, T. Jøssag, and P. Meakin. 1996. Avalanche dynamics in a pile of rice, *Nature*, **379**(6560), 49–52.

Frisch, U., and G. Parisi. 1985. Fully developed turbulence and intermittency. In *Turbulence and Predictability in Geophysical Fluid Dynamics*, M. Ghil, R. Benzi, and G. Parisi editors, North Holland, New York, 84–92.

Fujiwara, A., G. Kamimoto, and A. Tsukamoto. 1977. Destruction of basaltic bodies by high-velocity impact, *Icarus*, **31**, 277–288.

Gilbert, G. K. 1877. Report on the geology of the Henry Mountains, Rep. U.S. Geological Survey of the Rocky Mountains Regions, Government Printing Office, Washington DC.

Gilbert, G. K. 1909. The convexity of hilltops, *J. Geol.*, **17**, 344–350.

Gilbert, L. E. 1989. Are topographic data sets fractals?, *Pure Appl. Geophys.*, 241–254.

Gilbert, L. E., and A. Malinverno. 1988. A characterization of the spectral density of residual ocean floor topography, *Geophys. Res. Lett.*, **15**, 1401–1404.

Glansdorff, P., and I. Prigogine. 1971. *Thermodynamic Theory of Structure, Stability and Fluctuations*, Wiley Interscience, London.

Glock, W. S. 1931. The development of drainage systems: A synoptic view, *Geogr. Rev.*, **21**, 475–482.

Gore, A. 1990. *Earth in the Balance*, Pergamon, Boston.

Gould, S. J. 1989. *Wonderful Life*, W. W. Norton, Boston, MA.

Gould, S. J., and R. Eldredge. 1977. Punctuated equilibrium: The tempo and mode of evolution reconsidered, *Paleobiology*, **3**, 114–122.

Gould, S. J., and R. Eldredge. 1993. Punctuated equilibrium comes of age, *Nature*, **366**, 223–226.

Grassberger, P., and I. Procaccia. 1983. Characterization of strange attractors, *Phys. Rev. Lett.*, **50**, 346–351.

Gray, D. M. 1961. Interrelationships of watershed characteristics, *J. Geophys. Res.*, **66**(4), 1215–1223.

Gregory, K. J., and D. E. Walling. 1973. *Drainage Basin Form and Process*, Edward, London.

Grieger, B. 1992. Quaternary fluctuations as a consequence of self-organized criticality, *Physica A*, **191**, 51–56.

Grinstein, G., D. H. Lee, and S. Sachdev. 1990. Conservation laws, anisotropy and self-organized criticality in noisy nonequilibrium systems, *Phys. Rev. Lett.*, **64**(16), 1927–1931.

Groff, M. 1992. Implicationi idrologiche dells studio della funetione di-ampiezza in bacini reali, Thesis, Dipartimento di Ingegneria Civile ed Ambientale, Universita' di Trento.

Gupta, V. K., and O. J. Mesa. 1988. Runoff generation and hydrologic response via channel network geomorphology – Recent progress and open problems, *J. Hydrol.*, **102**, 3–28.

Gupta, V. K., and E. Waymire. 1983. On the formulation of an analytical approach to understand hydrological response and similarity at the basin scale, *J. Hydrol.*, **65**, 95–129.

Gupta, V. K., and E. Waymire. 1989. Statistical self-similarity in river networks parameterized by elevation, *Water. Resour. Res.*, **25**(3), 463–476.

Gupta, V. K., and E. Waymire. 1990. Multiscaling properties of spatial rainfall and river flow distributions, *J. Geophys. Res.*, **95**, 1999–2009.

Gupta, V. K., and E. Waymire. 1993. A statistical analysis of mesoscale rainfall as a random cascade, *J. Appl. Meteorology*, **32**, 251–267.

Gupta, V. K., E. Waymire, and C. T. Wang. 1980. A representation of an IUH from geomorphology, *Water Resour. Res.*, **16**(5), 885–862.

Gupta, V. K., E. Waymire, and I. Rodriguez-Iturbe. 1986. On scales, gravity and network structure in basin runoff. In *Scale Problems in Hydrology*, V. K. Gupta, I. Rodriguez-Iturbe, and E. F. Wood editors, Reidel, Dordrecht, pp. 159–184.

Gutemberg, B., and C. F. Richter. 1954. *Seismicity of the Earth and Associated Phenomena*, Princeton University Press, Princeton, NJ.

Gutjahr, A. L. 1989. Fast Fourier Transform for random field generation, Project Rep. for Los Alamos Grant to New Mexico, Contract no. 4 - r58 - 2690R.

Hack, J. T. 1957. Studies of longitudinal profiles in Virginia and Maryland, *U.S. Geol. Surv. Prof. Paper*, 294-B.

Hadley, R. F., and S. A. Schumm. 1961. Sediment sources and drainage basin characteristics in upper Cheyenne River Basin, *U.S.G.S. Water Supply Paper 1531-B.*

Hadwich, G., B. Merte', A. Hubler, and E. Luscher. 1990. Stationary dendritic structures in an electric field, *Herbsttagung der SPG/SSP*, **63**, 487–489.

Halsey, T. C., P. Meakin, and I. Procaccia. 1986a. Scaling structure of the surface layer of diffusion-limited aggregates. *Phys. Rev. Lett.*, **56**, 854–857.

Halsey, T. C., M. H. Jensen, L. P. Kadanoff, I. Procaccia, and B. I. Shraiman. 1986b. Fractal measures and their singularities: The characterization of strange sets, *Phys. Rev. A*, **33**(2), 1141–1151.

Hanks, T. C., and H. Kanamori. 1979. A moment-magnitude scale, *J. Geophys. Res.*, **84**, 2948–2980.

Hanks, T. C., R. C. Buckman, K. R. Lajoie, and R. E. Wallace. 1984. Modifications of wave-cut and faulting-controlled, *J. Geophys. Res.*, **89**, 5771–5790.

Havlin, S., and D. Ben-Avraham. 1987. Diffusion in disordered media, *Adv. Phys.*, **36**(6), 695–798.

Henderson, F. M. 1963. Some properties of the unit

hydrograph, *J. Geophys. Res.*, **68**(16), 4785–4793.

Henderson, F. M. 1966. *Open Channel Flow*, MacMillan, New York.

Hentschel, H. G. E., and I. Procaccia. 1983. The infinite number of dimension of fractals and strange attractors, *Physica D*, **8**, 435–444.

Hjemfelt, A. T. 1988. Fractals and the river length catchment-area ratio, *Water Resour. Bull.*, **24**, 455–459.

Holley, R., and E. C. Waymire. 1992. Multifractal dimensions and scaling exponents for strongly bounded random cascades, *Ann. Appl. Probability*, **2**(4), 819–845.

Horton, R. E. 1932. Drainage-basin characteristics, *EOS Trans. AGU*, **13**, 350–361.

Horton, R. E. 1945. Erosional development of streams and their drainage basins; hydrophysical approach to quantitative morphology, *Bull. Geol. Soc. Am.*, **56**, 275–370.

Hough, S. E. 1989. On the use of spectral methods for the determination of fractal dimension, *Geophys. Res. Lett.*, **16**(7), 673–676.

Howard, A. D. 1971a. Optimal angles of stream junction: Geometric, stability to capture, and minimum power criteria, *Water Resour. Res.*, **7**, 863–873.

Howard, A. D. 1971b. Simulation of stream networks by headward growth and branching, *Geogr. Anal.*, **3**, 29–50.

Howard, A. D. 1971c. Simulation model of stream capture, *Geol. Soc. Am. Bull.*, **82**, 1355–1363.

Howard, A. D. 1990. Theoretical model of optimal drainage networks, *Water Resour. Res.*, **26**(9), 2107–2117.

Howard, A. D. 1994. A detachment-limited model of drainage basin evolution, *Water Resour. Res.*, **30**(7), 2261–2285.

Howard, A. D., and M. J. Selby. 1994. Rockslopes. In *Geomorphology of Desert Environments*, A. D. Abrahams and A. J. Parsons editors, Chapman and Hall, London, pp. 123–172.

Howard, A. D., W. E. Dietrich, and M. A. Seidl. 1994. Modelling fluvial erosion on regional to continental scales, *J. Geophys. Res.*, **99**(B7), 13971–13986.

Huang, J., and D. L. Turcotte. 1989. Fractal mapping of digitized images: Application to the topography of Arizona and comparisons with synthetic images, *J. Geophys. Res.*, **94**, 7491–7498.

Huang, K. 1963. *Statistical Mechanics*, J. Wiley, New York.

Huber, G. 1991. Scheidegger's rivers, Takayasu's aggregates and continued fractions, *Physica A*, **170**, 463–470.

Huggett, R. J. 1988. Dissipative systems: Implications for geomorphology, *Earth Surf. Proc. Landforms*, **13**, 45–49.

Huse, A., and C. L. Henley. 1985. Pinning and roughening in domain walls in Ising systems due to random impurities, *Phys. Rev. Lett.*, **54**, 2708–2712.

Hwa, T., and M. Kardar. 1992. Avalanches, hydrodynamics and discharge events in models of sandpiles, *Phys. Rev. A*, **45**(10), 7002–7023.

Ikeda, S., G. Parker, and Y. Kimura. 1988. Stable width and depth of straight gravel rivers with heterogeneous bed materials, *Water Resour. Res.*, **24**, 713–722.

Ijjasz-Vasquez, E. J., I. Rodriguez-Iturbe, and R. L. Bras. 1992. On the multifractal characterization of river basins, Proceedings of the 23rd Binghampton Geomorphology Conference, 24–27.

Ijjasz-Vasquez, E. J., R. L. Bras, I. Rodriguez-Iturbe, R. Rigon, and A. Rinaldo. 1993a. Are river basins optimal channel networks?, *Adv. Water Resour.*, **16**, 69–79.

Ijjasz-Vasquez, E. J., R. L. Bras, and I. Rodriguez-Iturbe. 1993b. Hack's relation and optimal channel networks: The elongation of river basins as a consequence of energy minimization, *Geophys. Res. Lett.*, **20**(15), 1583–1586.

Ijjasz-Vasquez, E. J., R. L. Bras, and I. Rodriguez-Iturbe. 1994. Self-affine scaling of fractal river courses and basin boundaries, *Physica A*, **209**, 288–300.

Jackson, T. J., E. T. Engman, and F. R. Schiebe. 1993. Washita '92 experiment description. In *Hydrology Data Report*, T. J. Jackson and F. R. Schiebe editors, National Agricultural Water Quality Laboratory, **93**, pp. 1–5.

Jensen, M. H., L. P. Kadanoff, A. Libchaber, I. Procaccia, and J. Stavans. 1985. Global universality at the onset of chaos: Results of a forced Rayleigh–Benard experiment, *Phys. Rev. Lett.*, **55**, 2798–2801.

Jensen, M. H., G. Paladin, and A. Vulpiani. 1991. Multiscaling in multifractals, *Phys. Rev. Lett.*, **67**(2), 208–211.

Jin, C. 1992. A deterministic gamma-type GIUH based of path types, *Water Resour. Res.*, **28**, 479–486.

Johnson, D. 1987. More approaches on the travelling salesman guide, *Nature*, **330**(10), 525.

Kac, M. 1966. Can one hear the shape of a drum?, *Am. Math. Monthly* (Slaught Memorial Papers, no. 11), 1–23.

Kadanoff, L. P. 1995. A model of turbulence, *Physics Today*, **9**, 11–13.

Kadanoff, L. P., S. R. Nagel, L. Wu, and S. M. Zhou. 1989. Scaling and universality in avalanches, *Phys. Rev. A*, **39**, 6524–6533.

Kahane, J. P., and Peyrière. 1976. Sur certaines martingales de Benoit Mandelbrot, *Adv. Math.*, **22**, 131–145.

Kardar, M. 1985. Roughening by impurities at finite temperature, *Phys. Rev. Lett.*, **55**, 2923–2927.

Karlinger, M. R., and B. M. Troutman. 1985. Assessment of the instantaneous unit hydrograph derived from the theory of topologically random networks, *Water Resour. Res.*, **21**, 1693–1702.

Kardar, M., and Y. C. Zhang. 1987. Scaling of directed polymers in random media, *Phys. Rev. Lett.*, **58**, 2087–2090.

Kardar, M., G. Parisi, and Y. C. Zhang. 1986. Dynamic scaling at finite temperature, *Phys. Rev. Lett.*, **56**, 889–893.

Kashiwaya, K. 1987. Theoretical investigation of the time variation of drainage density, *Earth Surf. Proc. Landforms*, **12**, 39–46.

Kauffman, S. 1993. *The Origins of Order*, Oxford University Press, New York.

Kauffman, S. 1995. *At Home in the Universe*, Oxford University Press, New York.

Kauffman, S. A., and S. J. Johnsen. 1991. *J. Theor. Biol.*, **149**, 467–506.

Kennedy, J. F., P. D. Richardson, and S. P. Sutera. 1965. Discussion on geometry of river channels, *J. Hydraul. Div. Am. Soc. Civ. Eng.*, **91**(HY6), 332–334.

Kent, C., and J. Wong. 1982. An index of littoral complexity and its measurement, *Can. J. Fish. Aquat. Sci.*, **39**, 847–853.

Kenyon, P. M., and D. L. Turcotte. 1985. Morphology of a delta prograding by bulk sediment transport, *Geol. Soc. Am. Bull.*, **96**, 1457–1465.

King, L. C. 1967. *The Morphology of the Earth*, Oliver and Boyd, Edinburgh.

Kirkby, M. J. 1971. Hillslope process–response models based on continuity equation. In *Slopes: Form and Process*, M. J. Kirkby editor, Special Publication, 3, Institute of British Geographers, London, pp. 15–30.

Kirkby, M. J. 1976. Tests of the random model and its application to basin hydrology, *Earth Surf. Proc. Landforms*, **1**, 197–212.

Kirkby, M. J. 1993. Long term interactions between networks and hillslopes, In *Channel Network Hydrology*, K. Beven, and M. J. Kirkby editors, J. Wiley, New York, pp. 253–293.

Kirchner, J. W. 1993. Statistical inevitability of Horton's laws and the apparent randomness of stream channel networks, *Geology*, **21**, 591–594.

Kirkpatrick, S., G. D. Gelatt, and M. P. Vecchi. 1983. Optimization by simulated annealing, *Science*, **220**, 671–680.

Kirshen, D. M., and R. L. Bras. 1983. The linear channel and its effect on the geomorphologic IUH, *J. Hydrol.*, **65**, 175–208.

Kolmogorov, A. 1941. The local structure of turbulence in incompressible viscous fluid for very large Reynolds numbers, *Comptes Rendus, Dokl. Acad. Sci. USSR*, **30**, 301–305.

Koons, P. O. 1989. The topographic evolution of collisional mountain belts: A numerical look at the Southern Alps, New Zealand, *Am. J. Sci.*, **289**, 1041–1069.

Korcak, J. 1940. Deux types fundamentaux de distribution statistique, *Bull. Inst. Int. Stat.*, **30**, 295–307.

Korvin, G. 1992. *Fractal Models in the Earth Sciences*, Elsevier, New York.

Kramer, S., and M. Marder. 1992. Evolution of river networks, *Phys. Rev. Lett.*, **68**, 205-209.

Kurths, J., and H. Hertzel. 1987. An attractor in a solar time series, *Physica D*, **25**, 165–172.

La Barbera, P., and R. Rosso. 1987. The fractal geometry of river networks, *EOS Trans. AGU*, **68**, 1276.

La Barbera, P., and R. Rosso. 1989. On fractal dimension of streams networks, *Water. Resour. Res.*, **25**(4), 735–741.

La Barbera, P., and R. Rosso. 1990. On fractal dimension of streams networks, Reply to Tarboton et al., *Water. Resour. Res.*, **26**(9), 2245–2248.

La Barbera, P., and G. Roth. 1994. Invariance and scaling properties in the distributions of contributing area and energy in drainage basins, *Hydrol. Processes*, **8**, 125–135.

Landahl, M. T., and E. Mollo-Christensen. 1987. *Turbulence and Random Processes in Fluid Mechanics.*, Cambridge University Press, Cambridge.

Landau, L. D., and E. M. Lifschitz. 1959. *Fluid Mechanics*, Pergamon Press, Oxford.

Langbein, W. B. 1947. Topographic characteristics of drainage basins, *U.S. Geol. Surv. Prof. Paper*, 968-C.

Langbein, W. B. 1964. Geometry of river channels, *J. Hydraul. Div. Am. Soc. Civ. Eng.*, **90**(HY2), 301–312.

Langbein, W. B., and L. B. Leopold. 1964. Quasi-equilibrium states in channel morphology, *Am. Sci. J.*, **262**, 782–794.

Langbein, W. B., and S. A. Schumm. 1958. Yield of sediment in relation to mean annual precipitation, EOS *Trans. AGU*, **30**(6), 1076–1084.

Lapidus, M. 1989. Can one hear the shape of a fractal drum? Partial resolution of the Weyl–Berry conjecture, Proc. Workshop on Differential Geometry, Calculus of Variations, and Computer Graphics, MSRI, Berkeley, May 1988, Mathematical Sciences Research Institute Publications, Springer–Verlag, New York.

Lavallée, D., S. Lovejoy, D. Schertzer, and P. Ladoy. 1993. Nonlinear variability of landscape topography: Multifractal analysis and simulation. In *Fractals in Geography*, N. S. Lam and L. De Cola editors, Prentice Hall, Englewood Cliffs, NJ, pp. 158–192.

Leheny, R. L., and S. R. Nagel. 1993. A model for the evolution of river networks, *Phys. Rev. Lett.*, **71**(9), 1470–1473.

Leopold, L. B., and W. B. Langbein. 1962. The concept of entropy in landscape evolution, *U.S. Geol. Surv. Prof. Paper*, 500-A.

Leopold, L. B., and T. Maddock. 1953. The hydraulic geometry of stream channels and some physiographic implications, *U.S. Geol. Surv. Prof. Paper*, 252.

Leopold, L. B., and J. P. Miller. 1956. Ephemeral streams–Hydraulic factors and their relation to the drainage net, *U.S. Geol. Surv. Prof. Paper*, 282-A.

Leopold, L. B., and M. G. Wolman. 1957. River channel patterns: Braided, meandering and straight, *U.S. Geol. Surv. Prof. Paper*, 282-B.

Leopold, L. B., M. G. Wolman, and J. P. Miller. 1964. *Fluvial Process in Geomorphology*, Freeman, San Francisco.

Liao, K. H., and A. E. Scheidegger. 1968. A computer model for some branching type phenomena in hydrology, *Bull. Int. Ass. Sci. Hydrol.*, **13**(1), 5–13.

Lienhard, J. H. 1964. A statistical mechanical prediction of the dimensionless unit hydrograph, *J. Geophys. Res.*, **69**(24), 330–334.

Lin, S. 1965. Computer solutions for the travelling salesman problem, *Bell Syst. Tech. J.*, **44**, 2245–2258.

Lin, S., and B. W. Kernighan. 1973. An effective heuristic algorithm for the traveling salesman problem, *Oper. Res.*, **21**, 498–516.

Lovejoy, S. 1982. The area–perimeter relation for rain and cloud areas, *Science*, **216**, 185–187.

Lovejoy, S., and D. Schertzer. 1985. Generalized scale invariance in the atmosphere and fractal models of rain, *Water Resour. Res.*, **21**, 1233–1250.

Lovejoy, S., and D. Schertzer. 1988. Scaling, fractals and nonlinear variability in geophysics, *EOS Trans. AGU*, **69**, 143–145.

Lovejoy, S., and D. Schertzer. 1990. Multifractals, universality classes and satellite and radar measurements of cloud and rain fields, *J. Geophys. Res.*, **95**, 2021–2034.

Lovejoy, S., D. Schertzer, and P. Ladoy. 1986. Fractal characterization of inhomogeneous measuring networks, *Nature*, **319**, 43–44.

Lovejoy, S., D. Schertzer, and A. A. Tsonis. 1987. Functional box counting and multiple elliptical dimensions in rain, *Science*, **235**, 1036–1038.

Lowenherz, D. S. 1991. Stability and the initiation of channelized surface drainage: A reassessment of the short wavelength limit, *J. Geophys. Res.*, **96**, 8453–8464.

Lubowe, J. K. 1964. Stream junction angles in the dendritic drainage pattern, *Am. J. Sci.*, **262**, 325–339.

Luke, J. C. 1972. Mathematical models of landform development, *J. Geophys. Res.*, **77**, 2460–2470.

Luke, J. C. 1974. Special solutions for nonlinear erosion equations, *J. Geophys. Res.*, **79**, 4035–4044.

Malinverno, A. 1989. Testing linear models of sea-floor topography, *Pure Appl. Geophys.*, **131**, 139–155.

Mandelbrot, B. B. 1963. The variation of certain speculative prices, *J. Bus. U. Chicago*, **36**, 394–419.

Mandelbrot, B. B. 1967. How long is the coast of Britain? Statistical self-similarity and fractional dimension, *Science*, **156**, 636–638.

Mandelbrot, B. B. 1972. Possible refinement of the lognormal hypothesis concerning the distribution of energy dissipation in intermittent turbulence. In *Statistical Models and Turbulence*, M. Rosenblatt and C. Van Atta editors, Lecture Notes in Physics 12, Springer, New York, pp. 333–351.

Mandelbrot, B. B. 1974. Intermittent turbulence in self-similar cascades: Divergence of high moments and dimension of the carrier, *J. Fluid Mech.*, **62**, 331–358.

Mandelbrot, B. B. 1975. Stochastic models of the Earth's relief, the shape and the fractal dimension of the coastlines, and the number–area rule for islands, *Proc. Natl. Acad. Sci. USA*, **72**, 3825–3828.

Mandelbrot, B. B. 1977. *Fractals: Form, Chance and Dimension*, Freeman, San Francisco.

Mandelbrot, B. B. 1983. *The Fractal Geometry of Nature*, Freeman, New York.

Mandelbrot, B. B. 1985. Self-affine fractals and fractal dimension, *Physica Scripta*, **32**, 257–260.

Mandelbrot, B. B. 1990. Multifractal measures, especially for the geophysicist, *Pure Appl. Geophys.*, **131**(1/2), 5–42.

Mandelbrot, B. B., and J. W. Van Ness. 1968. Fractional Brownian motions, fractional noises and applications, *SIAM Rev.*, **10**, 422–437.

Mandelbrot, B. B., and V. R. Wallis. 1968. Noah, Joseph and operational hydrology, *Water Resour. Res.*, **4**, 909–918.

Mantoglu, A., and J. L. Wilson. 1988. The turning bands method for the simulation of random field using line generation by a spectral method, *Water Resour. Res*, **18**(5), 1379–1394.

Marani, A., R. Rigon, and A. Rinaldo. 1991. A note on fractal channel networks, *Water Resour. Res.*, **27**, 3041–3049.

Marani, M. 1994. Sulla funzione di ampiezza dei bacini idrografici naturali, *Quaderni di Informazione e di Studio*, 4, Centro Internazionale di Idrologia "Dino Tonini, " Universitá di Padova.

Marani, M., A. Rinaldo, R. Rigon, and I. Rodriguez-Iturbe. 1994. Geomorphological width functions and the random cascade, *Geophys. Res. Lett.*, **21**(19), 2123–2126.

Marder, S. P. 1993. Nonlinear models of river networks, Ph.D. Dissertation, University of Texas at Austin.

Mareschal, J. C. 1989. Fractal reconstruction of sea floor topography, *Pure Appl. Geophys.*, **131**, 197–210.

Maritan, A., A. Rinaldo, A. Giacometti, R. Rigon, and I. Rodriguez-Iturbe. 1996a. Scaling in river networks, *Phys. Rev. E*, **53**, 1501–1512.

Maritan, A., F. Colaiori, A. Flammini, M. Cieplak, and J.

R. Banavar. 1996b. Universality classes of optimal channel networks, *Science*, **272**, 984–986.

Mark, D. M. 1983. Relation between field-surveyed channel networks and map-based geomorphic measures, Inez, Kentucky, *Ann. Am. Geogr.*, **73**(3), 358–372.

Mark, D. M. 1988. Network models in geomorphology. In *Modelling in Geomorphological Systems*, M. G. Anderson editor, J. Wiley, New York, pp. 73–95.

Mark, D. M., and P. B. Aronson. 1984. Scale-dependent fractal dimensions of topographic surfaces: An empirical investigation with applications in geomorphology and computer mapping, *Math. Geol.*, **16**(7), 671–683.

Maslov, S., M. Paczuski, and P. Bak. 1994. Avalanches and $1/f$ noise in evolution and growth models, *Phys. Rev. Lett.*, **73**, 2162–2166.

Matsushita, M., and S. Ouchi. 1989. On the self-affinity of various curves, *Physica D*, **38**, 246.

Meakin, P. 1987a. Scaling properties for the growth probability measure and harmonic measure of fractal structures. *Phys. Rev. A*, **35**, 2234–2245.

Meakin, P. 1987b. Fractal aggregates and their fractal measures. In *Phase Transitions and Critical Phenomena*, C. Domb and J. L. Lebowitz editors, Academic Press, New York, pp. 141–160.

Meakin, P. 1991. Fractal aggregates in geophysics, *Rev. Geophys.*, **29**(3), 335–382.

Meakin, P., H. E. Stanley, A. Coniglio, and T. A. Witten. 1985. Surfaces, interfaces and screening of fractal structures, *Phys. Rev. A*, **32**, 2364–2369.

Meakin, P., A. Coniglio, H. E. Stanley, and T. A. Witten. 1986. Scaling properties for the surfaces of fractal and nonfractal objects: An infinite hierarchy of critical exponents, *Phys. Rev. A*, **34**, 3325–3340.

Meakin, P., J. Feder, and T. Jossang. 1991. Simple statistical models of river networks, *Physica A*, **176**, 409–429.

Mei, C. C. 1995. *Mathematical Analysis in Engineering*, Cambridge University Press, New York.

Meinhardt, H. A. 1976. Morphogenesis of lines and nets, *Differentiation*, **6**, 117–123.

Meinhardt, H. A. 1982. *Models of Biological Pattern Formation*, Academic, San Diego.

Melton, M. A. 1958. Geometric properties of mature drainage systems and their representation in an E_4 phase space, *J. Geol.*, **66**, 35–54.

Melton, M. A. 1959. A derivation of Strahler's channel ordering system, *J. Geol.*, **67**, 345–346.

Meneveau, C., and K. R. Sreenivasan. 1987. Simple multifractal cascade for fully developed turbulence, *Phys. Rev. Lett.*, **59**(13), 1424–1427.

Meneveau, C., and K. R. Sreenivasan. 1991. The multifractal nature of turbulent energy dissipation, *J. Fluid Mech.*, **224**, 420–484.

Merte', B., P. Gaitzsch, M. Fritzenwanger, W. Kropf, A. Hubler, and E. Luscher. 1988. Stable stationary dendritic patterns with minimal dissipation, Rapport de la Reunion d'automne de la SSP, 76–79.

Merte', B., G. Hadwich, B. Binias, P. Deisz, A. Hubler, and E. Luscher. 1989. Formation of self-similar dendritic patterns with extremal properties, *Helvetica Physica Acta*, **62**, 294–297.

Merte', B., J. Muller, R. Ruckerl, P. Hildebrand, and E. Luscher. 1990. Comparison of three different numerical methods to characterize the geometry of dendritic structures, *Fruhjahrstagung*, **63**, 821–822.

Mesa, O. J. 1986. Analysis of channel networks parametrized by elevation, Ph.D. Dissertation, Dept. of Civ. Eng., University of Mississippi.

Mesa, O. J., and V. K. Gupta. 1987. On the main channel length–area relationships for channel networks, *Water Resour. Res.*, **23**(11), 2119–2122.

Mesa, O. J., and E. R. Mifflin. 1988. On the relative role of hillslope and network geometry in hydrologic response. In *Scale Problems in Hydrology*, V. K. Gupta, I. Rodriguez-Iturbe, and E. F. Wood editors, Dordrecht, Holland, pp. 181–190.

Metropolis, N., M. Rosembluth, M. Teller, and E. Teller. 1953. Equations of state calculations by fast computing machines, *J. Chem. Phys.*, **21**, 1087–1096.

Miramontes, O., R. V. Solé, and B. C. Goodwin. 1993. Collective behaviour of random-activated mobile cellular automata, *Physica D*, **63**, 145–160.

Mock, S. J. 1971. A classification of channel links in stream networks, *Water Resour. Res.*, **7**, 1558–1566.

Monin, A. S., and A. M. Yaglom. 1972. *Statistical Fluid Mechanics*, MIT Press, Cambridge, MA.

Montgomery, D. R. 1991. Channel initiation and landscape evolution, Ph.D. Dissertation, 421 pp., University of California, Berkeley.

Montgomery, D. R., and W. E. Dietrich. 1988. Where do channels begin?, *Nature*, **336**, 232–234.

Montgomery, D. R., and W. E. Dietrich. 1989. Source areas, drainage density, and channel initiation, *Water Resour. Res.*, **25**, 1907–1918.

Montgomery, D. R., and W. E. Dietrich. 1992. Channel initiation and the problem of landscape scale, *Science*, **255**, 826–830.

Montgomery, D. R., and W. E. Dietrich. 1994. Landscape dissection and drainage-slope thresholds. In *Process Models and Theoretical Geomorphology*, M. J. Kirby editor, J. Wiley, New York, pp. 221–246.

Montgomery, D. R., and E. Foufoula-Georgiou. 1993. Channel networks source representation using digital elevation models, *Water Resour. Res*, **29**(12), 1925–1934.

Moore, I. D., and G. J. Burch. 1986. Sediment transport capacity of sheet and rill flow: Application of unit stream power theory, *Water Resour. Res.*, **22**(8), 150–1360.

Moore, I. D., E. M. O'Loughlin, and G. J. Burch. 1988. A contour-based topographic model for hydrological and ecological applications, *Earth Surf. Proc. Landforms*, **13**, 305–320.

Morisawa, M. E. 1964. Development of drainage systems on an upraised lake floor, *Am. J. Sci.*, **262**, 340–354.

Morse, T., and R. Feshback. 1967. *Mathematical Methods for Theoretical Physics*, vol. I, McGraw–Hill, New York.

Mosley, M. P. 1972. An experimental study of rill erosion, M.S. Thesis, Colorado State University, Fort Collins, CO.

Muir, J. 1873. A geologist's winter walk, *Overland Monthly*, **10**, 1873.

Muller, J. E. 1973. Re-evaluation of the relationship of master streams and drainage basins: Reply, *Geol. Soc. A. Bull.*, **84**, 3127–3130.

Murray, C. D. 1926. The physiological principle of minimum work. I, The vascular system and the cost of blood volume, *Proc. Natl. Acad. Sci. USA*, **12**, 207–214.

Nagatani. 1993. Kinetic growth transitions in a simple aggregation of charged particles with injection, *J. Phys. A*, **26**, 489–496.

Nakano, T. 1983. A fractal study of some rias coastlines in Japan, *Ann. Rep. Inst. Geosci. Univ. Tsukuba*, **9**, 75–80.

Nash, E. 1957a. The form of the instantaneous unit hydrograph, Proc. IAHS, Toronto, Tome III, 114–121.

Nash, E. 1957b. Systematic determination of unit hydrograph parameters, *J. Geophys. Res.*, **64**, 111–115.

National Research Council. 1991. *Opportunities in the Hydrologic Sciences*, National Academy Press, Washington DC.

Neill, W., and P. Murphy. 1993. *By Nature's Design*, Chronicle Books, San Francisco.

Newman, W. I., and D. L. Turcotte. 1990. Cascade model for fluvial geomorphology, *Geophys. J. Int.*, **100**, 433–439.

Nikora, V. I. 1991. Fractal structure of river plan forms, *Water Resour. Res*, **27**(6), 3569–3575.

Nikora, V. I. 1994. On self-similarity and self-affinity of drainage basins, *Water Resour. Res*, **30**(1), 133–137.

Nikora, V. I., and V. B. Sapozhnikov. 1993a. River network fractal geometry in computer simulation, *Water Resour. Res.*, **29**(10), 1327–1333.

Nikora, V. I., and V. B. Sapozhnikov. 1993b. Fractal geometry of individual river channel and its computer simulation, *Water Resour. Res.*, **29**(10), 3561–3568.

O'Callaghan, J. F., and D. M. Mark. 1985. The extraction of drainage networks from digital elevation data, *Comput. Vision Graphics Image Process.*, **28**, 323–344.

O'Loughlin, E. M. 1986. Prediction of surface saturation zones in natural catchments by topographic analysis, *Water Resour. Res.*, **22**(5), 794–804.

Ouchi, S., and M. Matsushita. 1992. Measurement of self-affinity on surfaces as a trial application of fractal geometry to landform analysis, *Geomorphology*, **5**, 115–130.

Paczuski, M., and P. Bak. 1993. Theory of the one-dimensional forest fire model, *Phys. Rev. E*, **48**, 321–332.

Paczuski, M., S. Maslov, and P. Bak. 1994. Field theory for a model of self-organized criticality, *Europhys. Lett.*, **27**(2), 97–102.

Paczuski, M., S. Maslov, and P. Bak. 1995a. Laws for stationary states in systems with extremal dynamics, *Phys. Rev. Lett.*, **74**, 4253–4256.

Paczuski, M., S. Maslov, and P. Bak. 1995b. Avalanche dynamics in evolution, growth and depinning models, *Phys. Rev. E.*, **53**, 414–418.

Paola, C. 1989. A simple basin-filling model for coarse-grained alluvial systems. In *Quantitative Dynamic Stratigraphy*, T. A. Cross editor, Prentice–Hall, Englewood Cliffs, pp. 363–374, NJ.

Paola, C., P. L. Heller, and C. L. Angevine. 1992. The large-scale dynamics of grain-size sorting in alluvial basins. 1, Theory, *Basin Res.*, **4**, 73–90.

Parker, G. 1978. Self-formed straight rivers with equilibrium banks and mobile bed. 2, The gravel river, *J. Fluid Mech.*, **89**, 127–146.

Parker, G. 1979. Hydraulic geometry of active gravel bed rivers, *ASCE J. Hydr. Eng.*, **105**, 1185–1194.

Parker, R. S. 1977. Experimental study of drainage basin evolution and its hydrologic implications, Ph.D. Dissertation, 353 pp., Colorado State University, Fort Collins, CO.

Patton, P. C., and S. A. Schumm. 1975. Gully erosion, Northwestern Colorado: A threshold phenomenon, *Geology*, **3**, 88–90.

Peano, G. 1890. Sur une courbe qui remplit toute une aire plane, *Matematische Annalen*, **36**, 157–160.

Peckham, S. 1995. New results for self-similar trees with applications to river networks, *Water Resour. Res.*, **31**(4), 1023–1030.

Peckham, S., and E. Waymire. 1992. On a symmetry of turbulence, *Comm. Math. Phys.*, **147**, 365–370.

Peitgen, H. O., and D. Saupe, editors, 1988. *The Science of Fractal Images*, Springer–Verlag, New York.

Peitgen, H. O., H. Jurgens, and D. Saupe. 1992. *Chaos and Fractals, New Frontiers of Science*, Springer–Verlag, New York.

Penck, W. 1953. *Morphological Analysis of Landforms*, MacMillan, London.

Pilgrim, D. H. 1976. Travel times and nonlinearity of flood runoff from tracer measurements on a small watershed, *Water Resour. Res.*, **12**(4), 487–496.

Pilgrim, D. H. 1977. Isochrones of travel time and distribution of flood storage from a tracer study on a small watershed, *Water Resour. Res.*, **13**(3), 587–595.

Press, W. H., S. A. Teulkosky, W. T. Vetterly, and B. P Flannery. 1992. *Numerical Recipes*, Cambridge University Press, New York.

Quinn, P., K. Beven, and O. Plachon. 1991. The prediction of flow paths for distributed hydrological modeling using digital terrain models, *Hydrol. Processes*, **5**, 59–79.

Rammal, R., C. Tannous, P. Breton, and A. M. S. Tremblay. 1985. Flicker $(1/f)$ noise in percolation networks: A new hierarchy of exponents, *Phys. Rev. Lett.*, **54**, 1718–1721.

Raup, D. M. 1991. *Bad Genes or Bad Luck*, W. W. Horton and Co., New York.

Rawls, W., D. L. Brakensiek, and K. E. Saxton. 1982. Estimation of soil properties, *Trans. ASAE*, **25**(5), 1316–1326.

Rényi, A. 1970. *Probability Theory*, North-Holland, Amsterdam.

Richards, K. 1988. Fluvial geomorphology, *Progr. Phys. Geogr.*, **12**(3), 433–456.

Richardson, R. L. 1961. The problem of contiguity: An appendix of statistics of deadly quarrels, *General Systems Yearbook*, **6**, 139–187.

Rigon, R. 1992. Il clima é scritto nella forma del reticolo idrografico?, *Istituto Veneto di Scienze, Lettere ed Arti*, Rapporti e studi, vol. CLI, 1–21.

Rigon, R. 1994. Principi di auto-organizzazione nella dinamica evolutiva delle reti idrografiche, Ph.D. Dissertation, Dipartimento di Ingegneria Civile e Ambientale, Universita' di Trento.

Rigon, R., A. Rinaldo, I. Rodriguez-Iturbe, E. Ijjasz-Vasquez, and R. L. Bras. 1993. Optimal channel networks: A framework for the study of river basin morphology, *Water Resour. Res.*, **29**(6), 1635–1646.

Rigon, R., A. Rinaldo, and I. Rodriguez-Iturbe. 1994. On landscape self-organization, *J. Geophys. Res.*, **99**(B6), 11971–11993.

Rigon, R., I. Rodriguez-Iturbe, A. Giacometti, A. Maritan, D. Tarboton, and A. Rinaldo. 1996. On Hack's law, *Water Resour. Res.*, **32**, 3367–3374.

Rinaldo, A., and A. Marani. 1987. Basin scale model of solute transport, *Water. Resour. Res.*, **23**(11), 2107–2118.

Rinaldo, A., and I. Rodriguez-Iturbe. 1996. The geomorphological theory of the hydrologic response, *Hydrol. Processes*, **10**, 803–829.

Rinaldo, A., A. Marani, and A. Bellin. 1989a. On mass response functions, *Water Resour. Res.*, **25**(7), 1603–1617.

Rinaldo, A., A. Marani, and A. Bellin. 1989b. A study of solute NO_3-N transport in the hydrologic response by a MRF model, *Ecological Modelling*, **48**, 159–191.

Rinaldo, A., R. Rigon, and A. Marani. 1991. Geomorphological dispersion, *Water. Resour. Res.*, **27**(4), 513–525.

Rinaldo, A., I. Rodriguez-Iturbe, R. Rigon, R. L. Bras, E. Ijjasz-Vasquez, and A. Marani. 1992. Minimum energy and fractal structures of drainage networks, *Water Resour. Res.*, **28**, 2183–2195.

Rinaldo, A., I. Rodriguez-Iturbe, R. Rigon, E. Ijjasz-Vasquez, and R. L. Bras. 1993. Self-organized fractal river networks, *Phys. Rev. Lett.*, **70**, 822–826.

Rinaldo, A., W. E. Dietrich, R. Rigon, G. K. Vogel, and I. Rodriguez-Iturbe. 1995a. Geomorphological signatures on varying climate, *Nature*, **374**, 632–636.

Rinaldo, A., G. Vogel, R. Rigon, and I. Rodriguez-Iturbe. 1995b. Can one gauge the shape of a basin?, *Water Resour. Res.*, **31**(4), 1119–1127.

Rinaldo, A., A. Maritan, A. Flammini, F. Colaiori, R. Rigon, I. Rodriguez-Iturbe, and J. R. Banavar. 1996a. Thermodynamics of fractal networks, *Phys. Rev. Lett.*, **76**, 3364–3368.

Rinaldo, A., A. Maritan, F. Colaiori, A. Flammini, J. R. Banavar, and I. Rodriguez-Iturbe, 1996b. On feasible optimality, Attie Memorie, Istituto Veneto di Scienze, Lettere ed Arti, Venice, vol. LXI, in press.

Robert, A., and A. G. Roy. 1990. On the fractal interpretation of the mainstream length–drainage area relationship, *Water Resour. Res.*, **26**, 839–842.

Rodriguez-Iturbe, I. 1993. The geomorphologic unit hydrograph. In *Channel Network Hydrology*, K. Beven and M. J. Kirkby editors, J. Wiley, New York, pp. 226–241.

Rodriguez-Iturbe, I., and J. B. Valdes. 1979. The geomorphologic structure of hydrologic response, *Water Resour. Res.*, **15**(6), 1409–1420.

Rodriguez-Iturbe, I., G. Devoto, and J. B. Valdes. 1979. Discharge response analysis and hydrologic similarity: The interrelation between the geomorphologic IUH and storm characteristics, *Water Resour. Res.*, **5**(6), 1435–1444.

Rodriguez-Iturbe, I., M. Gonzales-Sanabria, and R. L. Bras. 1982a. A geomorphoclimatic theory of the instantaneous unit hydrograph, *Water Resour. Res.*, **18**(4), 877–886.

Rodriguez-Iturbe, I., M. Gonzales-Sanabria, and G. Caamano. 1982b. On the climatic dependence of the IUH: A rainfall-runoff analysis of the Nash model and the geomorphoclimatic theory, *Water Resour. Res.*, **18**(4), 887–903.

Rodriguez-Iturbe, I., B. Febres de Power, M. H. Sharifi, and K. P. Georgakakos. 1989. Chaos in rainfall, *Water Resour. Res.*, **25**(7), 1667–1675.

Rodriguez-Iturbe, I., D. Entekhabi, and R. L. Bras. 1991. Nonlinear dynamics of soil moisture at climate scales. 1, Stochastic Analysis, *Water Resour. Res.*, **27**(8), 1899–1906.

Rodriguez-Iturbe, I., E. Ijjasz-Vasquez, R. L. Bras, and D.

G. Tarboton. 1992a. Power-law distributions of mass end energy in river basins, *Water Resour. Res.*, **28**(4), 988–993.

Rodriguez-Iturbe, I., A. Rinaldo, R. Rigon, R. L. Bras, and E. Ijjasz-Vasquez. 1992b. Energy dissipation, runoff production and the three-dimensional structure of channel networks, *Water Resour. Res.*, **28**(4), 1095–1103.

Rodriguez-Iturbe, I., A. Rinaldo, R. Rigon, R. L. Bras, and E. Ijjasz-Vasquez. 1992c. Fractal structures as least energy patterns: The case of river networks, *Geophys. Res. Lett.*, **19**(9), 889–892.

Rodriguez-Iturbe, I., M. Marani, R. Rigon, and A. Rinaldo. 1994. Self-organized river basin landscapes: Fractal and multifractal characteristics, *Water Resour. Res.*, **30**(12), 3531–3539.

Rodriguez-Iturbe, I., G. K. Vogel, R. Rigon, D. Entekhabi, and A. Rinaldo. 1995. On the spatial organization of soil moisture, *Geophys. Res. Lett.*, **22**(20), 2757–2760.

Rodriguez-Iturbe, I., R. Rigon, A. Maritan, W. E. Dietrich, and A. Rinaldo. 1996. Space/time dynamics of channel networks, preprint.

Rodriguez-Iturbe, I., G. Caldarelli, A. Maritan, A. Rinaldo, 1997. Evalutionary active self-organized critical landscapes, preprint.

Rosso, R. 1984. Nash model relation to Horton order ratios, *Water Resour. Res.*, **20**(7), 914–920.

Rosso, R., B. Bacchi, and P. La Barbera. 1991. Fractal relation of mainstream length to catchment area in river networks, *Water Resour. Res.*, **27**(3), 381–387.

Roth G., F. Siccardi, and R. Rosso. 1989. Hydrodynamical description of the erosional development of drainage patterns, *Water Resour. Res.*, **25**(2), 319–322.

Roy, A. G. 1983. Optimal angular geometry models for river branching, *Geogr. Anal.*, **15**, 87–96.

Roy, A. G. 1984. Optimal models of river branching angles. In *Models in Geomorphology*, M. A. Woldenberg editor, Allen and Unwin, Boston, pp. 269–280.

Ruhe, R. V. 1952. Topographic discontinuities of the Des Moines lobe, *Am. J. Sci.*, **250**, 46–50.

Russ, J. C. 1994. *Fractal Surfaces*, Plenum, New York.

Salvadori, G. 1992. Multifrattali stocastici: Teoria ed applicazioni, Ph.D. Dissertation, Universita degli Studi di Pavia.

Saupe, D. 1988. Discrete versus continuous Newton's method, *Acta Appl. Math.*, **13**, 59–80.

Scheidegger, A. E. 1964. Some implications of statistical mechanics in geomorphology, *IASH Bull.*, **9**(1), 12–16.

Scheidegger, A. E. 1967. A stochastic model for drainage patterns into an intramontane trench, *Bull. Ass. Sci. Hydrol.*, **12**, 15–20.

Scheidegger, A. E. 1979. The principle of antagonism in the Earth's evolution, *Tectonophys.*, **55**, 7–10.

Scheidegger, A. E. 1991. *Theoretical Geomorphology*, 3rd edition, Springer–Verlag, Berlin.

Schenck, H. 1963. Simulation of the evolution of drainage basin networks with a digital computer, *J. Geophys. Res.*, **68**(20), 5379–5745.

Schertzer, D., and S. Lovejoy. 1987. Physical modeling and analysis of rain and clouds by anisotropic scaling multiplicative processes, *J. Geophys. Res.*, **92**, 9693.

Schertzer, D., and S. Lovejoy, editors, 1989a. *Scaling, Fractals and Non-linear Variability in Geophysics*, D. Reidel, Hingham, MA.

Schertzer, D., and S. Lovejoy. 1989b. Generalized scale invariance and multiplicative processes in the atmosphere, *Pure Appl. Geophys.*, **130**, 57–81.

Scholz, C. H., and B. B. Mandelbrot editors. 1989. *Fractals in Gephysics*, Birkhauser, Basel.

Schoutens, J. E. 1979. Empirical analysis of nuclear and high-explosive cratering and ejecta. In *Nuclear Geoplosic Sourcebook*, Def. Nuclear Agency, Rep. DNA OIH-4-2, vol. 55, part 2.

Schroeder, M. 1991. *Fractal, Chaos and Power laws*, Freeman, New York.

Schumm, S. A. 1956. Evolution of drainage systems and slopes in badlands at Perth Amboy, New Jersey, *Geol. Soc. Am. Bull.*, **67**, 597–646.

Schumm, S. A. 1977. *The Fluvial System*, J. Wiley, New York.

Schumm, S. A., and H. R. Khan. 1971. Experimental study of channel patterns, *Nature*, **233**, 407–411.

Schumm, S. A., H. R. Khan, B. R. Winkley, and L. G. Robbins. 1972. Variability of river patterns, *Nature*, **237**, 75–76.

Schumm, S. A., M. P. Mosley, and W. E. Weaver. 1987. *Experimental Fluvial Geomorphology*, J. Wiley, New York.

Seginer, I. 1969. Random walk and random roughness models of drainage networks, *Water Resour. Res.*, **5**, 591–599.

Shapiro, A. M., and V. D. Cvetkovic. 1988. Stochastic analysis of solute arrival time in heterogeneous porous media, *Water Resour. Res.*, **24**, 1711–1718.

Sherman, T. F. 1981. On connecting large vessels to small, *J. Gen. Physiol.*, **78**(4), 431–453.

Shreve, R. L. 1966. Statistical law of stream numbers, *J. Geol.*, **74**, 17–37.

Shreve, R. L. 1967. Infinite topologically random channel networks, *J. Geol.*, **77**, 397–414.

Shreve, R. L. 1969. Stream lengths and basin areas in topologically random channel networks, *J. Geol.*, **77**, 397–414.

Shreve, R. L. 1974. Variation of mainstream length with basin area in river networks, *Water Resour. Res.*, **10**, 1167–1177.

Sinclair, K., and R. C. Ball. 1996. A mechanism for global optimization of river networks from local erosion rules, *Phys. Rev. Lett.*, **76**, 3359–3363.

Smalley, R. F., D. L. Turcotte, and S. A. Sola. 1985. A renormalization group approach to the stick-slip behavior of faults, *J. Geophys. Res.*, **90**, 1884–1900.

Smart, J. S. 1968. Statistical properties of stream lengths, *Water Resour. Res.*, **4**, 1001–1014.

Smart, J. S. 1969. Topological properties of channel networks, *Geol. Soc. Am. Bull.*, **80**, 1757–1774.

Smart, J. S. 1972. Channel networks, *Adv. Hydroscience*, **8**, 305–345.

Smart, J. S. 1973. The random model in fluvial geomorphology, *IBM Res. Tech. Rep. RC 4504*, 40 pp.

Smart, J. S. 1978. The analysis of drainage network composition, *Earth Surf. Proc. Landforms*, **3**, 129–170.

Smart, J. S., and V. L. Moruzzi. 1971a. Random walk model of stream network development, *IBM J. Res. Develop.*, **15**(3), 197–203.

Smart, J. S., and V. L. Moruzzi. 1971b. Computer simulation of Clinch Mountain drainage networks, *J. Geol.*, **79**, 572–565.

Smart, J. S., and C. Werner. 1976. Applications of the random model of drainage basin composition, *Earth Surf. Proc.*, **1**, 219–233.

Smart, J. S., A. J. Surkan, and J. P. Considine. 1967. Digital simulation of channel networks, *IAHS Publ. No. 75*, 87–98.

Smith, T. R. 1950. Standards for grading texture of erosional topography, *Am. J. Sci.*, **248**, 655–658.

Smith, T. R. 1974. A derivation of the hydraulic geometry of steady-state channels from conservation principles and sediment transport laws, *J. Geol.*, **82**, 98–108.

Smith, T. R., and F. P. Bretherton. 1972. Stability and the conservation of mass in drainage basin evolution, *Water Resour. Res.*, **8**(6), 1506–1529.

Snell, J. D., and M. Sivapalan. 1994. On geomorphological dispersion in natural catchments and the geomorphological unit hydrograph, *Water Resour. Res.*, **30**(7), 2311–2324.

Sneppen, K., P. Bak, H. Flyvbjerg, and M. H. Jensen. 1995. Evolution as a self-organized critical phenomenon, *Proc. Natl. Acad. Sci. USA*, **92**, 5209–5213.

Solé, R. V., and O. Miramontes. 1995. Information at the edge of chaos in fluid neural networks, *Physica D*, **80**, 171–180.

Solé, R. V., O. Miramontes, and B. C. Goodwin. 1993. Oscillations and chaos in ant societies, *J. Theor. Biol.*, **161**, 343–357.

Spivak, M. 1965. *Calculus on Manifolds*, Benjamin, New York.

Stanley, E. H. 1985. Fractal concepts for disordered systems: The interplay of physics and geometry. In *Scaling Phenomena in Disordered Systems*, R. Pynn and A. Skjeltop editors, Plenum, New York, pp. 49–69.

Stanley, E. H., and P. Meakin. 1988. Multifractal phenomena in physics and chemistry, *Nature*, **335**, 405–409.

Stanley, E. H., and N. Ostrowsky. 1986. *On Growth and Form*, M. Nijhoff, Dordrecht.

Stark, C. P. 1991. An invasion percolation model of drainage network evolution, *Nature*, **352**, 423–427.

Stauffer, D. 1985. *Introduction to Percolation Theory*, Taylor and Francis, London.

Stevens, P. S. 1974. *Patterns in Nature*, Little, Brown, and Co., Boston.

Strahler, A. N. 1950. Equilibrium theory of erosional slopes approached by frequency distribution analysis, *Am. J. Sci.*, **248**, 673–696.

Strahler, A. N. 1952. Hypsometric (area altitude) analysis of erosional topography, *Geol. Soc. Am. Bull.*, **63**, 1117–1142.

Strahler, A. N. 1957. Quantitative analysis of watershed geomorphology, *EOS Trans. AGU*, **38**, 912–920.

Strahler, A. N. 1964. Quantitative geomorphology of drainage basins and channel networks. In *Handbook of Applied Hydrology*, V. T. Chow editor, McGraw–Hill, New York, pp. 39–76.

Summerfield, M. A. 1991. *Global Geomorphology*, Longman, Singapore.

Sun, T., P. Meakin, and T. Jossang. 1994a. A minimum energy dissipation model for river basin geometry, *Phys. Rev E*, **49**, 4865–4872.

Sun, T., P. Meakin, and T. Jossang. 1994b. The topography of optimal drainage basins, *Water Resour. Res.*, **30**, 2599–2611.

Sun, T., P. Meakin, and T. Jossang. 1994c. A minimum energy dissipation model for drainage basins that explicitly differentiates between channel networks and hillslopes, *Physica A*, **210**, 24–47.

Sun, T., P. Meakin, and T. Jossang. 1995. Minimum energy dissipation river networks with fractal boundaries, *Phys. Rev. E*, **51**(6), 5353–5359.

Takayasu, H. 1989. Steady-state distribution of generalized aggregation systems with injection, *Phys. Rev. Lett.*, **63**, 2563–2568.

Takayasu, H. 1990. *Fractals in the Physical Sciences*, Manchester University Press, Manchester.

Takayasu, H., and H. Inaoka. 1992. New type of self-organized criticality in a model of erosion, *Phys. Rev. Lett.*, **68**, 966–969.

Takayasu, M., and H. Takayasu. 1989. Apparent independency of an aggregation system with injection, *Phys. Rev. A*, **39**, 4345–4347.

Takayasu, H., I. Nishikawa, and H. Tasaki. 1988. Power-law mass distribution of aggregation systems with injection, *Phys. Rev. A*, **37**, 3110–3117.

Takayasu, H., M. Takayasu, A. Provata, and G. Huber.

1991. Statistical models of river networks, *J. Stat. Phys.*, **65**, 725–740.

Tarboton, D. G., R. L. Bras, and I. Rodriguez-Iturbe. 1988. The fractal nature of river networks, *Water. Resour. Res.*, **24**(8), 1317–1322.

Tarboton, D. G, R. L. Bras, and I. Rodriguez-Iturbe. 1989a. Scaling and elevation in river networks, *Water. Resour. Res.*, **25**(9), 2037–2051.

Tarboton, D. G., R. L. Bras, and I. Rodriguez-Iturbe. 1989b. The analysis of river basins and channel networks using digital terrain data, Tech. Rep. no. 326, Ralph M. Parsons Lab., Massachusetts Institute of Technology, Cambridge, MA.

Tarboton, D. G, R. L. Bras, and I. Rodriguez-Iturbe. 1990. Comment on the fractal dimension of stream networks, by La Barbera and Rosso, *Water. Resour. Res.*, **26**(9), 2243–2244.

Tarboton, D. G., R. L. Bras, and I. Rodriguez-Iturbe. 1991. On the extraction of channel networks from digital elevation data, *Hydrol. Processes*, **5**, 81–100.

Tarboton, D. G., R. L. Bras, and I. Rodriguez-Iturbe. 1992. A physical basis for drainage density, *Geomorphology*, **5**, 59–76.

Taylor, G. I. 1921. Diffusion by continuous movements, *Proc. London Math. Soc. Ser.*, **A 20**, 196–211.

Taylor, G. I. 1954. The dispersion of matter in turbulent flow through a pipe. *Proc. R. Soc. London Ser. A*, **223**, 446–468.

Tel, T. 1988. Fractals, multifractals and thermodynamics, *Z. Naturforsch A. Phys. Sci.*, **43**, 1154–1174.

Tessier, Y., S. Lovejoy, and D. Schertzer. 1993. Universal multifractals: Theory and observation for rain and clouds, *J. Appl. Meteorology*, **32**, 223–250.

Tokunaga, E. 1978. Consideration on the composition of drainage networks and their evolution, Geograph. Rep. of Tokio Metropolitan University, no. 13.

Toy, T. J., and R. F. Hadley. 1987. *Geomorphology and Reclamation of Disturbed Lands*, Academic Press, Orlando, FL.

Tricot, C. 1989. *Rectifiable and Fractal Sets*, C. R. M., Montreal.

Tricot, C. 1995. *Curves and Fractal Dimensions*, Springer–Verlag, Berlin.

Tricot, C., J. F. Quiniou, D. Wehbi, C. Roques-Carmes, and B. Dubuc. 1988. Evaluation de la dimension fractale d'un graphe, *Rev. Phys. Appl.*, **23**, 111–124.

Troendle, C. A. 1985. Variable source area models. In: *Hydrological Forecasting*, M. G. Anderson and T. P. Burt editors, J. Wiley, New York, pp. 347–403.

Troutman, B. M., and M. R. Karlinger. 1984. On the expected width function for topologically random channel networks, *J. Appl. Prob.*, **22**, 836–849.

Troutman, B. M., and M. R. Karlinger. 1985. Unit hydrograph approximations assuming linear flow through topologically random channel networks, *Water Resour. Res.*, **21**(5), 743–754.

Troutman, B. M., and M. R. Karlinger. 1992. Gibbs' distribution on drainage networks, *Water Resour. Res.*, **28**(2), 563–577.

Troutman, B. M., and M. R. Karlinger. 1994. Inference for a generalized Gibbsian distribution on channel networks, *Water Resour. Res.*, **30**(7), 2325–2334.

Tsonis, A. A. 1990. Some probabilistic aspects of fractal growth, *J. Phys. A*, **20**, 5025–5028.

Turcotte, D. L. 1989. Fractals in geology and geophysics, *Pure Appl. Geophysics.*, **131**, 171–196.

Turcotte, D. L. 1992. *Fractals and Chaos in Geology and Geophysics*, Cambridge University Press, New York.

Uylings, H. B. M. 1977. Optimization of diameters and bifurcation angles in lung and vascular tree structures, *Bull. Math. Biol.*, **39**, 509–520.

Valdes, J. B., Y. Fiallo, and I. Rodriguez-Iturbe. 1979. A rainfall-runoff analysis of the geomorphological IUH, *Water Resour. Res.*, **15**(6), 1421–1434.

Van der Tak, L. D., and R. L. Bras. 1988. Stream length distributions, hillslope effects and other refinements of the geomorphologic IUH, Ralph M. Parsons Laboratory Tech. Rep. no. 301, Massachusetts Institute of Technology, Cambridge, MA.

Veneziano, D., G. E. Moglen, and R. L. Bras. 1995. Multifractal analysis: Pitfalls of standard procedures and alternatives, *Phys. Rev. E*, **52**(2), 1387–1398.

Vicsek, T. 1989. *Fractal Growth Phenomena*, World Scientific, Singapore.

Voss, R. F. 1986. Random fractal forgeries. In *Fundamental Algorithms for Computer Graphics*, A. Earnshaw editor, Springer–Verlag, Berlin, pp. 805–835.

Voss, R. F. 1988. Fractals in nature: From characterization to simulation. In *The Science of Fractal Images*, H. O. Peitgen and D. Saupe editors, Springer–Verlag, New York, pp. 21–70.

Wallace, R. E. 1977. Profiles and age of young fault scarps, north-central Nevada, *Geol. Soc. Am. Bull.*, **88**, 1267–1281.

Wang, S. X., and E. C. Waymire. 1991. A large deviation rate and central limit theorem for Horton ratios, *SIAM J. Disc. Math.*, **4**(4), 575–588.

Waymire, E. 1989. On the main channel length–magnitude formula for random networks: A solution to Moon's conjecture, *Water Resour. Res.*, **25**(5), 1049–1050.

Waymire, E. 1992. On network structure function computations. In *New Directions in Time Series Analysis*, D. Billinger et al. editors, Springer–Verlag, Berlin, pp. 365–373.

Wejchert, J. 1989. Optimally collecting networks, *Europhys. Lett.*, **9**(6), 503–508.

Werner, C. 1991. Several duality theorems for interlocking ridge and channel networks, *Water Resour. Res.*, **27**(12), 3237–3247.

Werner, C., and J. S. Smart. 1973. Some new methods of topological classification of channel networks, *Geogr. Anal.*, **5**, 271–295.

Whittle, P. 1962. Topographic correlation, power-law covariance functions and diffusion, *Biometrika*, **49**, 305–314.

Willgoose, G., R. L. Bras, and I. Rodriguez-Iturbe. 1989. A physically based channel network and catchment evolution model, Ralph M. Parsons Laboratory, Tech. Rep. no. 322, Massachusetts Institute of Technology, Cambridge, MA.

Willgoose, G. R., R. L. Bras, and I. Rodriguez-Iturbe. 1990. A model of river basin evolution, *EOS Trans. AGU*, **71**, 1806–1808.

Willgoose, G. R., R. L. Bras, and I. Rodriguez-Iturbe. 1991a. A coupled channel network growth and hillslope evolution model. 1. Theory, *Water Resour. Res.*, **27**, 1671–1684.

Willgoose, G. R., R. L. Bras, and I. Rodriguez-Iturbe. 1991b. A coupled channel network growth and hillslope evolution model. 2. Nondimensionalization and applications, *Water Resour. Res.*, **27**, 1685–1696.

Willgoose, G. R., R. L. Bras, and I. Rodriguez-Iturbe. 1991c. A coupled channel network growth and hillslope evolution model. 3. A physical explanation of an observed link area–slope relationship, *Water Resour. Res.*, **27**, 1697–1702.

Willgoose, G. R., R. L. Bras, and I. Rodriguez-Iturbe. 1991d. Results from a new model of river basin evolution, *Earth Surf. Proc. Landforms*, **16**, 237–254.

Willgoose, G. R., R. L. Bras, and I. Rodriguez-Iturbe. 1992. The relationship between catchment and hillslope properties: Implications of a catchment evolution model, *Geomorphology*, **5**, 21–38.

Witten, T. A., and L. M. Sander. 1981. Diffusion-limited aggregation, a kinetic critical phenomenon, *Phys. Rev. Lett.*, **47**, 1400–1404.

Wittmann, R., T. Kautzky, A. Hubler, and E. Luscher. 1991. A simple experiment for the examination of dendritic river systems, *Naturwissenschaften* **78**, 23–27.

Woldenberg, M. J., and K. Horsfield. 1986. Relation of branching angles to optimality for four cost principles, *J. Theor. Biol.*, **122**, 204.

Wolfram, S. 1988. *Mathematica: A System for Doing Mathematics by Computer*, Addison–Wesley, Reading, MA.

Wolman, M. G. 1955. The natural channel of Brandywine Creek, PA, USA, *U.S. Geol. Surv. Prof. Paper*, no. 271.

Wood, E. F. 1994. Scaling, soil moisture and evapotranspiration in runoff models, *Adv. Water Resour.*, **17**, 25–34.

Wyss, J. 1988. Hydrologic modelling of New England river basins using radar-rainfall data, M.S. Thesis, Massachusetts Institute of Technology, Department of Earth, Atmospheric and Planetary Sciences, Cambridge, MA.

Yang, C. T. 1971a. Formation of riffles and pools, *Water Resour. Res.*, **7**(6), 1567–1574.

Yang, C. T. 1971b. Potential energy and stream morphology, *Water Resour. Res.*, **7**(2), 311–322.

Author Index

Subject Index